Urban Operations

Plus

Engineer Operations

Department of Defense

Pentagon
Publishing

Field Manual
No. 3-06

Headquarters
Department of the Army
Washington, DC, 26 October 2006

Urban Operations

Contents

Distribution Restriction: Approved for public release; distribution is unlimited.

***This publication supersedes FM 3-06, 1 June 2003.**

Figures

Historical Vignettes

Preface

Doctrine provides a military organization with a common philosophy, a language, a purpose, and unity of effort. Rather than establishing a set of hard and fast rules, the objective of doctrine is to foster initiative and creative thinking. To this end, FM 3-06 discusses major Army operations in an urban environment. This environment, consisting of complex terrain, a concentrated population, and an infrastructure of systems, is an operational environment in which Army forces will operate. In the future, it may be the predominant operational environment. Each urban operation is unique and will differ because of the multitude of combinations presented by the threat, the urban area itself, the major operation of which it may be part (or the focus), and the fluidity of societal and geopolitical considerations. Therefore, there will always exist an innate tension between Army doctrine, the actual context of the urban operation, and future realities. Commanders must strike the proper balance between maintaining the capability to respond to current threats and preparing for future challenges.

PURPOSE

This manual provides the analytical tools for evaluating an urban operation to determine if the operation is necessary for overall mission success. It also provides the means to understanding and determining the impacts of the urban environment on military operations and provides information on managing, taking advantage of, and mitigating the effects of those impacts as appropriate. As such, this manual demonstrates how to apply the doctrinal principles in FM 3-0 to this unique environment.

SCOPE

Chapter 1 introduces theoretical and historical perspectives of urban operations that serve as the underlying basis for the rest of the manual. While this manual has incorporated lessons learned from recent and ongoing operations, it has deliberately taken a broad and varied historical perspective in order to remain relevant to future threats and circumstance differing from those that the Army currently faces. Chapter 2 discusses the characteristics of urban centers and populations as well as their impact on operations. It is unlikely that Army forces will ever operate in a benign urban environment; therefore, Chapter 3 discusses the varied nature of potential urban threats. Chapter 4 describes the effects of the urban environment on warfighting functions and tactics. An understanding of the complexities of the urban environment, the nature of the enemy as an adaptive, learning opponent, and the effects of the environment on warfighting capabilities and skills is essential to sound decision making. Chapters 5 and 6 discuss the potential costs of urban operations and risk reduction measures that the commander and his staff must consider early in their planning. These chapters also outline an urban operational framework and specific urban considerations that create the foundations necessary for successfully applying operational doctrine to an urban environment.

The second part of the manual (Chapters 7 – 10) discusses how urban operations are conducted and resourced. Urban operations include major offensive and defensive operations in urban environments as well as stability or civil support operations ranging from peace operations and counterterrorism to disaster relief and humanitarian assistance. For the different types of operations—offense, defense, and stability or civil support—the purpose, characteristics, organization, and considerations are discussed. However, commanders consider that most urban operations will normally involve the simultaneous execution of offense and defense with stability or civil support (although proportional emphasis will shift over time) and plan accordingly.

APPLICABILITY

This manual is intended for Army commanders and their staffs at the brigade level and above. It addresses full spectrum operations that Army units will execute in urban settings. However, users should also consult JP 3-06 for specific joint information. Additionally, users should be familiar with FM 3-06.1, FM 3-06.11, TC 90-1, and urban operations chapters, appendices, or sections found in other infantry, armor, combined arms, and proponent field manuals for the tactics, techniques, and procedures (TTP) and appropriate proponent information necessary to conduct tactical urban operations at the brigade level and below. This publication applies to the Active Army, the Army National Guard (ARNG)/Army National Guard of the United States (ARNGUS), and United States Army Reserve unless otherwise stated.

ADMINISTRATIVE INSTRUCTIONS

When this FM 3-06 was approved, doctrine was incomplete for incorporating stability operations and civil support operations in place of stability operations and support operations. This manual incorporates stability and reconstruction and civil support operations doctrine found in FM 1 and FMI 5-0.1; it will be revised to reflect the future updates of FM 3-0 and FM 3-07 and, when published, FM 3-28. (Note: Stability and reconstruction operations has been redesignated stability operations to comply with Department of Defense Directive 3000.05.)

Chapter 2 defines "city" according to a population size. However, in historical vignettes and accounts, the term "city" may be applied in its common usage without specific regard to size to maintain conformity with most other historical reports.

In this manual, the term "threat" is applied broadly to include an enemy force (conventional or unconventional), an armed belligerent in a peace operation, antagonistic or unfriendly elements of the civilian population, or some other hazardous condition in the urban environment that negatively influences mission accomplishment.

The term military operations on urban terrain (MOUT) is replaced by urban operations (UO). MOUT is an acronym from FM 90-10, which is superseded by this manual.

Otherwise, the glossary lists most terms used in FM 3-06 that have joint or Army definitions. Where Army and joint definitions are different, (Army) follows the term. Definitions for which FM 3-06 is the proponent manual (the authority) are marked with an asterisk (*). The proponent or amplifying manual for other terms is listed in parentheses after the definition.

The manual attempts to incorporate a broad range of historical vignettes into each chapter where the account supports the doctrinal line of reasoning. Two historical vignettes, however, were included as appendices (A and C) because of their longer lengths.

Unless this publication states otherwise, masculine nouns or pronouns do not refer exclusively to men.

This publication contains copyrighted material.

The proponent for this publication is the United States Army Training and Doctrine Command. Send written comments and recommended changes on DA Form 2028 (Recommended Changes to Publications and Blank Forms) directly to Commander, U.S. Army Combined Arms Center and Fort Leavenworth, Combined Arms Doctrine Directorate, ATTN: ATZL-CD, (FM 3-06), 201 Reynolds Avenue, Fort Leavenworth, KS 66027-2337. Send comments and recommendations by e-mail to webb-cadd@leavenworth.army.mil. Follow the DA Form 2028 format or submit an electronic DA Form 2028.

This page intentionally left blank.

This page intentionally left blank.

Chapter 1

The Urban Outlook

Today's security environment demands more from Army leaders than ever before. Army leaders must not only be able to lead Soldiers but also influence other people. They must be able to work with members of other Services and governmental agencies. They must win the willing cooperation of multinational partners, both military and civilian. But ultimately, the Army demands self-aware and adaptive leaders who can compel enemies to surrender in war and master the circumstances facing them in peace. Victory and success depend on the effectiveness of these leaders' organizations. Developing effective organizations requires hard, realistic, and relevant training.

FM 1

Given the prevalence of large cities throughout the world, Army forces will likely be required to conduct operations in, around, and over large urban areas. These operations will normally be in support of a joint force commander (JFC) conducting military operations pursuant to U.S. national security policy. This manual is designed to facilitate the planning and conduct of the full range and spectrum of land operations in a complex urban environment. Each urban environment and urban operation is unique; prescribing a specific doctrinal "solution" for each situation is impossible. Instead, this manual provides a framework to commanders and their staffs for understanding the urban environment, for analyzing and deciding whether or not to initiate urban operations (UO), and for applying operational doctrine to this complex environment. It also provides a broad base of historical vignettes and examples to help develop a refined analytical perspective and stimulate thought. The manual also includes some planning points and tactics and techniques to assist in preparing for and conducting UO. This information provides a foundation for approaching major UO. Combined with other joint and Army doctrine, this information also will help commanders and their staffs learn to adapt and succeed in this challenging environment.

THE PROSPECT OF URBAN OPERATIONS

1-1. The world is undergoing massive urbanization. Although exceptions exist, an overall trend of migration from rural to urban areas is occurring throughout the globe. (Australia, one of the world's most urbanized countries, is actually becoming less urbanized.) This trend is especially evident in developing nations. Combined with the exponential growth of the global population in the last quarter century, this migration has created massive urban areas that hold the centers of population, government, and economics in their respective regions. In many cases, rapid urbanization has overburdened already weak infrastructures, scarce resources, and a fragile economic base. As urbanization has changed the demographic landscape, potential enemies recognize the inherent danger and complexity of this environment to the attacker, and may view it as their best chance to negate the technological and firepower advantages of modernized opponents. Given the global population trends and the likely strategies and tactics of future threats, Army forces will likely conduct operations in, around, and over urban areas—not as a matter of fate, but as a deliberate choice linked to national security objectives and strategy, and at a time, place, and method of the commander's choosing.

Army Urban Operations

Army forces conduct UO either as one component of a larger operation or as a single operation focused totally on a specific urban environment. Major Army UO are often part of a joint and multinational effort requiring rigorous interagency and civil-military coordination that typically includes the full spectrum of military operations. Commanders of Army major operations must determine if UO are essential to mission accomplishment. If so, commanders must carefully integrate the operations into campaign planning to support the operational objectives of the JFC.

Army leaders conducting UO must—

- Understand the urban environment to determine decisive points.
- Shape the operation to set the conditions for success.
- Precisely mass the effects of combat power to thoroughly engage the decisive points that lead to centers of gravity.
- Continually consolidate gains essential to the retention of the initiative.
- Transition the urban area to the control of another force or agency or, ultimately, back to legitimate and functioning civilian control.

AN URBAN PERSPECTIVE

1-2. As a subset of all Army operations, UO are operations focused on an urban environment. UO include full spectrum operations—offensive, defensive, and stability or civil support—that may be executed, either sequentially or (more likely) simultaneously, during the conduct of a single urban operation (see figure 1-1). UO may be the sole mission of the commander or one of several tasks nested in a larger operation. Regardless of the types of operations conducted or whether the urban area is the single focus of the operation or only one component of a larger operation or campaign, the complex urban environment significantly affects the overall conduct of the mission.

1-3. When conceptualizing urban operations, commanders must understand two important terms: urban area and urban environment. The first is a subset of the second. An ***urban area*** **is a topographical complex where man-made construction or high population density is the dominant feature**. Focusing on urban areas means concentrating on the physical aspects of the area and their effects on weapons, equipment, line-of-sight, and tactics, techniques, and procedures. The ***urban environment*** **includes the physical aspects of the urban area as well as the complex and dynamic interaction and relationships between its key components—the terrain (natural and man-made), the society, and the supporting infrastructure—as an overlapping and interdependent system of systems.**

1-4. Importantly, commanders must also understand and consider that critical elements of the infrastructure may lie far beyond the area's physical confines. For example, the generating source providing power to the urban energy system is part of that system but may be located well outside of the urban area. Similarly, effects of the interaction between components of the infrastructure, located both inside and outside the urban area, extend well into smaller, neighboring urban areas and surrounding rural areas and often form their political, economic, and cultural focus. Understanding the total urban environment is essential to planning and conducting full spectrum urban operations.

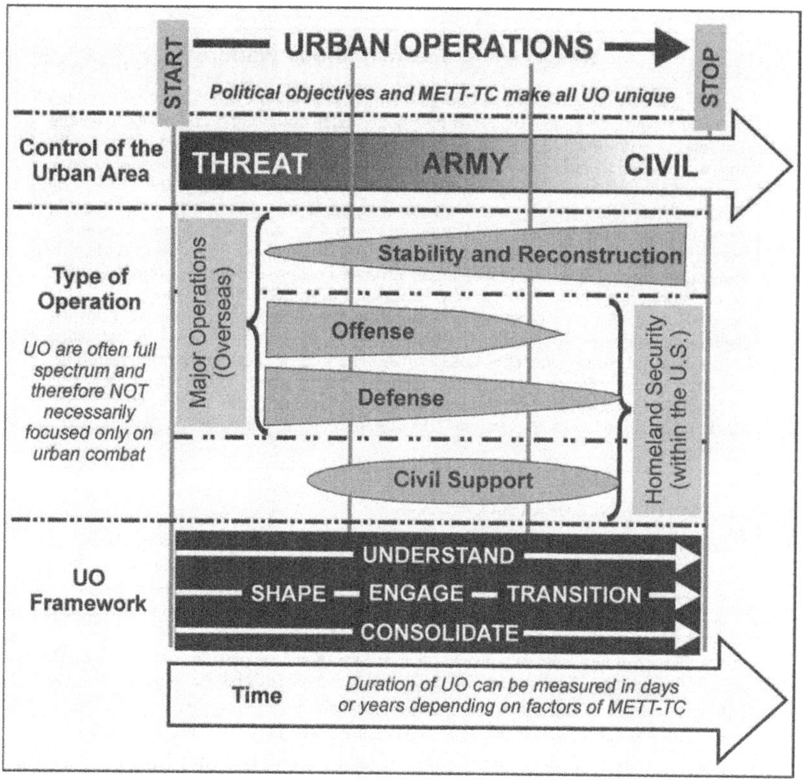

Figure 1-1. Full spectrum urban operations

HISTORICAL SIGNIFICANCE OF URBAN AREAS IN WARFARE

1-5. Urban areas always have been central to, or have significantly influenced, military operations. One of the first urban-centered battles was the siege of Troy at the beginning of Greek history. Moreover, much of the history of early Greece revolved around wars between its city-states or with Persia and centered on the conquest, siege, or blockade of cities. Five hundred years later, the Roman Empire replaced Greece as the dominant world power; although, urban areas remained central to the Roman method of warfare. Even Rome's history can be viewed as a microcosm of urban warfare over the past two thousand years. Though military operations within the physical confines of many of these historic urban areas were not the norm, the focus of these operations was their conquest or control.

1-6. Although Rome last saw combat in 1944, urban areas have been no less prominent in warfare since that time. Seoul in Korea, Beirut in Lebanon, Panama City in Panama, Grozny in Chechnya, Sarajevo in Bosnia-Herzegovina, Kabul in Afghanistan, and Baghdad in Iraq have been centers of conflict in the last 50 years. Urban areas, now more pervasive than ever before, will continue to be essential to successful operational and strategic warfighting. Today, armies cannot expect to execute major military operations without the influence of the urban environments within their area of operations.

Rome: A Microcosm of Urban Warfare

During two millennia, Rome has been the center of at least 12 battles. The Gauls lay siege to Rome first in 387 BC. That first siege lasted six months and ended after the barbarians burnt much of the city. The surviving patrician families paid a ransom for the withdrawal of Brennus' army. From 408 to 410 AD, the Goth leader, Alaric, successfully besieged Rome no less than three times. The Byzantine General Belisarius captured Rome twice from the Goths and withstood siege inside the city once between 536 and 549. Five hundred years later in 1084, Norman adventurer Robert Guiscard captured medieval Rome and sacked the city during a dispute between the Pope and the Holy Roman Empire. Forces of the Holy Roman Empire again stormed and captured the city to punish the Pope in 1527. During the Italian Revolution in 1849, a French army supporting the Pope captured the city from the Italian revolutionary army under Garibaldi. In 1944, the last military action took place in and around Rome when the U.S. Fifth Army captured the city from the retreating German army. Rome's turbulent history—fought over ethnic and religious differences, prestige, and military necessity—demonstrates the importance of urban areas in warfare and the various causes and combatants within this complex environment.

Strategic Importance of Urban Areas

1-7. Several reasons have attracted (and continue to attract) armies to combat in urban areas:

- A military force chooses to position itself in an urban area to capitalize on the perceived defensive advantages offered by the environment. In contrast, an opposing force, by analyzing the factors of the situation, determines that it must enter the urban area to attack and destroy its enemy (or devote essential combat power to their isolation).
- The urban environment's people (their allegiance and support), infrastructure, capabilities, or other resources have or can be of significant operational or strategic value.
- The urban area has significant symbolic importance.
- The urban area's geographical location dominates a region or avenue of approach.

1-8. Russia's 1994 experience in Chechnya illustrates an increasingly important motivation for conducting urban operations. The Chechen rebels, after failing to engage Russian forces outside the city, chose to turn Grozny into the main battlefield. Leaders of the defeated Chechen conventional forces recognized that fighting in the urban area provided them their best chance for success. The complexities of urban combat and the perceived advantages of defending an urban area mitigated their numerical and technological inferiority. The urban area provided the Chechens protection from fires, resources, interior lines, and covered and concealed positions and movement. Given such advantages offered by the environment, smaller or less-sophisticated military forces have similarly chosen to fight in urban areas either as a deliberate strategy or to escape certain destruction in open terrain.

1-9. Such advantages of operating in an urban environment also prompt forces to conduct an urban operation to facilitate a larger campaign plan and decisive battle in another location. The urban operation can focus the enemy on the urban area and allow other forces to conduct operations elsewhere. From a defensive perspective, an urban defense may gain time and space to reorganize forces in new defensive positions, to divert enemy forces from other critical tasks, or to prepare to conduct offensive operations. To some extent, these reasons motivated Soviet forces defending Leningrad and Stalingrad from the Germans in World War II. The stubborn defense permitted the Soviets to reorganize for later offensive operations. From an offensive perspective, an attack on an urban area may be a shaping operation used to divert enemy resources from the decisive operation that will follow.

1-10. Armies also fight in an urban area to obtain some critical feature or resource in the area, such as a port facility. The desire to control an important seaport and access to the Persian Gulf largely motivated the Iranian and Iraqi struggle for Basra in the 1980s. Earlier, in 1944, British forces fought German units in Arnhem for control of the Rhine River Bridge. Other key infrastructure of the urban environment may have operational or strategic significance and can compel military forces to attack or defend the area. As urban areas account for an increasing share of a country's national income, often generating over 50 percent of gross national product, the strategic implications for their control or influence become even greater.

1-11. Urban areas are often located on terrain that dominates a region or an avenue of approach. In these cases, offensive armies capture these areas to proceed with security to another objective. Conversely, defensive forces commonly defend the area to deny the area of operations. To illustrate, Cassino, Italy during World War II stood astride the critical highway approach up the Liri valley to Rome. The allies had to attack and capture the monastery to facilitate the allied offensive north. Cassino's location made bypassing virtually impossible. Likewise, in the early 1980's, Israeli army urban operations in Beirut were a result of its strategic location near the Israeli security zone; various Arab insurgent and terrorist groups used Beirut as a base for attacks against Israel. Beirut evolved as the major base of the Palestine Liberation Organization, a major opponent of Israel. Beirut's location made it a security threat to Israel and thus compelled several major Israeli operations in the urban area (see Appendix A).

1-12. Another reason for engaging in urban operations is the symbolic—historical, cultural, political, and even economic—importance of many urban areas. Often, capital cities—such as Rome, Paris, Seoul, Berlin, and Baghdad—are identified as the strategic centers of gravity of their respective nations. Possessing or threatening these urban areas may impact directly on the outcome of a conflict. The objective of Germany's wars with France in 1870 and 1914 was ultimately Paris. Napoleon's 1812 campaign had as its objective Moscow, as did Hitler's 1941 offensive into Russia. The objective of the Soviet 1945 offensive was Berlin, and the North Vietnamese 1975 offensive had as its objective the South's capital of Saigon. Still, history also reminds us that commanders must assess the sustainability and decisiveness of operations directed toward these "prestige" objectives. For example, in 1812, Napoleon captured Moscow but had to evacuate it within 30 days. He lacked supplies and shelter, failed to destroy the Russian Army, and failed to defeat the political will of the Czar and the people. Similarly, the North Korean occupation of Seoul during the Korean War was equally indecisive.

U.S. ARMY'S EXPERIENCE IN URBAN OPERATIONS

1-13. The U.S. Army has a varied history of conducting urban operations. The American Revolution saw the Army conduct several urban operations. These operations included the unsuccessful defense of New York, the successful attack on Trenton, and the decisive siege and attack on British forces at Yorktown. The Mexican War also had a successful assault on the fortified city of Monterey and the decisive siege of Mexico City. During the American Civil War, the armies, in the tradition of Napoleonic maneuver warfare, avoided urban areas and fought in the open. However, the opposing armies frequently made urban areas their objective because of their importance as railheads. Success in the siege of several key urban areas—Vicksburg, Atlanta, and Petersburg—contributed to the Northern victory.

1-14. Following the Civil War, the U.S. Army faced no large-scale urban combat for several generations. The Indian Wars, the Spanish-American War, the Philippine Insurrection, and even World War I did not require the Army to fight in large urban areas. Between the Civil War and World War II, the U.S. Army fought in several urban areas worldwide supporting U.S. commitments. These limited urban combat operations were small but essential parts of what are currently called urban stability operations. From 1900 to 1901, the Army provided public security for a sector of Peking, China of around 50,000 inhabitants. The Army conducted UO and, in the course of the operation, the 9th U.S. Infantry suffered 20-percent casualties while fighting in Tientsin. Punitive expeditions to places such as Siberia, Cuba, Philippines, Central America, and Mexico put the Army in various urban situations that required using military power, notably, the occupation and security of Vera Cruz, Mexico in 1914. In the context of these smaller-scale contingencies, UO became a staple of U.S. Army employment.

1-15. World War II forced the Army to grapple with the issues of large-scale urban combat almost immediately. In his 1941 defense of the Philippines, General MacArthur examined how to defend Manila.

Manila represented a large, modern, friendly urban area, which was the capital city of a close U.S. ally. Defending the urban area posed numerous challenges. Ultimately General MacArthur determined that he could best conduct its defense outside the city by defeating the enemy forces in combat on the invasion beaches or shortly after they landed. When Japanese forces defeated MacArthur's Philippine Army in a series of engagements, MacArthur had to decide how best to protect the friendly populace of Manila. He had two choices: abandoning the city or waging a costly defense that would likely result in the city's destruction, thousands of noncombatant casualties, and no operational advantage. He had little choice but to declare Manila an open city and move his forces to Bataan to wage an operational defense in the vain hope that a counteroffensive could relieve his isolated force. On 2 January 1942, Japanese forces entered Manila unopposed.

1-16. Had General MacArthur decided to defend Manila, his forces would have found scant doctrine in the Army regarding how to fight in an urban area. Doctrine for urban operations did not appear until early 1944, when faced with the possibility of fighting through the larger urban areas of Western Europe. At this time the U.S. Army published FM 31-50. This manual had the first formal discussion of how the Army viewed urban combat. It was based on the Army's limited experiences in the Mediterranean theater and the study of German and Soviet experiences on the Eastern front.

1-17. FM 31-50 emphasized a deliberate pace, individual and small unit initiative, the liberal use of direct and indirect firepower, and decentralized command and execution. It focused on the urban area (as opposed to the environment); however, it did include policies towards the noncombatants. The manual was also focused at the regimental combat team level. Complementing the doctrine of FM 31-50 was the 1944 operations manual, FM 100-5. This latter manual rightly emphasized the importance of combined arms actions and the need for extensive reconnaissance of prepared and defended cities. The Army successfully implemented this doctrine in several major instances of urban combat, most notably the capture of the first German city, Aachen, and hundreds of small-scale urban assaults on cities, towns, and villages across France, the Benelux, and Germany. Army forces also successfully employed this urban combat doctrine during the liberation of Manila in 1945.

1-18. The legacy of this era of Army operations was an effective tactical solution to urban offensive combat: isolate the urban area, seize a foothold, and expand the foothold block by block until occupying the entire urban area and destroying the enemy. The doctrine's emphasis on firepower kept friendly casualties to a minimum. Unfortunately, when enemy forces stoutly defended the urban area, the emphasis on firepower resulted in the area's virtual destruction and high casualties among noncombatants.

1-19. The doctrinal approach honed in World War II remained the accepted Army approach to urban combat to the century's end. The last successful implementation occurred during the Korean War and the final liberation of Seoul. The Vietnam conflict did not offer the Army opportunities or the requirement to practice urban combat or test and refine doctrine on a large scale. The largest urban battle, Hue, was a chaotic tactical battle that validated most of the historical lessons of urban combat without generating any new doctrinal insights for large-scale urban warfare.

1-20. From the mid-1950s through the 1990s, the Army conducted UO in the U.S. in support of civil authorities during civil unrest and anti-Vietnam protests. Some operations involved numerous active and Reserve Component forces engaged in restoring public order. The Detroit riots of 1967 and the Los Angeles riots of 1992 required the commitments of active and National Guard units. In 1968, the Army deployed over 35,000 troops to Washington D.C., Chicago, and Baltimore following the death of Dr. Martin Luther King, Jr.

1-21. In the 1970s and 1980s, Army doctrine predominantly focused on urban areas and successfully fighting a conventional ground war against Soviet and Warsaw Pact forces in Central Europe. FM 90-10 (1979) described how to conduct urban operations against Soviet forces in Germany. Its concepts were never tested other than in simulation, and its approach to urban combat was not substantially different from that practiced by the Army since World War II. Despite previous doctrine's admonition to avoid cities, the Army has had to fight in them in diverse circumstances.

1-22. In the 1990s and early 21st century, commanders and planners began to more carefully consider the impact of urban environments on the overall conduct of campaigns and major operations. Maneuver warfare characterized Operation Desert Storm in 1991; however, urban environments became prominent during subsequent operations in Somalia, the Balkans, and again in Iraq. (During Operation Iraqi Freedom in 2003, for example, Baghdad was considered as a strategic center of gravity for planning.) This was a result of adversaries recognizing the asymmetric advantages to be gained by fighting superior U.S. forces in an urban environment as well as a general worldwide trend of increasing urbanization. The evolving nature of Army operations in the urban environment was recognized in the Army doctrine of full spectrum operations. No longer could combat operations be considered exclusively as a separate phase of the major operation. Instead, full spectrum operations—combat and stability or civil support—would be conducted simultaneously. It also became clear to Army commanders that while one type of operation can often dominate for a period, simultaneous, full-spectrum urban operations were now the norm and an in-depth understanding of complex urban societies essential to overall success.

MODERN ARMY URBAN OPERATIONS

1-23. Modern urban operations will be full spectrum operations executed jointly and will often contain a multinational and interagency component. They will span the entire range of possible applications of military power. At higher echelons, these separations are often viewed as levels or scales of intensity. For the tactical units conducting urban operations, these divisions appear indistinct, as the intensity is often high despite where the operation falls within the range of military operations.

JOINT, INTERAGENCY, AND MULTINATIONAL

1-24. The urban environment is too multifaceted for a single-service, single-agency, or single-dimensional solution. Generating desired effects and avoiding unintended negative consequences in this complex environment requires careful integration of joint (and often multinational) forces and interagency capabilities throughout all phases of the operation. Effective interagency collaboration will help plan effects, supporting actions, and measures of effectiveness to ensure that military actions complement diplomatic, economic, and informational activities.

1-25. Joint urban operations (JUOs) in which Army forces are a major component will be primarily land operations. These operations may take place within the context of a joint campaign conducted by a joint force land component commander or a joint task force (JTF) commander. Or they may be an Army operation under an ARFOR commander who himself operates for a JFC, depending on the organization of the theater's joint command structure. In the later case, the JFC will manage joint issues in the urban area.

1-26. The JFC conducting JUOs will focus on effectively organizing his forces for UO and tasking them in accordance with their service capabilities. His guide for the conduct of the JUO will be the joint operational tasks described in JP 3-0. JP 3 06 will provide the JTF commander specific guidance regarding the conduct of joint operational tasks in the urban environment. Army commanders will execute tasks assigned by the JFC and advise him on using Army forces and capabilities. Army commanders will also ensure that Army UO are nested within the JFC's concept of operations. Also, the ARFOR commander will request support through the JFC from other service and functional commanders who have urban capabilities critical to the success of Army UO. See Appendix D for more information on joint capabilities in an urban environment.

1-27. The military conflict will not be an end to itself. It is inevitably a means of transition from a perceived unsatisfactory state of affairs to an improved end. Therefore, urban operations will require the careful orchestration of military and civilian capabilities to achieve success. An early identification of potential requirements will allow for the proper allocation of responsibilities between military and civilian agencies and a clear understanding of what is possible for each to achieve. Such understanding will preclude the establishment of false perceptions from all concerned, but particularly from the urban population and the public at large.

1-28. Interagency cooperation will include the Army as part of the Department of Defense, elements of the Department of State, and other various governmental and nongovernmental organizations. Urban operations demand the expertise and abilities of various organizations and a synergistic unity of effort if strategic goals aimed at bringing about a more satisfactory political, social, diplomatic, economic, and military situation are to be achieved. However, there will often be a gap in time before bringing many governmental capabilities to bear that the Army and other military forces—to include coalition partners—will have to fill. Routine interagency training, planning, and coordination will help to decrease that gap in time to its minimum. As it is likely that civilian organization leaders will be in charge during some phases of the overall operation, they should lead planning at those times. A careful survey of urban requirements and rapid communication to participating civilian organizations will also help speed civilian capabilities to where they are needed most.

Winning the Peace in Iraq:
The Requirement for Full Spectrum Operations

We found that if we concentrated solely on establishing a large security force and targeted counterinsurgent combat operations—and only after that was accomplished, worked toward establishing a sustainable infrastructure supported by a strong government developing a free-market system—we would have waited too long. The outcome of a sequential plan allowed insurgent leaders to gain a competitive advantage through solidifying the psychological and structural support of the populace.

Further, those who viewed the attainment of security solely as a function of military action alone were mistaken. A gun on every street corner, although visually appealing, provides only a short-term solution and does not equate to long-term security grounded in democratic process. Our observation was born not from idealism, but because it creates the essence of true security, protecting not only our soldiers, but Iraq, the region, and, consequently, our homeland.

Major General Peter Chiarelli
Commander, 1st Cavalry Division

FULL SPECTRUM OPERATIONS

1-29. Army forces will conduct full spectrum operations within urban areas. Army commanders conduct full spectrum operations abroad by executing offensive, defensive, and stability urban operations as part of an the joint, interagency, and multinational effort described above. Army forces within the United States and its territories conduct full spectrum operations by combining offensive, defensive, and civil support operations to support homeland security. The situation will mandate that one type of operation—offense, defense, stability, or civil support—dominates the urban operation. Depending on whether the operation is overseas or within the United States, however, commanders will often find themselves executing offensive and defensive operations and stability or civil support operations simultaneously. In fact, waiting until all combat operations are concluded before beginning stability or civil support operations often results in lost, sometimes irretrievable, opportunities. Unsurprisingly, the mission determines the dominant type of operation, with the other types of Army operations conducted to shape the AO for mission success. The dominant type of operation will vary between different urban areas even in the same campaign.

Offense

1-30. Against a large conventional enemy in a major urban area with a large civil population present, offensive operations will likely require a greater commitment of Army resources than in other environments. Urban offensive operations will also incur the greatest risks to Army forces and noncombatants. Within defensive and stability or civil support operations, forces may conduct tactical offensive UO, such as counterattacks to maintain the initiative or raids to eliminate elements disrupting the stability or civil support operation.

Defense

1-31. Defensive UO are generally conducted as a shaping operation within a larger major operation. These temporary operations often set conditions for successful offensive operations, stability operations, or civil support operations. Commanders often conduct defensive UO within other types of operations to protect essential facilities in the urban area, protect flanks against counterattack, prevent the breakout of isolated enemies, or protect valuable supply bases or vulnerable convoy routes. Army forces conducting defensive UO must creatively use the environment to enhance their combat power.

1-32. In UO, essential facilities will likely include urban ports and airfields required by sister services to support ground operations. Therefore, Army commanders will need to carefully plan integrated air and base defense operations with air and naval component commanders. Protection requirements increase dramatically as air and naval assets are most vulnerable when aircraft are on the ground and ships are docked at port. Without a carefully synchronized and integrated base defense plan, Army commanders risk sporadic or total loss of support by air and naval forces—forces essential to successful UO.

Stability

1-33. Stability operations in an urban environment require offensive and defensive operations, combined with other tasks unique to each stability operation. Army forces conduct urban stability operations for various reasons, including noncombatant evacuation operations, peace operations, or support to insurgencies (see Chapter 9). Urban stability operations will require an offensive capability to destroy any military capability that overtly threatens its objectives before that military threat can adversely affect the operation. Army forces employ defensive capabilities to safeguard themselves as well as secure critical places, populations, or infrastructure in the urban area. Commanders may also employ defensive capabilities to separate and protect one faction from another. Various stability operations will also require the distribution of food or aid and the protection or assistance of agencies conducting economic or humanitarian activities.

Civil Support

1-34. Army civil support operations in an urban environment aid other U.S. agencies and organizations in mitigating the consequences of natural and man-made disasters. In response to the disaster, civil support operations require the equipment, personnel, or organizational abilities of Army forces rather than the Army's combat capabilities. In a civil support mission, these capabilities often involve Army transportation, medical, quartermaster, or engineer forces. Unless conducted in conjunction with homeland defense, urban civil support operations will seldom require combat; however, commanders must determine if force protection threats exist that could hamper Army civil support operations. During homeland security operations, defensive and offensive capabilities will be required to defeat hostile armed forces.

PREPARING FOR FUTURE URBAN OPERATIONS

1-35. To operate successfully in a complex urban environment requires a thorough understanding of the urban environment and rigorous, realistic UO training. Training should be conducted by the complete combined arms team and cover full spectrum operations to include appropriate tactics, techniques, and procedures (see FM 3-06.11 and TC 90-1). It should also replicate—

- The psychological impact of intense, close combat against a well-trained, relentless, and adaptive enemy.
- The effects of noncombatants—including governmental and nongovernmental organizations and agencies—in close proximity to Army forces. This necessitates—
  - An in-depth understanding of culture and its effects on perceptions.
  - An understanding of civil administration and governance.
  - The ability to mediate and negotiate with civilians including the ability to effectively communicate through an interpreter.
  - The development and use of flexible, effective, and understandable rules of engagement.
- A complex intelligence environment requiring lower-echelon units to collect and forward essential information to higher echelons for rapid synthesis into timely and useable intelligence for all levels of command. Understanding the multifaceted urban environment necessitates a bottom-fed approach to developing intelligence (instead of a top-fed approach more common and efficient for open terrain and conventional threats). It also emphasizes the need for intelligence reach and a truly collaborative approach to the development and sharing of intelligence.
- The communications challenges imposed by the environment as well as the need to transmit large volumes of information and data.
- The medical and logistic problems associated with operations in an urban area including constant threat interdiction against lines of communications and sustainment bases.

Developing Joint, Interagency, and Multinational Operability

1-36. UO training extends from the individual Soldier to the joint level including the integration of conventional and special operations forces. Preparedness also includes enhancing interoperability in regards to urban multinational and interagency operations. Training, as well as campaign and contingency plan creation, should include significant and sustained participation by civilian agencies. Fruitful multinational and interagency relationships must be cultivated before the onset of operations, that is, before Soldiers and their coalition counterparts are making decisions—many with strategic implications and often when they are under fire. Joint, interagency, and multinational collaboration will help design effects, supporting actions, and measures of effectiveness necessary to ensure that military actions in urban environments complement the diplomatic, informational, and economic activities necessary to achieve strategic objectives.

Conducting Live, Virtual, and Constructive Training

1-37. Force preparedness mandates integrating the actual use of urban terrain, exercises at urban training sites, simulations, or any combination into tactical- and operational-level intra- and interservice training. This type of multi-faceted training will help commanders develop a better understanding of the complexity of the urban environment and enable them to execute missions across full spectrum operations. Careful use of these facilities will also allow Army forces to fully integrate urban operations within self-development, institutional, and operational training.

Ensuring Every Soldier is an Urban Warrior

1-38. In a complex urban environment, every Soldier—regardless of branch or military occupational specialty—must be committed and prepared to close with and kill or capture threat forces in an urban environment. Every Soldier must also be prepared to effectively interact with the urban area's noncombatant population and assist in his unit's intelligence collection efforts. Each urban operation will be unique and commanders—given an opportunity to surge and conduct mission-specific post-mobilization and in-theater training—will need to conduct an analysis of the tasks requiring emphasis or modification to fit the conditions of the operational environment. In UO, every soldier will likely be required to—

- Perform advanced rifle marksmanship to include advanced firing positions, short-range marksmanship, and night firing techniques (unassisted and with the use of optics).
- Operate their unit's crew-served weapons.
- Conduct urban reconnaissance and combat patrolling.
- Enter and clear buildings and rooms as part of an urban attack or cordon and search operation.
- Defend an urban area.
- Act as a member of a combat convoy (including specific drivers training).
- Control civil disturbances.
- Navigate in an urban area.

1-39. While not all-inclusive and necessarily urban-specific, other critical individual and collective UO tasks (often modified for the urban environment) might include—

- Conduct troop-leading procedures.
- React to contact, ambush, snipers, indirect fire, and improvised explosive devices.
- Establish an observation point, personnel or vehicle checkpoint, or roadblock.
- Secure a disabled vehicle or downed aircraft.
- Call for indirect fire and close air support.
- Create and employ explosive charges.
- Handle detainees and enemy prisoners of war.
- Treat and evacuate casualties.
- Accurately report information.
- Understand the society and culture specific to the area of operations.
- Use basic commands and phrases in the region's dominant language.
- Conduct tactical questioning.
- Interact with the media.
- Conduct thorough after-action reviews.

1-40. Commanders must understand that all Soldiers will require urban-specific equipment to conduct many urban-specific tasks. Critically, commanders cannot expect supporting forces to fight alongside and support ground maneuver units in an urban environment without the appropriate equipment and training.

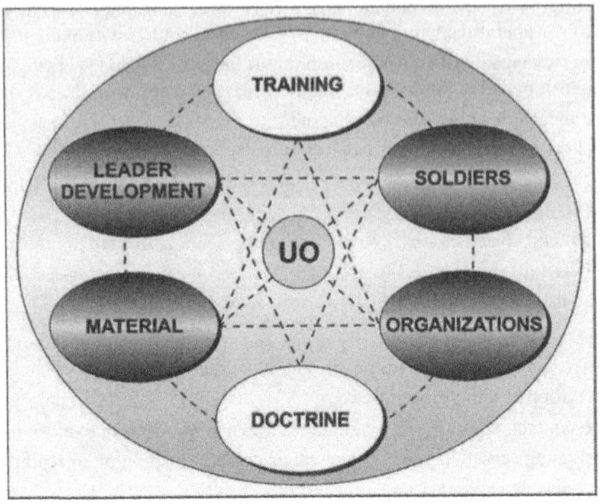

Figure 1-2. UO and DOTMLPF

Continual Adaptation and Innovation

1-41. Realistic UO training (as well as the conduct of real world operations) has the added benefit of identifying operational requirements and resultant changes necessary in our doctrine, organizations, training, materiel, leadership and education, personnel, and facilities (DOTMLPF) (see figure 1-2 and FM 100-11). While technology (materiel) and organizational changes are important, confident Soldiers (personnel) remain the decisive means for success. Technology and organizational changes will be a critical enabler to attain better understanding of the urban environment, enhance command and control on the noncontiguous battlefield, and achieve the agile, simultaneous, and precise effects required in UO. In the future, technology may lead to a radically new operational concept and approach to UO. Still, competent leaders and well-trained, disciplined, and adaptive Soldiers will remain the decisive means for the Army to succeed in this complex and multidimensional urban environment.

Chapter 2

Understanding the Urban Environment

From a planning perspective, commanders view cities not just as a topographic feature but as dynamic entities that include hostile forces, local population, and infrastructure. Planning for urban operations requires careful IPB, with particular emphasis on the three-dimensional nature of the topography and the intricate social structure of the population.

FM 3-0

Of all the environments in which to conduct operations, the urban environment confronts Army commanders with a combination of difficulties rarely found elsewhere. Its distinct characteristics result from an intricate topography and high population density. The topography's complexity stems from the man-made features and supporting infrastructure superimposed on the natural terrain. Hundreds, thousands, or millions of civilians may be near or intermingled with soldiers—friendly and enemy. This second factor, and the human dimension it represents, is potentially the most important and perplexing for commanders and their staffs to understand and evaluate. To this end, this chapter provides information essential to understanding the urban environment and conducting an effective intelligence preparation of the battlefield (see FM 2-01.3, FMI 2-91.4, and Appendix B).

Although urban areas possess general similarities, each environment is distinct and will react to and affect the presence and operations of Army forces differently. A tactical technique effective in one area may not be effective in another area due to physical differences, such as street patterns or the type of building construction. An Army policy popular with one urban group may cause resentment and hostility in another due to diverse cultural differences. All difficulties potentially exist, and they increase the complexity for Army forces operating in urban areas. These difficulties range from conventional military forces to disease and starvation (see Chapter 3) to a pervasive media—often acutely present in intricate combinations. Thus, commanders at all levels must make extraordinary efforts to assess and understand their particular urban environment to plan, prepare for, and execute effective urban operations (UO).

A COMPLEX ENVIRONMENT

2-1. Urban areas vary depending on their history, the cultures of their inhabitants, their economic development, the local climate, available building materials, and many other factors. This variety exists not only among urban areas but also within any particular area. The ever-changing mix of natural and man-made features in urban areas present commanders with some of the most difficult terrain in which to conduct military operations.

2-2. Although urban areas possess similar characteristics, no two are identical. The sprawl of Los Angeles, for example, bears little physical resemblance to New Delhi. Societal characteristics most significantly affect each area's uniqueness and complexity. While complex, information about the terrain, its potential effects on operations, and how it changes over time may be determined with some degree of certainty. However, the human dimension is much more difficult to understand and assess, particularly its effects on military operations. Like any environment, the side that can best understand and exploit the positive and mitigate the negative effects of the urban environment has the best chance of success.

2-3. Whether a large metropolis or a small village, each urban environment has identifiable components that constantly change and interact. This "system of systems" consists of the terrain, the society, and the infrastructure that links the two (see figure 2-1).

2-4. These systems are not separate and distinct categories, but rather overlapping and interdependent systems, acting dynamically with each other. Thoroughly analyzing these elements, along with the other factors of mission, enemy, weather, troops and support available, and time—

- Contributes to the accuracy of the commanders' situational understanding.
- Potentially lessens the number and cost of close combat engagements.
- Allows commanders to develop courses of action that apply appropriate resources against decisive points.

2-5. In stability operations and civil support operations, this understanding allows commanders to engage and dominate the decisive points critical to maintaining peace or restoring normalcy to the urban environment. Although each system is categorized into subordinate components or subsystems, commanders must often "step back" and visualize each system, the complex urban environment, and their area of operations (AO) as a unified whole. This "systems thinking" aids commanders in uncovering key relationships and intersections that can help reveal centers of gravity (COGs) and decisive points.

2-6. To comprehend the urban environment and its components to the fullest extent possible, commanders must carefully integrate and employ tactical reconnaissance forces, special operations forces (SOF)—to include psychological operations (PSYOP) and civil affairs units—and a myriad of other human intelligence (HUMINT) assets and regional, language, and cultural experts. The societal aspects and integrating infrastructure will challenge commanders' assessment and understanding. These aspects will also require greater dependence on nonmilitary and nongovernmental organizations (NGOs) and host-nation agencies for their information, knowledge, and expertise. This last consideration requires commanders to develop effective techniques and procedures for coordinating, interacting, and, to the greatest extent possible, synchronizing activities with these agencies.

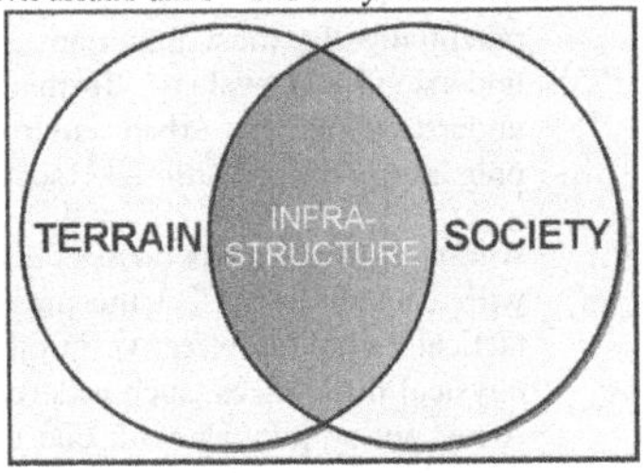

Figure 2-1. Keys to understanding the urban environment

URBAN TERRAIN

2-7. Although complex and difficult to penetrate with many intelligence, surveillance, and reconnaissance (ISR) assets, the terrain is the most recognizable aspect of an urban area. Truly understanding it, however, requires comprehending its multidimensional nature. The terrain consists of natural and man-made features, with man-made features dominating; an analysis considers both. Buildings, streets, and other infrastructure have varied patterns, forms, and sizes. The infinite ways in which these factors can intertwine make it difficult to describe a "typical" urban area. However, these various factors provide a framework for understanding the complex terrain in an urban area. Furthermore, man-made features significantly affect military systems and Soldiers, and thus tactics and operations. General effects on urban operations are discussed in this chapter. Specific effects on warfighting functions (see Chapters 4 and 10) and the spectrum of operations (see Chapters 7, 8, and 9) are interwoven throughout the manual.

MULTIDIMENSIONAL BATTLEFIELD

2-8. Urban areas present an extraordinary blend of horizontal, vertical, interior, exterior, and subterranean forms superimposed on the natural relief, drainage, and vegetation. An urban area may appear dwarfed on a map by the surrounding countryside. In fact, the size and extent of the urban area of operations is many times that of a similarly sized portion of undeveloped natural terrain. A multi-storied building may take up the same surface area as a small field, but each story or floor contains approximately an equal area as the ground upon which it sits. In effect, a ten-story building can have eleven times more defensible area than "bare" ground—ten floors and the roof. It is the sheer volume and density created by this urban geometry that makes UO resource intensive in time, manpower, and materiel.

2-9. Like natural disasters, UO can radically alter the physical characteristics of the urban terrain in ways not experienced in other environments. They may cause (either intentionally or not) uncontrollable fires or the loss of electricity. A power outage can cause flooding (especially in subsurface areas) by shutting down pumping stations. Entire buildings may be destroyed, eliminating reference points, leaving large piles of rubble, altering fields of fire, and making movement and transportation extremely difficult. Additionally, buildings and other urban structures, damaged but not destroyed, can become (or remain) effective obstacles and possible booby traps. Even without enemy exploitation, their weakened construction and unstable structure increase the risk of injury to Soldiers and civilians moving within them. (Engineer expertise will often be needed to determine whether the buildings can support occupation by Army forces or civilians.) Yet, even the total collapse of a building may not eliminate its defenders. Of additional concern, the likely presence of toxic industrial material (TIM) can create additional obstacles and health hazards.

2-10. Commanders in other environments normally address the depth, breadth, and height of their AO in terms of two areas: airspace and surface. In an urban environment, they broaden their scope to include supersurface and subsurface areas (see figure 2-2) that voluminously extend the commanders area of operations. Although spatially separated, each area may be used as an avenue of approach or mobility corridor, line of communications (LOC), or engagement area.

2-11. Supersurface and subsurface areas magnify the complexity of the urban physical environment. Commanders must consider activities that occur outside buildings and subterranean areas (the external space) as well as the activities happening unseen in buildings and subterranean systems (the internal space). This internal volume further challenges command, control, and intelligence collection activities and increases the combat power required to conduct UO. Commanders must develop methods and techniques to help themselves, their staffs, and their subordinate commanders and staffs to represent, visualize, and reference these multiple dimensions. Increasing the difficulty, such dimensions can change rapidly simply due to continued urban growth or, as described earlier, the effects of nature and UO themselves.

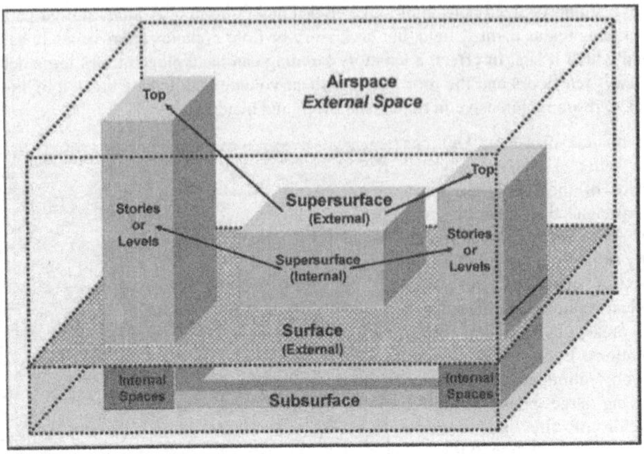

Figure 2-2. The multidimensional urban battlefield

Airspace

2-12. As in all other environments, aircraft and aerial munitions use the airspace as rapid avenues of approach in urbanized areas. Forces can use aviation assets for observation and reconnaissance, aerial attack, or high-speed insertion and extraction of Soldiers, supplies, and equipment. Some surface obstacles in an urban area, such as rubble, do not affect flight (though they may prevent the take-off and landing of aircraft). Buildings of varying height and the increased density of towers, signs, power lines, and other urban constructions, however, create obstacles to flight and the trajectory of many munitions (masking). Similarly, these obstacles can restrict a pilot's line of sight as well as physically limit low-altitude maneuverability in the urban airspace. Excellent cover and concealment afforded enemy gunners in an urban area increases aviation vulnerability to small arms and man-portable air defense systems (MANPADS), particularly when supporting ground forces. The potential for a high volume of air traffic (military and civilian) over and within urban airspace (including fixed-wing, rotary-wing, and unmanned aircraft systems) may become another significant hazard and necessitate increased airspace command and control measures.

Surface

2-13. Surface areas apply to exterior ground-level areas, such as parking lots, airfields, highways, streets, sidewalks, fields, and parks. They often provide primary avenues of approach and the means for rapid advance. However, buildings and other structures often canalize forces moving along them. As such, obstacles on urban surface areas usually have more effect than those in open terrain since bypass often requires entering and transiting buildings or radical changes to selected routes. Where urban areas border the ocean or sea, large lakes, and major rivers, the surface of these bodies of water may provide key friendly and threat avenues of approach or essential LOCs—a significant consideration for Army commanders. As such, amphibious, river-crossing, and river operations may be integral parts of the overall urban operation (see FM 3-05.212 and FM 3-90.12).

2-14. Larger open areas—such as stadiums, sports fields, school playgrounds, and parking lots—are often critical areas during urban operations. They can provide locations for displaced civilians, interrogation centers, and prisoner of war holding facilities. These areas also can afford suitable aircraft landing and pickup zones and artillery firing locations. They can provide logistic support areas and aerial resupply possibilities because they are often centrally located. Finally, large open areas (and immense or unusually shaped structures) within urban areas are often easier to see—especially from the air—and can serve as excellent target reference points from which to shift or control fires.

Supersurface

2-15. These areas include the internal floors or levels (intrasurface areas) and external roofs or tops of buildings, stadiums, towers, or other vertical structures. They can provide cover and concealment; limit or enhance observation and fields of fire; and restrict, canalize, or block movement. However, forces can move within and between supersurface areas creating additional, though normally secondary, avenues of approach. Rooftops may offer ideal locations for landing helicopters for small-scale air assaults and aerial resupply. First, however, engineers must analyze buildings for their structural integrity and obstacles. Such obstacles include electrical wires, antennas, and enemy-emplaced mines (although personnel may be inserted by jumping, rappelling, or fast roping from a hovering helicopter and extracted by hoist mechanisms). Some rooftops are designed as helipads. Roofs and other supersurface areas may also provide excellent locations for snipers; lightweight, handheld antitank weapons; MANPADS; and communications retransmission sites. They enable top-down attacks against the weakest points of armored vehicles and unsuspecting aircraft. Overall, elevated firing positions reduce the value of any cover in surrounding open areas and permit engagement at close range with less risk of immediate close assault. This area (and the subsurface area) requires commanders to think, plan, and execute ground operations vertically as well as horizontally. In this latter regard, UO share strong similarities with mountain operations (see FM 3-97.6).

Subsurface

2-16. Subsurface areas are below the surface level. They may serve as secondary and, in fewer instances, primary avenues of approach at lower tactical levels. When thoroughly reconnoitered and controlled, they offer excellent covered and concealed LOCs for moving supplies and evacuating casualties. They may also provide sites for caching and stockpiling supplies. Subsurface areas include subterranean areas such as subways, mines, tunnels, sewers, drainage systems, cellars, civil defense shelters, and other various underground utility systems. In older cities, they may include ancient hand-dug tunnels and catacombs. Both attacker and defender can use subsurface areas to gain surprise and maneuver against the rear and flanks of a threat and to conduct ambushes. However, these areas are often the most restrictive and easiest to defend or block. Their effectiveness depends on superior knowledge of their existence and overall design. Army commanders may also need to consider potential avenues of approach afforded by the subsurface areas of rivers and major bodies of water that border urban areas. This particularly applies when operating as part of a joint task force (JTF) task organized with SOF or when opposing a threat with similar capabilities.

MAJOR URBAN PATTERNS

2-17. Four major urban patterns (satellite, network, linear, and segment) can influence UO (see figure 2-3). Central to two of the patterns (satellite and network) is the hub or dominant urban area or pattern around which outlying urban areas or patterns radiate. (A segmented urban area, because it tends to be a larger urban area, can often be a hub.) In offensive and defensive operations, the hub serves as a pivot or strong point; as such, it can become a major obstacle to an attacker. If the attacker chooses to bypass the urban area (hub) located along his axis of advance without first isolating the area, he may expose his flank or LOC to attack from the hub as well as dependent urban areas or subordinate satellite patterns. Because the focus of stability and civil support operations is normally on people, commanders should understand the value and influence of the hub to the economic, political, or cultural well being of the surrounding area. Generally the larger the hub, the greater influence it has on satellite urban areas and surrounding rural areas. Commanders must remember that urban areas are not islands; all are connected to the surrounding rural (and other urban) areas through fluid and permeable boundaries and LOCs.

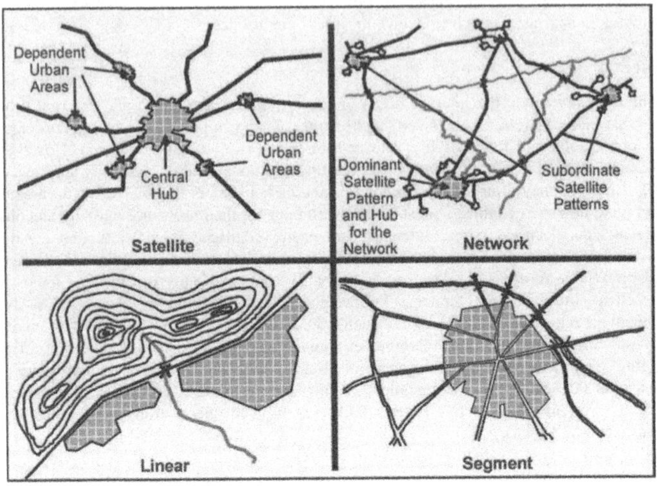

Figure 2-3. Major urban patterns

Satellite Pattern

2-18. This common pattern consists of a central hub surrounded by smaller, dependent urban areas. LOCs tend to converge on the hub. Outlying areas often support the principal urban area at the hub with means of reinforcement, resupply, and evacuation. In some instances, they may serve as mutually supporting battle positions. Commanders should consider the effects of the outlying urban areas on operations within the hub, and, conversely, the effects of operations within the hub on outlying urban areas. Information operations (IO), for example, targeted primarily at key leaders and other civilians located within the hub of a satellite pattern may subsequently influence civilians in outlying urban areas and achieve necessary effects without having to commit specific resources to these outlying areas.

Network Pattern

2-19. The network pattern represents the interlocking of the primary hubs of subordinate satellite patterns. Its elements are more self-sufficient and less supportive of each other, although a dominant hub may exist. Major LOCs in a network extend more than in a satellite pattern and take more of a rectangular rather than a convergent form. Its natural terrain may vary more than in a single satellite array. Operations in one area may or may not easily influence, or be influenced by, other urban areas in the pattern.

Linear Pattern

2-20. Potentially a subelement of the previous two patterns, the linear pattern may form one ray of the satellite pattern or be found along connecting links between the hubs of a network. Most frequently, this pattern results from the stringing of minor urban areas along a confined natural terrain corridor, such as an elongated valley, a body of water, or a man-made communications route. In offensive and defensive operations, this latter form of the linear pattern facilitates developing a series of strong defensive positions in depth, effectively blocking or delaying an attacking force moving along the canalized terrain.

Segment Pattern

2-21. When dominant natural terrain, such as a river or man-made features (canals, major highways, or railways), divides an urban area, it creates a segmented pattern. This pattern often makes it easier for commanders to assign areas of operations to subordinate commanders. However, this pattern may fragment operations and increase risk to an operation requiring mutual support between subordinate units. Still, the segmented urban areas may allow commanders to isolate threats more easily in these areas and focus operations within segments that contain their decisive points. Although an integral part of the whole (the urban area), each segment may develop distinct social, economic, cultural, and political characteristics. This social segmenting may benefit commanders faced with limited assets to influence or control the urban populace. After thoroughly analyzing the society, they may be able to focus IO and populace and resources control measures against only specific segments that affect decisive operations. Commanders may need only to isolate other segments or may need to just monitor for any significant changes in the attitudes, beliefs, or actions of the civilians located there.

LESSER STREET PATTERNS

2-22. Lesser patterns in the urban area result from the layout of the streets, roads, highways, and other thoroughfares. They evolve from influences of natural terrain, the original designer's personal prejudices, and the changing needs of the inhabitants. Street patterns (and widths) influence all warfighting functions; however, they greatly affect movement and maneuver, command and control, and sustainment. (In some portions of older Middle Eastern urban areas, the labyrinths of streets were designed only to allow two loaded donkeys to pass each other; tanks are too wide.) Urban areas can display any of three basic patterns and their combinations: radial, grid, and irregular (see figure 2-4).

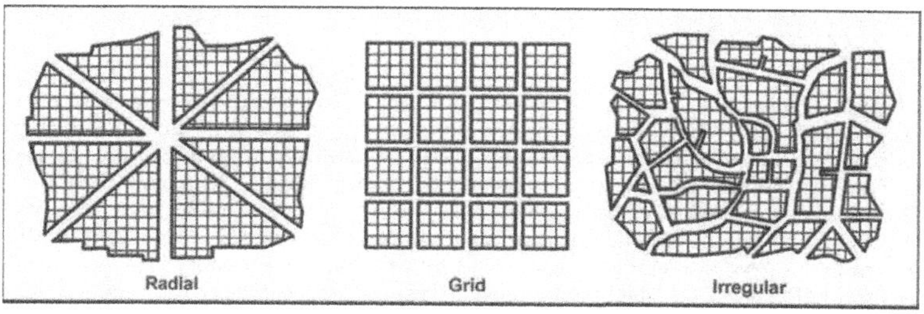

Figure 2-4. Basic internal street patterns

Radial

2-23. Societies of highly concentrated religious or secular power often construct urban areas with a radial design: all primary thoroughfares radiating out from the center of power. Urban areas with this design may signal an important historical aspect in the overall analysis of the urban society. Terrain permitting, these streets may extend outward in a complete circle or may form a semicircle or arc when a focal point abuts a natural barrier, such as a coastline or mountain. To increase mobility and traffic flow, societies often add concentric loops or rings to larger radial patterns. Unless commanders carefully plan boundaries, routes, and axes of advance, their subordinate units' movement or maneuver may be inadvertently funneled toward the center of urban areas with this pattern resulting in congestion, loss of momentum, and an increased potential for ambush or fratricide.

Grid

2-24. The most adaptable and universal form for urban areas is the grid or rectangular pattern: lines of streets at right angles to one another forming blocks similar to the pattern of a chessboard. A grid pattern can fill in and eventually take over an original radial pattern. Grid patterns often appear to ease the assignment of boundaries for subordinates units. However, commanders also consider how the natural terrain influences operations and the establishment of graphic control measures. They also consider the influence of the buildings and other structures lining these streets, such as their height and construction, before assigning boundaries and developing other control measures. Commanders should also consider the following when developing urban graphic control measures:

- Describing boundaries, phase lines, checkpoints, and other graphic control measures by easily recognizable features is as important in urban areas as elsewhere. While easily identifiable urban structures such as unusually tall or oddly-shaped buildings, cemeteries, stadiums, or prominent rail or highway interchanges can be useful references, available natural features are a better descriptor than man-made features that may be altered or unrecognizable. As an aid to air-to-ground coordination, commanders should select features that can be identified by both ground and air forces. Those that help in controlling ground forces may not be easily visible from the air and vice versa.
- Commanders should also consider whether a boundary along an easily-recognizable terrain feature, such as a river, will also be easy to identify by the threat who may seek to "find the seam" and exploit the likely control and coordination difficulties associated with boundaries, especially between higher-level units. This often requires commanders to carefully position a control measure away from the key feature to provide a designated subordinate force with the terrain and space necessary to control the feature. On the other hand, commanders working closely with local authorities during stability and civil support operations may not need to thoroughly understand the physical effect of street patterns on the assignment of boundaries as they might for combat UO. Instead, commanders may choose to assign boundaries overlaid on existing geopolitical boundaries used by local agencies to increase interoperability and aid in unity of effort.

IRREGULAR

2-25. In most urban areas, regardless of the original intent, plan, or vision, existing street patterns emerge from successive plans overlaid one on another. Some are well planned to fit with previous plans while others a haphazard response to explosive urban growth. The result may mix patterns. Urban engineers and planners may specifically design irregular patterns for aesthetic reasons (as in many suburban housing developments) or to conform to marked terrain relief. Irregular street patterns may alert commanders and analysts that the underlying natural terrain may exert greater influence over operations than in other portions of the urban area. Finally, irregular street patterns make the movement and maneuver of forces less predictable. However, a labyrinth of irregular or "twisting" street patterns may increase the possibility of fratricide particularly for units that are trained or accustomed only to grid patterns.

AN URBAN MODEL

2-26. Throughout the world, urban areas have similar form and function. In form, urban areas contain like characteristics, readily divisible into distinct sections or areas. Functionally, they tend to be the centers of population, finance, politics, transportation, industry, and culture. While urban areas may be modeled by several different means, figure 2-5 illustrates the general forms and internal functions. Some forms and functions may overlap. For example, high-rise buildings are located in core areas as well as in outlying areas and may be used for residential purposes. With the rapid urbanization associated with developing nations, the areas displayed in this urban model often manifest themselves less clearly there than in developed nations.

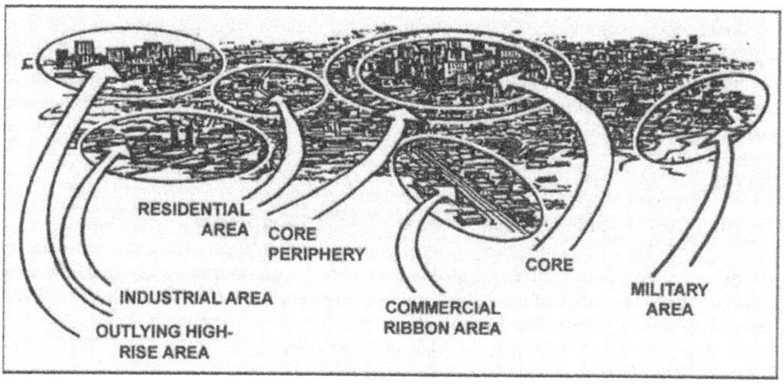

Figure 2-5. Urban functional zones

2-27. This analysis helps to determine, in general terms, potential advantages and disadvantages each portion of the urban area may have toward accomplishing the urban operation. However, construction materials and methods can vary drastically. Commanders and their staff will often need to identify specific building types and construction and understand weapons effects on them. If a commander desires precise effects, the chosen munitions or weapons system must be sufficiently accurate, capable of penetrating the target structure (without exiting the other side), and achieve effects within. Often noncombatants, critical infrastructure, or protected targets are in the vicinity. Commanders may need to determine if the surrounding walls or structures will sufficiently absorb or negate the blast or thermal effects of the weapon. Regardless, understanding the structure and composition of buildings and other structures in the urban AO may be necessary to allow commanders to determine the best means to accomplish the mission.

Core

2-28. The core is the heart of the urban area, the downtown or central business district. Relatively small and compact, it contains a large percentage of the urban area's shops, offices, and public institutions. Often, it houses the headquarters for commercial and financial activities and contains important cultural, historical, and governmental buildings. These activities prefer the core because of its accessibility. As the focal point of the transportation network, residents find the core the easiest part of the urban area to reach. It normally has the densest concentration of multistory buildings and subterranean features (underground parking garages, underground shopping centers, and basements).

2-29. High-rise buildings, varying greatly in height (possibly 50 stories above ground and four stories below ground), make up the cores of many of today's urban areas. Buildings routinely abut one another, with little or no setback from the sidewalks. Building height and density (except in outlying high-rise areas) often decreases from the core to the edge of the residential areas, while the amount of open areas frequently increases. Modern urban planning allows for more open spaces between buildings than found in the cores of older urban areas. Most core areas have undergone constant redevelopment resulting in various types of construction. Commonly, brick buildings abound in the oldest part of the core; framed, heavy-clad structures in the next oldest part; and a concentration of framed, light-clad buildings in the newest part. The outer edge of the core, the core periphery, has ordinarily undergone less change than the core resulting in buildings of uniform height (commonly two to three stories in towns and five to ten stories in larger urban areas).

2-30. Generally, offensive operations focused in core areas (even when effectively isolated) will require greater resources—particularly manpower, time, and information—than in many other parts of the urban area. Mounted maneuver often proves more difficult in core areas because of fewer open areas, buildings closer to the streets, and an increased density of civilian vehicles. Razed buildings in central core areas (especially high-rise buildings) become greater obstacles to mobility as they can collapse on and easily block thoroughfares. Rubble piles provide excellent covered and concealed positions for dismounted threat forces. Consequently, commanders often use more dismounted forces as part of their combined arms operations. Conversely, the core may be critical to urban defensive operations, particularly older areas of heavier construction that afford greater protection. Despite potential difficulties, the core area may be key to accomplishing many stability or civil support missions since it houses much of the human activity that occurs in the urban area.

Industrial Area

2-31. Industrial areas often develop on the outskirts of the urban areas where commercial transportation is easiest (along airfields and major sea, river, rail, and highway routes). The road networks in and around industrial areas are generally more developed and suitable for transportation assets. These areas will likely displace farther from the core and residential areas as urban planners recognize the potential threat of TIM. The dispersed pattern of the buildings provides sufficient space for large cargoes, trucks, and materiel handling equipment. These areas may provide ideal sites for sustainment bases and maintenance sites. While older heavier-clad structures may be found, new construction generally consists of low, large, flat-roofed factory and warehouse buildings with large parking areas and work yards. These structures generally have steel frame and lightweight exterior walls. Multistory structures usually have reinforced concrete floors and ceilings.

Toxic Industrial Chemical	Industrial/Commercial Uses
Ammonia	Commercial Refrigerant, Fertilizer and Food Production, Petroleum, Explosives, Other Chemicals
Arsine	Semiconductor Industry
Boron Trichloride	Organic Catalyst, Soldering Magnesium
Boron Trifluoride	Chemical Catalyst, Aluminum Refining
Carbon Disulfide	Industrial Solvent, Dry Cleaning, Agriculture, Petroleum, Electroplating
Chlorine	Potable Water, Disinfectants, Metal Treatment, Plastics & Rubber
Diborane	Plastics and Rubber
Ethylene Oxide	Industrial Alcohols, Fumigant, Industrial Sterilant
Fluorine	Uranium Processing, Rocket Fuel
Formaldehyde	Plastics, Fertilizers, Preservative/Corrosion Inhibitor, Fungicide and Germicide, Pesticide, Pharmaceuticals
Fuming Nitric Acid	Fertilizers, Explosives, Metal Processing, Pesticides, Rocket Fuel
Hydrogen Bromide	Chemical Industry, Pharmaceuticals
Hydrogen Chloride	Fabrics, Semiconductors
Hydrogen Cyanide	Pesticides, Other Chemicals, Pharmaceuticals, Electroplating
Hydrogen Fluoride	Glass Production, Chemical Catalyst
Hydrogen Sulfide	Metallurgy, Agricultural Disinfectant
Phosgene	Dyes, Pharmaceuticals, Herbicides & Insecticides
Phosphorus Trichloride	Metallurgy, Pesticides and Germicides, Gasoline Additive
Sulfur Dioxide	Paper, Food Processing, Ice Production, Disinfectant, Leather Processing
Sulfuric Acid	Fertilizers, Petroleum, Iron and Steel Production, Battery Electrolyte
Tungsten	Electronics, Other Chemicals

Figure 2-6. Some toxic industrial chemicals and their industrial or commercial uses

2-32. Toxic industrial chemicals and other TIM may be transported through an urban area (by rail, barge, truck, or pipeline) or found stored throughout. However, larger concentrations will exist in industrial areas, and their presence should concern Army forces operating near them. Some TIM are heavier than air and tend to settle in low-lying and subsurface areas.

2-33. Each year, over 70,000 different chemicals are produced, processed, or consumed globally. An estimated 25,000 commercial facilities around the world produce, process, or store chemicals that have a legitimate industrial use yet are also classified as chemical warfare agents. Many other chemicals (not classified as weapons) may still be sufficiently hazardous to pose a considerable threat to Army forces and civilians in urban areas as choking agents or asphyxiates, flammables or incendiaries, water contaminants, low-grade blister or nerve agents, or debilitating irritants. These chemicals can be released either accidentally or deliberately. For example, on 2 December 1984, nearly 40 tons of methylisocyanate used to produce pesticides leaked from a storage tank at Bhopal, India, killing thousands and injuring hundreds of thousands. Figure 2-6 contains a small sampling of other toxic industrial chemicals along with their industrial or commercial usage that commanders may encounter in an urban area. The most common chemicals that may pose an immediate risk to Army forces are highly toxic irritant gases such as ammonia, chlorine, hydrogen chloride, and sulfur dioxide.

2-34. Standard chemical defense equipment may not protect against (and chemical detection devices may fail to detect) many toxic industrial chemicals. Therefore, the risk to Soldiers operating near the chemicals may increase. Commanders must vigilantly identify these potential hazards, carefully consider them as part of their overall vulnerability analysis, factor the analysis into their risk assessment, and execute necessary contamination avoidance measures. (Local urban fire fighters may be a critical source of information for determining the likely locations of TIM.) Any assessment includes the chance that toxic industrial chemicals may be deliberately released by a threat to gain advantage or accidentally released by friendly actions (see FM 3-11.9, FM 3-11.14, FM 3-11.21, and the U.S. Department of Transportation's current version of the Emergency Response Guidebook).

Outlying High-Rise Area

2-35. High-rise areas consist of multistoried apartments, commercial offices, and businesses separated by large open areas, such as parking lots, parks, and individual one-story buildings. High-rise buildings are generally of framed, light-clad construction with thin walls of brick, lightweight concrete, or glass. The automobile, mass transit systems, and improved road networks encourage these areas to grow and function further from the urban core.

2-36. Similar to the urban core, units given the mission to clear these areas, or even portions therein, will need more resources—most notably personnel and time—to accomplish their mission. Commanders should consider courses of action that isolate these entire areas, multiple sections within these areas, or even individual buildings before assigning tasks. Without careful consideration and analysis, some tasks in these areas could unintentionally—but rapidly—drain a unit's resources or unhinge other portions of the major operation. When defending, commanders who can integrate these areas in the defense will present the attacker with similar resource problems and may be appropriate in a defense to delay. However, defending commanders must ensure that the defense is arranged so that this portion cannot be easily isolated and bypassed. Defensive positions in structures may require extensive reinforcement due to light-clad construction.

Residential Area

2-37. Residential areas can be found dispersed throughout the urban area; however, large suburban areas (or sprawl) normally form on the outskirts. Residential areas often consist of row houses or single-family dwellings set in a grid or ringed pattern in a planned development project. Yards, gardens, trees, and fences may separate the buildings in a residential area. In some areas of the world, residential areas may be located in high walled compounds with houses built right up to the edge of the street. Modern residential construction is more often of light-clad, framed wood construction, or brick, however; residential homes formed by poured or pre-cast concrete can be found throughout many parts of the world. The combined population of surrounding suburban areas often far outnumbers that of the urban area proper. Specific suburbs tend toward homogeneity based on ethnicity, religion, economics, or some other social aspect. Commanders must locate and analyze these areas to determine their impact on operations—often the most critical importance is the people located there (see the subsequent discussion in this chapter on the urban society).

2-38. In offensive and defensive operations, commanders should determine whether operations pose an unacceptable physical risk to civilians. If so, they may have to relocate civilians to a safer area, perhaps another residential area. If not, commanders may implement a "stay-put" policy for that area and attempt to isolate the effects of the operation from them. During civil support and stability operations, residential locations may be the initial focal point for operations since most of the permanent population is located there.

2-39. This area also contains an urban phenomenon known as shantytowns. These areas are commonly on unoccupied, low-value land in and around many urban areas in underdeveloped countries. Shantytowns may contain over 50-percent of the total urban population. They usually lack streets and public utilities. The lean-to structures tend to be irregularly laid out, connected by walking paths, and made of any scrap material available: lumber, brick, sheet metal, cardboard, cloth, or vegetation. The random arrangement of structures, the absence of formal street naming and numbering, and often the lack of easily identifiable buildings and terrain create challenges. These challenges include navigation, coordination, and the transmission of accurate information and intelligence. Depending on the operation, the temporary nature of the structures can also mean that mobility can be either more or less restricted than other sections of the urban area. A military force may easily knock down and traverse structures without affecting mobility at all. However, their destruction may cause unacceptable civilian casualties, in which case mobility becomes more restrictive as the narrow paths often do not accommodate vehicular traffic. Similarly, the makeshift materials inhibit weapons effects less than many other parts of the urban area built more solidly. A tank round, for example, may go much farther and injure many more noncombatants than in an area where the primary building material is stone. Regardless, commanders should consider the effects of their operations in this area, to include vehicles and weapons, as the weak structures increase the risk of fratricide, civilian casualties, and large, rapidly spreading fires.

Commercial Ribbon Area

2-40. Commercial ribbon areas are rows of stores, shops, and restaurants built along both sides of major streets that run through and between urban areas. These same types of areas often develop along the roads that connect one urban area to another (strip areas). The buildings uniformly stand two to three stories tall (about one story taller than the dwellings on the streets behind them).

Military Area

2-41. Fortifications and military installations may be found in or near urban areas throughout the world. Historically, they may have been the "seed" responsible for initiating the growth of the present-day urban area. Many countries possess long coastlines and borders with potentially hostile neighbors. To meet their defensive needs, they developed coastal and border defense works that include extensive subsurface facilities, many contiguous to urban areas. North Korea, for example, has built numerous hardened artillery, missile, and command and control facilities along both its coasts and the demilitarized zone which separates them from South Korea. While some fortifications may be inactive (particularly ancient fortifications), they can be rapidly activated and modified by threats (or Army forces) to accomplish their original role. Even if not activated, they may still serve as choke points and major obstacles to movement and maneuver.

2-42. Permanent-type fortifications can be made of earth, wood, rock, brick, concrete, steel-reinforced concrete, or any combination of the above. Some contemporary constructions have been built subsurface and employ heavy armor, major caliber weapons, internal communications, service facilities, and chemical, biological, radiological, and nuclear overpressure systems. Because they have been built specifically for military purposes, commanders and planners should carefully consider the effects of these military constructions on the conduct of UO.

URBAN SOCIETY

2-43. Although intricate, understanding the urban terrain is relatively straightforward in comparison to comprehending the multifaceted nature of urban society. UO often require Army forces to operate in close proximity to a high density of civilians. Even evacuated areas can have a stay-behind population in the tens of thousands. This population's presence, attitudes, actions, communications with the media, and needs may affect the conduct of operations. Homogeneity decreases drastically as the size of the urban area increases. Commanders must take into account the characteristics of a population whose beliefs and interests vary based on factors (see figure 2-7). Analysis and understanding of these societal factors is critical to a successful IO campaign and, thus, the entire operation.

2-44. Civilian populations continually influence, to varying degrees, operations conducted in an urban area. Thoroughly understanding these societal aspects and avoiding "mirror-imaging"—overlaying one's own values and thought processes on top of the person or group one is trying to understand—will help to accurately anticipate civilian actions and response. For example, while bribery and nepotism are not accepted norms in U.S. society, they may be customary practices in the foreign urban environment in which Army forces are operating. If commanders attempt to understand incidences of bribery and nepotism using their own societal norms and values as a basis, they are likely to misinterpret the significance of the events and implement an inappropriate course of action. Critically, however, the development of an in-depth understanding of the urban society does not mean that commanders and their Soldiers should adopt the culture of the foreign urban society. While understanding and respect of the urban area's culture is crucial to successful UO, Soldiers function best in dealing with the urban society when they act in accordance with American culture and values.

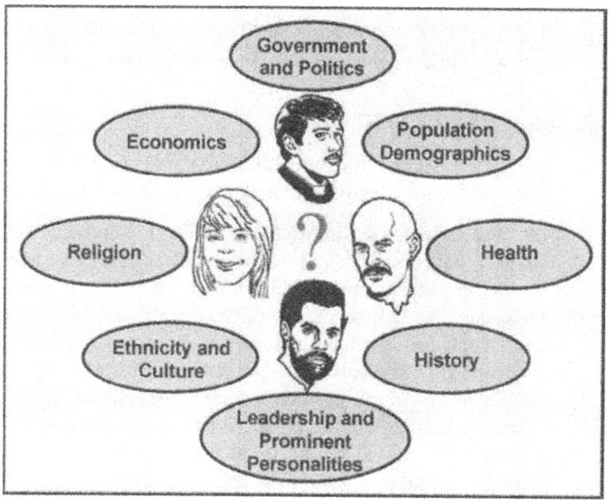

Figure 2-7. Key aspects of the urban society

POTENTIAL CENTER OF GRAVITY

2-45. A COG during an urban operation, particularly in stability and civil support operations, may be the civilian inhabitants themselves—specifically their behavior. However, supportive behavior is generally an advantage in any type of operation. Correspondingly, neutral behavior toward friendly forces is an advantage over hostile behavior. (However, a neutral population normally will not readily provide information to Army forces—often critical to understanding the urban environment and threat. In this light, neutrality can be seen as a disadvantage.) To influence or control their behavior, commanders must first understand the society's complex nature and character. Second, they must understand and accept that every military action (or inaction) may influence the relationship between the urban population and Army forces, and, by extension, mission success or failure. Lastly, they must understand that Army forces may play only a supporting (but essential) role as part of an integrated and synchronized multiagency effort focusing all aspects of national power—diplomatic, informational, military, and economic. With this awareness, commanders can take one or more actions:

- Coordinate and plan operations.
- Implement effective civil-military programs.

- Take the immediate action necessary to maintain support of a friendly populace, neutralize or gain the support of hostile or neutral factions, or do any combination of these activities to achieve precise effects and accomplish the mission.

Without this understanding, commanders increase the risk that their actions, particularly concerning the urban population, may not have the intended and necessary effects.

2-46. Although the factor of civil considerations takes on added significance in UO, it is just one that commanders evaluate. Sometimes, as in most urban counterinsurgency operations, it may be the most important factor to consider as a COG. At other times it may be the least important as to be almost negligible. Its importance is not constant; it changes over time (like all factors). At the beginning of the operation, civil considerations may not be essential to mission accomplishment, but as the operation progresses this factor's importance to success may increase. In other circumstances, the opposite may be true. Overall, commanders should consider three objectives regarding the civilians of the urban area:

- Minimize their interference with urban operations. In offensive and defensive operations this means moving them away from combat operations or establishing measures to shield them from its effects. In all operations, it often requires centralizing them in one or more locations—which may mean keeping them where they are.
- Maximize their support of Army, joint, and multinational forces and government agencies.
- Observe the necessary legal, moral, and humanitarian obligations.

General Population Size

2-47. Urban areas are commonly classified according to the general size of their population instead of landmass. Figure 2-8 lists one method of defining and categorizing urban areas by various population sizes.

2-48. These categories are useful to establish commonality and standardize terms that shape ideas, discussion, and concepts. Smaller populations usually suggest homogeneity among the inhabitants. Homogeneity can make consensus or compromise easier to achieve because fewer opposing viewpoints exist. Given this homogeneity, effects of change are more certain and often easier to determine. However, homogenous does not mean identical. If major social divisions do exist (either physical or ideological), commanders can more easily determine those divisions and their fundamental causes with smaller populations. Treating an urban population as a completely homogenous entity can lead to false assumptions, cultural misunderstandings, and poor situational understanding.

Category	Population
Village	3,000 or less.
Town	Over 3,000 to 100,000.
City	Over 100,000 to 1 million.
Metropolis	Over 1 million to 10 million.
Megalopolis	Over 10 million.

Figure 2-8. Urban areas by population size

2-49. As urban areas expand, the urban patterns begin to blur and the social complexity increases. For example, as satellite patterns continue to grow, the LOCs between a central hub and outlying urban areas may develop and begin to assume a linear urban pattern. Simultaneously, a hub and outlying urban areas may continue to expand until they merge into a single, large metropolis. On a larger scale, a network pattern can grow and unite as a single megalopolis. This growth physically unites smaller urban areas but cannot force conformity of needs and beliefs. It also increases the physical and social complexity of an urban area.

Group Size, Location, and Composition

2-50. Understanding how specific elements of the urban society affect operations (and vice versa), begins with analyzing their size, location, and composition (see figure 2-9). Because commanders must minimize civilian casualties, size and location (without regard to composition) are important initial demographic considerations. After determining the presence and numbers of civilians relative to decisive points, commanders can then decide whether civilian proximity and density represent a significant risk to their mission—refugees clogging LOCs, for example. If civilians are the primary focus of the operation, as in many stability and civil support operations, this same analysis may help to determine decisive points. In this analysis, commanders should consider that urban areas, on many levels, are in constant motion. The densities of circulating people and other traffic often vary according to cultural events or the time of day, such as religious holidays or sporting events and rush hours or market times. Therefore in planning urban operations, commanders must consider the timing, rhythms, or patterns of the population and their vehicular movements in the urban area. Identifying and understanding trends and patterns of activity (and disruptions to them) may provide critical information to commanders.

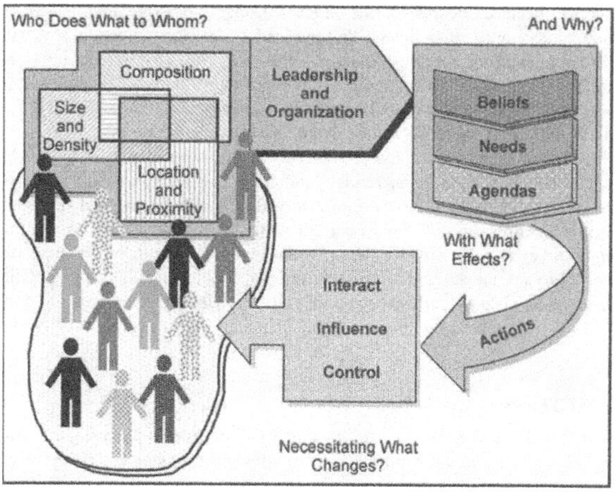

Figure 2-9. Simplified analysis of urban society

2-51. Commanders normally determine the composition of, or the identifiable groups or organizations within, the civilian urban population. Groups may be categorized by race, religion, national origin, tribe, clan, economic or social class, party affiliation, education level, union memberships, age, gender, occupation, or any other significant social demographic. Physical and ideological overlaps (and divisions) often exist between groups. Overlaps may provide early focus for analysis and suggest ways to affect more than one group simultaneously. In some cases, groups may have radically different ideologies but are (or can be) united by a single characteristic. Commanders must understand the intricacies of "who does what to whom." Such understanding furthers identifying the urban society's sources of power, influence (both formal and informal), and decisive points that hold the keys to controlling or protecting this potential COG. (See also the discussion of competing power structures in Chapter 3.) Commanders should have expert, detailed, and current knowledge and information to avoid developing simple formulas of social interaction that may actively mislead and add to a flawed course of action.

Leadership and Organization

2-52. Commanders must also understand how authority and responsibility is held or shared within and between each of the identified groups; they must understand leadership and the social hierarchy. For groups to exert meaningful influence, leadership provides vision, direction, and organized coherence. This leadership can be a function of personality as well as organization. Some groups depend on a charismatic leader to provide cohesion; although in some cultures, the spokesman is not the leader. Others de-emphasize individual leadership and provide redundancy and replacement in decision making. Others combine elements of both these types of leadership and organization. Based solely on personality, a leader may centralize power or, while still being in ultimate control, decentralize decision making and execution to subordinates. In contrast, a single person may head a group while a ruling council actually makes and executes policy. Groups centered on one leader (which may or may not be the officially designated leader) can often produce decisions and initiate actions rapidly but are vulnerable to disruptions if key personalities are removed or co-opted. Groups with shared or redundant leadership take longer to make decisions yet are more resistant to change and outside influence.

2-53. In UO, particularly stability operations, Army commanders and leaders at all levels will devote considerable effort to identifying and cultivating relationships with civilian leaders in their AO. This civilian leadership will include political, religious, tribal or clan, ethnic, and economic leaders. Commanders should consider that their attention toward and discussion with identified leaders may increase (or, some instance, decrease) the targeted leaders' prestige and power. While this may be intentional, commanders must often ensure that the leaders that they chose to deal with are the legitimate and accepted in the eyes of the urban population. Otherwise, they may further imbalance an already weak power structure and exacerbate the situation. In unique circumstances, commanders may need to identify and interact with the leadership of criminal organizations. These civilian leaders will be the means to affect change and allow the urban populace to understand and accept the purpose behind Army operations. As importantly, they will be the conduit for understanding the urban society's sentiments, perceptions, and reactions to Army operations. As such, reliable and trustworthy linguists that can not only interpret the language but can serve as cultural advisors to Army forces will be in great demand. Keeping the local population objectively informed not only of current Army operations but also of the intent and desired end state will often be a key task for commanders. A poor communications effort and a lack of understanding of intent may alienate a commander and cause significant problems in future negotiations. It is imperative that communications be clear and effective and that all Army leaders are aware of its implications. See FM 4-05.401 and FM 3-07 for the effective use of interpreters and the conduct of productive meetings and negotiations.

Interests and Actions

Me and Somalia against the world, me and my clan against Somalia, me and my family against the clan, me and my brother against my family, me against my brother.

Somali Proverb

2-54. Identifying and analyzing groups also helps commanders focus on specific segments of the urban society to determine their beliefs, needs, and agendas. It also helps commanders determine how those interests motivate groups to future action (or inaction)—previous patterns of activity are critical in this regard. This analysis seeks to determine why groups (and their leaders) act as they do. Commanders should consider political, economic, cultural, and religious factors in this analysis. These factors affect all groups to some extent and often provide the basis for their beliefs, needs (actual or perceived), and subsequent behavior. Size and location considerations also apply to each group to help determine to what extent its beliefs or ideologies, needs, and actions may impact the urban operation. However, size and proximity may not accurately indicate actual or potential capabilities. Individuals, small groups, and groups located some distance from the actual conduct of the urban operation may be able to influence large portions of the population. These individuals or groups may have a capability disproportionate to their size and proximity—especially against objectives that are not terrain oriented (as in the case of many stability operations).

Interaction, Influence, or Control

2-55. Commanders must cultivate an understanding of a group's—

- Size, location (and proximity to operations), and composition (to include leadership and organization).
- Interests.
- Capabilities.
- Potential actions (intent) and their effects—if any—on operations.

2-56. Then they can develop or modify courses of action as appropriate. Certain courses of action may be needed to improve the interaction between Army forces and civilians (and between other agencies) to accomplish common goals. Others may be needed to influence favorable support, stabilize neutral groups, or neutralize hostile groups. Still others may require more forceful means to control and protect—never to punish—civilians. Courses of action may include—

- Establishing buffer zones and restricted areas.
- Setting up checkpoints and roadblocks with other travel restrictions on people and goods.
- Screening civilians.
- Conducting negotiations (directly or as a mediator).
- Providing or protecting rations, water, and other critical resources.
- Restoring or improving specific, key infrastructure.
- Enforcing curfews.
- Inspecting facilities.
- Directing amnesty programs.
- Conducting internment and resettlement operations.
- Maintaining a "stay-put" policy.

2-57. Commanders must remember that many measures will require significant resources that may initially be beyond the capabilities of the Army force to impose and enforce. (Where possible, commanders should attempt to use host-nation security forces and local urban law enforcement to control activities.) Other elements of the environment, terrain and infrastructure, may fragment efforts and make it difficult to impose controls throughout the area. A careful assessment and understanding of the urban society's interests (beliefs, needs, and agendas) is essential before implementing any populace and resources control measures. Otherwise inappropriate controls, particularly if they are improperly perceived as punishment, may only aggravate the situation. Finally, an appropriate course of action may require no specific action towards the urban society. In most cases, training and discipline, grounded in cultural understanding and sensitivity, will help mitigate many potential adverse effects resulting from military-civilian interaction. (Commanders should review FM 3-19.15 and FM 3-19.40 for additional civilian control measures and considerations.)

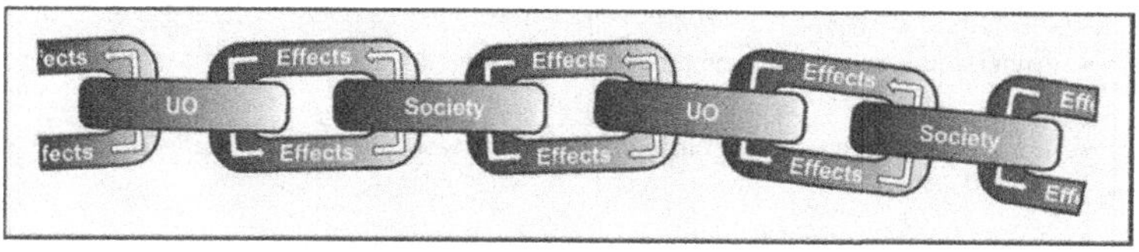

Figure 2-10. UO society chain of effects

A Chain of Effects

2-58. Since urban society is so dynamic and the relationship between various elements so complex, commanders must continually assess how their operations will affect the society and vice versa (see figure 2-10). Specifically, they assess how effectively their measures improve interaction with, influence of, and control over civilians. There is always a difference between intended and actual effects of a specific course of action. Nowhere is this more prominent than dealing with the urban society. This chain of effects frustrates understanding during UO. Therefore, commanders must continuously monitor these effects to make decisions and modifications while planning, preparing, executing, sustaining, and transitioning UO. (Commanders may conduct informal and formal polling activities to assess and understand changing civilian perceptions.) Initially certain aspects of the society, such as religion, may not affect the operation. However, if the threat successfully shapes the perceptions of the urban populace that Army forces are biased against their religion or religious beliefs (or at least critical segments are affected by propaganda), this element may become extremely important. In this instance, the urban commander may need to adjust his IO (to include PSYOP), public affairs (PA) activities, and CMO to counter this propaganda while diverting other combat power to control the populace. Overall, commanders must understand and account for second, third, and higher order effects of their actions and decisions.

Understanding the Effects of Unit and Soldier Actions in Iraq

During OPERATION IRAQI FREEDOM, an armor task force commander described his methodology as that of plotting and measuring everything. After a positive or negative event, he would have his staff evaluate all actions they had conducted before, during, and after the event. This would allow him to correlate activities with outcomes and develop tactics, techniques, and procedures for future success. While not all actions easily correlated to events, this methodological approach helped establish a base line of comparative success for his task force.

Some units assessed the Iraqi culture and developed practical guidelines or rules of interaction for their soldiers to follow as they conducted their missions. These guidelines encompassed a list of actions that their Soldiers should and should not do when dealing with the Iraqi populace. The guiding principle behind these actions was not only to "win the hearts and minds" of the Iraqi people but, more accurately, to establish legitimacy and credibility through appropriate Soldier actions that reflected a combined understanding of the mission and the Iraqi culture.

DO:

- Separate the men from the women and children during cordon and searches.
- Talk to the oldest or most senior male when civilian cooperation is needed.
- Bring something (for example, water, food, or replacement locks) to give residents to compensate for damages when the search ends up being a "dry hole."
- Learn Arabic phrases; say hello and greet people on the street.
- Use female Soldiers to search and talk to Iraqi females.
- Cordon and knock whenever possible instead of kicking in doors needlessly.
- Understand that in Iraq, women do not have the same social status as men.

DO NOT:

- Force a detainee's head to the ground or use your feet to hold someone to the ground.
- Stare at or touch Iraqi women.
- Show obvious disrespect to suspects in front of family members.
- Manhandle the Koran or other religious objects.
- Commit to anything; instead of agreeing to (or denying) an Iraqi's request, it is good practice to say "In'sha Allah" (If God wills it).

URBAN INFRASTRUCTURE

2-59. Urban infrastructures are those systems that support urban inhabitants and their economy. They form the essential link between the physical terrain and the urban society. During urban stability and civil support operations, restoration or repair of urban infrastructure will often be decisive to mission accomplishment. During urban combat operations, destroying, controlling, or protecting vital parts of the infrastructure may be a necessary shaping operation that can isolate a threat from potential sources of support. A threat force operating in an urban area may rely on the area's water, electricity, and sources of bulk fuel to support his forces. This is true particularly when his bases or facilities are physically located in or near the area. Isolating this threat from these sources may require him to generate his own electricity and transport his own water and fuel from outside the urban area. To transport supplies, the threat may rely on roads, airfields, sea or river lanes, and rail lines. Controlling these critical transportation nodes may prevent the threat from resupplying his forces. The control of key radio, television, and newspaper facilities may isolate him from the urban populace (another potential source of support).

INTERDEPENDENCE

2-60. Commanders must understand that destroying or disrupting any portion of the urban infrastructure can have a cascading effect (either intentional or unintentional) on the other elements of the infrastructure. Yet, they may be able to gain an operational advantage while minimizing unwanted and unintended effects. Commanders can control, seize, or secure an essential facility or structure by using precision munitions, electronic disruption of communications, or SOF and conventional ground forces. To gain this advantage, commanders will rely more on the expertise of Army engineer and civil affairs units; local urban engineers, planners, and public works employees; and others with infrastructure-specific expertise. After understanding the technical aspects of the area's systems and subsystems, commanders can then develop the best course of action.

A SYSTEM OF SYSTEMS

2-61. Hundreds of systems may exist. Each system has a critical role in the smooth functioning of the urban area. Simple or complex, all systems fit into six broad categories (see figure 2-11). Commanders should analyze key facilities in each category and determine their role and importance throughout all phases of the urban operation. As there is much overlap between infrastructure systems, this analysis considers each system individually and in relation to others to determine an appropriate course of action toward it.

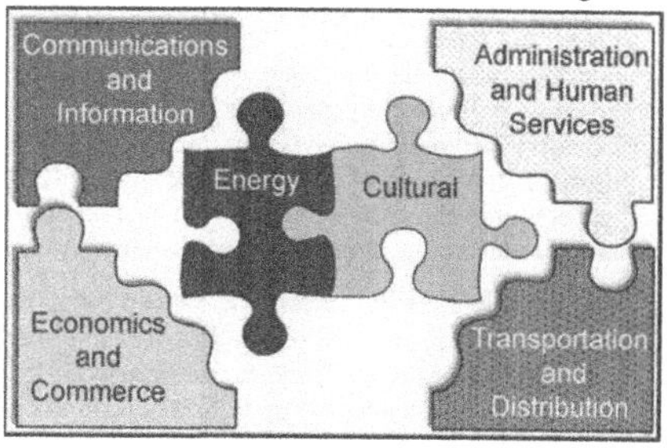

Figure 2-11. Urban infrastructure

A COMBINATION OF STRUCTURES AND PEOPLE

2-62. As depicted earlier in figure 2-1, each element of the infrastructure consists of both a physical (terrain) and human component. For example, the physical component of the electrical segment of the energy infrastructure consists of power stations, substations, a distribution network of lines and wires, and necessary vehicles and repair supplies and equipment. The human component of this same segment consists of the supervisors, engineers, linemen, electricians, and others who operate the system. Commanders must understand and recognize the physical and human components in their assessments.

POTENTIAL IMPACT ON FUTURE OPERATIONS

2-63. Destroying or incapacitating of any of these elements may impact future operations and inhabitants of the urban area. Destroying urban infrastructure during initial phases of an operation may require commanders to assume responsibility for repair, maintenance and clean up, and operation of those same facilities later. Although exceptions will exist, commanders cannot destroy or significantly damage the infrastructure of a foreign urban center during operations and expect the population to remain friendly to U.S. or allied forces. On the other hand, early repair or restoration of critical or essential infrastructure may improve civil-military relations, speed transition back to competent civilian authorities, and, overall, aid in successful mission accomplishment. Still, support from the urban society (albeit of increased importance in UO) is only one factor that commanders weigh while developing appropriate courses of action.

RESOURCE INTENSIVE

2-64. Requirements to protect, restore, or maintain critical infrastructure may divert substantial amounts of resources and manpower needed elsewhere and place additional constraints on subordinate commanders. Civilian infrastructure is often more difficult to secure and defend than military infrastructure. The potentially large and sprawling nature of many systems (such as water, power, transportation, communications, and government), make their protection a challenge. Yet, the infrastructure of an urban area may provide commanders with essential logistics and support. Therefore, the initial expenditure of time and other resources may be necessary to support concurrent or future operations. Legal considerations, however, may affect using the infrastructure and acquiring the urban area's goods and services. Commanders, their staffs, and subordinates (often down to the individual soldier) must know their limits concerning Army authority to commandeer civilian supplies or equipment to facilitate mission accomplishment (see the legal support discussion in Chapter 9). In stability and civil support operations, the safeguard or restoration of critical urban infrastructure for military or civilian use may be a decisive point in the overall operation.

2-65. Keys to understanding the magnitude of the resources and manpower required to restore the infrastructure are an initial infrastructure assessment and, as soon as practical afterward, a detailed infrastructure survey. An initial assessment provides the commander immediate feedback concerning the status of basic services needed to meet the urgent needs of the urban population. The systems assessed are based on mission, enemy, terrain and weather, troops and support available, time available, civil considerations, and the commander's vision of the overall end state. The infrastructure assessment, while typically performed by engineers, may be accomplished by, or in conjunction with, others with sufficient expertise to provide the type and quality of information required. These others may include civil affairs, medical, and chemical personnel. Those tasked with this assessment should routinely consult other Army and coalition forces and governmental and nongovernmental agencies currently operating in the urban area as well as the urban civilian leadership for their informed input.

2-66. While an infrastructure assessment functions to support the resolution of immediate challenges to urban reconstruction and restoration, it also provides the initial basis for determining the conditions for successful transition. However, commanders and planners must continually expand and refine their understanding. As a necessary follow-on, commanders initiate a detailed infrastructure survey. This survey is normally conducted by U.S. Army Corps of Engineers personnel assigned to forward engineer support teams. As with the assessment, the commander should incorporate other technical specialty personnel in the survey team to enhance the quality and accuracy of the product (see FM 3-34.250).

COMMUNICATIONS AND INFORMATION

2-67. This system is comprised of the facilities and the formal and informal means to transmit information and data from place to place. Understanding communication and information infrastructure of an urban area is important because it ultimately controls the flow of information to the population and the enemy. It includes—

- Telecommunications, such as telephone (to include wireless), telegraph, radio, television, and computer systems.
- Police, fire, and rescue communications systems.
- Public address, loudspeaker, and emergency alert systems.
- The postal system.
- Newspapers, magazines, billboards and posters, banners, graffiti, and other forms of print media.
- The informal human interaction that conveys information such as messengers, open-air speeches and protests, and everyday conversations.
- Other inventive informal means such as burning tires and honking horns.

2-68. Perhaps more than any other element of the infrastructure, communications and information link all the other elements in an interdependent "system of systems." It is a critical enabler that helps coordinate, organize, and manage urban activities and influence and control the urban society. Army commanders are acutely aware of the impact that a loss or degradation in communications has on their own operations. The urban environment experiences similar impacts to communication failures; however, urban governments and administrations are generally less prepared to deal with a collapsed communications and information infrastructure than are trained Army forces.

2-69. Militarily, a functioning urban communications and information system can serve as an alternate for both friendly and threat forces and can be easily secured with civilian, off-the-shelf technologies. Threats may make use of commercial systems intertwined with legitimate civilian users, making it unpalatable to prevent use of these assets. Forces can also use these systems to influence public opinion, gain intelligence information, support deception efforts, or otherwise support IO.

Increasing Impact of Computers

2-70. In many urban areas, computers link other elements of the urban infrastructure. They link functions and systems in the urban area and connect the area to other parts of the world. This latter aspect creates important implications for commanders of a major operation. Operations involving this cybernetic function may produce undesirable effects on a greater scale than initially intended. For example, commanders may be able to close or obstruct an urban area's banking system; however, this system may impact the international monetary exchange with unwanted or even unknown effects. The authority to conduct these types of IO will often be retained at the strategic level.

Pervasive Media

2-71. The media is central to the communications and information infrastructure and a critical operational concern. Compared to other operational environments (jungles, deserts, mountains, and cold weather areas), the media has more access to urban operations. This is due largely to airports, sea and river ports, and major road networks; ready access to power sources and telecommunications facilities; as well as access to existing local media structures. Hence, media presence may be pervasive and IO even more critical to success in UO than operations in many other environments.

A Complex Relationship

2-72. A complex relationship exists among information, the public, and policy formulation. Although the degree and manner in which public opinion shapes government policy are difficult to accurately determine, negative visual images of military operations presented by the media can change political objectives and, subsequently, military objectives. As important, media reporting can influence civilian activity in an urban AO to either the advantage or disadvantage of the commander.

Induce Cooperation Through Credibility

2-73. Commanders do not control the media; however, they monitor the flow of information that the news media receives and subsequently reports. Consequently, commanders should plan and execute PA operations that will induce cooperation between the media and Army forces. Successful relations between urban Army forces and the news media evolve from regular interaction based on credibility and trust. More information is usually better than less, except when the release of such information may jeopardize security and the success of the operations and threaten the safety of Soldiers. However, commanders cannot simply withhold information to protect the command from embarrassment. They consider media interests as part of the normal planning process and work to ensure that information presented to the news media is accurate, timely, and consistent with operations security. Since the media will likely arrive in the urban area before the conduct of operations, early deployment of PA assets may be critical. Commanders should synchronize PA activities with CMO and PSYOP. Such action eliminates duplicated effort and ensures a unity of purpose consistent with the IO concept of support (see Chapter 4).

2-74. Failure to provide sufficient information can hamper a commander's ability to conduct the mission. Commanders cannot refuse to deal with particular news media because they consistently report a negative image of Army forces and operations. Poor relationships with any media can result in inaccurate and even biased reporting. Such reporting can cause a public reaction that influences the ability to achieve operational objectives. During the Russian 1994-95 battle against Chechen separatists in Grozny, for example, the Russian military refused to communicate with reporters. The media reported primarily from the perspective of the Chechen rebels. This encouraged both local and international support for the rebels. It also allowed the Chechens, who lacked sophisticated information systems, to use the media to broadcast operational guidance to their forces. (During their second Chechnya campaign of 1999-2000, Russia learned this lesson well and the Russian view of the war dominated domestic public opinion.) On the other hand, successfully engaging the media can serve as a force multiplier. The Army's open and responsive interaction with the media during peacekeeping operations in Bosnian urban areas helped to explain the challenges and successes of Army forces in the Balkans to the public. This helped maintain domestic, international, and local political support for NATO operations and, with a successful command information program, helped maintain Soldiers' morale.

TRANSPORTATION AND DISTRIBUTION

2-75. This element of the infrastructure consists of—

- Networked highways and railways to include bridges, subways and tunnels, underpasses and overpasses, ferries, and fords.
- Ports, harbors, and inland waterways.
- Airports, seaplane stations, and heliports.
- Mass transit.
- Cableways and tramways.
- Transport companies and delivery services that facilitate the movement of supplies, equipment, and people.

Similar to communications and information, this facet provides the physical link to all other elements of the infrastructure.

2-76. Army forces deploying into a theater of operations depend on ports and airfields; seizure and protection of these critical transportation nodes may impact the projection of combat power. Once in theater, transportation and distribution systems in the urban area can contribute greatly to the movement of forces, maneuver, and logistic operations throughout the entire AO. Control of decisive points in this infrastructure may be important to the military operation and to the normal functioning of the urban area (and surrounding rural areas). Supplies traveling through the transportation and distribution system may be military-specific supplies (such as ammunition and repair parts) and supplies for both the military and urban population (such as food, medicine, oil, and gas). The system may also support the movement of military forces and the urban area's population (for which it was designed). Therefore, commanders of a major operation may have to develop innovative methods that limit the transit of threat supplies and reinforcements while facilitating the movement of their own resources and those of civilians. This last consideration attempts to minimize hardship and promote normalcy in the urban area and will increase in significance as the need for legitimacy increases.

2-77. Most urban areas (particularly in developing countries) have two forms of transportation and distribution systems that exist simultaneously: a formal system and an informal or paratransit system. Large organizations, bureaucracy, imported technology, scheduled services, and fixed fares or rates characterize formal systems. Low barriers to entry; family and individual entrepreneur organizations; adapted technology; flexible routes, destinations, and times of service; and negotiated prices characterize the informal system. The informal system is more decentralized and covers a much greater portion of the urban area than the formal system. The informal transportation and distribution system often includes a waterborne element, is more likely to function through turbulence and conflict, and can extend hundreds of kilometers beyond the urban area. Accordingly, commanders should understand both systems to establish effective movement control.

ENERGY

2-78. The energy system provides the power to run the urban area. It consists of the industries and facilities that produce, store, and distribute electricity, coal, oil, wood, and natural gas. This area also encompasses alternate energy sources, such as nuclear, solar, hydroelectric, and geothermal power. Energy is needed for industrial production and is therefore vital to economics and commerce. Among many other things, this system also provides the fuels to heat, cool, and light homes and hospitals, cook and preserve food, power communications, and run the transportation necessary to move people and their supplies throughout the urban area. Loss of an important energy source such as electricity or gasoline, especially for those accustomed to having it, will become an immense area of discontent that the commander of a major urban operation will need to quickly address. Therefore some threats, particularly terrorists and insurgents, may actively target this element of the urban area's infrastructure to erode support for civilian authorities and Army forces.

2-79. Sources of energy may be tens or hundreds of miles away from the urban area itself. Therefore, commanders may exert control without applying combat power directly to the urban area itself by controlling or destroying the source (power generation or refinement plant) or the method of distribution (pipelines or power lines). With electrical energy that cannot be stored in any sizable amount, the latter may be the best means as most major urban areas receive this energy from more than one source in a network of power grids. However, control may be as simple as securing a power station or plant and turning off switches or removing a vital component that could later be restored. On the other hand, lengthy pipelines and power lines may compound security and protection of this element of the infrastructure.

2-80. The number of nations that have invested in nuclear power and nuclear research is increasing. With this increase, the potential for Army forces to operate in urban areas that include (or are near) these facilities also increases. Damage to one of these facilities and potential radiation hazards will present special challenges to commanders of a major operation. To safeguard friendly forces and civilians, commanders will need to employ a blend of peacetime and tactical nuclear contamination avoidance principles (see FM 3-11.14).

ECONOMICS AND COMMERCE

2-81. This system encompasses—

- Business and financial centers to include stores, shops, restaurants, hotels, marketplaces, banks, trading centers, and business offices.
- Recreational facilities such as amusement parks, golf courses, and stadiums.
- Outlying industrial, mineral, and agricultural features to include strip malls, farms, food processing and storage centers, manufacturing plants, mines, and mills.

2-82. An essential aspect of this area during operations may be the political sensitivity of U.S. or allied industries investing and operating in a foreign country, particularly during stability operations. An enemy or a disgruntled civilian population may attack or disrupt commercial activities as a political statement against the United States or our allies. Food production facilities also may assist commanders in Army food services and may be an essential concern during relief operations. During long-term stability operations, visible, material, and tangible economic progress consisting of the creation or restoration (and protection) of businesses, agriculture, and overall jobs will often be critical to—

- Generating or maintaining the urban population's support to Army forces and operations.
- Reducing support to threat forces and operations to include eliminating civilians as a potential manpower pool for insurgent or terrorist organizations and activities.
- Lowering other hostile civilian activities such as protests and riots.
- Transitioning the urban area back to legitimate civilian responsibility and control.

2-83. This element of the infrastructure also consists of the production and storage of toxic industrial chemicals used in agriculture (insecticides, herbicides, and fertilizers), manufacturing, cleaning, and research (to include biological agents). Fertilizer plants may be of specific concern as they contribute to providing a key material in terrorist and insurgent bomb-making activities. A thorough analysis of this element of the infrastructure may also be essential to understanding how urban insurgencies are funded and supported. This helps commanders to understand the true organization of the insurgency as well as to suggest methods to isolate insurgents from their economic or financial support. In their overall assessment of this area of the infrastructure, commanders should also consider the activities and influence of criminal organizations or elements.

ADMINISTRATION AND HUMAN SERVICES

2-84. This wide-ranging system covers urban administrative organizations and service functions concerned with an urban area's public governance, health, safety, and welfare. Together, it encompasses—

- Governmental services that include embassies and diplomatic organizations.
- Activities that manage vital records, such as birth certificates and deeds.
- The judicial system.
- Hospitals and other medical services and facilities.
- Public housing and shelter.
- Water supply systems.
- Waste and hazardous material storage and processing facilities.
- Emergency and first-responder services such as police, fire, and rescue.
- Prisons.
- Welfare and social service systems.

CULTURAL

2-85. This system encompasses many organizations and structures that provide the urban populace with its social identity and reflect its culture. (This infrastructure system overlaps with many recreational facilities included under the economics and commerce infrastructure. For example, an urban society may radically follow soccer matches and teams. Hence, soccer stadiums relate to the society's cultural infrastructure.) Some of these facilities, particularly religious structures, will be protected targets and others may require security and law enforcement protection from looting and pilferage. However, commanders will need to quickly educate, inform, and continually remind the urban populace (and the media) that cultural infrastructure may lose its protected status when used by threats for military purposes. Cultural infrastructure may include—

- Religious organizations, places of worship, and shrines.
- Schools and universities.
- Museums and archeological sites.
- Historic monuments.
- Libraries.
- Theaters.

Restoring and Protecting Essential Services

2-86. Losing the support of essential elements of the infrastructure will have an immediate, destabilizing, and life-threatening impact on the inhabitants of the urban area. In stability and civil support operations, numerous parts of the administrative and human services and energy infrastructure often rise to critical importance before all other elements. Again, however, complete restoration of these essential services is often a lengthy, resource-intensive civil-military operation. (Following the end of major combat operations of OPERATION IRAQI FREEDOM, most units developed one or more of their logical lines of operations oriented along the restoration or improvement of urban infrastructure and essential services. An acronym used by many units to focus and track critical activities within this line of operation was sewer, water, electricity, and trash (SWET). Later, other units modified this acronym to SWEAT and then SWEAT-MS to include concerns for restoring academics [or schools], revitalizing medical facilities, and establishing security [police and host-nation security forces].) Of critical importance will be a simultaneous IO campaign that includes efforts to help ensure that the urban population develops realistic expectations about Army abilities to restore their essential urban infrastructure.

Understanding the Urban Environment: Paris – 1944

The summer of 1944 confronted German General Dietrich von Choltitz with a dilemma. As military commander of greater Paris, he was to eliminate French Resistance internal to the city while defending against approaching Allied units, missions for which he had insufficient forces. (Notably, General Eisenhower, the commander of the Allied forces, wanted to bypass Paris to sustain the offensive. Seizing the French capital would task his forces with the support of tens of thousands of civilians. Eisenhower was nevertheless ordered to capture the city. A political decision resulted in civilian assistance taking precedence over combat operations.)

Choltitz's situation was further complicated by Hitler's demand that he destroy the city, an action the general saw as needlessly destructive (and infeasible given his scant resources). Choltitz's seniors directed the preparation, and later the destruction, of Paris's 45 Seine River bridges. They were the only remaining crossing points over that waterway given Allied bombing of others outside the French capital. Premature destruction would trap German forces defending to their north, a second-order effect that Choltitz used to justify his disobedience of orders demanding the bridges' demolition.

The German general also recognized that some mission-critical elements were part of Paris's social rather than physical infrastructure: the leadership of the various resistance groups and the relationships between them. Choltitz understood that he lacked resources to defeat the many separate factions; he therefore chose the unorthodox (asymmetric) approach of accepting an intermediary's offer of a truce with these groups. Such an agreement provided some measure of the stability needed while Choltitz awaited promised reinforcements. Further, he realized that the resistance factions were by no means united in their goals. Communist elements sought a much different end than those looking toward a de Gaulle-led postwar government. A truce thus set the French Communists (who sought an uprising so as to legitimize their claims to power) against others trying to buy time until Allied forces arrived, forces that included Free French units supportive of de Gaulle.

Although his defense of the capital failed, Choltitz succeeded in harboring his available resources, reducing the effectiveness of the resistance organizations fighting his soldiers, and maintaining withdrawal routes for units north of the Seine. The German commander's analysis in support of these efforts was effective in part because of his insightful (1) identification of critical points that included elements of terrain, citizenry, and infrastructure; (2) understanding of the relationships between these parts; and (3) use of an asymmetric approach to address his lack of sufficient force to otherwise handle the urban densities that challenged him.

This page intentionally left blank.

Chapter 3

Understanding the Urban Threat

... [T]he United States could be forced to intervene in unexpected crises against opponents with a wide range of capabilities. Moreover, these interventions may take place in distant regions where urban environments, other complex terrain, and varied climatic conditions present major operational challenges.

Quadrennial Defense Review Report, 30 September 2001

As the strategic environment has become less stable, more uncertain, and more dangerous, Army forces must be trained and ready to address persistent and evolving urban threats. These threats range from regional conventional military forces, paramilitary forces, guerrillas, and insurgents to terrorists, criminal groups, and angry crowds. These threats can hide in plain sight and become indistinguishable from the noncombatant urban population that may help shield, protect, and sustain them. Although uncertain about events, Army forces can be clear about trends. Increasingly, the Army will face threats that severely differ in doctrine, organization, and equipment, are skilled at developing and adapting techniques to counter Army tactics, techniques, and procedures (TTP), and can fully interact with the three other components of the urban battlefield—terrain, society, and infrastructure. In urban operations, commanders must broaden their concept of the threat to include natural disasters, hunger and starvation, and rampant disease. Further, commanders must plan to contend with many passive urban threats, such as psychological illnesses and toxic industrial materials (TIM). These threats may be found in isolation, but most likely commanders will encounter them in various combinations. Moreover, each new threat will pose a different combination and likely have new capabilities that previous opponents lacked.

ASYMMETRICAL AND ADAPTIVE

3-1. An emphasis on asymmetry to offset U.S. military capabilities has emerged as a significant trend among potential threats and become an integral part of threat objectives and tactics (discussed below). Though, this is not to imply that all future urban threats will fight and operate asymmetrically. Some may be asymmetrical; others may not. Future threats may be a symmetric conventional military; others may be paramilitary, insurgent, or even a nonmilitary threat such as a criminal organization. Asymmetry results when one opponent has dissimilar capabilities—values, organization, training, or equipment—that the other cannot immediately counter. It is not a new concept. It naturally evolves from a sound mission, enemy, terrain and weather, troops and support available, time available, civil considerations (METT TC) analysis by an intelligent, freethinking, and adaptive threat. These asymmetric approaches will include the most advanced, commercially-available technology innovatively applied and mixed with crude, simple, and unsophisticated weapons and TTP. The success of these approaches often depends upon the predictability of the forces on which they are employed. Thus, Army commanders must consider whether their operations and activities lend themselves to pattern analysis by threat forces and, if so, they may need to vary their battle rhythm and take other steps necessary to become less predictable.

WEAPONS OF MASS DESTRUCTION

3-2. A chief asymmetric means of engaging the national power of the United States is to employ weapons of mass destruction (WMD) against the United States or its allies. These weapons can be used against military forces and include high-yield explosives as well as nuclear, biological, and chemical weapons. Operations in urban areas may require concentrating forces and may create a lucrative target for a threat that possesses fewer numbers and less equipment.

3-3. A threat's WMD use will adversely affect the Army's abilities to conduct urban operations (UO) to various degrees. For example, the intervening structures and the effects of urban microclimates complicate the ability to detect and identify radiological, chemical, or biological attacks from a standoff distance. Also, the individual Soldier's ability to recognize his leaders, understand oral and visual commands, and operate increasingly sophisticated equipment is difficult when wearing protective clothing and equipment—particularly if his training proficiency is low. Despite the increased challenges and complexity, Army forces have the best (or better) training and equipment necessary to respond to such an attack compared with most armies around the world or with any civilian organization.

3-4. Although initial casualties could be high, the public can accept military casualties before those of civilians. Therefore, threats may gain an initial tactical advantage but would achieve less asymmetric benefit by directly attacking Army forces. Instead, they may attempt to achieve a strategic advantage by employing WMD against U.S. or allied civilian populations. In doing so, threats hope to use political sensitivity to high civilian casualties to reduce popular support for the United States or its allies. The chance of these attacks occurring in an urban area increases because—

- The area facilitates weapon effects and camouflages delivery means.
- The dense civilian population ensures a high casualty rate.
- The attack (or even the threat of attack) often will receive more publicity and public attention.
- The urban infrastructure is especially vulnerable to WMD, (particularly the systems of the economics and commerce infrastructure located in large urban areas) and may have far-reaching national and global effects.

THREAT OBJECTIVES

3-5. The threat will seek to achieve several key objectives when opposing Army forces operating in an urban environment (see figure 3-1). These objectives focus more on how a threat might fight in an urban area rather than specifically who the threat might be or in what region of the world the conflict might occur. These objectives are more easily achieved in an urban environment due to—

- The high costs in time, material, and manpower involved in UO.
- The limiting effects of urban areas on many technological advantages.
- The proximity of airfields and ports to urban areas.
- The potential moral dilemmas created by exposing numerous civilians to harm or injury.
- The ability of the media to observe and report the threat's version of events.

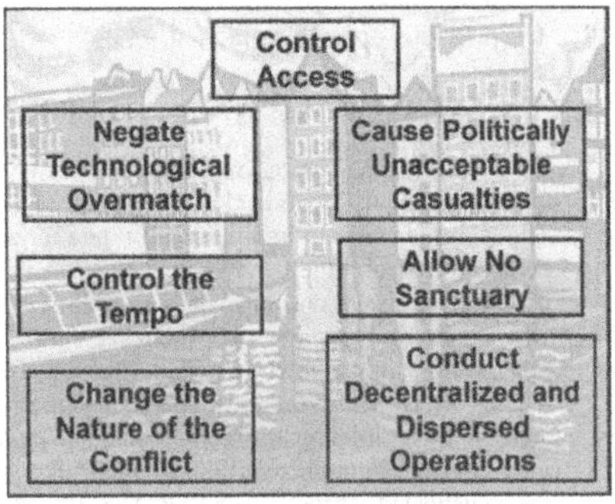

Figure 3-1. Threat objectives

3-6. These objectives complement and overlap each other; however, at their core is the threat need to defeat an enemy of superior numbers, technology, or both.

CONTROL ACCESS

> *... Iraq made no direct effort to impede the buildup in the theater. Planners should consider what might have happened if Iraq had attempted a strategy based on denying access to the region. Planners might also wonder what the outcome would have been if Iraq had attacked US forces in Kuwait before they were ready for the running start. The point is that the conditions in [US Central Command] in 2003 are unlikely to be replicated elsewhere.*
>
> *On Point: The United States Army in Operation IRAQI FREEDOM*

3-7. The Army may not be located where future conflicts are fought. Thus, the Army maintains the ability to rapidly project and sustain combat power over long distances and time spans. This capability demands that Army forces quickly gain and maintain control of seaports or aerial ports of embarkation or debarkation, particularly where the density of U.S. basing and en route infrastructure is low. Commanders gain control of these ports by unopposed (assisted or unassisted) or forcible entry operations. In either case, these phased-entry operations may present potential vulnerabilities, particularly—

- Unsuitable composition of initial or early entry forces lacking necessary combat power for immediate decisive operations.
- Initial command and control difficulties and an immature situational understanding.
- Weak initial port or base defenses without the necessary integration of naval and air component forces.
- Lack of developed logistic support.

3-8. Consequently, threats may attack during initial force projection operations to oppose, delay, disrupt, or otherwise manipulate the build-up of essential combat power into a theater of operations. These attacks may occur anywhere deploying Army forces are located, at overseas bases, at home stations, and even in military communities. Increasingly, deployment facilities such as airfields and ports exist as integral components of urban areas. Threats will invariably use the complex and concealing nature of these urban areas, coupled with the vulnerabilities, to create favorable conditions for their attacks.

NEGATE TECHNOLOGICAL OVERMATCH

3-9. Threats will always strive to force engagements at a time and place most advantageous to them. They may locate military forces and vital military capabilities in urban areas to achieve sanctuary from the effects of Army capabilities and make Army forces and systems more vulnerable to less-sophisticated weapons.

3-10. The clutter of the physical structures, electromagnetic radiation, and population diminishes Army capabilities. This clutter makes it difficult for Army forces to acquire and effectively engage targets at long ranges. In urban areas, the terrain often allows a threat to operate in closer proximity to friendly forces. Therefore, the threat may "hug" friendly forces to avoid the effects of high-firepower standoff weapon systems and degrade their ability to gain or maintain a thorough common operational picture. Additionally, this threat tactic attempts to inhibit friendly commanders from employing some weapon systems and munitions for fear of fratricide. Threats will also seek to hide and operate in culturally sensitive areas and sites, such as places of worship and schools, to establish temporary safe havens and refuge from Army forces operating within the urban environment.

CONTROL THE TEMPO

3-11. Threats will try to achieve a decisive advantage by setting and controlling the tempo. To prevent the Army's entry into theater, threats may try to create a high operational tempo to take advantage of the inherent weaknesses in power projection operations outlined earlier. As other efforts deny entry, threats may seize the initiative, achieve surprise, and exploit the tempo differential by attacking with heavy conventional forces potentially possessing greater firepower and more rapid ground mobility than the Army's initial-entry forces.

3-12. If they cannot deny entry or end the conflict quickly, threats may use any preparations made in the initial high-tempo period to prolong the event, aiming to degrade U.S. or allied commitment. (Some threats may conduct operationally planning in terms of decades rather than years.) The complex nature of the urban environment slows operations conducted in, around, and over these areas. Threats may maximize this characteristic by fighting (conventionally or unconventionally) for key urban complexes and infrastructure, forcing friendly forces to operate within these areas. If Army operations focus on one or more urban areas, the overall campaign slows. However, even when UO make up only one component of a much larger campaign, they may consume valuable resources needed for other operations and delay the entire campaign. (See also the discussion of Tempo in Chapter 7.)

Tempo

The battle for Aachen, Germany, in the fall of 1944, developed during the U.S. First Army's offensive to breach the Westwall fortifications. Aachen, the ancient capital of Charlemagne, had symbolic political and psychological significance for the Germans and Americans. Furthermore, it was the first city on German soil to face an assault by the Allies. Consequently, the symbolic importance of this first major battle in Germany ensured bitter resistance against American attackers. The Germans surrendered only after the city was destroyed. Expected to take a few days, instead, the battle took weeks. Although the Army had achieved a clear tactical victory, the German defense of Aachen cost the First Army valuable time and resources, and delayed the planned attack to the Rhine River.

Change the Nature of the Conflict

3-13. Threats may attempt to change the fundamental nature of the urban conflict to exploit ambiguous or tenuous political-military objectives. Many nations must gain and maintain domestic popular support to use their armies for political objectives. The threat may attempt to change the nature of the conflict by modifying its strategy and tactics, the environment, or any combination, ultimately hoping to reduce friendly popular support. For example, introducing an urban terrorist threat to U.S. civilians or Soldiers not directly engaged in operations changes the nature of the conflict. This type of threat may not have been an initial consideration, and this change may reverse public support for the operation. Another example, growing U.S. coalition combat power may cause the threat to switch from open maneuver warfare to urban insurgency operations to avoid decisive combat with superior forces and achieve a stalemate. Originally expecting a quick solution or victory, the political leadership may now envision a longer deployment with less chance of lasting success.

Cause Politically Unacceptable Casualties

3-14. Threat forces may gain an advantage against superior friendly forces by capitalizing on a perceived weakness of many Western nations: the inability to endure continuous losses or casualties for other than vital national interests or losses for which they are psychologically unprepared. A secondary U.S. interest may equate to national survival on the part of a threat. Therefore, the threat (particularly with fanatical leadership) may willingly sacrifice excessive amounts of money, equipment, and people (soldiers and civilians) to achieve victory. Threats may attempt to weaken U.S. resolve and national will to sustain the deployment or conflict by inflicting highly visible, embarrassing, and if possible, large losses on Army forces, even at the cost of similar losses to themselves. Many threat forces will use UO to inflict mass casualties and destroy premier Army weapon and information systems. The physical characteristics of the urban environment support these ambush techniques. Light infantry or insurgents with readily obtainable, hand-held antiarmor weapons can effectively attack armored vehicles and helicopters, no matter how sophisticated, in an urban area.

Allow No Sanctuary

3-15. Threats will attempt to deny Army forces safe haven anytime and anywhere. Terrorism and insurgency may be some of the tactics used to deny sanctuary to Army forces. They will attack Army forces anywhere, particularly while operating in urban areas where the fear from being attacked from any quarter is often greater. Threats may be or employ state-sponsored or independent terrorists, well equipped and motivated to accomplish their assigned missions.

3-16. Military buildings, facilities, and installations in urban areas are particularly vulnerable to high-yield explosive munitions as well as other clever means to create large explosions. The close-in nature of urban areas, large populations, and high volume of vehicle traffic provide a good environment for target reconnaissance, explosives positioning (conventional or high-yield), and cover for an attack. These attacks will likely be preceded by extensive, careful reconnaissance, necessitating a solid friendly counterterrorism, counterinsurgency, and counterintelligence effort.

Conduct Decentralized and Dispersed Operations

3-17. To a certain extent, decentralized and dispersed operations are an integral part of all threat objectives. However, this concept warrants separate emphasis as an objective since threat forces will likely place great significance on it on future urban battlefields. Both dispersed and decentralized approaches seek to reduce threat vulnerabilities to air power and precision-guided munitions (PGM) while increasing their agility, flexibility, and overall maneuverability in an urban environment.

3-18. Urban terrain tends to fragment and separate forces that operate in it. Threat forces recognize this characteristic, accept it, and make it work to their advantage. They conduct operations from dispersed urban locations to reduce their vulnerability to friendly decisive operations and massed firepower. Although separated, threat forces will attempt to retain the ability to assemble and mass quickly so to strike as opportunities present themselves. Once threat forces complete the operation, they will return to separate locations to avoid potential counterattack. The fluidity and seemingly disjointed appearance of these UO threats will challenge friendly efforts to conduct templating and pattern analysis. Ambushes (air and ground) will be used to deny friendly ground and air reconnaissance of their dispersed locations.

3-19. Dispersed operations normally depend on good command and control to achieve synchronization and massed effects. Threat forces also understand the debilitating effects of the urban terrain on communications and the execution of operations. When they cannot mass their forces or effects, they will depend on decentralized operations to achieve their objectives. They will operate autonomously, guided only by a higher authority's purpose and intent. These operations make them even less vulnerable to massed attacks and PGM as smaller threat forces do not present an objective or target that will allow friendly decisive operations. Again, pattern analysis and templating, although essential, will be extremely difficult. This objective may prolongs the conflict but is central to achieving the other threat objectives.

3-20. When the threat employs decentralized operations, is effectively isolated, or both, Army and coalition forces are much more likely to be able to share lessons learned and emergent TTP than a dispersed threat. Threat isolation, either imposed by Army forces or as a result of the threat's desire to remain concealed from Army intelligence, surveillance, and reconnaissance (ISR) capabilities, reduces the threat's ability to freely and easily exchange information and quickly adapt. As learning organizations, this allows Army forces to develop dominance over a threat's weapons and tactics for longer, albeit temporary, periods of time. This ability to rapidly train and learn is an advantage the Army and coalition forces must maintain while denying the same to future threats.

URBAN THREAT TACTICS

3-21. Urban areas provide a potentially casualty-producing and stress-inducing environment ideally suited for using specific urban tactics. Moreover, urban areas can provide threats with an unmatched degree of cover and concealment from friendly ISR and firepower systems. Understanding how these tactics will be used often requires a concurrent understanding of the threat's culture, customs, language, and philosophical ideals. While active urban threats may vary widely, many techniques will be common to all.

Figure 3-2 outlines a set of tactics available to potential threats opposing mission accomplishment in urban areas. Army forces may use many of these tactics, except those that violate the law, ethics, and morals, to defeat urban threats. Moreover, using asymmetric means is not the sole domain of the threat. Army commanders can also leverage capabilities, create conditions, and plan operations to develop their own asymmetric advantages to accomplish the mission.

USE THE POPULATION TO ADVANTAGE

3-22. Many urban areas may be too large to evacuate completely (if at all). Even if desirable, a military force may have no place to safeguard and secure the inhabitants. Therefore, future UO may see large segments of the populace remain. Offensive and defensive operations may be constrained not only by the terrain, but also by the presence of many civilians. Army forces involved in urban stability and civil support operations will certainly conduct missions in and amongst the residents. These residents may restrict operations and, when gathered in large numbers, may (even without initial hostile intent) present a critical force protection issue for the commander.

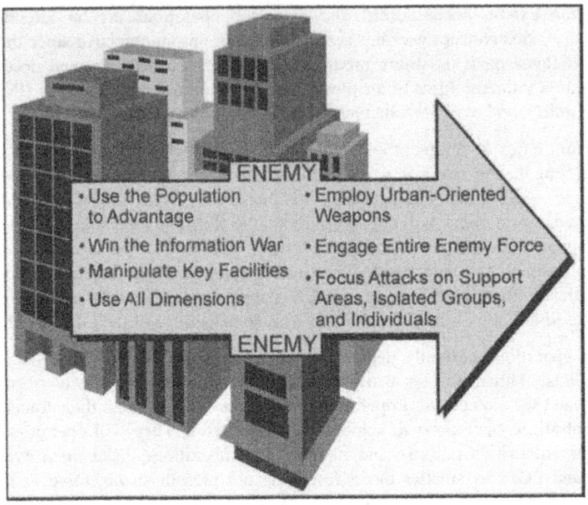

Figure 3-2. Urban threat tactics

Use as Key Terrain and Concealment

3-23. From the threat standpoint, the populace is similar to key terrain: the side that manages it best has an advantage. Threat forces may gain this advantage by using civilians as camouflage, concealment, and a means of deception. Guerrilla and terrorist elements may look no different from any other member of the community. Many foreign conventional and paramilitary troops—in addition to terrorists and insurgents—often have a "civilian" look. Western military forces originally adopted the clean-shaven and close-cut hair standards to combat disease and infection, but future opponents may not adhere to those standards. Instead, they may adopt grooming standards, civilian-looking clothing, and other "nonmilitary" characteristics to make themselves indistinguishable from the civilians. To avoid defeat or capture, threats will often throw down their weapons and attempt to blend into the urban population. Therefore, commanders must adopt effective populace and resource control measures to limit, identify, and track threat movements through the urban area of operation.

Identifying Threats from Noncombatants

During Russia's 1994-95 conflict with Chechnya, Russian forces had difficulty identifying Chechen guerrilla forces from Grozny's noncombatant population. Because their dress and ethnic appearance was identical to that of the urban populace, Chechen soldiers could freely walk around the city, suddenly disappear, and then abruptly reappear firing their weapons from basements, windows, or dark alleyways. To distinguish fighters from peaceful city dwellers, Russian forces began looking at men's shoulders to see if they were bruised (from firing weapons) and their forearms to see if there was burned hair or flesh (from the extraction of cartridges). They closely examined their clothing and smelled for gunpowder residue. To identify a Chechen artilleryman, Russian soldiers checked for glossy spots left by artillery and mortar rounds on the bends and cuffs of sleeves. They also turned pockets inside out to check for a shiny, silvery-leaden hue indicating the former presence of small arms ammunition. Russian forces also recognized a grenade launcher operator or mortar man from fibers and crumpled pieces of gun cotton on their clothing. U.S. Army commanders may need to develop similar, imaginative techniques and rapidly integrate technological means to identify and track threat forces such as—

- Military working dogs, ion spray, and electronic vapor tracers to locate explosives (or their former presence).
- Biometrics—measurable biological characteristics—such as fingerprints, voiceprints, facial scans, and retina scans.
- Electronic and chemical tagging devices.
- Monitoring hospitals for injuries that indicate participation in combat activities. For example, improper use of rocket-propelled grenade launchers may result in significant burns to the lower part of the body due to the back blast.

Gain Cover, Protection, and Increased Mobility

3-24. Threat forces may attempt to gain cover by using the urban inhabitants as human shields. With this increase in protection, they simultaneously increase their mobility. They recognize the Army's focus on developing and applying rules of engagement (ROE). They will take advantage of the restraining effects of international law and the Army ethical values to enhance their mobility in proximity to friendly positions. Knowing the Army's reluctance to cause noncombatant casualties and collateral damage, threats may operate in areas containing civilians and essential facilities to restrict the Army's use of massed or nonprecision firepower. For example, they may use school buses and ambulances to transport fighters and equipment about the urban battlefield. They may also employ "rent-a-crowds"—civilians paid or incited to demonstrate against military forces—armed only with sticks, stones, and Molotov cocktails (a potential asymmetric challenge). Threats can also hide military equipment within sensitive infrastructure—for example, caching weapons in places of worship, schools, or medical facilities. As Army and coalition forces potentially respond by targeting and firing upon these locations, threats will make use of these reactions for their possible propaganda value.

Make Moral Responsibilities a Weakness

3-25. Depending on their successes, threats may use these tactics and skillful information operations that attack national will and coalition sensitivities in an attempt to force the Army to establish more restrictive ROE. For example, a teenaged fighter coerced or enticed to fire a rifle or rocket-propelled grenade launcher can be quickly stripped of his weapon once dead and displayed at the local hospital as a child victim of coalition forces. Threat forces may also take advantage of the Army's moral responsibilities. By herding refugees into friendly controlled areas, threat forces try to make the civilians a burden on the Army's logistic and security resources. Similarly, some threats may attack to kill NGOs and force the remainder of them out of the area to achieve the same burden effect. At the same time, these threats hope to cause the population to lose confidence in the Army's (and potentially the government's) ability to protect and provide them essential services.

3-26. Threat forces, on the other hand, may not abide by international agreements, such as the Geneva conventions. Threats may not take prisoners unless they can be ransomed or made part of a local prisoner exchange. They may even execute friendly prisoners in front of the media to show their "strength" and, more importantly, to cause friendly forces to overreact and lose their legitimacy. Threat forces can then use such an overreaction to unite others with their cause. Pretending to be civilian noncombatants, they may also feign injury or pretend to need assistance (such as a disabled taxi) with the primary motive of luring Army forces into a deliberate ambush. While the threat may gain a short-term advantage by using ethical responsibilities as a weapon, the Army commander's strict adherence to moral principles will, in the long run, contribute greatly to threat defeat—particularly during long-term urban stability operations.

Acquire Intelligence and Logistic Support

3-27. Indigenous threat forces can normally use the local population for intelligence and logistic support far more effectively than can an alien army. Threat forces may manipulate local hires serving among U.S. Soldiers, such as those contracted by the Army for base operation purposes or translator duties. In addition, refugees moving through friendly controlled sectors may provide the threat with information on friendly dispositions, readiness, and intent. (Women and children may be specifically used for this purpose since U.S. Soldiers' cultural bias can create an incorrect perception that these noncombatants pose no threat to security.) Even friendly residents may become unwitting or unwilling informants, providing an enemy or a hostile with vital information on friendly activities, dispositions, and capabilities. However, a threat employing particularly cruel, abusive, or repressive measures may easily turn certain groups in the urban area against them, even when they share a common history, culture, and ethnicity with the civilians. This is more likely in those areas with higher population densities.

3-28. Threat forces may also seek to use some nongovernmental organizations (NGOs). They may try to obtain relief supplies either through the organizations' legitimate relief operations or as a target for theft. Some organizations may even be fronts for weapons, food, ammunition, money, and fighters. For example, during Russia's second conflict in Chechnya (1999-2000), documents purportedly found in Grozny by the Russians listed nations such as Sudan, Nigeria, Niger, and Ivory Coast as sending fighters to Chechnya under the guise of the International Islamic Relief Organization. (Chechen fighters also disguised themselves as Red Cross workers.) This deception increases the need for strict security and force protection measures, close coordination with NGOs operating in urban areas, and closer monitoring of suspect organizations' activities by civil affairs personnel.

WIN THE INFORMATION WAR

3-29. Threat forces will try to win the information war as much as they will directly oppose Army UO. Threat urban campaigns need not be tactical military successes. They need only to weaken legitimacy and make the opposition's campaign appear unpalatable to domestic, U.S., and world support. As a critical part of a threat's overall information operations, they will use the ever-present media to tell their story. Portable video cameras, commercial radios, and cellular telephones, available and easily concealed, will be as important to many threat actors as weapons and ammunition. Internet access, already firmly established in many urban areas, provides the means to easily disseminate threat propaganda, misinformation, and disinformation through web sites and electronic mail. The internet also allows threats to conduct their own distance learning and education. Threats will make good use of the instantaneity and global reach of television and radio broadcasts, particularly local and international media, to obtain information of immediate tactical value and broadcast their own operational and tactical guidance. Hackers, covered and concealed in the interior spaces of the urban area, may gain access to U.S. sites to manipulate and obtain information to the threat's advantage.

Information and the Media

The media coverage of the urban battle for Hue, South Vietnam, although only one of hundreds of different attacks of the Tet Offensive, affected the will of both the American people and their political leadership. On January 31, 1968, two North Vietnamese Army (NVA)/Vietcong (VC) regiments and two sapper battalions, moving rapidly and with the element of surprise, attacked and seized part of the walled city (Citadel) of Hue. It was the third largest city in South Vietnam, the former capital of a united Vietnam, the capital of Thua Thien province, and a spiritual and cultural center. Initially intending to hold the city for seven days, the NVA/VC retained portions of the city for approximately three weeks against determined U.S. and South Vietnamese attempts to retake it.

Hue marked a revolution in the coverage of war by modern media. It was the first time Americans could sit at home and watch an ongoing battle on the evening news. One of the most intense and savage battles of the Vietnam conflict, it was televised every evening for almost a month. Although the battle for Hue was a tactical victory for the United States, the North Vietnamese clearly achieved strategic success by searing the American consciousness with the high costs of urban warfare. Had U.S. leaders made winning the information war a central part of the overall campaign plan—for example, exposing the American people to the NVA's brutality by publicizing the civilian executions in Hue—civilian support for the war may have been bolstered and a different outcome achieved. See Chapter 7 for a more detailed account of the battle for Hue.

Manipulate Key Facilities

3-30. Threat forces will attempt to identify and quickly seize control of critical components of the urban area to help shape the operational environment. Urban telephone exchanges can provide threats with simple and reliable communications that can be easily secured with off-the-shelf technologies. Existing public address and loudspeaker systems can be used to incite riots or openly command and control dispersed threat forces. (Local language may provide threats with a measure of communications security.) Sewage treatment plants and flood control machinery can be used to implement WMD strategies or to make sections of the urban area uninhabitable. Media stations significantly improve the information operations abilities of the controlling force. Power generation and transmission sites provide means to control significant aspects of civilian society over a large area. Insurgents will attempt to destroy key facilities and critical infrastructure in order to shape the populace's perception that the existing government is incapable of providing for their safety and meeting their basic needs.

Use All Dimensions

3-31. Threats will think and operate throughout the depth, breadth, and height (including supersurface and subsurface areas) of the urban environment. However, threats will often choose to fight from within interior spaces rather than out in the streets and other open areas. Conventional lateral boundaries will often not apply as threat forces control some stories of the same building while friendly forces control others. The use of all three dimensions of an urban area makes identification, reporting, and targeting of enemy locations more difficult for forces accustomed and trained to acquire and engage targets in primarily two dimensions. Because of these target identification and acquisition difficulties, threats will likely use decoys in the urban area to cause erroneous assessments of their combat capabilities, strength, and disposition. They will also attempt to use these decoys to absorb expensive and limited precision-guided munitions as well as cause misallocation of other critical Army resources.

3-32. Roofs and other supersurface areas provide urban threats with excellent observation points and battle positions above the maximum elevation of many weapons. Shots from upper floors strike armored vehicles in vulnerable points. Basements and other subsurface areas also provide firing points below many weapons' minimum depressions and strike at another weakness in most armor. Sewers and subways may provide covered and concealed access throughout the area of operations.

EMPLOY URBAN-ORIENTED WEAPONS

3-33. Whether purpose-built or adapted, many weapons are more useful in an urban environment while others may have significant disadvantages. Urban threat weapons are much like the nature of urbanization and the urban environment: inventive and varied. Many threats will integrate widely available off-the-shelf technologies into their weapon systems and armed forces. However, sniper rifles and small, man-portable, fire-and-forget weapons and demolitions and other improvised explosive devices (IEDs), to include suicide and car bombs, will likely dominate the urban environment. Figure 3-3 lists examples of threat weapons favored in UO.

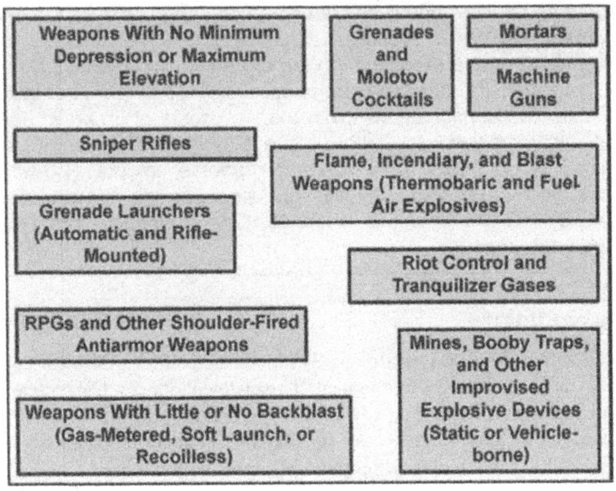

Figure 3-3. Favored threat weapons

ENGAGE ENTIRE ENEMY FORCE

3-34. Threats may attempt to keep all or significant portions of Army forces engaged in continuous operations to increase their susceptibility to stress induced illnesses. UO, by their nature, produce an inordinate number of combat-stress casualties. Continuous operations exacerbate this problem. Threat forces that employ this tactic will often maintain a large reserve or achieve respite by hiding among the civilian population to minimize the psychological impacts on their own forces.

3-35. To accomplish this key tactic, threat UO will likely involve decentralized maneuver, precision fires, and simultaneous operations involving unconventional and special purpose forces and guerilla, insurgent, and terrorist tactics. Threats may also attempt to bait Army forces by ambush or other means with a primary focus of attacking response or reaction forces. Overall, threat forces will take advantage of any exposed weakness in weapons, equipment, and TTP and they will cultivate the flexibility to engage in battles as favorable opportunities present themselves.

Focus Attacks on Support Areas, Isolated Groups, and Individuals

3-36. To supplement the previous tactic, threat forces will seek to target support areas, small groups, leaders and their headquarters, and unaware Soldiers. Their focus on resupply and convoy operations, casualty evacuation, and other sustainment activities, coupled with the compartmented terrain, navigational challenges, and multiple three-dimensional avenues of approach often makes these locations and Soldiers more susceptible to surprise raids and ambushes. Attacks on these areas and groups are conducted to erode the Army's ability to sustain UO, to inflict maximum casualties, and to induce psychological stress. These attacks can be mitigated by careful, regular evaluation of choke points and other restrictive terrain, regular awareness training for units and individuals operating in or transiting through potential incident-prone areas, well-planned and protected convoys, and thorough after-action analysis of incidents.

NEGATIVE EFFECTS OF URBANIZATION

3-37. Many urban areas are the engines for increased industrialism and economic growth as an expanding population provides the labor for manufacturing and service needs. However, rapid and inadequately planned growth can result in undesirable consequences. Uncontrolled urbanization may result in an infrastructure and economic base unable to support the growing population. A large transient, ill-housed, and idle population in a close geographic space may produce strife. Classes, cultures, ethnic groups, and races that might otherwise peacefully coexist can clash under the stress of survival. Uncontrolled urban growth has resulted in the negative effects listed in figure 3-4. In many urban stability and civil support operations, these may be the primary "threats" to mission accomplishment.

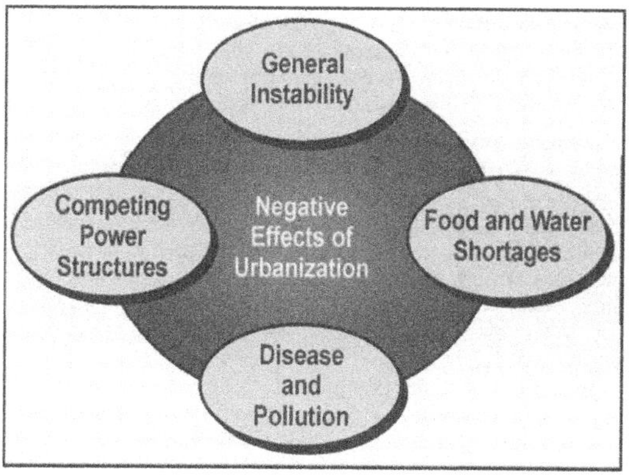

Figure 3-4. Negative effects of urbanization

3-38. Not all urban areas prevail as inherently unstable or hotbeds for unrest. Urban growth due to migration may remove sources of conflict, or it may provide the catalyst for violence. Commanders should recognize the possible effects of uncontrolled urbanization. During their intelligence preparation of the battlefield (IPB), they should determine if these conditions exist. Throughout mission analysis and the development of courses of action, commanders should consider the impact (if any) on their operations. At the same time, they must recognize that UO may create similar problems that may affect the current operation as well as the overall campaign.

General Instability

3-39. Urbanization can enhance stability by generating industrialization and economic growth resulting in more jobs, a higher overall standard of living, and an educated, relatively satisfied populace. However, the population dynamics associated with urbanization can also have an opposite, destabilizing effect. Radical population growth may create overcrowding and generate or aggravate resource and quality of life issues. Intense and destructive competition for employment, housing, and social status may develop in this climate of economic deprivation. The inability of some governments to handle these problems—

- Makes their urban areas potential sources of unrest.
- Increases the likelihood of the Army's involvement in stability operations.
- Complicates all operations conducted in such an urban environment.

Weak civil administrations—perhaps weakened by the effects of previous UO—have difficulty controlling their society, safeguarding their military armaments, and preventing their urban areas from serving as sanctuaries to terrorists, insurgents, and criminal organizations.

3-40. Urbanization in developing countries warrants more concern. Intense migration and growth, coupled with the forced closeness of people once separated by the rural countryside, may stress already struggling institutions, hasten conflict, and lead to overall instability. Commanders must understand that UO, depending on the operation, may either cause massive population movement out of or into urban areas.

3-41. Urban areas with a large youth population may also help to generate conditions for instability. Rural-to-urban migrants tend to be relatively young. In 1999, Cairo, for example, had more than 40 percent of its population younger than 15 years. Young urban populations generate enormous demands for social resources, primarily education and jobs. Even a strong urban economy may fold under the economic expectations of a tremendous influx of young migrants. Disorder and violence may result as hostiles (many nonstate actors) easily mobilize and manipulate the idle young to act politically and criminally. Urbanization and population growth are more dangerous when they combine to produce a cohort of young urban dwellers separated from traditional social controls, such as village elders and clan leaders.

Cultural and Religious Instability

The 1992 bombing of the Babri Masid Mosque in Ayodya, India, enflamed an already intense cultural and religious rivalry between Hindus and Muslims and led to rioting throughout many Indian urban areas. Of the 1,500 who died in conflicts and riots, almost 95 percent died in urban areas. The violence struck Ahmedabab and Bombay most seriously, with related acts of murder, gang rapes, and arson occurring months after the destruction of the mosque.

3-42. Ethnic, religious, and other social issues may become the vents for anger and frustration produced by the high tension of urban life. Major acts of violence and destruction, such as occurred in 1992 in India, can directly threaten a nation's security. Army forces may have to conduct large-scale, stability operations to promote peace and protect national interests. In these cases, all levels of command will be particularly concerned with maintaining impartiality and perceived legitimacy.

Food and Water Shortages

3-43. Rapid urbanization, primarily in developing nations, may lead to severe food shortages that could influence Army forces (or lead to their use). Such shortages may cause instability, massive migration, revolts, or increased support of armed opposition groups. Armed factions may target NGOs that supply aid as a means of furthering dissatisfaction among the populace. In effect, food may become a weapon. Deployed troops may need to provide or support humanitarian food aid networks to keep the humanitarian situation from escalating.

3-44. However, planners must be wary of the unintended consequences of well-intentioned humanitarian assistance operations. For example, providing free, safe food may alleviate starvation, but could also undercut the local agricultural system by reducing demand in the market. If the food is distributed through urban centers, urbanization could increase, further reducing the food supply and adding to the existing strains on the infrastructure. Unless they are trying to use them as a means to separate combatants from noncombatants, commanders should normally use centralized feeding centers as a last resort. Instead, Army forces should bring the food closest to the population to encourage civilians to stay in their homes. If safe areas or camps are created, they should be designed for use over as short a time as is feasible. The general rule should be to return the urban population to their homes as soon as possible. Army forces conducting domestic support or foreign humanitarian assistance operations that cannot maintain the safe food supplies may find the frustrations and hostility of the local population focused on them.

3-45. Water shortages (and quality) are becoming a serious problem in many regions. Sewage, industrial waste, other forms of pollution and deliberate contamination pose threats to existing water supplies. Commanders operating in an urban environment need to know the water supply origins and its treatment, purification, distribution, and vulnerabilities. Before beginning operations, commanders must know if they are providing water for the noncombatants as well as their own forces. Across the spectrum of operations, controlling and protecting a limited water supply is, or may become, an essential operational consideration during UO.

Food and Water Shortages

Countries as varied as Indonesia and Algeria exported their food surpluses only two generations ago but now import up to two-thirds of their basic staples. This cycle has resulted in many countries, which once exported agricultural products, facing the growing cost of imports to feed their urban populations. Estimates predict that in the 2010 timeframe, at least 65 countries (including 30 of Africa's 51 countries) may depend completely on food imports. For some countries, the outlook is even worse. The Democratic Republic of Congo (Zaire), once a net food exporter, faces severe malnutrition and utter starvation for a large portion of its population.

Other estimates predict that by 2025, 2.7 to 3.5 billion people will live in water-deficient countries. Even now in urban areas such as Jakarta, Indonesia, for example, urban authorities cannot provide the necessary water to support its growing population. Less than half of Jakarta's population is supplied by the urban area's water infrastructure; the remainder of the area's residents rely on water from wells and other groundwater supplies that lie only a few feet below the surface. These alternate sources need to be sufficiently boiled or residents risk increased disease. In the northern part of Jakarta, land is actually sinking as the urban area's groundwater supply is overused by its residents. Houses on low-lying areas have to be rebuilt every few years to keep them above sea level. Meanwhile, saltwater from the Java Sea is seeping into the land polluting the remaining groundwater and further exacerbating the situation.

DISEASE AND POLLUTION

3-46. Urban areas frequently spawn epidemics; therefore, widespread disease may pose a significant threat to Army forces that operate there. In many developing nations, rapid urbanization has occurred without a corresponding upgrade, expansion, or even development of adequate sewage and water systems. Some urban areas have only one toilet for every 750 people or more. In these areas, hundreds of thousands live much as they would in poor villages, yet so confined as to ensure high transmission rates for airborne, waterborne, sexually transmitted, and contact-transmitted microbes.

3-47. In urban areas lacking adequate trash and waste management infrastructure, insect-spread diseases proliferate. Mosquitoes that breed in polluted water, open water tanks, and irrigated urban gardens carry malaria and dengue fever—the leading causes of sickness and death from infectious disease in Latin America and Africa. The problem compounds with growing numbers of bacteria resistant to various antibiotics, a shortage of trained medical personnel, inadequate or insufficient medical facilities and supplies, and unclean agricultural and food-processing practices.

3-48. Pollution also creates critical health problems in developing areas and a potential health risk for intervening Army forces. For example, urban areas in China have recorded five to ten times the levels of sulfur dioxide found in the air of urban areas in the developed world. In parts of Poland, toxic waste has so polluted the land and water that ten percent of the babies have birth defects. Pollution may cause immediate health problems but more often, the insidious effects appear months or years after exposure. As discussed earlier, UO may contribute, either intentionally or unintentionally, to an increase in pollution. Destruction of industrial complexes that use, produce, and store hazardous material may produce toxic gas and smoke pollutants and create significant health concerns to exposed Soldiers and civilians. Soldiers forced to fight within the confines of subsurface areas such as sewers and drainage systems further increases their exposure to contaminants.

3-49. Commanders should initiate force health protection planning early, including analysis of the medical threat and other critical medical information requirements during the IPB process. A medical surveillance system should monitor the daily status of Army personnel throughout the operation. In preparation, all personnel must receive a predeployment medical examination. This exam establishes an accurate baseline health status of the force and ensures that Army forces do not introduce new diseases to an urban area, possibly exacerbating the situation. Conversely, Soldiers not immune to native viruses or possessing a weakened immune system due to continuous operations and the stress associated with UO may put Army forces at a significant disadvantage. An outbreak of plague during an operation would have an effect similar to a chemical or biological attack. The closer that Army forces operate to civilians (the humanitarian assistance operations conducted in Port-au-Prince, Haiti, and Mogadishu, Somalia, for example), the more probable that these situations may occur. (See Chapter 9.)

COMPETING POWER STRUCTURES

3-50. Many groups can exist that become strong enough to rival the power of the governing officials and eventually turn the urban area into a system of divergent and competing power structures. These groups can consist of insurgent forces, a merchant class or an economic elite, criminal organizations, or some other significant source of power such as religious organizations (see Chapter 10), clans, or tribes (figure 3-5). In the absence of a legitimate authority, armed factions headed by "warlords" may vie to fill the power void. Sometimes these groups or organizations, normally at odds with each other, may form alliances to achieve specific goals. Commanders must recognize, identify, and understand these alternate urban power bases and, if necessary, develop engagement strategies to neutralize or harness them to accomplish the Army mission. Four of these groups, not previously covered or covered in a subsequent chapter, are discussed in greater detail below.

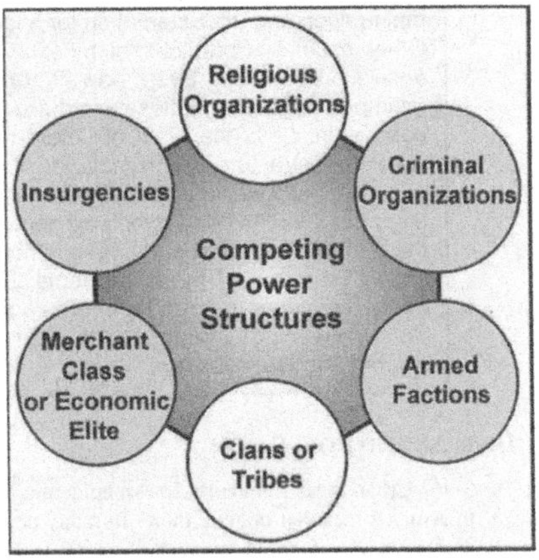

Figure 3-5. Competing power structures

Urban Insurgencies

3-51. As urban migration increases in the developing world, rural guerrillas appear to follow. This transition of insurgencies from rural to urban areas occurs because urban areas offer a rich field of targets for insurgent attacks. People immediately notice any disruption of urban infrastructure, thus having great propaganda value. A concentrated urban population whose allegiance and support are critical to the success of an insurgency is often more susceptible to propaganda, political organization, and terrorism. Insurgents can easily arrange mass demonstrations using available communications facilities, both overt and covert. Travel is effortless and large urban populations provide cover and concealment. On the whole, urban areas, particularly those suffering the effects of general instability outlined earlier, may provide a fertile environment for guerrillas to apply formerly rural insurgent strategies. However, even with a primarily rural-based insurgency, operations in urban areas will often be essential to the insurgents' strategy as they offer them distinct opportunities to disrupt, discredit, and demoralize the government (see FM 3-05.20 and FMI 3-07.22).

Merchant Class or Economic Elite

3-52. Urban areas normally possess a merchant class or economic elite as part of their social structure. In some urban areas, they may carry more power than the local or central state government. They may isolate themselves physically and socially from the sprawling poor yet wield enormous power over the country's political and economic activities. The degree of economic separation between the merchant class and the poor may be small but still socially or politically significant.

3-53. In a vastly impoverished area where the economy of the urban area is severely disrupted, the merchant class will often continue to operate and function and, as a result, achieve a measure of influence. To continue to operate under acute economic turmoil, they may form alliances in criminal organizations and secure loyalties within the government. Outside resources introduced into a crisis area (such as food, water, fuel, and pharmaceuticals) take on increased value, may replace currency as the medium for exchange, and often become the means to amass and hold wealth. One primary way to obtain wealth is to steal it.

3-54. In some turbulent situations that lead to the need for stability operations, commanders may harness the power of the merchant class as a force for peace and stability instead of one that uses crime to achieve economic goals. For example, in a relief situation, instead of competing with the merchant class by distributing food directly to the needy and possibly creating an environment of looting and black marketeering, it may be possible to monetize food. Food assistance from donor governments could be sold to merchants at an attractive price so they have a reliable source of supply. This could, in turn, create a healthy economic system and separate merchants from criminals and gangs.

Criminal Organizations

3-55. Organized criminal groups have grown common in urban areas; have also become an important part of the urban social structure (gangs for example); and can exert considerable influence on governments, people, and military forces conducting UO. These organizations can threaten the successful completion of urban operations both directly and indirectly. Criminals and criminal organizations may directly target Army forces, stealing supplies and extorting money or contracts. Criminals may also violently confront Army forces during mission execution. During many UO, particularly during or following combat, civil disturbances, or large natural disasters, looting (organized or unorganized) may become of critical concern.

For a portion of the population, crime may be the only reliable source of income and resources to support their families. Therefore, UO will often require a combined law enforcement and military response.

3-56. Some large criminal organizations relying on international connections often have better resources and equipment than their insurgent counterparts. (A prime example is drug traffickers.) Their large financial resources, long-reaching connections, and ruthlessness provide them the means to corrupt or intimidate local officials and government institutions. Their tactics parallel those of insurgents. They have developed an intuitive cultural understanding of slum neighborhoods and the ability to lure civilians into criminal activities. They have also mastered the management of mobs. They recruit teenagers and young adults in their efforts against rivals and authorities, just as insurgents muster armies from the youth of rural villages. In many developing nations, there exists an alliance between insurgents and organized criminal groups. In these alliances, the insurgents defend the criminals and the criminals fund the insurgents.

3-57. Excessive criminal activity can undermine the commander's efforts to establish a sense of security among the remainder of the urban populace. Commanders will rely on the ability of their military police (MP) units to assess and understand the potential influence that crime and criminal organizations may have on UO. MPs—in coordination with local police—will need to quickly identify law enforcement gaps to include dysfunctional police organizations and stations and inadequate prison systems. Urban police records may be an important intelligence database to assist in this assessment and may be particularly useful in counterterrorism and counterinsurgency operations. Success in ensuring that law and order is maintained or reestablishing law and order will directly affect the population's perceptions and support of Army operations. Also, MP intelligence operations will contribute immeasurably to the overall IPB effort and aid in understanding of the complexities of the urban society (see FM 3-19.50).

Crime and Criminal Organizations

Crime and poverty plague urban areas such as Rio de Janeiro, Brazil's second largest urban area, and would affect military operations conducted in their limits. Rio has some of the nation's highest negative urban indicators: the largest number of slum dwellers (1 million), the highest murder rate (1 of 700 residents per year), and the highest kidnapping rate (4 per week). In 1989, the homicide rate of the urban area was three times higher than New York City's, and the rate of urban violence has continued to rise. Therefore, law enforcement management may be a critical issue for Army forces operating in urban environments similar to that of Rio de Janeiro.

However, criminal elements or organizations may not always work against Army commanders. They can be co-opted or influenced to serve friendly objectives. For example, during World War II the U.S. Navy worked covertly with the Mafia in New York City to secure the New York harbor from German U-boats believed to be torpedoing ships there. The Mafia controlled most dock activities in New York harbor and was perfectly positioned to monitor other subversive waterfront activity. This capability provided needed information to the Navy for its counterintelligence and security tasks. New York civil authorities, therefore, agreed to permit a Navy-Mafia alliance to operate at the port for the greater good of the country. Although the Mafia was not the preferred ally of the Navy, it had the capability to protect U.S. ships and the interest (patriotism) to help in the war effort. In those circumstances, the temporary alliance worked (see also the civilian threat discussion in Appendix B).

Armed Factions

3-58. A characteristic of some stability operations has been the deterioration or complete collapse of political authority in the country or urban area in crisis. In some cases, "warlords" or the leaders of various armed factions have attempted to fill the power vacuum (see Appendix C for an example). These individuals often have no particular claim to legitimacy. Their power issues from their weapons, not necessarily from their political skills, human services provided, or popular consent (although they have some popular support to remain in their relative position of authority). Others, however, may already have a substantial base of support (for example, a religious base) and then seek to achieve their goals through violent means. In dealing with the leaders of these armed factions during stability operations, it may appear that there are three general courses of action. The first is to aggressively locate and target the leaders and their forces militarily. The second is to ignore them completely, and the third is to meet and work with them diplomatically. Depending on METT-TC, commanders may reduce some of the risks involved in these first three courses of action by combining the last two to create a fourth approach. Nevertheless, the technique chosen must clearly support political and military objectives and lead to the desired end state.

3-59. In some cases, the leaders of armed factions pose a direct threat to Army forces, the urban populace, and the overall accomplishment of the mission. Therefore, it may become necessary to target them and their forces for capture or destruction. Yet, commanders should generally avoid framing the success of a major operation on the death or capture of one or more individuals. Even with increased ISR capabilities, commanders may find an individual manhunt difficult or impossible to achieve. Instead, it may be better to seek to isolate warlords and their forces militarily, while simultaneously seeking to marginalize their influence through a combination of political, economic, and informational means. This does not necessarily mean that commanders must abandon the search if the death or capture the faction leader will contribute to achieving the mission. However, commanders must maintain the capability to strike with speed and precision if the threat leader's location is discovered in the clutter of the complex urban environment.

3-60. In some situations, commanders may attempt to marginalize warlords by ignoring them. However, refusal to acknowledge warlords may increase the threat to Army forces and NGOs. Their militias may attack Army forces to achieve recognition or simply due to misunderstanding or inherent friction between armed forces. On the other hand, dealing with them may provide legitimacy to the exclusion of other elements of the urban population such as professional groups (for example, doctors or teachers), religious leaders, and traditional clan or tribal chiefs—which may have a greater claim to legitimacy and better form the foundation for a reconstituted urban society.

3-61. A compromise between the last two options may offer the best chance for success. In this fourth course of action, commanders can generally recognize these warlords and work with them diplomatically to decrease the risk of armed conflict and civilian casualties; however, this recognition can be kept at staff levels to avoid bestowing any perceived legitimacy on them. Instead, commanders themselves may visibly meet the other elements of society that have a more legitimate claim to political, social, or economic leadership. Inevitably, commanders may need to meet with warlords. In those circumstances, clan or tribal elders, and others who represent traditional authority, should attend the meetings. Moreover, commanders should ask for and give deliberate consideration to their opinions above those of the leaders of the armed factions.

This page intentionally left blank.

Chapter 4

Understanding the Urban Environment's Effects on Warfighting Functions and Tactics

War is, above all things, an art, employing science in all its branches as its servant, but depending first and chiefly upon the skill of the artisan. It has its own rules, but not one of them is rigid and invariable. As new implements are devised, new methods result in its mechanical execution; but over and above all its mechanical appliances, it rests upon the complex factors of human nature, which cannot be reduced to formulas and rules.

Captain Francis V. Greene, 1883

Commanders of major operations should understand the potential effects that the urban environment may have on warfighting functions. They should also understand the possible effects that the urban environment may have on lower-level tactics to properly plan, prepare, and execute major operations that may include UO. Otherwise, commanders may ask their subordinates to achieve effects, accomplish objectives, or adhere to a timetable that is unsupportable due to the constraints imposed by the urban environment. Commanders and their staffs must do more than simply understand the impossible; rather, they must apply the art and science of warfighting to the urban environment and determine what it will take to make it possible.

WARFIGHTING FUNCTIONS

4-1. Understanding the potential effects of the urban environment on warfighting functions permits the urban commander to better visualize his operational environment. With this appreciation, he can conduct a more thorough assessment and thereby determine the most efficient and effective means of employing Army forces. The staff should be intimately familiar with effects in their area of expertise and use that knowledge to understand the problem and develop creative and innovative solutions to achieve their commander's intent.

WARFIGHTING FUNCTIONS

- Intelligence
- Movement and maneuver
- Fire support
- Protection
- Sustainment
- Command and control

INTELLIGENCE

4-2. The intelligence function facilitates understanding of the threat and the environment. The urban environment affects this critical function in many ways. Impacts of the environment on the intelligence function include degraded reconnaissance capability; more difficult IPB process; and an increased importance of credible HUMINT (including the contribution of local civilian liaisons), and an established intelligence reach capability. The Army forces' response to these effects can result in timely, accurate, and actionable intelligence that permits the effective application of other warfighting functions to the mission within the urban environment (see FM 2-0).

Degraded Reconnaissance and Surveillance Capability

4-3. The physical environment creates a major challenge to the intelligence function. The man-made construction in the urban areas provides nearly complete cover and concealment for threats. Although improving many sensor capabilities cannot penetrate the subsurface facilities and much of the space within supersurface areas. The mass of buildings can also defuse electronic signatures. Tall buildings shield movement within urban canyons from aerial observation except from directly overhead. Urban threats may be less technology dependent and may thwart some signals intelligence efforts simply by turning off their radios and using messengers. Threat forces will likely use elements of the civilian telecommunications infrastructure for C2. These systems may include traditional landline phones, cellular telephones, and computer-to-computer or Internet data communications. Most urban telecommunications systems use buried fiber or cables or employ modern digital signaling technology. Such systems are difficult to intercept and exploit at the tactical level.

4-4. From the above, it is evident that these characteristics make it more difficult for the intelligence function to use electronic means to determine threat dispositions and, in offensive and defensive UO, identify decisive points leading to centers of gravity. While the environment limits some typical collection methods, all enemy electronic and human activity creates some form of observable signature and exposes the enemy to potential collection. Seeking ways to take advantage of these vulnerabilities will provide the commander an information advantage over his opponent.

Challenging IPB Process

4-5. The sheer complexity of the environment also challenges the intelligence function. The intelligence function applies the IPB process to the urban environment in accordance with Army doctrine (see Appendix B). With more data points for the IPB process to identify, evaluate, and monitor, this application becomes more demanding. The human and societal aspects of the environment and the physical complexity primarily cause this difference. Relationships between aspects of the environment, built on an immense infrastructure of formal and informal systems connecting the population to the urban area, are usually less familiar to analysts. Thus, the urban environment often requires more specifically-focused intelligence resources to plan, prepare for, execute, and assess operations than in other environments.

4-6. Compounding the challenges is the relative incongruity of all urban environments. No two urban areas are alike physically, in population, or in infrastructure. Thus, experience in one urban area with a particular population and pattern of infrastructure does not readily transfer to another urban area. Any experience in UO is valuable and normally serves as a starting point for analysis, but the intelligence function cannot assume (and treat as fact) that patterns of behavior and the relationships in one urban area mirror another urban area. The opposite is as likely to hold true. The intelligence function will have to study each urban area individually to determine how it works and understand its complex relationships.

4-7. Each characteristic of the urban environment—terrain, society, and infrastructure—is dynamic and can change radically in response to UO or external influences. Civilian populations pose a special challenge to commanders conducting UO. Civilians react to, interact with, and influence to varying degrees Army forces. Commanders must know and account for the potential influence these populations may have on their operations. Intelligence analysts must revisit or continuously monitor the critical points looking for changes, relationships, and patterns.

4-8. The actions of Army forces will affect, positively or negatively, their relationship with the urban population and, hence, mission success. NGOs may deliberately or inadvertently influence civilians. The intelligence function can monitor and predict the reactions of the civil population. However, accurate predictive analysis of a large population requires specific training and extensive cultural and regional expertise.

Increased Importance of Human Intelligence

> ***HUMINT*** is the collection of foreign information—by a trained HUMINT collector—from people and multimedia to identify elements, intentions, composition, strength, dispositions, tactics, equipment, personnel, and capabilities. It uses human sources as a tool, and a variety of collection methods, both passively and actively, to collect information. FM 2-0

4-9. The intelligence function adjusts to the degradation of its technical intelligence gathering systems by increasing emphasis on HUMINT in UO. HUMINT operations may be the primary and most productive intelligence source in UO. In urban offensive and defensive operations, HUMINT gathers information from refugees, immigrants and former citizens (especially previous civil administrators), civilian contractors, and military personnel who have operated in the area. Credible intelligence of this type can help meet requirements, provide more detail, and alleviate some of the need to physically penetrate the urban area with reconnaissance forces. In many urban operations where HUMINT is the primary source of intelligence, acting on single-source reporting is a constant pitfall. Yet, situations may arise where commanders must weigh the consequences of inaction against any potential negative consequences resulting from acting on uncorroborated, single-source information. (See also the Human Capabilities discussion under the Urban ISR portion of Chapter 5.)

4-10. In urban stability operations, HUMINT identifies threats and monitors the intentions and attitudes of the population. A chief source of information contributing to the development of accurate HUMINT, particularly at the tactical level, is reconnaissance forces—especially small-unit dismounted patrols. Urban reconnaissance forces and patrols should be thoroughly and routinely debriefed by unit intelligence personnel to obtain information that aids in developing a clearer picture of the threat and the urban environment. Reliable and trustworthy HUMINT is particularly important in foreign internal defense, counterterrorism, and support to counterdrug operations. Leaders must organize intelligence resources appropriately, and learn and apply valuable techniques, such as pattern and link analysis (see FM 34-3). Additionally, Soldiers, as part of reconnaissance and patrolling training, should be taught to handle captured documents, weapons, material, and equipment as legal evidence much like military and civilian police. Proper "evidence" handling is often a critical intelligence concern in counterterrorism and counterinsurgency operations.

Developing Local Liaisons

4-11. Whenever Soldiers encounter the urban populace, the resulting interaction may become an important source of information the commander can use to answer questions about the threat and the urban environment. While military intelligence units are the primary collectors and processors of HUMINT, commanders are not likely to have enough trained HUMINT Soldiers to satisfy their requirements—particularly in a larger urban environment and during longer-term stability operations. Therefore, commanders may need to cultivate and establish local civilian associations to provide relevant information for decision making and to support the overall HUMINT effort.

4-12. Urban liaisons can be developed through positive civil-military interaction with the urban populace. Critical information may be acquired through interface with the urban leadership (both formal and informal), administration officials, business owners, host-nation support workers, inhabitants along a unit's patrol route, pedestrians at a checkpoint, civilian detainees, or any other human source willing to volunteer information to Army forces or who respond positively to tactical questioning. (Noncombatants are never coerced to provide information.) Commanders may also direct unit leaders to conduct liaisons with specific local leaders and key members of the community to obtain command directed information. Critical information may also come from other U.S. and coalition forces and intelligence organizations operating near or within the commander's AO. To this end, commanders should ensure that collectors operating in an urban area coordinate and deconflict activities and, if possible, outbrief subordinate, geographically-responsible commanders with any relevant information that may affect their current operations. Any relevant information obtained incident to civilian liaison activities should be routinely provided to intelligence staffs not only to gain assistance in verifying the credibility of the information but to share the information with all affected echelons and units.

4-13. NGOs operating in urban areas can also be especially beneficial resources for credible and relevant information about the urban environment. (However, they are generally not a good source for information about the threat since providing such information can violate their neutrality thereby making it difficult for them to achieve their humanitarian aid objectives.) During the 1999 fighting in Kosovo, for example, the Red Cross provided the most accurate figures regarding the number of Kosovar refugees, helping U.S. and other coalition forces to estimate the appropriate level of support required to handle their needs. In addition to a developed understanding of the current needs of the local urban populace, NGOs may also have—

- A network of influential associations.
- Historical archives.
- Extensive understanding of the urban infrastructure.
- Key knowledge of political and economic influences.
- A keen awareness of significant changes in the urban environment.
- Insight into the current security situation.
- Up-to-date web sites and maps.

4-14. While productive civilian associations may become long term, they should not be confused with HUMINT source operations. Only trained HUMINT personnel can recruit and task sources to seek out threat information. Information obtained from these societal connections is normally incidental to other civil-military relationships. For example, as part of infrastructure repair in an urban stability operation, a commander may be instrumental in obtaining a generator for a local hospital. Within the context of this relationship, the commander may develop a rapport with one or more of the hospital's administrators or health practitioners. These civilians may be inclined to provide valuable information about the threat and the urban environment—often on a continuing basis. In any civil-military relationship, however, commanders ensure that the information provided is not tied to promises of assistance or that such assistance is in any way perceived as a means to purchase civilian loyalty.

4-15. Commanders also understand that repeated interaction with any one individual may put that individual and his family in danger from threat forces. Before this potential danger becomes a reality, they should refer their civilian connections to trained HUMINT personnel for protection and continued exploitation. In addition to civilian protection considerations, commanders may also deem it necessary to turn their civilian associations over to trained HUMINT collectors anytime during the relationship if they consider the information that the contact is providing (or may provide) is credible, relevant, and—

- Provides essential threat information on a repetitive basis.
- Helps answer higher-level CCIR.
- Affects operations in another AO.
- Requires interrogation or monetary compensation to obtain.

However, turning a liaison over to trained HUMINT teams does not necessarily preclude maintaining a continued, albeit a more guarded, relationship with the individual.

4-16. In developing these civilian liaisons essential to understanding the urban environment, commanders must avoid the distinct possibility of conducting unofficial source operations by non-HUMINT Soldiers. While prohibited by regulatory guidance (see Defense Intelligence Agency Manual 58-11 for requirements and restrictions for source operations), such actions also run the additional risks of—

- Obtaining unevaluated information that cannot be crosschecked with and verified by other sources of information.
- Creating inequities that result from illegally rewarding contracts which can undermine HUMINT Soldiers who are constrained by intelligence contingency fund regulations.
- Disrupting ongoing HUMINT operations when different sources are seen to be treated differently by non-HUMINT Soldiers vice HUMINT Soldiers.
- Providing non-HUMINT and HUMINT Soldiers with the same information potentially leading to a false confirmation of information.

- Increasing the likelihood that untrained Soldiers may fall victim to a threat deception and misinformation.

Established Intelligence Reach

4-17. Understanding the complex urban environment, particularly the infrastructure and the society, will require more sources of information beyond a unit's organic intelligence capabilities. Therefore, commanders will have to make extensive use of intelligence reach to access information and conduct collaboration and information sharing with other units, organizations, and individual subject matter experts. Before deployment (and throughout the operation), units should establish a comprehensive directory of intelligence reach resources. These resources may include national, joint, Army, foreign, commercial, and university research programs. (Prior to deployment for OPERATION IRAQI FREEDOM, some units established contacts within the local community outside their bases such as police, fire department, and government officials that expanded their reach once in theater—particularly for information regarding civilian infrastructure and urban administration.) Once deployed, intelligence reach includes effective information sharing and collaboration among adjacent units, sister services, coalition partners, and other governmental and nongovernmental agencies operating in the area. Effective information sharing and collaboration requires common network analysis software and databases to be used among all Army forces and, if possible, other governmental agencies.

MOVEMENT AND MANEUVER

4-18. Army maneuver forces—infantry, armor, cavalry, and attack aviation—move to achieve a position of advantage. Entire urban areas or specific threat forces located within may be isolated from affecting other operations and then bypassed. However, when the situation requires entering the urban area to accomplish the mission—such as when bypassed urban threat forces interdict critical lines of communications (LOCs), the environment will significantly affect the Army's ability to move and maneuver (see figure 4-1). These negative effects include canalization, compartmentalization, and increased vulnerability. However, tactics and techniques equip Army forces to overcome these challenges and maneuver successfully. One tactic, effective combined arms task organization, includes an increased dismounted maneuver capability combined with armor and combat engineers, continuous operations, and technological enhancements.

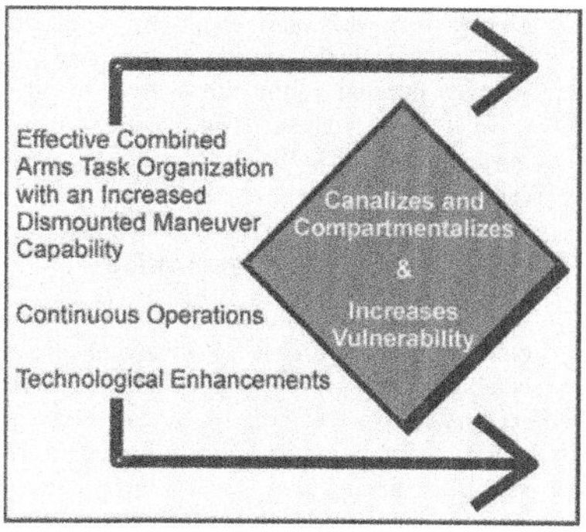

Figure 4-1. Urban maneuver challenges and means to overcome them

4-19. In all UO, mobility operations may allow civilian traffic and commerce to resume, letting the urban area return to some semblance of normalcy (often a critical objective). In stability operations, mobility often focuses on keeping lines of communications open and reducing the threat of explosive hazards to Soldiers and civilians. In civil support operations, mobility may focus on removing storm debris or reducing obstacles caused by destroyed property.

Canalization and Compartmentalization

4-20. The urban terrain will often canalize and compartment forces—and their fires—moving and maneuvering through it. Buildings pose obstacles to both mounted and dismounted movement, forcing units to be canalized along streets. The buildings also block movement between streets, thus compartmenting units. Fires are canalized into open and unmasked areas where vision is unobstructed, producing concentrated fire zones and areas, especially at road intersections and in front of defended positions. Hence, changing directions, repositioning committed forces, reinforcing forces in contact, bypassing threats, and maneuvering to the threat flank become extremely difficult. Units often breach obstacles to help solve this problem. Using helicopters to quickly move forces, both forward into contact and to rear areas as part of repositioning, also permits Army forces to overcome some terrain constraints.

4-21. The canalized and compartmented effects can slow movement and maneuver considerably. However, slowed movement also results from the physically demanding soldier tasks required in an urban environment. Soldiers operate dismounted across rubble and hard surfaces. Operating in three dimensions, they constantly move up the supersurface areas of building interiors and down into basements, cellars, and other subsurface areas. They breach many obstacles and use upper-body strength, ropes, and ladders to scale heights. The inability to see into the next room, floor, or building magnifies stress. The resulting fatigue slows the overall rate of Army force movement and maneuver.

Increased Vulnerability

4-22. The urban environment increases the vulnerability of Army forces executing movement and maneuver in offensive, defensive, and stability or civil support operations. Both the physical terrain and the urban population provide threat cover and concealment. Air movement and maneuver is vulnerable for many of the same reasons. In offensive or defensive operations, enemy forces can remain undetected in buildings and in position to ambush Army forces. Consistent with METT-TC, Army forces should clear buildings along maneuver routes prior to mounted movement along those axes. Failure to clear routes (and effectively mark cleared portions) may expose mounted movement to ambush at close range. Movement back across streets and obstacles may be difficult particularly if the element of surprise was essential in the initial crossing or breach. The same buildings also provide cover and concealment to enemy air defense capabilities—particularly man-portable air defense systems that can be fired from multiple positions hidden amongst the clutter of fires, lights, smoke and dust, and easily concealed and transported in civilian vehicles throughout the urban area. In all operations, but especially stability operations, civilians can conceal threat elements. The threat can then initiate offensive operations against Army forces from close range and where ROE will hamper applying combat power. Thus, maneuver through a dense population can be a high-risk operation.

Combined Arms Task Organization

4-23. Effective combined arms task organization ensures that forces are task organized with infantry—the essential building block for all organizations conducting UO. Infantry protects mounted elements as the combined arms unit moves and maneuvers through the urban area. (In some urban situations, mechanized infantry may not be able to provide dismounted support beyond support to its own vehicles—tanks may require the support of additional light infantry). The infantry destroys the enemy in buildings, bunkers, and subsurface areas where they cannot be defeated by mounted forces and prevents infiltration of threat forces back into hard-won urban terrain. Field artillery aids in dismounted and mounted (to include air) maneuver by suppressing known and suspected enemy positions with precision fires. Attack aviation make best use of their standoff capabilities (see later fire support discussion) and aircraft speed to conduct running and diving fires; in UO, hovering fire is generally avoided. Armored elements protect Soldiers from small arms fire and destroy or suppress enemy positions with precise, direct fire. Carefully protected artillery may also be used in this direct fire role. Armored forces and attack helicopters also can facilitate maneuver through shock action that can have a psychological effect, particularly against less well-trained threats and, in discrete instances, hostile crowds. (Although, commanders consider that the "intimidation" value of any method erodes quickly with its repetitive use.)

4-24. Combined arms also ensure that combat engineers support dismounted maneuver by assisting in covered and concealed maneuver through buildings and off exposed streets. In addition to combat engineers, explosive ordnance disposal teams, military police, chemical personnel, and other with essential expertise to conduct mobility missions significantly reduces mobility and maneuver challenges (see FM 3-34.2). Urban buildings are often obstacles to movement and mobility. Combat engineers, trained and equipped for UO, can turn these obstacles into an advantage by breaching them with "mouse holes" made by explosives, sledgehammers, bulldozers or armored vehicles, or high-strength (diamond or carbide-tipped) cutting devices. These breaches permit dismounted movement through buildings under both cover and concealment.

4-25. Combat engineers must also be trained and equipped to facilitate mounted mobility in the urban environment. Generally, buildings restrict mounted movement to the compartmented and canalized streets. Threats can block streets with roadblocks ranging from sophisticated log and concrete cribs reinforced with antitank and antipersonnel mines to the expedient use of cars, buses, and trucks to create obstacles. Combat engineers must be capable of breaching these obstacles to maintain the coherence of the combined arms team (mounted and dismounted). Combat engineers should be forward, often task organized down to platoon level, and have the expertise and equipment to rapidly reduce point obstacles. It even may be necessary that every armored vehicle (or section of two vehicles) be task organized with an associated engineer squad and vehicle. Because of the increased density and hardness of many urban building and construction materials, heavy engineer equipment (such as the D9 bulldozer) will be in great demand to accomplish mobility (and countermobility, and survivability) functions in an urban environment. However, commanders will need to consider increased protection requirements and the availability of equipment transport to move these slower-moving engineer assets around the urban battlefield.

4-26. A major difference of UO combined arms is in proportion and organization. Although based an accurate METT-TC assessment, UO will often require an increased proportion of dismounted infantry and engineer capabilities; armor may not be required in the same high numbers. As significant, genuine combined arms urban operations are required at lower tactical levels where small, well trained and led units will dominate. Commonly, company level will require true combined arms capability and may include combat engineers, military intelligence, reconnaissance, and artillery. Combined arms teams can then form at platoon and squad levels. Because of this, larger units will need more CA, military intelligence, and combat engineers than habitually attached for combat in more open or less restrictive terrain.

4-27. In determining the appropriate task organization, commanders consider a suitable span of control for subordinate commanders. They also consider the potential of dissipating a unit's combat power, capabilities, and synergy by breaking a unit up into smaller units in an attempt to ensure subordinate maneuver units have a complete combined arms capability. For example, an additional engineer battalion may be task organized to a brigade combat team. In turn, the brigade combat team may task organize this battalion into engineer companies under the control of their subordinate maneuver battalions. If this type of organization continues, maneuver companies may end up with an engineer platoon and, in the end; maneuver platoons each having an engineer squad. Ultimately, a combined arms capability may have been established at lower tactical levels but the parent maneuver unit (in this example, the brigade combat team) may have lost the ability to conduct larger engineer operations without having to re-task organize and potentially disrupt current operations and established relationships. As a guide, urban commanders may consider task organizing to create combined arms organizations at lower tactical levels when operations are predominately offensive or defensive, and bringing those assets back under their own control when the operation transitions to predominately stability or civil support operations.

Continuous Operations and Technology Enhancements

4-28. Two other means to improve Army forces' ability to move and maneuver in urban terrain is through continuous operations and the leveraging of technology, such as the Army's night operations capability. Historically, urban battles have been fought primarily during daylight because of technological limitations and fatigue. By utilizing night vision technologies, accurate situational understanding, a common operational picture (COP), training, and rotated units, Army forces can defeat threats who use the same soldiers in day and night operations and who are less well-equipped and adept at night operations. Night operations are also a means of mitigating the air defense threat against air maneuver. Continuous operations through night maneuver with fresh forces are challenging, but it can overcome many advantages that a stationary force has against maneuver in the urban environment. However, commanders should also consider that streetlights, fires, background illumination (as well as dark building interiors without ambient light), the increased heat absorption of many urban structures, and the skillful use of searchlights by threat forces may limit the effectiveness of night vision devices and make thermal imagery identification difficult.

Countermobility

4-29. Countermobility capabilities in urban terrain are also an essential consideration in all UO. In defensive operations, commanders use countermobility capability to control where the enemy moves in the urban area. Repositioning defensive forces in the urban area can be difficult and obstacles are essential to limiting the enemy's maneuver options. During offensive operations, countermobility protects exposed flanks and air assaulting forces from counterattack. In stability and civil support operations, countermobility operations may take the form of constructing barriers to assist in populace and resources control at critical urban locations.

FIRE SUPPORT

4-30. The fire support function includes the collective and coordinated use of several means to attack targets in the urban area (see Appendix D for joint capabilities). These means include target acquisition data, indirect fire weapons, rotary- and fixed-wing aircraft, offensive information operations (IO), and other lethal and nonlethal means. The urban environment affects these components of the fires function and their employment.

Target Acquisition

4-31. Target acquisition in an urban environment faces several challenges. First, forces have difficulty penetrating the urban environment's increased cover and concealment using sensors and reconnaissance. Acquiring targeting information and tracking targets throughout the depth of the urban area may prove challenging. Moving personnel or vehicular targets are normally easiest to acquire. However, the cover and concealment provided by urban terrain gives moving targets short exposure times requiring firing systems to act rapidly on targeting data. Targeting of opposing indirect fire units by acquisition radar may work more effectively in urban terrain because of the necessary high angles of indirect fire. The urban environment presents similar difficulties for battle damage assessment.

4-32. Targeting challenges are met by innovatively integrating reconnaissance capabilities. These capabilities include SOF, cavalry, unmanned aircraft systems, and aerial observers as well as the standard reconnaissance assets. More artillery systems may need to be used to ensure the responsiveness (rather than the weight) of fires. Positioning numerous artillery systems reduces the dead space (as discussed below) and permits units to establish more direct sensor-to-shooter links.

The Targeting Process

4-33. Heightened concerns for collateral damage will require that commanders pay particular attention to their targeting process. This process ensures that all available combat power, both lethal and nonlethal (including offensive IO), is effectively integrated and synchronized to accomplish the mission. Commanders ensure that techniques and procedures are in place, rehearsed, and understood by all members of their staffs. Additionally, the C2 system must be responsive and agile; otherwise, an elusive and adaptable threat will likely disappear before units can employ the appropriate weapon systems. In an urban area, even 10-digit grid coordinates may not be sufficient to accurately identify targets as buildings may be connected to each other—often throughout the entire block. Target locations, in addition to grid coordinates, may need to routinely include the street address, number of stories, shape, color, or any other distinguishing characteristic essential for ground and air forces to achieve targeting precision. A common urban reference system with graphics, reference points, and other control measures adequate for both ground and air forces may also help facilitate identification of targets and facilitate the rapid clearance of fires.

4-34. Greater concerns exist for the safety and health (environmental matters) of the urban populace and the protection of critical infrastructure and cultural structures. Hence, CA and staff judge advocates (see Chapters 5 and 10 respectively) will play a greater role for the expert advice they can provide regarding these elements of the urban environment. Nonetheless, all members of the staff ensure that operations minimize collateral damage. That responsibility does not end with identifying potential collateral damage; the goal, as always, is successful mission accomplishment. Again, staffs are guided by the commander's intent and work to develop courses of action that incorporate collateral damage concerns (short- and long-term) yet accomplish the mission. This requires a keen understanding of the legal issues and both friendly and enemy weapon systems' effects in an urban environment.

Urban Effects on Fire Support Systems

4-35. Both the physical and human components of the urban area affect how units use fire support weapon systems. The physical aspects of the urban environment, such as the heights and concentration of buildings, may cause significant masking and dead space (see figure 4-2). Intervening buildings that stand three or more stories tall hinder close indirect fire support. Tall buildings can potentially mask several blocks of area along the gun-target line of artillery. (For low-angle artillery fire, dead space is about five times the height of the building behind which the target sits.) The potential for collateral damage to adjacent buildings may also prevent engagement with artillery. Such damage might cause noncombatant and friendly troop casualties and unintentional (and unwanted) rubbling. Commanders can offset these effects by carefully placing artillery positions, repositioning artillery as targets change, and using mortars. Mortars have a steep angle of fall and short minimum ranges as a high-angle alternative to field artillery fire. (In comparison to artillery, dead space for mortar fire is only about one-half the height of the building.) For fixed-winged aircraft, precision munitions and weapons with low explosive yields and near-vertical impact angles resulting in bomb burial can significantly reduce collateral damage. Collateral damage concerns may also cause commanders to—

- Maintain approval authority for some sensitive or protected targets (churches or mosques, for example) at higher echelons of command.
- Restrict attacks to certain times of day.
- Give warning prior to an attack so that noncombatants can evacuate the area.
- Incorporate indigenous forces into the operation.
- Abort an attack unless the required level of precision effects can be achieved.
- Prepare specific branches and sequels to the IO plan to inform the populace why the collateral damage was justified. These plans may include filming or otherwise documenting the operation to thwart threat propaganda and claims of excessive collateral damage.
- Develop and rehearse detailed staff battle drills that address clearance of fires in an urban environment.

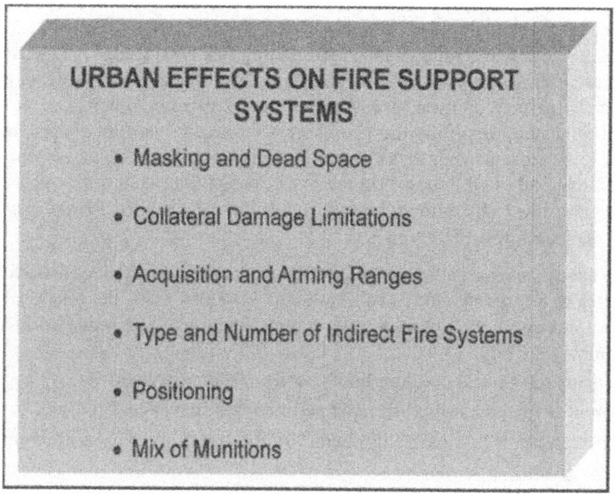

Figure 4-2. Urban effects on fire support systems

4-36. Vertical structures interrupt line of sight (LOS) and create corridors of visibility along street axes. The result is thereby shortened acquisition and arming ranges for supporting fires from attack helicopters and subsequently affected engagement techniques and delivery options. Pilots maintain LOS long enough to acquire targets, achieve weapons delivery solutions, and fly to those parameters. For example, tube-launched, optically tracked, wire-guided heavy antitank missile systems require 65 meters to arm. Similarly, the Hellfire missile requires at least 500 meters to reliably arm and stabilize on the intended target. Thus, attack helicopters firing from longer ranges actually improve the probability of a hit. Poor weather and heavy smoke and dust rising from urban fires and explosions may hinder target identification, laser designation, and guidance for rotary- and fixed-winged aircraft. Poor air-to-ground communications may also hinder effective use of airpower. The close proximity of friendly units and noncombatants requires units to agree on, thoroughly disseminate, and rehearse clear techniques and procedures for marking target and friendly locations. The ability for ground units to "talk-on" aircraft using a common reference system described earlier helps expedite aerial target acquisition (and helps mitigate potential fratricide). FM 3-06.1 details other aviation TTP in an urban environment.

4-37. The urban environment also affects the type and number of indirect fire weapon systems employed. Commanders may prefer high-angle fire because of its ability to fire in close proximity to friendly occupied buildings. Tactically, commanders may consider reinforcing units in UO with mortar platoons from reserve units. This will increase the number of systems available to support maneuver units. Unguided Multiple Launch Rocket Systems (MLRSs) may be of limited use in urban areas due to their exceptional destructive capabilities and the potential for collateral damage. However, commanders may use unguided MLRSs to effectively isolate the urban area from outside influence. Commanders may also employ field artillery systems as independent sections, particularly self-propelled systems, in the direct-fire role; decreasing volume and increasing precision of artillery fire helps minimize collateral damage. While discretely applying the effects of high-explosive and concrete-piercing munitions, these self-propelled systems take advantage of the mobility and limited protection of their armored vehicles.

4-38. The urban area may also affect the positioning of artillery. Sufficient space may not exist to place battery or platoon positions with the proper unmasked gun line. This may mandate moving and positioning artillery in sections while still massing fires on specific targets. Commanders must protect artillery systems, particularly when organized into small sections. Threats to artillery include raids and snipers. Therefore, maneuver and firing units will have to place increased emphasis on securing their positions and other appropriate force protection measures.

4-39. The mix of munitions used by indirect fire systems will change somewhat in urban areas. Units will likely request more precision-guided munitions (PGM) for artillery systems to target small enemy positions, such as snipers or machine guns, while limiting collateral damage. Currently, only conventional tube artillery, not mortars, has this capability. However, large expanses of polished, flat reflective surfaces common in urban areas may degrade laser designation for these munitions (as well as attack helicopter PGM). The vertical nature amplifies the geometrical constraints of many precision munitions. Remote designators need to be close enough to accurately designate but far enough away not to be acquired by the PGM during its flight path. PGMs based on the global positioning system (for instance, guided MLRS or the Air Force's joint direct attack munitions) or other optically guided PGMs may be more effective if urban terrain hinders laser designation.

4-40. The urban environment greatly affects the use of nonprecision munitions. Building height may cause variable time fuses to arm prematurely. Tall buildings may also mask the effects of illumination rounds. Units may choose not to use dual-purpose conventional munitions if (similar considerations apply to Air Force cluster bombs)—

- The enemy has several building floors for overhead protection.
- Dismounted friendly units need rapid access to the area being fired on.
- Large numbers of civilians will operate in the target areas soon after combat operations have ceased.

4-41. Depending on the building construction, commanders may prohibit or limit illumination, smoke, and other munitions because of fire hazards. In particular instances, they may specifically use them for that effect. Structure fires in an urban area are difficult to control and may affect friendly units. Conventional high-explosive munitions may work best against concrete, steel, stone, and other reinforced structures. When not used in the direct-fire role, a greater mass of indirect fire is often required to achieve desired effects. Commanders balance firepower and collateral damage since the rubbling caused by massive indirect fires may adversely affect a unit's ability to maneuver and provide a threat with additional cover and concealment.

4-42. Nonlethal weapons, munitions, and devices can help commanders maintain the desired balance of force protection, mission accomplishment, and safety of noncombatants by expanding the number of options available when deadly force may be problematic. As additional nonlethal capabilities are developed, they are routinely considered for their applicability to UO. In determining their use and employment, commanders, in addition to any previous experience at using these weapons, munitions, and devices, consider—

- **Risk**. The use of nonlethal weapons in situations where lethal force is more appropriate may drastically increase the risk to Army forces.
- **Threat Perspective**. A threat may interpret the use of nonlethal weapons as a reluctance to use force and embolden him to adopt courses of action that he would not otherwise use.
- **Legal Concerns**. Laws or international agreements may restrict or prohibit their use (see Chapter 10).
- **Environmental Concerns**. Environmental interests may also limit their use.
- **Public Opinion**. The apparent suffering caused by nonlethal weapons, especially when there are no combat casualties with which to contrast it, may arouse adverse public opinion.

PROTECTION

4-43. The protection function includes those tasks and systems that preserve the force so that commanders can apply maximum combat power. Preserving the force includes enhancing survivability and properly planned and executed air and missile defense as well as defensive IO (see IO discussion in Chapter 5) and chemical, biological, radiological, nuclear, and high-yield explosive (CBRNE) counterproliferation and consequence management activities (see CBRNE discussion in Chapter 9).

Survivability

4-44. Survivability in the urban environment is a significant force multiplier. Properly positioned Army forces can take advantage of the increased survivability afforded by the physical terrain. Even a limited engineer effort can significantly enhance the combat power of small Army forces. In stability operations, properly planned and constructed survivability positions can enable small groups of Soldiers to withstand the assaults of large mobs, sniping, and indirect fire. Well-protected support bases are often critically essential to minimizing casualties during long-term stability operations and can become a key engineer task.

4-45. While executing major combat operations or campaigns, in particular defensive operations, well planned and resourced engineer efforts can enhance the survivability characteristics of the urban area. These efforts, though still requiring significant time and materiel, can establish defensive strong points more quickly and with greater protection than can be done in more open terrain. Skillfully integrating the strong point into the urban defense greatly increases the overall effectiveness of the defense disproportionately to the number of forces actually occupying the strong point (see Chapter 8).

4-46. Commanders increase survivability by ensuring that all Soldiers have necessary protective equipment and are trained and disciplined in their use. In addition to standard equipment such as helmets, gloves, boots, and chemical protective overgarments, commanders should ensure, as necessary, availability of other protective equipment and material such as—

- Body armor.
- Goggles or ballistic eye protection.
- Knee and elbow protectors.
- Riot control equipment such as batons, face masks, and shields.

- Barrier material such as preformed concrete barriers, wire, sandbags, and fencing material.
- Up-armored or hardened vehicles.
- Fire extinguishers and other fire-fighting equipment.
- Immunizations.

4-47. The Army's urban survivability operations can become complex if the Army is tasked to support survivability operations for civilians. Such operations can range from constructing civil defense shelters or evacuating the population to assisting the population in preparing for or reacting to the use of weapons of mass destruction. However, Army forces are not organized or equipped to support a major urban area's requirements as well as its own mission needs. Normally, Army forces can render this type of support only as a focused mission using a unique, specially equipped task organization.

Air and Missile Defense

4-48. The air and missile defense protects the force from air surveillance and air and missile attack. This system uses—

- The careful massing of air and missile defense combat power at points critical to the urban operation.
- The proper mix of air defense weapon and sensor systems.
- Matched (or greater) mobility to the supported force.
- The integration of the air defense plan into the overall urban operation.
- The integration of Army systems with those of joint and multinational forces.

4-49. Properly planned and executed air and missile defense prevents air threats from interdicting friendly forces and frees the commander to synchronize maneuver and other elements of firepower. Even in a major combat operation or campaign, the enemy will likely have limited air and missile capabilities and so seek to achieve the greatest payoff for the use of these systems. Attacking Army forces and facilities promises the greatest likelihood of achieving results, making urban areas the most likely targets for air and missile attack.

Rotary- and Fixed-Winged Aircraft

4-50. Enemy rotary-wing aircraft can be used in various roles to include air assault, fire support, and combat service support. Some threats may use unmanned aircraft systems to obtain intelligence and target acquisition data on friendly forces. Increased air mobility limitations and targeting difficulties may cause enemy fixed-wing aircraft to target key logistics, C2 nodes, and troop concentrations outside the urban area, simultaneously attacking key infrastructure both in and out of the urban area.

Increased Missile Threat

4-51. The intermediate range missile capability of potential threats has increased to be the most likely air threat to an urban area. Urban areas, particularly friendly or allied, make the most attractive targets because of the sometimes-limited accuracy of these systems. By firing missiles at an urban area, a threat seeks three possible objectives:

- Inflict casualties and materiel damage on military forces.
- Inflict casualties and materiel damage on the urban population.
- Undermine the confidence or trust of the civil population (particularly if allied) in the ability of Army forces to protect them.

4-52. If facing a missile threat, commanders conducting UO work closely with civil authorities (as well as joint and multinational forces) to integrate the Army warning system with civil defense mechanisms. Similarly, Army forces may support urban agencies reacting to a missile attack with medical and medical evacuation support, survivor recovery and assistance in damaged areas, and crowd control augmentation of local police forces. Before such an attack, Army engineers might assist and advise urban officials on how to construct shelters.

Increased Security of Assets

4-53. When defending against an air or missile threat in a neutral or hostile urban environment, air defense assets are concerned with security. Separating air defense locations from high population and traffic centers, as well as augmenting these positions with defending forces, can prevent or defeat threat efforts to neutralize them. Additionally, increased density of UO means increased concentration of all friendly and enemy systems engaged in air and counter-air operations. This density may increase friend and foe identification challenges, air space management challenges, and the overall risk in the conduct of air operations. Finally, limited air defense assets, difficulties in providing mutual support between systems, potential mobility limitations, and other effects of the urban environment increase the need for (and effectiveness of) a combined arms approach to air defense (see FM 44-8).

SUSTAINMENT

4-54. The sustainment function incorporates support activities and technical service specialties, to include maximizing available urban infrastructure and contracted logistics support. It provides the physical means with which forces operate. Properly conducted, the sustainment function ensures freedom of action, extends operational reach, and prolongs endurance. Commanders conducting sustainment to support full spectrum operations must understand the diverse logistic requirements of units conducting UO. They must also understand how the environment (to include the population) can impact sustainment support. These requirements range from minimal to extensive, requiring Army forces to potentially provide or coordinate all life support essentials to a large urban population.

4-55. Commanders and staffs consider and plan for Army sustainment operations that are based in a major urban area. These operations are located in major urban areas to exploit air- and seaports, maintenance and storage facilities, transportation networks, host-nation contracting opportunities, and labor support. These operations are also UO. The commander gains additional factors to consider from basing the sustainment operation in an urban environment. See Chapter 10 for a detailed discussion of urban sustainment considerations.

COMMAND AND CONTROL

> *Fighting in a city is much more involved than fighting in the field. Here the "big chiefs" have practically no influence on the officers and squad leaders commanding the units and subunits.*
>
> Soviet General Vasili Chuikov
> during the 1942-43 Battle for Stalingrad

4-56. The command and control function is the related tasks and systems that support the commander in exercising authority and direction. The urban environment influences both the commander and his C2 system (which includes INFOSYS). The leader's ability to physically see the battlefield, his interaction with the human component of the environment, and his intellectual flexibility in the face of change all impact the mission. The C2 system faces difficulties placed on the tactical Internet and system hardware by the urban environment, by the increased volume of information, and by requirements to support the dynamic decision making necessary to execute successful UO.

Unity of Command

4-57. Although severely challenged, the principle of unity of command remains essential to UO. However, the number of tasks and the size of the urban area often require that Army forces operate noncontiguously. Noncontiguous operations stress the C2 system and challenge the commander's ability to unify the actions of his subordinates, apply the full force of his combat power, and achieve success. To apply this crucial principle in an urban environment requires centralized planning, mission orders, and highly decentralized execution. The method of C2 that best supports UO is mission command (see FM 6-0). Mission command permits subordinates to be innovative and operate independently according to clear orders and intent as well as clearly articulated ROE. These orders and ROE guide subordinates to make the right decision when facing—

- A determined, resolute, and adaptive threat.
- A complex, multidimensional battlefield.
- Intermittent or complete loss of communications.
- Numerous potentially hostile civilians close to military operations.
- The constant critique of the media and military pundits.

4-58. Decentralized execution allows commanders to focus on the overall situation—a situation that requires constant assessment and coordination with other forces and agencies—instead of the numerous details of lower-level tactical situations. Fundamentally, this concept of C2 requires commanders who can accept risk and trust in the initiative, judgment, and tactical and technical competence of their subordinate leaders. Many times, it requires commanders to exercise a degree of patience as subordinate commanders and leaders apply mental agility to novel situations.

Political and Media Impact

4-59. Commanders of a major operation consider how the need to maintain a heightened awareness of the political situation may affect their exercise of C2. A magnified political awareness and media sensitivity may create a desire to micromanage and rely solely on detailed command. Reliance on this method may create tactical leaders afraid to act decisively and with speed and determination—waiting instead for expected guidance from a higher-level commander. Threats may capitalize on this hesitation by conducting operations faster than Army forces can react. Mission orders that express the overarching political objectives and the impact of inappropriate actions, combined with training and trust, will decrease the need for detailed command. Leaders must reduce a complex political concept to its simplest form, particularly at the small-unit level. Even a basic understanding will help curtail potentially damaging political actions and enable subordinates to make the often instantaneous decisions required in UO—decisions that support military and political objectives.

Commander's Visualization

> *I heard small-arms fire and RPG explosions and felt shrapnel hit the vehicle.... Land navigation at this time was impossible; every time I tried to look out, I was thrown in a different direction.... At this time, I was totally disoriented and had not realized we were on our own.*
>
> Captain Mark Hollis
> "Platoon Under Fire"

4-60. Leaders at all levels need to see the battlefield to lead Soldiers, make effective decisions, and give direction. Sensors and other surveillance and reconnaissance assets alone cannot provide all the information regarding the urban environment that commanders will need. The focus of lead elements narrows rapidly once in contact with a hostile force limiting their assessment to the local area. Therefore, tactical commanders will not be able to observe operations from long, stand-off ranges. Their personal observation remains as critical in urban areas as elsewhere and helps to preclude commanders from demanding their subordinates accomplish a task or advance at a rate inconsistent with the immediate situation. In urban offensive and defensive operations, seeing the battlefield requires that commanders move themselves and their command posts forward to positions that may be more exposed to risk. Thus, commanders modify their C2 system capabilities to make them smaller, reduce their signature, and increase their mobility. Because of the greater threat to C2, security efforts may be more intense.

4-61. In stability operations, commanders often intervene personally to reassure the urban population and community and faction leaders about the intentions of Army forces. In these type operations, threats may attack leaders to gain the greatest payoff with the least expenditure of resources. Commanders carefully evaluate risk and potential benefits of such exposure. These risks however, cannot stop them from seeing the battlefield, personally intervening in situations as appropriate, and leading their Soldiers.

4-62. Commander's visualization also requires having an accurate understanding of friendly and enemy locations, detailed maps, other appropriate intelligence products, and INFOSYS that accurately depict the urban environment and help establish a COP. The reliability of these items is as important to planning major operations as it is to tactical-level operations. The commander of the major operation ensure that subordinate tactical-level commanders have the necessary products to achieve accurate situational understanding and dominate the urban environment as subordinate commands often lack the personnel or assets to develop these products. Frequently, satellite or aerial imagery is requested to compensate for the drastic changes that can occur due to UO, natural disasters, and outdated or imprecise maps. (Even maps developed and maintained by urban area's administrative activities may not be up-to-date. Extensive and continually expanding shantytowns, for example, may not be mapped at all. Maps may have even been purposefully distorted or critical detail intentionally omitted. The systems used to transliterate some languages such as Arabic and Chinese to Anglicized alphabets often result in the same location being spelled several—and frequently considerably different—ways. Maps may also assign names to features that are completely different than those used by locals to refer to them.)

4-63. Other critical intelligence products needed in the COP include overlays or gridded reference graphics. (Whenever possible, gridded reference graphics should conform to standard military grid reference system formats to reduce the probability of error when entering target coordinates into targeting systems that use global positioning systems.) These products should be developed and distributed to all participants prior to the UO. Overall, their focus should be on ease of reference and usefulness for all forces—ground and air (see Appendix B). Overlays and graphics can also portray important societal information or urban infrastructure, such as—

- Religious, ethnic, racial, or other significant and identifiable social divisions.
- Locations of police, fire, and emergency medical services and their boundaries or zones of coverage.
- Protected structures such as places of worship, hospitals, or other historical and culturally significant buildings or locations.
- Underground subway, tunnel, sewer, or water systems.
- Bridges, elevated roadways, and rail lines.
- Electrical generation (to include nuclear) and gas storage and production facilities and their distribution lines.
- Water and sewage treatment facilities.
- Telephone exchanges and television and radio stations.
- Toxic industrial material locations.

Mental Flexibility

4-64. Commanders conducting UO must remain mentally flexible. Situations can change rapidly because of the complexity of the human dimension. Typical of the change is a stability operation that suddenly requires the use of force. Commanders must be capable of quickly adjusting their mental focus from a noncombat to combat situation. Equally important is dealing with populations during combat operations. Consequently, commanders must also be capable of rapidly adjusting plans and orders for sudden stability or civil support tasks that emerge during or soon after a combat mission. In developing their vision, commanders must consider the second- and third-order effects of UO.

Information Systems

4-65. The urban environment also challenges INFOSYS that support the commander, especially communications. Urban structures, materials, densities, and configurations (such as urban canyons) and power constraints associated with man-portable radios significantly degrade frequency modulation (FM) communications. This causes problems at brigade-level and below where commanders rely heavily on constant FM radio contact with subordinates. Tactical communication problems might also cause an inability to maintain a COP, to give orders and guidance, to request support, or to coordinate and synchronize elements of the combined arms team. Communication problems in urban areas can prevent the achievement of information superiority and contribute directly to mission failure. In UO, allocating critical or high-value communication assets will be significant and essential to the main effort.

4-66. In an urban environment, units and staffs properly prepare for and mitigate the communication problems in urban areas (see figure 4-3). Adequate communications, in most cases, are ensured by—

- Training in and use of retransmission and relay sites and equipment, which may include unmanned aircraft systems (UAS).
- Airborne command posts, satellite communications, high-frequency radios, and other redundant communication platforms and systems.
- Careful positioning of commanders, command posts, and antennas to take advantage of urban terrain characteristics.
- Detailed communications analysis for movement from one AO to another due to the likely density of units operating in the urban environment.

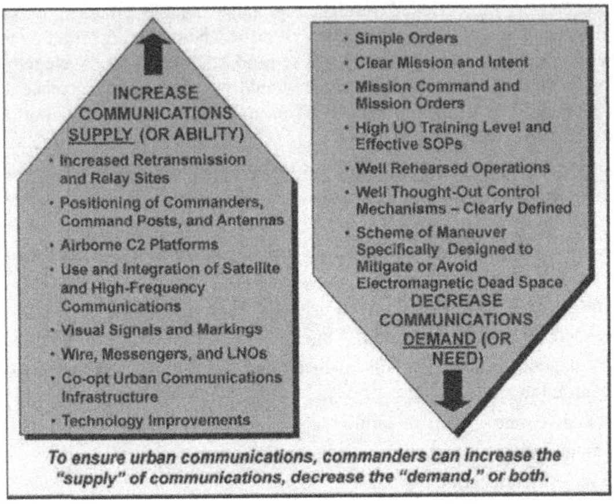

Figure 4-3. Methods to overcome urban communications challenges

4-67. Standing operating procedures (SOPs) for visual markings (both day and night) may assist in command and control. These SOPs indicate unit locations and other essential information. They coordinate with units across common boundaries. Given adequate consideration to limitations on multinational capabilities, these SOPs may assist in command and control and preclude fratricide incidents resulting from loss of FM communications. However, visual signals, including pyrotechnics, are less effective in buildings and enclosed spaces.

4-68. In defensive, stability, or civil support operations, positions often do not change as frequently as in offensive operations. Urban commanders then rely more on military wire (properly camouflaged amongst the civilian communications infrastructure), commercial communications, and messengers. Even in combat, some if not all of the urban area's organic communications structure remains intact for Army use. For example, every building may have one or more telephone distribution boxes that can control hundreds of individual telephone lines. Setting up wire communications using these points is relatively simple but, like all wire communications, is susceptible to wire-tapping. Cellular telephones can usually work well in urban areas; however, locating and destroying the repeater stations or other land-based elements of the cellular telephone system (or the effects of natural disasters) can easily disable them. (Cell phones may be a critical and only means to rapidly communicate with key civilian organizations and important community leaders.) Consequently, the C2 system may use these alternatives to FM communications but with proper operations and physical security procedures in place.

> **Example of Simple Communications Innovation:**
> **Israel's Six-Day War – 1967**
>
> In the 1967 battle for Jerusalem, the Israeli Defense Force placed flags on top of cleared buildings so that aircraft providing close air support could monitor the Israeli forward line of troops. They also used a spotlight during the night to mark specific buildings as close air support targets.

4-69. Units will likely use multiple means to communicate throughout the urban area. Hence, commanders emphasize proper operations security procedures (OPSEC) despite the level of security provided by the communications system. This emphasis helps to lessen the probability that Soldiers will inadvertently compromise essential information as they switch from one mode of communications to another (for example, from secure FM radio to unsecured cellular telephones or from classified to unclassified Internet domains).

4-70. Command posts above brigade-level ensure that they can communicate in an urban area without significant disruption. In stability and civil support operations, immediate and reliable communications between tactical and strategic levels may be necessary. Higher commanders anticipate that although the urban area does not significantly challenge their INFOSYS, the area may severely challenge systems at the lower tactical levels. For this reason, information flow from lower to higher may take longer. If the situation is not acceptable, the higher headquarters takes steps to mitigate it, such as increasing the number of liaison officers operating with units engaged in decisive operations. In some instances, the scheme of maneuver may be specifically designed to account for communications interference, propagation characteristics, and electromagnetic dead space. (However, this will require more time, resources, and a detailed urban communications IPB.)

4-71. Finally, urban areas can overload the INFOSYS with information. UO across the spectrum of operations can generate large volumes of information when crises threaten. This sheer volume can easily overwhelm UO commanders and command posts, and the information conduit connecting the two. Training prepares command posts to handle this volume of information and to filter the critical from the merely informative. Staffs must create products (visual or textual) that help their commanders understand the urban environment, not just present them information to know.

KEY TACTICAL CONSIDERATIONS

4-72. Commanders and planners of major UO must thoroughly understand the tactical urban battle. Especially, they must understand the effects of the environment on men, equipment, and systems. The complexity of urban environment changes and often compresses many tactical factors typically considered in the planning process. Figure 4-4 shows some of these compressed factors, however, commanders and their staffs should carefully review FM 3-06.11 for doctrine to support tactical urban operations.

Figure 4-4. Compressed tactical factors

TIME

4-73. The time available to think and act is compressed in urban combat operations. The tactical engagements that comprise battles and major UO are often quick and decisive; therefore, higher-level commanders require the ability to conduct battle command on the move so that their decision making remains correspondingly fast. The impact of decisions (or lack of) and the outcome of battle can occur in mere minutes. Often the amount of information and the number of decisions can overwhelm the overall ability of INFOSYS to respond. Commanders have little time to influence tactical actions with resources kept in reserve. Reserves and fire support assets are close to the point of decision so that they can respond in time to make a difference. The terrain causes C2 challenges that further inhibit commanders from responding quickly to changes in the situation. Small unit leaders receive training that emphasizes understanding the commander's intent so that they can recognize tactical opportunities and can act quickly to take advantage of them.

DISTANCES AND DENSITY

4-74. Distances in UO are compressed to correspond to the density of threat forces and noncombatants. In open terrain, squads, platoons, and companies may be able to control or influence thousands of meters of space. In UO, large buildings can absorb the efforts of several companies or battalions. Crowds of thousands can assemble in areas of a few hundred meters requiring correspondingly large forces for control. Maximum engagement ranges, as influenced by the urban terrain, are usually closer. Units may require field artillery for direct fire at targets ranging fewer than a hundred meters. Commanders and staffs understand the telescoping nature of the battlefield, the density of threat forces, and the density of noncombatants. In addition to the actual conduct of urban tactical operations, these factors will directly affect training, planning, force deployment, and strength.

4-75. Time-distance considerations are especially important throughout planning cycles. Though distances may be short, the physical nature of the environment can drastically change the planning factors for unit movements. The advance of a battalion may be measured in hundreds of meters per day. Thus, all time and distance calculations that relate to sequencing of forces, synchronizing combat power and other capacities, and making decisions require reevaluation based on the urban conditions.

4-76. Although it may appear counterintuitive, airspace above the urban area may be also be severely compressed as multiple fixed-wing, rotary-wing, and indirect fires (including multinational assets) compete for the same space. Increasing this density is the use of airspace by civilian aircraft and the proliferation of tactical UAS. Due to the potential for a high volume of air traffic, commanders and planers pay close attention to the integration and deconfliction of airspace over urban areas. Commanders should consider specific techniques and procedures for UAS which may include—

- Defining airspace control areas.
- Regulating flight times.
- Creating altitude restrictions.
- Including UAS on the air tasking order.

COMBAT POWER

4-77. The urban terrain can also compress combat power. This terrain increases the utility and effects of some weapons and systems, increasing overall combat power. One system that dramatically demonstrates this effect is the sniper. In open terrain, snipers slightly influence operations. In UO, snipers—well concealed, positioned, and protected—can take on significance disproportionate to their combat capability in other situations (see FM 23-10).

4-78. The density of ground combat power in a given size area is also increased because of the effect of the terrain on ranges. The complex terrain precludes standoff engagement from extended ranges by dispersed forces. Commanders often position weapon systems closer together and at shorter ranges to mass effects on the same target. Thus, commanders may position armored vehicles, which typically position themselves Hundreds of meters from friendly troops and other vehicles, within a few meters of each other to provide mutual support. Targets, which in open terrain are engaged at thousands of meters, are engaged in tens of meters on the urban battlefield.

4-79. The dense clutter of the urban environment also affects target acquisition. Systems, such as radar optimized for open terrain, will not be able to acquire targets as effectively. Decreased acquisition capability equates to diminished combat power. It may also require increasing the density of acquisition systems to compensate for reduced capability.

4-80. Finally, the density of combat power may also increase the vulnerability of Army forces. Many Army systems are protected from enemy systems at longer ranges. The number of enemy systems that can threaten Army forces at a short range increases dramatically. Lack of dispersal will make it more likely that multiple Army systems can be targeted by a single enemy threat.

LEVELS OF WAR

4-81. The levels of war are also compressed in the urban area. The tactical actions of individuals and small units can directly influence operational and even national and strategic objectives. Conversely, the decisions of the President can directly affect the conduct of tactical operations. UO have short cause and effect links between the tactical, operational, and strategic levels of operations. Because of the close media scrutiny of UO, the President can sometimes observe the actions of platoons in real time. For example, the media may film a platoon applying nonlethal force for crowd control. The President can view that film on the nightly news before the platoon even disengages from the action, much less reports formally through the various levels of command. If appropriate, the President can decide and direct the strategic and operational commanders to adjust ROE before the platoon has reported. Therefore, commanders at all levels must understand the urban environment's potential compressive effects on the levels of war. A major impact of these effects can be a lower tolerance for tactical errors and a greater need for detailed planning and precision in execution and weapons' effects (lethal and nonlethal).

DECISION MAKING

4-82. The nature of the urban environment compresses the time available to make decisions and increases the number of decisions to make. This is particularly true at the lower tactical levels. Units observing an urban AO face more potential unknowns than in other situations. A large structure presents many more potential firing positions that are observed than simpler terrain. Movement in one of those windows forces the soldier or unit to quickly make a decision regarding the nature of the target—deciding whether it is a threat or a noncombatant. Incorporating combatant and noncombatant discriminatory considerations into all live-fire training will improve Soldiers' ability to make these critical judgment decisions. Overall, lack of understanding regarding the urban environment requires commanders to rely more on analytic decision making, while greater understanding facilitates adaptability and allows commanders to make rapid and intuitive decisions while still accounting for second, third, and higher-order effects (see Figure 4-5 and FM 6-0). Realistically, however, all decisions made during urban operations will likely require a combination of analytic and intuitive decision-making abilities.

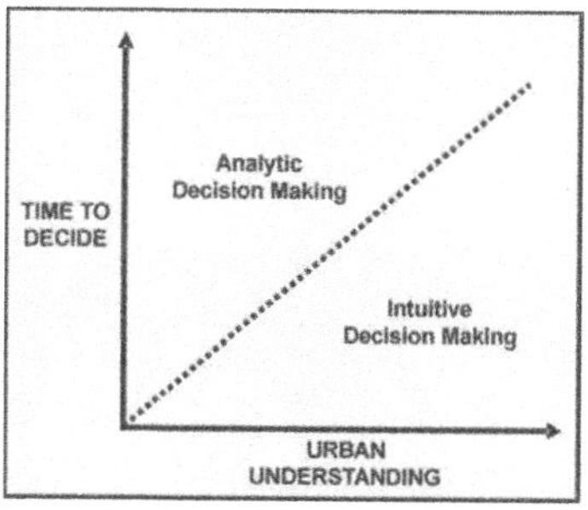

Figure 4-5. Urban understanding and decision making

This page intentionally left blank.

Chapter 5

Contemplating Urban Operations

We based all our further calculations on the most unfavorable assumptions: the inevitability of heavy and prolonged fighting in the streets of Berlin, the possibility of German counter-attacks from outside the ring of encirclement from the west and south-west, restoration of the enemy's defence to the west of Berlin and the consequent need to continue the offensive.

General of the Army, S. M. Shtemenko
describing operational level planning for taking Berlin
The Soviet General Staff at War

In any potential situation and in any area, Army commanders will likely need to assess and understand the relevance and impact of one or more urban areas on their operations. They will also need to determine whether full spectrum urban operations (UO) will be essential to mission accomplishment. UO may be the commander's sole focus or only one of several tasks nested in an even larger operation. Although UO potentially can be conducted as a single battle, engagement, or strike, they will more often be conducted as a major operation requiring joint resources. Such actions result from the increasing sizes of urban areas. Army commanders of a major urban operation must then ensure that UO clearly support the operational objectives of the joint force commander (JFC), requesting and appropriately integrating critical joint resources. Whether the urban operation is the major operation itself or one of many tasks in a larger operation, Army commanders must understand and thoroughly shape the conditions so subordinate tactical commanders can engage and dominate in this complex environment.

A ***major operation*** is a series of tactical actions (battles, engagements, strikes) conducted by various combat forces of a single or several services, coordinated in time and place, to accomplish operational, and sometimes strategic objectives in an operational area.

DETERMINING THE NECESSITY OF URBAN OPERATIONS

5-1. Early in planning, commanders of a major operation must address the necessity and feasibility of conducting operations in urban areas located throughout their areas of operations (AOs). Chapter 1 discussed strategic and operational considerations that compel forces to operate in urban areas. These reasons include the location and intent of the threat force; critical infrastructure or capabilities that are operationally or strategically valuable; the geographic location of an urban area; and the area's political, economic, or cultural significance. Additionally, humanitarian concerns may require the control of an urban area or necessitate operations within it. Several considerations exist, however, that may make UO unnecessary, unwarranted, or prohibitively inefficient. When determining whether to operate in an urban environment, commanders must consider the operational (and accidental) risks and balance them with mission benefits. The factors shown in Figure 5-1 highlight some measures to evaluate the risks associated with UO.

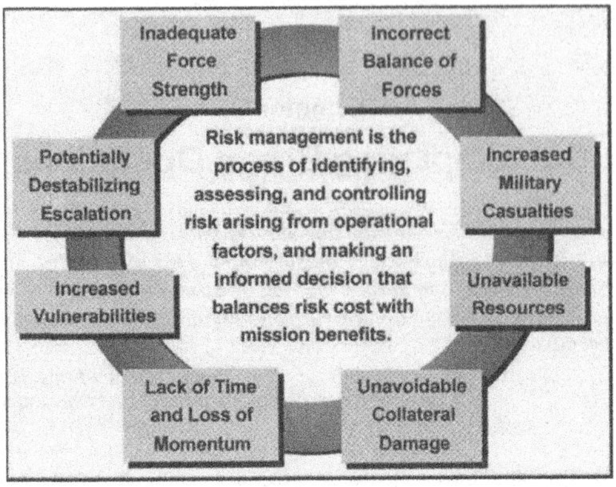

Figure 5-1. Risk management and the associated risks with urban operations

INADEQUATE FORCE STRENGTH

5-2. When facing prospective UO, commanders must consider if they have the necessary troops available to conduct the operation properly and within acceptable risk. Under normal circumstances, large urban areas require many forces merely to establish control. The New York City police department has over forty thousand officers and hundreds of support staff simply to conduct peacetime law enforcement. Major UO, particularly those that are opposed, will often require a significant number of forces. Offensive missions, for example, can require three to five times greater troop density than for similar missions in open terrain. If commanders lack sufficient force to conduct effective operations, they may postpone or consider not initiating those operations until they have the necessary strength. Commanders should also add to their analysis the requirements for troop strength elsewhere in the AO. Additionally, commanders must consider the number (and type) of forces required to make the transition from urban offensive and defensive to stability or civil support operations when determining overall force requirements—not just the number of forces required to realize objectives for major combat operations.

INCORRECT BALANCE OF FORCES

5-3. Along with force strength, commanders must consider the type and balance of forces available. This consideration includes an assessment of their level of training in urban operations. Generally, UO put a premium on well-trained, dismounted infantry units. The operational or tactical necessity to clear threat forces from the dense urban environment, hold hard-won terrain, and interact with the urban population greatly increase dismounted requirements. Therefore, Army forces conducting UO are often force tailored to include a larger infantry component. In addition, special operations forces (SOF) are invaluable in UO. SOF include psychological operations (PSYOP) and civil affairs (CA) forces. They should always be considered as part of the task organization.

5-4. UO must include combined arms to ensure tactical success in combat. Although masses of heavy forces may not be required, successful UO require the complementary effects of all Army forces. Even if an urban operation is unlikely to involve offensive and defensive operations, armor, combat engineers, and field artillery may be essential to mobility and force protection. In urban stability and civil support operations, successful mission accomplishment requires more robust CA organizations. They are also valuable in urban offensive and defensive operations. While commanders may have sufficient combat and combat support forces, they may lack enough sustainment forces to provide the support necessary to maintain the tempo. Again, commanders without balanced types of forces, to include their proficiency in operating in urban environments, should consider alternatives to UO or delaying UO until proper force types are trained and available in sufficient numbers.

INCREASED MILITARY CASUALTIES

5-5. Casualties in UO are more likely than in operations in other environments. In urban offense and defense, friendly and threat forces often engage at close range with little space to maneuver. The urban terrain provides numerous advantages to the urban defender; higher casualties occur among troops on the offensive, where frontal assaults may be the only tactical option. Conversely, defenders with limited ability to withdraw can also suffer high casualties when isolated and attacked. Casualties can be more difficult to prevent in urban stability operations because of the dense complex terrain, the close proximity of the urban population, and the possible difficulty in distinguishing friend from foe. The potential for high casualties and the subsequent need for casualty evacuation under difficult circumstances make the positioning and availability of adequate medical resources another important consideration. Additionally, high-intensity urban combat and the potential for increased stress casualties may require additional units to allow for adequate unit rotations so that Soldiers receive the rest they require.

5-6. Though casualties occur in all operations, commanders should recognize the likelihood of more casualties during large-scale or high-intensity UO. During the battle for Hue in 1968, for example, many company-size units suffered more than 60 percent casualties in only a few days of offensive operations. Commanders conducting urban stability operations must know the casualty risk and how it relates to national and strategic objectives. While a lower risk normally exists in stability operations than in offensive and defensive operations, just one casualty may adversely impact the success of the stability mission. A realistic understanding of the risk and the nature of casualties resulting from UO critically affect the decision-making process. If commanders assess the casualty risk as high, they must ensure that their higher headquarters understands their assessment and that the objectives sought within the urban area are commensurate with the anticipated risk.

UNAVAILABLE RESOURCES

5-7. Offensive and defensive operations in an urban environment put a premium on certain types of munitions and equipment. Forces may want to use vast amounts of precision munitions in the urban environment. At the tactical level, they will likely use more munitions than during operations in other environments. These munitions include—

- Grenades (fragmentation, concussion, stun, riot control, and smoke).
- Mortar ammunition (due to its rate of fire, responsiveness, and high-angle fire characteristic).
- Explosives.
- Small arms.

5-8. Soldiers need access to special equipment necessary to execute small-unit tactics effectively. In urban stability and civil support operations, this equipment may include antiriot gear, such as batons, protective clothing, and other nonlethal crowd-control devices. In urban offensive and defensive operations, special equipment can include sniper rifles, scaling ladders, knee and elbow pads, and breaching equipment such as door busters, bolt cutters, and sledgehammers. Soldiers can conduct UO with standard clothing and military equipment. However, failure to equip them with the right types and quantities of munitions and special equipment will make mission success more difficult and costly. When commanders consider whether to conduct UO, they must evaluate the ability of logistics to provide the necessary resources (see Chapter 10). Considerations must include the ability to supply all Soldiers regardless of branch or military occupational specialty with urban-specific equipment.

UNAVOIDABLE COLLATERAL DAMAGE

5-9. UO require an expanded view of risk assessment. When considering risk to Army, joint, and multinational forces, commanders must also analyze the risk to the area's population and infrastructure. This comprehensive analysis includes the second- and third-order effects of significant civil casualties and infrastructure damage. Collateral damage can influence world and domestic opinion of military operations and thus directly affect ongoing operations. It also influences the postconflict physical environment and attitudes of the population. Negative impressions of the civilian population caused by collateral damage can take generations to overcome. Destroying an urban area to save it is not a viable course of action for Army commanders. The density of civilian populations in urban areas and the multidimensional nature of the environment make it more likely that even accurate attacks with precision weapons will injure noncombatants. Unavoidable collateral damage of sufficient magnitude may justify avoiding UO, which, though it may be tactically successful, may run counter to national and strategic objectives.

LACK OF TIME AND LOSS OF MOMENTUM

5-10. Commanders conducting major operations must analyze the time required to conduct UO successfully. UO can be time consuming and can require larger quantities of resources. The density of the environment, the need for additional time to conduct a thorough reconnaissance, the additional stress and physical exertion imposed on Army forces operating in urban areas, and the potential requirements to care for the needs of the urban population consume time and slow momentum. Commanders cannot permit UO conducted as a shaping operation to divert resources from the decisive operation. Nor can they allow UO to interrupt critical time lines, unnecessarily slow tempo, or delay the overall operation. Threat forces may conduct UO with the primary purpose of causing these effects. Commanders must recognize that time generally works against political and military objectives and, hence, they must develop plans and operations to avoid or minimize UO that might unacceptably delay or disrupt the larger, decisive operation. Once commanders achieve major combat objectives, however, they will likely need to shift resources and focus on the urban areas that they previously isolated and bypassed.

INCREASED VULNERABILITIES

5-11. Commanders must weigh the potential for increased vulnerabilities when executing UO. The density of the environment makes protection (safety, field discipline, force protection, and especially fratricide avoidance) much more difficult. Forces operating in a large urban area increase their risk of isolation and defeat in detail. Joint capabilities, such as air power, may work less effectively to support a close urban battle than in some other environments. Thus, responding to unexpected situations or augmenting disadvantageous force ratios when applying joint capabilities is significantly more difficult. Although organized, trained, and equipped for success in any environment, the Army vulnerability to weapons of mass destruction (WMD) increases when forces concentrate to conduct UO. Commanders may consider not committing forces or limiting the size of a force committed to an urban area because of increased vulnerability to (and likelihood of) attack by WMD.

5-12. Fratricide avoidance is a matter of concern for commanders in all operations. The complex urban terrain and density of participating forces coupled with typical battlefield effects—smoke, dust, burning fires—and weather effects—fog, snow, rain, and clouds—immensely increase the potential for urban fratricide. Additionally, safety requirements make it difficult to fully replicate training conditions that allow leaders to become more aware of the conditions that contribute to fratricide. The effects of fratricide can be devastating to UO and spread deeply within the Army force. Critical effects include—

- Needless loss of combat power.
- Decreased confidence in leadership, weapons, and equipment. These lead to a loss in initiative and aggressiveness, failure to use supporting combat systems, and hesitation to conduct limited visibility operations.
- Disrupted operations and decreased tempo.
- General degradation of cohesion and morale.

5-13. Therefore, commanders should increase fratricide awareness and emphasis on prevention measures during UO. Causes can be procedural, technical, or a combination of the two and include—

- Combat identification failures due to poor situational understanding, lack of communication, failure to effectively coordinate, and short engagement ranges coupled with the need for quick reaction.

- Location errors involving either the target or enemy forces due to poor situational understanding.
- Inappropriate command and control and fire support coordinating measures; a failure to receive, understand, or adhere to these measures.
- Imprecise weapons and munitions effects such as, an antitank round that penetrates several walls before exploding near friendly forces.

Exacerbating these difficulties will be the likelihood of Army forces conducting operations with (or within proximity of) SOF, coalition forces, and indigenous security forces including local police. (During OPERATION IRAQI FREEDOM, the head of the Australian Defence Forces SOF in Iraq made it a point to carefully coordinate with U.S. Army forces—down to the company level—whenever his personnel would be in an American AO. He felt the danger of fratricide was a much greater risk than the loss of operational security.)

POTENTIALLY DESTABILIZING ESCALATION

5-14. In the urban environment, Army forces cannot avoid close contact with enemy forces and civilians that may potentially become hostiles. In urban stability and civil support operations, commanders should consider the chance of this contact escalating into confrontation and violence, which may become destabilizing. This consideration may delay, limit, or altogether preclude UO using Army forces.

CONSIDER ALTERNATIVES AND RISK REDUCTION MEASURES

5-15. Since UO are often high risk, commanders should consider courses of action that provide alternatives. When the objective of an urban operation is a facility, commanders should consider replicating that facility outside of the urban area. For example, a critical requirement for an airfield to sustain operations may lead commanders to consider UO to seize or secure one located in an urban area. However, if adequate resources exist (especially time and adequate general engineering support), Army forces may build an airfield outside of the urban area and eliminate the need to conduct the urban operation. Similarly, logistics over-the-shore operations may be an alternative to seizing a port facility. In some situations, the objective of UO may be to protect a political organization such as a government. Relocating the government, its institutions, and its personnel to a safer area may be possible. Commanders can also design an operation to avoid an urban area. For example, if an urban area dominates a particular avenue of approach, use a different avenue of approach. Using a different avenue of approach differs from isolating and bypassing because the entire operation specifically makes the urban area irrelevant.

5-16. If commanders execute UO, they must assess potential hazards, and then they develop controls to either eliminate or reduce the risks to Army forces. The first means to offset risk is always to ensure a thorough understanding of the urban environment and its effects on operations by all members of the force. Other measures to mitigate risk may include—

- Detailed planning to include thorough intelligence preparation of the battlefield and the development of appropriate branches and sequels.
- Integrated, accurate, and timely intelligence, surveillance, and reconnaissance (ISR).
- Clear missions and intent, which includes a well-articulated end state that looks beyond the cessation of combat operations.
- Sufficient reserves and rotation of forces.
- Vigilant physical security precautions to include increased use of barriers and other defenses, particularly when urban areas are used as support areas.
- Operative communications and other information systems (INFOSYS).
- Effective populace and resources control measures.
- Comprehensive and flexible rules of engagement (ROE) continuously reviewed to ensure they remain adequate for the situation.
- Sufficient control measures (which often include a common urban reference system) and standard marking and identification techniques that adequately consider limited visibility concerns for both air and ground forces. Measures should allow commanders to satisfactorily

control UO and minimize fratricide without unreasonably restricting subordinate commanders' ability to accomplish assigned missions. Commanders must ensure that all subordinate units thoroughly disseminate any approved nonstandard reference systems.

- Proper targeting procedures (including effective fire support coordinating measures and a streamlined legal review of targets), positive identification of targets, and controlled clearance of fires. The goal is achievement of precise (yet rapid) effects with both lethal and nonlethal means. In close air support, positive air-to-ground communications are essential to coordinate and authenticate markings.
- Well-synchronized information operations (IO) that begin before introducing Army forces into the urban environment and well through transition. Commanders should emphasize vigilant operations security (OPSEC) particularly when operating closely with the media, nongovernmental organizations (NGOs), and elements of the civilian population.
- Active and effective integrating, synchronizing, and coordinating among all forces, agencies, and organizations involved in the operation. Commanders should allow adequate planning and rehearsal time for subordinates.
- Responsive, sustainable, and flexible urban sustainment.
- Forces well trained in joint, interagency, multinational, and combined arms UO.
- The creation of adaptable, learning organizations. This requires thorough after-action analyses conducted during actual operations as well as after training exercises. In addition to official Army sites (such as the Center for Army Lessons Learned and the Battle Command Knowledge System), commanders must create a unit-level system to allow hard-won, lessons learned and tactics developed to be immediately passed on to other units and soldiers—even in the midst of an operation. This system may be technology-based, procedural, or both. (During OPERATION IRAQI FREEDOM, the 1st Cavalry Division developed an effective web-based knowledge network—called CAVNET—that allowed them to actively capture and share lessons learned among subordinate units.)

INTEGRATION INTO LAND OPERATIONS

5-17. The commander of the major operation, after determining that urban operations are required, will then integrate the urban operation into his overall operation. He does this by articulating his intent and concept for the urban operation to his subordinates. The commander of the major operation must also set the conditions for successful tactical urban operations by his subordinates. He should define ROE, focus ISR efforts, task organize his capabilities, ensure information superiority, design the operational framework, and initiate and sustain effective coordination with other agencies and organizations (see FM 6-0).

CONCEPT OF THE OPERATION

5-18. The commander's concept of the operation should address all operationally important urban areas in his AO. It should also articulate his vision of the urban operation through directions to his staff and subordinates. Subordinate commanders should address urban areas that the higher commander does not specifically address. The commander's concept should discuss each urban area in terms of task and purpose (see FM 5-0). The commander should also describes his vision of the situation's end state in terms of—

- The threat.
- The urban environment (terrain, society, and infrastructure).
- Friendly forces.
- The conditions necessary to transition control of urban areas within his AO to another agency or back to legitimate civilian control.

RULES OF ENGAGEMENT

5-19. National- or joint-level command authorities may develop urban-specific ROE. If not, Army commanders, as part of their assessment, must determine if urban-specific ROE are required for their situation and provide supplemental ROE. However, commanders must forward any conflicts or incongruities to their higher headquarters for immediate resolution.

5-20. Developing effective ROE relies on thoroughly understanding the national and strategic environment and objectives. It also relies on understanding how to conduct urban operations at the tactical level including weapons effects. For example, broad ROE may result in significant collateral damage and civilian casualties. Even in a major operation or campaign, significant collateral damage caused during UO can make postcombat operations difficult. Such damage may even change national and international public opinion or threaten the achievement of national and strategic objectives. In contrast, restrictive ROE can hamper tactical operations causing mission failure, higher friendly casualties, or both. ROE are often part of essential elements of friendly information (EEFI), protected to reduce the potential for threat exploitation. Even in a limited urban operation, ROE will frequently need to change as circumstances warrant. Therefore, commanders should plan ROE "branches" for anticipated changes in the operational environment.

5-21. In urban operations, ROE are flexible, detailed, and understandable. They should preclude the indiscriminate use of deadly force while allowing soldiers latitude to finish the mission and defend themselves. ROE should recognize that the urban area is not homogenous and may vary according to the key elements of the threat and environment: terrain, society, and infrastructure. To be effective, ROE are consistent throughout the force (an increased challenge in multinational urban operations), and soldiers are thoroughly trained and familiar with them.

Enemy Considerations

5-22. The nature of an urban enemy affects ROE as well. Commanders must consider the type of enemy weapon systems, the degree of defensive preparation, the ability to target enemy vulnerabilities with precision systems, and the ability to distinguish combatant from noncombatant.

Terrain Considerations

5-23. ROE may vary according to the terrain or physical attributes of an urban area. Physical factors may drive the ROE to preclude certain types of munitions. For example, if the construction of a portion of the area is sensitive to fire, then ROE may preclude using incendiary munitions in that area. The ROE may lift this prohibition when units move into areas of masonry construction. Toxic industrial chemicals or radiological contaminants in an industrial area may also affect ROE.

Societal Considerations

5-24. The societal or human dimension of the urban environment will often affect ROE the most. Commanders must base the ROE development on a thorough understanding of the civilian population and threat. They evaluate the loyalty of the population, its dynamic involvement in activities that affects the operation, and its size and physical location. A population that is present and supports Army forces will likely elicit more restrictive ROE than a hostile population actively supporting forces opposing the Army forces. A neutral population, not actively involved in behavior affecting Army forces, supports consideration of more restrictive ROE. In all cases, ROE conforms to the law of war. However, ROE may be much more restrictive than the law of war requires.

5-25. The location of the population also affects ROE. The evacuation or consolidation of noncombatants into controlled, safe areas may result in less restrictive ROE. A U.S. or allied population that remains in the urban area conducting routine business in and amongst Army forces during noncombat UO will normally require the most stringent ROE.

Infrastructure Considerations

5-26. Commanders must consider the urban infrastructure when developing ROE. Urban infrastructure vital to current or future Army operations may dictate that commanders adjust ROE to ensure that critical elements of the infrastructure remain intact during the conduct of operations. If Army forces conduct an urban operation to capture port facilities, for example, the ROE should address damage to the key facilities that are the objective of the operation.

RESOURCE ALLOCATION

5-27. Commanders of a major operation must ensure that subordinate tactical commanders have the resources necessary to conduct UO effectively. They assign appropriate forces to subordinate commanders tasked to conduct UO; support them with Army forces at the operational level; and request and coordinate their support by joint resources.

Task Organization

5-28. Task organizing subordinate units for urban operations will depend largely on the nature of the operation. Some units, however, are always part of the task organization to ensure the success of UO. Infantry, CA, aviation, military police, PSYOP, military intelligence, and engineers are units required for all urban operations across full spectrum operations. Other type forces—such as armor, artillery, and chemical—have essential roles in specific types of urban operations but may be less applicable across the spectrum of operations. Commanders and staffs of a major operation must understand their mission, the particular urban environment in which they operate, and the general effects of the environment across the warfighting functions to allocate the appropriate forces to their tactical commanders. See Chapter 4 for details.

Operational-Level Support

5-29. Commanders of a major operation should also support the tactical commander with forces remaining under their direct control. These forces can include Army SOF, such as CA, PSYOP, and Special Forces, ground and air cavalry, aviation, logistics, engineers, and communications support. These forces may not be under operational control of the supported command, but their efforts are synchronized and coordinated.

Coordinating and Requesting Joint Support

5-30. Commanders of a major operation will often provide forces to the JFC as well as receive assets. They will also coordinate for and integrate joint assets to support the tactical battle. These assets will usually include air support, such as close air support, tactical airlift, and aerial reconnaissance and surveillance. Intelligence support comes in the form of reach to strategic and national intelligence capabilities and to space-based systems. This reach to space assets provides reliable, robust long-range communications, environmental monitoring, ISR, positioning and navigation, and warning of enemy missile launch. Special operations capabilities can assist the tactical mission with special reconnaissance and direct action (including special operations aviation) against high-payoff targets. Joint resources also provide the Army forces augmentation by Marine ground forces. In coastal areas, Navy forces and Coast Guard elements assist Army forces with security, sealift, and fire support. Additionally, the potential agility, reach, and power projection provided by Navy forces and platforms, combined with their self-sustaining capability, can play a vital role in supporting UO. This is especially true when UO are conducted in the littorals, require forcible entry operations, and when host nation infrastructure is severely taxed or destroyed. Commanders of a major operation will coordinate with the JFC regarding available joint resources and their allocation. They then ensure that their efforts coordinate with and complement those of tactical Army forces in the urban area. Appendix D discusses the potential contribution of joint capabilities to Army UO.

URBAN ISR

5-31. Commanders at all levels require accurate and timely information to conduct assessments for successful urban operations. This is critical to planning and execution. Senior commanders have a large role in coordinating the urban ISR effort. National strategic sources (as well as open sources) provide most of the information that commanders and staffs require on the characteristics of the human dimension, the physical properties of the terrain, and the infrastructure. The general characteristics of these aspects of the urban environment do not change drastically over time, with one exception. Military operations or natural disasters can change physical characteristics radically. Analysts can obtain crucial information through diligent research of intelligence databases and open sources. However, the disposition and composition of the urban threat is time sensitive and not likely to be discovered through this type of investigation. Due to the effects of the urban environment, deceptive efforts may influence the threat more easily. The urban population is dynamic and critical aspects of this dimension must be updated or confirmed as a prelude to urban operations. Surveillance and reconnaissance (including geospatial assets) provide relevant information regarding threat dispositions, composition and the state of the population, and the specifics of the urban terrain. Successful urban operations depend on the successful conduct of urban reconnaissance. (See also the earlier discussion of effects on the intelligence and command and control warfighting functions in Chapter 4. While counterintelligence (CI) is an essential ISR component, it will be covered later in this chapter under IO.)

Challenges

5-32. The most significant challenge to urban ISR is physical. The physical organization and complexity of the urban terrain, both man-made and natural, challenges national strategic, operational, and tactical ISR capabilities. Commanders should understand the challenges when planning and allocating time and resources to their ISR efforts. They acknowledge that subordinate commanders will face similar challenges. Therefore, commanders consider subordinate capabilities, limitations, and needs when planning, requesting, allocating, and prioritizing ISR assets and capabilities.

Imagery Capabilities

5-33. A significant national, strategic, and operational ISR capability is imagery. Even at the tactical level, units employ unmanned aircraft systems (UAS) and vehicle and dismounted electro-optics, commercial cameras (still and video), and co-opt existing civilian surveillance systems to gain situational understanding. However, the structures of the urban area significantly degrade the information that imagery acquires and may make it susceptible to physical deception measures.

5-34. Imagery, however, is an excellent source regarding the arrangement and nature of many other physical aspects. It can be used to create updated mapping products and provide significant detail of major portions of the infrastructure. Imagery can also reveal what may be happening in structures through detailed study of patterns and other exterior indicators. Yet, the bulk of a skillful threat's forces, well positioned and concealed inside or underneath structures in the urban area, are largely immune from rapid detection by overhead imaging systems. The volume of civilian movement in an urban area will itself provide a degree of camouflage and increase the difficulty of employing pattern analysis. The success in 1999 of the Yugoslavian army concealing heavy forces when confronting NATO indicates the limits of these assets to penetrate an urban area. This leaves a requirement to supplement overhead imagery in the urban environment with a reliable vehicle-mounted or handheld imaging system. For example, thermal sights on a vehicle patrolling an urban street during limited visibility may note the hot engine of a vehicle on the side of the road, possibly indicating suspicious activity. Simply providing patrols with digital cameras or video cameras can also help provide needed information and fill in information gaps.

5-35. The requirement for high altitude and low oblique imagery in more open terrain has been supplemented by a growing need for detailed, close-up, and nearly vertical overhead shots of specific portions of the urban area. This forces collection managers and system operators to move expensive and scarce imagery collection systems closer to the target or to employ flight profiles that place them at greater risk from threat air defense systems. While gun-camera images from aircraft can provide a stand-off reconnaissance platform, UAS imagery may be one of the fastest, least risky methods by which commanders can conduct reconnaissance to update and verify current maps of the urban area and determine routes, obstacles such as damaged and destroyed buildings, and intact and destroyed bridges and other significant infrastructure. Increased UAS proliferation will require greater attention to airspace command and control procedures as well as the movement and protection of ground control stations.

5-36. As the media will likely be present in the urban environment, commanders may consider devoting intelligence resources (to include linguists) to monitoring the various local, national and international sources, particularly the medium of television. The media coverage of threat forces may provide important intelligence and relevant information of immediate tactical value. Commanders, however, should always be cognizant that threats may recognize that Army forces will be using this as a source of intelligence and, feeling no compulsion to be truthful, may use it for their own deception means.

Electronic Capabilities

5-37. The physical attributes of the urban area also diminish the effectiveness of electronic ISR capabilities. Buildings and other structures significantly disrupt radio communications in an urban area. Buildings not only make tactical radio communications difficult for the user, they also make them difficult to locate, intercept, and jam. The range and clarity of frequency modulation (FM) signals significantly diminish when antennas are located inside buildings or when buildings block line of sight between the source and receiving station. To mitigate this effect on signals intelligence means, detection capabilities often move closer to the transmission source. Thus, the threat's vulnerability to compromise by means of his FM and other wireless communications in an urban environment is much less than in many other environments.

5-38. Without losing tactical surprise and increasing risk, units cannot effectively use many electronic detection and surveillance capabilities until urban combat is imminent or perhaps already begun. Ground surveillance radars will have limited uses in urban areas because of the lack of wide-open spaces in which they best operate, but can be used along the periphery of urban areas to detect infiltration and exfiltration of threat forces. However, remote systems such as the remotely monitored battlefield sensor system and the platoon early warning device may have great utility in monitoring many of the numerous avenues of approach that cannot be kept under human surveillance due to manpower constraints. Among many urban applications, remote sensors and early warning devices can be used to monitor subsurface passageways, the entrances and exits to buildings, fire escapes, the perimeters of sustainment bases, and covert foot trails between urban and rural areas.

Human Capabilities

5-39. The limits on imagery and electronic ISR capabilities place a premium on human-based visual reconnaissance. Commanders have three types of human reconnaissance assets to augment electronic reconnaissance resources: special reconnaissance, conventional combat reconnaissance, and human intelligence (HUMINT) gathered by military intelligence from individuals and multimedia. The urban environment poses several challenges to these capabilities.

5-40. The urban area challenges special reconnaissance in several ways. First is the access to the urban area. Although avenues of approach may be numerous, concealed avenues of approach into a defended urban area may be limited and thoroughly covered. Air access is also more difficult because aircraft are detected more easily, airspace is smaller, drop and landing zones are limited or not secure, and more air defense systems probably exist. Still, special reconnaissance efforts to penetrate the urban area can be successful using unconventional techniques including high-altitude low-opening parachutes or underwater penetration.

5-41. Special reconnaissance then faces a second challenge: moving in and identifying targets in the urban area. Stealth movement in an occupied urban area is exceptionally difficult. Repositioning to new or alternate positions is also dangerous. The soldiers' ability to conceal themselves among the civil population can mitigate some of these challenges but includes inherent risks of a different nature. Also difficult is establishing observation positions that provide a field of view of several targets.

5-42. Lastly, special reconnaissance may face navigational and reporting challenges. Special reconnaissance's ability to locate themselves and communicate critical locations and routes are challenged by—

- Differences in language and numbering systems. Road signs, if they exist at all, may be removed or deliberately altered to confuse.
- Irregular street patterns.
- Outdated maps.
- Intervening structures that impede communications and global positioning systems.
- Changes to the landscape due to the effects of UO or natural disasters.
- Featureless shantytowns.

5-43. Conventional reconnaissance faces many of the same challenges as special reconnaissance. Conventional reconnaissance also may lack the advantage of surprise and the special equipment and training that provides special reconnaissance stealth capability. Conventional reconnaissance is not likely to operate undetected by the civilian population. Given the constraints discussed above on other sources, conventional reconnaissance units will likely begin their mission with much less information than they would have on threat dispositions in a less complex environment. Commanders may choose to have their reconnaissance elements fight for information in the urban area. While this high-risk option is more favorable under fluid conditions, it can be used at any time. It requires careful planning, rehearsal, and formulation of information requirements.

5-44. As previously indicated, human intelligence will be one of the most valuable sources for information regarding the situation inside an urban area. HUMINT may take advantage of the proximity and large numbers of potential informants to gather information about threat activities and capabilities. It is especially valuable because it can address all elements of the environment. HUMINT sources can describe political and religious nuances that may be significant to commanders. Such information is useful for insights regarding the human dimension but extremely difficult to obtain from other means. This intelligence also can describe the infrastructure uncovering essential details of how the infrastructure functions. Obtaining good HUMINT requires skilled interrogators and linguists—often in large numbers. Commanders must know and account for some of the possible shortcomings of HUMINT:

- It is susceptible to the influence of the threat; the threat can threaten and influence the source as well as local interpreters. Commander may need to consider that people may not want to talk through a local interpreter because they are providing information they might not want others to know; anonymity may be essential. On the other hand, a local interpreter may provide community access. Yet, those trusted by the locals may not be truthful with Army forces. Conversely, those truthful with Army forces may not be trusted by the indigenous population particularly if these interpreters come from another country, region, or demographic group. In such cases, commanders may need more than one interpreter.
- It is limited by the accuracy of the source and interpreter's perceptions. A local interpreter may have difficulty understanding and translating military topics or the intricacies of urban infrastructure. Some local translators may want to please both participants, so they may shade the conversation, telling each side what they want to hear. Finally, accuracy may be affected by the collector's own perceptions if not firmly grounded in an understanding of social, cultural, and other local conditions.
- Multiple echelons collecting from a given area may inadvertently collect from the same sources and thereby give the false impression of independent sources confirming the same data. To help mitigate this possibility, commanders should ensure that all HUMINT sources are properly recorded on intelligence source registers.
- It may not be timely. The process of identifying and cultivating a source (particularly in an environment where most civilians support threat forces), gathering information, analyzing the information, and providing the intelligence to commanders can be extremely time consuming.
- Some informants may come from unscrupulous or sordid elements of the urban society and may also have their own agenda. They may attempt to use protection afforded them by their relationship with Army forces to conduct activities (even atrocities) that will compromise

political and military objectives. In some cases, a local interpreter's own agenda may influence his translation. (For example, a local schoolteacher used as a translator may report to the collector that the source believes that repairing schools in their urban area is critical, when actually the source may have said that potable water is most needed.)

Conducting Urban ISR

5-45. To be successful, ISR efforts (national to tactical level) are exceptionally comprehensive and synchronized. Moreover, success necessitates the integration of all ISR sources into operational and tactical planning. This requires that ISR assets deploy and execute early, diversify, properly focus, and integrate under a comprehensive ISR plan. Successful ISR also requires flexibility to adapt to the operational and tactical needs of the commander (see Figure 5-2). Commanders must ensure that the appropriate intelligence architecture (including robust links to joint and multinational elements) is established and tested prior to execution of ISR effort.

Early Deployment

5-46. One of the first requirements for effective urban ISR is the early deployment and employment of assets. The complex urban terrain presents a significant challenge. Commanders should consider that ISR assets will normally take longer to gather data amid the complexity.

5-47. Limited national, strategic, and operational imagery intelligence (IMINT) and signals intelligence (SIGINT) capabilities should be considered and quickly requested. If they are approved, they are tasked and deployed or repositioned to begin urban ISR operations. This takes time. Fortunately, spacing the ISR effort over time permits the analysis of the information or data as it is received. Such time also permits subsequently refining the ISR effort before all assets are committed.

5-48. SOF or conventional units will require significantly more time to execute reconnaissance missions and maintain an acceptable survivability rate. Urban reconnaissance operations require additional time for stealthy insertion into the urban area. IMINT and SIGINT capabilities are used to identify possible locations of high-value targets and corresponding observation positions; this helps minimize time-consuming and high-risk repositioning in the urban area. Again, reconnaissance units may require extensive time to observe from observation positions for indicators of threat activity and disposition and identify patterns.

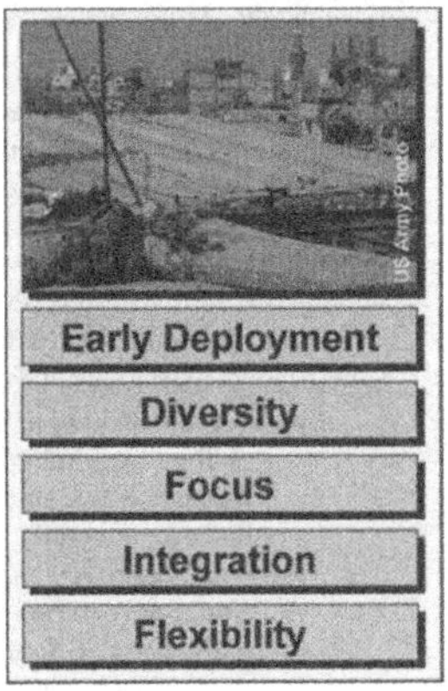

Figure 5-2. Urban ISR considerations

5-49. As conventional combat forces prepare to commit to the urban area, conventional reconnaissance precedes their actions. Conventional reconnaissance will often be a slow and methodical effort. Such forces may need time to reconnoiter the interior of structures for snipers and other small threat teams. They may also need time to deploy and destroy snipers and small delaying elements and to breach harassing obstacles. If necessary, they may need time to mass the combat power necessary to fight through enemy security and continue the reconnaissance.

Diversity

5-50. No single ISR capability can solve the riddle of the urban defense. In order to gain an accurate common operational picture of the complex urban terrain necessary to focus combat power on decisive points, commanders must employ diverse ISR capabilities. These capabilities will each contribute pieces of relevant information to permit identifying operational objectives and leveraging tactical combat power to achieve those objectives quickly. Higher-level commanders should understand that tactical reconnaissance capabilities alone often cannot provide all the tactical information required for success at lower echelons.

5-51. Using diverse capabilities challenges the threat's ability to conduct deception and otherwise defeat the friendly ISR effort. A threat that focuses on minimizing his vulnerability to overhead imagery (for example, satellites) may increase his reliance on communications and thus his vulnerability to SIGINT. At the same time, he may decrease his ability to detect the actions of ground reconnaissance units. Conversely, a threat that actively campaigns to detect ground reconnaissance may make himself more vulnerable to SIGINT and IMINT. Moreover, as Army forces enter urban areas, threat activity and susceptibility to SIGNINT often increases.

5-52. Diverse capabilities also facilitate the tactical ISR effort. Tactical reconnaissance units often consist of small dismounted teams and small combined arms teams with a dismounted element and an armor-protected mounted element. Combat engineers with their technical expertise and breaching capability are essential to the combined arms reconnaissance effort. The teams' movements are synchronized and coordinated with other assets, such as UAS and air cavalry reconnaissance. These teams use several movement techniques including infiltration, with the primary objective of conducting zone reconnaissance along key axes that support brigade and battalion actions against decisive points. To accomplish this mission, reconnaissance reconnoiters the proposed routes and alternate approaches. This supports deception and contingency planning. (Reconnaissance of alternate routes and objectives also applies to aerial reconnaissance; helicopters and UAS are not invisible. They can alert a threat to impending operations.) Infiltration of dismounted reconnaissance is made easier when a threat focuses on combined arms reconnaissance teams. Aerial reconnaissance, such as air cavalry and UAS, provides early warning of threat elements to ground reconnaissance, identifies obstacles and ambush sites, and helps select the routes for ground reconnaissance. Air elements may also reduce the mobility of counterreconnaissance forces.

Focus

5-53. Another key to successful ISR is the ability to focus the assets on commander's critical information requirements (CCIR). This focus begins with mission analysis and the commander's initial planning guidance. It is incrementally refined throughout planning and execution as each ISR effort provides information and permits more specific focus in subsequent efforts. The size and complexity of the urban environment require that the ISR effort center strictly on decisive points or centers of gravity (COGs). Therefore, the overall ISR effort will have two major focuses. The first is to uncover and confirm information on the decisive points and COG. The second is to determine the approaches (physical, psychological, or both) leading to the decisive points. The first focus will likely drive ISR in support of major operations. The second focus will likely provide the impetus for tactical ISR efforts. For example, special operations reconnaissance might focus on a major command center that controls the entire urban area and that is one of the higher-echelon command's CCIR. Tactical reconnaissance might focus on the nature of the defense along a particular avenue of approach to the objective.

Integration

5-54. Another important aspect of urban ISR is integration. All reconnaissance capabilities provide both distinctive information as well as information that confirms and adds to that coming from other sources. Essential to urban ISR is the link between all of these sources, either directly or through an integrating headquarters.

5-55. ISR operations must be vertically and horizontally linked. Vertical links ensure that ISR operations among the various levels of command are complementary and that the information flow between these levels is rapid. Horizontal links ensure that forces operating in close proximity (particularly adjacent units), where areas of interest overlap, can rapidly share results of their individual ISR efforts. Together, this helps ensure that all Army forces share a common operational picture and permits the greatest flexibility and survivability of ISR resources.

5-56. ISR operations also are integrated into the planning system, especially the targeting process. As part of targeting, positioned reconnaissance and surveillance elements may become the trigger and terminal control for applying precision fires when appropriate and after considering the risks of compromise of the position or platform. ISR operations and plans must also be synchronized with the air tasking order, as many surveillance platforms are coordinated on the same timeline.

Flexibility

5-57. The urban ISR effort must be more flexible than in other operations. This flexibility permits the ISR effort to meet unforeseen circumstances and to deal with the challenges of the urban environment. As indicated previously, the urban environment is particularly difficult to penetrate. The practical effects of this characteristic are that—

- The initial ISR effort may not be as successful as in other operations.
- More intelligence requirements may be discovered later while executing ISR operations than otherwise.
- The threat may be more successful in active counterreconnaissance because of the concealment advantages of the urban environment (hiding in structures as well as among the urban population).

Therefore, tactical and operational commanders should consider requesting greater than usual ISR support from higher headquarters. Higher headquarters should be proactive in augmenting units conducting urban operations with additional ISR assets. Additionally, ISR assets remaining under the control of the higher headquarters must respond more quickly to the CCIR of supported commanders. Sequencing reconnaissance missions over time provides flexibility by creating uncommitted reconnaissance assets.

5-58. Time sequencing of ISR assets is essential to flexibility. It makes ISR assets more survivable and allows the intelligence cycle to mature the CCIR. It also creates a ready ISR capability to augment committed forces in critical areas if required or diverts them around centers of threat resistance. If not required, the original ISR tasks can be executed as envisioned in planning. Cueing allows a high-value ISR asset to be able to respond to multiple targets based on an ongoing assessment of the overall reconnaissance effort and the changing CCIR. Redundancy permits the effort to overcome line of sight restrictions, the destruction of an ISR asset, and the ability to combine ISR resources to create combat power if required. Maximizing the ISR effort requires applying all available ISR assets to support the urban operation. Additionally, assets—such as air defense artillery and field artillery radars and engineer reconnaissance teams—are integrated into the ISR effort. In urban operations, units will also commit infantry and armor elements (plus their organic reconnaissance elements) into the tactical reconnaissance effort. These units increase the dismount capability and the ability of reconnaissance elements to fight for information and fight through security zones.

INFORMATION OPERATIONS

5-59. Information operations are an integral part of all Army operations and a critical component in creating and maintaining information superiority. The information environment is the sum of individuals, organizations, or systems that collect, process, and disseminate information; it also includes the information itself. In UO, the information environment is extremely dense due to the proliferation of INFOSYS and widespread access to those systems. In urban operations, commanders must consider how the urban environment, particularly the human component, uniquely relates to executing IO. They must also anticipate the threat's information operation campaign and preclude successful operations on its part. Overall in regard to the urban society, commanders must convince the urban inhabitants of the inevitable success of UO, the legitimacy of Army actions, and the beneficial effects that Army success will eventually generate.

5-60. Because urban operations are likely to be joint, interagency, and multinational, commanders must guard against various forms of information fratricide as a multitude of actors plan and conduct IO in the urban area. For example, the lead governmental agency's key IO objective may be to establish the legitimacy of newly formed civilian authorities. In an attempt to develop close relationships with the civilian populace, Army commanders may continue to work closely with traditional, informal leaders to the exclusion of the new authority. These actions, while they are often conducted out of practical and immediate necessity, may run counter to the lead agency's goal. Overall, Army commanders must nest their IO campaign objectives and themes within those of the lead agency, aggressively coordinate with other governmental agencies and coalition partners, and synchronize activities down to the tactical level to prevent working at odds and avoid information fratricide. However, the process must not be so centralized and rigid that subordinate units lose the flexibility necessary to develop their own products and themes essential to addressing their own AO and the particular situation that confronts them.

5-61. IO are executed using core and supporting elements and related activities (see Figure 5-3 and FM 3-13). The elements of IO are employed in either an offensive or defensive role. Many aspects of IO are not affected differently in an urban environment from any other environment. The following sections outline some IO considerations unique to urban operations.

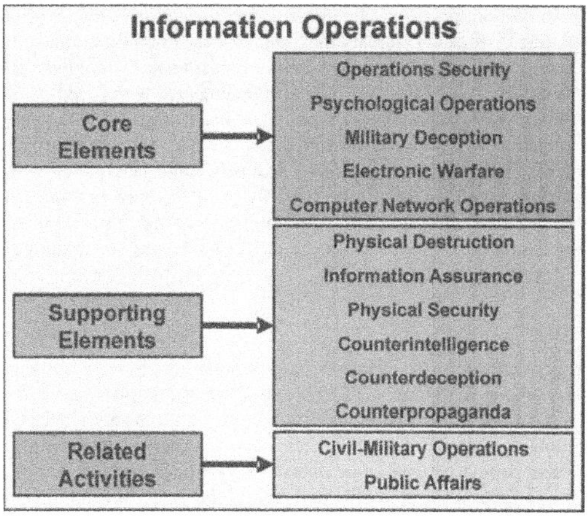

Figure 5-3. IO elements and related activities

Operations Security

5-62. In the urban environment, Army forces can leverage existing urban infrastructure, including the communications and information infrastructure, to enhance Army operations. The danger in integrating these systems is violating OPSEC. Commands ensure that Army forces use only approved systems and proper safeguards exist. Commands also supervise subordinate units for inadvertent breaches of OPSEC policies when using existing urban systems. Finally, established OPSEC procedures must be reasonable and practical. Overly complicated and stringent procedures increase the likelihood that unintentional OPSEC violations may occur as Army forces seek to rapidly accomplish the mission.

5-63. Of particular concern are computers (see also the discussion of Computer Network Operations below). During longer-term UO, commanders will constantly be upgrading and improving the quality of life for their soldiers through the establishment of well-resourced forward operating and sustainment bases. This will likely include the provision of unclassified access to the Internet and e-mail allowing Soldiers to communicate with friends and families around the world. E-mail addresses, facsimile numbers, cell phone numbers, photographs, and other sensitive (but unclassified) information can be valuable sources of threat information. Soldiers must be periodically trained and reminded of everyday procedures to combat potential threat exploitation of these unclassified sources (such as not opening e-mails from unknown sources as they may harbor Trojan programs and viruses).

5-64. To compound concerns, commanders will also establish protected systems to disseminate tactical lessons learned to individual soldiers. This creates the potential that Soldiers can inadvertently release classified information about ongoing operations to threat intelligence systems. In addition to OPSEC training and persistent awareness, commanders must establish procedures to shutdown unclassified Internet and e-mail access immediately, both before and after critical events, to mitigate potential OPSEC violations. (Commanders may also need to randomly shutdown Internet and e-mail access as part of their overall OPSEC measures to preclude such information blackouts serving as a potential warning of impending operations.) Technical surveillance countermeasures—the identification of technical collection activities conducted by threat intelligence entities—will be an important CI service to identify this and other technical OPSEC concerns.

5-65. The close proximity of Army operations to a civil population, particularly in stability and civil support operations, makes Army activities themselves an additional OPSEC concern. Hostile civilians or other threats integrated into the urban population may have more chances to observe Army activities closely. Such observations can provide insight into tactics, techniques, and procedures (TTP) and expose operational vulnerabilities. However, threats may coerce even friendly civilians to provide a threat with EEFI. Therefore, commanders in an urban environment must ensure that civilians cannot observe critical TTP. Any observable patterns and TTP vary and are supplemented with deception efforts. Physical security is increasingly important in urban areas to control civilians' access. Although many urban operations require close coordination with NGOs, commanders should screen information provided to these organizations to protect EEFI. Release of EEFI to NGOs is controlled and done with full recognition and understanding of potential consequences—the benefits must far outweigh the risks involved. (Even the best-intentioned NGO might inadvertently compromise security. For example, by revealing that a commander has closed a particular route the next day, the NGO may unwittingly reveal the objective of a pending offensive operation.)

Psychological Operations

5-66. PSYOP aim to influence the behavior and attitude of foreign audiences, both military and noncombatant, in the urban environment. PSYOP are a force multiplier and contribute in many ways to mission success (see FM 3-05.30). Effective PSYOP focuses on transmitting selected messages to specific individuals and groups in order to influence their actions. Their ability to influence the attitudes and disposition of the urban population cannot be overstated. While the complexity of the societal component of the urban environment can make PSYOP challenging, the urban society also offers many options and resources. Potentially, PSYOP (with other political and economic actions) may help limit or preclude the use of military force in urban areas. In some circumstances, UO may be relevant to the major operation only in terms of their psychological effect.

5-67. The positive influence created by PSYOP is often useful in developing an effective HUMINT capability particularly in an urban area where many civilians actively or passively support the threat. Persuading and influencing a few to support friendly forces may pay great dividends. These few supporters may allow Army forces to penetrate the urban area and obtain essential information. Such information can apply to threat capabilities, threat intentions, and even the urban environment itself.

5-68. PSYOP, combined with other elements of offensive IO, aid in isolation of a threat—a critical shaping action for any urban operation. For example, commanders may use PSYOP to inform civilians about new food distribution points located away from urban combat operations. This action supports the UO fundamental of separating combatants from noncombatants and helps to further isolate the threat (both physically and psychologically) from the civilian populace. Aside from projecting a positive image of friendly forces over threat forces, PSYOP can also isolate the threat by identifying and exploiting ethnic, cultural, religious, and economic differences between the elements of the civilian populace and threat forces as well as the differences among supportive and unsupportive civilian factions. The complexity of the urban environment enables quick changes in opinion or attitude. Commanders must continually evaluate the results of PSYOP for mission relevance.

Military Deception

5-69. Urban operations present numerous challenges to tactical commanders; however, higher-level commanders may help to mitigate some challenges through effective military deception Commanders can use military deception efforts designed to mislead threat decision makers as to friendly force disposition, capabilities, vulnerabilities, and intentions. Military deception actions may allow commanders to achieve tactical surprise or improve relative combat power at a selected location. For example, allowing the threat to observe certain activities on a selected avenue of approach may cause the threat to shift his forces (and effort) to the area perceived to be threatened. (This movement may also aid in determining the overall disposition of threat forces and intentions.) Repositioned forces or effort to activities or locations that are not decisive to the achievement of friendly objectives, combined with other IO designed to overwhelm threat information and intelligence systems, may create the force and tempo differential necessary to achieve success. Commanders should tailor urban deception plans to the specific urban area, paying close attention to the societal characteristics of the target population.

Electronic Warfare

5-70. Electronic warfare (EW) includes all actions that use electromagnetic or directed energy weapons to control the electromagnetic spectrum or to attack a threat. Conducting EW in urban areas seeks to achieve much the same results as in other environments. A major consideration in urban areas is collateral effects on portions of the urban infrastructure that rely on the electromagnetic spectrum for service. Thus, precision is a major factor in planning for EW operations. For example, EW attacking a threat's television broadcasts avoids affecting the television broadcasts of neutral or friendly television. Likewise, EW attacking military communications in a large urban area avoids adversely affecting the area's police and other emergency service communications. Urban offensive and defensive operations will have the least restrictions on EW operations while urban stability may have significant constraints on using EW capabilities.

Computer Network Operations

5-71. Computer network operations (CNO) include computer network attack (CNA), computer network defense (CND), and computer network exploitation (CNE). CNO are not applicable to units at corps and below. Echelons above corps (EAC) units will conduct CNA and CNE. If tactical units require either of these network supports, they will request it of EAC units.

Computer Network Defense

5-72. In urban operations, CND will require extreme measures to protect and defend the computers and networks from disruption, denial, degradation, or destruction. The nature of the urban environment and configuration of computer networks provides the threat with many opportunities to interdict local area networks (LANs) unless monitored by military forces. LANs controlled by military forces are normally more secure than the civilian infrastructure. Commanders should prepare for opportunities by the threat to insert misinformation.

Computer Network Attack

5-73. Considerations regarding the execution of CNA in urban operations are similar to those of EW: CNAs that do not discriminate can disrupt vital civilian systems. However, possible adverse effects on the civilian infrastructure can be much larger—potentially on a global scale. In the short term, CNAs may serve to enhance immediate combat operations but have a debilitating effect on the efficiency of follow-on urban stability operations. Because of these far-reaching effects, tactical units do not execute CNA. CNA is requested of EAC units. EAC units will receive all requests from lower echelons, carefully consider second- and third-order effects of CNA, and work to ensure its precise application.

Computer Network Exploitation

5-74. CNE consists of enabling operations and intelligence collection to gather data from target or adversary automated INFOSYS or networks. Tactical units do not have the capability for CNE. CNE contributes to intelligence collection at EAC. In UO, CNE will be centrally controlled.

Physical Destruction

5-75. Physical destruction includes those actions—including direct and indirect fires from air, land, sea, space, and Special Forces—taken with, to augment, or supplement IO actions. Like many other IO elements, major concerns with employing physical destruction in UO are precision and follow-on effects. Thus, commanders using physical destruction to support IO must adhere to the same constraints as all other urban fires.

Information Assurance

5-76. Information assurance in UO takes on an added dimension. As with other operations, availability of information means timely, reliable access to data and services by authorized users. In UO, the timeliness of information may be restricted because structures block the transmission waves. The need for retransmission facilities may overwhelm the signal community. The reliability can be questioned because of the blockage between units and communications nodes. Unauthorized users may intercept the communications and input misinformation or disinformation. Commanders must protect the integrity of all information from unauthorized changes, including destruction. INFOSYS with integrity operate correctly, consistently, and accurately. The authentication of information may be accomplished by sophisticated electronic means. However, it is more likely that communications-electronics operating instructions authentication tables will authenticate the information. Commanders should consider the confidential nature of all information in UO and establish procedures to protect the information from unauthorized disclosure. Of additional concern, the density of the infrastructure in urban areas may inhibit receipt by the intended individual or unit. Therefore, commanders must develop alternative means to determine if the message was received.

Physical Security

5-77. Physical security consists of those actions and measures to safeguard personnel; to prevent unauthorized access to equipment, installations, material, and documents; and to safeguard them against espionage, sabotage, damage, and theft. The joint, interagency, and multinational nature of urban operations increases the need for physical security measures, especially with organizations that do not emphasize physical and operational security within their own organizations. The potential for locating sustainment bases in or near urban areas and the use of indigenous personnel on these facilities to accomplish a multitude of tasks mandates that commanders carefully consider the effectiveness of their physical security measures. In some stability operations where Army forces must recruit and train security or police forces, and then conduct joint operations and patrols with them, commanders must examine how best to establish close working relations while maintaining essential physical (and operations) security.

Counterintelligence

5-78. The urban environment, particularly in stability operations, is ideal for espionage, other intelligence activities, sabotage, or assassination. Threats can approach, conduct reconnaissance, and escape under the concealment offered by the urban environment. The dense environment and electromagnetic clutter serves to hide technological collection means. To counter threat intelligence collection activities, the Army employs CI forces. They provide analysis of the threat's human, imagery, signals, and measurement and signature intelligence capabilities to develop countermeasures against them (see FM 34-60).

5-79. In conjunction with military police, engineers, medical personnel, and other as required, CI help create urban threat and vulnerability assessments necessary in developing comprehensive force protection measures and evaluating their effectiveness. CI personnel can also supplement the screening of linguists and other critical host-nation personnel employed by Army forces. As Soldiers operate in close proximity to the urban society, threats will attempt to use three, time-tested tools to influence them: money, sex, and drugs. Periodic CI refresher training remains crucial to decrease the effectiveness of these threat tactics.

Counterdeception

5-80. In UO, threat forces can easily accomplish deception operations. The force that controls the area above and below ground will have freedom of movement. Deception aimed at friendly commanders will cause them to deploy combat power at the wrong place and the wrong time. Counterdeception by friendly commanders will identify and exploit threat attempts to mislead friendly forces. Counterdeception is difficult. Cultures of certain rhetoric and actions are more predisposed to deception than others. Knowing a threat's previous deception methods is important. Dismissing tactical indicators because they conflict with preconceptions may allow a hostile deception operation that plays on the preconception to succeed.

Counterpropaganda

Four hostile newspapers are more to be feared than a thousand bayonets.

Napoleon Bonaparte

5-81. Because propaganda is aimed at both combatants and noncombatants, UO are especially concerned with its use. Propaganda can rapidly and dramatically affect the attitudes of the urban population and will probably occur (or increase significantly) after urban operations have begun. Thus, it can create situations in the human dimension of the environment quite different from those discovered in the pre-operations assessment. Counterpropaganda is, therefore, essential to urban operations. To negate, deflect, or destroy the threat's propaganda capability, counterpropaganda requires—

- Monitoring and evaluating the effectiveness the threat's propaganda efforts.
- Determining innovative methods using all Army force capabilities, especially PSYOP and PA units.

5-82. In particular, humanitarian assistance, when coupled with an effective information operations campaign, can be a powerful counterpropaganda tool to sway public support. For example, media coverage of Army forces directly providing or coordinating NGO humanitarian efforts to feed, shelter, and provide medical care to the urban population can help counter threat efforts to discredit Army forces as an evil occupying force. This type of coverage can be used to turn opinion against the threat and could also prove an effective tool in gaining favorable international support.

Public Affairs

5-83. A closely related activity to IO is public affairs (PA). PA influences urban operations by transmitting information through the media to internal (urban Army forces as well as the urban civilian populace) and external audiences. At higher levels of command, PA can help maintain popular national support for the urban operation by clarifying the links between strategic goals and operational objectives. At both the operational and tactical levels, it links Army units, the urban inhabitants, the U.S. and international public, and the media. The media will often focus on the negative actions that are ongoing in an AO; hence, the urban commander's challenge will be to publicize his unit's good news stories. Commanders should not wait for media inquiries and interview requests. Instead, they should aggressively plan public affairs to get the Army's story out to the media and, ultimately, the public.

5-84. PA can help determine potential media issues that may influence planned UO. It can also aid commanders in understanding the impact of UO on the environment (particularly its citizens) and other agencies and organizations operating in the urban area. PA also helps to counter rumors, uncertainty, fear, loneliness, confusion, and other factors that cause stress (to both soldiers and civilians) and undermine effective UO. If the populace does not understand the mission or the Army forces' capabilities, false expectations may be created that Army forces may not be able to meet. PA can help prepare the American public for the possibility of high casualty rates. Overall, PA supports urban commanders in their goals to achieve information superiority and preserve public support.

5-85. The density of information sources and reporters in UO ensures that all Army activities will be subject to media and public scrutiny. Many reporters will congregate in cities for their own comfort and take advantage of established communications networks. Urban areas are densely populated and, together with Army forces and NGOs operating there, will present the greatest number of human-interest stories.

The local urban or host-nation media, however, will often have their own agendas developed over a longer period of time. This local media may also have a greater influence over the urban population than the international media. The indigenous media may not follow international norms. Commanders are responsible for understanding the media (particularly the local media), its role, and its potential for negative or positive influence.

5-86. Commanders cannot allow themselves to be intimidated by the media. Media sources are going to get and report their story one way or another. Therefore, commanders should support open and independent reporting and grant access to their units as early and as far forward as the situation permits. Successful units often go beyond simply accommodating the media to actually integrating them into their units. In the future, embedding of media may become the standard for reporting Army operations. Other considerations for improving urban media relations include:

- Encouraging the media to report on all ongoing aspects of the unit's urban operations—not exclusively on combat operations. This may create a unit responsibility to protect journalists that participate in covering suggested activities.
- Hosting media luncheons or other events for key journalists. This provides commanders the ability to get the whole story out to the press from their unit's perspective. To gain media interest and unit credibility, this will often require key leadership participation.
- Identifying Soldiers that speak the local and international journalists' native language. Even if journalists speak English as a second language, they may overlook or misunderstand important points due to language subtleties. Additionally, people tend to be friendlier to those that speak their own language.
- Identifying selected Soldiers at lower tactical levels to act as PA representatives for their unit as an additional duty. However, commanders must provide them with appropriate training and resources to accomplish specific PA tasks.
- Writing articles for local newspapers that emphasize the unit's good news. Depending on the situation, subordinate units may be tasked to write stories about the good accomplishments in their AO. Commanders, with appropriate OPSEC considerations, may hire locals to assist in this effort. However, commanders must take extra precautions to ensure that these efforts are not perceived as covert propaganda or as undermining free and independent news reporting.

5-87. PA does not distort, direct, or manipulate information. Its effectiveness stems directly from establishing and maintaining credibility with the urban population and media. PA protects the overall integrity of the Army as an institution by ensuring that the Army is recognized as the most reliable source for official information among all other competing sources. Commanders must synchronize PA with the core and supporting elements of IO (particularly PSYOP and counterpropaganda) to ensure that all Army sources send only one message. Urban commanders must plan for the media and integrate PA into their decision-making and (through IO) targeting processes. Similar to lethal targeting, commanders should consider target audiences and coverage areas for the various new agencies and mediums. They also pay attention to print media circulation statistics and television and radio broadcast times. For example, an urban populace may tend to read morning newspapers vice late editions and watch or listen to nightly news instead of morning or daytime programming. Additionally, low literacy rates among the urban population may indicate that television and radio are better mediums to use to influence the urban population than print media such as newspapers, magazines, leaflets, and posters.

5-88. The PA principles listed in Figure 5-4 and addressed in FM 46-1 summarize PA. They serve as useful guides toward planning and executing PA operations regardless of the environment. However, the principles of "practice security at the source," and "truth is paramount" particularly apply to the urban environment. The compartmented nature of most UO impede commanders' and their PA officers' ability to be at all places where the media will likely be. Friendly inaccuracies, intentional or not, will help fuel the threat information campaign. Therefore, all soldiers are trained, provided with clear and understandable PA guidance, and prepared to accurately communicate to the civilian media. The keys at all levels are understanding, prepared acceptance, and truthfulness tempered with an essential concern for OPSEC.

- Truth is Paramount
- If News is Out, It is Out
- Public Affairs Must be Deployed Early
- Not All News is Good News
- Practice Security at the Source
- Media are Not the Enemy
- Telling Our Story is Good for the Army
- Soldiers and Families Come First

Figure 5-4. Public affairs principles

Civil-Military Operations

5-89. Civil-military operations (CMO) are critical and will be an essential, implied task of any urban operation. They are included here as a closely related activity of IO. Commanders use CMO to establish, maintain, influence, or exploit relations to achieve operational objectives. These relations are among military forces, civilian authorities and organizations (both governmental and nongovernmental), and the civilian population. CMO range from support to combat operations to assisting in establishing political, economic, and social stability. In some instances, commanders may need to be prepared to assume temporary responsibility for the functions and capacities of the urban government. Like PA, effective CMO is based on establishing and maintaining credibility and trust with the urban populace and civilian organizations operating in the urban environment. Because of the potential importance of CMO, commanders may consider placing an assistant commander, executive officer, or some other senior leader to monitor, guide, liaise with other units and organizations, and, overall, ensure that the appropriate command emphasis is placed on these activities.

Civil Affairs

5-90. Civil affairs (CA) units and teams will be critical during UO. While any military force can conduct CMO, CA units are specifically organized, trained, and equipped to conduct activities in support of CMO. They have experience in planning and conducting CMO, a regional focus (which includes enhanced cultural awareness and language training), civilian technical expertise that parallel common urban government and administration functions, and experience in negotiation and mediation. Such experience ensures relevant support to commanders conducting urban operations. (As it is likely that there will not be enough CA forces to cover all requirements, commanders should survey their Soldiers to determine the whether any possess civilian expertise that may be beneficial to CMO.) CA units organize their skills into six functional specialty areas (see Figure 5-5). Commanders should use these skills, often unfamiliar to most military personnel, to—

- Develop their situational understanding of the urban environment (particularly the infrastructure and society).
- Plan CMO to support UO.
- Achieve many of the fundamentals of UO.

Note: To ensure optimal integration, commanders must establish clear command and support relationships between CA and maneuver forces.

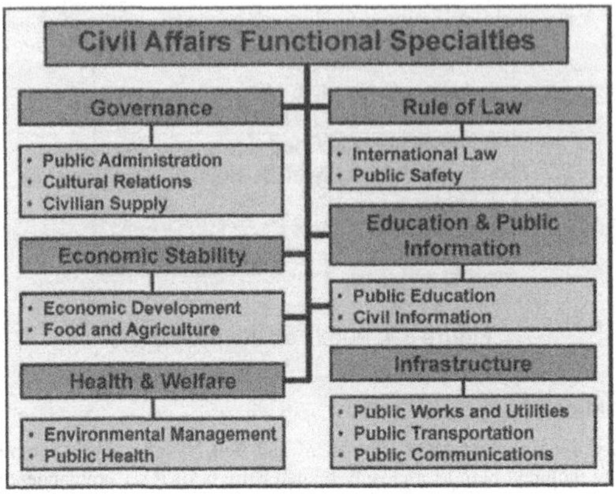

Figure 5-5. Civil affairs functional specialties

5-91. In addition to providing essential information for understanding the urban environment, CA personnel and activities help shape the battlefield, engage a civil problem, consolidate gains, and transition to a legitimate civil authority. Specifically, CA teams and CMO help urban commanders—

- Minimize civilian interference with UO and the impact of urban operations on the populace and infrastructure. CA personnel can help establish and run a civil-military operations center (CMOC), discussed later in this chapter, to coordinate UO with civilian agencies (both governmental and nongovernmental), other services, and multinational partners.
- Provide advice and assistance to restore or rehabilitate portions of the infrastructure.
- Plan, supervise, and execute necessary populace and resources control measures (in close coordination with military police units) until no longer required or the urban operation is completed.
- When requested or when military necessity or legitimate directives require, establish all or portions of the urban civil administration.
- Determine available supplies and services in the urban area and if necessary assist in negotiating their acquisition. CA also help commanders assess the capability, dependability, and willingness of urban sources to provide and sustain identified needs as well as to calculate the impact of using them on other aspects of the urban environment.
- In conjunction with the SJA, fulfill the Army's responsibilities toward the urban population under international, host-nation, and U.S. law.
- Plan and conduct the transition of control for the urban area or operation to another military or civilian governmental or nongovernmental organization or agency.

5-92. As part of CMO, stability or civil support operations will be critical as commanders begin to restore essential services to meet the emergency needs of a potentially devastated urban population. Besides repairing or restoring vital services and other critical infrastructure, these and other projects may also stimulate the urban area's economy and provide work for the unemployed. (Employed workers are less likely to be co-opted by terrorists, insurgents, and criminal gangs.) As such, they can be a critical stabilizing tool for commanders as they attempt to transition control and responsibility to other agencies or to legitimate civilian control. These projects may include—

- Providing clean water.
- Restoring waste management functions (less-extensive sewer repairs and trash collection and disposal activities).
- Improving medical care.
- Rehabilitating schools.
- Assisting in the establishment of local governments.
- Improving security and emergency services (police, fire, and rescue).
- Providing electrical generators for hospitals, clinics, and factories.
- Restoring irrigation to surrounding farmland.
- Repairing important cultural infrastructure.

5-93. As part of the initial planning process, CA units conduct an *area assessment*, which can provide commanders with essential information about the environment (see FM 41-10). Commanders should integrate this initial assessment into the overall urban-focused IPB process. To help analyze civil considerations in any environment, commanders and staffs can consider many characteristics such as areas, structures, capabilities, organizations, people, and events (ASCOPE). These characteristics easily align with an urban area's three main characteristics of terrain, society, and infrastructure; and, like them, they are overlapped and interdependent (see FM 6-0). Of these, the category of events is a critical addition to urban characteristics previously categorized and discussed. Significant spikes (and lulls) to threat and civilian activity can be expected to occur in conjunction with important religious, historic, and political events (for example elections). Therefore, commanders should determine, track, and pay particular attention to these events when planning operations and managing transitions (see also the Key Event Chart discussion in Appendix B, paragraph B-43).

5-94. Similar to PA operations, CMO are related to information operations. The nature of CMO and the need for CA personnel to develop and maintain close relationships with the urban population put CA personnel in a favorable position to collect information. CA personnel work daily with civilians, their equipment, and their records that may be prime sources of information. If used correctly, CA personnel can complement the intelligence collection process necessary to understand the dynamic societal component of the urban environment and detect significant changes. However, CA personnel are not, and cannot appear as, intelligence agents; otherwise, it will undermine their ability to interact with the civilian community. Examples of information available to CA units include government documents, libraries, and archives; files of newspapers and periodicals; industrial and commercial records; and technical equipment, blueprints, and plans.

5-95. Overall, CA expertise is critical to a commander's understanding of the complexities of the infrastructure and societal components of the urban area. These components (together with the terrain or physical component of the urban area) interconnect. CA forces help identify and understand the relationships and interactions between these urban components. From this understanding, commanders can anticipate how specific military actions affect the urban environment and the subsequent reactions. CA personnel help commanders predict and consider the second- and third-order effects and reactions as well as the long-term consequences. Understanding these long-term consequences helps in achieving stability and ensuring a smoother and speedier transition of the urban area back to civilian responsibility and control. Oppositely, an unconsidered short-term solution may, in the end, turn out to be the solution requiring the greatest amount of time. Poor solutions may exacerbate the situation. Others may contribute little to conflict resolution and waste precious resources (particularly time) until a more reasoned approach is formulated—one that should have been the first solution.

Civil-Military Reconstruction Efforts

5-96. Reconstruction efforts and activities, an integral part of CMO, should complement NGO efforts and are conducted in concert with larger urban area and nationwide projects often headed by the U.S. Army Corps of Engineers (USACE) and the U.S. Agency for International Development (USAID). For example, extensive, large-scale sewer projects or major repairs to the urban area's electrical infrastructure would be left to the USACE and USAID. When planning and conducting urban reconstruction, commanders should—

- **Keep expectations realistic.** With so much to do, it is easy for Army forces to become overcommitted. Commanders must focus on the achievable, build off successes, and create constructive forward process.
- **Seek opportunities.** While commanders think long-term, they should look for catalytic investment of resources. Minimal resources provided at the right time can generate significant activities resulting in increased momentum toward the desired end state.
- **Align resources with local needs.** In analyzing the merit of a project, commanders must check to determine if it is supported and desired by the urban populace. Local participation in the process reinforces ownership and acceptance and eventual transition of responsibility.
- **Balance speed and quality.** Commanders must balance the achievement of rapid response with quality construction and repairs.
- **Ensure efforts do not undermine locals.** Army reconstruction efforts should not undermine the growth and legitimacy of local institutions and authorities.
- **Integrate in the targeting process.** Commanders should routinely include approval of civil-military reconstruction products as part of the information operation's contribution to the overall targeting process.

INTEGRATION OF CONVENTIONAL AND SPECIAL OPERATIONS FORCES

5-97. One important Army and joint resource that commanders of a major operation can use to influence urban operations is SOF. Several types of these forces exist (including the CA forces discussed above), each with unique and complementary capabilities. They can be extremely valuable in UO for their ability to execute discrete missions with a higher degree of precision than conventional forces, to provide information, and to enhance cultural understanding. However, the challenges of using SOF include command and control; integration; coordination with conventional forces that will normally command, control, and conduct the bulk of UO tasks; and the imprudent inclination to use SOF forces for conventional purposes. The density and complexity of UO make close coordination and synchronization of conventional forces and SOF essential to mission success. The nature of the environment dictates that both forces will work in close proximity to each other; the separation in space and time between SOF and conventional forces will often be much less in urban areas than in other environments. Overall, the nature of the environment demands a synergistic combination of capabilities to achieve effects on the threat and mission success.

5-98. Successfully integrating SOF occurs with proper integration into, or coordination with, the command structure of the force conducting the UO. SOF within a theater (less CA and tactical PSYOP) ordinarily falls under joint command and control. Therefore, the commander of the major operation responsible for an urban area, if he is not a JFC, will have to coordinate through the JFC to integrate SOF capabilities into the UO. Examples of critical coordination elements include:

- Boundaries.
- No-fire areas. (If assigned, SOF liaison officers (LNOs) must be routinely included in targeting meetings.)
- Coordination points.
- The exchange of key intelligence and information.
- Requirements to support personnel recovery contingencies.

Conventional and Special Forces Integration

On 20 April 2003 during OPERATION IRAQI FREEDOM (OIF), an Army Special Forces (SF) operational detachment alpha (ODA) was patrolling through villages and secondary roads east-southeast of An Najaf. A local Iraqi approached the team and volunteered the location of a senior Ba'ath Party official in the nearby town of Ghamas. Unfortunately, he did not provide an exact location but rather an approximate location and the name of the family dwelling. The ODA coordinated with their company commander and nearby elements of the 101st Airborne Division to conduct a raid on the residence at dawn the next day. Joint planning, preparation, and rehearsals among the units occurred throughout the night.

At 0430, the raid convoy departed its base en route to Ghamas. The small task force consisted of an assault team, a security team, a command and control element, and three blocking forces composed of Soldiers from a scout platoon and antiarmor company of the 101st. At 0515, the convoy hit the release point outside Ghamas. The 101st vehicles moved to their blocking positions at three bridges surrounding the town. The remainder of the task force moved into the town. An interpreter quickly found a local guide who knew where the house was located. He took the SF team to a walled two-story dwelling with a courtyard. The security team isolated the objective and provided overwatch while the assault team forcibly seized the house and apprehended three adult males. Among them was Abd Hamden, the target of the raid and a senior Ba'ath Party official from Baghdad.

After a tactical interrogation, one of the men provided the location of a Fedayeen major nearby. The SF and conventional force raided this house minutes later but only found the major's relatives. As the Soldiers left the town with their captives, they were cheered by the locals throughout the town. Although only one small example, the extensive coordination between SOF and conventional forces during urban operations was a tremendous success during OIF.

SYNCHRONIZATION OF ACTIVITIES WITH OTHER AGENCIES

5-99. The population density of the urban environment, its economic and political importance, and its life-supporting infrastructure attracts many types of organizations. These organizations include—

- Other U.S. governmental agencies.
- International governmental organizations.
- Allied and neutral national governments.
- Allied and coalition forces.
- Local governmental agencies and politicians.
- NGOs.

5-100. Even in a major operation or campaign, many organizations operate in the area as long as possible before combat or as soon as possible after combat. Therefore, coordination with these organizations sharing the urban AO will be essential to achieve synchronization; however, effective synchronization is challenging, time consuming, and manpower intensive. The staffs of larger headquarters (divisions or higher) normally have the breadth of resources and experience to best conduct the necessary synchronization. They can effectively use or manage the organizations interested in the urban area and mitigate their potential adverse effects on UO. By taking on as much of the synchronization requirements as possible, the operational headquarters permits its tactical subordinates to remain focused on accomplishing their tactical missions. The higher headquarters should assume as much of the burden of synchronization as possible. However, the density of the urban environment will often require that smaller tactical units coordinate and synchronize their activities with other agencies and the local civilian leadership (formal and informal) simply because of their physical presence in the units' AOs. In urban stability and civil support operations, mission accomplishment will require effective civil-military coordination and synchronization activities and measures at all levels as either a specified or implied task.

Civil-Military Operations Centers

5-101. To coordinate activities among the varied agencies and organizations operating in an urban area and the local population, urban commanders can establish a CMOC. The CMOC synchronizes Army activities and resources with the efforts and resources of all others involved (see FM 41-10). This can be particularly important in stability and civil support operations where combat operations are not the dominant characteristic of the major operation. CMOCs can be established at all levels of command. Hence, more than one CMOC may exist in an AO, particularly large urban areas. CMOCs may be organized in various ways and include representatives from as many agencies as required to facilitate the flow of information among all concerned parties. Commanders must still ensure that force protection and OPSEC requirements are not compromised. Effective CMOCs can serve as clearinghouses for the receipt and validation of all civilian requests for support, can aid in prioritizing efforts and eliminating redundancy, can decrease the potential for inappropriate displays of wealth by one or more of the participating organizations, and, most importantly, can reduce wasting the urban commander's scarce resources.

Liaison Officers

5-102. LNOs—sufficiently experienced and adequately trained in liaison duties and functions—are necessary to deal with the other agencies that have interests in the urban area. Army LNOs work with the lead agency or other organizations (including local civilian agencies such as police) that the commander has identified as critical to mission success. Together they work to rapidly establish unity of effort and maintain coordination, often before a CMOC is established. The additional coordination afforded by the physical presence of LNOs within these organizations may be required even after the CMOC is fully functional. When commanders lack enough LNOs to meet requirements, they should prioritize and often assign a single LNO to several organizations. That LNO will then share his time and presence to those organizations based on the situation and his commander's guidance.

Commander's Personal Involvement

5-103. Overall, establishing a close relationship with other agencies and the urban civilian population will often be a major, positive factor in successful mission accomplishment, particularly in urban stability operations. Despite internally established command and control relationships, commanders that develop a direct and personal relationship with the leaders and staff of other agencies can often avoid conflict, win support, foster trust, and help eliminate the "us versus them" mentality that can frustrate cooperation among Army forces and civilian organizations.

Chapter 6

Foundations for Urban Operations

Utilities such as electricity and water are as much weapons of war as rifles, artillery pieces or fighter aircraft. . . . In the case of Manila, where there was a noncombatant, civilian population of one million in place, it was the attacker's aim to capture the utilities which the defender planned to destroy.

The Battle for Manila

Commanders conducting major urban operations (UO) use their ability to visualize how doctrine and military capabilities are applied within the context of the urban environment. An operational framework is the basic foundation for this visualization. In turn, this visualization forms the basis of operational design and decision making. To accurately visualize, describe, and direct the conduct of UO, commanders and their staffs must understand the basic fundamentals applicable to most UO.

URBAN OPERATIONAL FRAMEWORK

6-1. Army leaders who have an urban area in their area of operations (AO) or are assigned missions in an urban area follow an urban operational framework. They identify the portion of the urban area essential to mission success, shape the area, precisely mass the effects of combat power to rapidly dominate the area, protect and strengthen initial gains without losing momentum, and then transition control of the area to another agency. This framework divides into five essential components: understand, shape, engage, consolidate, and transition. These five components provide a means for conceptualizing the application of Army combat power and capabilities in the urban environment.

6-2. The urban operational framework assists commanders in visualizing urban operations. This framework is simply an aid to the commander. It is not sequential, nor is it a planner's tool for phasing an operation. Commanders should combine the urban operational framework with—

- The principles of war.
- The tenets of Army operations.
- The components of operational design.
- Considerations for stability operations and civil support operations.
- Sustainment characteristics.
- Running estimates.
- Commander's critical information requirements (CCIR).
- Each commander's experience.

The framework contributes to the visualizing, describing, and directing aspects of leadership that make commanders the catalysts of the operations process (see Figure 6-1). In the same manner, the urban operational framework contributes to the overall operations process (see FM 3-0).

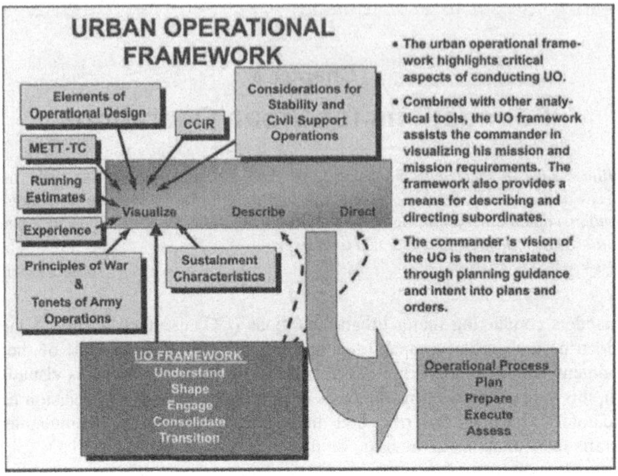

Figure 6-1. The urban operational framework and battle command

UNDERSTAND

6-3. Understanding requires continuous assessment—throughout planning, preparation, and execution—of the current situation and progress of an operation, and the evaluation of it against measures of effectiveness to make decisions and adjustments. Commanders use visualization, staff officers use running estimates, and all use the intelligence preparation of the battlefield (IPB) process to assess and understand the urban environment. Commanders and staffs begin to understand by observing and then collecting information about the situation. They observe and learn about the urban environment (terrain, society, and infrastructure), and other factors of METT-TC—mission, enemy, weather, troops and support available, and time available. They use intelligence, surveillance, and reconnaissance means; information systems (INFOSYS); and reports from other headquarters, services, organizations, and agencies. Then they orient themselves to the situation and achieve situational understanding based on a common operational picture (COP) and continuously updated CCIR. Largely, the ability to rapidly and accurately achieve a holistic understanding of the urban environment contributes to the commanders' abilities to seize, retain, and exploit the initiative during UO.

Disproportionately Critical

6-4. The Army operations process requires continuous assessment and understanding; it precedes and guides every activity. In UO, however, in-depth understanding is disproportionately critical for several reasons. First, each urban environment is unique. Other environments can be studied and their characteristics quantified in a general manner with accuracy. This is fundamentally not true of different urban areas. The characteristics and experience in one urban area often have limited value and application to an urban area elsewhere. This characteristic sets UO apart from operations in other environments.

Extremely Dynamic

6-5. The urban environment is also extremely dynamic. Either deliberate destruction or collateral damage can quickly alter physical aspects of the urban environment. The human aspect is even more dynamic and potentially volatile. A friendly civil population, for example, can become hostile almost instantaneously.

These dynamics (combined with initial difficulty of understanding and describing this unique environment) make it difficult for commanders and staffs to initially develop and maintain a COP and establish situational understanding. Furthermore, public reaction to media coverage of the urban operation and political changes influence national security strategy and objectives. Such changes can affect the basic nature of an operation, especially after it has commenced. Anticipating these potential effects and developing appropriate branches and sequels based on an accurate understanding often determines how quickly commanders can achieve the desired end state.

Risk Assessment

6-6. As in any environment, UO pose both tactical and accident risks. However, the level of uncertainty, ambiguity, and friction can often be higher than that of many other environments. Such challenges increase the probability and severity of a potential loss due to the presence of the enemy, a hostile civilian group, or some other hazardous condition within the urban environment (see Determining the Necessity of Urban Operations in Chapter 5). Therefore, commanders must—

- Identify, assess, and understand hazards that may be encountered in executing their missions.
- Develop and implement clear and practical control measures to eliminate unnecessary risk.
- Continuously supervise and assess to ensure measures are properly understood, executed, and remain appropriate as the situation changes.

6-7. Risk decisions are commanders' business. However, staffs, subordinate leaders, and even individual Soldiers must also understand the risk management process and continuously look for hazards at their level or within their area of expertise. Any risks identified (with recommended risk reduction measures) must be quickly elevated to the appropriate level within the chain of command (see FM 3-100.12 and FM 5-19).

Complex and Resource Intensive

6-8. The urban environment is the most complex of all the environments in which the Army conducts operations. It is often comprised of a diverse civil population and complex, ill-defined physical components. A sophisticated structure of functional, social, cultural, economic, political, and informational institutions unites it. Thus, the analysis to understand the environment is also complex and time and resource intensive. The nuances of the urban environment can take years to uncover. Hence, constant analysis of the environment requires greater command attention and resources. Accurately understanding the environment is a prerequisite to shaping it, and both understanding and shaping activities are crucial to effectively engage the elements of the urban environment critical to success.

SHAPE

6-9. Shaping operations, part of all Army operations, are essential to successful UO. They set the conditions for decisive operations at the tactical level in the urban area. Rapid action, minimum friendly casualties, and acceptable collateral damage distinguish this success when the AO is properly shaped. Failure to adequately shape the urban AO creates unacceptable risk. The commander of a major urban operation has several resources with which to begin shaping the AO. Important capabilities include—

- Fires.
- Information operations.
- Special operations capabilities.
- The maneuver of major subordinate units.

6-10. Critical urban shaping operations may include actions taken to achieve or prevent isolation, understand the environment, maintain freedom of action, protect the force, develop cooperative relationships with the urban population, and train Army forces for sustained UO.

Isolation

6-11. Isolation of an urban environment is often the most critical component of shaping operations. Commanders who's AO includes operationally significant urban areas often conduct many shaping operations to isolate, or prevent isolation of, those areas from other parts of the AO. Likewise, commanders operating in the urban area focus on isolating decisive points and objectives in the urban area or averting isolation of points that are critical to maintaining their own freedom of action. Isolation is usually the key shaping action that affects UO. It applies across full spectrum operations. Most successful UO have effectively isolated the urban area. Failure to do so often contributed to difficult or failed UO. In fact, the relationship between successful isolation and successful UO is so great that the threat often opposes external isolation actions more strongly than operations executed in the urban area (or critical areas within). In some situations, the success of isolation efforts has been decisive. This occurs when the isolation of the urban area compels a defending enemy to withdraw or to surrender before beginning or completing decisive operations. In UO that are opposed, Army forces attempt to isolate the threat three ways: physically, electronically, and, as a resultant combination of these first two, psychologically (see Figure 6-2).

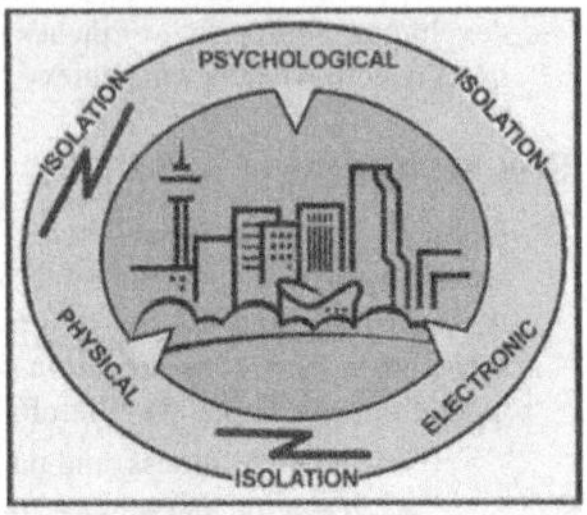

Figure 6-2. Urban isolation

Physical Isolation

6-12. In offensive UO, physical isolation keeps the threat from receiving information, supplies, and reinforcement while preventing him from withdrawing or breaking out. Conversely, a defending Army force attempts to avoid its own physical isolation. Simultaneously, this force conducts operations to isolate the threat outside, as they enter, or at selected locations in the urban area. Physical isolation can occur at all levels. In many situations, the commander of a major combat operation may attempt to isolate the entire urban area and all enemy forces defending or attacking it. At the tactical level, forces isolate and attack individual decisive points often using a cordon technique. In stability operations, physical isolation may be more subtly focused on isolating less obvious decisive points, such as a hostile civilian group's individual leaders. In many operations, isolation may be temporary and synchronized to facilitate a decisive operation elsewhere. To effectively isolate an urban area, air, space, and sea forces are often necessary additions to the capabilities of ground forces.

Electronic Isolation

6-13. Electronic isolation is achieved through offensive information operations (IO). Electronic warfare (particularly two of its components: electronic warfare support and electronic attack) and computer network attack are critical to electronic isolation (see FM 3-13 and the Information Operations discussion in Chapter 5). At the operational level, offensive IO aims to quickly and effectively control the information flow into and out of an urban area. This isolation separates the threat's command and control (C2) system in the urban area from its operational and strategic leadership outside the urban area. Offensive IO also focuses on preventing the threat from communicating with civilians through television, radio, telephone, and computer systems. At the tactical level, IO aim to isolate the threat's combat capability from its C2 and leadership within the urban area, thus preventing unity of effort within the urban area. Defensive IO can prevent isolation of friendly forces defending in an urban area.

Psychological Isolation

6-14. Psychological isolation is a function of physical actions, electronic warfare, and other forms of IO, especially military deception and psychological operations. Psychological isolation denies the threat political and military allies. It separates the enemy or hostile civilian group from the friendly population, nongovernmental organizations (NGOs) operating in the urban area, and from political leaders who may consider supporting Army forces. Psychological isolation destroys the morale of individual enemy soldiers or hostile civilians. It creates a feeling of isolation and hopelessness in the mind of the threat. It undermines the threat leadership's ability to think, plan, decide, and act. As such, it undermines the confidence of the threat in their leadership. On the other hand, IO, as well as the disciplined conduct of Army personnel, can help to forge legitimacy for Army operations. In stability operations, psychologically isolating the threat can result in the friendly urban population and NGOs positively supporting Army operations.

Other Shaping Actions

6-15. Other shaping actions can include the proper sequencing and deployment of forces, reconnaissance operations, and force protection. These actions contribute equally to the success of any urban operation. Commanders must understand how the urban environment affects their ability to accomplish these shaping actions. However, civil-military operations (CMO), another closely related activity of IO, are important to shaping the urban operational environment for decisive operations. The specific civil-military task can vary greatly and may include affecting a cooperative relationship with the civil political system, protecting portions of the civil population or infrastructure, or establishing refugee camps or safe areas for noncombatants. This is most true in stability and civil support operations. Successful CMO also can contribute to the psychological isolation of the threat. (See Civil-Military Operations in Chapter 5 for a more detailed discussion.)

Training and Education

6-16. Finally, Army commanders must consider that critical shaping actions often occur prior to the urban operation in the form of professional education, and home-station and in-theater training. Commanders can enhance training through joint, interagency, multinational, and combined arms exercises and effective rehearsals.

ENGAGE

6-17. In UO, Army forces engage by appropriately applying their full range of capabilities against decisive points leading to centers of gravity. Successful engagement takes advantage of the Army force's superior training, leadership, and, within the constraints of the environment, equipment and technology. Successful engagement also requires the establishment of necessary levels of control and influence over all or portions of the urban environment until responsibilities can be transferred to other legitimate military or civilian control. Engagement may range from the overwhelming, yet precise, application of combat power to defeat a threat, to large-scale humanitarian assistance operations, to unobtrusive advice and assistance to the urban security forces. It may require brief engagement in a relatively small portion of an urban area or with small segments of the population to long-term engagement throughout multiple, large urban areas and with vast sectors of the population. The commander's ability to engage in a specific urban operation depends, as always, on the situation and the assigned mission.

Offense: Attack Decisive Points

6-18. In offensive UO, forces successfully engage by striking at the enemy's center of gravity using multiple offensive actions from unexpected directions and throughout all dimensions. Army forces aim to engage identifiable decisive points. Successful efforts against decisive points lead to effects on the center of gravity. The center of gravity will differ in each offensive situation. It may be an individual enemy leader, the enemy's combat power, the enemy's communications capability, or a physical structure of cultural, political, or economic significance.

Defense: Deny Vital Functions and Critical Infrastructure

6-19. In defensive UO, engage translates into denying the enemy control of the vital functions and critical infrastructure of the urban area. Forces achieve this by leveraging the defensive advantages of the urban terrain, defending essential areas in depth, using economy of force in nonessential areas, controlling the enemy direction of attack with natural and man-made obstacles, and retaining the initiative through counterattacks.

Stability and Civil Support: Apply Innovation and Imagination

6-20. The ability to engage in urban stability operations hinges on the type of stability operation commanders execute. In a noncombatant evacuation operation, forces limit engagement to finite geographic areas and times. In contrast, a peace operation may require engagement of a large urban area for an extended time. In this operation, engage is defined as using the array of Army capabilities to create specific conditions among the belligerents. Thus, the techniques used for engagement in stability operations vary according to the situation and as situations mature during long-term operations.

6-21. In urban civil support operations, engagement is accomplished by innovative and subtle application of Army capabilities. Since Army forces usually support other agencies that lead the operation, effective engagement results from carefully and discretely applying Army capabilities to the tasks assigned by the lead agency. In a humanitarian relief situation, Army forces may be tasked to transport supplies in the urban area. Successful engagement in this activity then becomes the goal of Army forces and may be achieved by providing, managing, and protecting transportation assets and clearing and maintaining sufficient lines of communications (LOCs).

CONSOLIDATE

6-22. Army forces consolidate to protect and strengthen initial gains and ensure retention of the initiative. Consolidation includes actions taken to eliminate or neutralize isolated or bypassed threat forces (including the processing of enemy prisoners and civilian detainees) to increase security and protect LOCs. It includes the rapid repositioning and, as necessary, reorganization of forces and intelligence, reconnaissance, and surveillance (ISR) assets. Consolidation may also include activities in support of the civilian population such as relocating displaced civilians, reestablishing law and order, humanitarian assistance and relief operations, and restoration of key urban infrastructure. A significant consolidation effort may be the reduction or elimination numerous explosive hazards resulting from previous urban combat operations. During consolidation, commanders often coordinate with NGOs to establish unity of effort in the execution of stability or civil support operations. Consolidation begins with the initiation of the urban operation and follows each critical activity.

TRANSITION

6-23. When planning UO, commanders ensure that they plan, prepare for, and manage transitions. Transitions are movements from one phase of an operation to another or a change in responsibility for all or portions of the urban environment from one unit, organization, or agency to another. They involve significant changes in the type of operation, concept of the operation, mission, situation, task organization, forces, resource allocation and support arrangements, or C2. Transitions occur in all operations, but in UO they occur with greater frequency and intensity, are more complex, and often involve agencies other than U.S. military organizations. For example, a successful attack may transition to a defend mission that includes not only defense tasks but also stability tasks. All operations will likely include a transition of responsibility for some aspect of the urban environment to (or back to) a legitimate civilian authority. Unless planned and executed effectively, transitions can reduce the tempo of the urban operation, slow its momentum, and cede the initiative to the threat.

Mental and Physical Preparation

6-24. Transitions occur as conditions warrant. They can be carefully planned and controlled, or they can be quick and dramatic, such as the swift transformation of a stability operation into offense or defense—and back again. Units prepare mentally and physically to address rapid transitions. Accordingly, plans include branches and sequels that address anticipated or possible transition points. When the dominant type of operation changes from an offense to stability or civil support, the types of units originally conducting the UO may no longer be appropriate. A large mobile reserve may permit increased flexibility to react to unplanned transition requirements. Operations or units in one part of an urban area may transition before operations or units in a different part of the same urban area. This will require commanders to exercise careful command and control and execute various types of operations and associated tasks simultaneously.

Transition to Legitimate Civilian Authorities or Agencies

6-25. Combat operations are not an end to themselves. They are inevitably a means of transition, an activity for moving from an unsatisfactory state of affairs to an improved end (as determined by national objectives). In UO, a distinct aspect of transition is the requirement to quickly and efficiently transition the major portions of Army responsibilities to civil agencies. Some tasks to which units will transition are not traditional combat tasks but rather stability tasks more closely associated with CMO. In stability or civil support operations this is often a near-term critical mission objective. In these operations, commanders aim to alleviate the circumstances requiring Army forces and ensure that other civilian agencies assume the functions provided by Army forces. Following combat operations, civilian agencies are assisted and encouraged to quickly resume specific support activities—such as providing sanitary services, food services, law enforcement, and health services—because of their high demand on Army resources.

6-26. Transitioning responsibility for the urban environment (or selected portions) to legitimate civilian agencies or authorities requires commanders to make an accurate assessment and develop an in-depth understanding of the civilian organization's ability to accept and handle the responsibility. Commanders consider not only resources available but whether the organization has the appropriate leadership and technical expertise as well as a vision in consonance with U.S. objectives. Commanders should avoid forcing a transition only to alleviate the burden on Army forces. On the other hand, they may need to encourage acceptance of responsibility while continuing to provide "behind-the-scenes" advice, assistance, and resources. Transitioning responsibility prematurely may alleviate the commander's immediate resource concerns but may, in the longer term, negatively impact the achievement of strategic objectives.

Clearly Visualize and Describe the End State

6-27. Army UO conclude when Army forces depart and have no further mission requirements in the urban area. At the outset, commanders visualize and describe the intended end state of a unit's execution of UO. Commanders then clarify and update this visualization as the political or strategic situation is refined or changes. Importantly, commanders must create a command climate and procedures that allow their staffs and subordinate commanders the ability to ask questions that will enable all to share in an accurate perception of the commander's visualization. This enables subordinate units to identify likely transitions and ensures that current operational planning takes into account second- and third-order effects. As long as an active Army AO contains an urban area, some type of urban operation will exist. After urban combat successfully ends, combat forces may move on. Support forces conducting sustaining operations may then occupy the area and continue to conduct a different form of UO.

Applying the Urban Operational Framework: Panama – 1989

The U.S. conducted OPERATION JUST CAUSE in December 1989 to remove the illegal ruler of Panama, Manuel Noriega, and to restore that country to a democracy. It also conducted the operation to ensure the safety of a substantial number of U.S. personnel as well as the security of U.S. interests in Panama. The major focus of JUST CAUSE was in Panama City, the country's capital. Most operations occurred in this large urban area, one of the numerous smaller urban areas, or the urban-like military bases. These bases increased the AO and were directly linked to operations in the capital city. This successful operation illustrates how commanders can apply the urban operational framework to visualize, describe, and direct the critical aspects of urban operations.

Understand

The synchronization achieved during the operation may have obscured the challenges faced in the initial assessment process in Panama. However, it was not as simple as it may have seemed. Using the framework of the urban environment, U.S. forces required details of the physical characteristics of the environment, the infrastructure, and the human dimension including the capabilities of the Panamanian military.

Because Army forces had a long history in Panama, commanders clearly understood the physical challenges and layout of critical urban areas (see Figure 6-3), particularly Panama City. They also understood how the infrastructure in each urban area functioned and which parts was a key element to success. Examples of key portions of the infrastructure included the Madden Dam, which controlled the water flow through the Panama Canal, and the Cerro Azul television tower, which was the main Panamanian broadcast tower.

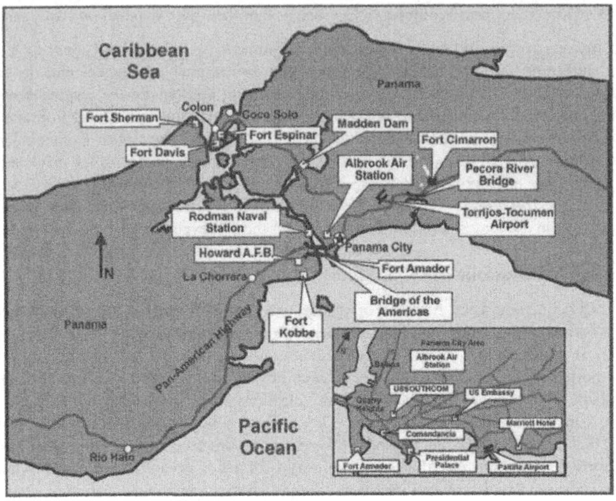

Figure 6-3. Panama

Collecting information and developing intelligence on the human elements of the urban environment was critical to operational success and a challenge. Because the Panamanian Defense Force (PDF) had traditionally been an ally of the United States, Army forces did not have a systemic database that adequately depicted their order of battle and their true capabilities. Additionally, much of the situation in Panama was colored in political terms making it more difficult for traditional military sources to evaluate the status of PDF forces. For example, Army planners needed to know if PDF military units (when faced with a formidable U.S. force) would fight at all for Noriega and if they did fight, how hard and long would they fight. The answers depended largely on their political loyalty to Noriega and on the individual loyalty of the unit officers to the Panamanian president. Thus, Army commanders needed to understand the military characteristics of PDF units and their political affiliations and tendencies.

Because transition from combat to noncombat tasks would be critical to achieving all objectives, particularly the restoration of democracy, Army forces also needed an accurate assessment of the political opposition to Noriega—including that opposition's capabilities and vulnerabilities. Again, Army forces were required to make assessments outside those needed solely for combat operations. Ultimately, assessing the political opposition's vulnerabilities led to assigning Army units to protect them throughout the operation so that they could serve as a foundation for a new democratic government.

Finally, the commander's assessment included an evaluation (often subjective) of the attitudes and disposition of the Panamanian people. Human intelligence (HUMINT) was the primary source of information on the population. Army forces had good access to the population because of their close proximity and historical ties to Panama. Many Soldiers were married to Panamanians, and the Army had total access to local media and to prominent individuals.

National imagery and special operations forces (SOF) also contributed to the ability of Army forces to assess the urban environments of JUST CAUSE. All units executing operations had detailed satellite photos of objective areas. Additionally, key objectives were placed under SOF surveillance well in advance. This surveillance revealed unit readiness, vulnerabilities, detailed disposition, and other patterns critical to mission success. The combination of the two capabilities allowed units to plan and achieve the synchronization necessary for such a complex urban operation.

Shape

During OPERATION JUST CAUSE, commanders conducted numerous shaping operations to establish the conditions for the decisive operations. Many operations were designed to control information, such as an assault on the Azul television tower identified during the assessment of the infrastructure. Planners designed many shaping operations to isolate various garrisons and PDF units. An example of tactical isolation was the plan for the Pacora River Bridge to prevent reinforcements from reaching the garrison at Torrijos-Tocumen Airport.

Operational isolation was achieved through the Ranger Regiment's and 82nd Airborne Division's assault on targets at Rio Hato in the west and Fort Cimarron in the east. These actions in conjunction with the securing of Maddam Dam had the primary objective of isolating Panama City. They were also the largest of the major actions occurring during OPERATION JUST CAUSE. The airborne assault was also the largest airborne operation conducted by U.S. forces since World War II. This large-scale shaping operation demonstrates that shaping operations are critical to mission success and can be more resource intensive than the actual operations that achieve domination.

Engage

U.S. Army forces achieved successful engagement in OPERATION JUST CAUSE by establishing unchallenged military control over Panama City and eliminating Noriega's capability to challenge that control. Toward this end, the operation attacked two decisive points. The first was the assault on the PDF headquarters located in Panama City: the Comandancia. The second was the operation undertaken to locate and seize Noriega himself.

Three battalions of task force (TF) Bayonet (5-87th Infantry, 1-508 Infantry [Airborne], and 4-6th Infantry [Mechanized]) executed the attack on the Comandancia and Fort Amador. They were also tasked to protect the American Embassy in downtown Panama City. To execute these missions, they moved from various staging areas located throughout the city to their assigned objectives using air assault, mounted, and dismounted approaches. The ground movement through the city proved to be the most difficult and hazardous part of the mission due to the vulnerability of the troops in their armored personnel carriers and trucks. The dismounted movement was slower than the mounted movement but allowed the Soldiers greater cover and concealment.

The strongest opposition to TF Bayonet occurred at the Comandancia. Elements of three PDF companies and two public order companies held out for three hours. The troops moving to Comandancia were subject to a large volume of sniper fire, and in the assault, unidentified indirect fire caused significant casualties among the mechanized forces. TF Bayonet forces, supported by airborne armored reconnaissance vehicles and Hellfire missiles from Apache helicopters, captured the

Comandancia. Commanders noted in particular the precision of the supporting fires from attack helicopters. The assault by fire from supporting AC 130 gunships severely damaged much of the reinforced Comandancia building.

Simultaneously, SOF attacked several targets where Noriega might be located. These initial attacks were unsuccessful. However, many subsequent actions neutralized Noriega's influence and eventually resulted in his apprehension on 3 January 1990. These actions included the well-organized and relentless manhunt conducted by SOF units, the isolation of Panama City itself, population control efforts, sophisticated IO, and cooperation with other U.S. agencies.

Consolidate and Transition

OPERATION JUST CAUSE demonstrated the vital need for a thought-out plan that adequately addresses consolidation and the transition from combat to noncombat before commanders initiate operations. Normally in complex UO, commanders cannot leave the details of transition until after the operation has begun without unacceptable risk to overall mission accomplishment. The follow-on stability operation, OPERATION BLIND LOGIC (later renamed OPERATION PROMOTE LIBERTY) began 24 hours after the initial assault and thus both operations were occurring simultaneously. This simultaneity of different types of operations is typical in major operations conducted in a large urban area. The stability operation involved more time than the combat operation and continued well after the close of OPERATION JUST CAUSE and after most of the major combat units had redeployed. It involved significant resources without the same level of risk to U.S. forces as the combat operations.

Civil affairs (CA) were a dominant part of the transition from combat to stability operations. The 96th Civil Affairs battalion was central to this operation. CA forces established a civil police force, emergency food distribution, property protection, production and distribution of newspapers, cleanup of the city, and building support for a new civil government. Most tasks were coordinated through Army CA forces and executed by other Army forces under the supervision of CA.

IO were also a major aspect of affecting a stable transition and successful postcombat operations. These operations built support for the U.S. operation among the population. They emphasized that the U.S. conflict was with Noriega and not the Panamanian people and that the U.S. forces would depart as soon as a new Panamanian government could take over.

Other U.S. agencies played critical roles in stability operations in Panama. The U.S. Drug Enforcement Agency and Justice Department were important to the negotiations that led to Noriega's capture. The U.S. State Department helped to negotiate for Noriega and develop military policies and plans during stability operations. The American Embassy advised commanders regarding the large diplomatic community that existed in Panama City.

FUNDAMENTALS OF URBAN OPERATIONS

6-28. UO often differ from one operation to the next. However, some fundamentals apply to UO regardless of the mission, geographical location, or level of command. Some of these fundamentals are not exclusive to urban environments. Yet, they are particularly relevant to an environment dominated by man-made structures and a dense noncombatant population (see figure 6-4). Appendix A provides an historical example of how many of these fundamentals applied to an actual conflict situation. Vitally, these fundamentals help to ensure that every action taken by a commander operating in an urban environment contributes to the desired end-state of the major operation.

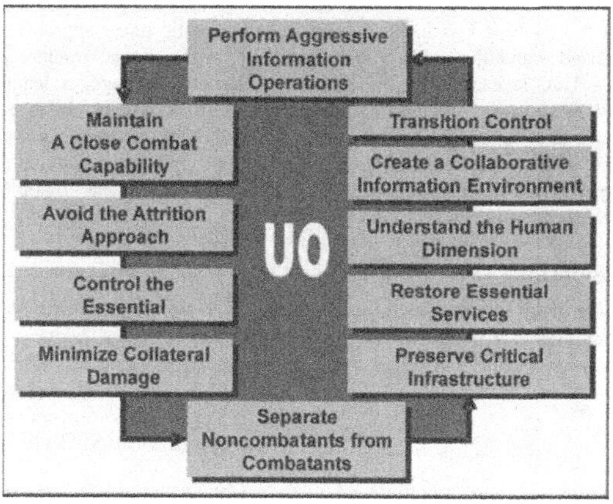

Figure 6-4. Fundamentals of urban operations

PERFORM AGGRESSIVE INFORMATION OPERATIONS

6-29. Information operations aimed at influencing non-Army sources of information are critical in UO across the spectrum of operations. Because of the density of noncombatants and information sources, the media, the public, allies, coalition partners, neutral nations, and strategic leadership will likely scrutinize how Army forces participate in UO. The proliferation of cell phones, video cameras, Internet capability, and media outlets ensure close observation of the activities of Army forces. With information sources rapidly expanding, public information of Army operations will often be available faster than the internal military INFOSYS can process it. Army forces should aggressively integrate IO into every facet and at all levels of the operation to manage perceptions and mitigate unintended consequences. Under media scrutiny, the actions of one Soldier may have significant strategic implications. IO aim to make the information accurate; placed in the proper context of the Army's mission; and available to all interested parties: the public, the media, and other agencies. Ensuring that the urban population understands how they and their urban area fit within the commander's vision, intent, and end state will be a critical task.

MAINTAIN A CLOSE COMBAT CAPABILITY

6-30. Close combat is required in all offensive and defensive UO. This core capability must also be present and visible in urban stability operations and may be required in urban civil support operations. Close combat in any urban operation is resource intensive, requires properly trained and equipped forces, and has the potential for high casualties. However, the ability to decisively close with and destroy enemy forces as a combined arms team remains essential. Potential threats respect strength and, instead, seek to attack ill-prepared or complacent Army forces—not necessarily forces that are better protected or armored, but those less likely or capable of inflicting casualties against them. In stability and civil support operations, a lack of respect and fear of Army forces can hinder recovery as much as the ill-advised use of forces. Hence, all Soldiers, regardless of branch or military occupational specialty, must be properly equipped and trained to fight in an urban environment. This ability allows Army forces to deter aggression, compel compliance, morally and physically dominate a threat and destroy his means to resist, and terminate urban conflicts on the Army commander's terms.

Avoid the Attrition Approach

6-31. Previous Army doctrine was inclined towards a systematic linear approach to urban combat. This approach emphasized standoff weapons and firepower. Army force structure does not support this approach towards UO. It can result in significant collateral damage, a lengthy operation, and an inconsistency with the political situation and strategic objectives. Enemy forces that defend urban areas want Army forces to adopt this approach because of the likely costs in resources. Commanders should only consider this approach to urban combat as an exception and justified by unique circumstances. Instead, commanders should seek to achieve precise, intended effects against multiple decisive points that overwhelm a threat's ability to react effectively.

Control the Essential

I'm talking about attacking those things from which the regime draws power but being very careful about it so that we don't get large bodies of young Americans caught up in house-to-house Berlin, World War II-type scenario.

Lt. Gen. William C. Wallace
Commander, V Corps
OPERATION IRAQI FREEDOM

6-32. Many modern urban areas are too large to be completely occupied or even effectively controlled without an enormous force. Therefore, Army forces focus their efforts on controlling only the essentials to mission accomplishment. At a minimum, this requires control of key terrain. Key terrain is terrain whose possession or control provides a marked advantage to one side or another. In the urban environment, commanders determine key terrain based on its functional, political, economic, or social significance. A power station or a place of worship may be key terrain.

6-33. All principles of war can apply to UO. The principle of mass and the principle of economy of force (in addition to the principle of unity of command discussed in the previous chapter) are particularly important in guiding UO and providing mission focus. Army forces mass combat power only to control those requirements essential for mission success. This permits conservation of combat power. It also implies economy of force and associated risk in those areas where Army forces choose not to exercise control.

Minimize Collateral Damage

6-34. Forces should integrate precision fires, IO, and nonlethal tactical systems consistent with mission accomplishment while decreasing the potential for collateral damage. Commanders develop unique rules of engagement (ROE) for each urban operation and provide necessary firepower constraints. IO and nonlethal systems may compensate for some restrictions, especially in stability or civil support operations. Commanders continually assess the short- and long-term effects of operations and firepower on the population, infrastructure, subsequent missions, and national and strategic objectives. They must also consider what, if any, provisions should be made to amend or address potential collateral damage. Overall, commanders must balance restraint and precision with speed and overwhelming combat power. By avoiding unnecessary harm to all elements of the urban environment, commanders retain the moral high ground and help sustain legitimacy for their operations. Minimization of collateral damage allows civilians to continue to provide for their own needs or the rapid return of the urban area to civilian self-sufficiency.

Separate Noncombatants from Combatants

6-35. Promptly separating noncombatants from combatants (psychologically and physically) may make the operation more efficient and diminish some of the threat's potential asymmetrical advantages. This separation also may reduce restrictions on the use of firepower, enhance force protection, and strip the threat from its popular support base. This important task becomes more difficult when the threat is an unconventional force that can mix with civilians.

6-36. In more recent operations, threats have sought to integrate their military capabilities as closely as possible into the civilian population and infrastructure. In these conditions, commanders must increase their efforts to discriminate between the two. Soldiers managing violence in this setting require the highest level of individual and organizational discipline and judgment. Soldiers will require the mental agility to separate their aggression toward threats from the noncombatant civilian population. The training, effort, and command emphasis in this area is as important as fully successful results. Such efforts strongly impact national and international perceptions of the operation.

Preserve Critical Infrastructure

6-37. Commanders should analyze the urban area to identify critical infrastructure. They should attempt to preserve and protect the critical elements for postcombat sustainment operations, stability or civil support operations, or the overall health and well-being of the indigenous population. Urban areas remain in the AO after combat operations have ceased. Postcombat UO are unavoidable. Different from simply avoiding collateral damage, Army forces may have to initiate actions to prevent an enemy or a hostile civilian group from removing or destroying critical infrastructure and assets. This may include cultural infrastructure such as religious and historical places. In some cases, preserving the infrastructure and the urban society's sources of economic and cultural wealth may be the assigned objective of the urban operation.

Restore Essential Services

6-38. Army forces plan to restore essential services that may fail to function before or during an operation. Essential services include power, food, water, sewage, medical care, and security and law enforcement. When planning for and conducting Army UO, units can use nonlethal and less destructive munitions and capabilities to keep potentially vital infrastructure intact. Initially, Army forces may be the only force able to restore or provide essential services and commanders must plan accordingly. Failure to do so can result in serious health problems for the civilians, which can affect the health of Army forces and negatively impact overall mission success. Army forces seek to transfer responsibility for providing essential services to other agencies, NGOs, or the local government as quickly and effectively as possible. Despite the potential causes for the failure or destruction of essential services, commanders also seek to ensure that civilians continually perceive Army restoration activities as assistance rather than an Army requirement. Otherwise, the civilian populace may be slow to accept or resume responsibility for their urban area.

Understand the Human Dimension

Do not try to do too much with your own hands. Better the Arabs do it tolerably than that you do it perfectly. It is their war, and you are to help them, not to win it for them. Actually, also, under the very odd conditions of Arabia, your practical work will not be as good as, perhaps, you think it is. It may take them longer and it may not be as good as you think, but if it is theirs, it will be better.

T. E. Lawrence
"Twenty-Seven Articles"

6-39. Commanders carefully consider and manage the perceptions, allegiance, and morale of the civilians. Their assessment of the environment needs to accurately identify the attitudes of the people toward Army forces. Operational guidance to subordinates including ROE, protection, sustainment operations, and fraternization—is based on this assessment. Commanders expect and consider the demographic variance in the attitudes of an urban population. They cannot inadvertently apply Western cultural norms to a non-Western urban population. Commanders can only make reliable assessments based on a thorough understanding and appreciation of the local society and their culture. Developing this ability also requires the additional ability to effectively share information among all echelons of command.

6-40. Sound policies, proper discipline, adequate consideration for local culture, and rapid engagement of local urban leaders will positively affect the attitudes of the population toward Army forces. Additionally, well-conceived and executed IO will enhance the position of Army forces relative to the urban population. Even during high-intensity urban combat, heightened awareness of and sensitivity toward the civilians can lead to a better postcombat situation than if civil considerations were unobserved or diminished in importance. An improved postcombat situation enhances transition. As the environment of conflict becomes more complex, the human dimension (and associated moral aspects) takes on greater importance and may have the greatest potential for affecting the successful outcome of UO. Therefore, the human aspect and development of cultural acuity creates a discrete planning factor.

CREATE A COLLABORATIVE INFORMATION ENVIRONMENT

6-41. The complexity of the urban environment, particularly the human dimension, requires rapid information sharing—providing and receiving—from the national level to the tactical level, among Army headquarters at each echelon, with sister services and coalition partners, and with participating governmental and (at appropriate times) nongovernmental agencies. The analysis of urban information into the relevant intelligence necessary to refine and deepen a commander's understanding of the urban environment and its infrastructure of systems also demands collaboration among the various information sources and consumers. Therefore, commanders seek to establish streamlined procedures, develop commonality among databases, and make best use of existing information systems to disseminate and receive the necessary intelligence and relevant information for subordinates and partner organizations and agencies to exercise effective leadership and decision making and establish synergistic unity of effort in this multifaceted environment.

TRANSITION CONTROL

6-42. Because UO are resource intensive, commanders should plan to end them as quickly as possible, yet consistently with successful mission accomplishment. (Depending on the factors of METT-TC, successful transition may take a few days or many years to achieve.) The end state of all UO transfers control of the urban area to another agency or returns it to legitimate civilian control and responsibility. Rapid transition releases Army resources for use elsewhere and improves the civilian morale and disposition toward Army forces. This requires the successful completion of the Army force mission and a thorough transition plan. The transition plan may include returning control of the urban area to another agency a portion at a time as conditions permit. A successful transition plan should consider early alignment of military capabilities with existing urban governmental and administrative organizations, agencies, structures, and districts. Transition planning must be conducted before the onset of operations and continually adjusted as the situation develops.

Chapter 7

Urban Offensive Operations

... Capture Suez City "provided it does not become a Stalingrad situation."

Order to the Adan Armored Division
prior to its 1973 attack on Suez City
On the Banks of the Suez

Offensive urban operations (UO) are one of the most challenging operations that military forces can undertake. Campaigns and wars have sometimes hinged on their success or failure. Costly in resources, even when successful, they are not lightly entered into. Once engaged, they are executed rapidly and decisively. For reasons already discussed, threat forces defending in UO may gain advantages from the environment while Army force capabilities may diminish. Despite the challenges, Army forces conduct successful urban offensive operations by combining the Army's existing offensive doctrine with a thorough understanding of the environment.

PURPOSE OF URBAN OFFENSIVE OPERATIONS

7-1. Like all offensive operations, urban offensive operations are designed to impose the will of commanders on the threat. The urban offense often aims to destroy, defeat, or neutralize a threat force. However, the purpose may be to achieve some effect relating to the population or infrastructure of the urban area. Army forces may conduct offensive operations to secure a port or a communications center, to eliminate a threat to a friendly government or the urban population, or to deny the threat use of urban infrastructure. No matter the purpose, commanders must use a combined arms approach for successful urban offensive operations.

OFFENSIVE CHARACTERISTICS APPLIED TO URBAN OPERATIONS

7-2. All offensive operations, including those in urban areas, contain the characteristics of surprise, concentration, tempo, and audacity (see FM 3-0). Commanders consider and incorporate these characteristics in their offensive UO plans.

SURPRISE

7-3. Army forces can achieve offensive surprise at two levels: operational and tactical. In urban offensive operations, operational surprise can be decisive. The goal is to attack the urban area before the threat expects it, from a direction he doesn't expect, or in a manner he doesn't expect. In UO, this requires an attack against urban areas that the threat believes will provide sanctuary from the technological advantages of Army forces. Usually, urban areas that meet this criterion are not easily accessible. Army forces launch an attack against these urban complexes in different ways: through a vertical assault using airborne or air assault forces, through an amphibious assault, or through a penetration followed by a rapid and deep advance. All three attacks aim to achieve surprise and to deny the threat time to prepare and establish a defense. Surprise in a major urban operation prevents a threat from falling back to occupy prepared positions in and around an urban area.

7-4. At lower tactical levels, forces achieve surprise by attacking asymmetrically. An asymmetric method attacks the threat so he cannot respond effectively. This may be achieved by using special operations forces

(SOF) against a threat prepared for a conventional attack, by attacking decisively with heavy forces when the threat expects an effort by light forces or SOF, or by leveraging Army forces' extensive information operations (IO) capability. Offensive IO—primarily using IO elements of deception, electronic warfare, and operations security (OPSEC)—can help achieve surprise at all levels (see Chapter 4). Attacking at night surprises the threat and maximizes the Army forces' training, command and control (C2), and technological advantages. Attacking from unexpected or multiple directions achieves surprise by leveraging Army information systems (INFOSYS) and superior synchronization of combat power and capabilities.

CONCENTRATION

7-5. In UO, the attacking force creates a major advantage by concentrating the effects of combat power at the point and time of its choosing. The area and its compartmented effects naturally disperse and dissipate combat capability. Often commanders must position considerable amounts of stay-behind forces to protect hard-fought gains. The environment also hinders repositioning forces rapidly. Such effects can work equally against defending and attacking forces. However, in a well-prepared defense, the defender often has the advantage of interior lines. The defender can reinforce or reposition forces more quickly using covered and concealed routes (such as, sewers, tunnels, or prepared holes made in walls). Successful UO need synchronized air and ground maneuver with overwhelming effects from fires at decisive points on the urban battlefield. To achieve proper synchronization and precise effects, commanders should consider the unique time and distance relationships set by the environment.

TEMPO

7-6. Tempo is the rate of military action; it differs between UO and operations in more open terrain. Often, the primary purpose of the threat's urban defense is to take advantage of that differential to disrupt the rapidity and overall tempo of the Army's major operation. Additionally, the complexity and the potential for increased risk in an urban environment may invoke a cautious and methodical response on the part of commanders and their staffs thereby exacerbating tempo differences. While careful and meticulous preparation and planning are required in UO, commanders conducting major operations that include urban areas must strive to maintain an active tempo in offensive operations through synchronization of combat power and anticipation of threat reactions. A high tempo allows Army forces to achieve surprise and quickly gain positions of advantage. Controlling operational tempo and not allowing the different tempo of UO to adversely affect other operations is a challenge that commanders of major operations must overcome.

Operational Context of Urban Operations: Brittany Ports – 1944

The plan for the invasion of Normandy, France, in June 1944 was meticulously developed. The plan not only addressed the invasion itself, but also contained detailed planning for the campaign to follow. A major concern of the detailed campaign planning was logistics. To address this critical concern, and specifically the problem of ports to supply the allied armies once ashore, the pre-invasion planning called for the major ports of the French province of Brittany—Brest, Lorient, and Saint Nazaire—to be objectives of General Patton's Third Army, once it was activated.

Early August 1944, almost two months after the successful Normandy invasion, the operational situation significantly differed from that envisioned by the D-Day planners. General Montgomery's Twenty-first Army Group was still fighting in the Bocage of Normandy. In contrast, General Bradley's Twelfth Army Group had just achieved a major breakthrough at Saint Lo, secured the Cotentin Peninsula, and reached the city of Avranches. Here was a decision point. Bradley and Eisenhower had to decide whether to adhere to the original plan and turn west with Patton's forces to secure the peninsula or to take advantage of the breakout at Saint Lo and turn east to exploit the disruption of the German defenses.

Ultimately they reached a compromise. General Middleton's VIII Corps was tasked to secure the peninsula, and the bulk of Patton's Army, three Army corps, was turned northeast to exploit the operational collapse of the main German defenses. (See figure 7-1.)

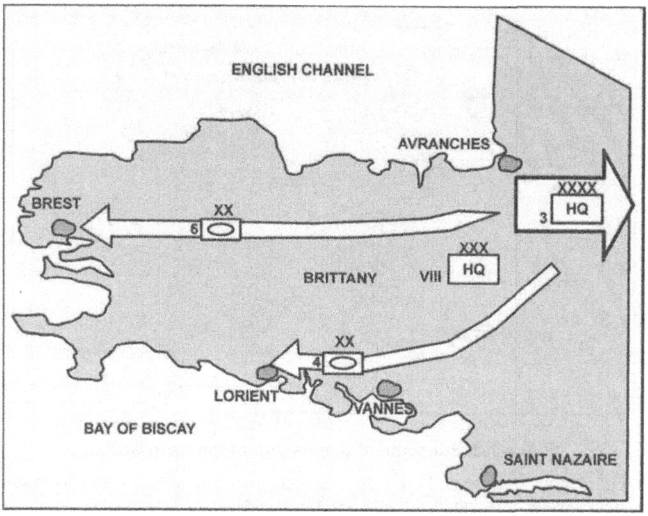

Figure 7-1. Initial attack in Brittany

Middleton's corps sprinted into the peninsula with the 4th and 6th Armored Divisions leading the way. However, poor communications, disagreements between commands, and contradictory orders caused the corps to hesitate before pushing the two divisions to continue to exploit toward the ports. The result: the 6th Armored Division missed an opportunity to seize Brest against light resistance by one day. The 4th Armored Division, after capturing the smaller port of Vannes, was also frustrated on the approaches to Lorient. The American reaction to the inability to rapidly seize the ports demonstrated an understanding of changing circumstances. The 6th Armored Division turned the attack at Brest to the 8th Infantry Division and then relieved the 4th Armored Division at Lorient. The 4th Armored was moved to rejoin the rest of Third Army exploiting to the east and north. Ultimately Brest fell to VIII Corps on 19 September after a 43-day siege by three infantry divisions. The victory yielded 36,000 German prisoners of war (POWs). However, the German defense and demolitions of the port left the port without an impact on the logistic situation of the allies. Brest cost the U.S. Army almost 10,000 casualties and the commitment of significant supplies. The experience convinced commanders to surround and bypass the other major Brittany ports. Lorient and Saint Nazaire remained under German control, deep in allied territory, until the war ended ten months later (see figure 7-2).

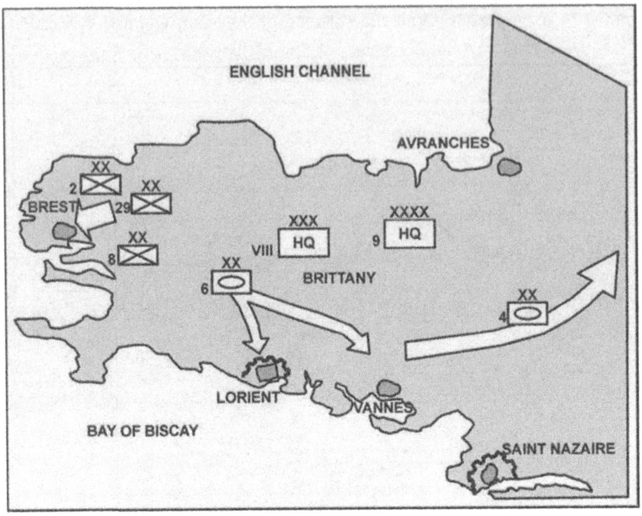

Figure 7-2. Subsequent disposition of forces in Brittany

The operational lessons of the Brittany campaign are numerous. First, commanders are responsible to continually assess assumptions and decisions made during planning based on the changing circumstances of the battlefield. This includes the planning decision to conduct urban offensive operations. When the allies arrived at the Brittany Peninsula, the focus of the operational maneuver was no longer securing logistics facilities but exploiting the breakthrough at Saint Lo and the disintegrating the German defense. The bulk of Third Army then was turned to the north and east rather than west into the peninsula.

The Brest experience also demonstrates that the costs of urban offensive operations are continually assessed against the operational value of the objective. This lesson was applied to the cities of Lorient and Saint Nazaire. The cities were never seized from the Germans because their logistic value failed to warrant the required resources to take them. German retention of the ports had no major adverse effect on the overall campaign.

Another lesson is that commanders cannot allow urban operations to disrupt the tempo of other offensive operations. One German goal of defending the ports was to disrupt the rapid tempo of the U.S. exploitation. They failed to achieve this goal because General Bradley continued the exploitation with the bulk of Third Army and executed the original plan with only a single corps.

Finally, commanders cannot allow emotion to color their decision to conduct or continue UO. The failure of 6th Armored Division to seize Brest rapidly caused some commanders to believe that Brest had to be captured because the prestige of the Army was committed to the battle. Costs of the continuing combat operations to seize Brest were significant. These resources might have been better committed elsewhere in the theater.

7-7. Tactical tempo is also important in urban combat. Because of the complex terrain, defending forces can rapidly occupy and defend from a position of strength. Once Army forces initiate tactical offensive operations, they cannot allow the threat to set the tempo of the operation. Instead, attacking forces seek to maintain a high tempo of operations. However, the tactical tempo of urban operations differs from operations in other terrain. Not necessarily slow, it requires a careful balance of preparation, speed, and security. In terms of unit fatigue, resource consumption, and contact with the threat, the tempo of most urban offensive operations may be rated as very high. On the other hand, in distances traveled and time consumed to achieve objectives, the tempo of many urban offensive operations might be rated as slow. The urban battlefield's density concentrates activity and consumes resources in a relatively small area. The lack of terrain seized or secured is not to be construed to mean a low tempo in the battle. Although the tempo may seem excruciatingly slow at higher levels of command and exceeding fast at lower, tactical levels, in reality, the natural tempo of urban operations is not faster or slower than other types of operations, merely different. Creating and operating at a tempo faster than an opponent can maintain, however, can favor forces which are better led, trained, prepared, and resourced.

7-8. A high tactical tempo in urban offensive operations challenges logisticians to provide for the increased consumption of munitions and degrades soldiers' physical capabilities. Commanders must anticipate these challenges and develop the means and abilities to overcome them. In the past, these challenges forced commanders to conduct urban offensives cyclically. They used night and other periods of limited visibility to resupply, rest, and refit forces. The environment influenced the tempo of their operations. This type of "battle rhythm" resulted in the forces spending each new day attacking a rested threat that was in a well-prepared position.

7-9. Army forces must maintain the tempo. Offensive operations continue even during darkness. Moreover, Army forces increase the tempo of operations at night to leverage the limited visibility capabilities, increased situational understanding, training, and INFOSYS that give an advantage to Army forces in all environments. To overcome the physical impact of the environment on soldiers, commanders can retain a large reserve to rotate, continuing offensive operations at night. The force that fights in daylight becomes the reserve, rests, and conducts sustaining operations while another force fights at night. Army forces can then maintain the tempo of operations and leverage technological advantages in urban offensive combat.

7-10. Tempo in UO does not necessarily mean speed. Offensive operations balance speed, security, and adequate firepower. Commanders must plan for the complex tactical environment and the requirements to secure flanks and airspace as the operation progresses. Mission orders allow subordinate units to make the most of tactical advantages and fleeting opportunities.

AUDACITY

7-11. Audacity is a simple plan of action, boldly executed. Superb execution and calculated risk exemplify it. In an urban attack, a thorough understanding of the physical terrain can mitigate risk. The terrain's complexity can be studied to reveal advantages to the attacker. Audacity can also be embodied in an operation by inventively integrating and coordinating the direct action tasks of SOF throughout the operation. Combining SOF actions with conventional attacks can asymmetrically unhinge a threat's defensive plan. Well trained Soldiers confident in their ability to execute urban offensive operations foster audacity.

URBAN OFFENSIVE OPERATIONS AND BATTLEFIELD ORGANIZATION

7-12. Urban offensive operations, like all operations, are arranged using the overall battlefield organization of decisive, shaping, and sustaining operations. Each operation is essential to the success of an urban offensive, and usually two or more of these operations occur simultaneously. Decisive operations are attacks that conclusively determine the outcome of UO. These attacks strike at a series of decisive points and directly lead to neutralizing the threat's center of gravity. Shaping operations in urban offensive operations create the conditions for decisive operations. In UO, much of the shaping effort focuses on isolation, which is critical in both major operations and tactical battles and engagements. Sustaining operations in urban offensive operations ensure freedom of action. They occur throughout the area of operations (AO) and for the duration of the operation.

Decisive Operations

7-13. A tactical commander fights decisive urban combat, whereas commanders conducting a larger major operation influence urban combat by setting the conditions for tactical success. Higher commanders may directly influence urban offensive operations by operational maneuver, by coordinating joint fires, by closely coordinating conventional forces, or with SOF.

7-14. Tactical urban offensive operations quickly devolve into small-unit tactics of squads, platoons, and companies seizing their objectives. The compartmented effect of the terrain and the obstacles to command and control of small units, especially once they enter close combat inside buildings or underground, often restricts the higher commander's ability to influence operations. Commanders can influence the actions of subordinates by clearly identifying the decisive points leading to the center of gravity; using mission orders (as discussed in Chapter 5); developing effective task organizations; synchronizing their decisive, shaping, and sustaining operations; and managing transitions.

7-15. Like all operations, successful decisive operations in UO depend on identifying the decisive points so the forces can destroy or neutralize the threat's center of gravity. Seizing a key structure or system that makes the threat's defense untenable; interdicting a key resupply route that effectively isolates the threat force from his primary source of support; or isolating the threat so that his force can no longer influence friendly activity may be more effective than his outright destruction.

7-16. Commanders must select the right subordinate force for the mission and balance it with appropriate attachments. Higher commanders do not direct how to organize the small tactical combined arms teams, but they must ensure that subordinates have the proper balance of forces from which to form these teams. Successful urban offensive operations require small tactical combined arms teams. Urban offensive operations require abundant infantry as the base of this force. However, successful urban combat requires a combined arms approach adjusted for the conditions of the environment. In urban offensive operations, it will normally not be a question of whether include armor and mechanized elements but, rather, how best to accomplish this task organization. Precision-capable artillery systems generally support urban operations better than unguided rocket artillery.

7-17. Divisions entering urban combat may require additional resources. These resources include military intelligence support in the form of linguists, human intelligence (HUMINT) specialists, and unmanned aircraft systems (UAS). Engineering assets will be at a premium; the task organization of a task force executing the decisive operation may require a one-to-one ratio of engineer units to combat units. Corps and higher engineering support may be necessary to meet these requirements and to repair vital and specialized infrastructure. A tailored and dedicated support battalion or group may need to assist in providing anticipated support to a displaced and stressed civil population. Finally, divisional CA units may require augmentation to deal with NGOs and civilian government issues.

7-18. Successfully conducting decisive operations in the urban environment requires properly synchronizing the application of all available combat power. Army forces have a major advantage in the command and control of operations. Commanders can use this advantage to attack numerous decisive points simultaneously or in rapid succession. They also use it to attack each individual decisive point from as many directions and with as many different complementary capabilities as possible. Commanders must completely understand urban environmental effects on warfighting functions to envision and execute the bold and imaginative operations required. Significantly, these operations require that C2 systems account for the mitigating effects of the environment as execution occurs.

7-19. Properly synchronized actions considerably enhance the relative value of the combat power applied at the decisive points. They present to the threat more requirements than he has resources with which to respond. Synchronized IO and multiple maneuver actions paralyze the threat's decision-making capacity with information overload combined with attacks on his C2 systems. Additionally, well-synchronized actions limit the time the threat has to make decisions and forces him into bad decisions. In the urban environment, these effects are enhanced because C2 systems are already strained, poor decisions are harder to retrieve, and units that do not react effectively are more easily isolated and destroyed.

SHAPING OPERATIONS

7-20. Shaping operations that support the urban attack separate into those focused on isolating the threat and all others. Army forces isolate the threat to ensure successful urban offensive operations. Depending on the threat reaction to isolation efforts and the nature of the threat center of gravity, this task may become decisive. Other shaping operations include those common to all offensive operations and others unique to urban operations. Unique urban shaping operations may include securing a foothold in a well-fortified defensive sector, securing key infrastructure, or protecting noncombatants. Because of the nature of UO, shaping operations may consume a much larger proportion of the force than during other operations and may take place both inside and outside the urban area (see Applying the Urban Operational Framework: Panama in Chapter 5). By successfully isolating a threat force, the force needed to conduct the decisive operation may be relatively small.

SUSTAINING OPERATIONS

7-21. Commanders conducting urban offensive operations must ensure security of the sustaining operation and bases; in many situations, protecting lines of communications (LOCs) and sustaining operations may be the greatest vulnerability of the attacking force. Those supporting an urban offensive are tailored to the urban environment and are well forward. Ideally, the supporting forces closely follow the combat forces and move within or just outside the urban area as soon as they secure an area. Operating in the urban area during offensive operations allows the sustaining operation to take advantage of the defensive attributes of the environment for security purposes.

7-22. Counterattacks against sustaining operations may take the form of special operations activities aimed at LOCs leading to or within the urban area. Choke points—such as bridges, tunnels, and mountain passes—are vulnerable to these attacks and may require combat forces to protect them. Threat forces attack the LOC to blunt the Army's combat power advantage in the urban area.

7-23. Attacks against the LOC into the urban area may also attempt to isolate the attacking Army forces from its sustainment base. Isolated forces in an urban area are greatly disadvantaged. Commanders must plan and aggressively execute strong measures to protect their LOC, even if it requires reduced combat power to execute their offensive operation.

7-24. Sustaining operations anticipate the volume and unique logistics requirements of urban operations. Specialized individual equipment—such as grappling hooks, ladders, and pads—is identified and provided to troops in quantity before they are needed. Forces stockpile and distribute their attacking units' special munitions requirements including small arms, explosives, and grenades of all types, precision artillery munitions, and mortar ammunition. Forces also supply transport to move the resources rapidly forward, both to and through the urban environment. Sustaining operations cannot rely on "operational pauses" to execute their tasks. Commanders must plan to continuously supply resources and capabilities to the most forward combatants as offensive operations advance.

7-25. Sustaining operations also anticipate the growth of sustainment requirements as Army forces secure and take responsibility for large portions of the urban area. The success of Army urban offensive operations will often uncover the civil population in former threat occupied areas. It may attract the civil population from sections of the urban area where the Army is not operating to areas occupied by Army forces. Rural populations may migrate to the urban area as the result of successful Army offensive operations.

7-26. Army forces may be required to take initial responsibility to provide for the urban population. This consideration is integrated into logistics planning and organization from the start of the planning process. To be successful and efficient in such a situation, sustainment planning includes Army civil affairs (CA) specialists and local government representatives. It also integrates and consults with the international community and nongovernmental organizations (NGOs) that might augment or supplement Army sustainment capabilities.

FORMS AND TYPES OF URBAN OFFENSE

7-27. Traditional forms of offensive maneuver include envelopment, turning movement, infiltration, penetration, and frontal attack. Traditional types of offensive operations are movement to contact, attack, exploitation, and pursuit. These traditional forms and types listed apply to urban combat. Some have greater application to an urban environment than others do. Moreover, success will belong to commanders who imaginatively combine and sequence these forms and types throughout the depth, breadth, and height of the urban battlefield. This is true at the lowest tactical level and in major operations.

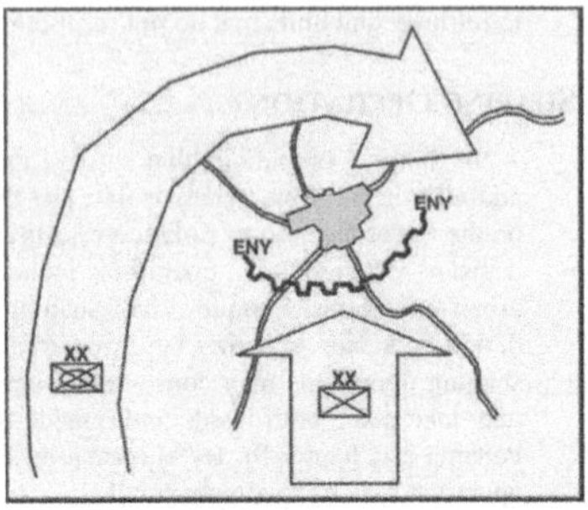

Figure 7-3. Envelopment isolates an urban area

FORMS OF OFFENSIVE MANEUVER

Envelopment

7-28. The envelopment is the ideal maneuver for isolating threat elements in the urban area or isolating the area itself. A deep envelopment effectively isolates the defending forces and sets the conditions for attacking the urban area from the flank or rear. Yet, enveloping an objective or threat force in the urban area is often harder since achieving speed of maneuver in the environment is so difficult (see figure 7-3). Vertical envelopment, however, works effectively if Army fires can effectively suppress or neutralize the threat air defense.

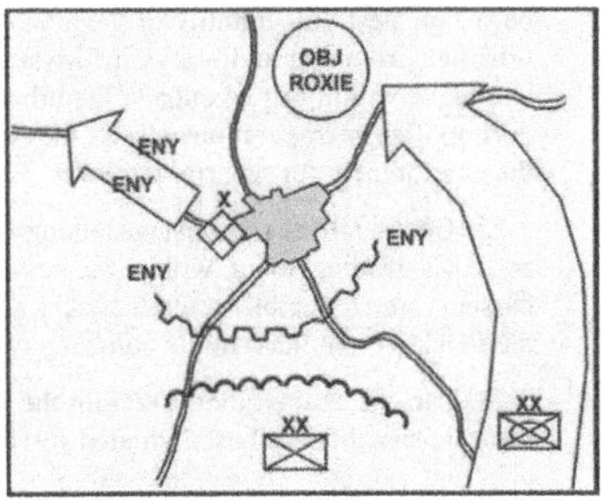

Figure 7-4. Turning movement

Turning Movement

7-29. Turning movements can also be extremely effective in major operations (see figure 7-4). By controlling key LOCs into the urban area, Army forces can force the threat to abandon the urban area entirely. These movements may also force the threat to fight in the open to regain control of LOCs.

Infiltration

7-30. Infiltration secures key objectives in the urban area while avoiding unnecessary combat with threat defensive forces on conditions favorable to them (see figure 7-5). This technique seeks to avoid the threat's defense using stealthy, clandestine movement through all dimensions of the urban area to occupy positions of

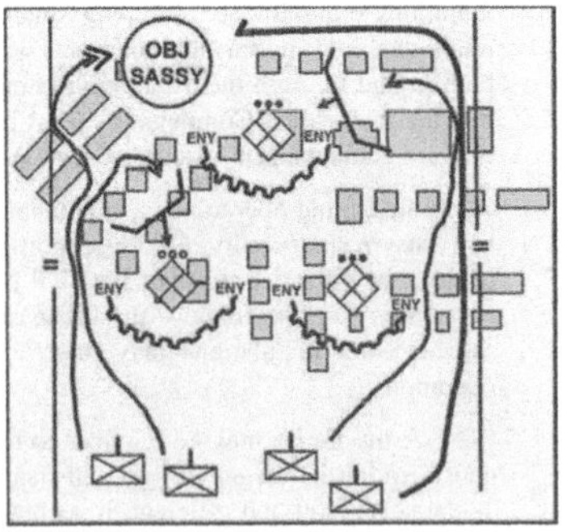

Figure 7-5. Infiltration

advantage in the threat's rear (or elsewhere).It depends on the careful selection of objectives that threaten the integrity of the threat's defense and a superior common operational picture (COP). Well-planned and resourced deception operations may potentially play a critical role in masking the movement of infiltrating forces. The difficulty of infiltration attacks increases with the size and number of units involved. It is also more difficult when Army forces face a hostile civilian population. Under such circumstances, infiltration by conventional forces may be impossible. Armored forces are generally inappropriate for infiltration operations. However, they may infiltrate large urban areas if the threat is not established in strength and had insufficient time to prepare defenses.

Penetration

7-31. Penetration is often the most useful form of attack against a prepared and comprehensive urban defense (see figure 7-6). It focuses on successfully attacking a decisive point or on segmenting or fragmenting the defense—thereby weakening it and allowing for piecemeal destruction. (The decisive point may be a relatively weak or undefended area that allows Army forces to establish a foothold for attacks on the remainder of the urban area.) Ideally in urban combat, multiple penetrations in all dimensions are focused at the same decisive point or on several decisive points simultaneously. In urban combat, the flanks of a penetration attack are secure, and resources are positioned to exploit the penetration once achieved. Although always a combined arms team, rapid penetrations are enhanced by the potential speed, firepower, and shock action of armored and mechanized forces.

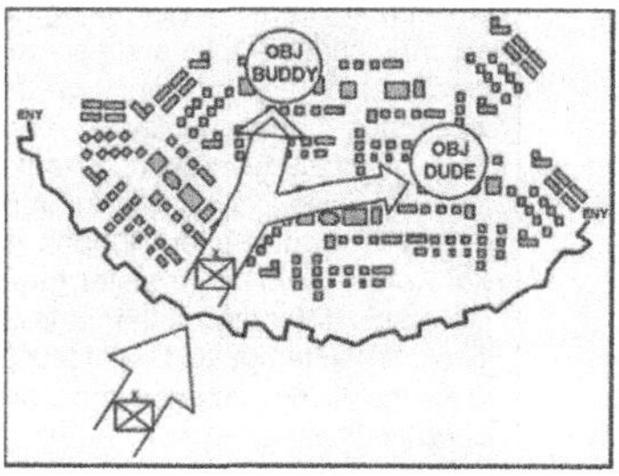

Figure 7-6. Penetration

7-32. Importantly, commanders must consider required actions and resources that must be applied following a successful (or unsuccessful) penetration. A penetration may result in the rapid collapse and defeat of the threat defense and complete capitulation. On the other hand, success may cause threat forces to withdraw but leave significant stay-behind forces (or disperse into the urban population as an insurgent-type force) which may necessitate methodical room-to-room clearance operations by significant dismounted forces. Additionally, securing portions (or all) of an urban area requires occupation by Army forces to prevent re-infiltration of threat forces thereby further increasing manpower requirements. Based on the factors of the situation, commanders may conclude that over time methodical clearance operations conducted from the outset, while frontloading time and resource requirements, will be less costly than a penetration followed by systematic clearing operations.

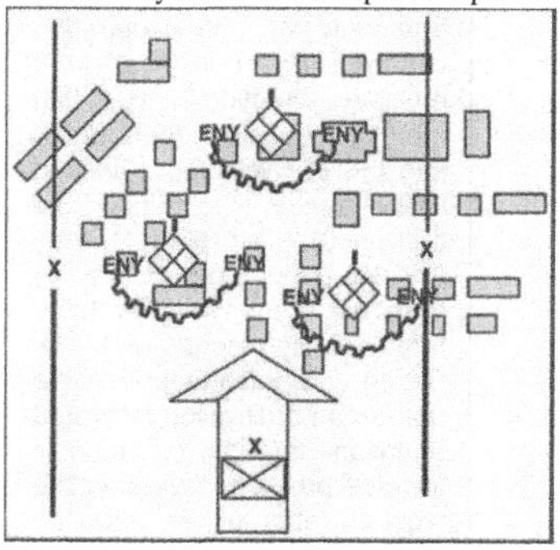

Figure 7-7. Frontal attack

Frontal Attack

7-33. For the commander of a major operation, the frontal attack is generally the least favorable form of maneuver against an urban area unless the threat is at an obvious disadvantage in organization; numbers, training, weapons capabilities, and overall combat power (see figure 7-7). Frontal attacks require many resources to execute properly, risk dispersing combat power into nonessential portions of the area, and risk exposing more of the force than necessary to threat fires. In urban offensive combat, forces most effectively use the frontal attack at the lowest tactical level once they set conditions to ensure that they have achieved overwhelming combat power. Then the force of the frontal attack overwhelms the threat with speed and coordinated and synchronized combat power at the point of attack. The assigned frontage for units conducting an attack on an urban area depends upon the size and type of the buildings and the anticipated threat disposition. Generally, a company attacks on a one- to two-block front and a battalion on a two- to four-block front (based on city blocks averaging 175 meters in width).

Forms of Attack in the Urban Offense: Metz – 1944

In November 1944, the U.S. Third Army launched its final effort to take the French city of Metz from the defending Germans. This was the Army's third attempt. The first attempt had been a surprise, mounted attack. This was followed by a series of piecemeal infantry assaults on the surrounding fortresses. Finally, a deliberate effort was made to take the city in a coordinated effort by XX Corps.

The initial failures stemmed from a shortage of resources on the U.S. side, to include fuel and units—especially infantry. This added to the ad-hoc nature of the first two efforts. The third effort, though more deliberate, was still constrained by resources. XX Corps could only muster three nearly full-strength infantry divisions to attack the German's defending with four under-strength divisions.

The third attempt to take Metz demonstrates how a corps operates with multiple divisions using various forms of attack to achieve its objective in urban offensive operations.

The opening phase of the Metz battle had attacks by the 90th and 5th Infantry Divisions to envelop the city from the north and south (see figure 7-8). This isolated the city and ensured the garrison could not escape nor be reinforced. The garrison was under orders from Hitler not to conduct a breakout and to resist to the last man; thus, German forces strongly counterattacked both wings of the envelopment to prevent isolation.

The second phase of the operation was the penetration of the city defenses from multiple directions. The 5th Infantry Division penetrated into the city with the 10th and 11th Infantry Regiments from the south. The 95th Infantry Division penetrated into the city from the north with the 377th Infantry Regiment and TF Bacon. Simultaneously, the 95th Infantry Division infiltrated battalions through the string of fortress positions guarding the western approaches into the city, isolated, and bypassed these positions with its other two regiments.

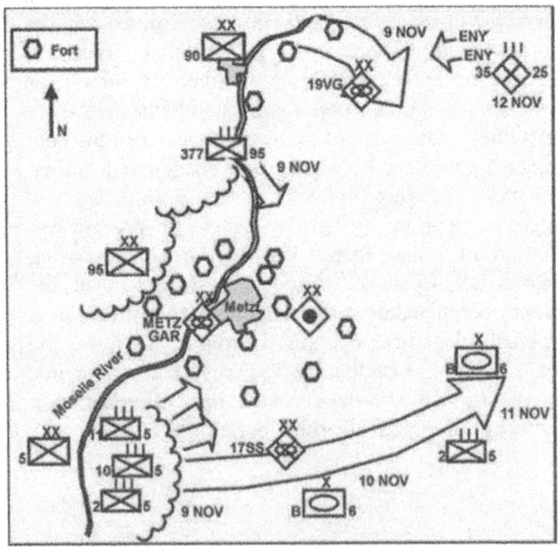

Figure 7-8. Metz envelopment

The final reduction of the defense was a series of battalion frontal attacks, which took place against the last remaining strongholds within the city. Even in these final engagements, however, the infantry battalions isolated, bypassed, and then attacked from the flanks and rear whenever possible (see figure 7-9).

The city was declared secured on 19 November. However, at that point more than a half-dozen of the fortresses had yet to be reduced. The 95th Infantry Division, after a four-day rest, quickly moved forward to rejoin the still rapidly advancing corps forward elements. Elements of the 5th Infantry Division remained in siege posture around the remaining strong points, the last of which surrendered on 19 December 1944 when it ran out of food. U.S. forces made no efforts to attempt to assault these bypassed fortresses although extensive psychological operations (PSYOP) were used.

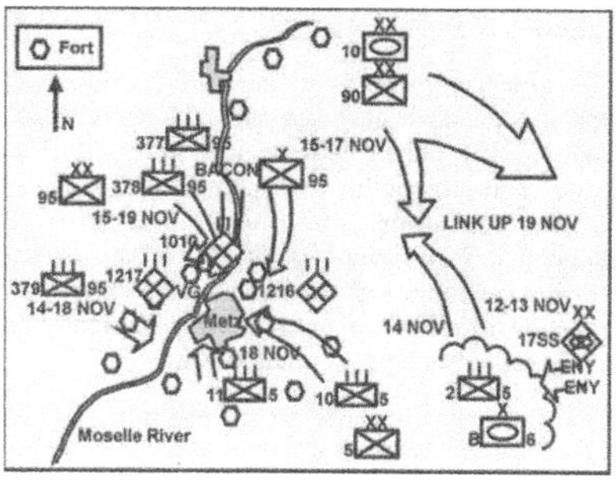

Figure 7-9. Metz final assault

TYPES OF OFFENSIVE OPERATIONS

Movement to Contact

7-34. In an urban area where the threat situation is vague, Army forces will often conduct a movement to contact to establish or regain threat contact and develop the situation. A movement to contact in an urban area occurs as both sides try to establish their influence or control over a contested urban area. The situation determines whether the movement to contact or its specific technique, the search and attack, is appropriate. A conventional force-oriented movement to contact will likely take place when friendly and threat conventional forces attempt to establish control simultaneously. Initially, neither side is defensive. The friendly force aims to quickly locate and fix the threat while establishing control of the urban area and its key infrastructure. The search and attack technique works well when a smaller threat has established a noncontiguous defense in an urban area. This operation is characterized by the friendly point defense of key infrastructure, robust reconnaissance, and rapidly concentrated combat power to fix and defeat or destroy threat resistance once located.

7-35. A meeting engagement often results from the movement to contact. It occurs when a moving force that is partially deployed for battle collides with and engages a threat at an unexpected time and place. In a meeting engagement in an urban area, the unit that reacts most quickly and decisively will likely win. Rapid and accurate decision making depends heavily on understanding the nature of the urban area and its impact on operations. Thus, in a meeting engagement, commanders should quickly assess the impact and role of all components of the urban environment (terrain, infrastructure, and society) on the operation. To this end, responsive reconnaissance is important. This permits accurate decision making regarding where to attack, where to defend, and how to allocate resources. Situational understanding enhanced by robust digital INFOSYS that provide an accurate COP also facilitates the rapid reaction of Army units and a synchronized response. This reaction and response allow Army forces to seize the initiative and dominate the threat.

Attack

7-36. The attack is the most common and likely offensive operation that Army forces will conduct in an urban environment. Commanders conducting major operations and commanders of large tactical units usually execute deliberate attacks. In the urban environment, units larger than battalion-size rarely conduct hasty attacks. Hasty attacks are common below company level as units use their initiative to take advantage of tactical opportunities. However, larger units will conduct hasty attacks when threat defenses are disrupted or unprepared, to take advantage of an unexpected situation, and to prevent the threat from establishing or re-establishing a coherent defense.

Exploitation

7-37. Exploitation follows a successful attack to disrupt the threat in depth. Commanders of major operations should consider focusing exploitation attacks on urban areas. A threat defeated in an attack will attempt to rally units, reinforce with reserves, and reorganize his defense. With its information and communications capability, transportation network, and defensive attributes, the urban area is the natural focal point to reestablish a disrupted defense. By establishing urban centers as the objectives of the exploitation, commanders deny the threat the sanctuary needed to reorganize and reestablish his defense. The exploitation focuses on the urban area as well as on the remnants of the threat. A successful exploitation to seize an urban area works efficiently because the attack preempts the defense and denies the threat the full advantages of urban terrain.

7-38. Commanders conducting exploitation must acknowledge the vulnerability of their forces to counterattack and ambush in urban areas. An urban area provides ideal cover and concealment to hide threat reserves, reinforcements, or reorganized forces. Constrictions of routes into and through the urban area make exploitation forces a potentially dense target and limit maneuver options. Robust and well-coordinated reconnaissance, tactical dispersal, and use of advance guard security forces protect against this threat (see Defensive Combat Power: Suez City vignette in Chapter 8).

Pursuit

7-39. The pursuit is designed to destroy threat forces attempting to escape. It focuses on the threat and not on urban areas. When conducting a pursuit, Army forces move through undefended urban areas and, if possible, bypass those in which threat forces successfully take refuge. The threat will likely attempt to use urban areas to disrupt the pursuit and permit the threat main body to escape. Commanders can prevent escape by denying the threat the time to establish forces in urban areas that cannot be bypassed. The agility of Army aviation forces for attack, reconnaissance, and transportation is essential to execute a successful pursuit around and through urban areas.

URBAN OFFENSIVE CONSIDERATIONS

7-40. The urban operational framework (understand, shape, engage, consolidate, and transition) provides a structure for developing considerations unique to urban offensive operations. The considerations vary depending on the situation and scale of the operation. Some considerations applicable to major operations that include an urban area will also be considerations at the tactical level focused in the urban area. However, no set rules exist. All urban operations are unique. Issues addressed at the operational level in one situation may be addressed in a new situation only at the tactical level. Under the right circumstances, a consideration may become an operational issue, a tactical issue, or a combination of the two. The following identifies some planning and execution issues that commanders conducting major operations should address.

UNDERSTAND

7-41. The first requirement, and a continuing requirement throughout the conduct of urban operations, is the assessment and understanding of the situation. Commanders should base this understanding on detailed information regarding the particular urban area. Since the threat will likely dominate or control most of the urban area during the planning phase of offensive operations, achieving an accurate understanding of the urban environment will be difficult. A comprehensive intelligence, surveillance, and reconnaissance (ISR) effort in support of a rigorous intelligence preparation of the battlefield (IPB) process overcomes this obstacle.

Integrated Intelligence, Surveillance, and Reconnaissance

7-42. The commander of a major operation that includes an urban area, unlike his subordinate commanders, can target reconnaissance deep into his AO and area of interest. This begins the application of ISR resources against the urban area that may lead to decisive ground operations. This ISR effort and the understanding it supports continue as long as the urban area remains in the AO. Commanders of major operations initially direct ISR assets on those information requirements that support determining whether or not to conduct urban offensive operations. Once decided, ISR resources shift to support the planning and execution of the operation in the urban area.

7-43. The first resources that a senior commander can use are national and strategic sensors. He requests them through the appropriate joint force commander. The commander aggressively pursues full use of these systems to begin building an initial database for analyzing the significant aspects of the terrain; key infrastructure considerations; the status and disposition of the population; and the size, type, disposition, and intentions of threat forces in the area.

7-44. Simultaneously, multiple intelligence sources contribute to the database. The sources collect, process, store, display, and disseminate the relevant information on large urban areas through open and classified resources. These information sources include—

- Historical research.
- Travel brochures that include cultural information and recent maps.
- Classified debriefings of diplomats, businesses, DOD personnel, and allies.
- Military maps and special geospatial products of the urban area.
- Previous intelligence assessments of the country, government, and population.
- Reachback to appropriate economic, political, cultural, and infrastructure subject matter experts not in the commanders AO.

The gathering and analysis of HUMINT plays a critical part of this process and assists commanders in understanding ethnic, cultural, religious, economic, political, and other societal and infrastructural facets of the environment.

7-45. As the intelligence and the national reconnaissance and surveillance efforts progress, commanders will insert, if available and feasible, Special Forces reconnaissance assets into the urban environment. These elements will seek to confirm or deny the information received from imagery intelligence (IMINT), signals intelligence (SIGINT), and HUMINT sources. Among many factors, using SOF depends on their availability, the particular urban area, the area's ethnic composition, and the relationship between the urban population and the threat. Other joint operational reconnaissance and surveillance assets that higher-echelon commanders may have available might include the Joint Surveillance Target Attack Radar System, Guard Rail targeting aircraft, UAS, and space-based systems.

7-46. The commander's staff will use all sources of information—IMINT and SIGINT sensors, HUMINT, historical research, and reconnaissance—to refine his ability to understand the urban environment. Digitally linking subordinate commanders with information sources helps to develop a COP essential to their situational understanding of the urban environment. The IPB process guides this assessment. As operations progress, additional reconnaissance and surveillance assets may become available. These may include UAS, long-range reconnaissance and surveillance units, counterfire radar, and air and ground cavalry. As these assets are employed, they are linked into the net of sources sharing information and further refine a common situational understanding of the environment.

Focused Assessment Efforts

7-47. In urban offensive operations, the tactical commander's assessment focuses on defeating the threat in the urban area within the constraints of the environment. Toward this end, identifying and assessing decisive points to attack is a commander's priority assessment task. Some unique aspects of the urban environment also require the focus of the commander's assessment efforts. These include the character of the urban defense, collateral damage considerations, and the effects of the environment on friendly and threat courses of action.

Character of the Urban Defense

7-48. To be both efficient and effective, Army urban offensive operations focus on what is decisive. Decisive points for an urban attack depend primarily on the mission within the urban area. They can vary widely in composition and size. Since commanders only focus on the essential, they may determine the decisive point to be a single building or a limited sector of an urban area. It could be an entire system within the urban infrastructure such as communications and information, or a limited subsystem of the transportation and distribution infrastructure such as a single airfield. Sometimes what is decisive in the urban area is the threat military capability, but even this large an objective, when carefully analyzed, may not require destruction of all threat forces or control of the entire urban area. Decisive points relate directly to the threat's center of gravity and to mission success. Some decisive points related to the urban threat's center of gravity may be physically located outside the urban area.

7-49. To gain specifics on threat dispositions within the urban area requires reconnaissance capability to see into the depths of the area and the intelligence capability to determine the threat's likely defensive course of action. With this information, commanders can determine decisive points and apply Army combat power discretely against them. Effective urban offensive operations require detailed situational understanding of an area of interest that extends well beyond the perimeter of the urban area.

7-50. Commanders see throughout the depth of the urban area using several actions (see figure 7-10). First, they evaluate sensor data and imagery. This guides targeting of special reconnaissance. Simultaneously, HUMINT is conducted using any persons who might know the urban area and threat. This includes civilians (allies, aides, neutrals, obstacles, and hostiles) and POWs. Finally, tactical conventional reconnaissance assets including reconnaissance forces, aviation, artillery radar, signals intelligence, and UAS are directed at the urban area. All these sources and data are linked through digital INFOSYS to provide commanders and their subordinates with improved situational understanding and a COP.

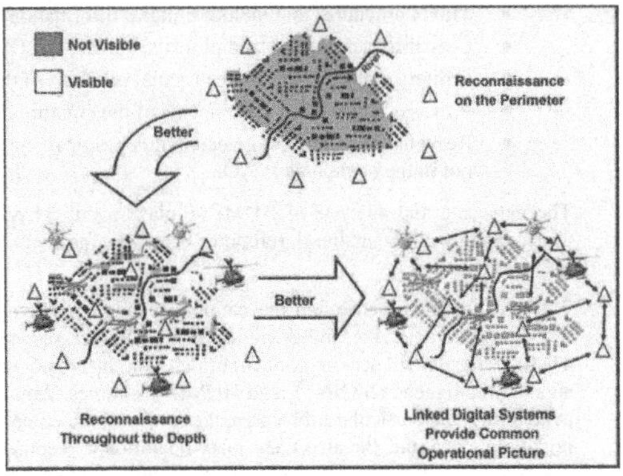

Figure 7-10. Required urban reconnaissance capabilities

Collateral Damage Considerations

7-51. Commanders must also assess the collateral damage risks that their operations may include. This assessment helps to initially determine the viability of a course of action. However, commanders should reassess their courses of action at frequent intervals in urban offensive operations based on known information to determine if the original evaluations remain valid. This reassessment minimizes potential collateral damage from a change in mission or a change in a course of action. Many aspects of the environment can change during mission execution.

The Environment's Effects on Courses of Action

7-52. The urban environment's unique aspects can significantly impact the course of action chosen by Army forces and the threat. Commanders must assess these effects in planning, but they must also verify and monitor these effects as forces execute offensive missions. In particular, commanders will want to confirm the civilian population's locations, beliefs, and actions and to monitor any changes. They will need to validate terrain considerations and monitor the effects of any changes due to rubble and other damage to structures. In urban terrain, dead space, cover, and concealment can only be identified physically and may change considerably as operations are executed.

SHAPE

7-53. Commanders of major operations have a primary contribution to urban operations: the planning and conduct of effective shaping operations that set the conditions for subordinate tactical success. In urban operations, isolation will be a critical condition. Effective isolation will require persistent, continuous surveillance and reconnaissance, innovative use of fires and maneuver (including effective force allocation decisions), and well-established sensor-to-shooter links. These efforts—combined and synchronized with SOF direct actions, IO that minimize noncombatant influences, and necessary shaping attacks (particularly the seizure of a foothold)—establish the conditions necessary for the subsequent offensive domination of the area.

Isolation is Essential

7-54. One key to success in the history of urban operations has been the effective isolation of the threat force (see figure 7-11). This applies today and equally well to major urban offensive operations as it does to smaller-unit attacks. This isolation not only denies access to the urban area from outside but also contains threat forces within. In a modern metropolis or megalopolis, this can appear a daunting task. Operational isolation requires dominating all physical and electronic contact between the threat in the urban area and supporting threat forces outside the urban area. This does not necessarily require physically encircling the urban area, but it does require that Army forces be able to exert control over the area's entire perimeter, as well as decisive points within. For a sprawling urban area, successful isolation may require the commitment of a large amount of resources.

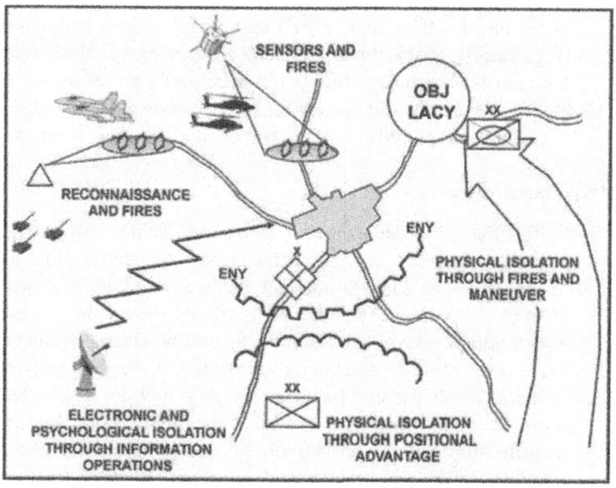

Figure 7-11. Shaping through isolation

7-55. Successful isolation of the urban area depends as much on the nature of the threat as it does on any other factor. A conventional threat in a large urban area may be much easier to isolate than an insurgent threat in a much smaller urban area. The forces needed in the former situation may be less than those needed in the latter. The more the characteristics of the threat are conventional in nature generally the easier it will be to isolate him using standard combat methods and equipment. Isolating a more unconventional force requires many of the same techniques as used against conventional forces. However, it also requires a much greater ability to simultaneously conduct offensive IO, to integrate CA units and civil-military operations (CMO), and to work with allies, nongovernmental organizations, and local authorities. Fundamentally, isolating a less conventional threat puts increased emphasis on separating combatants from noncombatants.

Offensive Isolation Objectives

7-56. Isolation seeks to achieve two primary objectives with respect to defeating a threat's urban defense:

- Weaken the overall coherence of his defense.
- Limit or manipulate his maneuver options.

Isolating the threat in the urban area from external support, as well as isolating him from sources of support within the urban area, weakens his overall defense. The defense is weakened through a combination of attrition (the threat cannot replace his losses) and the diversion of his combat power from the defense to operations to counter the isolation effort. Isolation can also prevent the threat from shifting his forces to reinforce decisive points in the urban area or to conduct counterattacks.

7-57. Commanders may choose not to isolate the urban area completely—or at least make it appear so to threat forces. Instead, they may afford the threat an apparent means of escape, create the conditions for its use through effective fire and maneuver against the defenders, and then destroy the threat through various ambush methods. While friendly forces may be able to move undetected to appropriate ambush sites, it is more likely that this technique will necessitate rapidly mobile air and ground forces moving along carefully chosen routes through the urban area. Commanders must consider maintaining the ability to complete the isolation of the urban area to prevent reinforcement and escape of urban threat forces—particularly if the ambush attack does not achieve desired effects.

Persistent Surveillance

7-58. Persistent surveillance of the urban area is essential to all types of actions used to isolate an urban area and as complete as resources will allow. Surveillance of the urban area relies on either reconnaissance forces or sensors continuously observing or monitoring urban avenues of approach. This network of ISR assets updates the commander's situational understanding and provides the means to quickly identify and, if necessary, attack threat elements as they move. However, particularly with sensors, commanders must consider that each detection is not necessarily an enemy to be attacked. Noncombatant activity clutters the environment making it easier for threats to disguise themselves and increases the burden (and the number of resources required) on Army forces to distinguish friend from foe.

Fires and Maneuver

7-59. Fires and maneuver may be used to achieve isolation, either singly or in combination. (As always, effective obstacles, monitored by sensors or observation, are integral to any isolation technique.) First, attacking forces can pre-position themselves along avenues of approach to deny entry and exit through positional advantage. Relying primarily on this method of isolation, particularly around a large urban area with multiple avenues of approach, can be overly resource intensive. Instead, the pairing of fires and maneuver provides attacking commanders more flexibility and allows them to isolate several avenues of approach with fewer resources. Highly mobile attack helicopters, operating outside threat-controlled portions of the urban area, are ideal for this purpose. Inside threat-controlled areas, it is more difficult to identify, eliminate, or effectively suppress the air defense threat. The threats may have numerous man-portable air defense weapons and enhanced effects of small arms used for air defense. Therefore, the risk to using this equipment may outweigh the potential benefits. However, mobile ground units—such as an air assault (subject to the same air defense threat considerations as attack aviation), armored, or mechanized forces—can also rapidly move to attack and destroy a threat moving in or out of an urban area. Potential disadvantages of the combined, fires and maneuver, option are that—

- Critical assets, on standby and dedicated to isolation efforts, may be unavailable for other missions.
- The attacking force may not locate the threat in time to complete its mission (an inherent risk to any attack).

7-60. Another alternative relies on indirect or joint fires alone to destroy the threat force. Its disadvantage is that fires alone often cannot completely destroy or stop a determined force from moving into or out of an urban area. Although targets and avenues of approach will require continual surveillance, it is usually a less resource-intensive option than those that include maneuver. It also does not normally require fires assets to remain on standby to accomplish the mission. However, fires must be able to reliably and quickly respond. For Army field artillery units and naval gunfire, the units must also be in range, which requires careful positioning. A skilled threat can avoid interdiction fires by using the geometry of the area to identify gaps due to obstructing terrain or the firing unit's range limitations. It can also use concealment and weather to avoid observation. However, effective sensor-to-shooter links throughout the urban battlefield will reduce the threat's ability to hide (see figure 7-12). A resolute threat may risk significant losses to fires to prevent isolation or may attempt to use noncombatants as a shield. Ultimately, commanders should use innovative combinations of all techniques discussed. Some units will physically block key avenues of approach. Surveillance will monitor less important routes and avenues. Artillery fires, joint fires, and maneuver units will then respond to the results of surveillance depending on the circumstances.

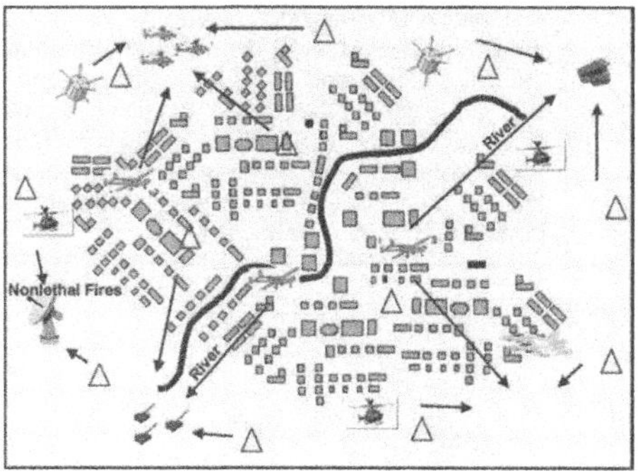

Figure 7-12. Critical sensor-to-shooter links

Threat Reactions

7-61. The reaction of the threat to the effects of isolation will depend on his mission, morale, force structure, and overall campaign plan. The threat may recognize isolation actions early and withdraw from the urban area before isolation is completed instead of risking destruction. On the other hand, the threat, based on a different or flawed assessment (perhaps a perception shaped by the Army force commander), may choose to—

- Continue to defend (or hide) and conduct local ambushes and counterattacks.
- Attack to break into the urban area or infiltrate forces and supplies in.
- Attack to break out of the urban area or infiltrate forces out.
- Execute any combination of the above (see figure 7-13).

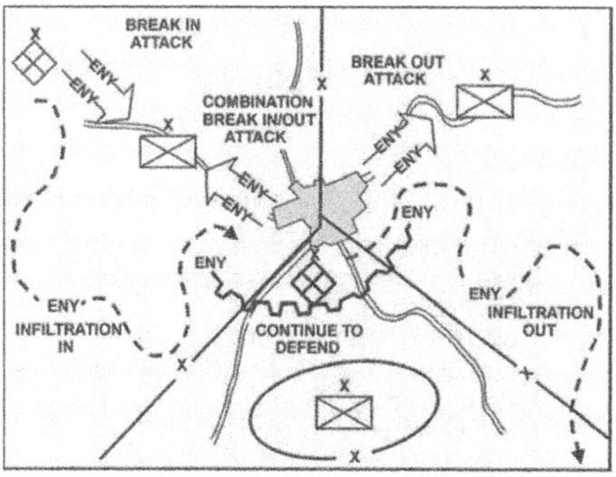

Figure 7-13. Reactions to isolation

7-62. Attacking commanders must consider how the threat leadership's subsequent actions will affect the continuance of overall offensive operations. They deliberate many considerations, to include—

- The allocation of more forces to the shaping operations to isolate the urban area.
- The allocation of more combat power to achieve rapid penetration and seizure of objectives to take advantage of developing threat dispositions in the urban area.

Isolating an Urban Area: Hue, Vietnam – 1968

On 31 January 1968, the 4th and 6th North Vietnamese Army (NVA) regiments and attached NVA and Vietcong (VC) sapper battalions attempted to seize control of Hue from the South Vietnamese Army's (ARVN) 1st Division as part of the North Vietnamese Tet Offensive. The attack, which was launched with complete surprise, successfully established temporary control over most of the city and facilitated the occupation of strong defensive positions within the city's ancient fortress known as the Citadel. However, it did not lead to the capture of the headquarters of the ARVN 1st Division located in the Citadel nor the U.S. military assistance command—Vietnam compound located in the southeast part of the city.

Both ARVN and U.S. forces swiftly reacted to the NVA attack. Within 24 hours, ARVN infantry and cavalry units counterattacked to recapture the Citadel. South of the Perfume River, the 1st Marine Division began attacking to clear the southern portion of the city (see figure 7-14).

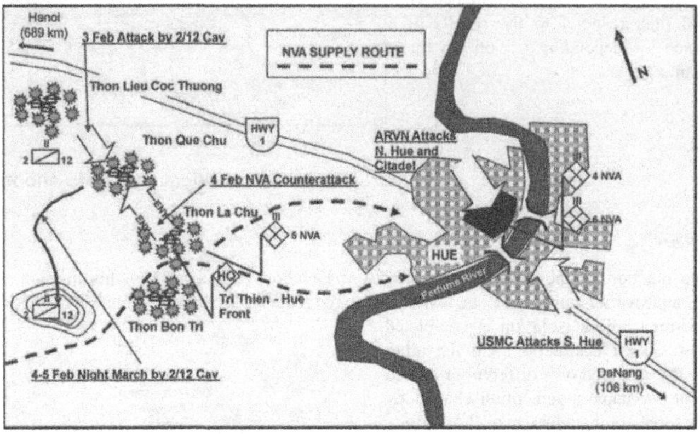

Figure 7-14. Initial attack to isolate Hue

On 2 February, the first Army unit was committed to Hue. The 2-12th Cavalry (an infantry battalion) was ordered to attack southeast along Highway 1 and clear enemy interdicting that route. On 3 February, 2-12th Cavalry began their attack moving along the south side of the highway. The plan was to clear a series of small hamlets in succession en route to the city itself. The first hamlet, Thon Lieu Coc Thuong, was cleared easily. Moving toward the second hamlet, Thon Que Chu, the battalion encountered a strong enemy force that was well entrenched. Air and artillery supported the battalion as it attacked and captured the hamlet against strong resistance. The battalion dug in for the night and prepared to resume the attack against the third hamlet, Thon La Chu, on 4 February.

Unknown to the 2-12th Cavalry, Thon La Chu was the headquarters of the Tri Thien-Hue Front, and it was defended by the NVA's 5th Infantry Regiment. Also important, the hamlet sat astride the NVA's primary supply route to the regiments fighting in Hue. The NVA was determined that 2-12th Cavalry's attack would not succeed; the NVA was determined to destroy the threat to its LOCs.

As the 2-12th Cavalry prepared to resume the attack on 4 February, the NVA fiercely counterattacked with all three battalions of the 5th NVA Regiment. It soon became apparent that 2-12th Cavalry could not continue to attack. As the day continued and the NVA pressure steadily increased, the battalion began to lose its position in Thon Que Chu. To avoid destruction, the battalion broke contact.

As darkness fell on 4 February, 2-12th Cavalry finally broke contact with the NVA. However, instead of retreating north back to its start line, the battalion moved west and then south to good defensive high ground deeper in the NVA AO. Eleven hours later the battalion was set in its new defensive position.

At dawn on 5 February, 2-12th Cavalry was established 5 kilometers west of Hue. The battalion soon observed enemy forces and supplies moving toward Hue. From its high ground position, the battalion directed artillery and air strikes against the NVA forces. By its bold move to bypass the 5th NVA Regiment, the battalion now held excellent position to direct fires on the primary NVA supply line into Hue. These fires were the first step toward isolating the NVA in Hue.

The fires controlled by the 2-12th Cavalry shut down the NVA LOCs into Hue during the daytime. However, under the cover of darkness supplies and reinforcements were still entering the city (see figure 7-15). It became apparent that the isolation of the NVA in Hue would require the capture of Thong La Chu. The problem facing American forces was concentrating combat power against the NVA. All U.S. units at this time were actively engaged against the numerous NVA attacks that constituted the NVA's 31 January Tet Offensive.

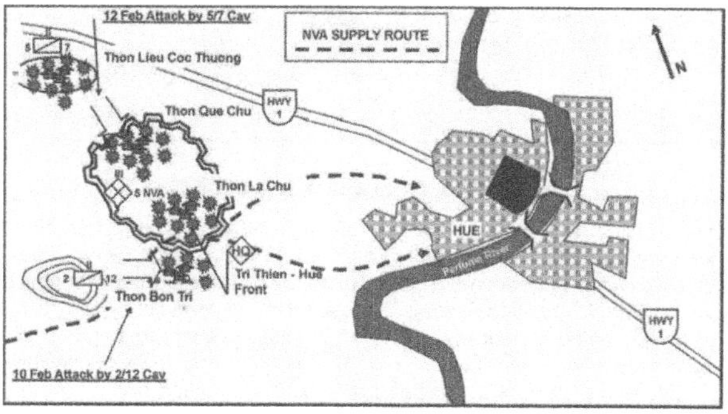

Figure 7-15. Subsequent attack to isolate Hue

The first additional American unit was not available until 12 February when the 5 7th Cavalry attacked Thong Que Chu much like the 2-12th Cavalry had attacked previously. The 5-7th Cavalry had even less success against the alert 5th NVA. The 5-7th Cavalry was forced to occupy defensive positions in Thon Lieu Coc Thuong and await the build up of combat power before it could continue to attack. In the interval, 2-12th Cavalry had moved off the high ground and captured the hamlet of Thon Bon Tri, south of the 5th NVA Regiment.

On 21 February, the 1st Cavalry Division had moved enough resources to the area to launch an effective attack to isolate Hue (see figure 7-16). In addition to the 5-7th and 2-12th Cavalry, the 1-7th Cavalry arrived in the AO and the 2-501st Airborne Infantry of the 101st Airborne Division was attached. On 21 February, after a combined artillery, air, and naval gunfire bombardment, the four battalions attacked the Thon La Chu stronghold. Elements of the 5th NVA Regiment were either destroyed in place or fled northeast. The next day resistance in Hue was noticeably lighter. U.S. Marine and ARVN units began the last phase of fighting to recapture the Imperial Palace. On 26 February, the North Vietnamese flag was removed from the Citadel and the ARVN I Corps declared the city secured.

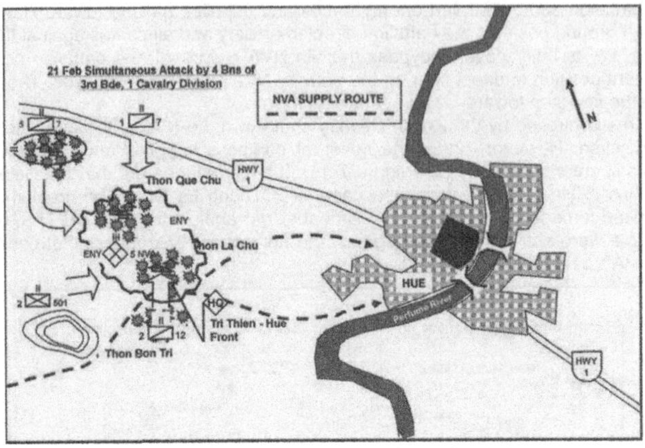

Figure 7-16. Final attack to isolate Hue

> The actions of the 1st Cavalry Division forces northwest of Hue demonstrated the importance and the difficulty of isolating an enemy fighting in an urban area. Isolating Hue was difficult not only because of the dispersion and surprise with which the Tet Offensive caught U.S. forces, but also because of the tenacity of the NVA. At least one-third of the combat power of the NVA in the Hue AO was focused on maintaining access to the city.
>
> Although only one aspect of this urban battle, Hue's isolation had an immediate and important, if not decisive, impact on operations. It not only resulted in restriction and then elimination of supplies and reinforcements, but it also immediately impacted the conduct of the defending NVA forces. Isolation caused an immediate drop in NVA morale and changed the nature of the defense. Once the enemy was isolated from external support and retreat, the objective of the NVA in the city changed from defending to avoiding destruction and attempting to infiltrate out of the city.

Civilian Reactions

7-63. Commanders must also consider the potential effects on and reactions and perceptions of the population living in the urban areas that they choose to isolate and bypass—either as a direct effect or as a response of the threat force being isolated. Isolation to reduce the threat's ability to sustain itself will likely have similar (and worse) effects on the civilian population remaining in the isolated area. (If food and water are in short supply, threat forces may take from noncombatants to satisfy their needs, leaving civilians to starve.) Isolation may also create a collapse of civilian authority within an urban area as it becomes apparent that the military arm of their government is suffering defeat. Due to their isolation, elements of the population may completely usurp the governmental and administrative functions of the former regime and establish their own local control, or the population may lapse into lawlessness. Returning later, Army commanders may find that these self-governing residents are proud of their accomplishments and, in some instances, less willing to allow Army forces to assume control since they may be perceived as having done nothing to earn that privilege. Alternatively, as witnessed in some urban areas during OPERATION IRAQI FREEDOM in 2003, a power vacuum may lead to intra-urban conflicts among rival factions coupled with general public disorder, looting, and destruction of the infrastructure (see also the discussion of Competing Power Structures in Chapter 3).

Direct Action by Special Operations Forces

7-64. Although SOF in urban offensive operations will likely conduct essential reconnaissance, they also have a direct action capability to shape the offensive operation (see figure 7-17). Special Forces and Rangers can use direct action capabilities to attack targets to help isolate the urban area or to directly support decisive actions subsequently or simultaneously executed by conventional forces. Successful attacks against urban infrastructure, such as transportation or communications centers, further the area's physical and electronic isolation. Direct action against command centers, logistics bases, and air defense assets can contribute to the success of conventional attacks by destroying or disrupting key threat capabilities. Direct action can also secure key targets such as airports, power stations, and television stations necessary for subsequent operations. Direct action by Special Forces and Rangers in these operations can help achieve precision and reduce potential damage to the target or noncombatant casualties.

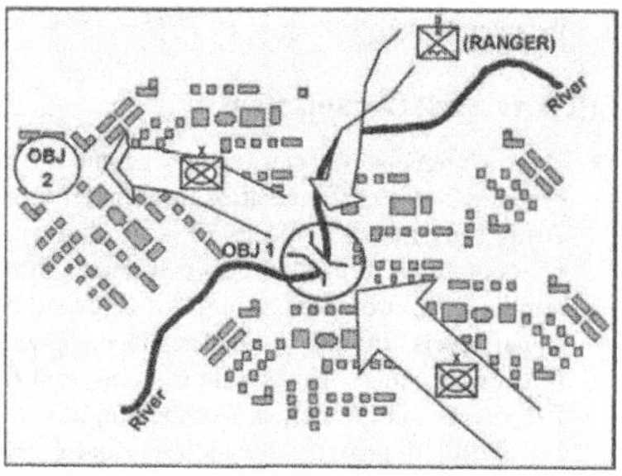

Figure 7-17. Coordination of SOF and conventional capabilities

Information Operations

7-65. Regardless of how Army forces physically isolate the urban area, they combine physical isolation with offensive IO to electronically isolate the threat and undermine his morale. Electronic isolation will cut off communications between forces in the urban area from their higher command to deny both from knowing the other's status. IO combined with isolation may persuade the threat's higher command or leadership that its forces located in the urban area are defeated. Thus, the command or leadership's intentions to break through to the besieged threat forces may be affected. PSYOP can undermine the morale of the threat in the area and reinforce electronic isolation and perceptions of abandonment. IO can be used to reduce any loyalty the civil population may have to the threat. IO can also ensure that civilians have the information that minimizes their exposure to combat and, as a result, overall noncombatant casualties. In addition, IO aim to deceive the threat regarding the time and place of Army force operations and intentions.

Detailed Leader Reconnaissance

7-66. Army commanders must clearly see the urban environment to understand the challenges facing their brigades, battalions, companies, platoons, and squads. Urban terrain can be deceptive until viewed from the soldier's perspective. Commanders are responsible to intimately know the conditions to allocate resources effectively to subordinate units. Often, particularly at battalion level and above, commanders will not be able to command and control dispersed forces from positions forward, but be forced by the terrain to rely on semifixed command posts. Detailed leader reconnaissance of the AO by commanders, their staff, and their subordinates before the mission can compensate for this challenge. This reconnaissance will give commanders a personal feel for the challenges of the terrain and will facilitate more accurate planning and better decision making during operations.

Mission Orders

7-67. Before contact, commanders can mitigate some terrain challenges to effective C2 using mission orders. Subordinates have mission orders to take advantage of opportunities before C2 systems can adversely impact the environment. To see the battle and provide effective and timely direction, tactical leaders will follow closely behind units as they assault buildings, floors, and rooms. Thus, only the most mobile INFOSYS can accompany tactical leaders into combat, and they will suffer the degrading effects of the environment. Mission orders permit rapid and decisive execution without commanders intervening at battalion level and above. Higher-level commanders can facilitate mission orders through their subordinates by articulating their desired end state, clearly stating their intent, and building flexibility into the overall plan.

Effective Task Organization

7-68. Commanders can shape urban offensive operations through effective and innovative task organization. Combined arms, often starting with an infantry base, are essential to success and may be the Army's asymmetric means of defeating an urban threat. Urban attacks will quickly break down into noncontiguous firefights between small units. To achieve the tactical agility for mission success in this nonlinear environment, many Army capabilities are task organized down to the company, platoon, and squad levels. Infantry provides the decisive capability to enter buildings and other structures to ensure threat destruction. Tanks, gun systems, and fighting vehicles provide additional speed and mobility, direct firepower, and protection. Combat engineers provide specialized breaching and reconnaissance capability. Field artillery provides the indirect (and if necessary, direct) firepower. Such mobility and firepower create the conditions necessary for the dismounted infantry to close with and destroy a covered threat in an urban defense. When a threat skillfully uses the urban area to limit ground maneuver, vertical envelopment or aerial attack using precision-guided munitions from Army aviation may circumvent his defenses and achieve necessary effects. Generally, ground systems used within the urban area will not be able to operate independently from dismounted infantry. The infantry will be required to protect armor and mechanized systems from close antiarmor weapons, particularly when those weapons are in well-prepared positions located throughout the urban area but especially on rooftops and in basements.

7-69. In urban offensive operations, direct fire support can be critical. Armor vehicle munitions types do not always achieve decisive effects against some urban structures. In some cases, field artillery high explosive munitions work better than armor for direct fire support of infantry. Large caliber (105 or 155mm) high explosives directly fired at a structure often produce a more severe shock effect than tank and fighting vehicle cannon and machine guns produce. Artillery is also able to achieve higher elevation than armor and engage threats located at greater heights.

7-70. However, commanders must view artillery as not just a weapon but a weapon system. As such, artillery should normally be placed under tactical control (TACON) of maneuver commanders, such as a platoon of three guns TACON to a company or a battery to a battalion, not just one gun to a company or other maneuver unit. Self-propelled artillery has some of the mobility characteristics of armor; however, it provides minimal ballistic protection from fragmentation for the crew. Although these systems seem formidable, they provide less crew protection than a Bradley fighting vehicle, for example, and contain large amounts of onboard ammunition and propellant. They are susceptible to catastrophic destruction by heavy automatic weapons, light cannon, and antitank fire. Therefore, infantry units must carefully secure and protect these systems (even more so than armored vehicles) when employed in urban offensive operations, particularly when forward in the direct fire role.

Creative Task Organization: Using Artillery in the Direct Fire Role

Task organizing artillery to permit its use in a direct fire role demonstrates the innovative task organization required for urban operations. The following provides three historical examples of task organizing and using field artillery for a direct fire role.

In 1944, U.S. Army units of the 1st Infantry Division were assigned to attack and seize the German city of Aachen. The city's internal defense included bunkers designed to serve as air raid shelters. These positions, buildings of stone, seemed impervious to direct fire tank weapons, demolitions, and small arms. To reduce the positions, the 1st Infantry Division relied on the artillery's direct fire.

Field artillery used this way had physical and psychological effects on the defenders. The 26th Regimental Combat Team's history of the battle describes the German reaction to the artillery pieces:

The chief shock to the defenders, Colonel Wilck (Aachen defense commander) said, came from the self-propelled 155s and tanks. The colonel spoke with considerable consternation of the 155mm self-propelled rifles. A shell from one of the guns, he said, pierced three houses completely before exploding and wrecking a fourth.

The 26th Infantry Regiment also described how the artillery, one piece attached to each assaulting infantry battalion, helped the infantry to penetrate buildings.

With solid blocks of buildings comprising most of the city, there wasn't any easy way to get at the Germans in the buildings. The eight-inch gun solved the problem. Beginning on the eastern outskirts the gun would plow a round into the side of the built up block of buildings at about ground level. One shell would usually open an entrance into the first tier of floors, i.e. the first building. Then several more shells were fired through the first hole. Thus a tunnel would be rapidly made all the way to the next cross street. Soldiers could then rush the newly formed entrance, clear the upper floors with hand grenades and rifles and then move on to the next building to repeat the process. When a block or square, was thus completely cleared of Germans—soldiers, skulkers, or even snipers—the next square was treated in the same way, working forward square by square, right and left, thereby avoiding nearly all exposure in the streets.

In 1982, Israeli forces invaded southern Lebanon to destroy base camps of the Palestine Liberation Organization (PLO). This operation involved significant fighting in urban areas including major operations in Beirut. Artillery, firing in a direct fire role, played a major part of the tactical solution. Artillery was particularly effective in the 33-day siege of Beirut. During this siege, Israeli forces used artillery in its traditional role as well as in the direct fire role.

The Israeli army was committed to a policy of disproportionate response during the Beirut siege. When fired on with small arms, crew-served weapons, tanks, or indirect artillery, the Israeli forces responded with intense, high-caliber direct and indirect fire from tanks and artillery positioned around the city. Many firing positions were on heights to the south and southwest that dominated much of the city. These positions had almost unrestricted fields of view. Israeli artillery fired from these positions directly into high-rise buildings concealing PLO gunners and snipers. The artillery, using direct fire, destroyed entire floors, collapsed floors on top of each other, and completely removed some upper floors. Such a response, as in Aachen in 1944, had as much a psychological impact as it did a physical impact on the PLO defenders.

In the early hours of 20 December 1989, the United States launched OPERATION JUST CAUSE. One of this operation's objectives was removing the Panamanian dictator, Manuel Noriega. U.S. forces carefully planned using all fires before the operation to minimize casualties and collateral damage. Part of this detailed fire planning called for applying artillery in a direct fire role.

The Panamanian Defense Force's (PDF) 5th Rifle Company based at Fort Amador was one of the key objectives of U.S. forces at the start of hostilities. This unit was high priority because it was the closest PDF unit to Noriega's headquarters, the Comandancia. U.S. forces expected the 500-man company to react first to

> OPERATION JUST CAUSE by reinforcing the defense of the Comandancia. It also posed a threat to U.S. military dependents housed at Fort Amador.
> To quickly neutralize this force, the United States assembled a three-company force composed of A, B, and headquarters elements of 1-508th Infantry (Airborne), supported by 105mm towed howitzers of 320th Field Artillery and M113 armored personnel carriers. The howitzers and the personnel carriers were covertly pre-positioned at the fort. At approximately 0100, helicopters transported the two airborne rifle companies into position. The howitzers then suppressed any personnel in the PDF-controlled buildings on Fort Amador while demonstrating the firepower of the U.S. task force. They used direct fire into the PDF barracks. The impact of the 105mm high explosives and .50-caliber fire from the M113s convinced the PDF infantry to give up after token resistance. Following the direct fire, U.S. infantry assaulted and cleared the dozen PDF buildings, finding that most occupants had fled or surrendered. For more details of OPERATION JUST CAUSE, see Applying the Urban Operational Framework: Panama in Chapter 5.
>
> ---
>
> The three examples cited indicate the importance of the situation-dependent and innovative task organization of artillery and its use in the direct fire role. Using artillery helps overcome some challenges of offensive operations in the urban environment, and it has an important psychological effect on a defending threat. Such task organization takes a traditional tool of a higher-level tactical commander and uses it to directly influence the company-level battle. This philosophy of task organization can be applied to other types of forces—not just artillery. PSYOP teams, interpreters, CA specialists, armor, engineers, and reconnaissance teams may require task organization different from traditional organization. The compartmented urban environment drives the requirements for these assets lower in the tactical scheme than in open operations. Consequently, commanders must understand and account for more of these assets for UO than for operations in less restrictive environments.

7-71. Army aviation may also be inventively task organized. It can support urban operations with lift, attack, and reconnaissance capabilities. Tactical commanders down to company may use all these capabilities to positively influence ground close combat. Army attack and reconnaissance aircraft can provide flank security for attacking ground forces. Attack aircraft may also provide direct fire support to individual platoons or squads. Lift may move entire battalions as part of brigade operations, or it may move single squads to a position of advantage (such as a roof) as part of a small unit assault. Army aviation can assist with C2 by providing airborne retransmission capability, airborne command posts, and the confirmed status and position of friendly forces. However, Army aviation is a limited and high-value asset; commanders should review its use in innovative task organizations. It is particularly vulnerable to urban air defense threats unless used over terrain secured by ground forces. From these positions, aircraft can use enhanced sensors to conduct reconnaissance and use precision weapons with standoff capability.

Shaping Attacks

7-72. In a large urban area, the defending threat cannot be strong everywhere. Shaping operations can also take the form of attacks against vulnerable positions to force the threat to maneuver and redeploy his forces in the urban area. This prevents him from merely defending from prepared positions. Forcing the threat to move negates many of the defensive advantages of urban terrain, confirms his dispositions, exposes vulnerable flanks, and permits target acquisition and engagement with precision standoff fires.

7-73. A critical shaping operation in urban offensive operations is usually an initial attack to seize a foothold. Once Army forces establish this foothold, they accrue some of the defensive advantages of urban terrain. From this protected location, Army forces continue offensive operations and have a position of advantage against neighboring threat defensive positions.

ENGAGE

7-74. Commanders may employ several methods to decisively engage elements of the urban area during offensive operations. These include—

- Rapid maneuver.
- Appropriate use of SOF.
- Precise application of fires and effects.
- Proper balance of speed and security.

7-75. None is unique to UO. Their effective execution, however, allows Army commanders to dominate in this challenging environment by effectively using resources with the least amount of collateral damage. Overall, decisive engagement results from urban offensive operations when forces achieve the objective of the assigned mission and establish preeminent control over the necessary terrain, population, and infrastructure. Largely, the Army commander's ability to engage is based on superior situational understanding and the correct application of unit strengths to the challenges found in the urban environment.

Rapid and Bold Maneuver

US Army Photo

7-76. Commanders of major operations may have or create the opportunity to seize an urban area with rapid and bold maneuver. Such maneuver requires striking while the area remains relatively undefended—essentially preempting an effective defense. This opportunity occurs when the urban area is well to the rear of defending threat forces or before the onset of hostilities. Under such conditions, an attack requires striking deep behind threat forces or striking quickly with little time for the threat to make deliberate preparations. Attacks under such conditions may entail significant risk; however, the potential benefit of audacious offensive operations may be well worth possible losses. Three potential ways to accomplish such attacks (and their combinations) are:

- Airborne or air assault.
- Amphibious assault.
- Rapid penetration followed by an exceptionally aggressive exploitation, for example, a heavy force using shock, armor protection, and mobility.

7-77. Commanders should analyze all potential urban operations to seek an opportunity or advantage to apply rapid and bold operational maneuver to the task. Using operational maneuver to avoid urban combat against an established threat defense potentially marks a significant operational achievement and can have decisive strategic consequences. Just influencing the threat's morale can positively affect all future operations. However, commanders must evaluate the challenges of such a course of action. These challenges may include—

- Sustaining the operation.
- Avoiding isolation and piecemeal destruction.
- Successfully conducting shaping attacks.
- Achieving the necessary tactical, operational, and strategic surprise.

7-78. Commanders can also build on the shaping effects of isolating the urban area internally and externally by attacking urban decisive points from multiple directions. They can attack multiple decisive points either simultaneously or in a systematic, synchronized manner. This complicates the threat's situational understanding of the urban environment, further impedes his decision making, and allows Army commanders to dictate the tempo.

Bold Operational Maneuver to Seize an Urban Area: Inchon and Seoul, Korea – 1950

In August 1950, UN forces in Korea were desperately attempting to stave off defeat by establishing a viable defense in southern Korea. This defense, known as the Pusan Perimeter, was the focus of the world's attention. The commander of U.S. Forces Far East, General Douglas MacArthur, was not focused on the U.S. Eighth Army in Pusan. Instead, he focused on how to dramatically and decisively transition to the offense—the recapture of the South Korean capital of Seoul.

Seoul was more than just the South Korean political capital, although that was an important consideration. Seoul was the key to the Korean transportation system, the hub of the national communications system, and the center of the country's economy and culture. It was also strategically placed astride maneuver corridors to the north, south, and northeast. It had been Korea's capital since 1394 when King Taejo Yi Sung-Gye established it as the center of the Chosun dynasty. It had been captured by the surprise communist invasion of the south in June 1950, and the key to success in Korea was its recapture. The question was how.

The obvious military solution to the Korean situation in August 1950 was to build military strength within the Pusan Perimeter and then, when sufficiently strong, to breakout and counterattack north and recapture Seoul. The disadvantage of this course of action was that the counterattack would be through hundreds of miles of mountainous terrain, through several other major cities, and across numerous rivers. The North Korean People's Army (NKPA) would be able to fall back from one mountain defense and one river line to another and would have weeks or months to prepare the defenses of Seoul.

MacArthur's solution was to delay a conventional counterattack, boldly maneuver by sea deep into the flanks of the NKPA, land on the Korean west coast at Inchon, and rapidly seize Seoul before the enemy could react (see figure 7-18). This course of action had numerous disadvantages: achieving surprise; assembling a landing force trained in amphibious operations; few good landing sites; supplying a large force once landed; and needing to simultaneously continue to defend and prepare a counterattack with the Eighth Army from Pusan. MacArthur understood the challenges of the operation but believed that the potential value compensated for the risk.

The potential value of a rapid assault on Seoul through a deep amphibious operation was immense. The operation would trap the bulk of the NKPA in South Korea and facilitate its destruction. It would also capture Seoul before the NKPA could prepare a defense. The quick recapture would immeasurably impact the morale of UN forces and South Korea. MacArthur felt the payoff of success justified the risks and challenges of the operation.

On 15 September, elements of the 1st Marine Division landed in Inchon taking the NKPA completely by surprise. On D+1 they were joined by elements of the 7th Infantry Division, both divisions under the control of U.S. X Corps. Naval gunfire and Marine close air support flying from Navy aircraft carriers supported the landings. Seven days later, the 5th Marine Regiment had battled 25 miles to the outskirts of Seoul and began attacking the city from the north. Three days later, on 25 September, the 1st Marine Division and elements of the 7th Infantry Division had defeated the 10,000 defenders of the NKPA's 18th Rifle Division.

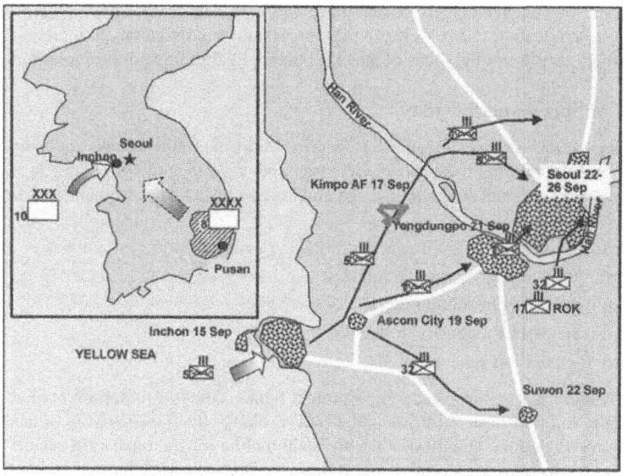

Figure 7-18. Inchon-Seoul campaign, September 1950

> Most of MacArthur's predictions for the Inchon-Seoul campaign were accurate. Superb staff work, excellent logistics operations, and unmatched training permitted the UN forces to assemble a trained landing force, land it over one of the most difficult shores in the history of military amphibious operations, and keep it supported. More importantly, the operation achieved total strategic and operational surprise. Seoul was recaptured quickly and, although resistance was fierce, the NKPA could not react fast enough to influence the outcome. In conjunction with the Eighth Army counterattack from Pusan, 75 percent of the NKPA was destroyed. The Inchon-Seoul campaign was an important turning point in the war, and had not the Chinese People's Army intervened two months later, it would have been the decisive campaign of the war.

Appropriate Use of Special Operations Forces

7-79. Sometimes Army forces can decisively engage (instead of simply shape) the urban area using the direct action capability of SOF. When the threat fails to develop a comprehensive defense and does not possess large, capable conventional forces, then Army forces can achieve operational surprise. Commanders, by synchronizing conventional and SOF effects, may actively control offensive operations to dominate the area. Importantly, however, conventional ground forces must be available to quickly assume the mission because SOF units acting as the primary striking force have limited logistic capability to sustain long-term operations. OPERATION JUST CAUSE offers several examples of this type of synchronization (see Applying the Urban Operational Framework: Panama in Chapter 6).

Precise Application of Fires and Effects

7-80. Precisely applied fires and the massed effects of combat power characterize successful urban attacks. The fires can be direct fire from combined heavy or light ground teams; direct or indirect fires from supporting Army aviation standing off from the target and any possible air defense threat; precision indirect fires from conventional tube artillery; or direct and indirect fires from supporting joint assets including Air Force and naval assets. All efforts strive to reduce collateral damage around the point of attack, consistent with mission success. Forces use fires to deny the threat the ability to maneuver in the urban area and to destroy the threat when he attempts to maneuver. When the threat exposes himself by moving, the environment no longer protects him, and fires can effectively engage him. Overall, precise fires and effects demonstrate the power of Army forces and help the urban population understand that only legitimate military targets are the focus of attacks, potentially building public support of Army UO.

Proper Balance of Speed and Security

7-81. Attacking units balance speed and security. Forces secure flanks as units advance, control dominating terrain (buildings), evacuate civilians, and keep the integrity and synchronization of the combined arms team. Obstacles are anticipated and rapidly breached. Commanders can choose avenues of approach to—

- Provide cover and concealment for following aviation and support units.
- Permit travel by all classes of vehicles.
- Easily defend from counterattack.
- Avoid nonessential centers of threat resistance.
- Avoid population concentrations.

7-82. Army aviation is a critical resource to protect flanks. Another important resource is engineers who seal off surface and subsurface entries and avenues along the flanks of the attack. Finally, as in all offensive operations, ground and air cavalry are ideal mobile forces to perform security in an economy of force role along flanks allowing decisive forces more freedom of maneuver.

CONSOLIDATE

7-83. Commanders, at all levels, must consolidate to strengthen their position during urban offensive operations without loss of momentum. They take the steps necessary to make permanent any temporary battlefield successes while maintaining relentless pressure on threat forces. Consolidation provides security and protection, facilitates reorganization, and allows forces to prepare for counterattack.

Repositioning of Forces

7-84. Following seizure of the objective, commanders normally consolidate by adjusting and repositioning forces. While UO are likely to be noncontiguous, commanders reposition ISR assets (including observation posts and reconnaissance patrols) to maintain contact with the threat, to establish contact with nearby friendly units and ensure that there are no exploitable gaps or seams, and to help maintain freedom of action. Physical occupation of the terrain (as well as continued reconnaissance) will provide commanders with a fuller understanding of the urban environment. With this enhanced understanding, commanders may need to adjust boundaries and other control measures to better adapt to the effects of urban terrain features such as canals, subway tunnels, raised roadways, and tall buildings. As necessary, commanders may also need to reposition communications assets and C2 facilities to enable subsequent operations.

7-85. Immediately after the conduct of successful urban operations, units must also be alert to the potential for rapid and violent counterattacks. Defenders may launch a quick counterattack to regain terrain before the offensive forces have consolidated and fully assumed the defensive advantages of the urban terrain. Delaying a counterattack in UO, even for a few minutes, permits the environment's advantages to shift to the successful attacker. Thus, attacking units anticipate this reaction and reposition forces such that they are prepared to defeat it.

7-86. As necessary, commanders should reposition armor and artillery (and other fire support assets) to account for the changing situation and battlefield geometry. Commanders should also consider consolidating and repositioning armored and artillery forces in positions—either inside or outside the urban area—to add significant combat power to a hasty defense, to defeat threat counterattacks, or to allow for quick resumption of the attack. If integrated into a hasty defense inside the urban area, these forces will require continued infantry protection. As a mobile counterattack force positioned inside the urban area, armored forces will also require careful selection of attack positions and counterattack routes. Damage to buildings and infrastructure may limit maneuver and the use of direct and indirect weapon systems.

Repositioning forces outside the urban area may contribute to strengthening or reestablishing the isolation of the urban area and will make better use of range and standoff capability, allow them to take full advantage of their speed and mobility, and decrease their need for additional protective support.

Eliminating Pockets of Resistance

7-87. Focusing on controlling the essential and attacking decisive points means that attacking units will have bypassed some elements of the threat's defense. Within the bounds of the initial plan and current situation, commanders should consider whether it is necessary to devote the time and resources to immediately clear and eliminate remaining threat forces and pockets of resistance with all or parts of the attacking force or leave the task to follow-on forces. In their deliberations, commanders should determine if remaining threat forces are capable of consolidating and mounting effective counterattacks before follow-on units can engage the remaining threat and take advantage of their dispersion and disunity. As another part of their deliberations, commanders should also consider if remaining threat forces will likely (and immediately) endanger—

- The urban inhabitants.
- Lines of communications.
- Critical resources within the urban area necessary to support the inhabitants or subsequent operations.

Reorganization

7-88. Reorganization includes all measures taken by the commander to maintain combat effectiveness or return to a specified level of combat capability. As necessary, these actions include—

- Redistributing or cross-leveling supplies and equipment until resupply can be accomplished.
- Replacing key personnel and combining units or crews to form mission-capable ones.
- Integrating replacement soldiers and systems into the unit and matching operational systems with available crews.
- Recovering, treating, and evacuating casualties, enemy prisoners of war, and damaged equipment.
- Conducting training and disseminating critical lessons learned.
- Other actions to reestablish unit cohesion.

TRANSITION

7-89. Effective transitioning allows commanders to continue full spectrum operations in the urban area and elsewhere in the AO without unnecessary delays. Commanders can accomplish this by thorough planning—including appropriate branches and sequels (revised as the situation changes)—that gives adequate consideration to post-offensive organizational, training, psychological, and civilian requirements. If properly prepared, commanders can anticipate rather than react to potential mission changes.

Early and Concurrent Transition Planning

7-90. Commanders can ensure smooth transitions of urban offensive operations by planning for post-offensive operations early. Based on the mission envisioned, they determine which subordinates and what type of force structure to use. Post-offensive missions, like all urban operations, encompass full spectrum operations. At the successful conclusion of offensive operations, Army forces transition to some type of stability or civil support operation. Commanders may leave the subordinate unit in place to execute the new mission, may reorganize the subordinate unit for the mission, or relieve the unit that just completed offensive operations with a new unit.

Changes to Task Organization

7-91. Commanders should consider the organization of forces following offensive operations. Hostile civilians may require significant combat forces or military police forces to maintain stability. On the other hand, friendly civilians may require a minimum of military police or combat forces, but significant logistic support. Commanders must carefully consider the urban situation before deciding how to use combat forces that recently participated in a high-intensity offensive operation.

Training and Psychological Considerations

7-92. Many Army combat tasks may not support follow-on stability or civil support operations without considerable modification (if at all). Often, noncombat skills—not normally part of a unit's mission essential task list such as negotiating or mediating skills—will be required. However, the greatest modification required applies to each soldier's mental outlook. Forces that transition directly from combat to stability operations may not be psychologically prepared for a rapid and drastic change of mission. Commanders cannot expect troops who have just completed high-intensity offensive operations to rapidly adjust and exercise the sensitivity and judgment required in most stability operations. This especially applies if the population is hostile to Army forces. If possible, combat forces assisting in stability operations, particularly in hostile civilian situations, should not have had recent experience in high-intensity urban operations and they should have trained for the mission.

Return to Civil Agencies

7-93. Commanders of major operations also have the critical role of transitioning aspects of the urban offensive operation to civil, allied, NGOs, and other agencies as appropriate. Transition planning is detailed and aims to quickly return as much civil control of the area as is feasible after the attack. Beyond local civil control, outside civil agencies and NGOs assume tasks as completely and as rapidly as possible. These organizations are consulted and integrated into the planning process as early as possible. Commanders should begin planning for transition simultaneously with planning for offensive operations. They consider the feasibility of relinquishing control of urban areas to civil government, law enforcement, or NGOs even before completing offensive operations. During the conduct of urban operations, these transition operations are closely synchronized with the execution of the attack.

Transition to a New Mission

7-94. In urban offensive operations, like other offensive missions, the change in mission after a successful urban attack may be to a hasty defense or a continuation of offensive operations outside the area. However, in urban offensive operations the mission will just as likely rapidly change to a civil support or stability mission. This is particularly true if the unit has had special training and is task organized for urban operations. Transition to stability or civil support operations is often accompanied by a transition in roles from supported to supporting.

7-95. Even more challenging than transition at the end of the mission is transition during the conduct of the mission. Soldiers may have a difficultly transitioning from stability or civil support to offense and defense, and back again multiple times during an urban offensive operation. Soldiers may apply the tactics, techniques, and procedures of urban offensive operations directly to the stability or civil support missions with potentially disastrous results. Commanders need to segregate missions in time and space, and, if sufficient forces exist, by unit. To this end, commanders should consider permanently designating specific units to conduct civil-military and humanitarian support tasks and avoid rapid mission changes that rotate units (particularly at company level and below) between violent and nonviolent tasks. However, commanders may not have that luxury and may need to rely heavily on preparatory training (including the inculcation of Army values) and strong unit leadership to mitigate potential difficulties.

Chapter 8

Urban Defensive Operations

Generally, a modern city magnifies the power of the defender and robs the attacker of his advantages in firepower and mobility. A city can ingest an invading army, paralyze it for weeks on end, and grind it down to a state of ineffectiveness.

"Military Operations on Urbanized Terrain:
The 2d Battalion, 26th Infantry, at Aachen, October 1944"

The skillful defense of an urban area can decisively affect a campaign. The urban area offers many advantages to defending forces. An adroit defender can use the advantages of the urban environment to negate combat power disparities, blunt the tempo of an attack, attrit threat forces, and sap the morale of attacking troops. The defender gains an opportunity to concentrate resources, reconstitute attrited units, and transition to the offense. A successful defense of an urban area can also deny the threat vital resources. Defense in the urban environment is an essential Army capability and can significantly affect the outcome of entire campaigns and the achievement of national objectives.

PURPOSE OF URBAN DEFENSIVE OPERATIONS

8-1. Army forces defend urban areas for various reasons: defeating a threat attack, buying time, economizing forces, protecting an ally's political institutions and economic infrastructure, protecting an urban population, shaping conditions for decisive offensive operations, and shaping conditions for executing stability or civil support operations. During force projection operations, urban areas may be used as initial lodgment areas that Army commanders may need to defend at the outset until they build sufficient combat power. Usually two or more of these purposes apply to the urban defense. Urban defensive operations provide commanders great opportunities to turn the environment's characteristics to the advantage of Army forces. Urban areas are ideal for defensive operations and greatly enhance the combat power of defending units.

DEFENSIVE CHARACTERISTICS APPLIED TO URBAN OPERATIONS

8-2. There are five general characteristics of the successful defense: preparation, security, disruption, massing effects, and flexibility. All apply to the successful urban defense and to the higher commander supporting a subordinate defending in the urban area.

PREPARATION

8-3. The urban area suits the defense since the area's physical characteristics naturally enhance the combat power of defending units. These characteristics include protection, obstacles, and concealment. Urban terrain provides superb defensive positions with minimum preparation. With deliberate preparation, urban defensive positions can rapidly become strong points.

8-4. One primary characteristic of urban terrain that enhances the defense is protection. With little or no advance preparation, buildings, subsurface structures, and walls protect Soldiers from direct and indirect fire, interdict indirect fire, limit observation, and limit engagement ranges (requiring Soldiers to be skilled at combat in close quarters and quick-fire techniques). Nearly all buildings provide some ballistic protection from direct and indirect fire. Mason and stone buildings with basements and cellars can protect Soldiers from most fires except the largest caliber or tonnage bomb. Minimal additional preparation turns them into formidable, defensive strong points.

8-5. Buildings in urban areas, because of their height and close proximity, also can protect Soldiers by masking them from indirect fire. The height of a building may interdict the flight path of an artillery round, rocket, missile, or bomb at a point short of the intended target. Masking protects static defending forces and protects forces moving along routes bordered with tall buildings that form urban "canyons". These protected routes can be used for sustainment, counterattacks, and maneuver.

8-6. Structurally significant buildings in an urban area can create major obstacles to maneuver. These obstacles immediately canalize maneuver into existing streets and routes without any preparation by the defense. These obstacles then become kill zones for well-positioned and sited defensive forces. Minimal obstacle construction as point obstacles blocking streets and routes can further restrict the maneuver options of the attacking force. Rubble from structures collapsing into streets after fires (intentional or unintentional) can also block routes.

8-7. Buildings also conceal the location, disposition, and intent of the defense. They limit visual observation to the external perimeter of the urban area. They degrade radar and electronic position identifiers and decrease the utility of overhead imagery. The physical aspect of the urban environment greatly enhances the defense by degrading the opposition's intelligence, surveillance, and reconnaissance (ISR) capabilities. Buildings can conceal static defensive positions and the maneuver of defensive forces in the urban area. Although the environment constrains defensive mobility in much the same manner as offensive mobility, the defender has the time and opportunity to conduct careful reconnaissance and select and prepare routes. This gives the defender the ability to move reserves, maneuver counterattack forces, and plan sustainment without observation. Careful preparation provides the defender a mobility advantage over attacking forces.

SECURITY

8-8. The urban area can be an advantage or a disadvantage to the security of defending forces. This largely depends on the nature of the human dimension of the environment. If the population is evacuated or allied with Army forces, then the environment may assist in the security of defending Army forces. However, if the population is present and hostile, then the environment may make security difficult.

8-9. The physical aspects of the urban environment, uninfluenced by the human dimension, may assist in the security of defending Army forces. The combat power of small security forces manning observation posts is greatly enhanced. Forces can more easily restrict and monitor avenues of approach for threat reconnaissance. Defending forces positioned mostly in structures are difficult to locate.

8-10. The physical aspects of the environment may also present some security challenges, primarily with observation. The compartmented terrain limits the field of observation from any one point. The defense may require more security forces to adequately observe the mounted and dismounted avenues to prevent infiltration. Threat forces that successfully infiltrate will be more difficult to locate. These forces will gain numerous hide positions for small reconnaissance units in complex terrain and the effect the terrain has in masking electronic signatures.

8-11. Friendly civilians in the urban area can help identify threat forces attempting to conduct reconnaissance. Civilian activity will also help to mask defense preparations. However, a hostile element of the population may pass intelligence information to the threat. They may assist threat reconnaissance to infiltrate the urban area or provide guides, manpower, or resource support for threat forces. Commanders should take measures to ensure strict control of hostile populations. If resources permit, commanders may consider removing potentially hostile civilians from the area.

DISRUPTION

8-12. The urban environment's attributes assist defending Army forces to disrupt the attacker. It does this through compartmentalization, inhibiting command and control (C2), and facilitating counterattacks.

8-13. The physical aspects of the urban area force the attacking threat into compartmented urban canyons that make mutual support between attacking threat columns difficult. Shifting resources from one portion of the threat attack to another is also difficult. Physically, the urban area disrupts tactical communications making synchronization of combat power difficult.

8-14. The urban terrain hinders the mobility capabilities of the defense. However, careful planning, preparation, and rehearsals can facilitate more rapid movement of larger forces. Defending forces can assemble counterattacks undetected, move them along covered and concealed routes, and achieve surprise at the point of the counterattack. Attacking forces, using the compartmented terrain, often leave forward elements in position to be isolated or expose long and vulnerable flanks to friendly counterattack and interdiction.

MASSING EFFECTS

8-15. The urban environment allows defenders to better protect their centers of gravity and decisive points. The restrictive terrain reduces the attacker's maneuver options. Defenders can position forces in protected and mutually supportive positions oriented on deadly engagement areas. Relatively few well-positioned defenders can generate significant combat power. Without the positional advantage and the corresponding protective effects of the terrain, attacking forces often mass numbers to achieve the necessary combat power.

8-16. Knowing the complex terrain permits defending forces to plan engagement areas that maximize the effects of their combat power. Defending forces can remove fences, walls, rooftops, and even entire buildings to facilitate fields of fire and unmask indirect fire flight paths. Forces carefully choose firing positions for indirect fire systems so that flight paths travel between buildings into engagement areas. By leveraging this knowledge of the terrain, numerically inferior defenders can synchronize devastating fires on offensive forces that are forced by terrain and reinforcing obstacles to mass in confined spaces where fires can have the greatest effect.

FLEXIBILITY

8-17. Defensive flexibility results from detailed planning and mental agility. Commanders develop defensive flexibility by ensuring that plans adequately address branches and sequels that include alternate and subsequent positions and emphasize counterattack options. The urban area facilitates defensive flexibility because the urban terrain can be quickly adapted for defensive operations with little or no preparation. The effect is similar to having multiple, prepared positions on nearly every possible approach. The urban area can also permit rapid, covered movement on interior lines. This permits swift movement to and occupation of strong defensive positions with little or no preparation. The defense also has more flexibility since defenders often know and better understand the urban terrain's effects on operations. Normally, defenders will not get lost as easily, will know complex lines of sight and masking effects, and will best understand the ballistic characteristics of individual structures.

8-18. Mental agility allows commanders to see that the best urban defense may actually be to defend outside of the area. Such a defense mitigates the danger to the urban population and potentially reduces collateral damage. It takes advantage of Army long-range engagement capabilities and denies the threat the opportunity to "hug" Army forces or noncombatants as protection from fires. This defense may be appropriate when Army forces have enough resources to defend more open terrain, when time permits deploying extensive obstacles and constructing protected positions, and when natural terrain such as river obstacles aids the defense.

URBAN DEFENSIVE OPERATIONS AND BATTLEFIELD ORGANIZATION

8-19. Urban defensive operations are organized within the overall battlefield organization of sustaining, shaping, and decisive operations. The success of urban defense depends on each operation, but commanders must synchronize these simultaneous operations as one action. Sustaining operations in defensive urban operations (UO) ensure freedom of action. Critically, urban sustaining operations ensure security of the lines of communications and establish effective movement control. Shaping operations in defensive UO create the conditions for decisive operations. Shaping operations vary greatly depending on the type of defense. For example, in a mobile defense the shaping operation may be the fixing force. In contrast, in an area defense the fixed defense may be the decisive operation. In the urban defense, decisive operations focus on accomplishing the commander's mission. The decisive operation may not defeat the threat's main effort, and it may not prevent threat occupation of large portions of the urban area if those tasks are not essential to mission accomplishment. For example, if the defense's objective is to protect a critical communications node, then, depending on the commander's overall intent, threat actions to secure an airfield elsewhere may not be important.

TYPES OF URBAN DEFENSE

8-20. Commanders can view urban area defensive operations two ways: as conducting a major defensive operation with an urban area in their area of operations (AO) and as defending entirely in an urban area.

AREA DEFENSE

8-21. At the operational level, an area defense may include both urban areas and open maneuver areas. The most common defense in an urban area and the most suitable for the characteristics of this distinct environment is the area defense. As a defensive operation, the area defense concentrates on denying threat forces access to designated terrain for a specific time rather than destroying the threat outright. Although an area defense in an urban area does not directly seek to destroy or defeat attacking threat forces, as an objective it does aim to force culmination of the threat's attack. The urban area defense often works effectively to exhaust threat resources and shape conditions for a transition to offensive operations. The urban area may also be used as a strong point to force threat movement in a different direction or to fix threat forces as part of a large, mobile defense taking place in the AO outside the urban area.

MOBILE DEFENSE

8-22. A mobile defense can operate in an urban area but only under specific conditions. It focuses on destroying or defeating the threat through a decisive attack by a striking force. It requires the defender to have greater mobility than the attacker. To shape a mobility advantage, the urban defender effectively uses the terrain and correctly task organizes his forces' mobility. The principles of applying the mobile defense in the urban area remain the same: a small fixing force stops the threat and limits any ability to maneuver while a striking force quickly maneuvers and counterattacks to destroy him.

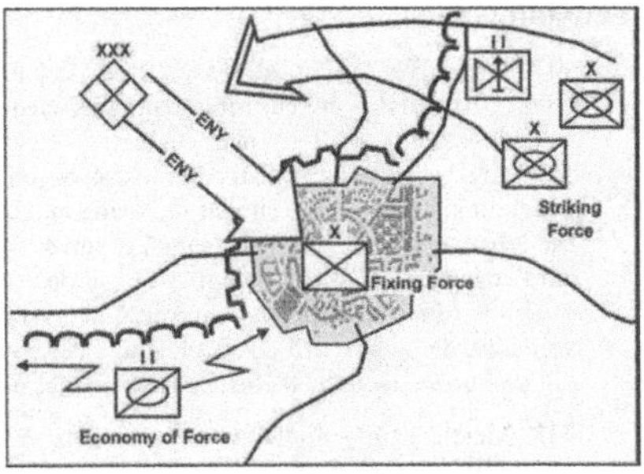

Figure 8-1. An urban area incorporated into a larger mobile defense

8-23. One key to executing a mobile defense in the urban area is to entice a threat force into the depths of the urban area where it begins to lose mobility options. A well-placed fixing force augmented with man-made obstacles and taking advantage of the naturally constrictive terrain can stop a much larger force. If the attacking force is largely mounted and armored, its mobility in the urban area may be reduced to less than that of dismounted infantry. In addition, if the attacking force's movement into the urban area is mounted and rapid, the commander's situational understanding also diminishes. Then the striking force, consisting of dismounted infantry forces, can execute the counterattack with surprise from multiple directions and dimensions (subsurface, surface, supersurface, and airspace). Man-portable antiarmor weapons—firing from flanks and top down and supported by precision indirect fires from both organic and joint systems—can rapidly destroy the threat.

8-24. From the perspective of commanders of the major operation, the urban environment can help defending forces achieve a mobility advantage over an attacker in a broader sense. Defending commanders can attempt to shape the battlefield so that the attacker commits significant resources into an urban area, where his maneuver capabilities are reduced (see figure 8-1). A disproportionately small defending force, which relies on the defensive combat power advantages of the urban environment, reduces and fixes the attacker's maneuver capabilities. Other defending forces mass outside the urban area then strike the threat with a combined mobility and firepower advantage.

Urban Defense in a Major Operation: Stalingrad – 1942 to 1943

The German and Soviet battle for Stalingrad in late 1942 and into early 1943 illustrates how a tactical urban area defense integrates into a larger mobile defense. Stalingrad was a battle fought on a huge scale: it involved army groups on both sides and thousands of square kilometers. Though the city was relatively small, it remained the focus of both German and Soviet offensive and defensive operations during the six-month battle.

In the summer of 1942, the Germans launched a strategic offensive in southern Russia. Its goal was the valuable oil fields of the Caucasus. German forces turning south into the Caucasus exposed a vulnerable flank to Soviet forces positioned between the Don and Volga Rivers. For the German Caucasus operation to succeed, it had to destroy Soviet forces between the Don and Volga, establish a good defensive line, and capture Stalingrad. This city would anchor the German defense and interdict the critical flow of supplies from the Caspian Sea via the Volga River into central Russia. Stalingrad, by virtue of its name, also had important political and cultural value to the Germans and Soviets.

The opening phases of the German offensive were successful; German forces—the 6th Army and 4th Panzer Army—entered the outskirts of Stalingrad in late August 1942 (see figure 8-2).

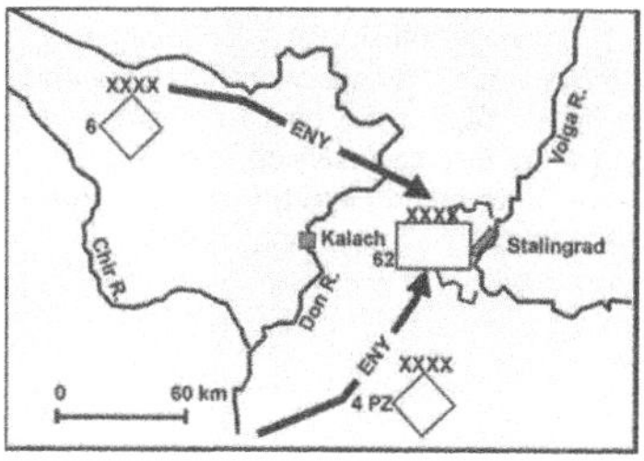

Figure 8-2. German attacks to seize Stalingrad

After a month of intense fighting, the Germans possessed nearly 90 percent of the city. At this point, the 6th Army commanded all German forces in and around Stalingrad. The Soviet 62nd Army's defense was reduced to a front only a few hundred meters deep and a couple of kilometers long on the banks of the Volga. The Soviet defenses hinged on fortress-like concrete industrial buildings and the fanatical bravery and tenacity of soldiers and civilians fighting in the city's remains (see figure 8-3 on page 8-6).

Beginning in mid-September, the Soviet command began looking at how to convert the defense of Stalingrad into an operational opportunity. During October and November, the 62nd Army held on to its toehold in Stalingrad. While maintaining the defense of the 62nd Army, the Soviets secretly began to build up strength on both flanks of the German 6th Army. The Germans increased their vulnerability by securing the German 6th Army's flanks with less capable Romanian, Hungarian, and Italian armies. Also, the 6th Army moved powerful German divisions into the city and rotated with German divisions that were exhausted by urban combat.

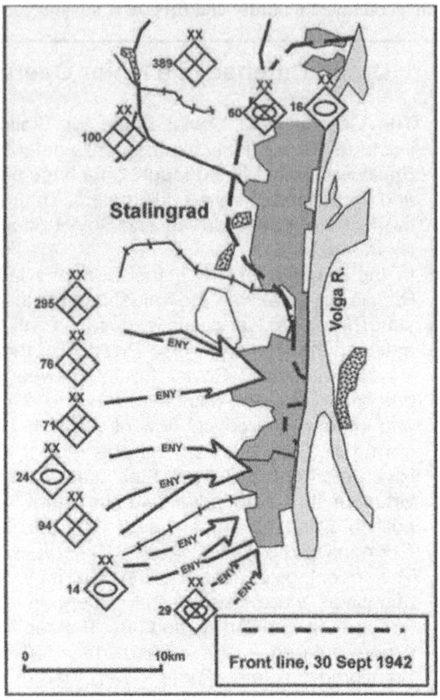

Figure 8-3. German attacks to seize Stalingrad, September 1942

On 19 November, the Soviets launched Operation Uranus that attacked two Romanian armies with seven Soviet armies. Simultaneously, the 8th Russian Army attacked to aid the 62nd Army in further fixing the German 6th Army. Within five days, the Soviet armies of the Don Front, Southwest Front, and Stalingrad Front met near the city of Kalach and sealed the fate of the German 6th Army's 300,000 troops in Stalingrad (see figure 8-4).

On the third day of the Soviet offensive, when encirclement seemed inevitable but not yet complete, the 6th Army commander asked permission to withdraw from the trap. The German high command denied permission believing that the Army could be supplied by air and then a renewed offensive could break through to the city. On 12 December, the German LVII Panzer Corps launched an offensive north to break through to Stalingrad. This offensive made progress until another Soviet offensive on 16 December forced its cancellation. This ended any hope of recovering Stalingrad and the 6th Army. On 31 January 1943, the 6th Army surrendered after sustaining losses of almost two-thirds of its strength. The Soviets took over 100,000 prisoners.

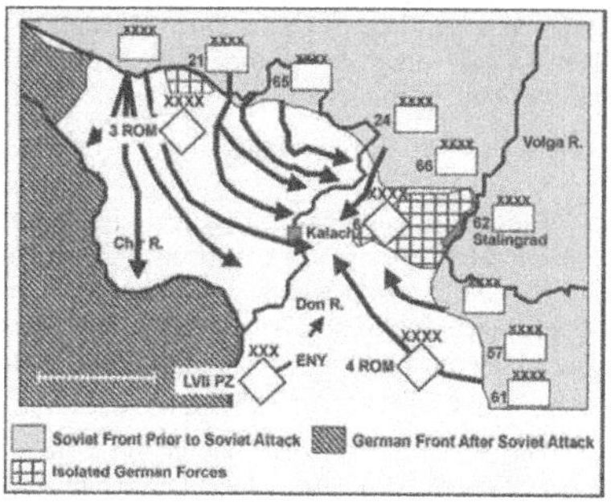

Figure 8-4. Soviet attacks trap German 6th Army

Many lessons emerge from the successful defense of Stalingrad. Tactically, the defense showed how using the terrain of a modern industrial city wisely could increase the combat power of an inferior, defending force and reduce the maneuver options of a mobile, modern attacking force. Another element in the Soviet's tactical success was the Germans' inability to isolate the defenders. The Germans never threatened the Soviet supply bases east of the Volga and, despite German air superiority, the Soviets continuously supplied and reinforced the 62nd Army across the Volga River. Also, Soviet artillery west of the river was able to fire in support of Soviet forces and was never threatened with ground attack.

At the operational level, the Soviets demonstrated a keen understanding of using an urban area within the context of a mobile defense. The 62nd Army's stubborn area defense of Stalingrad drew the bulk of the German combat power into the urban area where they were fixed by a smaller and quantitatively inferior defending force. This allowed the Soviets to build combat power outside the urban area. The Soviets set the conditions for a mobile defense by positioning powerful Soviet armor forces in open terrain outside the urban area against quantitatively inferior German allied forces. In OPERATION URANUS, the mobile defense's strike force destroyed the enemy outside the urban area and trapped the greater part of the best enemy formations inside the urban area. The trapped units were then subjected to dwindling resources and extensive psychological operations, further isolated into pockets, and defeated in detail.

RETROGRADE

8-25. A retrograde involves organized movement away from the threat. Retrograde operations include withdrawals, delays, and retirements. These defensive operations often occur in an urban environment. The urban environment enhances the defending force's ability to conduct retrograde operations successfully (see figure 8-5).

8-26. The cover and concealment afforded by the urban environment facilitates withdrawals where friendly forces attempt to break contact with the threat and move away. The environment also restricts threat reconnaissance, which is less able to detect friendly forces moving out of position, and presents excellent opportunities for deception actions. Finally, a small security force's ability to remain concealed until contact in the urban environment significantly slows threat attempts to regain contact once Army forces have broken contact and begun to move.

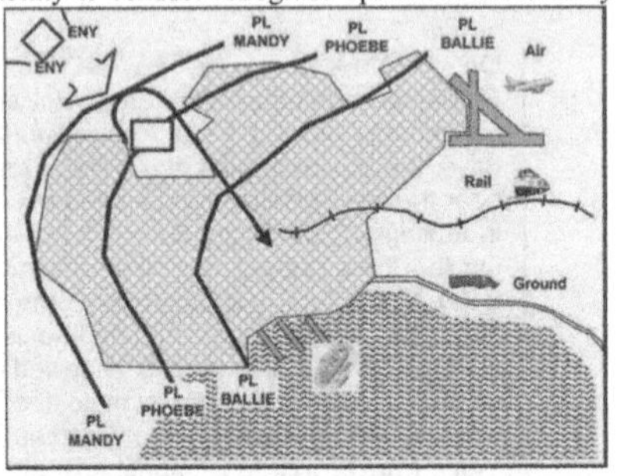

Figure 8-5. Retrograde through an urban area

8-27. The urban environment's natural cover and concealment, as well as the compartmented effects, facilitates delays. Delays can effectively draw the threat into the urban area for subsequent counterattack or as an integral part of a withdrawal under threat pressure. Delaying units can quickly displace from one covered and concealed position to another; the repositioning options are vast. Compartmented effects force the attacking threat to move on well-defined and easily interdicted routes and limit the threat's ability to flank or bypass delaying positions.

8-28. The urban area's transportation and distribution network facilitates retiring forces that are not in contact. Properly used, the urban transportation system can quickly move large forces and associated resources, using port facilities, airfields, railheads, and well-developed road networks.

URBAN DEFENSIVE CONSIDERATIONS

What is the position about London? I have a very clear view that we should fight every inch of it, and that it would devour quite a large invading army.

Winston Churchill
War in the Streets

8-29. The urban operational framework—understand, shape, engage, consolidate, and transition—provides structure to developing considerations for defensive operations. The considerations can vary depending on the level of war at which the operation is conducted, the type of defense, and the situation. Most issues discussed may, in the right circumstances, apply to both commanders conducting major UO and commanders at lower tactical levels of command.

UNDERSTAND

8-30. The commander defending in the urban area must assess many factors. His mission statement and guidance from higher commanders help him focus his assessment. If the mission is to deny a threat access to port facilities in an urban area, the commander's assessment will be focused much differently than if the mission is to deny the threat control over the entire urban area. The mission, enemy, terrain and weather, troops and support available, time available, civil considerations (METT TC) structure guides the commander's assessment. Of these, the impacts of the threat and environment—to include the terrain, weather, and civil considerations—are significant to the commander's understanding of urban defensive operations.

The Threat

8-31. In the urban defense, a key element is the commander's understanding of the threat. One of his primary concerns is to determine the attacker's general scheme, methodology, or concept. Overall, the attacker may take one of two approaches. The most obvious would be a direct approach aimed at seizing the objectives in the area by a frontal attack. A more sophisticated approach would be indirect and begin by isolating Army forces defending the urban area. Innumerable combinations of these two extremes exist, but the threat's intentions toward the urban area will favor one approach over another. The defending Army commander (whose AO includes but is not limited to the urban area) conducts defensive planning, particularly his allocation of forces, based on this initial assessment of threat intentions. This assessment determines whether the commander's primary concern is preventing isolation by defeating threat efforts outside the area or defeating a threat attacking the urban area directly. For the higher commander, this assessment determines how he allocates forces in and outside the urban area. For the commander in the urban area, this assessment clarifies threats to sustainment operations and helps shape how he arrays his forces.

The Environment's Defensive Characteristics

8-32. A second key assessment is the defensive qualities of the urban environment. This understanding, as in any defensive scenario, is based on mission requirements and on a systemic analysis of the terrain in terms of observation and fields of fire, avenues of approach, key terrain, obstacles, and cover and concealment (OAKOC). It is also based on potential chemical, biological, radiological, nuclear and fire hazards that may be present in the urban area. This understanding accounts for the unique characteristics of urban terrain, population, and infrastructure as discussed in Chapter 2.

8-33. Generally, units occupy less terrain in urban areas than more open areas. For example, an infantry company, which might occupy 1,500 to 2,000 meters in open terrain, is usually restricted to a frontage of 300 to 800 meters in urban areas. The density of building in the urban area, building sizes and heights, construction materials, rubble, and street patterns will dictate the actual frontage of units; however, for initial planning purposes, figure 8-6 provides approximate frontages and depths for units defending in an urban area.

UNIT	Frontage (Blocks*)	Depth (Blocks*)
Battalion	4 – 8	3 – 6
Company	3 – 4	2 – 3
Platoon	1 – 2	1
*Average block is 175 meters		

Figure 8-6. Approximate defensive frontages and depths

SHAPE

8-34. Commanders of a major operation shape the urban battle according to the type of defense they are attempting to conduct. If conducting an area defense or retrograde, they use shaping actions like those for any defensive action. Important shaping actions that apply to all defensive UO include—

- Preventing or defeating isolation.
- Separating attacking forces from supporting resources.
- Creating a mobility advantage.
- Applying economy of force measures.
- Effectively managing the urban population.
- Planning counterattacks.

Preventing or Defeating Isolation

8-35. Failure to prevent isolation of the urban area can rapidly lead to the failure of the entire urban defense. Its importance cannot be overstated. In planning the defense, commanders must anticipate that the threat will attempt to isolate the urban area. Defensive planning addresses in detail defeating threat attacks aimed at isolation of the urban area. Commanders may defeat this effort by allocating sufficient defending forces outside the urban area to prevent its isolation. Defensive information operations (IO) based on deception can also be used to mislead the threat regarding the defensive array in and outside the urban area. Such information can convince the threat that a direct attack against the urban area is the most favorable approach.

8-36. If the threat has successfully isolated the urban area, commanders of a major operation have several courses of actions. Three options are an exfiltration, a breakout attack by forces defending the urban area, or an attack by forces outside the urban area to relieve the siege. A fourth option combines the last two: counterattacks from both inside and outside the urban area to rupture the isolation (see breakout operations in FM 3-90). Time is critical to the success of either operation. Commanders should plan for both contingencies to ensure rapid execution if necessary. Delay permits threat forces surrounding the urban area to prepare defenses, reorganize their attacking force, retain the initiative, and continue offensive operations. The passage of time also reduces the resources of defending forces and their ability to breakout. Therefore, commanders and staff of a major operation must vigilantly avoid isolation when Army forces are defending urban areas in their AO.

Separating Attacking Forces from Supporting Resources

8-37. Commanders of the major operation primarily use fires and IO for separating in space and time threat forces attacking the urban area from echelons and resources in support. The purpose of this shaping action is the same as for any conventional area defense. It aims to allow the defending forces to defeat the threat piecemeal as they arrive in the urban area without support and already disrupted by deep fires and IO against information systems. This separation and disruption of the threat also sets the conditions for a mobile defense if commanders choose to execute that type of defense. These operations also prevent the threat commander from synchronizing and massing his combat power at the decisive point in the close battle.

8-38. If the urban area is part of a major mobile defense operation, the urban defense becomes the fixing force. Commanders can shape the defense to encourage the threat to attack into the urban area. They can lure the threat using a combination of techniques depending on the situation. They may make the urban area appear only lightly defended while other alternative courses of action appear strongly defended by friendly forces. Placing the bulk of the defending forces in concealed positions well within the urban area and positioning security forces on the periphery of the urban area portray a weak defense. In other situations, the opposite is true. If the urban area is an important objective to the threat, friendly forces can make the urban area appear heavily defended, thus ensuring that he commits sufficient combat power to the urban area to overwhelm the defense. Both cases have the same objective: to cause a major commitment of threat forces in the urban area. Once this commitment is made, the mobile defense striking force attacks and defeats the threat outside the urban area. This isolates the threat in the urban area and facilitates its destruction.

8-39. In the urban tactical battle, many shaping actions mirror those in all defensive operations. The size and complexity of the urban area prevent defending forces from being strong everywhere; shaping operations designed to engage the threat on terms advantageous to the defense have particular importance. Shaping actions include reconnaissance and security operations, passages of lines, and movement of reserve forces prior to their commitment. In addition, shaping operations critical to urban defense include mobility and countermobility operations, offensive IO, economy of force operations, and population management operations.

Creating a Mobility Advantage

8-40. In urban terrain, countermobility operations can greatly influence bringing the threat into the engagement areas of defending forces. Countermobility operations—based on understanding the urban transportation system, design, and construction characteristics—can be unusually effective (see Chapter 2). Demolitions can have important implications for creating impassable obstacles in urban canyons as well as for clearing fields of fire where necessary. Careful engineer planning can make the already constrictive terrain virtually impassable to mounted forces where appropriate, thus denying the threat combined arms capabilities. Countermobility operations in urban terrain drastically increase the defense's ability to shape the attacker's approach and to increase the combat power ratios in favor of the defense. As with all aspects of UO, countermobility considers collateral damage and the second- and third-order effects of obstacle construction.

8-41. Well-conceived mobility operations in urban terrain can provide defending forces mobility superiority over attacking forces. This is achieved by carefully selecting routes between primary, alternate, and subsequent positions, and for moving reserves and counterattack forces. These routes are reconnoitered, cleared, and marked before the operation. They maximize the cover and concealment characteristics of the terrain. Using demolitions, lanes, and innovative obstacles denies the defense of these same routes.

Applying Economy of Force Measures

8-42. Economy of force is extremely important to effective tactical urban defense. A megalopolis is too large and too easily accessible for defending forces to be strong everywhere. Economy of force enables the defending force to mass effects at decisive points. Forces used in an economy of force role execute security missions and take advantage of obstacles, mobility, and firepower to portray greater combat power than they actually possess. They prevent the threat from determining the actual disposition and strength of the friendly defense. If, contrary to expectations, they are strongly attacked, their mobility—stemming from a mounted maneuver capability, planning, and an intimate knowledge of the terrain—allows them to delay until reserves can meet the threat. Security forces in an economy of force role take position in parts of the urban area where the threat is less likely to attack.

Defensive Combat Power: Suez – 1973

At the end of October, the Israeli Army was in the midst of effective counterattack against the Egyptian Army. The Israelis had success attacking west across the Suez Canal. Their armored divisions were attempting to achieve several objectives, to include destroying Egyptian air defense sites and completing the encirclement of the Egyptian 3rd Army, which was trapped on the canal's east side.

To completely encircle the Egyptian 3rd Army, the Israelis had to seize all possible crossing sites to it from the canal's west bank and the Red Sea. Also, as international negotiations towards a cease-fire progressed, the Israeli government wanted to capture as much Egyptian territory as possible to improve their negotiating position after hostilities.

Consequently, the Israeli Adan Armored Division was tasked to seize the Egyptian Red Sea port of Suez on the morning of 24 October. A cease-fire was to begin at 0700, and the Israeli intent was to be decisively engaged in the city by that time and then consolidate their position as part of the cease-fire compliance.

The Adan Division plan to seize Suez was a two-part operation. Each of the division's armored brigades would have a role. The 460th Brigade would attack west of the city and complete the city's encirclement. Simultaneously, the 217th Brigade would attack in columns of battalions through the city to seize three key intersections in the city. This was in accordance with standard Israeli armored doctrine for fighting in an urban area. The 217th Brigade would seize its objectives through speed, firepower, and shock action. Once the objectives were seized, infantry and armored teams would continue attacking from the secured objectives to mop up and destroy pockets of resistance. The Israeli commanders expected to demoralize the defending Egyptians—two infantry battalions and one antitank company—by this rapid attack. The armored division commander was specifically advised by his commander to avoid a "Stalingrad" situation.

The attack got off to an ominous beginning as mist greatly inhibited a scheduled aerial bombardment in support of the attack. The 217th Brigade began its attack without infantry support and was quickly stopped by antitank missiles and antitank fire. Infantry was quickly integrated into the brigade and the attack resumed.

At the first objective, the Israelis encountered their first problems. A withering barrage of small arms, antitank missiles, and antitank fire hit the lead tank battalion, including direct fire from SU-23 anti-aircraft guns. Virtually all the officers and tank commanders in the tank battalion were killed or wounded, and several tanks were destroyed. Disabled vehicles blocked portions of the road, and vehicles that turned on to secondary roads were ambushed and destroyed. The battalion, however, successfully fought its way through the first brigade objective and on to the final brigade objective.

Hastily attached paratroop infantry in company strength were next in column following the tanks. They were traveling in buses and trucks. As the lead tank battalion took fire, the paratroopers dismounted, and attempted to secure adjacent buildings. The tank battalion's action of fighting through the objective caused the paratroopers to mount up and also attempt to move through the objective. Because of their soft skinned vehicles the paratroopers were unable to remain mounted and again dismounted, assaulted, and secured several buildings that they could defend. Once inside the buildings, the paratroopers found they were cut off, pinned down, and unable to evacuate their considerable casualties, which included the battalion commander. The paratroopers were on the initial brigade objective but were unable to maneuver and were taking casualties.

A second paratroop company also dismounted and quickly became stalled in house-to-house fighting. The brigade reconnaissance company in M113 personnel carriers brought up the rear of the brigade column and lost several vehicles and was also unable to advance.

By 1100 the Israeli attack culminated. Elements of the 217th Brigade were on all three of the brigade's objectives in the city. However, the armored battalion, which had achieved the deepest penetration, was without infantry support and under severe antitank fire. Both paratroop companies were isolated and pinned down. In addition, an attempt to link up with the paratroopers had failed. At the same time, the civilian population of the city began to react. They erected impromptu barriers, ambushed isolated Israeli troops, and carried supplies and information to Egyptian forces.

The Israeli division commander ordered the brigade to break contact and fight its way out of the city. The armored battalion was able to fight its way out in daylight. The paratroop companies were forced to wait until darkness and then infiltrated out of the city carrying their wounded with them. Israeli casualties totaled 88 killed and hundreds wounded in addition to 28 combat vehicles destroyed. Egyptian casualties were unknown but not believed to be significant.

The fight for Suez effectively demonstrates numerous urban defensive techniques. It also vividly demonstrates the significant effect on defensive combat power of the urban environment.

The Egyptian defense demonstrates how the compartmented urban terrain restricts the mobility and the massing of firepower of armored forces. Trapped in column on the road, the Israelis were unable to mass fire on particular targets nor effectively synchronize and coordinate their fires. The short-range engagement, also a characteristic of urban combat, reduced the Israeli armor protection and eliminated the Israeli armor's ability to keep out of small arms range. Thus, hand-held antiarmor weapons were more effective in an urban area. Additionally, Egyptian small arms and sniper fire critically affected Israeli C2 by successfully targeting leaders.

The Egyptian defenders effectively isolated the mounted Israelis by defending and planning engagement areas in depth. The Egyptians synchronized so that they engaged the entire Israeli force simultaneously. This forced the Israelis to fight in multiple directions. It also separated the Israeli infantry from the armor and prevented the formation of combined arms teams necessary for effective urban offensive operations.

Suez also demonstrated how civilians come to the advantage of the defense. After the battle was joined, the population—by threatening isolated pockets of Israelis and building barricades—helped prevent the Israelis from reorganizing while in contact and hindered the Israelis breaking contact. The population was also a valuable source of intelligence for the Egyptians and precluded covert Israeli movement in daytime.

Suez shows the ability of a well-placed defense in depth to fix a superior force in an urban area. Despite the Israeli commander's caution to avoid a "Stalingrad," the Israeli division, brigade, and battalion commanders were quickly trapped and unable to easily break contact. Even a successful defense on the perimeter of the city would not have been nearly as effective, as the Israelis would have easily broken contact once the strength of the defense was recognized.

Another key to the success of the Egyptian defense was the Israelis' inadequate reconnaissance. While the Israelis knew the approximate size of the defending forces, they had no idea of the Egyptian dispositions. In this case, time prevented adequate reconnaissance. Key to a successful defense is adequate security to obscure defense dispositions, which permits surprise and shock effect.

The Suez defense was a decisive defeat of elite Israeli forces by regular infantry units inferior in training, morale, and numbers. Total disaster was averted only because of the professionalism of the Israeli armored forces and paratroopers that permitted them to continue to fight and eventually exfiltrate the urban trap. The Israeli forces thus escaped total destruction. Suez strongly demonstrates how the enhancing effects of the urban environment on defensive combat power are significant enough to permit inferior regular forces to defeat elite formations. Since the 1973 Suez battle, U.S. forces in Mogadishu, Somalia, and Russian forces in Grozny, Chechnya have faced similar urban defensive ambushes.

Effectively Managing the Urban Population

8-43. Another way to shape the urban defensive battle is population management. In most cases, defending force commanders are in the urban area before combat. This time gives them the chance to manage civilians. Consequently, they can better manage and protect the population (a legal requirement) and gain more freedom of action for their forces.

8-44. Managing the civilians during the defense is a function of the size, disposition, and needs of the population and the resources available to the commander. Requesting higher support or coordinating with nongovernmental organizations and the local civil leadership for support may make up shortages of resources. Resources devoted to population management are carefully weighed against availability, military mission requirements, and possible collateral damage affecting tactical, operational, or strategic success. It may prove impractical to evacuate an urban area's population; still, commanders should attempt to create and move most civilians to protected areas. Moving the population allows defending forces to more liberally apply fires, emplace obstacles, and relieve combat units and support units of requirements to continue life support for civilians while executing combat operations. Overall, effective civil-military operations can turn a friendly (or a neutral) population into an effective force multiplier providing support to every warfighting function.

Planning Counterattacks

8-45. Counterattacks are also an important tool in shaping the battlefield for defensive success. Counterattacks as a shaping tool have two applications: retaining the initiative and separating forces. However, opportunity for effective counterattacks will be brief and, therefore, timing will be critical. If conducted too soon, the counterattack may expend resources required later; if conducted too late, it may not be effective. Commanders should understand the effect of the urban environment on time-distance relationships; otherwise, the timing of the attack may be upset and the operation desynchronized. Additionally, commanders should develop plans beyond the counterattack to exploit potential success.

ENGAGE

8-46. Engaging the urban area in a defensive operation requires decisively defeating the threat's attacks. Defensive forces use the terrain to their advantage, employ precision supporting fires, and use direct fire from protected positions aligned against carefully selected avenues of approaches and kill zones. The combat power of the defense augmented by shaping actions and the characteristics of urban terrain force culmination of the threat attack. Like urban offensive operations, effective engagement in urban defensive operations typically results from successful actions at the tactical level of war. These actions include—

- Performing aggressive ISR.
- Creating depth.
- Executing an effective obstacle plan.
- Conducting coordinated counterattacks.

Performing Aggressive ISR

8-47. ISR efforts of the defender are focused initially on identifying relevant information about the location and nature of the threat's main effort. Once identified, the defender's ISR focus shifts to assessing the rate at which the threat attack moves to its culminating point. Indicators of culmination may be physical fatigue of Soldiers, a breakdown in C2 capability, difficulty providing logistics support, or the increasing time required to reorganize small units to attack. When that culmination is achieved, friendly forces counterattack before the threat has a chance to transition to a hasty defense.

Creating Depth

8-48. Depth in the defense is the key to forcing the threat to culminate. The urban defense cannot allow itself to be penetrated nor permit forward elements to be destroyed. The defense is designed with the greatest depth possible. Defending forces weaken the threat to the fullest extent possible by attack from each position but not permit themselves to be destroyed by fires or close assault. Instead, as threat combat power builds up against individual positions, the use of mission orders permits subordinate leaders to disengage on their own initiative and move on preplanned routes to subsequent positions. Positions are designed to be mutually supporting—withdrawing from one position to a subsequent one while supporting positions cover by fires. The attacker is constantly forced to deploy and reorganize without being able to achieve decisive effects against the defender.

Executing an Effective Obstacle Plan

8-49. Obstacles in the urban defense are designed to break up the threat's combined arms capability. Separating dismounted forces from mounted forces disrupts the cohesion of the attacker and reduces his combat power. It also exposes his individual elements to the effects of asymmetric counterattack. The leading threat dismounted force can be effectively counterattacked by a friendly combined arms element while the threat armored force remains vulnerable to antiarmor attack by dismounted forces.

Conducting Coordinated Counterattacks

8-50. The counterattack is one of the key actions of the urban defense. However, commanders do not counterattack unless there is a reasonable chance of success. As the attacker moves into the depth of the urban area, his forces become fatigued, attrited, and increasingly disorganized. He likely also creates an increasingly long and exposed flank. At all levels, forces defending in urban terrain look for opportunities to counterattack. As the offensive force reaches the culmination point where it can no longer continue to attack with the available forces, the defensive commander executes a planned and coordinated counterattack. The counterattack aims to regain the initiative and to make the threat fight in multiple directions. Infiltration using superior knowledge of the terrain (including supersurface and subsurface capabilities) and the skillful use of stay-behind forces permits attacking the threat throughout the depth of his formations. Small-scale counterattacks focus on C2 and sustainment capabilities. These counterattacks can set the conditions for a deliberate attack leading to the ultimate destruction of the attacking threat force.

CONSOLIDATE

8-51. Consolidation is as important to urban defensive operations as to offensive operations. Many of the same consolidation considerations for the urban offense apply equally well to the defense. Commanders should reinforce or reposition maneuver forces and fire support assets on the urban battlefield based on weaknesses uncovered during rehearsals and opportunities discovered during actual execution. While maximizing the many advantages of the urban defense, the commander of the urban defense aggressively seeks ways of attriting and weakening enemy forces before they enter the urban area and close combat is initiated. Commanders should combine the static and mobile elements of the defense to strengthen their positions in relation to the enemy while seeking every opportunity to transition to urban offensive operations. As in urban offensive operations, commanders should conduct any necessary reorganization actions that they were unable to accomplish during execution.

TRANSITION

8-52. Transitions in urban defensive operations occur at all levels. As with offensive operations, commanders of major operations should address which units are assigned to continue to operate in the area after defensive operations have ceased. In defensive UO, this task is not as challenging as an occupation mission during urban offensive operations. The psychology of troops defending an urban area differs from those attacking into it. Defending forces become accustomed to the environment, having experience in the environment before combat. In terms of training, it is easier for follow-on missions to be assigned to a unit that has successfully defended the urban area. This course of action takes advantage of the defending unit's experience in the area and its relationships with other agencies—agencies that were operating alongside the units before and possibly during the defense. In defensive operations, regardless of the civilians' attitudes, policies regarding that population are established before the successful defense, and the command likely has experience executing operations with civil authorities and other agencies. Thus, these relationships are neither new nor as significant an issue as in offensive operations. Therefore, commanders must be prepared to execute various stability or civil support operations or use a successful defense to springboard into more decisive offensive operations elsewhere in the commanders' AO.

Transition Emphasis to Stability or Civil Support Operations

8-53. At the end of a successful urban defense, operational commanders should generally expect civil authority, control, and jurisdiction to increase. Additionally, the civil population may be anxious to return. Defensive combat will require virtually complete military control of the urban area; however, after the successful defense, a rapid transition will occur from military control to civil or joint military and civil control afterward. Although the full spectrum of Army operations will often be conducted simultaneously in UO, commanders will transition emphasis to stability or civil support operations. Important transition tasks will include demilitarizing munitions, clearing obstacles, and searching for isolated threat pockets of resistance. Conclusion of the defensive operations will also require transition to joint civil-military tasks, such as evaluating structures for safety, restoring essential services, and possibly creating joint law enforcement. Commanders of major operations, using a civil-military operations center and G9, should anticipate these requirements and begin early preparations to ensure a smooth, successful transition.

Transition to Offensive Operations

8-54. Units that have successfully defended the urban area may then transition to offensive operations. A rapid transition to offensive operations will require identification, preparation, and training of units designated to assume missions as the defending units leave the urban area. This preparation emphasizes continuity of policies and relationships already established. A relief in place occurs. The new occupying units provide not only a continuity of policy, but also a continuity of attitude toward the urban area, its population, and its institutions.

Chapter 9

Urban Stability and Civil Support Operations

The fundamental shared aims between stability and civil support operations and any operation conducted in an urban environment are the increased significance and influence of the civil population and nonmilitary organizations. Often, no military victory is to be achieved. The center of gravity for these operations normally cannot be attacked through military means alone; the Army (and the military in general) is often but one tool supporting a larger, civil-focused effort. When conducted, military defeat of an armed adversary is only a phase in a larger campaign, the overall objectives of which will be to bring about a more satisfactory political, social, diplomatic, economic, and military situation than existed prior to the conflict. Without a tightly coordinated civil-military effort, overall success will be difficult or impossible to achieve. Commanders who can understand and cope with the complexities of stability or civil support operations gain insights that directly apply to executing any urban operation. Urban stability or civil support operations may complement urban offensive and defensive operations, or may dominate the overall operation. Army forces may need to conduct offensive and defensive operations to defend themselves or destroy urban threats seeking to prevent the decisive stability or civil support mission. During hostilities, urban stability operations may keep armed conflict from spreading, encourage coalition and alliance partners, and secure the civilian population's support in unstable urban areas (and the surrounding rural areas under their influence). Following hostilities, urban stability operations may provide a secure environment for civil authorities to rebuild. Urban civil support operations can range from transporting, feeding, and sheltering the population made homeless as a result of man-made or natural disasters to conducting civil search and rescue and providing urgent medical care.

PURPOSE OF URBAN STABILITY AND CIVIL SUPPORT OPERATIONS

9-1. Army forces conduct stability and civil support operations to deter war, resolve conflict, promote peace, strengthen democratic processes, retain U.S. influence or access abroad, assist U.S. civil authorities, and support moral and legal imperatives. Stability operations promote and sustain regional and global stability. Civil support operations meet the urgent needs of designated U.S. civilians until civil authorities can accomplish these tasks without military assistance. Nearly every urban operation will involve some type or form of stability operation or civil support operation combined and sequenced with offensive and defensive operations.

CHARACTERISTICS OF URBAN STABILITY AND CIVIL SUPPORT OPERATIONS

9-2. Worldwide urbanization, migration trends from rural to urban areas, and more centralized populations in urban areas increase the chance that Army forces will conduct stability or civil support operations in or near urban areas. Simply put, many people live in urban areas, and their welfare will be the primary reason for conducting these operations. Urban areas that serve as economic and government centers (the ideal location for U.S. and allied embassies) are often the focal point for many threat activities. Therefore, Army forces may need to conduct stability operations in these foreign cities to counter those threats. Additionally, urban areas may contain the resources and infrastructure to support both types of operations, regardless of whether the overall focus is in urban or rural areas. Repairing or restoring the infrastructure may be a critical task in accomplishing either type of operation. Supported governmental and nongovernmental agencies are not as logistically self-sufficient as the Army. As such, these agencies will often need to center their operations in and around urban areas to use the area's infrastructure to support themselves and their objectives. These agencies may require military protection to accomplish their missions. Figure 9-1 lists some defining characteristics of these wide-ranging operations.

- Long or Short Duration
- Joint and Interagency
- Unilateral or Multinational
- Increased Civil-Military and Legal Considerations
- Greater Potential for Ambiguity
- Increased Constraints Necessitating More Restrictive ROE
- Amplified Need for Cultural and Political Sensitivity

Figure 9-1. Characteristics of stability and reconstruction and support operations

9-3. Stability and civil support operations are diverse, varied in duration, and unilateral or multinational. Stability operations are conducted overseas as part of a campaign or major operation, while civil support operations are conducted within the United States as part of homeland security. Like all urban operations (UO), they are usually joint. Unlike urban offensive and defensive operations, they are more often interagency operations and require more restrictive rules of engagement (ROE). The multiplicity of actors involved usually increases the scope and scale of required coordination and communication. In urban stability or civil support operations, adverse conditions arising from natural or man-made disasters or other endemic conditions—such as human suffering, disease, violations of human rights, or privation—will significantly modify the urban environment. Unresolved political issues and tenuous agreements, difficulties discriminating combatants from noncombatants or between parties of a dispute, and the absence of basic law and order all serve to complicate an already complex and uncertain environment. Civil-military and legal considerations take on added significance in all urban operations (see Civil-Military Operations and Legal Support in Chapters 5 and 10 respectively), but even more so in urban stability and civil support operations. Finally, recognizing, defining, and achieving the desired end state is often more difficult than in offensive and defensive operations.

9-4. Overall, commanders of major operations involving urban stability and civil support operations should not expect clear guidance. They must learn, adapt, and live with ambiguity. They cannot expect to operate in a political vacuum (even commanders at the tactical level); rather, they should expect to work alongside both governmental and nongovernmental leaders and organizations. Commanders conducting UO should not expect an easily identifiable enemy located across a clearly demarcated line. In fact in many peace operations, they and their Soldiers must resist the need to have an enemy—difficult at best when one side or another (or both) may be sniping at them. Commanders should also expect changing and additional missions and tasks, without being allowed to use every means at their disposal to carry out those missions. Many tasks required may be ones for which their units have never, or rarely, trained. Finally, commanders should expect to show restraint with a keen sensitivity to political considerations and to alien cultures, either or both of which they might find confusing or even repugnant.

URBAN STABILITY AND CIVIL SUPPORT OPERATIONS AND BATTLEFIELD ORGANIZATION

9-5. Each urban stability operation or civil support operation will be distinct from one another. These operations differ even more when applied to a specific urban area. Due to the complexity of the environment, commanders must carefully arrange their forces and operations according to purpose, time, and space to accomplish the mission. In most UO the terrain, the dense population (military and civilian), and the participating organizations will further complicate this arrangement.

9-6. The support and assistance that Army forces will provide during these operations is only temporary (although it may be of long duration) and this facet requires particular emphasis. Commanders should plan and execute both of these types of operations with that essential consideration always in mind. Eventually, the government and administration—either foreign or domestic—must secure and support their population by themselves. Therefore, commanders must envision and set the conditions that allow for the transition of control and responsibility to shift to legitimate civilian authorities. While commanders may only provide assistance and support based on specific and well planned civilian requests, more often commanders will need to determine requirements in collaboration with competent civilian authorities and agencies or, in some cases, with little or no initial civilian assistance at all. Ultimately, transition planning should occur as an integral part of the overall operational planning and include collaboration with appropriate civilian agencies and organization as early as possible to ensure a seamless transition to civilian control without major setbacks and loss of forward momentum.

STABILITY OPERATIONS

9-7. Stability operations establish, sustain, and exploit security and control over foreign areas, populations, and resources. Urban areas will be decisive to accomplishing many types of stability operations because urban areas are the centers of population, culture, economy, and government. Much of the support provided by Army forces will aim to assist local, regional, or national governments to restore essential services and infrastructure and reestablish civil order and authority. The location of civilian authorities in urban areas will, by necessity, be a dominating factor in accomplishing the mission. As importantly, many stability operations—enforcing peace in Bosnia for example—will require interacting with, influencing, controlling, or protecting all or parts of the civilian population. Assessing, understanding, and gaining the support of civilians in key economic, cultural, or political urban areas may influence surrounding regions (smaller urban areas and the rural countryside) and may be decisive to achieving overall stability objectives.

9-8. Stability operations may involve both coercive and cooperative actions. They are conducted in situations in which legitimate civil authority is unable to provide the necessary security and control for the urban population as a result of—

- Deliberate operations to influence a regime change.
- Offensive or defensive operations or natural disasters resulting in ineffective civil authorities, conditions beyond the capabilities of the foreign urban government, or both.

CIVIL SUPPORT OPERATIONS

9-9. Civil support operations are an integral part of homeland security which aim to protect U.S. citizens and infrastructure from conventional and unconventional threats and mitigate the effects of man-made and natural disasters. (The other part to homeland security is homeland defense.) Army forces conduct civil support operations by providing Army resources, expertise, and capabilities in support of a lead agency—which may be to Department of Defense—to address the consequences of man-made or natural accidents beyond the capabilities of local or regional U.S. civilian authorities. Such a situation could result from a chemical, biological, radiological, nuclear, and high-yield explosive (CBRNE) incident, hurricane, flood, or civil disturbance in a domestic urban area.

CBRNE Incidents

9-10. CBRNE incidents can be disastrous and are of particular concern for both stability and civil support UO. In urban areas, the potential for catastrophic loss of life and property is enormous. The Army categorizes CBRNE incidents separately from other natural and man-made disasters because it has specific expertise with these weapons. CBRNE incidents usually result from a military or terrorist threat (adding a law enforcement dimension to the disaster). Other urban CBRNE considerations include the following:

- Subways and other subsurface areas offer ideal areas for limited chemical or biological attacks. Nuclear attack (and high-yield explosives) can produce tragic results due to the effects of collapsing structures, flying debris, and fires. Dispersion patterns are affected by the urban terrain and are more difficult to predict and monitor. Large-scale incidents may produce hundreds of thousands of casualties, but even a limited attack may require evacuating and screening large numbers of civilians. Requirements for medical support, basic life support, and, if necessary, decontamination may quickly overwhelm the Army force's capabilities even with augmentation.
- Similar to natural disasters, panic and disorder may accompany the CBRNE event. Fleeing civilians may clog elements of the transportation and distribution infrastructure. Physical destruction may also affect other components of the infrastructure of critical and immediate concern, such as energy and administration and human services (water, sanitation, medical, fire fighting, and law enforcement). Because all elements of the infrastructure may be affected, the overall recovery time may be lengthened and the effects broadened to include much of the surrounding area. The effects of a single urban CBRNE event potentially could be felt nationally or globally.

Decisive Operations

9-11. In urban stability operations, decisive operations may take many years and include multiple actions before achieving the desired end state. This particularly applies to the strategic and operational levels. Oppositely, decisive operations involved in an urban civil support operation for mitigating or reducing disease, hunger, privation, and the effects of disasters normally achieve faster results. However, any operation that attacks the underlying cause and seeks to prevent or relieve such conditions will usually take longer. In urban areas, establishing law and order to protect critical infrastructure and the inhabitants from lawlessness and violence is often critical and also the decisive operation.

Shaping Operations

9-12. Shaping operations establish and maintain the conditions for executing decisive operations. In urban stability or civil support operations, shaping operations always include information operations (IO) that influence perceptions and maintain legitimacy. Often, various participants, and their potentially divergent missions and methods, are involved. Army commanders must coordinate their planning and efforts (early and continuously) to ensure that their decisive, shaping, or sustaining operations are not working against other agencies' efforts and operations—agencies that may have the lead role in the operation. Thus, a critical shaping operation may be to establish the coordination to help develop a common purpose and direction among agencies, particularly those that may experience continuous personnel turnover during the conduct of a lengthy operation. In some instances and with some organizations and agencies, particularly nongovernmental organizations (NGOs), genuine unity of effort may not be achievable; however, recognizing the differences in aims and goals will allow Army commanders to conduct operations with less friction. Commanders should actively request and include NGOs and appropriate governmental agencies in mission readiness exercises or any other training for stability or civil support operations.

SUSTAINING OPERATIONS

9-13. Sustaining operations enable decisive and shaping operations and include logistics, rear area and base security, movement control, terrain management, and infrastructure development. Sustainment bases, especially those located in urban areas, become an attractive target for hostile civilians; therefore, commanders must actively and aggressively protect these bases as well as lines of communications (see Chapter 10).

CONSIDERATIONS FOR URBAN STABILITY AND CIVIL SUPPORT OPERATIONS

In wars of intervention the essentials are to secure a general who is both a statesman and a soldier; to have clear stipulations with the allies as to the part to be taken by each in the principal operations; finally, to agree upon an objective point which shall be in harmony with the common interests.

Lieutenant General Antoine-Henri,
Baron de Jomini

9-14. The urban operational framework (understand, shape, engage, consolidate, and transition) provides a structure for developing considerations unique to urban stability and civil support operations. Many considerations presented in urban offensive and defensive operations apply to urban stability and civil support operations, particularly those that address how to understand the urban and overall operational environment. Because the situations in which stability and civil support operations normally occur share strong similarities with any urban environment, many of these considerations are closely linked to the urban fundamentals presented in Chapter 5. Taken together, commanders will often find them useful in conducting UO throughout full spectrum operations. Appendix C has a historical example of how to apply the urban operational framework to a stability operation.

UNDERSTAND

9-15. In urban stability or civil support operations, commanders must carefully assess and understand the political dimension of the operational environment, as well as their role and the media's part in managing information. These operations are inherently tied to the exercise of diplomatic power. All operations in urban areas are often the focus of the media and thus gain considerable public and political attention. Therefore, military objectives in urban stability or civil support operations are more directly linked with political objectives. The relationship between the levels of war—strategic, operational, and tactical—is often closer than in urban offensive and defensive operations. Military objectives must be carefully nested within political objectives. Commanders must ensure that the ways and means to accomplish their objectives, to include security and force protection measures, will hold up to media scrutiny and are appropriate for the situation and environment. All levels of command must understand the link between political and military objectives, to include a basic understanding at the Soldier level. One uncoordinated, undisciplined, or inappropriate action, even at the lowest level, could negate months or years of previous, disciplined effort. Commanders must balance security and force protection measures with mission accomplishment. Ineffective measures can put Soldiers at too great a risk and jeopardize the mission. Conversely, overly stringent measures may make it difficult for forces to interact with the population closely—essential in many of these operations. Finally, commanders will need a thorough assessment of the governmental and nongovernmental organizations and agencies that will be operating in or near urban areas that fall within their area of operations.

Political and Military Objectives

9-16. Commanders must translate political objectives into military objectives that are clear and achievable (clear tasks and purposes) and can lead to the desired end state. (In most stability operations, defeat of a threat is not the ultimate desired end state.) Political objectives may be vague making it difficult for commanders to conduct their mission analysis. This applies to tactical- and even operational-level commanders, unskilled at higher level, strategic political-military assessments. Each type of stability operation or civil support operation is distinct, often unfamiliar to the executing unit, and unique to the specific situation. These factors often make it difficult to confidently determine the specific tasks that will lead to mission success. Therefore, commanders must also establish measures of effectiveness that aid in understanding and evaluating progress and help gauge mission accomplishment.

Note: Commanders should consult the US Agency for International Development's (USAID) Field Operations Guide for Disaster Assessment and Response when conducting their assessments and developing measures of effectiveness for many urban relief operations.

9-17. These measures should be quantifiable (in some circumstances, a qualitative assessment may be most appropriate) and link cause with effect. They help determine the changes required and are essential to the assessment cycle required for urban stability or civil support operations. In a humanitarian relief operation to aid the starving, commanders could determine that the decisive effort is delivering safe food to the urban area. To judge success or effectiveness, they could determine that the appropriate measure is the number of food trucks dispatched daily to each distribution site; the more trucks, the more effective the efforts. However, this measure must correlate with the overarching measure of effectiveness: decline in the mortality rate. If no significant decrease in deaths due to starvation occurs, they may need to reassess and modify the tasks or measures of effectiveness. A better measure may be to track the amount of food consumed by those in need instead of simply counting the number of trucks dispatched. Areas around which measures of effectiveness can be formed for many stability or civil support operations (including the example above) and which will help return most societies to some degree of normalcy and self-sufficiency include:

- Restoring law and order.
- Decreasing morbidity and mortality rates.
- Securing safe food and water.
- Restoring critical infrastructure.
- Resettling the population.
- Reestablishing economic activity.

Commanders can often develop measures of effectiveness to address these areas in terms of establishing or restoring security and providing logistics. (Figure 9-2 provides example measures of effectiveness from the strategic to tactical levels for a possible stability operation.)

9-18. Political objectives are fluid and modified in response to new domestic and international events or circumstances. Thus, assessment is continuous, and commanders must adjust their own objectives and subsequent missions accordingly. In urban stability or civil support operations, commanders often develop military objectives that support or align with the objectives of another agency that has overall responsibility for the urban operation. In this supporting role, commanders may receive numerous requests for Soldier and materiel assistance from the supported agency and other supporting agencies operating in the urban area (to include elements of the urban population). With such unclear lines of authority and areas of responsibility, they must ensure that the tasks, missions, or requested Army resources fall clearly in the intended scope and purpose of the Army's participation in the operation. They must not develop or execute missions based on inadequate or false assumptions, misinterpreted intent, or well-meaning but erroneously interpreted laws or regulations by any organization, to include even the lead agency. When missions appear outside their scope, commanders must quickly relay their assessment to their higher headquarters for immediate resolution. The commander's goal is not to limit or slow military participation but to contribute as intended and in consonance with political objectives and the law.

Security and Force Protection Measures

9-19. Commanders must plan for and continually assess the security of their forces operating in an urban area as well as constantly review protection measures. Establishing a robust intelligence—particularly human intelligence (HUMINT)—network that can determine the intentions and capabilities of the threat and the urban populace is the basis for establishing force protection for Army forces operating in the urban environment. However, many such operations, particularly stability operations, require extra time to forge a lasting change. Over time, and particularly in peacetime when objectives center on helping others and avoiding violence, even the complex urban environment may seem benign. Without continued, aggressive command emphasis, Soldiers may become lulled into complacency. Also, during periods of transition or transfer of authority from one unit or organization to another, departing Soldiers' focus often shifts to redeployment activities and away from force protection concerns. It is usually then that Army forces are most vulnerable to terrorist and insurgent tactics, such as bombings, kidnappings, ambushes, raids, and other forms of urban violence.

Measures of Effectiveness: The Need to Measure Progress

Strategic Level Criteria	Operational Level Criteria	Tactical Level Criteria
Accountable to the American people for defining and measuring progress towards defeating terrorism and meeting national security goals. Examples: • Prevention of the insurgency from receiving aid or re-sources from other international groups. • A functioning national government. • Amount of international support and aid to reconstruction. • Number of nations contributing manpower to coalition forces.	Accountable to the strategic level for measuring operational success and providing linkage to strategic goals. Examples: • Host-nation security forces trained and equipped. • Denial of the merging of insurgent forces with terrorist groups. • Amount and distribution of— • Electricity. • Liquid propane gas. • Gasoline. • Functioning provincial governments.	Responsible to the operational level for measuring tactical success and providing linkage to operational goals. Examples: • Number of insurgent forces in the AO neutralized. • Reduced attacks on coalition forces in the AO. • Reduced civilian-on-civilian violence in the AO. • Host-nation security force recruitment goals met. • Host-nation security force training goals met. • Number of reliable human intelligence walk-ins. • Amount of unexploded explosive ordnance cleared. • Functioning neighborhood and district advisory councils.

Figure 9-2. Example measures of effectiveness

9-20. Although force protection will not ensure successful urban stability or civil support operations, improper assessment and inadequate force protection measures can cause the operation to fail. In either operation, keeping a neutral attitude toward all elements of the urban population, while maintaining the appropriate defensive posture, enhances security. One threat principle discussed in Chapter 3 was that threats would seek to cause politically unacceptable casualties. An improper threat assessment and a lapse in security at the tactical level could result in casualties. That result could affect strategy by influencing domestic popular support and subsequently national leadership decisions and policy.

9-21. Emphasizing security and force protection measures does not mean isolating Soldiers from contact with the urban population. On the contrary, commanders should balance survivability with mobility according to the factors of METT-TC—mission, enemy, terrain and weather, troops and support available, time available, civil considerations. Survivability measures—such as sandbagging, hardening, or fortifying buildings and installations, particularly where large numbers of Soldiers are fed and billeted—will be necessary and require considerable command attention. On the other hand, mobility operations are essential in preserving freedom of action and denying a threat the opportunity to observe, plan, and attack urban forces. Continual Army presence in the urban area may provide the urban population a sense of security and allow Soldiers to develop a detailed knowledge of the "patterns of life" in their assigned area of operation (AO). Armed with this knowledge, they can detect the absence of the normal or the presence of the abnormal that might indicate a potential threat. Overall, mission degradation and increased risk to the force can result if force protection measures prevent Army forces from conducting prudent missions and establishing an active and capable presence.

Assessment of Security and Force Protection: Belfast, Northern Ireland

Since 1969, Belfast has significantly affected the British military campaign for stabilizing the area. British operations in Belfast illustrate the difficulty of balancing the security and protection of forces with maintaining the stabilizing presence necessary to uphold law and order, minimize violence, and control the urban population.

British successes in protecting Belfast's infrastructure and government facilities from terrorist attacks compelled various terrorist cells, especially the Irish Republican Army and the Provisional Irish Republican Army, to attack more military targets.

At the time, British soldiers and bases presented relatively unprotected targets to these factions, and attacks against them solidified their legitimacy as an "army." In response, British commanders implemented extreme security and force protection measures—from ballistic protection vests and helmets to fortress-like operational bases and large unit patrols. These protection measures successfully decreased the violence against British soldiers in Belfast. However, they also decreased the soldiers' interaction with the population and their ability to stabilize the city. The large patrols protected the soldiers, but also inhibited effective saturation of neighborhoods. These patrols, coupled with fortress-like bases and bulky protective clothing, created an "us-versus-them" mentality among civilians and soldiers.

As force protection increased and stabilizing effects decreased, the terrorists were provided more targets of opportunity among the civilians and infrastructure. British commanders reassessed the situation, identified this "see-saw" effect, and adapted to strike a better balance between force protection and effective presence patrols. For example, British forces switched to four-man patrols to enable greater mobility and wore berets instead of helmets to appear less aggressive.

While the above historical example illustrates a critical aspect of stability and civil support operations, it subtly reveals another important lesson for Army commanders. Many of our future coalition partners and assisting governmental and nongovernmental organizations will bring key insights to UO (particularly dealing with

> urban societies) based on their historical and operational experience. Army commanders and their staffs should seek out and remain open to others' unique understandings that are based on hard-earned and well-analyzed lessons learned.

Participating Organizations and Agencies

9-22. Across the spectrum of urban operations, but more so in these operations, numerous NGOs may be involved in relieving adverse humanitarian conditions. Dense populations and infrastructure make an urban area a likely headquarters location for them. (In 1994 during OPERATION UPHOLD DEMOCRACY, for example, over 400 civilian agencies and relief organizations were operating in Haiti.) Therefore, commanders should assess all significant NGOs and governmental agencies operating (or likely to operate) in or near the urban area to include their—

- Functions, purposes, or agendas.
- Known headquarters and operating locations.
- Leadership or senior points of contact (including telephone numbers).
- Communications capabilities.
- Potential as a source for critical information.
- Financial abilities and constraints.
- Logistic resources: transportation, energy and fuel, food and water, clothing and shelter, and emergency medical and health care services.
- Law enforcement, fire fighting, and search and rescue capabilities.
- Refugee services.
- Engineering and construction capabilities.
- Other unique capabilities or expertise.
- Previous military, multinational, and interagency coordination experience and training.
- Rapport with the urban population.
- Relationship with the media.
- Biases or prejudices (especially towards participating U.S. or coalition forces, other civilian organizations, or elements of the urban society).

Commanders can then seek to determine the resources and capabilities that these organizations may bring and the possible problem areas to include resources or assistance they will likely need or request from Army forces. These organizations will be critical to meeting the population's immediate needs and minimizing the effects of collateral damage or disaster. However, commanders should consider whether a close relationship with any of these organizations will compromise the organization's appearance of neutrality (particularly threat perceptions) and adversely affect their ability to assist the population.

SHAPE

9-23. Commanders conduct many activities to shape the conditions for successful decisive operations. In urban stability and civil support operations, two activities rise to the forefront of importance: aggressive IO and security operations.

Vigorous Information Operations

9-24. IO, particularly psychological operations (PSYOP) and the related activities of civil affairs (CA) and public affairs, are essential to shape the urban environment for the successful conduct of stability or civil support operations. Vigorous IO can influence the perceptions, decisions, and will of the threat, the urban population, and other groups in support of the commander's mission. IO objectives are translated to IO tasks that are then executed to create the commander's desired effects in shaping the battlefield. These operations can isolate an urban threat from his sources of support; neutralize hostile urban populations or gain the support of neutral populations; and mitigate the effects of threat IO, misinformation, rumors, confusion, and apprehension. Developing effective measures of effectiveness is essential to a good IO campaign strategy. One of the most valuable methods for obtaining data for use in this process is face-to-face encounters with targeted audiences by unit patrols and HUMINT, PSYOP, and CA teams. A valuable technique may be to conduct periodic, unbiased surveys or opinion polls of the civilian population to determine changes in their perceptions and attitudes.

Security Operations

Protecting Civilians and Critical Infrastructure

9-25. Security for NGOs and civilians may also be an important shaping operation, particularly for stability and civil support operations. Commanders may need to provide security to civil agencies and NGOs located near or operating in the urban area so that these agencies can focus their relief efforts directly to the emergency. Commanders may also need to protect the urban population and critical infrastructure to maintain law and order if the urban area's security or police forces are nonexistent or incapacitated, or the urban area's security situation has undergone drastic change (such as the result of a natural disaster) and security or police forces require additional augmentation. (See also the discussion of the legal aspects of Civilians Accompanying the Force in Chapter 10.)

Preserving Resources

9-26. Just as forces may be at risk during urban stability or civil support operations, so may their resources. In urban areas of great need, supplies and equipment are extremely valuable. Criminal elements, insurgent forces, and people in need may try to steal weapons, ammunition, food, construction material, medical supplies, and fuel. Protecting these resources may become a critical shaping operation. Otherwise, Army forces and supporting agencies may lack the resources to accomplish their primary objectives or overall mission.

Prioritize Resources and Efforts

9-27. During UO, commanders will always face limited resources with which to shape the battlefield, conduct their decisive operations, and accomplish their objectives. They must continually prioritize, allocate, and apply those resources to achieve the desired end state. To this end, they may develop an order of merit list for proposed projects and constantly update it over time. To some degree, the local urban population will usually need to be protected and sustained by Army forces. Hence, commanders must tailor their objectives and shape their operations to achieve the greatest good for the largest number. Commanders first apply the urban fundamental of preserving critical infrastructure to reduce the disruption to the residents' health and welfare. Second, they apply the urban fundamental of restoring essential services, which includes prioritizing their efforts to provide vital services for the greatest number of inhabitants possible. In operations that include efforts to alleviate human suffering, the criticism for any participating organization is likely to be there is not enough being done or Army forces are not being responsive enough. Therefore, commanders must develop clear measures of effectiveness not only to determine necessary improvements to operational plans but also to demonstrate their Soldiers' hard work and sacrifice and U.S. commitment to the operation.

Improve the Urban Economy

9-28. When conducting reconstruction and infrastructure repair, commanders should consider using such activities to simultaneously improve the urban economy. Hiring local civilians and organizations to do reconstruction work helps satisfies urban job requirements, may inspire critical elements of the urban society to assume responsibility for the success or failure of urban restoration efforts, and may potentially reduce threat influence by diminishing their civilian sources of aid. Hiring indigenous personnel for short-term projects does not replace the need for long-term economic planning and the development of stable jobs. The overall reconstruction effort must be guided by the commander's vision of the end state. However, the commander's guidance and intent should be broad and expansive enough to allow responsive decision making by subordinate commanders based on their analysis of the urban population's needs in their assigned AO.

ENGAGE

If there is any lesson to be derived from the work of the regular troops in San Francisco, it is that nothing can take the place of training and discipline, and that self-control and patience are as important as courage.

Brigadier General Frederick Funston
commenting on the Army's assistance following
the 1906 San Francisco earthquake and fire

9-29. The focus of the Army is warfighting. Therefore, when Army commanders conduct many urban stability and civil support operations, they must adjust their concept of success. Commanders will most often find themselves in a supporting role and less often responsible for conducting decisive operations. They must accept this supporting function and capitalize on the professional values they have instilled in each Soldier, particularly the sense of duty to do what needs to be done despite difficulty, danger, and personal hardship. Commanders must also put accomplishing the overall mission ahead of individual desires to take the lead—desire often fulfilled by being the supported rather than supporting commander. In many stability and civil support operations, success may be described as settlement and compromise rather than victory. Yet, the Army's professionalism and values—combined with inherent adaptability, aggressive coordination, perseverance, and reasonable restraint—will allow Army forces to engage purposefully and dominate during complex urban stability or civil support operations.

Adaptability

9-30. Adaptability is particularly critical to urban stability or civil support operations because these operations relentlessly present complex challenges to commanders for which no prescribed solutions exist. Commanders often lack the experience and training that provide the basis for creating the unique solutions required for these operations. Since the primary purpose for the Army is to fight and win the nation's wars, the challenge then is to adapt urban warfighting skills to the unique stability or civil support situation.

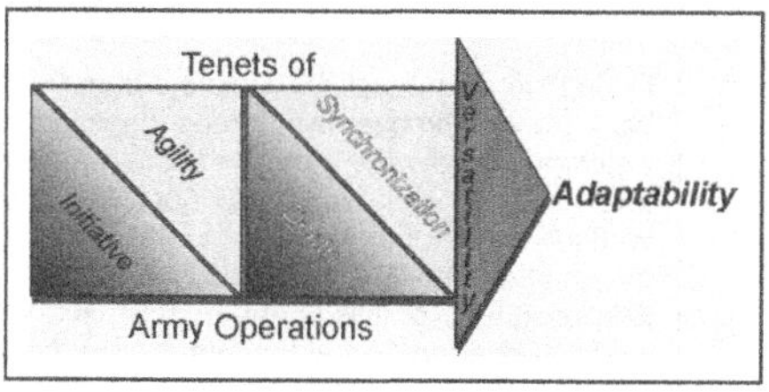

Figure 9-3. Adaptability

9-31. Doctrine (joint and Army) provides an inherent cohesion among the leaders of the Army and other services. Still, Army commanders conducting urban stability or civil support operations will often work with and support other agencies that have dissimilar purposes, methods, and professional languages. Army commanders must then capitalize on three of the five doctrinal tenets of Army operations: initiative, agility, and versatility (see figure 9-3 and FM 3-0). Commanders must bend as each situation and the urban environment demand without losing their orientation. They must thoroughly embrace the mission command philosophy of command and control addressed in FM 6-0 and Chapter 5 to encourage and allow subordinates to exercise creative and critical thinking required for planning and executing these UO. Commanders must also be alert to and recognize "good ideas," and effective tactics, techniques, and procedures, regardless of their source—other services, coalition partners, governmental and nongovernmental organizations, and even the threat—and adapt them for their own purposes in UO.

9-32. Adaptability also springs from detailed planning that carefully considers and realistically accounts for the extent of stability and civil support operations. Although no plan can account for every contingency and completely eliminate the unexpected, good plans—which include detailed civil considerations—provide platforms from which to adjust course more readily. Adequate planning allows commanders not only to react more effectively, but also to be forward-thinking and take actions that favorably steer the course of events.

Aggressive Coordination and Synchronization

9-33. In urban stability and civil support operations, the increased number of participants (both military and nonmilitary) and divergent missions and methods create a significant coordination and synchronization challenge. Significant potential for duplicated effort and working at cross-purposes exists. The success of UO often depends on establishing a successful working relationship with all groups operating in the urban area. The absence of unity of command among civil and military organizations does not prevent commanders from influencing other participants not under his direct command through persuasion, good leadership, and innovative ideas.

9-34. Commanders may consider establishing, as necessary, separate organizations for combat operations and for stability operations or civil support operations to increase coordination and enhance local, NGO, and international support. Further, aligning the unit or subordinate units with NGOs may contribute to establishing popular legitimacy for the operation and place greater pressure on threat forces. In some instances, commanders may consider organizing part of their staff around government, administrative, and infrastructure functions that mirror the urban area in which their forces are operating. Development of a mirror urban area organization may give greater legitimacy to the urban government or administration and ease transition of responsibility once the end state is achieved. Commanders must be mindful, however, that local groups seen as allying themselves with Army or coalition authorities will likely experience pressure to demonstrate their independence as established dates for redeployment or other critical events approach. In some instances that demonstration of independence may be violent.

Civil Support and Coordination with Civilian Authorities: Los Angeles – 1992

During the spring of 1992, Soldiers from the 40th Infantry Division, California National Guard were among the forces deployed to Los Angeles County to assist the California Highway Patrol, Los Angeles County Sheriffs, and civilian law enforcement. They worked to quell the riots that were sparked by the "not guilty" verdicts concerning four police officers who, following a lengthy high-speed chase through Los Angeles, were accused of brutally beating Rodney King.

Successful accomplishment of this civil support operation was attributed to the exercise of strong Army leadership and judgment at lower tactical levels, particularly among the unit's noncommissioned officers. An essential component of combat power, it was especially critical in executing noncontiguous and decentralized operations in the compartmented terrain of Los Angeles. As important, however, was the clear understanding that Army forces were to support civilian law enforcement—and not the other way around. The 40th Infantry Division aligned its area of operations with local law enforcement boundaries and relied heavily on police recommendations for the level at which Soldiers be armed (the need for magazines to be locked in weapons or rounds chambered).

One incident emphasized the need for coordination of command and control measures with civilian agencies even at the lowest tactical levels. To civilian law enforcement and Army forces, the command "Cover me" was interpreted the same: be prepared to shoot if necessary. However, when a police officer responding to a complaint of domestic abuse issued that command to an accompanying squad of Marines, they responded by immediately providing a supporting base of fire that narrowly missed some children at home. However, the Marines responded as they had been trained. This command meant something entirely different to them than for Army Soldiers and civilian law enforcement. Again, coordination at all levels is critical to the success of the operation (see also the vignette in Appendix B).

9-35. In the constraints imposed by METT-TC and operations security (OPSEC), commanders should seek to coordinate all tactical stability operations with other agencies and forces that share the urban environment. Importantly, they should seek to coordinate appropriate information and intelligence sharing with participating organizations. Commanders must strive to overcome difficulties, such as mutual suspicion, different values and motivations, and varying methods of organization and execution. Frequently, they must initiate cooperative efforts with participating civilian agencies and determine where their objectives and plans complement or conflict with those agencies. Commanders can then match Army force capabilities to the needs of the supported agencies. In situations leading to many urban civil support operations, confusion may initially make it difficult to ascertain specific priority requirements. Reconnaissance and liaison elements—heavily weighted with CA, engineers, and health support personnel—may need to be deployed immediately to determine what type of support Army forces provide. Overall, aggressive coordination will foster trust and make unity of effort possible in urban stability or civil support operations where unity of command is difficult or impossible to achieve.

Perseverance

9-36. The society is a major factor responsible for increasing the overall duration of urban operations. This particularly applies to urban stability operations where success often depends on changing people's fundamental beliefs and subsequent actions. Modifying behavior requires influence, sometimes with coercion or control, and perseverance. The urban population often must be convinced or persuaded to accept change. Necessary change may take as long as or longer than the evolution of the conflict. Decades of problems and their consequences cannot be immediately corrected. Frequently, the affected segments of the urban society must see that change is lasting and basic problems are being effectively addressed.

9-37. In most stability operations, success will not occur unless the host nation, not Army forces, ultimately prevails. The host urban administration must address the underlying problem or revise its policies toward the disaffected portions of the urban population. Otherwise, apparent successes will be short lived. The UO fundamental of understanding the human dimension is of paramount importance in applying this consideration. After all Army forces, particularly commanders and staff of major operations, understand the society's history and culture, they can begin to accurately identify the problem, understand root causes, quickly engage and assist key civilian leadership, and, overall, plan and execute successful Army UO.

Reasonable Restraint

9-38. Unlike offensive and defensive operations where commanders seek to apply overwhelming combat power at decisive points, restraint is more essential to success in urban stability and civil support operations. It involves employing combat power selectively, discriminately, and precisely (yet still at decisive points) in accordance with assigned missions and prescribed legal and policy limitations. Similar to the UO fundamentals of minimizing collateral damage and preserving critical infrastructure, restraint entails restrictions on using force. Commanders of major operations should issue or supplement ROE to guide the tactical application of combat power. Excessively or arbitrarily using force is never justified or tolerated by Army forces. Even unintentionally injuring or killing inhabitants and inadvertently destroying their property and infrastructure lessens legitimacy and the urban population's sympathy and support. Collateral damage may even cause some inhabitants to become hostile. In urban stability or civil support operations, even force against a violent opponent is minimized. Undue force often leads to commanders applying ever-increasing force to achieve the same results.

9-39. Although restraint is essential, Army forces, primarily during urban stability operations, must always be capable of decisive (albeit in some circumstances, limited) combat operations. This is in accordance with the UO fundamental of maintaining a close combat capability. This capability must be present and visible, yet displayed in a nonthreatening manner. A commander's intent normally includes demonstrating strength and resolve without provoking an unintended response. Army forces must be capable of moving quickly through the urban area and available on short notice. When necessary, Army forces must be prepared to apply combat power rapidly, forcefully, and decisively to prevent, end, or deter urban confrontations. Keeping this deterrent viable requires readiness, constant training, and rehearsals. It also requires active reconnaissance, superb OPSEC, a combined arms team, and timely and accurate intelligence.

Restraint: An Najaf, Iraq – 2003

On 31 March 2003, during OPERATION IRAQI FREEDOM, The 101st Airborne (Air Assault) Division transitioned from isolating An Najaf—among the holiest places for Shia Muslims in Iraq and the home of one of their leading holy men, the Grand Ayatollah Ali Hussein Sistani—to an attack to clear Iraqi fighters from the town. The decision to attack was made in response to increasing attacks against 101st forces situated near An Najaf and due to alarming reports that the Fedayeen were killing Iraqi family members to force adult males to fight coalition forces. Since An Najaf had an airfield, control of the town would also allow the 101st to obtain needed hardstands for their aircraft.

Ultimately, the division used two brigades to attack and clear the town. The 2nd Brigade Combat Team (BCT) attacked from the north and the 1st BCT—responsible for the initial penetration—attacked from the southwest. Using tanks, infantry fighting vehicles, and light infantry, the commander of the 1st BCT formed effective combined arms teams supported by artillery and air that successfully fought their way into the city.

On 3 April 2003, Soldiers of TF 2-327 Infantry turned a street corner to face a group of civilian men blocking their way and shouting in Arabic, "God is great." The crowd quickly grew into hundreds as they mistakenly thought the Soldiers were trying to capture Sistani and seize the Imam Ali Mosque. Someone in the crowd threw a rock at the Soldiers, which started a hailstorm of rocks; even the battalion commander was hit on the head, chest, and the corner of his sunglasses.

Based on information from a Free-Iraqi Fighter accompanying the battalion and his own observations during the attack, the commander of TF 2-327 believed that most of the people in An Najaf neither wanted to fight nor obstruct his unit's efforts. Consequently, he ordered his Soldiers to position themselves on one knee, smile, and point their weapons toward the ground. At this gesture, many of the Iraqis backed off and sat down, enabling the commander to identify the true troublemakers. He identified eight. In case these agitators were to produce weapons and start to shoot, the commander wanted to make sure that the remainder of the crowd would know from where the shooting would originate. Next, he ordered his Soldiers to withdraw to allow the tension to subside. With his own rifle pointed toward the ground, the battalion commander bowed to the crowd and led his Soldiers away. When tempers had calmed, the Grand Ayatollah Sistani issued a decree (fatwa) calling on the people of Najaf to welcome the American Soldiers.

The attack on An Najaf by the 101st provides an excellent example of well-led Soldiers capable of understanding cultural differences and possessing the discipline and mental flexibility to transition from the aggressive mindset required for high-intensity urban combat to the restraint essential for the initiation of stability operations. In his weekly radio address, President George W. Bush commented: "This gesture of respect helped defuse a dangerous situation and made our peaceful intentions clear."

CONSOLIDATE

9-40. As urban operations will often be full spectrum, many of the consolidation activities necessary to secure gains in urban offensive and defensive operations will be applicable to urban stability and civil support operations. However, the greatest obstacles to attaining strategic objectives will come after major urban combat operations. Therefore, emphasis will appropriately shift from actions to ensure the defeat of threat forces to those measures that address the needs of the urban population, manage their perceptions, and allow responsibility to shift from Army forces to legitimate indigenous civilian control (or the intermediate step to other military forces, governmental agencies, and organizations).

Continued Civilian and Infrastructure Protection

9-41. Following urban offensive or defensive operations, there will likely be a need to secure and protect the civilian population and much of the civilian infrastructure from the civilians themselves. After having minimized collateral damage and preserved critical infrastructure, commanders must implement measures to preclude looting and destruction of critical and essential infrastructure by the urban population and civilian-on-civilian violence. This may be as relatively simple as allowing the urban police force to return to work or may be as difficult as hiring, vetting, and training an indigenous police force. In the latter case commanders may need to determine—

- Number and operability of police stations.
- Responsibility for recruiting, hiring, training, and equipping the urban security or police force. The accountable unit, organization, or agency will need to consider a vetting process, suitable salaries and wages, and appropriate training standards.
- The appropriate responses toward those civilians who threaten, oppose, or harm the new police force.

9-42. Alternatively, civilian security firms (from inside or outside the urban area or country) can provide supplemental protection until indigenous police forces are fully functional. Further, the commander of the major urban operation may plan to manage expected instability primarily with Army forces. Often, this will require larger numbers of infantry, military police, and dismounted forces. Other populace and resource control measures such as curfews may also assist in civilian (including NGOs) and infrastructure protection. Previous shaping operations aimed at improving the local economy can also assist in this regard.

Resolute Legitimacy

9-43. Closely linked to the restraint described previously is legitimacy or the proper exercise of authority for reasonable purposes. Achieving or maintaining legitimacy during urban stability or civil support operations is essential in gaining and maintaining the support of the urban population. Commanders can ensure legitimacy by building consent among the population, projecting a credible force, and appropriately using that force. Perceptions play a key role in legitimacy, and skillful IO can shape perceptions. Commanders must send messages that are consistent with the actions of their forces. Generally, the urban population will accept violence for proper purposes if that force is used impartially. Perceptions that force is excessive or that certain groups are being favored over others can erode legitimacy and generate resentment, resistance, and, in some situations, violent acts of revenge.

9-44. In stability operations against an elusive insurgent threat, commanders may need to explain to urban residents why damage was necessary, apologize, or make near-instant restitution for some unsuccessful Army operations that may have been planned based on inaccurate or incomplete intelligence. (Soldiers may even make it a point to thank homeowners for allowing the search of their homes during cordon and search operations.) During urban operations, a single Soldier's misbehavior can significantly degrade a commander's ability to project an image of impartiality and legitimacy. Fortunately, disciplined Soldiers can contribute immeasurably to gaining and maintaining legitimacy, mitigating ill will, or otherwise winning the urban population's trust and confidence. In stability operations, the greater fight will often be to win the battle of perceptions and ideas instead of one to seize terrain and triumph over an enemy. Inconsistencies in message and behavior will provide threats with raw material for their propaganda and precipitate doubt in the minds of the urban populace who might otherwise support Army objectives.

Transition

9-45. Commanders of major operations are the focal point for synchronizing tactical stability operations or civil support operations with strategic diplomatic and political issues. They are also the critical links between national intelligence resources and the tactical commander. Because strategic, diplomatic, and political changes can quickly transition the type of urban operation, they must keep subordinate tactical commanders abreast of changes in intelligence, policy, and higher decisions. The potential to rapidly transition to urban combat operations emphasizes the need to maintain the capability to conduct close, urban combat. Failure to recognize changes and transition points may lead to UO that do not support the attainment of the overall objective and needlessly use resources, particularly Soldiers' lives. Therefore, Army forces on the ground in an urban stability operation must be more aware of the strategic environment than the threat and the civilian population, each of whom will have their own means of monitoring the national and international situation.

Legitimate and Capable Civilian Control

I met immediately with the [special operations forces liaison officer] to discuss how we would establish a new government that would be able to make instant progress. We decided to select a city council with the following departments: commerce, public works, social services, health, emergency services, education, public relations, and agriculture. A city council director would lead the council, and each department would also have a director in charge of its respective services. A critical step in selecting the city council was ensuring that there was a balance on the council between Altun Kupri's three major ethnic groups. An ethnic imbalance on the council or perceived favoritism would have decreased the legitimacy of the Coalition forces' efforts in the city and served as fuel for the fire of competition for power between the Turkomen and Kurds.

Captain Jeffrey B. Van Sickle
"Stability Operations in Northern Iraq"

9-46. Commanders must maintain or enhance the credibility and legitimacy of the government and police of the urban area and, in the case of stability operations, of the host nation's military forces operating there. In accordance with the urban fundamental of transitioning control, urban commanders must conclude UO quickly and successfully, often to use assets elsewhere in their area of operations. This entails returning the control of the urban area back to civilian responsibility as soon as feasible. The host nation's military and the urban area's leadership and police are integrated into all aspects of the urban stability or civil support operations to maintain their legitimacy. They must be allowed (or influenced) to take the lead in developing and implementing solutions to their own problems. This requires commanders to transition from "leading from the front" to "leading from behind" in an advisory and assistance role. Effective transition to civilian control and responsibility requires commanders at all echelons who understand the basic operation of civil governments and the administration and management of key urban infrastructure.

9-47. If the host nation's leadership, military, and police are not up to the task, commanders can take steps to increase its capabilities through training, advice, and assistance by CA units or by other nongovernmental or governmental organizations and agencies. Sometimes, new leadership and a restructured police force may be required, particularly when corrupt and no longer trusted by the population. This candid assessment of the urban leadership's ability to govern, protect, and support itself is made early in the planning process. Only then can commanders ensure that resources and a well thought-out and coordinated plan (particularly with civilian organizations and agencies) are available for a speedy transition. IO will be paramount in these instances to ensure that the urban population sees the training and rebuilding process itself as legitimate. Throughout urban stability or civil support operations, commanders shape the conditions to successfully hand over all activities to urban civilian authorities.

Longer-Term Commitment

Maintain the Focus

9-48. Many stability operations often require perseverance and a longer-term U.S. commitment requiring operational endurance and the establishment of a sustainable battle rhythm. Lengthy operations will also require a transitional rotation of Army units into the area of operations to continue the mission. Considerations for these transitions are similar to a relief in place and battle handover (see FM 3-90) combined with considerations for deployment and redeployment. FM 41-10 contains comprehensive guidance on transition planning and coordination activities applicable to UO. In addition to any threat considerations, planning for urban transitions between units will often include emphasis and understanding of—

- Formal and informal civilian leadership and relationships.
- Government institutions and administrative functions.
- On-going reconstruction projects.
- The urban economy.
- Participating governmental and nongovernmental organizations and established relationships and cooperation activities, particularly information sharing.
- Significant key events affecting (or likely to affect) operations.
- Significant cultural lessons learned as a result of the outgoing unit's operations.

9-49. The commander of the major operation must ensure that the incoming unit also understands the political and strategic objectives behind the tasks that they will be required to accomplish. Otherwise, the new unit may begin to plan operations that are similar to those conducted by the previous unit without achieving the desired end state or accomplishing the mission.

Transition Trust

9-50. If units are successful in establishing and cultivating relationships with the urban populace and their leadership, and have built an effective level of trust, their anticipated departure from the AO as part of a unit rotation schedule may have detrimental effects on working relationships. As a result, local leaders may develop deep anxieties about the replacement force and their ability to "be as good" as the current command. Therefore, incumbent forces must take steps to identify successful and ongoing efforts and activities in which the new unit can continue to achieve success. In some instances, the current unit may delay a project or activity whose scheduled completion is close to the date of relief or the transfer of authority so as to allow the new unit immediate success. In other instances, projects or activities can be backward planned to ensure that the completion date is scheduled soon after the new unit arrives. The goal for the outgoing unit is to ensure that the new unit is readily accepted and that both units execute a seamless relief in place.

Modify Objectives to Match the Current Environment

9-51. While it is important that incoming commanders create a seamless transition with their outgoing counterparts, they must make full use of the opportunity to review the current political, strategic, and local urban environment to determine potential modifications to logical lines of operations and unit objectives. Otherwise, units may fall into the trap of executing a series of six-month to one-year rotations that do not significantly contribute to solidifying the conditions required to ultimately transition the urban environment back to legitimate civilian control. In operations that involve longer-term commitments, the potential exists for the same unit to conduct two (or more) rotations back into the same AO. Commanders must resist the desire to simply continue with previous lines of operations that, at the time, were deemed successful. The urban environment will change in response to Army actions and this change will necessitate subsequent operational adaptations.

Provide an End State Not an End Date

9-52. Though it may not be feasible, due to political constraints, commanders of major urban operations should fight the public desire to establish a fixed transition timetable. Instead, they should plan urban stability operations based on an attainable end state and not on a specified withdrawal or end date. Progress toward strategic (and operational) goals is susceptible to many changes and delay, particularly in coalitions which will likely be a part of most future urban operations. As such, an end date will only serve to unduly constrain commanders in the attainment of their urban operational objectives. Having set and announced a firm timetable, American and coalition publics will expect withdrawal to commence as scheduled and will not easily be convinced of the need to stay longer even if the situation changes drastically. In general, the public can understand and accept a logical and clearly articulated end state as long as commanders can demonstrate adequate progress based on well developed measures of effectiveness. As important, a timetable allows threats to adjust their plans to the anticipated withdrawal date. Instead of yielding to relentless diplomatic, informational, military, and economic pressure backed by firm political resolve, threats will likely exercise strategic and operational patience.

This page intentionally left blank.

Chapter 10

Urban Sustainment

Sustainment activities exist to enable the Army to initiate and sustain full spectrum operations. Inseparable from decisive and shaping operations, logistics is the central component of sustaining operations and provides the means for commanders to generate and maintain combat power and extend the operational reach of the force. In offensive and defensive operations, sustainment activities are not by themselves likely to be decisive or shaping, yet significantly contribute to those operations. In stability and civil support operations, when the critical objectives may be restoring the infrastructure and the welfare of civilians, sustainment activities will often be decisive. Their success will allow Army forces to dominate this complex environment. However, like all urban operations (UO), sustainment activities affect and are affected by the environment. The urban terrain, infrastructure, and existing resources, coupled with supportive civilians, may facilitate sustainment activities. In contrast, a poorly designed or damaged infrastructure and a hostile population may severely hamper sustainment activities. In the latter case, critical Army resources required elsewhere in the area of operations (AO) may be diverted to repair facilities and control and support the inhabitants of the urban areas.

URBAN SUSTAINMENT CHARACTERISTICS

10-1. Fundamental sustainment characteristics (see figure 10-1) guide prudent sustainment planning regardless of the environment. They provide commanders an excellent framework to analyze and help develop urban sustainment requirements, assess the impact of the environment on the provision of logistics, and gauge the effectiveness of urban sustainment activities.

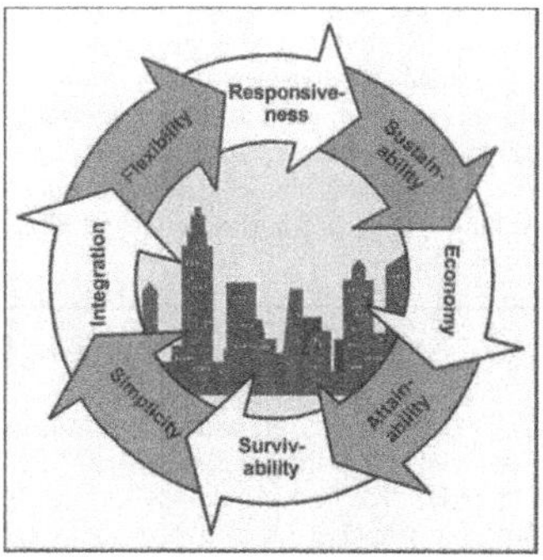

Figure 10-1. Sustainment characteristics

Responsiveness and Sustainability

10-2. UO require responsiveness and sustainability to establish and maintain the tempo necessary for success. Responsiveness—providing the right support in the right place at the right time—is the essential sustainment characteristic. It requires that sustainment commanders and planners accurately forecast urban operational requirements. Continuous urban operations will drain personnel, equipment, and supplies at rates vastly different than in other environments. Therefore, sustainability—the ability to maintain continuous support throughout all phases of the operation—will be a significant concern. Anticipation is critical to both responsiveness and sustainability. It requires that sustainment commanders and planners comprehend the potential effects that components of the urban environment (terrain, infrastructure, and society) may have on operations and sustainment, either benefiting or impeding UO. Effective urban operational and sustainment planning cannot be accomplished separately. Operational and sustainment planners, as well as sustainment operators, must be closely linked to aid in synchronizing and attaining responsiveness and sustainability. The key is often the right balance between efficiency and effectiveness.

Economy and Attainability

10-3. A thoughtful assessment and understanding of the urban environment can also help determine how specific urban areas can contribute to or frustrate the achievement of economy and attainability. Economy is providing the most efficient support at the least cost to accomplish the mission. Attainability means generating the minimum essential supplies and services necessary to begin operations. If available, obtaining support in the AO costs less than purchasing the supplies outside the area and then transporting them there. Critical resources may be available in urban areas to support the operation. However, relying on sources outside the established military sustainment system may create conflict with other sustainment characteristics. A strike by longshoremen, for example, may shut down port operations (at least temporarily) lowering responsiveness and sustainability.

Survivability

10-4. Survivability is being able to protect support functions from destruction or degradation. Commanders often choose to locate sustainment functions in an urban area because the buildings may better protect and conceal equipment, supplies, and people. Urban industrial areas and airports are frequently chosen as support areas because they offer this protection as well as sizeable warehouses, large parking areas, and materials handling equipment (MHE). Such areas facilitate the storage and movement of equipment and supplies. They may also provide readily available water, electricity, and other potentially useful urban resources and infrastructure. However, these areas may also contain toxic industrial materials (TIM) (see Chapter 2). These materials and chemicals in close proximity to support areas may unjustifiably increase the risk to survivability. Sustainment activities in any environment will always be targeted by threat forces. Furthermore, sustainment activities located in any type of confined urban area can offer lucrative targets for terrorists, insurgents, or even angry crowds and mobs. Therefore, no sustainment activity should be considered safe from attack during UO. (During OPERATION JUST CAUSE, Panamanian paramilitary forces and deserters even attacked marked ambulances). Although host-nation support may include assets to assist in defending sustainment units, bases, and lines of communications (LOCs), sustainment commanders must carefully consider if adequate protection measures can ensure survivability. The sustainment commander's greatest challenge to force protection may be complacency born of routine.

> **Base Security: Tan Son Nhut, Vietnam – 1968**
>
> Colonel Nam Truyen was the commander of the 9th Vietcong Division who planned and conducted the attack on the U.S. airbase at Tan Son Nhut during the 1968 Tet Offensive. He had previously entered the airbase during the 1967 Christmas cease-fire using forged identity papers to conduct his own personal and highly thorough reconnaissance.

UO increase the likelihood that the indigenous population will be utilized as a workforce to supplement Army sustainment activities. Hence, strict operations security will be paramount. Commanders will need to carefully screen potential workers to determine their relative trustworthiness. Human intelligence (HUMINT) and counterintelligence (CI) assets will likely be required to assist in this effort. However, higher-level commanders should seek to take on as much of this burden as possible so that lower-level, tactical commanders can use their limited intelligence assets to accomplish other critical intelligence collection and analysis activities.

All commanders with base protection responsibilities should implement random, changing force protection measures that focus on civilians—even those civilians who have, over time, earned a perceived measure of trust. (Civilians can be co-opted and coerced by threat forces.) As such, all civilian activity that seems out of the ordinary should be promptly investigated regardless of whether a specific force protection measure is in effect. Although leaders and Soldiers seek to establish mutual trust and respect between themselves and the local populace, commanders should normally err on the side of protecting their Soldiers.

SIMPLICITY

10-5. Simplicity is required in both planning and executing sustainment activities in this complex environment. Developing standard procedures among the Army, other services, and especially civilian governmental and nongovernmental agencies, liaison and open channels of communication, simple plans and orders, and extensive rehearsals contribute immeasurably to attaining this necessary characteristic.

INTEGRATION

10-6. The need for sustainment integration increases in UO due to its joint and multinational nature and greater numbers of other governmental and nongovernmental organizations (NGOs) and agencies operating in or near urban areas. More NGOs will likely exist because urban areas often contain most of a region's population. Most NGOs focus on people. Army forces and other military and nonmilitary groups must cooperate and coordinate their actions. Much of their coordination will revolve around sustainment activities. Cooperation and coordination will take advantage of each group's sustainment capabilities, help to avoid duplicated effort (contributing to economy), and create sustainment synergy. It will also help to curtail competition for the same urban resources and assist in developing a unified list of priorities. Such coordination will help ensure that other operations by one force or agency will not disrupt or destroy portions of the urban infrastructure critical to another's sustainment operations and the overall mission.

10-7. Success in UO requires combined arms and innovative task organizations. As such, UO will often result in the integration of heavy and light units necessitating a simultaneous integrated approach to sustainment. Gaining units must integrate forces into their maintenance systems and carefully consider potentially increased resource requirements. (For example, armor units consume greater amounts of Class III—bulk and petroleum—than a similar-sized light unit.) When integrating into a light unit, armored and mechanized units should bring as much of their critical parts and supplies as possible and immediately integrate into the gaining unit's logistical reporting system to ensure adequate resupply.

FLEXIBILITY

10-8. Lastly, commanders must develop flexibility. Although they and their staffs must thoroughly understand the urban environment essential to planning sustainment operations, they cannot possibly anticipate every eventuality. Urban commanders must possess the ability to exploit fleeting opportunities. Knowledge of the environment, particularly its infrastructure, can aid in developing innovative solutions to sustainment acquisition and distribution problems. Flexibility enables sustainment personnel to remain adaptive and responsive to the force commander's needs. Key to maintaining flexibility is the constant assessment of the sustainment situation and the readiness to modify or change procedures to adapt to the current conditions confronting Army forces operating within the urban environment.

10-9. Maneuver and sustainment commanders should consider and prioritize these characteristics as they visualize UO. Each characteristic does not affect every operation and urban area in the same way. The sustainment characteristics seldom exert equal influence, and their importance varies according to mission, enemy, terrain and weather, troops and support available, time available, civil considerations (METT-TC). Like the principles of war, commanders must not ignore the potential impact of sustainment characteristics and how their influence changes as the operation evolves (see FM 4-0).

LOGISTICS PREPARATION OF THE THEATER

[Sustainment] planning accounts for increased consumption, increased threats to lines of communications, and anticipated support to noncombatants...Urban operations place a premium on closely coordinated, combined arms teams and carefully protected [sustainment activities]. Urban operations are [logistic]-intensive, demanding large quantities of material and support for military forces and noncombatants displaced by operations.

FM 3-0

10-10. A thorough logistics preparation of the theater (LPT) is critical for an adaptable UO support plan. Sustainment planners conduct the LPT to understand the situation from a sustainment perspective and determine how best to support the maneuver commander's plan. However, decisions that impact the urban environment and the political situation require a combined operational and sustainment perspective. Sustainment planners must understand the urban environment, the fundamentals of UO, the proposed course of action, and the urban environment's effects on sustainment (as well as the other warfighting functions). Such knowledge allows the planners to develop a detailed estimate of support requirements. A thorough LPT helps commanders determine the most effective method of providing adequate, responsive support to meet support estimates while minimizing the sustainment footprint. Overall, it helps tie together UO requirements with acquisition and distribution. As with all operations, but particularly in a dynamic urban environment, this assessment process is continuous since requirements will change as the urban operation unfolds and matures.

SUPPORT TO INTELLIGENCE

10-11. The LPT resembles and runs parallel to the intelligence preparation of the battlefield (IPB). Products generated under IPB may be useful in the sustainment analysis. Conversely, the LPT may contribute to the IPB by identifying critical resources and infrastructure and assessing their potential to influence (positively or negatively) the operational plan. This information may warrant a course of action that includes offensive or defensive operations to seize, secure, or destroy those critical resources. In UO initially planned for other than sustainment reasons, the information may require altering the plan or imposing additional constraints to protect the identified resources. While these resources may or may not be critical to current operations, they are usually important to set or shape the conditions necessary for Army forces to consolidate and transition to subsequent missions or redeploy. This close relationship between IPB and LPT underscores the need to quickly and continuously involve sustainment personnel for their sustainment expertise and perspective in planning UO.

10-12. As sustainment units are among the most dispersed and omnipresent in any AO, sustainment Soldiers are potentially valuable intelligence collectors. Soldiers working in and around urban sustainment bases and executing frequent convoys through urban areas quickly become very familiar with the urban terrain and the routine associated with that terrain. As with all Soldiers, sustainment personnel must be incorporated into a comprehensive human intelligence collection system.

URBAN SUSTAINMENT INFORMATION

10-13. Figure 10-2 illustrates that a thorough analysis of the key components of urban areas in the commander's AO provides the data for an accurate LPT and subsequent UO sustainment plan (see Chapter 2 and Appendix B). Analyzing the urban terrain and infrastructure helps to determine—

- Geographic influences on consumption factors and on the provision of support (weather, climate, and topography).
- The availability of supplies, such as safe food, potable water, petroleum, electrical energy, barrier material, and compatible repair parts.
- The location of facilities, such as warehouses, cold-storage sites, water treatment facilities, manufacturing plants, hospitals, hotels for billeting, and waste treatment facilities.
- Transportation information, such as seaport and harbor facilities, airfields, rail and road networks, traffic flow patterns, choke points, and control problems.
- Locations and accessibility of maintenance facilities and equipment, and machine works for the possible fabrication of parts.
- The available general skills among the urban population, such as linguists, drivers, mechanics, MHE operators, longshoremen, and other vital trade skills.

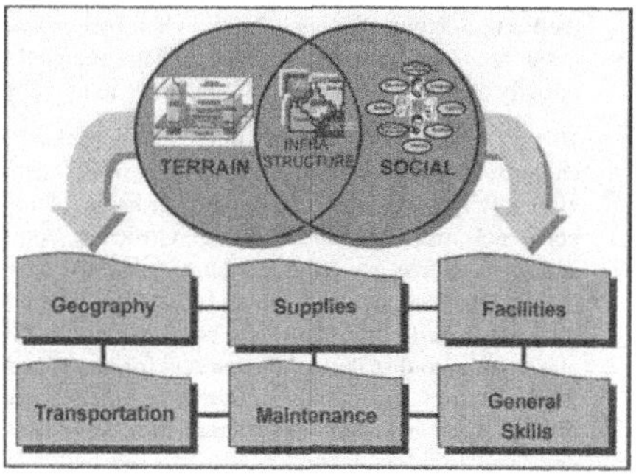

Figure 10-2. Urban environment and essential elements of sustainment information

POTENTIAL RESTRICTIONS

10-14. Commanders must be aware of restrictions that apply to the use of some non-U.S. resources. Security and requirements for U.S. national control dictate that only U.S. assets may perform certain services and functions. Therefore, some foreign urban-area capabilities, even if abundantly available, may not be used. These might include—

- Command and control of medical supply, service, maintenance, replacements, and communications.
- Triage of casualties for evacuation.
- Treatment of chemical, biological, radiological, nuclear, and nuclear (CBRN) casualties, as well as the decontamination of U.S. equipment, personnel, and remains.
- Identification and burial of deceased U.S. personnel.
- Veterinary subsistence inspection.
- Law and order operations over U.S. forces and U.S. military prisoner confinement operations, as well as accountability and security of enemy prisoners of war in U.S. custody.

URBAN SOCIETAL CONSIDERATIONS

10-15. As in all aspects of UO, the urban society is a critical element of the LPT analysis. Sustainment planners must assess whether they can acquire and use urban resources and infrastructure without overly disrupting the urban society and their environment. If the resources are only sufficient for the urban area's inhabitants (and dependent populations in outlying areas), and the facilities cannot increase production to accommodate the needs of Army forces, then commanders may not be able to rely on those resources to support their operations. In fact, the opposite may be true. The effects of UO on the inhabitants, particularly during offensive and defensive operations, may place increased burdens on the Army's resources. For example, commanders may require civilians to evacuate all or parts of an urban area to safeguard them from the effects of planned combat operations. This precautionary measure may require Army forces to support civilian evacuees with temporary housing, food, water, clothing, and medical and sanitary facilities. It will also require carefully planned evacuation routes to ensure that the movement does not interfere with sustainment of the fight.

10-16. Sustainment planners must take into account the potential effects that the provision of supplies to the urban population may have on the operational environment (provided by Army forces or not). For example, an embargo on petroleum products was lifted after Army forces entered Haiti in the 1990s and a less repressive regime assumed governmental control. Few civilian vehicles had moved on the streets of the Port-au-Prince, the capital city, when fuel was scarce and the movement of mounted forces through the urban area was virtually unimpeded. This changed almost overnight once fuel supplies became available, initially making it very difficult for forces to move where required (unless transported by helicopter).

10-17. Sustainment planners must also consider the urban society's ability to restore their own facilities and provide for themselves (if necessary with assistance from Army forces). Throughout this analysis, civil affairs (CA) units can advise and assist in identifying and assessing urban supply systems, services, personnel, resources, and facilities. Critically, commanders must understand that purchasing local goods and services may have the unintended negative consequence of financially sustaining the most disruptive and violent factions in the area. Army forces must seek to purchase urban products and services that will not contribute to prolonging the conflict or crisis. In many stability or civil support operations, they must also attempt to distribute the contracts for goods and services purchased locally as fairly as possible among urban factions and ethnic groups to maintain impartiality and legitimacy. As part of their coordination efforts, commanders should attempt to achieve the cooperation of relief agencies and other NGOs in this endeavor.

10-18. Finally, sustainment planners must also identify potential threats and increased protection requirements that the urban society (criminals, gangs, and riotous mobs) may present, particularly when support units and activities are positioned in urban areas. The disposition or allegiance of the urban population is also important to consider. The infrastructure of an urban area may exhibit great potential to support the sustainment efforts of Army forces, but if the population is hostile or unreliable, use of the facilities and resources may be unfeasible.

Urban Support Areas and Bases

10-19. A major influence on the operation plan and its subsequent execution is often the proper identification and preparation of support areas and bases. The LPT helps commanders determine the need, advantages, and disadvantages of using urban areas in the AO as bases from which to provide support and conduct distribution operations. Ideally, these areas should support reception, staging, onward movement, and integration operations. They should also allow easy sea and air access, offer adequate protection and storage space, increase the throughput of supplies and equipment, and be accessible to multiple LOCs. Consequently, commanders often establish support areas and bases near seaports and airports that are part of a larger urban area. However, threats recognize the Army's need for ports and airfields and may devote substantial resources and combat power to defend them. Army commanders also face the challenge of integrating airfield and port defensive operations with air and naval component commanders to deny threats the ability to conduct stand-off attacks against aircraft and ships. Only an integrated, joint approach can ensure persistent air and sea support– essential not only to sustainment but to firepower as well—is not degraded by the threat. Therefore, planners may determine during the LPT that the risks of seizing or establishing urban lodgment areas may be too high (see Chapter 4). Instead, they may recommend building an airfield, conducting logistics over-the-shore operations, or constructing sustainment bases in more isolated—but more easily secured—locations.

Overall Assessment

10-20. As shown above, the LPT process and analysis help to determine if urban areas in the AO—

- Are suitable as areas and bases for support.
- Can contribute sufficient quantities of and are a dependable source for resources for the overall operation.
- May additionally drain the supported commander's resources.

10-21. The results of this process often serve as a basis for reviewing requirements for civilian contract support and host-nation support and for developing sustainment input into time-phased force and deployment data. This chapter focuses on the effects urban areas may have on accomplishing sustainment functions and related activities, particularly when sustainment units and activities are in urban areas.

ACHIEVING FORCE AGILITY

10-22. To maintain the responsiveness necessary to support full spectrum operations against a wide array of threats and operational environments, sustainment forces must remain agile, possessing the mental and physical agility to transition within or between types of operation with minimal augmentation, no break in contact, and no significant additional training. Responsiveness, flexibility, and economy are key characteristics that enable sustainment forces to support a dynamic combat force; they represent the capability of sustainment forces to provide the resources to initiate, sustain, and extend UO. Agile sustainment forces are characterized by modular force designs, the ability to tailor sustainment organizations for the supporting mission, and the ability to conduct split-based operations. These sustainment aspects are critical to achieving the agility necessary for Army forces to succeed in UO.

MODULAR DESIGN

10-23. Sustainment units are structured as modular organizations, or must possess the ability to modularize in order to conduct decentralized sustainment operations required in UO. At the company level, modularity enables each major company sub-element to assume a cross-section of the company's total capabilities, allowing commanders to employ individual modules to provide a support function, while the rest of the unit remains operational. Modularity enhances responsiveness while enabling support forces to project capabilities as far forward as possible into urban areas, yet with the minimum forward footprint necessary to sustain operations.

FORCE TAILORING

10-24. Sustainment force tailoring refers to the process of determining and deploying the right mix of capabilities to support the force or mission. Sustainment commanders may task organize specific capabilities into functional elements in order to maximize effectiveness and efficiency, and to minimize the sustainment footprint. During UO, the sustainment commander can tailor the support element required to accomplish a specific mission or task, thereby mitigating the risk associated with deploying a larger, more robust capability package forward into the urban area.

SPLIT-BASED OPERATIONS

10-25. Split-based operations refer to performing certain sustainment administrative and management functions outside the area of operations. During UO, Soldiers and civilians can perform personnel, materiel, and distribution management functions without deploying forward into the urban area if the information systems are adequate. This is essential to minimizing risk to support forces and the sustainment footprint in the urban area, and still ensures that all support requirements are met.

SUSTAINMENT FUNCTIONS

- Supply and Field Services
- Maintenance Support
- Transportation Support
- Force Health Protection
- Explosive Ordnance Disposal Support
- Human Resources Support
- Financial Management Support
- Religious Support
- Legal Support to Operations
- General Engineering Support

Figure 10-3. Sustainment functions

SUSTAINMENT FUNCTIONS

10-26. Sustainment consists of eleven functions that must be carefully planned, managed, and synchronized to provide responsive and efficient support to maneuver commanders (see figure 10-3).

Similar to the components of the urban environment, particularly its infrastructure, they overlap and are interdependent. The success of one function depends on the success of several others. Like urban infrastructure, they have two components: a physical component (supplies, equipment, and facilities) and a human component (the Soldiers and civilians who execute these functions). Like city mayors, commanders must plan, manage, and synchronize these functions to provide responsive and efficient sustainment to UO.

10-27. Commanders and planners should consider two essential aspects when addressing these sustainment functions. One aspect looks outward and one looks inward. The first aspect is how these functions can best support full-spectrum UO—the outward analysis. The second aspect is how the urban area affects the conduct of sustainment functions, particularly when those functions are located or performed in an urban area—the inward analysis.

SUPPLY AND FIELD SERVICES

10-28. The supply function involves acquiring, managing, receiving, storing, protecting, maintaining, salvaging, and distributing all classes of supply (except Class VIII) required to equip and sustain Army (and joint) forces. For UO, commanders may need to decide early in the planning cycle whether to stockpile resources forward or to rely on distribution-based sustainment management to satisfy requirements. Some specialized items identified below may not be available through the normal military supply system and may take sustainment personnel much longer to obtain or fabricate. Operation planners must quickly identify the special equipment and increased supply requirements for UO to give sustainment personnel the time necessary to acquire them.

10-29. Field services are those essential services required to enhance a Soldier's quality of life. They consist of clothing exchange, laundry and shower support, textile repair, mortuary affairs, preparation for aerial delivery, food services, billeting, and sanitation. The urban commander determines the need and priority of each service after careful METT-TC analysis. Some facilities such as shower, laundry, and cold storage may be available in the urban area through theater support contracting or provided by external support contracts such as the logistics civilian augmentation program (LOGCAP). Additionally, requirements to care for the urban population will increase requirements for field services immensely. In some circumstances, most notably urban stability or civil support operations, adequately protected field service units or activities will be critical and may be the only support provided.

Increased Urban Supply Requirements

Water

10-30. Water production and distribution will become more difficult for sustainment forces during UO. (Many parts of an urban area can become virtual deserts if public water supplies are cut off or contaminated due to deliberate military action or natural disasters.) Consequently, sustainment managers will be called upon to find more efficient and effective means of moving and distributing potable water, especially when bulk storage of water within the urban area may not be possible or feasible. In sustainment bases that do not have access to urban or other water sources, the transportation and storage of non-potable water may be as great a concern as potable water. Unless adequate storage is available for both potable and non-potable water, units may be forced to rely solely on potable water for all bulk requirements to mitigate health risk. Due to the compartmented nature of many urban operations, planners may need to ensure that additional containers for water (and fuel) are available to support dispersed stock at small-unit level, and that man-portable systems are available in sufficient quantities to ensure timely replenishment.

Class II

10-31. Clothing and individual equipment exposed to the urban environment (concrete, glass, and steel) experience wear and damage at accelerated rates. Units will also require items such as rope, grappling hooks, crowbars, ladders, chain saws, elbow- and kneepads, special vehicle and personnel body armor, fire-fighting equipment, packboards, carpentry tools, and other specialized items to conduct UO. Planners must recognize this and plan accordingly to ensure operations are not interrupted due to equipment shortfalls.

Class III (Bulk)

10-32. With the potential exception of aviation fuel, bulk Class III requirements for UO generally decrease at the maneuver unit level. However, increased fuel requirements for engineer and power-generation equipment attached to or operating with forward units may offset these decreases. Generally, fuel sources within the urban area (refinement facilities, gas stations, garages, and airfields) should not be utilized unless tested and approved for use by a fuel laboratory.

Class IV

10-33. Requirements for construction and barrier material (and the power tools to cut it) during UO will likely far exceed the capability of the sustainment force to provide. Commanders must prioritize distribution, defensive projects, and construction and repair efforts in order to maximize the efficient use of available Class IV. Every effort should be made to obtain supplemental Class IV material from the local economy, as long as that effort does not adversely affect operations or the restoration and improvement of critical urban infrastructure. Of note, specialized, prefabricated road barriers will also be in great demand to block roads, create checkpoints, reinforce defensive positions, and protect headquarters and sustainment activities.

Class V

10-34. Urban combat operations will often increase requirements for Class V. Ammunition consumption rates for urban environments have been as much as five to ten times greater than operations in other environments. Urban offensive and defensive operations will often increase overall requirements for certain types of ammunition (crew-served and small arms ammunition, mortars, antitank rounds, grenades, mines, and demolitions) while others may typically decrease (tank, artillery, attack helicopter). Since forces operating on urban terrain may not be capable of storing ammunition, an established system of distribution is vital to ensuring a constant flow of ammunition necessary to sustain continuous operations and retain the initiative.

Class IX

10-35. Requirements for repair parts generally decrease in UO, with the notable exception of those systems that experience increased ammunition consumption. The heavy use of crew-served and small arms, mortar, and antitank systems in the urban area causes wear and failure at significantly increased rates. Additionally, wheeled-vehicle tires and armored-vehicle track pads and road arms, road wheels, center guides, wedge bolts, and track blocks may wear rapidly on urban streets, rubble, and debris. Based on the tempo of operations, it may be more feasible to replace whole sets of tracks rather than have crews replace individual track pads. Hence, commanders may consider resourcing all wheeled vehicles with spare tires and creating containerized and rapidly transportable sets of track and track parts for speedy delivery to urban sustainment bases. On the whole, planners should consider the use of repair parts packages specifically configured for UO that can be transported by tailored, modular sustainment elements that directly support operations within the urban environment. During long-term stability operations, outgoing units may leave large portions of their prescribed load list in theater to ensure that relieving units have adequate supplies to maintain the tempo of operations.

Storage and Distribution

10-36. UO present challenges to storage and distribution that may not be experienced outside of the urban environment. Forces operating within these areas may not be static and, therefore, cannot store equipment and resources beyond what they can transport organically. Even forces conducting operations from secure locations within the urban area may not be able to store materiel, necessitating a continuous, coordinated distribution system to ensure forces are adequately resourced. During UO, sustainment forces may need to utilize unit or throughput distribution techniques to provide the level of support required to sustain the force.

10-37. Urban commanders must also consider the benefits of stockpiling resources forward against relying on the Army's distribution-based sustainment system. Stockpiling brings supplies close to urban forces and helps ensure supplies are immediately available to support potentially increased consumption rates. However, this method may burden the support structure that moves, handles, and protects large quantities of resources often on a repetitive basis. The Army's normal distribution-based sustainment system reduces this burden significantly; however, available transportation assets (sea, air, and ground) impact delivery response times. These transportation assets often combine military assets overlaid on the host nation's (and urban area's) transportation and distribution infrastructure. Although the civilian infrastructure may initially support the Army's distribution system, later effects of UO, such as destruction of equipment and facilities or loss of civilian workers, may degrade the system. Moreover, Army forces may have to share these assets with other military, multinational force, or civilian organizations participating in the urban operation, as well as with the indigenous civilian population at large. This shared system may put at risk the timely delivery of critical supplies to Army forces.

10-38. Sustainment planners must understand the urban environment and its effects on the proposed method of distribution. They must also understand how urban operations (to include sustainment UO) may affect the urban environment. Storing bulk fuel or ammunition in or near an urban area, for example, may increase the risk of fire and explosive hazard to civilians and Army forces. With this awareness, planners present the commander with an estimate that considers both risks and benefits. Depending on the particular area and other METT-TC factors, they may recommend one method of distribution then transition later to another or a combination of methods. Preconfigured resupply loads delivered as close as possible to where they are needed is vital to responsiveness. Only an agile distribution-based sustainment system will allow Army forces to be responsive and operationally effective across full spectrum operations in an urban environment.

Food Preparation

10-39. The daily feeding standard for Soldiers operating in urban areas remains the same: three quality meals per Soldier. UO creates an increased energy demand on Soldiers, requiring a caloric intake of about 5,000 calories each day. Since producing and delivering prepared meals to forward elements may be difficult or even impossible due to the urban environment, the innovative use of meal supplements can be critical to meeting the dietary requirements of the force. Energy drinks and bars, rich in electrolytes and carbohydrates, are an essential supplement to prepared meals and should be provided in quantities that support or exceed the requirements of the forces operating in the urban area.

10-40. Food (and water) may be available in the urban area; however, local sources must be tested, carefully monitored, and medically approved before consumption. Garbage disposal may be an important consideration in the urban area. Improper trash disposal may leave a signature trail (particularly during urban defensive operations) that may degrade sanitary conditions leading to increased disease and nonbattle injuries (DNBIs). Commanders must understand that food operations, if not properly positioned and secured, can become a focal point for the urban population. Strict policy regarding distribution and control of any Class I supplies (including waste products) will be enforced. Black marketers will be attracted to Army food service activities as well (even during offensive and defensive operations).

Water Purification

Vigilant Monitoring

10-41. Urban areas will often have a ready source of water to support the urban inhabitants and its infrastructure. However, this water may not be suitable for U.S. and allied forces (though the urban population may have developed immunity toward its microorganisms). The higher concentration of TIM in urban areas compounds this problem, as ground water is highly susceptible to chemical contamination, even supplies located miles away from the source of contamination. U.S. Soldiers should be trained and cautioned against using water from an urban area (to include ice and bottled water) until preventive medicine and veterinary personnel can determine its quality. When water quality is unknown, commanders ensure use of tactical water purification equipment that will upgrade it to the Army's water quality standards. Even if initial testing indicates that urban water sources are safe for Army forces, personnel must continuously monitor the water quality. However, Army water purification, storage, and transportation requirements for UO can be greatly reduced if the existing urban water supply can be integrated into sustainment operations. An early assessment of the feasibility of this course of action is critical to sustainment planning.

Potentially Greater Requirements

10-42. Offensive and defensive UO are often intense and can produce more casualties, including civilians. Consequently, medical facilities, already consumers of large volumes of water, may require even more water. Water purification, particularly in the urban areas of developing nations and during urban disaster relief operations, will be a critical and constant concern for Army forces.

Possible Key Terrain

10-43. Sites that can control the water of the urban area may be key terrain, providing not only a resource for Army forces, but also a means to control the threat, the civilians, or both. These sites may be the sources of the water—the river, lake, reservoir, or storage tanks—or the means that process and transport the water—pipelines, pumping stations, or treatment facilities. Many sites may be outside the urban area, as many large urban areas draw water from distant sources. The seizure of a pumping station or pipeline may make it possible for commanders to control water supplies without expending resources required to enter the urban area. To preserve critical infrastructure, commanders may increase security to protect these locations from contamination or destruction. Engineers also may need to restore, maintain, or operate existing water facilities damaged by the threat or disaster and to drill new wells and construct new water facilities.

Mortuary Affairs

10-44. Mortuary affairs provide the necessary care and disposition of deceased personnel. It supports the Army across the spectrum of operations. It may directly and suddenly impact (positively or negatively) the morale of the Soldiers and the American public and may influence relations with the civilian population in the AO. It can also affect the health of Soldiers and the urban populace. Overall, commanders must carefully plan evacuation routes and temporary collection and internment sites, adhering to local customs and traditions to lessen potentially negative consequences.

10-45. Units are responsible for recovering the remains of their own fatalities and evacuating them to the closest mortuary affairs collection point, usually located at the nearest support area. Urban governmental authorities, family, and representatives of appropriate nongovernmental organization are responsible (and should always be relied on to the greatest extent possible) for recovering and handling indigenous civilian remains. However, because of the sheer density of noncombatants in UO, commanders may also find themselves responsible for civilian remains. Deaths of civilians under Army control, such as urban evacuees and refugees at Army-operated sites, often obligate the Army to care for their remains including medical certification and records of death. High-intensity urban combat may result in civilian deaths, and health concerns will require Army forces to deal with civilian remains expeditiously. Commanders should consult local religious leaders, the Staff Judge Advocate (SJA), CA personnel, and chaplains to verify that they are abiding by law and customs. Overall, commanders must ensure that forces treat all deceased, including civilians, with dignity and respect (another important aspect of adhering to the urban fundamental of understanding the human dimension).

Aerial Delivery

10-46. Aerial delivery is the movement by fixed- or rotary-wing aircraft and delivery by the use of parachute or sling load of Soldiers, supplies, and equipment. As a vital link in the distribution system, it contributes to flexibility and provides the capability of supplying the force even when urban ground LOCs are disrupted. Forces use aerial delivery to deliver supplies and equipment when not possible or timely by other means. During urban stability or civil support operations, it is used extensively to move supplies to meet the urgent needs of a population in crisis. In all UO, delivery aircraft are highly vulnerable to small arms, rockets, and air defense systems. A threat may further decrease an already limited number of urban drop zones (aircraft may be able to avoid air defense systems, but ground forces may not be able to secure the drop zone and retrieve the cargo). Equipment and supplies transported by helicopter sling-load lessen the latter disadvantage. There are usually more available sites to deposit sling loads, such as rooftops (engineers may need to determine if they have sufficient structural integrity), parking lots, and athletic fields. However, load instability during flight may restrict a helicopter's airspeed and maneuver capabilities making it more vulnerable to small arms and man-portable air defense systems. For these reasons, aerial delivery of supplies in UO may be much less efficient than in many other environments.

MAINTENANCE SUPPORT

10-47. Maintenance entails keeping materiel in operating condition, returning it to service, or updating and upgrading its capability. It includes recovering and evacuating disabled equipment; replacing forward; performing preventive maintenance checks and services (PMCS); increasing battle damage assessment and repair (BDAR); and determining potential maintenance resources in the urban area.

Rapid Recovery Essential

10-48. Disabled vehicles produce obstacles to movement during UO, blocking narrow thoroughfares and creating vulnerable targets for threat forces. This makes rapid recovery operations essential to avoid clogging limited LOCs and mounted avenues with vehicle evacuation operations. Hence, maintenance personnel should be proficient in hastily securing unit maintenance collection sites near the damaged equipment, ensuring route security to and from the disabled vehicle, and methods of egress under fire. The increased potential for the task organization of armored units into smaller attachments during UO (often platoon-sized or smaller) will strain limited recovery assets and increase the need for sections to be well trained and resourced (to include adequate tow-bars and chains) for self-recovery operations. Units must maintain centralized and responsive control over these potentially critical recovery assets and position them as far forward as the situation allows.

Replace Forward

10-49. One of the guiding maintenance principles is to replace forward and fix rear. Maintenance activities, with a forward focus on system replacement, task and use the distribution and evacuation channels to push components and end items to the sustainment level for repair. However, the urban environment may make distribution and evacuation difficult or even impossible. Repairing equipment as far forward as possible will often be vital to maintaining the tempo necessary to succeed in UO. When recovery is required, equipment is moved only as far back as the point where repairs can be made. When selecting a maintenance site, commanders consider:

- Security and force protection measures.
- Sufficient area around equipment for lift or recovery vehicles.
- Availability of existing maintenance facilities or garages.

10-50. The unforgiving urban terrain will invariably increase damage to man-portable weapons and equipment, particularly electronic equipment sensitive to jarring. More frequent jarring can also translate into more frequent calibration requirements. Although by definition man-portable weapons and equipment are easier to transport, evacuating these systems may prove as difficult as evacuating vehicles and larger, heavier equipment. Therefore, field support maintenance support teams (MSTs) will frequently need to repair equipment at (or as near as possible to) the point where it was damaged. Equipment operators are responsible to properly diagnose the fault or damage. Such action ensures that the correct repair parts and maintenance personnel are sent forward to complete necessary repairs. In UO, particularly offensive and defensive operations, units may need to replace rather than repair equipment. Subsequently, sustainment personnel may need to plan for increased replacement of what might normally be repairable equipment, as well as plan for increased repair parts for man-portable items.

Add-On Protection Increases Wear

10-51. To increase protection against small arms, mines, rocket-propelled grenades, and lightweight antiarmor weapons, units may need to attach additional armor and improvised wire-mesh screens and cages to both wheeled and tracked vehicles operating in an urban environment. (Units can also sandbag vehicles to achieve a degree of increased protection.) These modifications, however, increase wear on brakes, springs, suspension, and tires, all of which are already vulnerable to the increased amount of debris present in the urban environment. Over time, this wear will begin to increase failure rates in major assemblies, such as engines, transmissions, and transfers. To offset the increase in wear and, subsequently repair parts requirements, commanders conducting UO should consider emphasizing these aspects during daily operator PMCS and increasing the frequency of scheduled services.

Increased Battle Damage Assessment and Repair

10-52. In UO, operators, crews, MSTs, and recovery teams execute BDAR far more than in other environments. BDAR quickly restores minimum essential combat capabilities for a specific mission (normally of short duration) or allows the equipment to self-recover by expediently fixing or bypassing components. Commanders may need to authorize supervised battlefield cannibalization and controlled exchange when units lack critical parts or cannot bring them forward. Sustainment commanders should consider the use of BDAR kits, specifically designed for the systems they support. The use of these kits, though providing only temporary repair, will extend the availability of combat systems until full repair capability is accessible.

Potential Urban Maintenance Resources

10-53. Although urban areas can complicate maintenance, they may contribute to this sustainment function. Analyzing the urban area in the commander's AO may reveal potential sources of parts, tools, equipment, and facilities, and the expertise necessary to fix equipment, repair components, and fabricate critical parts. Sites of potential value may include automotive repair shops, foundries, car dealerships, parts stores, junkyards, scrap-metal dealers, and machine shops. Urban areas may also serve as key sources for parts and facilities (and theater support contractor personnel) to repair automation and network communication equipment.

Host Nation Repair Operations: OPERATION IRAQI FREEDOM

During operations in Mosul, Iraq in the summer months of 2003, elements of the 2nd Brigade Combat Team (BCT), 101st Airborne Division (Air Assault) suffered from a decline in combat power that outpaced the ability of the distribution-based sustainment system to provide repair parts. As readiness across the BCT threatened to degrade operational capability, sustainment personnel from the brigade's forward support battalion (FSB)—drawing on lessons learned from the 1990-1991 Gulf War—scoured the local populace for repair parts, facilities, and expertise in an attempt to offset the losses suffered due to supply shortfalls.

Initially, local retailers provided limited repair parts and facilities to the FSB, but the effect was significant and readiness rates began to stabilize. As time passed, however, the continued wear on equipment operating in the urban environment, coupled with the harsh summer weather, began to take a toll on reparable components and major assemblies. With limited availability in theater, and order-ship times as long as 60 days, requisitions would not be filled before the effect on combat fleet readiness reached a critical point.

Recognizing a trend that threatened the success of the BCT's urban operations in Mosul, the FSB established a contract maintenance facility in the brigade support area, used its contingency contracting officer to hire a group of local mechanics certified through General Motors, and opened the first host nation repair facility in Iraq. The group repaired major assemblies and reparable components and systems, maintaining a failure rate consistent with major CONUS rebuild facilities. The effect was immediate: during the first four months of the program, a full two-thirds of the combat platforms repaired and returned to service operated with components or assemblies rebuilt in the host nation facility.

Over time, the program was expanded to support other divisional units in northern Iraq. This effort provided maneuver commanders with responsive maintenance support and demonstrated the effect of sustainment flexibility and sustainability in urban operations. Furthermore, this action met the division commander's intent of infusing much-needed funds into the local economy while strengthening American-Iraqi relations in the region.

TRANSPORTATION SUPPORT

10-54. Transportation supports the concept of the urban operation by moving and transferring units, Soldiers, equipment, and supplies. Transportation incorporates military, commercial, host-nation, and urban area capabilities to establish a flexible system that expands to meet the needs of the force. Transportation includes movement control, terminal operations, and mode operations.

10-55. Urban areas are often critical to transportation operations. These areas may serve as a lodgment or support area for entry of Army forces and sustainment supplies. The existing transportation and distribution infrastructure may be essential to reception, staging, and onward movement. Theater support contracts, as well as host-nation support agreements, may greatly increase the ability of Army forces to use the urban area's facilities, which may include docks, airfields, warehouses, and cargo handling equipment. Urban support may also include skilled urban workers, such as longshoremen, truck drivers, and MHE operators.

Urban Terminals

10-56. In addition to serving as major seaports and aerial ports of debarkation, urban areas may provide additional terminals in the AO. Forces may use these terminals for further staging, loading, discharging, and transferring the handling between various inland transportation modes and carriers (motor, air, rail, and water). These urban terminals—with synchronized movement control—permit commanders to rapidly shift transportation modes and carriers. This increases flexibility and ensures the continued forward movement of equipment and supplies to support ongoing operations. Movement control, particularly in urban areas, relies heavily on support from military police in their maneuver and mobility support role. Without this support, urban LOCs may become congested, hinder movement and maneuver, and degrade force effectiveness (see FM 3-19.4). Urban commanders may need to establish multiple checkpoints, roadblocks, and traffic control points; restrict selected roads to military traffic; and reroute movement to unaffected road networks when civil support and refugee control operations compete for available routes. Again, military police operations are critical in this regard and will require continuous, close coordination with urban civilian police (if available).

Obstacles to Ground Transportation

10-57. Although urban areas can contribute to transportation operations, rubble and other damage can become obstacles to ground movement. Even in an undamaged urban environment, road and bridge weight restrictions, barriers and medians, steep embankments, large drainage systems, and other urban structures may limit transportation operations. Urban route maintenance, to include reinforcing bridges and constructing bypasses, may become a priority task for engineer units. Bypassed pockets of resistance and ambushes pose a constant threat along urban supply routes and routes in close proximity to urban areas. Urban convoy and resupply operations will often require increased security in the form of continuous route security operations, regular (daily, if necessary) mine and improvised explosive device (IED) clearance operations, numerous observation posts, attack helicopter support, a larger, more mobile tactical combat force, and adequate communications capabilities among convoy vehicles and with supporting forces and higher headquarters. While the level of security and force protection may vary, convoy operations are never considered an "administrative" operation (see FM 4-01.45). Security needs may increase manpower requirements for sustaining operations and potentially reduce resources from decisive operations. Moving critical supplies may require heavily armed convoys and armored vehicles. Drivers must be well trained in convoy operations, rehearsed, and alert. They must recognize and avoid potential mines and IEDs (stationary and vehicle-borne) and be able to react rapidly to ambushes. Additionally, commanders may need to institute deception measures such as randomly closing and opening supply routes to confuse threat pattern analysis. As before, aerial resupply can alleviate problems due to ground obstacles, but the air defense threat and proximity of threat forces may preclude their routine use.

Population Effects

10-58. The ability of Army forces to use vital urban transportation facilities depends largely on the civilians and the threat. The civilian population can affect the transportation system if they do not support the goals of Army operations. Urban transportation systems—such as ports, railroads, and rivers—require many specialists to operate. Without these specialists, the system's utility is degraded and may not function at all. In urban stability or civil support operations, Army forces will share the system with civilians and other agencies. Civilian authorities may refuse to allow Army forces to use any portions of an urban area's transportation system. Negotiating for access to that system under these circumstances then becomes a command priority.

Threat Effects

10-59. The threat can significantly affect urban transportation systems. Many are composed of smaller subsystems. Each subsystem is vulnerable to attack, which in turn often shuts down the whole system. A large canal system, for example, may have entrance and exit facilities, the canal itself, a means to pull the vessel along such as a locomotive engine, and the civilians that run each of these subsystems. Both an attacker and defender understand the components of the particular transportation system. If important to current or subsequent operations, defending forces then develop plans and allocate forces to protect these subsystems. Attacking forces, on the other hand, often avoid collateral damage to the system, while simultaneously preventing enemy destruction of the facilities.

FORCE HEALTH PROTECTION

10-60. Force health protection (FHP) consists of all services performed, provided, or arranged to promote, improve, conserve, or restore the mental or physical well-being of personnel in the Army and, as directed, for other services, agencies, and organizations. FHP conserves the force by—

- Preventing DNBIs and controlling combat stress.
- Clearing the urban area of casualties.
- Providing forward medical treatment and hospitalization and en route care during medical evacuation.
- Providing required veterinary, dental, and laboratory services.
- Ensuring that adequate Class VIII supplies, medical equipment, and blood are available.

10-61. FHP operations minimize the effects of wounds, injuries, disease, urban environmental hazards, and psychological stresses on unit effectiveness, readiness, and morale. Effective UO require acclimated Soldiers trained in specific urban tactics, techniques, and procedures. FHP helps maintain the health of urban forces, thereby conserving that trained manpower. Historically, urban combat operations have generated three- to six-times greater casualty rates than operations in any other type environment. (Greater enemy wounded and a need to provide care for them will also exacerbate this situation.) FHP operations that maintain the health and readiness of Army forces reduce the strain on the replacement and evacuation systems. Such care allows Soldiers to concentrate on the task at hand instead of the increased risks associated with UO.

10-62. As part of the overall LPT, commanders and medical planners must analyze and continuously assess the urban area. They must identify medical threats, required medical resources, and the quality and availability of medical facilities and resources (to include civilian medical personnel). This assessment prevents duplicated services and permits more effectively and efficiently organized medical resources.

10-63. FHP analysis also identifies available hospitals, clinics, medical treatment facilities, and medical supplies and equipment (including production facilities) in the urban area. It may also identify NGOs capable of providing medical services and supplies. (However, stringent federal regulations, standards of medical care, and a need for unavailable advanced technologies may limit their use by Army forces. But, this analysis will still be crucial in determining whether urban medical facilities can support civilian requirements.) FHP personnel keep abreast of the operational situation and its impact on FHP and the urban population. Peace operations, for example, may rapidly transition to high-intensity offensive and defensive operations requiring medical support able to handle potential mass casualty scenarios which may include civilians.

Care of Civilians

10-64. During combat operations, the military normally does not provide injured civilians with medical care, instead relying on civilian medical personnel to provide this service. However, based on METT-TC and requirements under Geneva Conventions, commanders may need to recover, evacuate, and treat numerous civilians (particularly in urban civil support or stability operations) until local civilian medical personnel and facilities can be reconstituted and supplied. In urban civil support operations involving weapons of mass destruction, the primary focus of Army support may be FHP.

10-65. If commanders provide civilian medical support, they adhere to the UO fundamental of transitioning control and transferring responsibilities for medical care of civilians to another agency or into civilian hands as soon as is practical. In UO, commanders consider and address the medical treatment of civilians (enemy and friendly) early in the planning process. Any initial assessment or survey teams therefore contain medical representatives. Of significant concern, commanders may need to ensure that adequate supplies and equipment are on hand to treat the urban population's children as combat health service support is normally focused on the treatment of adult Soldiers. Additionally, follow-on military care for civilians after treatment at a combat support hospital is often nonexistent or extremely limited to a case-by-case basis necessitating the quick restoration of civilian hospital capabilities. Finally, military health practitioners must also assess and consider cultural factors related to civilian medical treatment. (For example, male doctors may not be able to examine or treat indigenous women.) To alleviate confusion and preclude avoidable negative repercussions and unrealistic expectations, commanders may need to develop specific medical policies, directives, and standing operating procedures. These "medical rules of engagement" (ROE) ensure that subordinates know how much medical care they may provide to—

- The urban population.
- Other host-nation and third-country civilians.
- Coalition and host-nation forces.
- Contractor personnel.

10-66. Commanders should also consider that medical aid to the local urban population and other civilians can be a powerful influential tool. Properly planned and executed, its potential benefits for Army forces include enhanced security and intelligence collection created by increasing the willingness of civilians to interact with Soldiers. These effects cannot be left to happenstance. They should be a conscious part of the information operations campaign nested within the overall UO plan. Consideration must also be given, however, to mitigating the potential negative consequences involved in the inevitable reduction or curtailing of medical assistance by Army FHP personnel. As with any aspect of UO, providing care without considering the longer-term consequences could result in far greater disadvantages than benefits.

Disease and Nonbattle Injury

10-67. DNBI is a major medical threat during all operations and presents additional risk during UO. Urban pollution hazards and potential exposure to TIM may increase the risk to soldier health. Some urban areas, particularly those in developing countries, are already large sources of communicable diseases, such as tuberculosis, cholera, typhus, hepatitis, malaria, dengue, and acquired immune deficiency syndrome (AIDS). Other factors that increase the number of disease vectors and the potential for the transmission of infectious diseases might include—

- Physical damage or deterioration of urban infrastructure—such as electricity, water, and sewage services.
- Industries that use or produce hazardous materials.
- Increased habitats for rodents and other pests due to rubble, debris, and uncollected trash.
- Inability for the threat or local authorities to recover and dispose of human remains and animal carcasses.
- Greater threat of animal diseases transmittable to humans due to larger numbers of animals wounded or injured by the effects of UO.
- Greater susceptibility of urban civilians to infectious diseases—increasing the potential of person-to-person disease transmission—due to declining public health capabilities and nutrition levels.

10-68. The density of the environment extends these urban medical risks to Army forces. Commanders should establish a medical epidemiological surveillance system early to assess the health of the force and promptly identify unusual or local occurrences that may signal preventive medicine problems or the influence of biological or chemical agents. These potential hazards, particularly the release of TIM, will influence the type of medical supplies needed by medical personnel and will also necessitate critical planning and preparation for potential mass casualties (civilian as well as military).

10-69. Adhering to the UO fundamental of preserving critical infrastructure will likely prevent surges in DNBI. Preventive medicine personnel must identify the diseases and recommend control and preventive measures. In urban areas, they may also conduct civilian health screening, health education, and immunization programs. Medical screening of military personnel, particularly multinational forces, may be required to prevent introducing new diseases (especially drug-resistant strains) into an urban area. A new disease may tax the medical system and introduce a new medical problem into an area already in crisis. Lastly, field sanitation training (to include a general emphasis on hand washing; training in the use of barrier protection such as latex gloves when rendering care to any person; and animal, rodent, and pest control), equipment, and supplies are part of overall preventive medicine measures and considerations.

Combat Stress

10-70. Stress occurs in every operation and type of environment; while limited stress levels are beneficial, too much is harmful. Controlled combat stress can invoke positive, adaptive reactions such as loyalty, selflessness, and heroism. On the other hand, uncontrolled combat stress can result in negative, harmful behavior and combat stress casualties. Such behaviors and casualties—battle fatigue, misconduct stress behaviors, and post-traumatic stress disorder—can interfere with the unit's mission. Physical and mental factors leading to combat stress result from the environment and the deliberate enemy actions aimed at killing, wounding, or demoralizing Soldiers. However, many stressors are also generated from the Soldier's own leaders and mission demands.

Combat Stress: Chechnya – 1994 to 1996

Russia's 1994-1996 conflict with Chechnya, a republic in the southwestern part of the Russian Federation, produced an increased number of psychological trauma and combat stress casualties. One medical survey found 72 percent of the soldiers screened had some sort of psychological disorder symptoms. Of that, 46 percent exhibited asthenic depression (a weak, apathetic, or retarded motor state). The other 26 percent exhibited psychotic reactions such as high states of anxiety, excitement, or aggressiveness and a deterioration of moral values or interpersonal relations. The statistics showed more troops experienced combat stress disorders than during their 1980s war in Afghanistan. One primary difference was that in Chechnya, Russian forces conducted combat mostly in cities rather than in mountains, valleys, and other rural areas.

Combat always invokes fear in soldiers. However, poor training and planning, uncertainty in their cause, and urban populations that resented their presence exacerbated the psychological climate for the Russian forces in Chechnya. Acts of subversion and terrorism by Chechen guerrillas kept the Russians in a constant, high state of readiness and anxiety; the soldiers viewed every civilian—young or old, male or female—as a potential enemy. This psychological pressure was not simply a by-product, but an objective of information operations and a prime reason for taking the fight into the close confines of Chechnya's urban areas. Torture and mutilation of prisoners; immediate execution of captured pilots; imitative electromagnetic deception (Chechens mimicked Russian radio transmissions and directed Russian close air support against their own forces); and propaganda to convince civilians that Russia's actions had a religious bias against Muslims and Islam were conducted to exert intense, unremitting psychological pressure on Russian forces—with great success.

The characteristics of urban areas combined with Chechen insurgent activities and information operations, civilians that did not welcome foreigners, and an unpopular and poorly supported conflict with an open-ended mission reinforced the need for a national will to strengthen and support the urban fight. These characteristics also reinforce the need for clear objectives, proper leadership and training, and available medical assets able to properly diagnose and treat combat stress casualties.

10-71. During UO, compartmented urban combat leads to physical isolation while difficulties in transmitting radio signals lead to communications isolation; together, they create an overwhelming sense of loneliness. Snipers, mines, and IEDs combined with the closeness and high intensity of urban combat contribute to an unremitting fear of attack from any quarter that further increases stress casualties. Additionally, seeing and perhaps accidentally inflicting casualties on civilians (especially women and children) increase battle fatigue. If civilians are hostile or a threat uses the population as cover and concealment, then the potential for misconduct stress behaviors often increases. Urban areas may provide temptations for looting, alcohol and substance abuse, black marketeering, and harmful social interactions; these temptations may increase misconduct stress behaviors. Leaders can prevent or rapidly identify, successfully manage, and treat stress-related casualties (see FM 22-51 and FM 6-22.5) and prevent misconduct stress behaviors as well as potential violations of the law of war. To reduce the number of stress casualties, commanders must—

- Provide tough, realistic training in urban combat (and combat stress management).
- Plan for rotation of forces and adequate rest and relaxation opportunities.
- Establish effective ROE.
- Create unit cohesion.
- Ensure adequate mental health support is available.
- Provide strong, capable leadership. Leaders, forward on the battlefield, will determine whether the necessary discipline perseveres.

Evacuation

10-72. Transportation restrictions may preclude the medical evacuation (MEDEVAC) of urban casualties. MEDEVAC by air ambulance remains the preferred method; however, it may not be feasible. The threat from small arms fire, rocket-propelled grenades, and other shoulder-fired weapons may prohibit the employment of helicopters. Aeromedical evacuation can also be hampered by the lack of landing zones due to narrow streets, heavily damaged buildings, and barricaded or booby-trapped roofs. If the employment of air ambulances is not feasible, a ground evacuation operation will be necessary. In UO, ground ambulances should often be—

- Hardened to improve the safety of the crew and patients as well as to increase the survivability of the evacuation platform.
- Field-sited as close to the operation as is tactically and physically possible.
- Maneuverable in tight spaces and narrow streets.

10-73. Evacuation by ground ambulance may have their own set of obstacles to surmount such as rubble; debris; barricades; mines, booby-traps, and IEDs; damaged or destroyed vehicles, and flooded or snow-filled streets. If enough ground ambulances are unavailable due to increased casualties or the tactical situation does not allow, units will need to execute casualty evacuation (CASEVAC) operations.

10-74. In UO, CASEVAC may entail evacuation by litter and the expedient use of a vehicle of opportunity or other military vehicle not dedicated for medical evacuation. Based on the threat situation, CASEVAC may also be limited to armored vehicles, conducted only during limited visibility, or both. For evacuation by litter, units may require more litter bearers to move the injured to a point where they can be further evacuated by ground or air ambulance. Lengthy evacuation routes require more litter bearers, as multiple litter relay teams will be necessary to conserve energy and expedite evacuation. However, unless augmented or relieved of this responsibility by another unit, these litter bearers will come from the casualties' own unit thereby diminishing the unit's strength levels necessary to accomplish its primary mission. Instead, tactical commanders may need to wait until a lull in or a cessation of the battle occurs. Depending on projected casualty rates, higher-level commanders may augment units with additional personnel to perform evacuation or may assign maneuver units this mission. Evacuation personnel will require specific training in urban evacuation techniques (moving casualties from subsurface and supersurface levels to and along the surface level).

10-75. Overall, urban commanders at all echelons must develop detailed casualty evacuation plans that include the establishment of treatment facilities as far forward as the situation allows. Engineers are critical to clear routes for medical resupply and evacuation. Based on the threat situation, additional armored vehicles may need to accompany ambulances in a multi-vehicle convoy not only to and from the patient pickup point but also through to a higher level of care. Army forces may also need to plan for and acquire unique capabilities, equipment, and skills now typically seen in civilian urban search-and-rescue teams to clear debris and search for urban casualties. For example, specially trained dogs may play a vital role in locating victims among urban rubble.

Treatment

10-76. First aid training will have increased significance in UO. The compartmented nature of UO, transportation restrictions, communications difficulties, and the low density of trauma specialists (combat medics) may limit the urban casualty's initial treatment administered by nonmedical personnel or to self-treatment measures. To mitigate this risk, units identify and train combat lifesavers (CLS) to perform in the absence of trauma specialists. Since the likelihood of Army forces performing UO continues to increase, commanders strive to meet or exceed Army standards for the number of fully trained CLS required for their specific unit. CLS (and trauma specialist) training must be realistic, hands-on, and conducted at appropriate intervals to ensure CLS can perform lifesaving tasks to include starting intravenous lines. This increase in self, buddy, and CLS care, as well as longer delays in evacuation, may also increase requirements for additional first aid, CLS, and trauma specialist-carried supplies. (Ultimately, commanders may have to consider weighting the main effort with greater overall FHP support for units conducting UO than for units conducting operations in other environments.)

10-77. The increased potential for delayed evacuation during UO mandates that Army trauma specialists be skilled in prolonged casualty care. During the 3-4 October 1993 battle in Somalia, for example, seven trauma specialists managed 39 casualties for more than 14 hours before they could be evacuated. Evacuation delays significantly increase the potential for infection; therefore, trauma specialists should also be trained to administer antibiotics on the battlefield.

10-78. FHP personnel must be capable of recognizing and treating injuries due to IEDs and incendiary or thermobaric weapons (fuel-air explosives)—favored, urban-oriented threat weapons (see Chapter 3). Thermobaric weapons explode; create a cloud of volatile gases, liquids, or powders; and then ignite, creating an immense fireball consuming oxygen and creating enormous overpressure. When employed in an urban structure, the blast wave or overpressure of any explosive is greatly amplified. Injuries resulting from these weapons are massive burns, broken or crushed bones, concussions and traumatic brain injuries, missile injuries, and internal injuries. To counter this threat, medical personnel must be aware of the capabilities of these blast weapons, wounding patterns, and treatment methods.

10-79. The increased use of body armor during UO will help prevent penetrating chest and abdomen wounds. As such, FHP personnel should expect more groin, pelvis, head, and extremity injuries. Furthermore, when fighting soldiers that are known to use body armor, a threat (particularly snipers) can be expected to target the head and face more often than other anatomic areas resulting in more head injuries. Medical personnel must also anticipate and be prepared to treat an increase in hearing loss due to the firing of weapons (particularly recoilless) in enclosed spaces with little ventilation. Medical personnel (and engineers) should also be trained on how to properly remove rubble that has fallen on a Soldier or civilian so as not to increase the extent of his injuries. Lastly, FHP personnel will need to research and develop unique treatments for injuries due to an ever-increasing arsenal of nonlethal weapons.

Explosive Ordnance Disposal Support

10-80. Explosive ordnance disposal (EOD) support provides the capability to neutralize domestic or foreign conventional explosive hazards which include unexploded explosive ordnance (UXO), booby traps, IEDs, captured enemy ammunition (CEA), and bulk explosives. Such material and devices threaten military operations and military and civilian facilities, equipment, and personnel. Explosive hazards create a much greater risk during UO than operations in any other environment. Confined spaces, hard surfaces, and more personnel (both Soldiers and civilians) in the vicinity may magnify the detonating effects of explosives. Dense terrain makes all types of explosive hazards more difficult to locate. Moreover, terrorists select urban settings for their booby traps and IEDs to potentially kill and destroy more, thus gaining greater visibility for their message.

10-81. 10-81. EOD units detect, mark, identify, render safe, and dispose of explosive hazards. Importantly, EOD specialists will work closely with unit intelligence personnel to conduct explosive forensics—technical intelligence collection and exploitation—to help identify the makers of the devices as well as their ever-evolving methods of construction, placement, concealment, and detonation. Chemical reconnaissance with their array of chemical sensors will also work closely with EOD units to reduce the uncertainty of the chemical threat. This integrated forensic approach is a proactive method to take the advantage away from a terrorist or insurgent threat that seeks to employ these devices as a primary tactic.

10-82. Urban operations will increase EOD specialists' role as advisors and instructors on explosive hazards, protection measures, and disposal techniques. They will advise and train other Army forces, other services, multinational partners, and civilian authorities. EOD specialists will also work closely with public affairs and psychological operations personnel to increase public awareness of the horrific effects of these munitions and devices. This education process is designed not only to teach the urban population to identify and avoid explosive hazards, especially UXO, booby traps, and IEDs, but to gain civilian assistance in reporting their locations. (In this regard, rewards for information leading to the discovery of unexploded ordnance may be beneficial. However, the largest reward should normally be reserved for information leading to the arrest or capture of bomb makers.)

10-83. In an urban environment where there are large amounts of UXO, IEDs, and CEA, the sheer amount of explosives may overwhelm the capabilities of available EOD personnel. Because of their training in demolitions and explosives, commanders will often use combat engineers to augment the destruction and clearance effort. Engineers are normally given responsibility for UXO clearance and disposal while EOD teams use their technical skills for IEDs. Commanders may also consider embedding EOD expertise within engineer units to provide the knowledge necessary to determine the correct amount and type of explosive necessary to detonate UXO and to understand potential risks and effects of detonation of various munitions. For example, EOD specialists would better understand the second-order effects and risk reduction measures necessary to detonate a large missile containing numerous submunitions.

10-84. Unfortunately, there is no single device or technique that can eliminate every explosive hazard encountered. Clearance operations are dangerous and time-consuming. Based on METT-TC, commanders may consider the following options when dealing with these explosive hazards:

- Report, mark, and bypass the hazard.
- Employ tactical breaching procedures.
- Self-extract to alternate routes or positions.
- Restrict routes or positions from further use.
- Clear specific areas or positions, or conduct an extensive route clearance.
- Lastly, accept risk of casualties and continue with the assigned mission.

10-85. Effective tactics, techniques, and procedures (TTP) for addressing explosive hazards must be captured and shared with friendly forces and coalition partners. Commanders recognize, however, that adaptive threats will continually modify their own techniques to counter the unit's response. See FM 3-34.119, FM 20-32, FM 3-100.38, FM 4-30.51, and FM 4-30.16.

HUMAN RESOURCES SUPPORT

10-86. Human resources support (HRS) encompasses the following functions: manning the force, personnel support, and personnel services. These activities include personnel readiness; replacement operations; personnel accounting; personnel information management; casualty operations; essential personnel services; postal operations; morale, welfare, and recreation; and band support provided to Soldiers, their families, Department of Army civilians, and contractors.

10-87. Successful UO require HRS functions. Clearly, HRS focuses on caring for the needs of people: Soldiers and civilians who support them. Since a critical component of the urban environment is the population, these activities, when required, may also support them. For example, HRS elements, in conjunction with mortuary affairs, military police, and CA, may provide administrative support to assist in the accountability of displaced persons and civilian internees and the establishment of other populace and resources control measures. This support may include the provision of identification and registration cards, travel permits, and other documents, and administrative support to casualty operations. Human resources managers may also assist with community activities and coordination with the Red Cross. If necessary, a postal platoon may assist urban officials in training and reestablishing civilian postal operations. Overall, this type of assistance and support helps to care for the immediate needs of the urban population and, ultimately, speed transition back to legitimate civilian responsibility and control.

10-88. In offensive and defensive UO, HRS personnel will need to account for more casualties and more frequent reconstitution. In all UO, success relies on training individual replacements. Urban combat requires Soldiers skilled in specific urban combat TTP. They understand the societal aspects of the urban population and have training in crowd control to avoid escalating potentially volatile situations. Training these replacements prior to their actual deployment or while in the replacement system frees urban commanders from having to do so. It also helps reduce Soldier (and supporting civilian) isolation, anxiety, and fear associated with urban areas and operations.

Financial Management Support

10-89. Money and its management are often critical to the success of UO. As such, financial management support makes resources available when and where they are needed and assist the urban commander in maintaining fiscal responsibilities. Financial management operations are necessary to conduct the contracting necessary to restore and rebuild infrastructure and provide real-time information, accounting, and services. Financial management operations ensure that urban operational policies and procedures adhere to laws and regulations, develop command resource requirements, and leverage appropriate fund sources to meet them.

10-90. UO will likely include other U.S. military services, governmental agencies, and contractors; multinational forces; NGOs; and various elements of the host nation. So many actors complicate financial accounting, resource management, and cost capturing. Yet, these activities are vital to accomplishing UO and maintaining legal requirements. Army forces will receive support from and provide support to these participants. Financial managers provide crucial expertise to the urban commander on the obligation authority. They also furnish the documentation necessary to obtain and pay for local goods and services using contract or commercial vendor services. However, commanders (with financial management assistance) must assess the economic impact of UO on the economy. An appropriate analysis includes how well the urban economics and commerce infrastructure can support the deployed force and how an influx of U.S. currency may affect the overall economy either positively or negatively.

Labor, Reward, and Solatia Programs

10-91. Financial management elements can also provide joint pay support and non-U.S. pay support for host-nation employees and day laborers supporting UO, as well as for civilian internees. These units can support reward programs and solatia (financial compensation). Commanders can use reward programs to purchase weapons, radios, information, and other items from the urban inhabitants. However, they must use caution when paying for critical human intelligence. First, they must ensure that multiple collection agencies in the command avoid paying the same source for the same information and interpreting these repeated inputs as validation that information is accurate and reliable. Second, they must establish a price or pay scale so subordinate units (particularly in multinational UO) equally compensate their sources for information. Commanders may also need to make solatia to alleviate grief, suffering, and anxiety resulting from injuries and property or personal loss resulting from some aspect of UO. At other times, commanders make solatia to meet cultural expectations of the urban population. These payments are nominal in amount and made without an admission of liability by the Army. Significantly, they may help maintain the legitimacy of the operation and preserve the support of the urban populace.

Financial Support to Civil-Military Projects

> *...[I]n an endeavor like that in Iraq, money is ammunition. In fact, depending on the situation, money can be more important than real ammunition—and that has often been the case in Iraq since early April 2003 when Saddam's regime collapsed and the focus rapidly shifted to reconstruction, economic revival, and restoration of basic services. Once money is available, the challenge is to spend it effectively and quickly to rapidly achieve measurable results. This leads to a related observation that the money needs to*

> *be provided as soon as possible to the organizations that have the capability and capacity to spend it in such a manner.*
>
> Lieutenant General David H. Petraeus
> Commander, 101st Airborne Division (Air Assault) and, later,
> Commander, Multi-National Security Transition Command-Iraq
> during OPERATION IRAQI FREEDOM

10-92. Other financial resources separate from operational funds may be appropriated to support civil-military operations (CMO) projects during stability operations. (During OPERATION IRAQI FREEDOM, these funds were referred to as commander's emergency response program [CERP] funds.) These funds are typically disbursed as bulk currency to tactical units, where the funds are allocated against lower-cost, high-impact projects that support the immediate needs of civil authorities and their citizens. As such, commanders must be prepared to receive, account, secure, and properly disburse these resources in accordance with established procedures. Commanders have an inherent responsibility for ensuring the proper use and disbursement of these funds and must execute appropriate judgment, fiscal responsibility, and supervisory oversight of any resources appropriated in support of CMO. While commanders establish appropriate guidelines to accomplish this fiscal responsibility, they must provide their subordinate commanders—who have the need, capacity, and capability to put these funds to use—reasonable flexibility in how they spend at least a portion of the money, so that they can address emerging needs within their AO.

Contracting Expertise

10-93. Due to the nature of UO, tactical-level units will be likely required to plan and initiate contracts not only for reconstruction projects but to supplement their own sustainment needs. Therefore, higher-level commands must be prepared to assist tactical-level units with contracting expertise. This expertise is required to overcome difficulties in delineating the scope of work, defining construction standards, and establishing viable delivery dates. For small projects and purchases, a standard contractual format may be sufficient. However, for large projects, commanders must provide their subordinates with knowledgeable contracting personnel, who can write contracts with clearly-understood specifications, create purchase orders that specify quality and delivery times, and help monitor work for quality control and assurance. Payment guidelines and standards among all services and members of a coalition are essential to eliminate competition and potential overpayment. Translators will be also essential to accurately translate requirements and standards in addition to trusted local agents to assist in selecting quality engineering and construction firms and workers. Overall, subordinate commanders need adequate contracting support and expertise to meet civilian needs as well as their expectations for timeliness and quality. Without it, commanders may risk losing critical opportunities to positively influence civilian perceptions. (See also Legal Support to Operations below.)

Religious Support

10-94. Religious support entails providing and performing religious support operations for commanders to protect the free exercise of religion for Soldiers, family members, and authorized civilians. This includes personal delivery of rites, sacraments, ordinances, pastoral and spiritual care, and religious education. Such support also consists of:

- Advising the commander on matters of religion, morals, and morale as affected by religion.
- The impact of local religions on threat courses of action, the urban society, and, subsequently, the mission.
- The ethical impact of command decisions.

Moral and Ethical Climate

10-95. Chaplains can help commanders sense the moral and ethical climate in Army units and understand potential moral dilemmas associated with planned UO. The urban environment affects Soldiers' mental health and can increase combat stress casualties, especially misconduct stress behaviors. Chaplains are an important part of combat stress support. The chaplain's presence and faith can sustain Soldiers throughout periods of great trauma. Religious support contributes to the total well being of Soldiers and aids in their return to combat readiness. Chaplains help bolster Soldiers' moral and ethical behaviors through spiritual fitness training so that Soldiers may better cope with ambiguous moral and ethical situations. Chaplains can help identify ethical concerns before they become critical command problems. To this end, their observations help develop and modify ROE, which—due to ambiguity or overly restraining rules—may be part of the problem.

Influence of Local Religions

10-96. Chaplains advise commanders on matters of religion as it affects the Soldiers within their units. They also explain the influences of local religions on threat courses of action and the urban populace and, subsequently, their potential effects on Army forces and UO. Religion is a crucial aspect of assessing the societal component of the urban environment. Understanding the major tenets and concepts of the religions and the impact of faith on civilians' lives may help commanders understand what motivates the threat and the urban populace. (Historically, when religion was a critical factor in armed conflict, the level of lethality has been significantly higher.) To understand threat strategy and tactics, commanders understand their ideology, which, in many cases, involves strong religious beliefs. This understanding also helps commanders to appreciate the inhabitants' attitudes toward other races, religions, and cultures and to identify unacceptable kinds of social interaction (particularly between Soldiers and civilians). Failure to recognize and respect religious beliefs can rapidly erode the perceived legitimacy of the mission and increase the length of the operation. In addition to its influence on threat courses of action, a thorough analysis of the urban environment also includes the degree of influence religion and religious leaders have on the area's government and economy.

Legal Support to Operations

10-97. Legal support provides operational law support in all legal disciplines (including military justice, international law, civil law—composed of contract, fiscal, and environmental law, claims, and legal assistance). This support assists in command and control, sustainment, and HRS of UO. Legal considerations are important in any operation; they take on added significance during UO. They form the foundation for establishing ROE and are critical in the targeting process. (For example, with the increased likelihood that the threat will use the media in an attempt to categorize Army attacks as unlawful, legal support will be critical in determining the protected status of targets.) Adequate legal instruction and training may help subordinate commanders and their Soldiers better understand some aspects ROE and, with this understanding, help them make the rapid decisions often required in a complex urban environment.

10-98. Legal considerations also affect how units acquire goods and services from urban areas and provide support to other agencies and organizations operating in an area. The environment's complex nature requires commanders and their staffs to review and closely consider applicable legal constraints when developing and executing friendly courses of action. Most urban areas have a highly developed legal system nested within county, provincial, state, and national systems. The SJA support to commanders addresses this urban system in conjunction with other legal systems and their potential to affect, positively or negatively, the success of UO.

10-99. International, host-nation, and U.S. law and other regulatory guidelines may vary in their applicability by time and place; actions permissible in one situation may be prohibited in another. These exceptions and complexities increase requirements for SJA, often working with CA personnel, to identify and resolve technical legal issues. Therefore, the SJA must actively advise and participate in all aspects of UO from predeployment training and initial planning through transition and redeployment. FM 27-100 contains detailed legal guidelines affecting UO.

International and Host-Nation Law

10-100. International law consists primarily of agreements, treaties, and customary law to include the law known as the law of war (see FM 27-10). The law of war consists of four general principles applicable when conducting any operation but requiring particular attention during UO. Figure 10-4 lists and describes the four principles: military necessity, discrimination (or distinction), unnecessary suffering (or humanity), and proportionality.

10-101. International law may affect urban operational issues, such as the right of entry, base operations, use of urban infrastructure, and overflight and landing rights. Status-of-forces agreements (SOFAs) exist or can be negotiated to resolve legal issues, such as the status of Soldiers (and, as necessary, contractors) operating in foreign areas to include criminal and civil jurisdictions, taxation concerns, and claims for damages and injuries. Unless a SOFA or other convention exists, Soldiers operating in foreign urban areas have the same legal status as tourists; they are subject to the laws and judicial process of the host nation. (During armed conflict, however, Soldiers always have the rights afforded to them by the Geneva Conventions). Commanders are responsible for understanding the international and host-nation agreements and laws that influence foreign UO. If local law hinders the operation, commanders may be able to inform the local U.S. diplomatic mission and request that it negotiate a solution.

10-102. Commanders may encounter civilian resistance groups whose actions may range from providing enemies with sustainment support to actively fighting Army and coalition forces. Members of such resistance groups must be dealt with in accordance with applicable provisions of the law of war. Commanders should seek legal guidance concerning detention and disposition of persons participating in various acts harmful to friendly forces and detrimental to the mission.

U.S. Law

10-103. Commanders conducting UO must also comply with U.S. law whether it is in the form of a statute, executive order, regulation, or other directive from a federal branch or agency. U.S. law influences UO by governing the acquisition of supplies and services for Army forces, regulating the assistance that can be rendered to foreign nations, and controlling intelligence activities. The Posse Comitatus Act, for example, makes it a crime for Army forces to enforce civil law. Similarly, portions of the Foreign Assistance Act prohibit Soldiers from performing law enforcement activities in foreign urban areas. However, circumstances—expressly authorized by the Constitution, acts of Congress, and other exceptions to these statutes—exist that allow the Army to support civilian law enforcement. Although not nearly all-inclusive, the above demonstrates how U.S. law further complicates urban operations, particularly stability operations, and increases the need for proactive SJA advice and counsel in all facets of UO.

Military Necessity: The principle that justifies the use of measures not forbidden by international or domestic law necessary to rapidly achieve military objectives.

Discrimination: The principle of distinguishing between combatants (who may be attacked) and noncombatants (who may not be attacked).

Unnecessary Suffering: The principle that prohibits the use of weapons, projectiles, or other materials in a manner calculated to cause superfluous injury and unnecessary suffering.

Proportionality: The principle that injury to persons and damage to property incidental to military action, in the circumstances ruling at the time, must not be excessive in relation to the concrete and direct military advantage anticipated.

Figure 10-4. General principles of the law of war

Legal Aspects of Nonlethal Force

10-104. Nonlethal capabilities can augment the means of deadly force and extend urban firepower options. They enhance the commander's ability to apply force in proportion to the threat and to allow discrimination in its use. The range of nonlethal capabilities includes offensive information operations, smoke and obscurants, irritants (such as chemical riot control agents), nonpenetrating projectiles, high-pressure water devices, and directed-energy weapons. These continually expanding capabilities give commanders more options to confront situations that do not warrant deadly force but require Soldiers to employ overwhelming decisive power. However, nonlethal capabilities are subject to the same legal constraints as lethal force (in fact, some nonlethal capabilities can cause serious injury and death, particularly if not employed properly) and undergo the same legal review. Like lethal force, nonlethal capabilities show military necessity, distinguish between combatants and noncombatants, distinguish between military objectives and protected property, are used proportionally, and do not result in unnecessary suffering.

10-105. Of special note, commanders cannot employ chemical herbicides or riot control agents as a method of warfare without prior Presidential approval. However, the Secretary of Defense and regional combatant commanders have authority to use herbicides and riot control agents in limited circumstances during armed conflict and in peacetime operations. Considering the complexity of this issue, commanders of major operations must streamline the approval process so that subordinate commanders do not discount use of these agents simply because of the high-level approval authority. Subordinate commanders must, however, be able to anticipate and clearly justify their potential use in UO. Also, any approval does not mean that these agents must be employed.

Civilians Accompanying the Force

10-106. LOGCAP personnel and other civilian support contractors provide a variety of sustainment functions for the Army. These functions may range from providing unskilled labor, transportation support, and health care to technical support of sophisticated equipment and weapons systems. Commanders must ensure that civilians providing support in their AO are not placed in positions of jeopardy, but if they are, these civilians must understand the risks that they assume when they engage in activities that might be misconstrued as direct or active participation in hostilities.

10-107. This last caution indicates an important legal consideration for commanders. They must try to ensure that threat forces can distinguish civilian support contractors from combatants and thereby remove any justification for them to be targeted or attacked. (As Army forces seek to work closely with the members of governmental and nongovernmental agencies and organizations in urban stability operations, commanders may need to take steps necessary to mitigate the risk of attack caused by their involvement with Army forces.) In the event that threat forces capture civilian support contractors, this same concern helps to ensure that, as lawful civilians accompanying the force, they will be afforded prisoner of war status. As the role of contractor support to Army operations continues to expand, commanders must stay abreast of changes in the law (to include evolving interpretative distinctions) as it affects the use of contractors on the urban battlefield. Other important considerations for commanders with civilian contractors supporting their force are—

- **Minimize Exposure to Harm**. Civilian contractors should be assigned duties at locations that minimize their exposure to harm. Commanders will need to periodically review these locations since they are likely to change as the urban operation evolves and transitions.
- **Prohibit the Performance of Combat Functions**. Civilian support personnel should not be working in situations that involve combat operations where they might be conceived as combatants. As an example, they should not perform force protection for Army forces such as fortification construction and checkpoint security.
- **Prohibit the Use of Arms**. Civilians should not be armed, unless approved by the combatant or service component commander on a limited, by-exception basis for personal defense.
- **Provide Identification Cards**. Civilians should be provided, and carry on their person, a Geneva Convention identification card identifying them as civilians authorized to accompany military forces and entitling them to be treated, if captured, as prisoners of war.
- **Restrict the Wear of Military Uniforms**. Civilians should not normally wear distinctive U.S. military uniforms unless the combatant or service component commander authorizes them to do so. Regardless of their clothing, however, civilians will wear a symbol that establishes their civilian status.

General Engineering Support

10-108. General engineering support will be essential during UO. This support helps assess, construct, maintain, and restore essential LOCs and urban facilities to sustain Army forces, the urban population, or both. Using civilian resources and investing Army general engineering resources requires careful consideration by commanders and staff planners. Since all elements of the urban infrastructure interconnect, general engineering support touches each category to some degree. Figure 10-5 illustrates how urban-specific, general engineering tasks align primarily with the transportation and distribution, energy, and administration and human services components of the urban infrastructure. These engineering tasks are significant and readily apply to UO. The last two, providing fire fighting support and waste management, have not been previously addressed and require more specific consideration.

Construct, Maintain, or Restore

- Roads and Highways
- Over-the-Shore Facilities
- Ports
- Railroad Facilities
- Airports and Heliports
- Fixed Bridges
- Electric Power Facilities
- Petroleum Pipelines and Storage Facilities
- Water Facilities

and Provide

- Mobile Electric Power
- Construction Support
- Fire Fighting Support
- Waste Management

Figure 10-5. General engineering support tasks

The Use or Investment of Resources

10-109. During urban offensive and defensive operations, Army engineer units accomplish tasks to sustain or improve movement and mobility, protection, and sustainment of U.S. and allied forces. These units should maximize the existing urban facilities, host-nation support, civilian contractors, and joint engineer assets. However, commanders should consider how using urban facilities to support military forces may negatively affect the population. On the other hand, construction and repair may benefit both Army units and the urban inhabitants. Restoring the urban transportation network not only improves military LOCs, but may also allow needed commerce to resume. Repairing urban airfields or ports increases throughput capabilities for military supplies, facilitates medical evacuation operations to the support base, accelerates needed relief efforts, and allows international commerce to proceed. Commanders may first invest resources and conduct general engineering tasks to restore facilities for civilian use. Such actions stem future drains on operational resources or facilitate later transition of control back to civilian authorities. For example, repairing police stations, detention facilities, and marksmanship ranges may help urban governments reestablish law and order after completing urban offensive or defensive operations. During stability or civil support operations, the focus of general engineering will often be to support and assist the urban population rather than Army forces.

Fire Fighting Support

10-110. Fire protection and prevention, as well as fire fighting, takes on added significance during UO, particularly offensive and defensive operations. Most ordnance produces heat and flame. This, coupled with an abundance of combustible material (buildings, furniture, gasoline, oil, and propane), poses a serious risk to Soldiers, civilians, and the urban operation. Large shantytowns can exacerbate this problem. In highly combustible areas, commanders may even need to limit or preclude the use of small-arms tracer ammunition. Fire threats to urban areas can be categorized as—

- **Isolated Fires.** These are restricted to a single structure or a specific area within a structure.
- **Area Fires.** These fires consume two or more structures and may extend to encompass an entire block. Generally, streets will serve as firebreaks and help contain the fire within a single block.
- **Firestorms.** These are the most violent and dangerous fires capable of rapidly consuming large areas by creating windstorms and intense heat. Firestorms are often inextinguishable until they have consumed all available combustible materials.

- **Explosive Hazards**. These are present in areas containing fuels and chemicals as well as military explosive hazards.

10-111. When analyzing the administration and human services component of the infrastructure, commanders must determine the adequacy of existing civilian fire fighting support. A deteriorated or nonexistent infrastructure that cannot support the urban area will likely fail to handle the increased risk due to military operations. Commanders may need to provide fire fighting teams to support their own forces and civilians.

10-112. A military force task organized with multiple fire fighting teams (even with maximum use of available civilian fire-fighting assets) will only be able to fight some fires in the AO. Water distribution systems damaged during operations, chemical and other TIM, and hostile activities will further complicate and limit fire fighting capabilities. Commanders must develop priorities for equipment, facility, and infrastructure protection. All Soldiers need training in fire prevention and initial or immediate response fire fighting. Such training includes planning covered and concealed movement, withdrawal, and evacuation routes. Soldiers should be trained to identify and remove ignition and fuel sources and provided additional fire fighting material such as extinguishers, sand, and blankets (see FM 5-415 and the U.S. Department of Transportation's current version of the Emergency Response Guidebook).

Waste Management

10-113. Management of all forms of waste, particularly human, putrescible (such as food), and medical, may become a critical planning consideration for Army forces. This particularly applies if the urban waste management infrastructure was previously inadequate or damaged during UO or natural disasters, the Army force is operating in the urban area for an extended period of time, and a significant number of the urban population remains. Failure to adequately consider this aspect, possibly coupled with the decaying remains of humans and animals and an inadequate or tainted water supply (such as may be caused by flooding), may create unacceptable sanitary and hygiene conditions and subsequently increase DNBIs as well as civilian casualties.

Appendix A

Siege of Beirut: An Illustration of the Fundamentals of Urban Operations

The [Israeli Defense Forces] IDF had neither the strategy nor the experience nor the configuration of forces to fight and sustain a house-to-house campaign in Beirut.

Richard A. Gabriel
Operation Peace for Galilee: The Israeli-PLO War in Lebanon

OVERALL STRATEGIC SITUATION

A-1. In 1982, Israel launched OPERATION PEACE FOR GALILEE designed to destroy the Palestine Liberation Organization (PLO) presence in southern Lebanon. On 1 June, the IDF launched a massive assault across the border into southern Lebanon. The Israeli attack focused on the PLO, but the operations quickly involved major ground and air combat between Israel and Syrian forces.

A-2. In the first few weeks, Israeli forces quickly pushed back both the Syrians and the PLO. However, except for some PLO forces isolated in bypassed urban areas, such as Tyre and Sidon, most of the PLO fell back into Beirut (see Figure A-1). By 30 June, Israeli forces had reached the outskirts of southern Beirut, occupied East Beirut, isolated the city from Syria and the rest of Lebanon, and blockaded the sea approaches to the city. Even so, with most of the PLO intact inside and with significant military and political capability, the Israelis had yet to achieve the objective of OPERATION PEACE FOR GALILEE. The Israeli command had to make a decision. It had three choices: permit the PLO to operate in Beirut; execute a potentially costly assault of the PLO in the city; or lay siege to the city and use the siege to successfully achieve the objective. The Israelis opted for the latter.

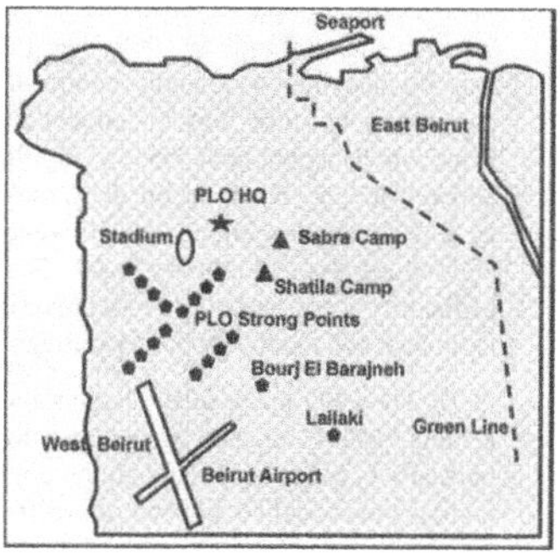

Figure A-1. The city of Beirut

Israeli Military Position

A-3. The Israelis had an excellent position around Beirut. They occupied high ground to the south and west, virtually dominating the entire city. Israeli naval forces controlled the seaward approaches to Beirut. The Israelis' position was also strong defensively, capable of defeating any attempt to break out of or into the city from northern Lebanon or Syria. The Israeli air force had total and complete air superiority. The Israelis controlled the water, fuel, and food sources of West Beirut. Although the PLO forces had stockpiles of food and supplies, the Israelis regulated the food, water, and generating power for the civil population.

A-4. Despite the superior positioning of Israeli forces, the IDF faced significant challenges to include the combat power of the PLO, Syria, and other threats in Beirut. Israeli doctrine and training did not emphasize urban operations. Additionally, Israel was constrained by its desire to limit collateral damage and friendly and noncombatant casualties. Organizationally, the Israeli army was not optimized to fight in urban terrain.

Armor and self-propelled artillery formations dominated the Israeli forces, and most Israeli infantry was mechanized. The Israeli forces had only a few elite formations of traditional dismounted infantry.

PLO MILITARY POSITION

A-5. Despite being surrounded and cut off from support, the PLO position in Beirut offered numerous advantages in addition to the characteristic advantages of urban defense. The PLO had long anticipated an Israeli invasion of southern Lebanon; it had had months to prepare bunkers, obstacles, and the defensive plan of Beirut and other urban areas. Approximately 14,000 Arab combatants in West Beirut readied to withstand the Israeli siege. This was done with the advice of Soviet, Syrian, and east European advisors. The preparation included stockpiling essential supplies in quantities sufficient to withstand a six-month siege. Also, the PLO fighters integrated into the civil populations of the urban areas. Often their families lived with them. The civil population itself was friendly and provided both information and concealment for PLO forces. PLO fighters were experienced in urban combat and knew the urban terrain intimately. PLO forces had been involved in urban fighting against Syrian conventional forces and Christian militias in Beirut several years prior to the Israeli invasion. Finally, the organization of the PLO—centered on small teams of fighters armed with machine-guns and antitank weapons, and trained in insurgent, hit-and-run tactics—was ideally suited to take maximum advantage of the urban environment.

ROLE OF CIVILIANS

A-6. Various ethnic and religious groups make up the civil population of southern Lebanon. However, West Beirut's population was heavily Palestinian and Lebanese. The civil population of West Beirut was between 350,000 and 500,000. The Palestinian population supported the PLO. The Lebanese population may be described as friendly neutral to the Israelis. Although unhappy under Palestinian dominance, this population was unwilling to actively support Israel. The civilian population was a logistic constraint on the PLO, which would have become significant had the siege lasted longer. The civilians in West Beirut were an even larger constraint on the Israelis. The presence of civilians significantly limited the ability of the Israelis to employ firepower. However, the Palestine refugee camps located in West Beirut were both civilian centers and military bases. The Israeli constraints on artillery and other systems against these parts of the city were much less restrictive than in other parts of West Beirut where the population was mostly Lebanese and where fewer key military targets existed.

A-7. The PLO knew of the Israeli aversion to causing civil casualties and purposely located key military centers, troop concentrations, and logistics and weapons systems in and amongst the population—particularly the refugee Palestinian population in the southern part of West Beirut. Tactically, they used the civilians to hide their forces and infiltrate Israeli positions.

A-8. The friendly Palestinian population provided intelligence to the PLO while the friendly Lebanese population provided intelligence for the IDF. Throughout the siege, the IDF maintained a policy of free passage out of Beirut for all civilians. This policy was strictly enforced and permitted no weapons to leave the city. Some estimates are that as many as 100,000 refugees took advantage of this policy.

INFORMATION OPERATIONS

A-9. The siege of Beirut involved using information operations (IO) to influence the media. PLO information operations were aimed at controlling the media and hence the international perception of the operation. This was done by carefully cultivating a select group of pro-PLO media years before hostilities even began. Once hostilities started, only these media sources were permitted to report from the besieged portions of the city, and they were only shown activities that portrayed the IDF negatively. The IDF did not vigorously counter the PLO plan. In fact, the IDF contributed to it by limiting media access to their activities. The PLO information operations had a successful impact. The international community was constantly pressuring the Israeli government to end hostilities. This put pressure on the IDF to conduct operations rapidly and to limit firepower and casualties.

CONDUCT OF THE URBAN OPERATIONS

A-10. The siege of Beirut began 1 July (see Figure A-2). By 4 July, Israeli forces occupied East Beirut, the Green Line separating East and West Beirut, and dominating positions south of the airport. IDF naval forces also controlled the sea west and north of Beirut. On 3 and 4 July, IDF artillery and naval fire began a regular campaign of firing on military targets throughout West Beirut. On 4 July, the IDF cut power and water to the city.

A-11. From 5 to 13 July, the Israeli fires continued to pound PLO targets in West Beirut. The PLO gave one significant response, firing on an Israeli position south of the city and causing several casualties. On 7 July, reacting to international pressure, the IDF returned power and water to West Beirut's civil population. On 11 July, the IDF launched its first attack, probing the southern portion of the airport with an armored task force (see Figure A-3). The PLO repulsed this attack and destroyed several IDF armored vehicles.

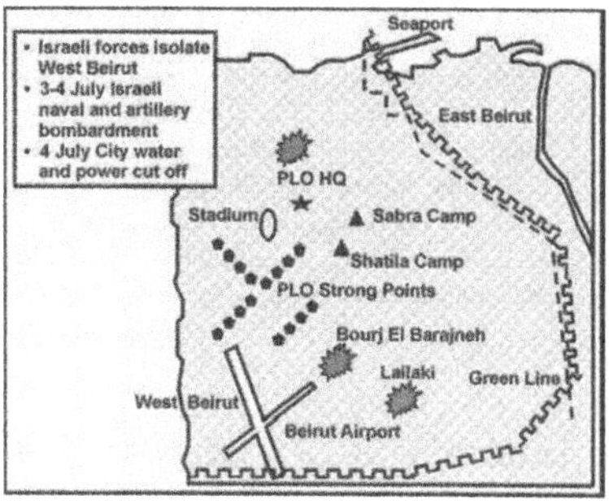

Figure A-2. Initial conduct of the urban operation

A-12. On 13 July, both sides entered into a cease-fire that lasted until 21 July. They began negotiations, mediated by international community, to end the siege. The PLO used this period to continue to fortify Beirut. The Israelis used the time to train their infantry and other arms in urban small unit tactics in Damour, a town the Israeli paratroopers had captured.

A-13. The cease-fire ended on 21 July as PLO forces launched three attacks on IDF rear areas. The Israelis responded with renewed and even more vigorous artillery, naval, and air bombardment of PLO positions in the city. The IDF attacks went on without respite until 30 July. On 28 July, the IDF renewed its ground attack in the south around the airport (see Figure A-4). This time IDF forces methodically advanced and captured a few hundred meters of ground establishing a toehold.

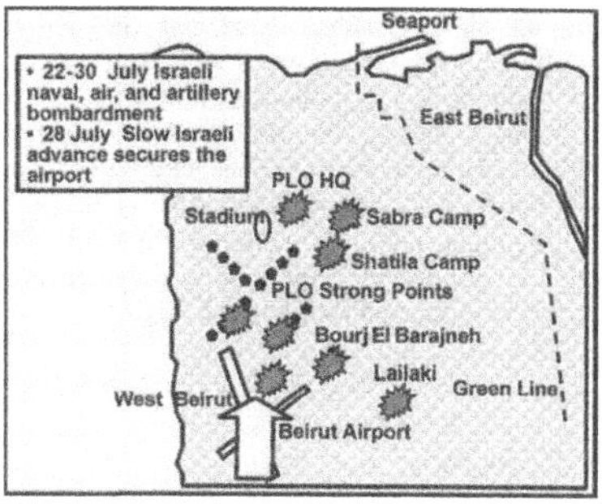

Figure A-3. Israeli probe of PLO defenses

A-14. The Israeli bombardment stopped on 31 July. However, on 1 August the IDF launched its first major ground attack, successfully seizing Beirut airport in the south (see Figure A-5). Israeli armored forces began massing on 2 August along the green line, simultaneously continuing the attack from the south to the outskirts of the Palestinian positions at Ouzai. On 3 August, the Israeli forces continued to reinforce both their southern attack forces and forces along the green line to prepare for continuing offensive operations. On 4 August, the IDF attacked at four different places. This was the much-anticipated major Israeli offensive.

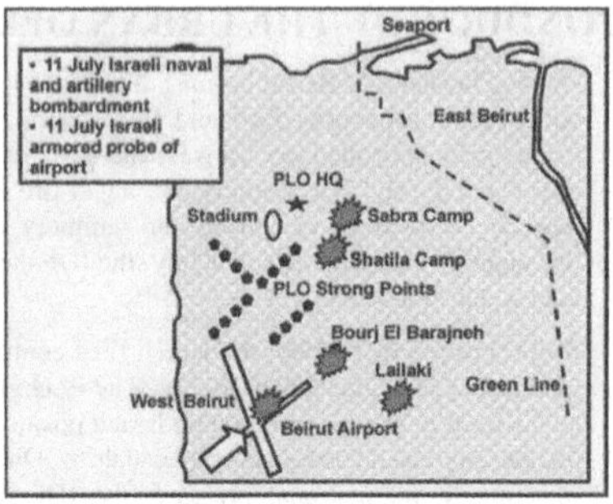

Figure A-4. Initial Israeli attack

A-15. The Israeli attack successfully disrupted the coherence of the PLO defense. The southern attack was the most successful: it pushed PLO forces back to their camps of Sabra and Shatila and threatened to overrun PLO headquarters. Along the green line the IDF attacked across three crossing points. All three attacks made modest gains against stiff resistance. For this day's offensive, the Israelis suffered 19 killed and 84 wounded, the highest single day total of the siege, bringing the total to 318 killed. Following the major attacks on 4 August, Israeli forces paused and, for four days, consolidated their gains and prepared to renew the offensive. Skirmishes and sniping continued, but without significant offensive action. On 9 August, the IDF renewed air and artillery attacks for four days. This activity culminated on 12 August with a massive aerial attack that killed over a hundred and wounded over 400—mostly civilians. A cease-fire started the next day and lasted until the PLO evacuated Beirut on 22 August.

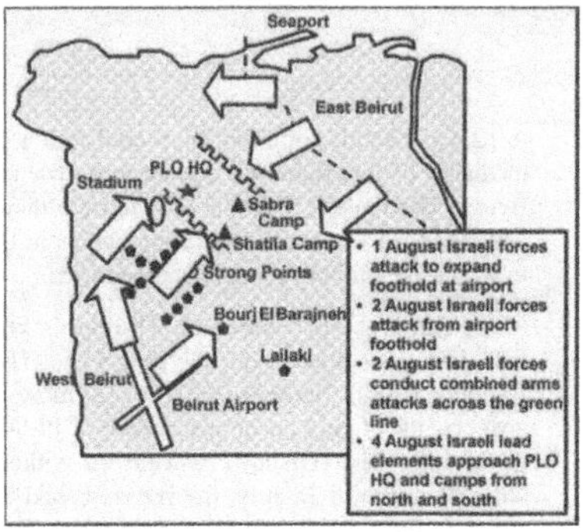

Figure A-5. Final Israeli attack

LESSONS

A-16. The Israeli siege of West Beirut was both a military and a political victory. However, the issue was in doubt until the last week of the siege. Military victory was never in question; the issue in doubt was whether the Israeli government could sustain military operations politically in the face of international and domestic opposition. On the other side, the PLO faced whether they could last militarily until a favorable political end could be negotiated. The answer was that the PLO's military situation became untenable before the Israeli political situation did.

A-17. This favorable military and political outcome stemmed from the careful balance of applying military force with political negotiation. The Israelis also balanced the type of tactics they employed against the domestic aversion to major friendly casualties and international concern with collateral damage.

PERFORM AGGRESSIVE INFORMATION OPERATIONS

A-18. The PLO devoted considerable resources and much planning on how to use IO to their best advantage. They chose to focus on media information sources as a means of influencing international and domestic opinion.

A-19. The PLO's carefully orchestrated misinformation and control of the media manipulated international sentiment. The major goal of this effort was to grossly exaggerate the claims of civilian casualties, damage, and number of refugees—and this was successfully accomplished. Actual casualties among the civilians were likely half of what the press reported during the battle. The failure of the IDF to present a believable and accurate account of operations to balance PLO efforts put tremendous pressure on the Israeli government to break off the siege. It was the PLO's primary hope for political victory.

A-20. In contrast to the overall weak performance in IO, the IDF excelled in psychological operations. IDF psychological operations attacked the morale of the PLO fighter and the Palestinian population. They were designed to wear down the will of the PLO to fight while convincing the PLO that the IDF would go to any extreme to win. Thus defeat was inevitable. The IDF used passive measures, such as leaflet drops and loudspeaker broadcasts. They used naval bombardment to emphasize the totality of the isolation of Beirut. To maintain high levels of stress, to deny sleep, and to emphasize their combat power, the IDF used constant naval, air, and artillery bombardment. They even employed sonic booms from low-flying aircraft to emphasize the IDF's dominance. These efforts helped to convince the PLO that the only alternative to negotiation on Israeli terms was complete destruction.

Maintain a Close Combat Capability

A-21. The ground combat during the siege of Beirut demonstrated that the lessons of tactical ground combat learned earlier in the twentieth century were still valid. Small combined arms teams built around infantry, but including armor and engineers, were the key to successful tactical combat. Artillery firing in direct fire support of infantry worked effectively as did the Vulcan air defense system. The Israeli tactical plan was sound. The Israelis attacked from multiple directions, segmented West Beirut into pieces, and then destroyed each individually. The plan's success strongly influenced the PLO willingness to negotiate. Tactical patience based on steady though slow progress toward decisive points limited both friendly and noncombatant casualties. In this case, the decisive points were PLO camps, strong points, and the PLO headquarters.

A-22. The Israeli willingness to execute close combat demonstrated throughout the siege, but especially in the attacks of 4 August, was decisive. Decisive ground combat was used sparingly, was successful and aimed at decisive points, and was timed carefully to impact on achieving the political objectives in negotiations. The PLO had hoped that their elaborate defensive preparations would have made Israeli assaults so costly as to convince the Israelis not to attack. That the Israelis could successfully attack the urban area convinced the PLO leadership that destruction of their forces was inevitable. For this reason they negotiated a cease-fire and a withdrawal on Israeli terms.

Avoid the Attrition Approach

A-23. The Israelis carefully focused their attacks on objects that were decisive and would have the greatest impact on the PLO: the known PLO headquarters and refugee centers. Other areas of West Beirut were essentially ignored. For example, the significant Syrian forces in West Beirut were not the focus of Israeli attention even though they had significant combat power. Selectively ignoring portions of the urban area allowed the Israelis to focus their combat power on the PLO and limit both friendly casualties and collateral damage.

Control the Essential and Preserve Critical Infrastructure

A-24. The Israeli siege assured Israeli control of the essential infrastructure of Beirut. The initial Israeli actions secured East Beirut and the city's water, power, and food supplies. The Israelis also dominated Beirut's international airport, closed all the sea access, and controlled all routes into and out of the city. They controlled and preserved all that was critical to operating the city and this put them in a commanding position when negotiating with the PLO.

Minimize Collateral Damage

A-25. The Israeli army took extraordinary steps to limit collateral damage, preserve critical infrastructure, and put in place stringent rules of engagement (ROE). They avoided randomly using grenades in house clearing, limited the use of massed artillery fires, and maximized the use of precision weapons. With this effort, the Israelis extensively used Maverick missiles because of their precise laser guidance and small warheads.

A-26. The strict ROE, however, conflicted with operational guidance that mandated that Israeli commanders minimize their own casualties and adhere to a rapid timetable. The nature of the environment made fighting slow. The concern for civilian casualties and damage to infrastructure declined as IDF casualties rose. They began to bring more field artillery to bear on Palestinian strong points and increasingly employed close air support. This tension underscores the delicate balance that Army commanders will face between minimizing collateral damage and protecting infrastructure while accomplishing the military objective with the least expenditure of resources—particularly soldiers. ROE is but one tool among many that a commander may employ to adhere to this UO fundamental.

Understand the Human Dimension

A-27. The Israelis had a noteworthy (although imperfect and at times flawed) ability to understand the human dimension during their operations against the PLO in Beirut. This was the result of two circumstances. First, the PLO was a threat with which the Israeli forces were familiar after literally decades of conflict. Second, through a close alliance and cooperation with Lebanese militia, the Israelis understood a great deal regarding the attitudes and disposition of the civil population both within and outside Beirut.

Separate Noncombatants from Combatants

A-28. Separating combatants from noncombatants was a difficult but important aspect of the Beirut operation. The Israelis made every effort to positively identify the military nature of all targets. They also operated a free passage system that permitted the passage of all civilians out of the city through Israeli lines. The need to impose cease-fires and open lanes for civilians to escape the fighting slowed IDF operations considerably. Additionally, Israeli assumptions that civilians in urban combat zones would abandon areas where fighting was taking place were incorrect. In many cases, civilians would try to stay in their homes, leaving only after the battle had begun. In contrast, the PLO tied their military operations closely to the civilian community to make targeting difficult. They also abstained from donning uniforms to make individual targeting difficult.

A-29. Earlier in OPERATION PEACE FOR GALILEE when the IDF attacked PLO forces located in Tyre, Israeli psychological operations convinced 30,000 Lebanese noncombatants to abandon their homes and move to beach locations outside the city. However, the IDF was subsequently unable to provide food, water, clothing, shelter, and sanitation for these displaced civilians. IDF commanders compounded the situation by interfering with the efforts by outside relief agencies to aid the displaced population (for fear that the PLO would somehow benefit). Predictably, many civilians tried to return to the city complicating IDF maneuver and targeting—that which the separation was designed to avoid. IDF commanders learned that, while separation is important, they must also adequately plan and prepare for the subsequent control, health, and welfare of the noncombatants they displace.

Restore Essential Services

A-30. Since essential services were under Israeli command, and had been since the beginning of the siege, the Israelis had the ability to easily restore these resources to West Beirut as soon as they adopted the cease-fire.

Transition Control

A-31. In the rear areas of the Israeli siege positions, the Israeli army immediately handed over civic and police responsibility to civil authorities. This policy of rapid transition to civil control within Israeli lines elevated the requirement for the Israeli army to act as an army of occupation. The Israeli army believed the efficient administration of local government and police and the resulting good will of the population more than compensated for the slightly increased force protection issues and the increased risk of PLO infiltration.

A-32. Upon the cease-fire agreement, Israeli forces withdrew to predetermined positions. International forces under United Nations (UN) control supervised the evacuation of the PLO and Syrian forces from Beirut. These actions were executed according to a meticulous plan developed by the Israeli negotiators and agreed to by the PLO. Israeli forces did not take over and occupy Beirut as a result of the 1982 siege (an occupation did occur later but as a result of changing situations).

SUMMARY

A-33. The Israeli siege of West Beirut demonstrates many of the most demanding challenges of urban combat. Apart from Israel's poor understanding of the strategic influence of the media on operations (a lesson that they learned and subsequently applied to future operations), the IDF's successful siege of Beirut emerged from their clear understanding of national strategic objectives and close coordination of diplomatic efforts with urban military operations. A key part of that synchronization of capabilities was the understanding that the efforts of the IDF would be enhanced if they left an escape option open to the PLO. This option was the PLO's supervised evacuation that occurred after the siege. Although the PLO was not physically destroyed, the evacuation without arms and to different host countries effectively shattered the PLO's military capability. Had Israel insisted on the physical destruction of the PLO in Beirut, the overall operation might have failed because that goal may not have been politically obtainable in view of the costs in casualties, collateral damage, and international opinion.

This page intentionally left blank.

Appendix B

An Urban Focus to the Intelligence Preparation of the Battlefield

Maneuvers that are possible and dispositions that are essential are indelibly written on the ground. Badly off, indeed, is the leader who is unable to read this writing. His lot must inevitably be one of blunder, defeat, and disaster.

Infantry in Battle

The complexity of the urban environment and increased number of variables (and their infinite combinations) increases the difficulty of providing timely, relevant, and effective intelligence support to urban operations (UO). Conducted effectively, however, the intelligence preparation of the battlefield (IPB) allows commanders to develop the situational understanding necessary to visualize, describe, and direct subordinates in successfully accomplishing the mission.

IPB is the systematic process of analyzing the threat and environment in a specific geographic area—the area of operations (AO) and its associated area of interest (see figure B-1). It provides the basis for intelligence support to current and future UO, drives the military decision-making process, and supports targeting and battle damage assessment. The procedure (as well as each of its four steps) is performed continuously throughout the planning, preparation, and execution of an urban operation.

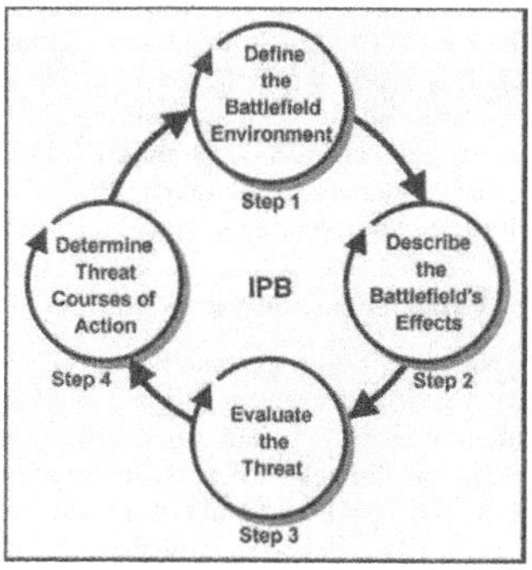

Figure B-1. The steps of IPB

UNAFFECTED PROCESS

B-1. The IPB process is useful at all echelons and remains constant regardless of the operation or environment. However, an urban focus to IPB stresses some aspects not normally emphasized for IPBs conducted for operations elsewhere. The complex urban mosaic is composed of the societal, cultural, or civil dimension of the urban environment; the overlapping and interdependent nature of the urban infrastructure; and the multidimensional terrain. This mosaic challenges the conduct of an urban-focused IPB. There is potential for full spectrum operations to be executed near-simultaneously as part of a single major operation occurring in one urban area with multiple transitions. Such multiplicity stresses the importance of a thorough, non-stop IPB cycle aggressively led by the commander and executed by the entire staff. Overall, the art of applying IPB to UO is in properly employing the steps to the specific operational environment. In UO, this translates to analyzing the significant characteristics of the environment, the role that its populace has in threat evaluation, and understanding how these affect the planning and execution of UO. FM 2-01.3 details how to conduct IPB; FM 34-3 has the processes and procedures for producing all-source intelligence. This appendix supplements the information found there; it does not replace it.

Increased Complexity

B-2. Uncovering intricate relationships takes time, careful analysis, and constant refinement to determine actual effects on friendly and threat courses of action (COAs). These relationships exist among—

- Urban population groups.
- The technical aspects of the infrastructure.
- The historical, cultural, political, or economic significance of the urban area in relation to surrounding urban and rural areas or the nation as a whole.
- The physical effects of the natural and man-made terrain.

B-3. A primary goal of any IPB is to accurately predict the threat's likely COA (step four—which may include political, social, religious, informational, economic, and military actions). Commanders can then develop their own COAs that maximize and apply combat power at decisive points. Understanding the decisive points in the urban operation allows commanders to select objectives that are clearly defined, decisive, and attainable.

Reducing Uncertainty and Its Effects

B-4. Commanders and their staffs may be unfamiliar with the intricacies of the urban environment and more adept at thinking and planning in other environments. Therefore, without detailed situational understanding, commanders may assign missions that their subordinate forces may not be able to achieve. As importantly, commanders and their staffs may miss critical opportunities because they appear overwhelming or impossible (and concede the initiative to the threat). They also may fail to anticipate potential threat COAs afforded by the distinctive urban environment. Commanders may fail to recognize that the least likely threat COA may be the one adopted precisely because it is least likely and, therefore, may be intended to maximize surprise. Misunderstanding the urban environment's effect on potential friendly and threat COAs may rapidly lead to mission failure and the unnecessary loss of Soldiers' lives and other resources. A thorough IPB of the urban environment can greatly reduce uncertainty and contribute to mission success.

Training, Experience, and Functional Area Expertise

B-5. Not all information about the urban environment is relevant to the situation and mission—hence the difficulty and the reason for IPB and intelligence analysis. Although it may appear daunting, institutional education, unit training, and experience at conducting intelligence support to UO will improve the ability to rapidly sort through all the potential information to separate the relevant from merely informative. (This applies to any new or difficult task.) The involvement and functional expertise of the entire staff will allow commanders to quickly identify the important elements of the environment affecting their operations. Fortunately, IPB as part of the entire intelligence process (plan, prepare, collect, process, and produce) is comprehensive enough to manage the seemingly overwhelming amounts of information coming from many sources. Accomplished properly, it allows commanders to recognize opportunities often without complete information.

B-6. As in any operational environment, tension exists between the desire to be methodical and the need to create the tempo necessary to seize, retain, and exploit the initiative necessary for decisive UO. Quickly defining the significant characteristics of the urban environment requiring in-depth evaluation (not only what we need to know but what is possible to know) allows rapid identification of intelligence gaps (what we know versus what we don't know). Such identification leads to information requirements and will drive the intelligence, surveillance, and reconnaissance (ISR) plan (how will we get the information we need). Commanders must carefully consider how to develop focused priority information requirements (PIR) to enable collectors to more easily weed relevant information from the plethora of information. Commanders can then make better decisions and implement them faster than urban threats can react.

AMPLIFIED IMPORTANCE OF CIVIL (SOCIETAL) CONSIDERATIONS

B-7. The Army focuses on warfighting. The experiences in urban operations gained at lower echelons often center on the tactics of urban offensive and defensive operations where the influences of terrain and enemy frequently dominate. At higher echelons, the terrain and enemy are still essential considerations, but the societal component of the urban environment is considered more closely and throughout the operational process. Moreover, the human or civil considerations gain importance in civil support or stability operations regardless of the echelon or level of command. In addition to the echelon and the type of operation, a similar relationship exists between the key elements of the urban environment and other situational factors. These factors can include where the operation lies within the range of operations or the level of war and the conventional or unconventional nature of the opposing threat. Figure B-2 graphically represents the varying significance of these elements to urban IPB. Overall, population effects are significant in how they impact the threat, Army forces, and overall accomplishment of strategic and operational goals.

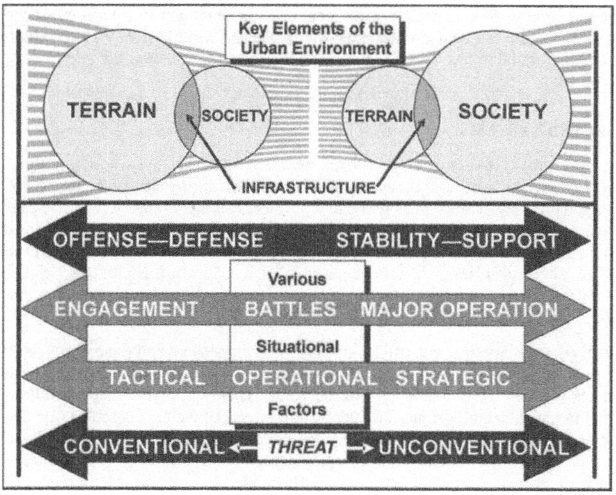

Figure B-2. Relativity of key urban environment elements

B-8. Describing the battlefield's effects—step two in IPB—ascribes meaning to the characteristics analyzed. It helps commanders understand how the environment enhances or degrades friendly and threat forces and capabilities. It also helps commanders understand how the environment supports the population. It also explains how changes in the "normal" urban environment (intentional or unintentional and because of threat or friendly activities) may affect the population. Included in this assessment are matters of perception. At each step of the IPB process, commanders must try to determine the urban society's perceptions of ongoing activities to ensure Army operations are viewed as intended. Throughout this process, commanders, staffs, and analysts cannot allow their biases—cultural, organizational, personal, or cognitive—to markedly influence or alter their assessment (see FM 34-3). This particularly applies when they analyze the societal aspect of the urban environment. With so many potential groups and varied interests in such a limited area, misperception is always a risk.

SIGNIFICANT CHARACTERISTICS

B-9. Urban intelligence analysis must include consideration of the urban environment's distinguishing attributes—man-made terrain, society, and infrastructure—as well as the underlying natural terrain (to include weather) and the threat. Because the urban environment is so complex, it is often useful to break it into categories. Then commanders can understand the intricacies of the environment that may affect their operations and assimilate this information into clear mental images. Commanders can then synthesize these images of the environment with the current status of friendly and threat forces and develop a desired end state. Then they can determine the most decisive sequence of activities that will move their forces from the current state to the end state. Identifying and understanding the environment's characteristics (from a friendly, threat, and noncombatant perspective) allows commanders to establish and maintain situational understanding. Then they can develop appropriate COAs and rules of engagement that will lead to decisive mission accomplishment.

B-10. Figures B-3, B-4, and B-5 are not intended to be all-encompassing lists of urban characteristics. Instead, they provide a starting point or outline useful for conducting an urban-focused IPB and analysis that can be modified to fit the specific operational environment and meet the commander's requirements. Commanders and staffs can compare the categories presented with those in the civil affairs area study and assessment format found in FM 41-10 and the IPB considerations for stability and civil support operations found in ST 2-91.1.

INTERCONNECTED SYSTEMS

B-11. Since the urban environment comprises an interconnected "system of systems," considerations among the key elements of the environment will overlap during an urban intelligence analysis. For example, boundaries, regions, or areas relate to a physical location on the ground. Hence, they have urban terrain implications. These boundaries, regions, or areas often stem from some historical, religious, political, administrative, or social aspect that could also be considered a characteristic of the urban society. Overlaps can also occur in a specific category, such as infrastructure. For instance, dams are a consideration for their potential effects on transportation and distribution (mobility), administration and human services (water supply), and energy (hydroelectric).

B-12. This overlap recognition is a critical concern for commanders and their staffs. In "taking apart" the urban environment and analyzing the pieces, commanders and staffs cannot lose perspective of how each piece interacts with any other and as part of the whole. Otherwise, their vision will be shortsighted, and they will fail to recognize the second- and third-order effects of their proposed COAs; the actual end state differing dramatically from the one envisioned by the commander. The increased density of combatants and noncombatants, infrastructure, and complex terrain means that a given action will likely have unintended consequences—positive or negative. Those consequences will be more widely felt and their impact will spread in less time than in other environments. These unintended results may have important strategic and operational consequences. The multiple ways these dynamic urban elements and characteristics combine make it necessary to approach each urban environment as a unique challenge for intelligence analysis.

URBAN TERRAIN AND WEATHER

Terrain

B-13. Earlier admonitions that civil considerations are more closely considered in UO do not necessarily mean that consideration for urban terrain is de-emphasized. In every urban operation, terrain and its effects on both threat and friendly forces must be assessed and understood. Then commanders can quickly choose and exploit the terrain (and weather conditions) that best supports their missions. Effective terrain analysis thoroughly assesses urban structures as well as the ground on which they stand (see figure B-3 and FM 5-33). An analysis of urban terrain first considers broader urban characteristics and effects and then progresses to a more detailed examination.

Natural Terrain

B-14. The natural terrain features beyond the urban area and beneath urban structures significantly influence unit operations. They dictate where buildings can be constructed, the slope and pattern of streets, and even the broad urban patterns that develop over longer periods of time, thereby influencing a unit's scheme of maneuver. The military aspects of terrain—observation and fields of fire, avenues of approach, key terrain, obstacles, and cover and concealment (OAKOC)—remain critical to the analysis of natural terrain in, under, and around urban areas. Fortunately, commanders and their staffs are normally accustomed to this type of analysis.

Forms and Functions

- Cores
- Industrial areas
  - Toxic industrial material production and storage facilities
  - Standard signs and markings for toxic chemicals
- Outlying high-rise areas
- Residential areas and shantytowns
- Commercial ribbon areas
- Forts and military bases

Broad Urban Patterns

- Types
  - Satellite
  - Network
  - Linear
  - Segment
- Dominant or central hub (if any)
- Area covered (square miles)

Street Patterns

- Basic Types
  - Radial
  - Grid
  - Irregular (planned and unplanned)
- Variations
  - Rayed
  - Radial-Ring
  - Contour-forming
  - Combined
- Widths

Construction and Placement

- Construction
  - Mass or framed
  - Light or heavy clad
  - Material (dirt, wood, stone, brick, cinder block, concrete, steel, and glass)
  - Density and thickness (roofs, floors, and interior and exterior walls)
  - Load bearing walls and columns
  - Height (floors)
  - Doors, windows, fire escapes, and other openings
  - Interior floor plan (including crawl spaces, elevators, and stairs)
- Placement
  - Random
  - Close-orderly block
  - Dispersed
- Ownership

Military Aspects of Terrain: OAKOC

- Observation and fields of fire
  - Smoke (fire), dust (explosions), and flying debris
  - Rubble
  - Engagement ranges (including minimum safe distances and backblast factors) and obliquity/angle (ricochets)
  - Elevation and depression considerations
  - Lasers and reflective concerns
- Avenues of approach (mobility corridors)
  - Airspace
  - Surface
  - Supersurface
  - Subsurface
- Key terrain
  - Landmarks
  - Buildings of significant cultural, social, political, historical, or economic significance
- Obstacles
  - Rubble and vehicles
  - Steep embankments
  - Medians
  - Inadequate bridges and overpasses (destroyed, weight-restricted, or narrow)
  - Tunnels and underpasses (destroyed or narrow)
  - Mines and roadside improvised explosive devices
  - Masking of fires
  - Burning buildings or other fire hazards
  - Rivers and lakes
- Cover and concealment
  - Building protection
  - Weapon penetration (single shot and multiple rounds) considerations
  - Rubble and vehicles

Figure B-3. Significant urban terrain characteristics

Analysis of an Urban Area's Underlying Terrain: Mitrovica, Kosovo – 1999

An urban area's underlying terrain provides many clues into its history, economy, society, and current situation. Mitrovica, Kosovo is an illustrative example. The Ibar River creates a natural line of communications through the middle of the city as well as an obstacle that bisects the urban area. This bisection naturally divides the two resident ethnic groups: Albanians and Serbs. The separation became significant at both the strategic and tactical levels during 1999 deployments to Kosovo. Army forces had to ensure that the Orthodox Church located south of the Ibar was accessible to Serbs residing in the north. North Atlantic Treaty Organization (NATO) peacekeepers built a footbridge across the river that allowed reliable, safe passage. The natural feature separating the two groups assisted NATO troops in maintaining stability in the region.

Man-Made Terrain

B-15. Building composition, frontages, placement, forms and functions, size, floor plans, and window and door locations affect maneuver, force positioning, and weapons deployment considerations. Angles, displacement, surface reflection, and antenna locations influence command and control. Structures also influence ISR operations. The increased density and volume created by man-made structures increases how much information commanders and their staffs collect and assess as well as the number of forces required. Building materials and construction will also influence force structures to include weapons and equipment required. Heavily constructed buildings combined with hot and cold extremes may affect target identification for thermal sights. Thick walls, for example, may make combat vehicle identification difficult by distorting hotspots. Additionally, the increased use of heaters and warming fires may clutter thermal sights with numerous hotspots. The ability to maneuver through the urban dimensions—airspace, supersurface, surface, and subsurface—and shoot through walls, ceilings, and floors also creates increased psychological stress. The physical characteristics of man-made terrain can also be analyzed using OAKOC.

Weather

B-16. Weather and its effects are often considered when examining the military aspects of terrain. Military aspects of weather include temperature (heat and cold), light conditions, precipitation (cloud cover, rain, snow, fog, and smog), and wind. Their military effects during UO are similar to any operational environment (see FM 34-81 and FM 34-81-1). Extremes of heat and cold affect weapon systems and the Soldiers that operate them. The extra luminescence provided by the ambient light of an urban area, unless controlled, may affect night vision capabilities and the ability of the Army to "own the night." Precipitation affects mobility and visibility. Smog inversion layers are common over cities. An inversion layer may trap smoke and chemicals in the air and increase ground and air temperatures to the detriment of Soldiers' health. (If the conditions are severe enough, it might require the use of protective masks.) Changes in temperature as a result of air inversions can also affect thermal sights during crossover periods of warm to cold and vice versa. Winds, which may increase as they are funneled through urban canyons, may—

- Increase other weather effects (for example, wind chill).
- Decrease visibility (blowing debris, sand, rain, and snow).
- Spread radiation, biological, and chemical hazards.
- Adversely affect low-altitude air mobility and airborne ISR platforms.

B-17. Commanders also analyze weather for its potential effect on civilians and civilian infrastructure as well as Soldiers and military equipment. Rain might create sewage overflow problems in urban areas with collapsed sewage infrastructure and in refugee camps, increasing disease and even creating panic. Rain and flooding may also make some subsurface areas impassable or extremely hazardous to civilian and military forces alike. Chemical agents and TIM may be washed into underground systems creating toxic hotspots. These effects become more pronounced as chemical agents or TIM are absorbed by brick and unsealed concrete sewers. Other weather effects on UO can include—

- Heavy snowfall in an urban area that may paralyze area transportation and distribution infrastructure, hindering the urban administration's ability to provide vital human services (police, fire fighting, medical, and rescue). Heavy rains and flooding may have similar effects especially on poorly designed and constructed roads or roads that have been damaged by tracked vehicles.
- Extreme hot and cold weather climates that increase the dependence (and military significance) of many elements of the infrastructure. For example, the energy infrastructure may be critical; without it, civilians may not be adequately cooled or heated or they may not be able to cook their food.

- In urban areas located in tropical regions, it can rain at the same time each day during the wet season. Threat forces may attack during these periods knowing aircraft will have difficulty responding. Bad weather also reduces the effectiveness of surveillance, direct and indirect fire, and logistic support.
- Inclement weather may preclude demonstrations or rallies by threats. Good weather may mean a maximum turnout of civilians for events such as festivals, sporting events, and other social, cultural, or religious activities.
- Severe weather may affect psychological, civil-military and humanitarian assistance operations. Heavy rains and severe dust storms may disrupt leaflet drops, construction projects, food and water distribution, and medical and veterinary assistance programs.

URBAN SOCIETY

B-18. This manual shows that societal considerations take on added importance in UO. Critical to operational success is knowing and understanding which groups live in an urban area, what relationships exist among them, and how each population group will respond to friendly and threat activities. Often determining any of this is very difficult. Cultural acuity is also essential in helping commanders and their staffs to view the urban area as the residents view it. The demographics presented depict what conditions exist, while the other categories help to explain the root causes or why conditions exist (see figure B-4 on page B-8). Other categories besides basic demographics that are important to gain this understanding include health, history, leadership, ethnicity and culture, religion, and government and politics.

From Cultural Awareness to Understanding and Competence

B-19. This societal understanding must go beyond a superficial awareness of gestures and taboos to developing a thorough understanding of the organizing principles that make the urban society hold together and function as a coherent entity—or not. In UO, particularly longer-term stability operations, commanders use this knowledge and insight to design and plan operations that work effectively within that society. For example in OPERATION IRAQI FREEDOM, Army units, as they shifted to long-term stability operations, were required to recruit, organize, equip, and train indigenous security forces in order to be able to transition security responsibility from U.S. forces to the burgeoning Iraqi government. Army commanders were successful—to a point. The first units that they built and trained attained a remarkable level of unity and tactical ability in a relatively short period of time. The newly formed units effectively fought alongside Army forces. However when prodded into conducting operations with minimal Army supervision, many Iraqi units lost unity and coherence. Army commanders initially failed to account for a basic Iraqi societal consideration: their society is patriarchal—each member or organization in Iraqi society always seeks a patron or sponsor to support them and provide for their needs. When Iraqi units trained and operated with Army forces, this relationship was satisfied. Operating on their own, however, Iraqi commanders and units lost this patronage without an equivalent Iraqi replacement. In hindsight, the establishment of an effective and more robust ministry of defense, either as a precursor or simultaneously with the establishment of tactical units, may have provided the Iraqi security forces the patronage that their culture required.

Resource the Societal Analysis

B-20. Aside from friendly and threat forces, the society is the only thinking component of the urban environment able to rapidly impact the urban operation. (Even people going about their daily routines can unwittingly hamper the mission.) Urban residents create conditions for restrictive rules of engagement, increase stress on Soldiers and logistic capabilities, and confuse threat identification (see Threat Considerations in this appendix). Demographic, health, safety, ethnic, and cultural concerns will be essential considerations in most UO. Other situational factors—the mission, enemy, and time available—dictate the balance between the level of detail and analysis to support the overall urban operation with the level of detail that commanders and their staffs can achieve. However, an analysis that fails to devote enough time and resources to understanding the urban society can find large elements of the population turned against the Army force. Detailed analysis of the urban society will help save time and ISR resources. While not necessarily a resource saver in the short term, a thorough understanding of the urban society and its culture ultimately contributes to faster achievement of strategic goals.

Population Demographics

- General population size
  - Village
  - Town
  - City
  - Metropolis
  - Megalopolis
- Group size based on race age, sex, political affiliation, economics, religion, tribe, clan, gang, criminal activities, or other significant grouping
  - Significant US or allied populations
  - Distribution, densities, and physical boundaries and overlaps
  - Majority, minority, and dominant groups
- Increasing or decreasing migration trends
  - Dislocated civilians
- Nongovernmental organizations (NGOs)
  - Local
  - National
  - International
- Languages (distribution, dialects, relationship to social structure)
- Educational levels and literacy rates
- Crime rates
- Birth and death rates
- Labor statistics and considerations
  - Skilled and unskilled
  - Imported and exported
  - Unemployment
  - Standard wages and per capita income
  - Workday and workweek norms

Health

- Diseases
- Nutritional deficiencies
- Local standards of care
- Pollution and environmental hazards (air, water, food, and soil)
- Health workers (types, numbers, and degree of skill)

History

- General and for a specific group
  - Internal or external
  - Recent conflicts
- Relationship with US, allies, and other participating multinational forces
- Applicable international treaties
- Status-of-forces agreements
- Antagonists/protagonists
- Heroes
- Events, facts, and dates considered important or celebrated
- Urban area's historical importance

Leadership and Prominent Personalities

- Identification, location, and prioritization of influential leaders (exploitation, evacuation, protection, etc.)
- Affiliation (ethnic, religion, military, government, industry, criminal, or entertainment)
- Education attained
- Organization and distribution of power
- Associations among different leaders and groups

Ethnicity and Culture

- Values, moral codes, taboos, and insults (verbal and non-verbal)
- Attitudes towards age, sex, and race (including same-sex interaction)
- Role of the clan, tribe, or family
- Biases between ethnic groups
- Privacy and individuality
- Recreation, entertainment, and humor
- Fatalism or self-determination
- Exchanges of gifts
- Displays of emotion
- Lines of authority
- Dating and marriage
- Greetings, leave-takings, and gestures
- Visiting practices
- Alcohol and drug use
- Important holidays, festivals, sporting, or entertainment events
- Eating and dietary practices
- Significance of animals and pets
- Urban-rural similarities and differences
- Driving habits
- Clothing

Religion

- Sects, divisions, and overlaps
- Religious biases and problems
- Relationship and influence on government, politics, economics, and education
- Impact on ethnic and cultural beliefs
- Key events or celebrations (daily, weekly, monthly, or annually)
- Funeral and burial practices

Government and Politics

- Present and past forms
- Organization and powers (executive, legislative, judicial, and administrative divisions)
- Scheduled elections and historical turnouts
- Degree of control over the population
  - Identification required
  - Border-crossing procedures
- Relationship with US or multinational governments, national governments, and criminal elements
- Political factions and boundaries
- Political traditions
- Grievances
- Censorship
- Nepotism and other clan, tribal, or social ties
- Civil defense and disaster preparedness (organization, plans, training, equipment, and resources)
  - Evacuation routes
- Legal system
  - System of laws
  - Applicable treaties
  - Courts and tribunals
  - Procedures
  - Records (birth and deeds)
- Property control
- Monetary system (formal and informal)
- Domestic and foreign trade
  - Taxation and tariffs
  - Customs requirements
  - Rationing and price controls
  - Economic performance and contribution to gross national product
- Economic aid
- Perception of relative deprivation
- Trade unions
- Competition with the black market and organized crime

Figure B-4. Significant urban societal characteristics

Transportation and Distribution

- Water
  - Shipyards and other port and harbor facilities
  - Inland waterways, canals, and locks
  - Offshore pipeline berths
  - Cargo storage and handling
  - Types and number of ships, boats, and ferries
  - Dams
- Streets and roads
  - Bridges and fords
  - Over- and underpasses
  - Raised embankments, tunnels, culverts, and other subterranean features (widths and clearances)
  - Parking areas (surface, subsurface, and supersurface)
  - Weight restrictions
  - Traffic light operations
  - Traffic patterns
  - Widths
  - Surface materials
- Rail
  - Lines
  - Terminals
  - Switchyards and junctions
  - Subways, bridges, elevated rail lines, and underpasses (clearances)
  - Track gauges
  - Types and number of rolling stock
  - Electrification
- Air
  - Airfields and runways (including capabilities)
  - Heliports and helipads (including rooftop)
  - Types and number of aircraft
  - Cargo storage and handling
- Trucking companies and delivery services
- Available material-handling equipment
- Rush hour and market time considerations
- Seasonal (weather) effects
- Rubble effects
- Impact of dislocated civilians and migration patterns
- Likely population congregation points
- Identifiable primary and alternate lines of communication

Economics and Commerce

- Industries
  - Types and locations
  - Important companies (including US or allied)
  - Military production facilities
- Sources of raw materials
- Use of toxic industrial materials and biological agents
  - Agriculture (insecticides, herbicides, and fertilizers)
  - Manufacturing
  - Cleaning
  - Research
- Food types, quantities, and sources
  - Requirements and availability
  - Storage and processing
  - Cleanliness standards
- Stores, shops, restaurants, hotels, and strip malls
- Recreation facilities
  - Outdoor and amusement parks
  - Stadiums and other sports facilities
- Machine shops
- Brick and lumber yards
- Banking and investment institutions

Administration and Human Services

- Police and fire protection
  - Headquarters, station, and key facilities locations
  - Organization and strengths
  - Equipment
  - Functions, authority, and jurisdictional boundaries
  - Contract guard services
- Welfare and public assistance
  - Monetary assistance
  - Orphanages
  - Elderly care facilities
- Water supply systems
  - Water sources and storage (lakes, reservoirs, cisterns, pools, and public baths)
  - Water treatment and quality
  - Pumping stations and other distribution methods (trucks, bottles)
  - Hydrant locations
- Snow removal capabilities
- Street light operations
- Health facilities
  - Hospitals
  - Emergency medical services
  - Mental institutions
  - Medical supplies and equipment
  - Research and pharmaceutical buildings
  - Blood banks
- Governmental buildings
  - Embassies
  - Capitol building
  - Legislative, judicial, and ministry buildings
  - Hall of records
- NGOs providing assistance
- Effects of military control measures on providing vital human services
- Waste and sanitation
  - Types (solid, sewage, and toxic)
  - Collection, processing, and disposal
  - Dumps or landfills
  - Drainage systems

Cultural

- Religious buildings and places of worship
- Shrines, monuments and other historical structures
- Schools
- Museums
- Theaters
- Libraries

Energy

- Types
  - Electric
  - Oil
  - Coal
  - Natural gas
  - Nuclear
  - Solar
  - Hydroelectric
  - Geothermal
- Facilities
  - Production and processing
  - Storage
- Distribution
  - Pipelines (above and below ground)
  - Power lines (overhead and underground)
  - Water, rail, and road
  - Potential hazards

Communication and Information

- Print media
  - Newspapers, periodicals, and pamphlets
  - Billboards and posters
  - Postal facilities
- Telephone facilities
  - Wire or wireless
  - Facsimile machines
- Telegraph facilities
- Radio facilities
- Police, fire, and rescue systems
- Security systems
- Television facilities
- Computers and the Internet
- Antennas, towers, relay stations, and lines (surface and subsurface)
- Integration of space-based capabilities
- Public forums and speech
- Low-technology media (car horns, drums, graffiti, and burning fires)
- Key media organizations and reporters
  - Local
  - International
  - US

Figure B-5. Significant urban infrastructure characteristics

URBAN INFRASTRUCTURE

B-21. Functional and analytical interdependence is readily apparent when examining urban infrastructures (see figure B-5 on page B-9). They are composed of physical structures or facilities and people. Hence, much of the analysis conducted for terrain and society can apply when assessing the urban infrastructure. For example, commanders, staffs, and analysts could not effectively assess the urban economic and commercial infrastructure without simultaneously considering labor. All aspects of the society relate and can be used to further analyze the urban work force since they are a sub-element of the urban society. Similarly, the OAKOC aspects used to evaluate terrain may also apply to the urban infrastructure, especially considerations of key terrain. Depending on the mission, elements of the urban infrastructure may become key terrain. While infrastructure analysis is not an engineer-specific task, Army engineers may provide the general and geospatial engineering expertise necessary to better understand the urban terrain and infrastructure.

THREAT CONSIDERATIONS

> *... [T]he adversaries of freedom ... send arms, agitators, aid, technicians and propaganda to every troubled area.... [S]ubversives and saboteurs and insurrectionists . ... [possess] the power to conscript talent and manpower for any purpose. and long experience in the techniques of violence and subversion....It is a contest of will and purpose as well as force and violence—a battle for minds and souls as well as lives and territory.*
>
> John F. Kennedy
> Message to Congress, 5 May 1961

B-22. Chapter 3 outlines the instability and uncertainty of the strategic environment. Commanders, staffs, and analysts identify and analyze the threat's composition, strength, disposition, leadership, training, morale, weapons and capabilities, vulnerabilities, internal logistics and external support, doctrine (if any), strategy or modus operandi, and tactics. The threat can take a variety of forms:

- Conventional military forces.
- Paramilitary forces.
- Guerrillas and insurgents.
- Terrorists.
- Militia or special police organizations.

B-23. A general study of guerrilla and insurgent tactics, techniques, and procedures may prove beneficial to many types of operations regardless of the actual composition or type of threat forces. Insurgent strategies and tactics may work especially well in this complex environment and will likely be a part of any threat COA. Particularly, commanders should understand how a threat might restrict itself by the laws of land warfare and similar conventions, or exploit the use of these conventions to its own gain. Commanders can refer to FM 31-20-3 and FMI 3-07.22 for more information. For many of the above threats, no doctrinal templates may exist. Intelligence staffs will need to identify and track common threat patterns and tactics, techniques, and procedures as they develop and change during the course of the operation. Understanding culture will be critical to determining threat motives that drive decisions. Comman415081ders, staffs, and analysts must evaluate, update (or create), and manage threat databases early (and continuously) during the operational process.

URBAN ENVIRONMENTAL THREATS

B-24. While threats vary, they share a common characteristic: the capability and intent to conduct violence against Army forces to negatively influence mission accomplishment. These threats are often the most recognizable for forces trained for warfighting—these are often the enemy. In order for Army units to be able to conduct full spectrum urban operations, commanders broaden their concept of the threat when analyzing the urban environment's terrain, societal, and infrastructure characteristics. This analysis includes many environmental dangers (potentially affecting both sides of a conflict as well as noncombatants) such as—

- Natural disasters (earthquakes, tornados, hurricanes, volcanic eruptions, fires, floods, tidal waves or tsunamis, and heavy snows).
- Food shortages.
- Water shortages (both quantity and quality).
- Rampant disease.
- Pollution and toxic industrial materials.

B-25. A critical difference between the latter forms of threat and the former is the lack of intent to do harm. The latter may stand alone as threats, or these conditions may be created, initiated, or used by the enemy or a hostile civilian group as a weapon or tool. Threat analysis includes identifying and describing how each relevant characteristic of the area of operations can hinder mission accomplishment. This analysis, particularly during stability or civil support operations, will require extensive coordination and cooperation with urban civil authorities, law enforcement, and numerous governmental and nongovernmental organizations.

CIVILIANS

B-26. In a major combat operation or campaign where offensive and defensive operations are conducted against a conventional enemy, threat identification is more readily accomplished. However, adaptation may be necessary to even further broaden the threat concept to include specific elements of the urban society and, in some instances, nongovernmental organizations (NGOs) and other civilian agencies working in the urban area. In many stability operations, this modification can account for opposing armed forces that are not an enemy but are a threat to the mission. As discussed in Chapter 9, Army forces in many stability or civil support operations must avoid classifying or thinking of these threats as the enemy. Further, commanders ensure that Soldiers do not shift the animosity that they feel toward urban threats onto the civilian population. In OPERATION IRAQI FREEDOM, for example, some Soldiers began to refer to Iraqi civilians as "hajjis" in reference to the annual pilgrimage—the Hajj—that forms one of the tenets or pillars of Islamic faith. Those who have completed the pilgrimage are entitled to add the phrase al-Hajj or hajji (pilgrim) to their name. However, the Soldiers were not using the term in its proper context but rather to denigrate. In response, commanders prohibited the inappropriate use of this term as one of many measures to stress and develop cultural understanding and respect for the people that Army forces were there to assist.

Need for More Accurate Categories

B-27. Urban commanders recognize that threat evaluation is not solely a straightforward assessment of the capabilities of a known, armed enemy. Instead, they develop a system of categories that mitigates potential situational uncertainty and better indicates the level of threat (or utility) that civilian groups pose for Army forces conducting UO. Categories can range from a simple construct—threat, neutral, or friendly—to a more refined one (see figure B-6 on page B-12). In most UO, a more intricate approach is necessary to detect and monitor shifts in key relevant relationships. While this analysis is important in any urban operations, it is especially critical when Army forces are opposed by urban insurgent and guerrilla forces.

B-28. In any system, commanders consider that the classification of a group is relative to the perspective from which it is viewed. This is an especially important consideration in multinational and interagency UO. During operations in Somalia, for example, U.S. Army forces may have viewed a particular clan as a hostile element. The United Nations' Italian contingent, with their colonial background in the region, may have considered the same clan as neutral or even an ally.

Shifting and Overlapping Groups

B-29. Commanders recognize that no system of categorization will precisely classify any given group; no system can reflect the overall nature and complexity of the urban society. A single group may fit in a particular category. It may also have components in two or more categories simultaneously. Often, it can shift among categories during an operation. A given group may have individuals in it who have interests identical to or different from that group and these individual interests may change over time.

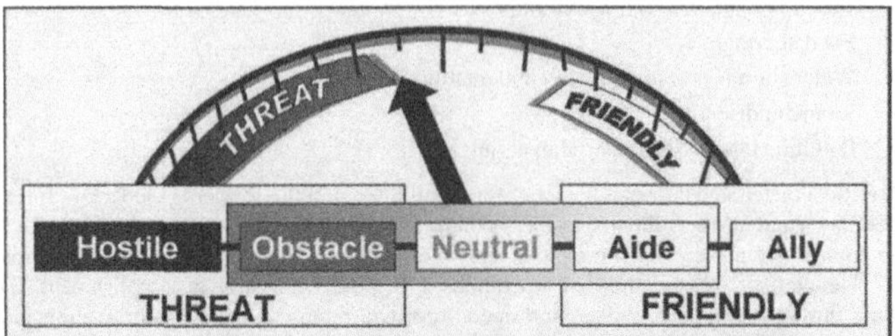

Hostile: An element (*individual, group, or organization*) that has the intent to negatively influence mission accomplishment by supporting enemy forces (or another hostile element) or exploiting friendly vulnerabilities but is not engaging in combat operations—otherwise, they lose any civilian status and become an enemy force.

Obstacle: An element whose presence or activities serve to unintentionally impede, disrupt, or constrain friendly operations by imposing additional losses in personnel, time, equipment, and resources. Its interests may or may not be compatible with friendly objectives.

Neutral: An element whose presence, activities, and overall interests do not conflict with friendly, enemy, or hostile elements.

Aide: An element whose *present* intentions are—either wittingly or unwittingly—compatible with friendly force objectives.

Ally: An element whose *fundamental* interests and intent align with friendly force objectives.

Figure B-6. A refined approach to civilian allegiance

B-30. A peace enforcement operation illustrates the varying nature of groups. An identified criminal group might be classified as an obstacle to the commander's mission because its illegal activities impede unit progress. Its compelling interest, however, is to make money rather than resist the actions of friendly forces. In the same operation, one of the armed belligerents may be intent on disrupting the peace process and would be, therefore, classified as a hostile. (Again, they are not an enemy unless the belligerents are engaged in planned combat operations against the peacekeeping force. Commanders should consider, however, that under international law, if any civilian takes a direct or active role in hostilities, that civilian can be subject to attack.) The belligerent force may finance the criminal organization to assist in further obstructing the peace mission. In this case, the criminal organization moves from being an obstacle to being a hostile.

B-31. This classification effort, therefore, is not a one-time undertaking; commanders must constantly review and update it. Groups or individuals can be influenced into assisting either the friendly or opposing force. People will also act opportunistically, shifting support and alliances as perceived advantages arise. Even seemingly passive and law abiding members of the urban society may conduct themselves in unexpected ways given the right conditions (mob violence, for example).

> **Shifting Civilian Interests and Intent**
>
> Among other applicable lessons (see also the vignette in Chapter 9), the Los Angeles riots of 1992 illustrate how urban population groups can shift their relative positions due to changing conditions in an urban AO. Several gangs exist in the Los Angeles area. Usually, these gangs are hostile to one another. During the riots, however, several rival gangs formed a "united front" against what was seen as a larger obstacle to their own interests: law enforcement. As a result, the hostile gangs became one another's aides during that time.

Similarities, Differences, Capabilities, and Vulnerabilities

B-32. Similarities and differences in interests and interdependencies between groups are often focal points for analysis (and the allocation of ISR assets). They may indicate how commanders may influence, coerce, or align civilian interests and intentions with mission objectives. Simultaneously, commanders should consider an analysis of the civilian element's (individual, group, or organization) capability to influence the accomplishment of friendly objectives. They should also consider civilians' vulnerabilities and dependencies. If a civilian group's fundamental interests align with friendly objectives and this group has the intent to assist friendly forces, it is clearly an ally. However, with limited or no capability, a specific group will not help move the commander any closer to achieving his desired end state and accomplishing the mission. Then the commander would normally limit the resources expended on this group to those necessary for maintaining their commitment to common goals and objectives.

B-33. In contrast, commanders may provide a group with resources to enhance or create the capability to assist in mission accomplishment. They may do this if they felt that the assistance gained (or reduction to threat support) exceeds the potential diminishment of their own force's capabilities from losing those same resources. Commanders would also consider the group's dependencies (such as food, infrastructure, and protection) and overall vulnerability to outside influence. If vulnerable to friendly influence or control (understanding urban societal considerations and matters of perception are critical in this regard), then forces are likely susceptible to enemy or hostile manipulation. Even if commanders can generate extra resources (and not significantly affect their own force's capabilities), they should still conduct this same cost-benefit analysis to determine which civilian group (if any) should receive resources.

Greatest Potential Nearer the Center

B-34. The most critical population sectors often are those nearest the middle of the spectrum shown in figure B-6, particularly if their capabilities (or potential) significantly degrade or enhance mission accomplishment. These are the obstacle, neutral, or aide categories. If their beliefs and interests can be adequately understood, commanders may have great chance to influence the population segment and significantly contribute to mission accomplishment. In the fight for information superiority, persuading a neutral population to become an aide may provide critical information required about the urban environment and threat forces operating within.

Political and Strategic Concerns

B-35. The aide category may be of significant political or strategic concern. An aide group may be invaluable for accomplishing intermediate objectives but become a vulnerability to accomplishing a larger stability operation. (Even an urban offensive or defensive operation is likely to transition to a stability operation.) For example, commanders may provide resources to a criminal organization to assist in defeating insurgent forces during urban combat operations. Once these forces have been subdued, the interests (monetary gain and defeat of the threat) linking friendly forces with this criminal element disappear. What may remain is a criminal organization with more power than a reconstituted or newly established law enforcement agency and a truly destabilizing force. This also illustrates the second- and third-order thinking that will be required of commanders and their staffs during UO.

URBAN INTELLIGENCE TOOLS AND PRODUCTS

B-36. Developing tools and products to assist with situational understanding is not the task of intelligence sections alone. Commanders and all elements of their staffs must develop and adapt products and tools suited to their particular requirements. Listed below are some of the analytic tools and products that may help meet those requirements in a complex environment (see figure B-7 on page B-14 and FMI 2-91.4). Standard tools and products include: modified combined obstacle overlays and doctrinal, situation, event, and decision support templates or matrices. In addition to these standards aids, staffs and analysts may develop or produce other innovative tools to assist commanders in their situational understanding of the complex urban environment. These tools may contain overlapping information as different types of information are compared to determine patterns among them. Staffs and analysts may also initiate requests for products (or information) from their higher headquarters or other agencies with the technical means or control over assets when the capability lies outside the Army force's means. The tools that may be developed or requested include—

- Imagery.
- Three-dimensional representations.
- Infrastructure blueprints.
- Hydrographic surveys.
- Psychological profiles.
- Matrices, diagrams, or charts.
- Various urban overlays.

B-37. There are many software applications available to the Army that can be used to conduct intelligence analysis and create many of the relevant products described above. These applications currently range from such programs as the Analyst Notebook and Crimelink which have link analysis, association matrix, and pattern analysis software tools to the Urban Tactical Planner developed by the Topographic Engineering Center as an operational planning tool. The focus of the following information, however, is the various types of tools that can be developed and used to understand the urban environment rather than the constantly evolving and changing software or hardware that may be used to create them.

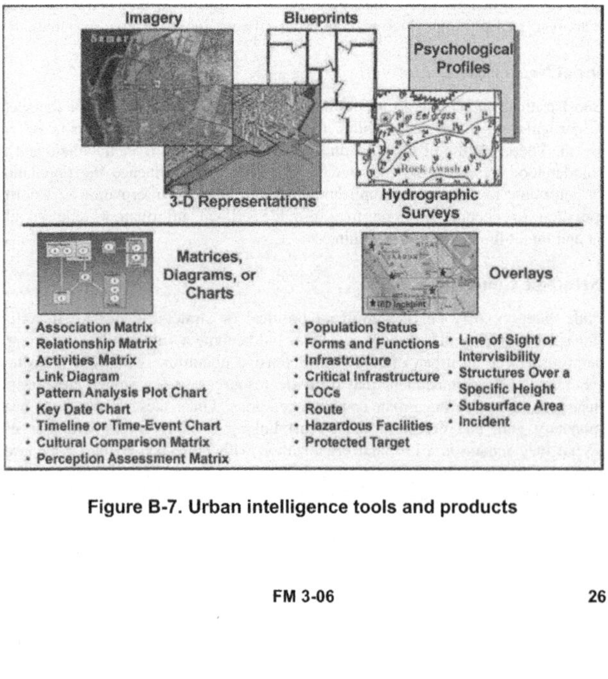

Figure B-7. Urban intelligence tools and products

Imagery

B-38. Recent satellite imagery or aerial photography will be required for most types of UO. Such images clarify vague and inaccurate maps and other graphic representations. Satellite assets provide responsive data input into the geographic information systems (GIS). (The National Geospatial Intelligence Agency [NGA] is one important source of imagery and map data.) GIS will often form the basis for creating the three-dimensional representations and the various overlays described below. Frequently updated (or continuous real-time) satellite or aerial imagery may be required for detailed pattern analysis and maintaining accurate situational understanding. For example, imagery taken during an area's rainy season may appear significantly altered during the summer months. Finally, digital, hand-held still and video cameras, particularly at the tactical level, will also be critical in developing and attaining situational understanding in a complex urban environment.

Three-Dimensional Representations

B-39. Often, physical or computer-generated (virtual) three-dimensional representations may be required to achieve situational understanding. These models or computer representations include specific sections of the urban area or specific buildings or structures. Such detail is particularly important for special operating forces and tactical-level units. These units require detail to achieve precision, increase the speed of the operation, and lessen friendly casualties and collateral damage.

Infrastructure Blueprints

B-40. Urban police, fire, health, public utilities, city engineer, realty, and tourist agencies, and other urban organizations often maintain current blueprints and detailed maps. Such documents may prove useful to update or supplement military maps or to clarify the intricacies of a specific infrastructure. They may prove critical in operations that require detailed information to achieve the speed and precision required for success. Without such detail, analysts determine interior configurations based on a building's outward appearance. That task generally becomes more difficult as the building size increases.

Hydrographic Surveys

B-41. Many urban areas are located along the world's littoral regions and major rivers. Therefore, commanders may need current hydrographic surveys to support amphibious, river crossing, riverine, and sustainment operations.

Psychological Profiles

B-42. Psychological profiles analyze how key groups, leaders, or decision makers think or act—their attitudes, opinions, and views. They include an analysis of doctrine and strategy, culture, and historical patterns of behavior. The degree to which the attitudes, beliefs, and backgrounds of the military either reflect or conflict with the urban populace's (or civilian leadership's) core values is extremely important in this analysis. Psychological profiles help to assess the relative probability of a threat (or noncombatant group) adopting various COAs as well as evaluating a threat's vulnerability to deception. These profiles are derived from open-source intelligence as well as signals and human intelligence.

Matrices, Diagrams, and Charts

B-43. Matrices, diagrams, and charts help to identify and understand key relationships among friendly and threat forces and other significant elements of the urban environment. While similar, each looks at information in a different way to uncover hidden patterns and connections useful in understanding the complex urban environment, particularly the threat and the urban society. These tools and products might include—

- **Association Matrix**. The association matrix helps identify the nature and relationship between individuals. Association matrices also help to identify those personalities and associations needing a more in-depth analysis necessary to establish the degree of relationship, contacts, or knowledge between individuals. Threat organizational structure and functions are uncovered as connections between personalities are made.

- **Relationship Matrix**. Relationship matrices are an extension of association matrices described above. They are used to examine the relationship between groups and organizations—threat forces, friendly forces, NGOs, the media, and the various elements of the urban population. Urban terrain and infrastructure may also be included as analytical factors to help expose relationships. These matrices graphically depict how human elements of the urban area interact with the physical elements to promote or degrade mission success. A significant relationship matrix may be a comparison of cultural perspectives—ideology, politics, religion, acceptable standards of living, and mores—between urban population groups and Army (and multinational) forces to help understand and accurately predict a civilian element's actions. The relationships discovered may suggest ways that commanders can shape the environment. For example, conflicting ideals or issues between two disparate threat forces may be emphasized to cause threats to focus their resources against each other rather than Army forces. Conversely, stressing common ideals identified between opposing noncombatant organizations may cause these groups to unite and work harmoniously toward a goal reflecting their commonalities.
- **Activities Matrix**. An activities matrix connects individuals to organizations, professions, events, activities, or addresses. Information from this matrix (combined with information from association and relationship matrices) also helps to link personalities, uncover the structure of an organization (threat or otherwise), and recognize differences, similarities, and dependencies for possible exploitation.
- **Link Diagram**. This tool graphically depicts many of the relationships and associations described above—people, groups, events, locations, or other factors deemed significant in any given situation. It helps commanders and analysts visualize how people and other factors are interrelated to determine key links.
- **Pattern Analysis Plot Chart**. This chart depicts the times and dates of a selected activity (such as ambushes, sniper and mortar attacks, bombings, and demonstrations) to search for patterns of activity for predictive purposes as well as to discern intent. This analysis can be conducted using a time-event wheel as well as by plotting events on maps using multiple historical overlays (analog or digital). (Compare with Incidents Overlay covered later in this chapter.) During pattern analysis, commanders and analysts consider not only what is occurring but also what is not occurring (or ceases to occur).
- **Key Event Chart**. In many urban operations, particularly stability operations, key holidays, historic events, and other significant cultural or political events can be extremely important for commanders to understand. More than just dates, these charts depict what can be expected to happen on each particular event. In Bosnia, for example, weddings are often held on Fridays and celebratory fire is a common occurrence. Understanding this cultural phenomenon could reduce collateral damage and accompanying civilian deaths.
- **Timeline or Time-Event Chart**. Timelines are a list of significant historical dates with relevant information and analysis that provide the commander with a record of past activities necessary to understand current operational conditions. A timeline may highlight a specific feature of the present situation such as population movements or political shifts or outline the general chronological record of the urban area, perhaps highlighting the activities of a certain population sector. Understanding past events also helps commanders predict future reactions to proposed COAs. Similarly to link diagrams, time-event charts display large amounts of information in a small space. These charts may help commanders discover larger-scale patterns of activity and relationships.
- **Cultural Comparison Matrix**. Commanders must avoid ethnocentricity—assuming that only one cultural perspective exists (often their own) and using that as the single lens with which to view the situation. To avoid this common obstacle to an accurate understanding of the urban society, commanders use this matrix to compare local ideology, attitudes, beliefs, and acceptable standards of living with associated U.S. and Army norms and values. This matrix not only highlights differences but also similarities.

- **Perception Assessment Matrix**. All of the above tools contribute to developing the perception assessment matrix; they allow commanders to see the urban environment from the perspective of its inhabitants. The perception assessment matrix is another tool that uses this refined perspective and the framework outlined in the matrix to predict the urban population's perception of a proposed COA. Although perceptions are not actions, they—more than reality—drive decision-making and resultant civilian activity. (As an example, legitimacy is a critical factor in developing many COAs that affect the civilian population, particularly during stability operations. This legitimacy must be viewed from the perspective of the inhabitants and not based on the Army commander's own culturally-shaped perceptions. Moreover, the legitimacy of an operation may need to be considered from other perspectives as well. These might include those of the American public, coalition partners and their publics, and those of neighboring states in the region.) This same matrix can later be used as a foundation to track and gauge the effectiveness of the chosen COA with favorable perception as one criterion of success. Commanders must be wary that perceptions are not fixed. They may change based on factors that the commander may not be able to control or influence. Therefore, commanders must continually monitor perceptions for deviation.

Various Urban Overlays

B-44. Staffs can produce various map overlays. These overlays depict physical locations of some aspect critical to the planning and conduct of the urban operation. Given adequate lead time, NGA can produce many overlays as an integrated map product (including satellite imagery). These overlays can include the—

- **Population Status Overlay**. This tool depicts the physical location of various groups identified by any significant social category such as religion, tribe, or language. During offensive and defensive operations, it may simply be where significant numbers of people are "huddled" or located throughout the battlefield. Population dispersal can vary significantly through the day, particularly at night, and must be considered as part of the overall analysis leading to the development of this tool.
- **Forms and Functions Overlay**. Based on the urban model, this overlay depicts the urban core or central business district, industrial areas, outlying high-rise areas, commercial ribbon areas, and residential areas, to include shantytowns.
- **Infrastructure Overlay**. This overlay is actually a series of overlays. It depicts identifiable subsystems in each form of urban infrastructure: communications and information, transportation and distribution, energy, economics and commerce, administration and human services, and cultural. Each subsystem can be broken down into more detail. Infrastructure data may be used to develop three other overlays—
  - **Critical Infrastructure Overlay**. This tool displays specific elements of the urban infrastructure that, if harmed, will adversely affect the living conditions of the urban society to the detriment of the mission. These elements may include power generation plants, water purification plants and pumping stations, and sewage treatment plants. This information could be coded as part of the overall infrastructure overlay.
  - **Lines of Communications (LOCs) Overlay**. The LOCs overlay highlights transportation systems and nodes, such as railways, road, trails, navigable waterways, airfields, and open areas for drop zones and landing zones. It also includes subsurface areas and routes such as sewage, drainage, and tunnels and considers movement between supersurface areas. The LOCs overlay and the route overlay (below) consider traffic conditions, times, and locations, to include potential points where significant portions of the urban population may congregate.
  - **Route Overlay**. This overlay emphasizes mobility information to assist commanders and planners in determining what forces and equipment can move along the urban area's mobility corridors. Pertinent data includes street names, patterns, and widths; bridge, underpass, and overpass locations; load capacities; potential sniper and ambush locations (which may be its own overlay); and key navigational landmarks such as major roads and highways, rivers and canals, cemeteries, bridges, stadiums, and churches. The structures over a specific height

overlay and subsurface overlay may assist in its development. As with the LOCs overlay, commanders, planners, and analysts think in all dimensions.

- **Line of Sight or Intervisibility Overlay**. This product creates a profile view (optical or electronic) of the terrain from the observer's location to other locations or targets. It can show trajectory or flight-line masking as well as obstructed or unobstructed signal pathways.
- **Structures over a Specific Height Overlay**. This level of detail may also be critical to communications, fires, and Army airspace command and control (air mobility corridors especially low-level flight profiles). Incorporated as part of this overlay, it may include floors or elevations above limitations for particular weapon systems at various distances from the structure.
- **Subsurface Area Overlay**. As an alternate to the building or structure height overlay, this product provides the locations of basements, underground parking garages, sewers, tunnels, subways, naturally-occurring subterranean formations, catacombs, and other subsurface areas. Similar to elevation "dead spaces," this overlay may show areas that exceed depression capabilities of weapon systems and potential threat ambush locations—again, affecting maneuver options.
- **Urban Logistic Resources Overlay**. This product identifies the locations of urban logistic resources that may contribute to mission accomplishment. It may contain specific warehouse sites, hospitals and medical supply locations, viable food stores, building material locations, fuel storage areas, car or truck lots, maintenance garages, and appliance warehouses. (NGO locations, taken from an NGO relationship matrix, may be an essential, overlapping element of this overlay.)
- **Hazardous Facilities Overlay**. This overlay identifies urban structures with known or suspected chemical, biological, or radiological features, such as nuclear power plants, fertilizer plants, oil refineries, pharmaceutical plants, and covert locations for producing weapons of mass destruction. These locations are critical to maneuver and fire planning.
- **Protected Target Overlay**. This overlay depicts terrain that should not be destroyed or attacked based on restrictions due to international, host-nation, or U.S. law and subsequent rules of engagement. These may include schools, hospitals, historical or other culturally significant monuments, and religious sites. This overlay may incorporate no-fire areas, such as special operations forces locations, critical infrastructure, logistic sources, and hazardous sites that must be protected as part of the commander's concept of the operation.
- **Incident Overlay**. Similar to the pattern analysis plot chart, this product depicts the location of different threat actions and types of tactics employed to uncover recurring routines, schemes, methods, tactics, or techniques and overall threat interests, objectives, or the desired end state.

B-45. The above IPB tools and products constitute a small sampling of what staffs and analysts can produce. They are limited only by their imaginations and mission needs (not all tools presented above may be relevant or necessary to every operation). Many products can be combined into a single product or each can generate further products of increasing level of detail. This is similar to transparent overlays positioned one atop another on a map. As discussed earlier, technology—software and hardware—will allow for more urban data to be combined, compared, analyzed, displayed, and shared—ultimately allowing scaleable products that can expand and contract to meet the intelligence requirements of any echelon of command down to the individual Soldier. The challenge will remain, however, to provide timely, accurate, complete, and relevant information in an understandable, usable, and share-capable form—share-capable among different levels of command as well as among multinational partners and various governmental and (as required) nongovernmental organizations—without overloading the commander.

Appendix C

Operations in Somalia: Applying the Urban Operational Framework to Stability Operations

It's impossible for an American mother to believe that a Somali mother would raise children to avenge the clan.

Major General Thomas M. Montgomery

GENERAL SITUATION

C-1. Following decades of political unrest and the fall of Somali dictator Siad Barre, a civil war broke out as 14 clans vied for power. The resulting nation composed of hostile social factions was held together by weak political alliances—none strong enough to unite and lead the country to national reconciliation. An ongoing drought led to famine and compounded the ethnic tensions and political instability. This volatile situation rapidly led to a phased U.S. involvement (see figure C-1). Army forces combined, sequenced, and proportionally emphasized the different types of operations to accomplish changing political objectives. Throughout all operations in Somalia, urban areas were critical to achieving mission success.

SOMALI OPERATIONS

INITIAL UN RESPONSE

C-2. The United Nations (UN) initially responded to requests for assistance from international relief organizations by sending supplies and other forms of humanitarian aid to Somalia. However, widespread looting, fighting between gangs, and other lawlessness prevented supplies from reaching the hungry and sick. Only 20 percent of the food entering the country reached the people who needed it. An estimated 25 percent of Somalia's 6 million people died of starvation or disease. In April 1992, the UN issued Security Council Resolution (UNSCR) 751 which authorized 50 unarmed observers, but the action had little effect. Under continuing pressure for additional measures to ensure the delivery of supplies and relief, the UN authorized 500 armed peacekeepers (furnished by Pakistan and transported by U.S. sea- and airlift) to protect humanitarian workers. The battalion's limited mission, designated UN Operations in Somalia I (UNOSOM I), encompassed safeguarding the unloading of ships and providing convoy security.

PROVIDE RELIEF (UNOSOM I)

C-3. In July 1992, the UN requested an increased airlift of supplies and the United States quickly responded. U.S. Central Command (USCENTCOM) activated joint task force (JTF) OPERATION PROVIDE RELIEF. Based on careful mission analysis, USCENTCOM limited the JTF's actions to—

- Deploying a humanitarian assistance survey team to assessing relief requirements.
- Providing an emergency airlift of supplies.
- Using Air Force cargo aircraft for daily relief sorties into Somalia.

C-4. USCENTCOM restricted the sorties to flying during daylight hours and to locations that would provide a permissive and safe environment. In mid-September 1992, the United States prudently expanded its role by stationing the amphibious ready group, USS Tarawa, offshore to provide support to the Pakistani security battalion and to provide security for U.S. airlift operations. The 11th Marine expeditionary unit (MEU) was on board the USS Tarawa to rapidly respond to any change in mission (see Appendix D for a description and the capabilities of a MEU).

Operation	Dates	UN Security Council Resolution	Relative Proportionality Between Types of Operations
PROVIDE RELIEF (UNOSOM I)	Aug 92 – Dec 92	UNSCR# 751 24 Apr 92	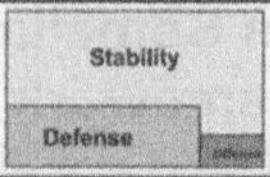
RESTORE HOPE (UNITAF)	Dec 92 – May 93	UNSCR# 794 3 Dec 92	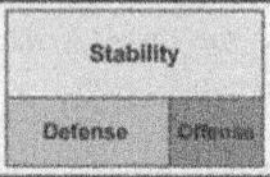
CONTINUED HOPE (UNOSOM II)	May 93 – Mar 94	UNSCR# 814 26 Mar 93	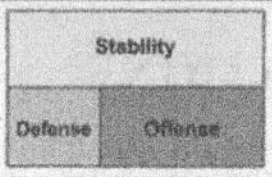

Figure C-1. Relative proportionality between types of operations

RESTORE HOPE (UNITAF)

C-5. By November 1992, the magnitude of the task, UN organizational deficiencies, and a continued lack of security precluded delivery of sufficient supplies to the needy. Notably, a ship laden with relief supplies was fired on in the harbor at Mogadishu, forcing its withdrawal before the supplies could be brought ashore, and a Pakistani peacekeeper was shot when his car was hijacked. Subsequently, the United States offered to provide forces and lead a UN-sponsored operation to reopen the flow of food to where it was needed most. In December 1992, the UN issued UNSCR 794, which authorized member states "to use all necessary means to establish a secure environment for humanitarian relief operations in Somalia" and demanded all "factions in Somalia immediately cease hostilities." To allay concerns of colonialism by a number of African countries, the UN Secretary-General was given oversight of the operation. The resolution also required soldiers to be withdrawn once order was restored; however, it provided no exit strategy. As clearly as possible, the USCENTCOM mission statement (below) for OPERATION RESTORE HOPE reflected the UN mandate:

> *When directed by the [President or the Secretary of Defense], [Commander, USCENTCOM] will conduct joint/combined military operations in Somalia to secure the major air and sea ports, key installations and food distribution points, to provide open and free passage of relief supplies, provide security for convoys and relief organization operations, and assist UN/[nongovernmental organizations] in providing humanitarian relief under UN auspices. Upon establishing a secure environment for uninterrupted relief operations, [Commander, USCENTCOM] terminates and transfers relief operations to UN peacekeeping forces.*

C-6. Mogadishu was the largest port in the country and the focal point of previous humanitarian relief activities of nongovernmental organizations (NGOs). It was also the headquarters of the coalition of 20 nations and over 30 active humanitarian relief organizations. As such, Mogadishu became the entry point for the operational buildup of the multinational force known as Unified Task Force (UNITAF) and the key logistic hub for all operations in Somalia. UNITAF immediately gained control over the flow of relief supplies into and through Mogadishu and stabilized the conflict among the clans. In less than a month, UNITAF forces expanded control over additional ports and interior airfields. They secured additional distribution sites in other key urban areas in the famine belt to include Baidoa, Baledogle, Gialalassi, Bardera, Belet Uen, Oddur, Marka, and the southern town of Kismayo (see figure C-2). With minimal force, the U.S.-led UNITAF established a secure environment that allowed relief to reach those in need, successfully fulfilling its limited—yet focused—mandate.

CONTINUED HOPE (UNOSOM II)

C-7. In March 1993, the UN issued UNSCR 814 establishing a permanent peacekeeping force, UNOSOM II. However, the orderly transition from UNITAF to UNOSOM II was repeatedly delayed until May 1993. (The UN Secretary-General urged the delay so that U.S. forces could effectively disarm bandits and rival clan factions in Somalia.) This resolution was significant in two critical aspects:

- It explicitly endorsed nation building with the specific objectives of rehabilitating the political institutions and economy of Somalia.
- It mandated the first ever UN-directed peace enforcement operation under the Chapter VII enforcement provisions of the Charter, including the requirement for UNOSOM II to disarm the Somali clans. The creation of a peaceful, secure environment included the northern region that had declared independence and had hereto been mostly ignored.

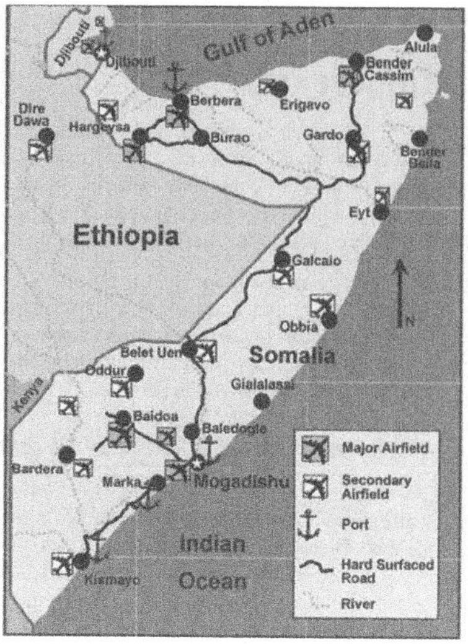

Figure C-2. Map of Somalia

C-8. These far-reaching objectives exceeded the limited mandate of UNITAF as well as those of any previous UN operation. Somali clan leaders rejected the shift from a peacekeeping operation to a peace enforcement operation. They perceived the UN as having lost its neutral position among rival factions. A more powerful clan leader, General Mohammed Farah Aideed (leader of the Habr Gidr clan), aggressively turned against the UN operation and began a radio and propaganda campaign. This campaign characterized UN soldiers as an occupation force trying to recolonize Somalia.

C-9. The mounting crisis erupted in June 1993. Aideed supporters killed 24 Pakistani soldiers and wounded 57 in an ambush while the soldiers were conducting a short-warning inspection of one of Aideed's weapons arsenals. UNSCR 837, passed the next day, called for immediately apprehending those responsible and quickly led to a manhunt for Aideed. The United States deployed 400 rangers and other special operations forces (SOF) personnel to aid in capturing Aideed, neutralizing his followers, and assisting the quick reaction force (QRF), composed of 10th Mountain Division units, in maintaining the peace around Mogadishu.

PHASED WITHDRAWAL

C-10. On 3 October 1993, elements of Task Force (TF) Ranger (a force of nearly 100 Rangers and SOF operators) executed a raid to capture some of Aideed's closest supporters. Although tactically successful, 2 helicopters were shot down, 75 soldiers were wounded, and 18 soldiers were killed accomplishing the mission. The U.S. deaths, as well as vivid scenes of mutilation to some of the soldiers, increased calls to Congress for withdrawing U.S. forces from Somalia. The President then ordered reinforcements to protect U.S. Forces, Somalia (USFORSOM) as they began a phased withdrawal with a 31 March deadline. The last contingent sailed from Mogadishu on 25 March, ending OPERATION CONTINUED HOPE and the overall U.S. mission in Somalia.

C-11. Although U.S. forces did not carry out the more ambitious UN goals of nation building, they executed their missions successfully, relieving untold suffering through humanitarian assistance with military skill and professionalism. Operations in Somalia occurred under unique circumstances, yet commanders may glean lessons applicable to future urban stability operations, as well as civil support operations. In any operations, commanders balance changing mission requirements and conditions.

UNDERSTAND

C-12. Although accomplished to varying degrees, U.S. forces failed to adequately assess and understand the urban environment, especially the society. Somali culture stresses the unity of the clan; alliances are made with other clans only when necessary to elicit some gain. Weapons, overt aggressiveness, and an unusual willingness to accept casualties are intrinsic parts of the Somali culture. Women and children are considered part of the clan's order of battle.

C-13. Early in the planning for OPERATION RESTORE HOPE, U.S. forces did recognize the limited transportation and distribution infrastructure in Mogadishu. The most notable was the limited or poor airport and harbor facilities and its impact on the ability of military forces and organizations to provide relief. Therefore, a naval construction battalion made major improvements in roads, warehouses, and other facilities that allowed more personnel, supplies, and equipment to join the relief effort faster.

UNDERSTANDING THE CLAN (THE HUMAN DIMENSION)

C-14. During OPERATION RESTORE HOPE, the UNITAF worked with the various clan leaders as the only recognized leadership remaining in the country. The UNITAF was under the leadership of LTG Robert B. Johnston and U.S. Ambassador to Somalia, Robert Oakley. In addition, UNITAF forces also tried to reestablish elements of the Somali National Police—one of the last respected institutions in the country that was not clan-based. This reinstated police force manned checkpoints throughout Mogadishu and provided crowd control at feeding centers. Largely because of this engagement strategy, the UNITAF succeeded in its missions of stabilizing the security situation and facilitating humanitarian relief. Before its termination, the UNITAF also worked with the 14 major Somali factions to agree to a plan for a transitional or transnational government.

C-15. The UN Special Representative of the Secretary-General, retired U.S. Navy Admiral Jonathon Howe, worked with the UNOSOM II commander, Turkish General Cevik Bir. During OPERATION CONTINUED HOPE, Howe and General Bir adopted a philosophy and operational strategy dissimilar to their UNITAF predecessors. Instead of engaging the clan leaders, Howe attempted to marginalize and isolate them. Howe initially attempted to ignore Aideed and other clan leaders in an attempt to decrease the warlord's power. Disregarding the long-established Somali cultural order, the UN felt that, in the interest of creating a representative, democratic Somali government, they would be better served by excluding the clan leadership. This decision ultimately set the stage for strategic failure.

THREAT STRATEGY AND TACTICS

C-16. During OPERATION RESTORE HOPE, U.S. forces also failed to properly analyze and understand their identified threat's intent and the impact that the urban environment would have on his strategy, operations, and tactics. The UN began to view eliminating Aideed's influence as a decisive point when creating an environment conducive to long-term conflict resolution. Aideed's objective, however, remained to consolidate control of the Somali nation under his leadership—his own brand of conflict resolution. He viewed the UN's operational center of gravity as the well-trained and technologically advanced American military forces, which he could not attack directly. He identified a potential American vulnerability—the inability to accept casualties for an operation not vital to national interests—since most Americans still viewed Somalia as a humanitarian effort. If he could convince the American public that the price for keeping troops in Somalia would be costly, or that their forces were hurting as many Somalis as they were helping, he believed they would withdraw their forces. If U.S. forces left, the powerless UN would leave soon after, allowing Aideed to consolidate Somalia under his leadership.

VULNERABILITY AND RISK ASSESSMENT

C-17. U.S. forces failed to assess, understand, and anticipate that Aideed would adopt this asymmetric approach and attack the American public's desire to remain involved in Somalia. By drawing U.S. forces into an urban fight on his home turf in Mogadishu, he could employ guerrilla insurgency tactics and use the urban area's noncombatants and its confining nature. Such tactics made it difficult for the U.S. forces to employ their technological superiority. If U.S. forces were unwilling to risk harming civilians, his forces could inflict heavy casualties on them, thereby degrading U.S. public support for operations in Somalia. If, on the other hand, the U.S. forces were willing to risk increased civilian casualties to protect themselves, those casualties would likely have the same effect.

C-18. However, an assessment and understanding of the Somali culture and society should have recognized the potential for Aideed's forces to use women and children as cover and concealment. Accordingly, the plan should have avoided entering the densely populated Bakara market district with such restrictive rules of engagement. As legitimacy is critical to stability operations, TF Ranger should have been prepared and authorized to employ nonlethal weapons, to include riot control gas, as an alternative to killing civilians or dying themselves.

C-19. U.S. forces also failed to assess and understand the critical vulnerability of their helicopters in an urban environment and the potential impact on their operations. TF Ranger underestimated the threat's ability to shoot down its helicopters even though they knew Somalis had attempted to use massed rocket-propelled grenade (RPG) fires during earlier raids. (Aideed brought in fundamentalist Islamic soldiers from Sudan, experienced in downing Russian helicopters in Afghanistan, to train his men in RPG firing techniques). In fact, the Somalis had succeeded in shooting down a UH-60 flying at rooftop level at night just one week prior to the battle. Instead, TF Ranger kept their most vulnerable helicopters, the MH-60 Blackhawks, loitering for forty minutes over the target area in an orbit that was well within Somali RPG range. The more maneuverable AH-6s and MH-6s could have provided the necessary fire support. Planning should have included a ready ground reaction force, properly task organized, for a downed helicopter and personnel recovery contingency.

C-20. Information operations considerations apply throughout the entire urban operational framework; however, operations security (OPSEC) is critical to both assessment and shaping. OPSEC requires continuous assessment throughout the urban operation particularly as it transitions among the spectrum of operations. As offensive operations grew during OPERATION CONTINUED HOPE, U.S. forces did little to protect essential elements of friendly information. Combined with the vulnerability of U.S. helicopters, Aideed's followers used U.S. forces' inattention to OPSEC measures to their advantage. The U.S. base in Mogadishu was open to public view and Somali contractors often moved about freely. Somalis had a clear view both day and night of the soldiers' billets. Whenever TF Ranger would prepare for a mission, the word rapidly spread through the city. On 3 October 1993, Aideed's followers immediately knew that aircraft had taken off and, based on their pattern analysis of TF Ranger's previous raids, RPG teams rushed to the rooftops along the flight paths of the task force's Blackhawks.

SHAPE

C-21. One of the most critical urban shaping operations is isolation. During OPERATION CONTINUED HOPE, U.S. forces largely discounted other essential elements of friendly information and did not establish significant public affairs and psychological operations (PSYOP) initiatives. In fact, Army forces lacked a public affairs organization altogether. Consequently, Aideed was not isolated from the support of the Somali people. This failure to shape the perceptions of the civilian populace coupled with the increased use of lethal force (discussed below) allowed Aideed to retain or create a sense of legitimacy and popular support. Ironically, the focus on Aideed helped create a popular figure among the warlords through which many noncompliant, hostile, and even neutral Somalis could rally behind.

C-22. During OPERATION RESTORE HOPE, Aideed conducted propaganda efforts through "Radio Aideed"—his own radio station. UNITAF countered these efforts with radio broadcasts. This technique proved so effective that Aideed called MG Anthony C. Zinni, UNITAF's director of operations, over to his house on several occasions to complain about UNITAF radio broadcasts. General Zinni responded, "If he didn't like what we said on the radio station, he ought to think about his radio station and we could mutually agree to lower the rhetoric." This approach worked.

ENGAGE

C-23. The complexity of urban operations requires unity of command to identify and effectively strike the center of gravity with overwhelming combat power or capabilities. Complex command and control relationships will only add to the complexity and inhibit a commander's ability to dominate and apply available combat power to accomplish assigned objectives. Stability operations as seen in Somalia required commanders to dominate within a supporting role and, throughout, required careful, measured restraint.

UNITY OF COMMAND (EFFORT)

C-24. During OPERATION RESTORE HOPE, UNITAF successfully met unity of command challenges through three innovations. First, they created a civil-military operations center (CMOC) to facilitate unity of effort between NGOs and military forces. Second, UNITAF divided the country into nine humanitarian relief sectors centered on critical urban areas that facilitated both relief distribution and military areas of responsibility. Third, to establish a reasonable span of control, nations that provided less than platoon-sized contingents were placed under the control of the Army, Marine Corps, and Air Force components.

C-25. On the other hand, during OPERATION CONTINUED HOPE, UNOSOM II command and control relationships made unity of command (effort) nearly impossible. The logistic components of USFORSOM were under UN operational control, while the QRF remained under USCENTCOM's combatant command—as was TF Ranger. However, the USCENTCOM commander was not in theater. He was not actively involved in planning TF Ranger's missions or in coordinating and integrating them with his other subordinate commands. It was left to TF Ranger to coordinate with the QRF as needed. Even in TF Ranger, there were dual chains of command between SOF operators and the Rangers. This underscores the need for close coordination and careful integration of SOF and conventional forces (see Chapter 4). It also emphasizes overall unity of command (or effort when command is not possible) among all forces operating in a single urban environment.

C-26. Following TF Ranger's 3 October mission, the command structure during OPERATION CONTINUED HOPE was further complicated with the new JTF-Somalia. This force was designed to protect U.S. forces during the withdrawal from Somalia. JTF-Somalia came under the operational control of USCENTCOM, but fell under the tactical control of USFORSOM. Neither the JTF nor USFORSOM controlled the naval forces that remained under USCENTCOM's operational control. However unity of effort (force protection and a rapid, orderly withdrawal) galvanized the command and fostered close coordination and cooperation among the semiautonomous units.

MEASURED RESTRAINT

C-27. During OPERATIONS PROVIDE RELIEF and RESTORE HOPE, U.S. forces dominated within their supporting roles. Their perseverance, adaptability, impartiality, and restraint allowed them to provide a stable, secure environment. Hence, relief organizations could provide the food and medical care necessary to reduce disease, malnourishment, and the overall mortality rate. However, during OPERATION CONTINUED HOPE, U.S. operations became increasingly aggressive under the UN mandate. However, peace enforcement also requires restraint and impartiality to successfully dominate and achieve political objectives. The increased use of force resulted in increased civilian casualties, which in turn reduced the Somalis' perception of U.S. legitimacy. As a result, most moderate Somalis began to side with Aideed and his supporters. Many Somalis, without effective information operations to shape their perceptions, felt that it was fine for foreign military forces to intervene in their country to feed the starving and even help establish a peaceful government, but not to purposefully target specific Somali leaders as criminals.

CONSOLIDATE AND TRANSITION

C-28. Across the spectrum of conflict, Army forces must be able to execute full spectrum operations not only sequentially but, as in the case of operations in Somalia, simultaneously. OPERATION PROVIDE RELIEF began primarily as a stability operation in the form of foreign humanitarian assistance. Later this operation progressed to include peacekeeping (another stability operation), defensive operations to protect

UN forces and relief supplies, and minimum offensive operations. As operations transitioned to OPERATION RESTORE HOPE, it became apparent that while foreign humanitarian assistance was still the principal operation, other operations were necessary. More forceful peacekeeping, show of force, and arms control (all forms of stability operations), and offensive and defensive operations grew increasingly necessary to establish a secure environment for uninterrupted relief operations. In the final phase of U.S. involvement during OPERATION CONTINUED HOPE, major changes to political objectives caused a major transition in the form of the stability operation to peace enforcement with an even greater increase in the use of force, offensively and defensively, to create a peaceful environment and conduct nation building. Finally, the TF Ranger raid demonstrated the need to maintain a robust, combined arms force capable of rapidly transitioning from stability operations (primarily peace operations) to combat operations.

SUMMARY

C-29. OPERATIONS PROVIDE RELIEF and RESTORE HOPE were unquestionably successes. Conversely, during OPERATION CONTINUED HOPE, the 3-4 October battle of Mogadishu (also known as the "Battle of the Black Sea") was a tactical success leading to an operational failure. TF Ranger succeeded in capturing 24 suspected Aideed supporters to include two of his key lieutenants. Arguably, given the appropriate response at the strategic level, it had the potential to be an operational success. After accompanying Ambassador Oakley to a meeting with Aideed soon after the battle, MG Zinni described Aideed as visibly shaken by the encounter. MG Zinni believed Aideed and his subordinate leadership were tired of the fighting and prepared to negotiate. Unfortunately, the U.S. strategic leadership failed to conduct the shaping actions necessary to inform and convince the American public (and its elected members of Congress) of the necessity of employing American forces to capture Aideed. The President was left with little recourse after the battle of Mogadishu but to avoid further military confrontation.

C-30. Despite this strategic failing, the operational commanders might have avoided the casualties, and any subsequent public and Congressional backlash, had they better communicated among themselves and worked with unity of effort. Recognizing the separate U.S. and UN chains of command, the UN Special Representative, along with the USCENTCOM, USFORSOM, and TF Ranger commanders, should have established the command and control architecture needed. This architecture would have integrated planning and execution for each urban operation conducted. These commanders failed to "operationalize" their plan. They did not properly link U.S. strategic objectives and concerns to the tactical plan. The TF Ranger mission was a direct operational attempt to obtain a strategic objective in a single tactical action. Yet, they failed to understand the lack of strategic groundwork, the threat's intent and capabilities, and the overall impact of the urban environment, to include the terrain and society, on the operation. Such an understanding may not have led to such a high-risk course of action but instead to one that de-emphasized military operations and emphasized a political solution that adequately considered the clans' influence.

This page intentionally left blank.

Appendix D

Joint and Multinational Urban Operations

[Joint force commanders] synchronize the actions of air, land, sea, space, and special operations forces to achieve strategic and operational objectives through integrated, joint campaigns and major operations. The goal is to increase the total effectiveness of the joint force, not necessarily to involve all forces or to involve all forces equally.

JP 3-0

As pointed out earlier, Army forces, brigade size and larger, will likely be required to conduct operations in, around, and over large urban areas in support of a joint force commander (JFC). The complexity of many urban environments, particularly those accessible from the sea, requires unique leveraging and integration of all the capabilities of U.S. military forces to successfully conduct the operation. This appendix discusses many of these capabilities; JP 3-06 details joint urban operations.

PURPOSE

D-1. In some situations, a major urban operation is required in an inland area where only Army forces are operating. Army commanders determine if the unique requirements of the urban environment require forming a joint task force (JTF) or, if not, request support by joint capabilities from the higher joint headquarters. Sometimes the nature of the operation is straightforward enough or the urban operation is on a small enough scale that conventional intraservice support relationships are sufficient to meet the mission requirements.

D-2. Most major urban operations (UO), however, require the close cooperation and application of joint service capabilities. A JTF may be designated to closely synchronize the efforts of all services and functions in an urban area designated as a joint operations area (JOA). If a large urban area falls in the context of an even larger ground force area of operations, a JTF dedicated to the urban operation may not be appropriate. These situations still require joint capabilities. In such cases, the responsible JFC designates support relations between major land units and joint functional commands. The major land units can consist of Army forces, Marine Corps forces, or joint forces land component command. The joint functional commands can consist of the joint special operations task force, joint psychological operations task force, or joint civil-military operations task force.

D-3. This appendix describes the roles of other services and joint combatant commands in UO. It provides an understanding that enables Army commanders to recommend when to form a JTF or to request support from the JFC. It also provides information so commanders can better coordinate their efforts with those of the JFC and the commanders of other services or components conducting UO. Lastly, this appendix describes some considerations when conducting UO with multinational forces.

SISTER SERVICE URBAN CAPABILITIES

D-4. Army forces conducting UO rely on other services and functional joint commands for specialized support in the urban environment. These capabilities are requested from and provided through the commanding JFC. Army forces request the assets and capabilities described in this annex through their higher headquarters to the joint command. The JFC determines if the assets will be made available, the appropriate command relationship, and the duration of the support. Army forces prepare to coordinate planning and execution with other services and to exchange liaison officers. These capabilities can greatly increase the Army's ability to understand, shape, engage, consolidate, and transition within the context of UO.

Air Force

D-5. Air Force support is an important aspect of the Army force concept for urban operations. Air Force elements have a role to play in UO across the spectrum of operations.

D-6. Air Force intelligence, surveillance, and reconnaissance (ISR) systems contribute significantly to understanding the urban environment. These ISR systems include the E-8 Joint Surveillance Target Attack Radar System (JSTARS) (see figure D-1), U2 Dragon Lady, RC-135 Rivet Joint, or RQ-4 Global Hawk and RQ/MQ-1 Predator unmanned aerial systems. Air Force ISR systems can provide vital data to help assess threat intentions, threat dispositions, and an understanding of the civilian population. These systems also can downlink raw information in real-time to Army intelligence processing and display systems, such as the common ground station or division tactical exploitation system.

Figure D-1. USAF E-8 JSTARS platform

D-7. Air interdiction (AI) can be a vital component of shaping the urban operational environment. Often, AI of the avenues of approach into the urban area isolates the threat by diverting, disrupting, delaying, or destroying portions of the threat force before they can be used effectively against Army forces. AI is especially effective in major combat operations where restrictions on airpower are limited and the threat is more likely to be a conventionally equipped enemy. In 1991 during OPERATION DESERT STORM, AI helped prevent the Iraqi 5th Mechanized Division from reaching Khafji.

D-8. Precision-guided munitions (PGMs) delivered from combat aircraft can positively and directly affect the conduct of Army close combat actions in the urban area. Special munitions designed to penetrate reinforced concrete can provide unique support to land forces executing UO. Although laser-guided bombs are extremely accurate with an effective laser lock-on, environmental factors such as smoke, haze, and weather can adversely affect successful laser designation. If the launching aircraft can achieve a successful laser designation and lock-on, these weapons have devastating effects, penetrating deep into reinforced concrete before exploding with great force. If launched without a lock-on, or if the laser spot is lost, these weapons are unpredictable and can travel long distances before they impact. However, continual advancements in tactics and weapons systems, such as joint direct attack munition (JDAM) weapons, help decrease the adverse effects of these environmental factors. (For example, in 2003 during OPERATION IRAQI FREEDOM, the Air Force supported the Army's attack on An Najaf with JDAM-equipped F 16s and B 1s during a blinding sandstorm. With near zero visibility, JDAM weapons provided significant fire support, allowing Army forces to complete their mission and continue the attack north.) When PGMs are unavailable, general-purpose bombs from 500 to 2,000 pounds can also prove effective. However, there may be risk factors associated with their use such as collateral damage due to decreased accuracy, threat exposure to the delivering aircraft due to lower delivery altitudes, or reduced weapons effectiveness due to shallower penetration angles. (See FM 3-06.1 for a more detailed description of various aircraft weapons and capabilities.)

D-9. In addition to shaping and engaging the UO through firepower, commanders can use Air Force capabilities to improve and augment the urban transportation and distribution infrastructure. Air Force units can repair or improve airfields, revitalize civil aviation maintenance facilities, manage air-delivered cargo, and control civil and military air traffic. These latter capabilities particularly enhance urban stability and civil support operations. These capabilities may even be decisive. Air-delivered cargo and air traffic management, for example, were the decisive factors in U.S. forces' successful resistance of the Soviet blockade of Berlin in 1948. In OPERATION RESTORE HOPE, from December 1992 to May 1993, Air Force operations in Mogadishu were critical to airlifting and staging supplies and forces. The Air Force determined the ultimate success of the humanitarian assistance operation (see Appendix C).

D-10. In unique situations, such as the Berlin Blockade and OPERATION RESTORE HOPE, exercising Air Force urban capabilities may be the decisive action of the operation. Most often, however, Air Force capabilities will play a shaping role, sustaining role, or both in joint urban operations because of the inherent requirement to occupy terrain and interface with the population.

MARINE CORPS

D-11. The Marine Corps can assault across water obstacles into a defended urban environment. This capability is an invaluable tactical and operational tool. The mere threat of this capability can divert many enemy forces from other avenues of approach and obscure the true nature of an attack. The impact of the threat of amphibious assault was vividly demonstrated during OPERATION DESERT STORM where embarked Marine Forces diverted several Iraqi divisions to defensive positions along the coast and near Kuwait City.

D-12. The presence of Marine amphibious equipment, apart from Marine infantry, provides Army forces unique capabilities. In UO, the amphibious operation is often not an assault from the sea, but rather an assault river crossing. In 1950, the 7th U.S. Infantry Division used amphibious tractor support from the 1st Marine Division to conduct an assault river crossing of the Han River into downtown Seoul. See FM 3-31.1.

D-13. The worldwide deployment of Marine air-ground task forces (MAGTFs) enables a short notice response into any urban areas accessible from the sea. Typically, a deployed MAGTF is a Marine expeditionary unit (special operations capable) (MEU[SOC]). The MEU(SOC) can perform forcible entry operations, seize lodgments, and may execute these tasks anticipating reinforcement by Army or joint forces. They are also well positioned and equipped to rapidly reinforce Army forces already deployed in theater. The special-operations-capable training that these units accomplish before deploying includes urban warfare training and contributes to their value in UO. The MEU(SOC) is relatively small (its core unit is a Marine infantry battalion), is forward deployed, and has a wide spectrum of organic capabilities. It is an important asset in crisis stability and civil support operations.

D-14. The Marine expeditionary brigade is the MAGTF between a MEU and a Marine expeditionary force (the largest MAGTF). All MAGTFs include an aviation combat element, which is particularly skilled at providing accurate and timely close air support. The responsiveness and accuracy of Marine close air support aptly suits it to UO where the lethality of combat and the close range of engagements demand accurate and responsive fires.

D-15. Marine forces that conduct UO work well in littoral urban areas because of their unique relationship with naval forces and thus their capability to closely integrate land and sea operations. A supporting arms liaison team may be attached to Army forces at battalion level to provide ship-to-shore communications and coordination for naval gunfire support.

NAVY

D-16. Many major urban areas are accessible from the sea. Therefore, Army commanders should understand how sea power can influence and support UO. The Navy brings several major capabilities to UO. These include naval gunfire support, naval air support, and port and coastal security. In addition, the Navy can offer secure command post sites and meeting areas; provide limited emergency shelter and critical care for Soldiers and, as necessary, civilians; and store and transport critical supplies ashore providing flexibility in meeting support requirements. The ability to provide support from "over the horizon" not only enhances force protection but also the legitimacy of urban and host nation authorities during transition.

D-17. Naval surface fire support (which includes naval gunfire support) particularly applies to forcible entry operations in littoral urban areas. It provides an initial indirect fire support capability until Army forces land ashore. However, naval gunfire support lacks a precision munitions capability. In defensive operations, naval gunfire support can also add major fires, especially during retrograde operations through an urban area. Naval gunfire support of the 3rd Infantry Division was the primary fire support for the last two days of the X Corps retrograde operation out of the port of Hungnam, North Korea, in December 1950.

D-18. Destroyers and cruisers, which mount one and two 127mm MK45 lightweight gun systems respectively (see figure D-2), are capable of providing naval gunfire support. This gun system can provide a rate of fire of 20 rounds per minute per gun to a range of approximately 23 kilometers. Destroyers, cruisers, and attack submarines are capable of precision strikes with Tomahawk cruise missiles. Depending on the location of the urban area, naval air can provide responsive aviation support to UO alleviating the need for fixed-wing aviation bases.

Figure D-2. USN MK45 lightweight gun system

D-19. Due to its flat trajectory, terrain masking affects naval gunfire more than field artillery. Naval gunfire also results in large range probable errors (the dispersion pattern of the naval gun is roughly elliptical with the long axis in the direction of fire). Hence, coverage of targets such as roads and airfields is most effective when the gun-target line (GTL) coincides with the long axis of the target. Very close supporting fire can be delivered when the GTL is parallel to the front line of troops. Oppositely, a GTL perpendicular to the front trace can endanger friendly forces. Within the limits of hydrographic conditions, the ship can maneuver to achieve a better GTL, but ship movement also makes it difficult to adjust fire. Overall, naval and air threats, bad weather, and large range probable errors make naval gunfire difficult and can cause cancellation of supporting fires.

D-20. Water terminals located in urban areas are usually the debarkation points for the bulk of Army forces as well as a joint force theater logistics requirement. Army forces are responsible for water terminal operations while naval capabilities protect these strategic and operationally vital facilities.

D-21. Naval coastal warfare (NCW) is the responsibility of the JFC and is often exercised through the Navy component commander. He may assign a naval coastal warfare commander for an appropriate geographic area. NCW includes coastal sea control, port security, and harbor defense. While coastal sea control is conducted in the environment of the open seas, port security and harbor security include the urban environment. Port security is the safeguarding of vessels, harbors, ports, waterfront facilities, and cargo from internal threats. It includes destruction, loss, or injury from sabotage or other subversive acts; accidents; thefts; or other causes of similar nature. The Navy's role in protecting essential urban infrastructure is often key. This task is also important when executing stability operations if the threats against urban infrastructure will likely be unconventional.

D-22. Harbor defense protects harbor approaches, harbors, anchorages, and ports from external threats. Harbor defense focuses on the conventional defense of port infrastructure. It is a task appropriate in a major operation or campaign and often includes port security as a subtask. The JFC executes NCW (focused on harbor defense and port security) using combined Navy capabilities including surface warfare, aviation, and naval special operations. See also discussions in this appendix on transportation command and special operations command.

COAST GUARD

D-23. The Coast Guard, like the Navy, can significantly influence the conduct of UO when the urban area is accessible from the sea. The U.S. Coast Guard (USCG) is the federal authority for port security and harbor defense of domestic facilities. When directed by the President, the USCG can augment the Navy in operations overseas. Historically, the entire USCG was under Navy control during both World Wars I and II. USCG elements deployed overseas and operated under Navy control during the Vietnam War and during OPERATIONS DESERT SHIELD and DESERT STORM.

D-24. The USCG uses surface warfare systems and aerial reconnaissance systems to conduct its missions. Its air systems are unarmed, whereas its surface systems are armed for self-defense and law enforcement operations. The USCG is experienced and adept at supporting other agencies, local governments, and law enforcement. Its capabilities can best support Army UO in civil support operations or the civil support aspects of an operation. The USCG works effectively against an unconventional threat and threats with on-water capability.

URBAN FUNCTIONAL COMBATANT COMMAND CAPABILITIES

D-25. Three commands provide urban functional combatant command capabilities. Transportation Command manages the global transportation system. Strategic Command integrates plans for and employs space forces ad assets. Special Operations Command is responsible for providing special operations forces.

TRANSPORTATION COMMAND

D-26. U.S. Transportation Command (USTRANSCOM) provides strategic air, land, and sea transportation for the Department of Defense to deploy, employ, sustain, and redeploy U.S. military forces worldwide. USTRANSCOM provides global transportation management, using an integrated transportation system across the spectrum of operations through its transportation component commands (TCCs). The TCC consists of Air Mobility Command (AMC), Military Sealift Command (MSC), and Surface Deployment and Distribution Command (SDDC).

D-27. During urban operations, USTRANSCOM, through its TCC, can provide common-user terminal services in support of strategic transportation movements to a theater of operations. AMC provides common-user airlift, air refueling, and aeromedical evacuation services. It is the worldwide aerial port manager and, where designated, the operator of common-user aerial ports of embarkation and aerial ports of debarkation. MSC provides common-user sealift services between seaports of embarkation (SPOEs) and seaports of debarkation (SPODs). SDDC provides common-user ocean terminal service and, where designated, serves as the single port manager at SPOEs and SPODs. The urban transport system, in the form of railheads, ports, and airfields, is integral to many urban operations across the spectrum of Army operations and often the objective of Army UO. Army forces planning, preparing, and executing urban operations engage the supported geographic commander of a combatant command or his component commands in all aspects of conducting UO when the objective is transport related.

D-28. In offensive operations, commanders use transportation expertise to identify the urban transportation infrastructure (both the terrain and social [human] aspects) that is secured or that can affect current and future operations. This analysis also includes second- and third-order support systems. USTRANSCOM is consulted regarding the degree of acceptable damage that the system can sustain and still meet mission requirements. During execution, USTRANSCOM units may integrate into the operation so they can begin operating the transportation systems as early as possible. During the Inchon landing of September 1950, Army forces had begun rail operations on D+1 and port operations under way by D+3 of the forcible entry.

D-29. In defensive and stability or civil support operations, USTRANSCOM units safeguard and prevent disruption of the transport system by the conduct of defensive operations or stability tasks.

USTRANSCOM advises Army commanders of the impact of defensive tactics, techniques, and procedures (TTP) on USTRANSCOM operations as well as security requirements for USTRANSCOM facilities. The JFC provides guidance to deconflict any issues that may arise from these potentially divergent missions and tasks.

STRATEGIC COMMAND

D-30. The United States Strategic Command (USSTRATCOM) is the combatant command that integrates and synchronizes space operations, information operations, integrated missile defense, global command and control (C2); ISR, global strike, and strategic deterrence to provide. A unified source for greater understanding of specific threats around the world and the means to rapidly respond to those threats including weapons of mass destruction. USSTRATCOM operates assigned forces through its service component commands: U.S. Army Space Command/Army Forces Strategic Command; Air Force Space Command, and Naval Net Warfare Command.

D-31. USSTRATCOM provides critical space systems that offer global coverage and potential for real-time and near real-time support to urban operations. Space systems are unconstrained by political boundaries. Commanders can use space systems during peacetime or times of crisis to monitor an urban area before inserting friendly forces. Space systems enhance operations and assist commanders in overcoming some of the physical challenges of the urban environment. These enhancement operations include ISR; positioning and navigation; environmental monitoring; and communications. Space-based imagery and sensors are important ISR capabilities that contribute to situational understanding throughout the depth, breadth, and height of the urban area. ISR systems can provide route and target information for mission planning, locate presurveyed missile launch sites, detect camouflage, assess threat operations and movements, and warn of hostile acts and reconnaissance. Positioning and navigation systems also assist situational understanding through links to digital information systems (INFOSYS) while assisting tactical navigation in ambiguous terrain found in some urban areas. These enhancements can include precise location and position information for urban fires, ingress and egress routes, and rendezvous coordination. Environmental monitoring systems can provide weather and ionospheric information needed to assess weapon selection, air routes, ground and water trafficability, and communications. Communications systems provide secure, survivable links between elements of Army and joint forces to disseminate plans, orders, and warnings. These systems may form a critical link in the INFOSYS that transmit data to assessment centers and intelligence to key decision makers. However, some of the same environmental influences and degradation as ground-based systems may affect space-based systems. Commanders should refer to FM 3-14 for more information on space support to Army operations.

SPECIAL OPERATIONS COMMAND

D-32. U.S. Special Operations Command (USSOCOM) exercises combatant command of all active and reserve special operations forces stationed in the United States. USSOCOM also provides trained and combat-ready special operations forces (SOF) to the geographical combatant commanders and, when directed by the President or the Secretary of Defense, command-designated special operations.

D-33. SOF provide commanders with capabilities critical to success in the urban environment. The density of this environment in both space and time requires the careful integration of SOF and conventional forces. Army forces conducting UO have a clear and unambiguous C2 relationship with the SOF in the urban area to ensure coordination, massing of effects, and unity of effort.

D-34. Each service has unique special operations (SO) capabilities. For example, SOF can identify and seize or destroy key terrain or infrastructure in denied areas; secure or capture key personnel; identify supportive civilians (aides and allies) who can facilitate future infiltrations and the cache of critical supplies; counter urban insurgencies; and conduct unconventional warfare in enemy-held urban areas. SOF can also emplace covert sensors and surveillance devices, provide clandestine intelligence collection, and provide target acquisition information or emplace beacons (for weapons guidance or navigation) in the highly restrictive terrain of the urban environment. In multinational UO, they can provide coalition support teams with trained, culturally aware, language proficient, military liaison personnel with organic communications connectivity. Army leaders understand the SO capabilities available to ensure that they request the right support for Army UO and to ensure unity of effort within the urban JOA.

Army SOF

D-35. The Army provides five types of SOF units to USSOCOM. Two of these, psychological operations (PSYOP) and civil affairs, are discussed in Chapter 4.

Ranger Forces

D-36. The U.S. Army 75th Ranger Regiment can rapidly deploy light infantry forces from company through regimental size. Rangers specialize in direct action (DA) SO missions and focus on airfield seizure and raids. Typically, Ranger units turn over their objectives to conventional units upon mission completion. The Ranger DA capability especially applies to UO because of the many critical infrastructures that often prompt Army forces to engage in UO. Ranger capabilities are ideal for seizing critical facilities to preempt their defense or destruction. The size and combat power of Ranger units permit execution of offensive and defensive operations against enemy conventional units for periods of limited duration; austere combat service support capabilities limit the Ranger regiment's ability to sustain combat action without extensive augmentation.

Special Forces

D-37. U.S. Army Special Forces Command (Airborne) trains and prepares Army Special Forces (SF) to deploy and execute operational requirements for geographical combatant commanders. SF units are small and capable of extended operations in remote and hostile locations. SF units execute seven basic missions: foreign internal defense (FID), unconventional warfare (UW), counterproliferation of weapons of mass destruction, special reconnaissance (SR), DA, counterterrorism, and information operations. Important collateral activities consist of coalition support, combat search and rescue, counterdrug activities, humanitarian demining activities, humanitarian assistance, security assistance, and special activities.

D-38. SR, DA, and coalition support are particularly important in UO. SF units may physically penetrate an urban area to conduct SR to determine threat strengths, dispositions, and intentions. In some situations, SF units, due to their language and cultural training, can gather accurate information regarding the disposition and attitudes of the population. SF special reconnaissance can also determine or verify the functional status of urban infrastructure as well as conduct target acquisition, area assessment, and poststrike reconnaissance.

D-39. SF teams can execute terminal guidance and with qualified joint tactical air controllers can control operations for Army and joint precision fires. Although these teams possess limited organic combat power for their short-duration, DA missions, they can bring significant effects to bear against high-payoff targets. Yet, SF units have no capability to conduct a sustained defense of such targets.

D-40. SF units can advise, train, and assist urban indigenous movements already in existence to conduct unconventional warfare and possibly accompany these groups into combat. The upsurge in urban insurgency and terrorism has caused worldwide concern since it is not confined to developing countries. Present day dissident groups are well aware of this situation. These groups realize that to be successful, they must center the insurgent activities on the major cities or political center of their countries.

Special Operations Aviation Forces

D-41. Army special operations aviation provided by the 160th Special Operations Aviation Regiment operates primarily to support SO missions. They execute insertion, extraction, and resupply missions to support SO. Similar to conventional aviation, this capability may be more vulnerable to concealed air defenses when operating over hostile or unsecured urban terrain than in many other environments.

Navy SOF

Sea-Air-Land Teams

D-42. Navy sea-air-land teams (SEALs) specialize in water approaches to targets. They operate in small, squad-size teams and have many of the same capabilities as Army Special Forces (see above). Navy SEALs do not typically have the cultural and language training of Army Special Forces. Their capability to insert from sea gives them a unique ability to penetrate into urban areas that are accessible from the sea.

Special Boat Units

D-43. Special boat units (SBUs) employ, operate, and maintain various surface combatant craft (see figure D-3) to conduct and support naval and joint special operations, riverine warfare, and coastal patrol and interdiction. The SBU can infiltrate and exfiltrate forces; provide small-caliber gunfire support; and conduct coastal patrol, surveillance, harassment, and interdiction of maritime lines of communications. These units are ideal in UO that include ports and rivers. They can assist in port security, conduct river patrols, and participate in harbor defense. They are well suited for preventing sea infiltration by unconventional threats.

Figure D-3. USN MK V special

Operations CraftAir Force SOF

D-44. The Air Force has three primary SOF elements: special operations air units, battlefield Airmen, and combat aviation advisory units.

Special Operations Wings and Groups

D-45. Air Force special operations aviation elements operate both fixed-wing and rotary-wing aircraft (which include variants of both C-130s and MH-53s). Air Force SOF conduct core missions which include specialized air mobility, precision aerospace fires, aerospace surface interface, personnel recovery operations, specialized refueling, combat aviation advisory, information operations, and PSYOP. As a result, Air Force SOF conduct infiltrations, exfiltrations, resupplies, close air support of SOF ground elements, electronic warfare, and aerial C2 support.

D-46. The AC-130 U and AC-130 H model aircraft (see figure D-4) are designed specifically to provide close air support to ground SOF. However, they will be an invaluable asset to any ground force operating in an urban environment. Hence, Army commanders seek opportunities to train with AC-130 gunships prior to the conduct of UO. AC-130s are armed with one 40mm autocannon and one 105mm howitzer. The AC-130 U is also armed with one 25mm autocannon. This aircraft has night capability and is extremely accurate. Its fires are responsive and can be decisive in close urban combat. Its cannon and howitzer are accurate enough to concentrate fire onto a single spot to create a rooftop breach that allows fire to be directed deeper into the building. The aircraft can provide excellent covert illumination with its infrared spotlight. The AC-130 is very vulnerable to air defense systems, which friendly forces must suppress or destroy to effectively use this system. (During OPERATION DESERT STORM in 1991, an AC-130 was shot down over Kuwait.)

Battlefield Airmen

D-47. Battlefield Airmen consist of special tactics teams (STTs) and special operations weather teams (SOWTs). Combat controllers and pararescue personnel constitute STTs. STTs are specially tailored to meet mission criteria and may vary from a small three-man team to a larger twenty-five-man element. STTs support the UO Army commanders by—

- Performing air-land-sea personnel recovery operations.
- Providing terminal attack control or guidance.
- Establishing and operating navigational aids and beacons.
- Providing liaison to ground commanders.
- Providing visual flight rules and limited instrument flight rules air traffic control.
- Providing positive control of the terminal objective area aviation environment during SOF operations.

Note: SOWTs are normally attached to Army SOF to provide weather observation and limited tactical forecasting.

Combat Aviation Advisory (CAA) Operations

D-48. CAA operations are part of an interagency process aimed toward freeing and protecting the host nation's population from subversion, lawlessness, and insurgency. Specifically, CAA operations are tailored to assess, train, advise, and assist foreign aviation forces in air operations employment and sustainability. CAA operations support geographical combatant commanders throughout the spectrum of conflict primarily by facilitating the availability, reliability, safety and interoperability, and integration of friendly and allied aviation forces supporting joint, interagency, and multinational forces and organizations. CAA operations provide assistance in the interrelated areas of FID, coalition support, UW, and humanitarian and disaster relief. CAA operations also include a liaison role in coalition support.

MULTINATIONAL CONSIDERATIONS

D-49. Army UO in foreign urban areas will often be joint and likely have a multinational component. When properly executed, integrating multinational forces into UO greatly enhances the operation's military (as well as political) effectiveness. Properly integrating multinational forces into UO requires a thorough understanding of both the urban environment and the nature of individual national forces. This understanding includes the political, cultural, and historical characteristics of the other national forces. Such understanding also includes the national force's doctrine and military capabilities, strengths, and weaknesses. In UO, these considerations are critical because these factors will alter the urban population's attitude toward multinational forces and the behavior of such forces as they interact with the urban population. Combining this understanding with effective C2 and an equal understanding of the urban environment results in effective multinational UO (see FM 100-8).

CULTURAL COMPATIBILITY

D-50. When evaluating and assigning UO tasks, Army commanders also consider the degree of interaction with the civilian population. A national force from a Muslim-majority country may work better with a civilian population that is also Muslim than a force having a different religion. In such a situation, the national force with the same religion as the urban population may be assigned tasks that require close relations with civilians. Army forces, in this situation, may be assigned tasks that are more remote from the population. National forces that have a national history of animosity to the civil population (or that sympathize with antagonists within the civilian population) are not used in tasks requiring diplomacy and close cooperation with the civilians or government.

Military Capabilities and Restrictions

D-51. Army forces are responsible to understand the military capabilities of national forces with which they work. Some national forces, as part of their normal capabilities, are adept at police functions that enable them to operate with little training in a law enforcement role. Other national forces specialize in small-unit, light infantry patrolling. These forces may be ideal in a stability operation. In contrast, a national force composed of conscripts and trained primarily in conventional warfare techniques may best work as a reinforcing force or may require extensive training before mission execution in an urban environment or a stability situation. Army commanders also consider the type of weapon systems with which participating multinational forces are equipped; they may be more or less effective in an urban environment. For example, some countries may still possess the Vulcan antiaircraft gun system (or similar weapon) that can be very useful in urban offensive and defensive operations.

D-52. Commanders in multinational UO must also understand nation-specific rules of engagement (ROE). Some nations will severely limit their participating military's use of lethal and nonlethal fires. On the other hand, other nations may not be as constrained as Army units. Ideally, commanders attempt to create ROE that are the same across the coalition but recognize that this may not be achievable. Hence, in an environment where minimization of collateral damage will be paramount, Army commanders will need to carefully consider national limitations on the employment of forces when planning and executing UO. Relatedly, commanders take into account each nation's ability to achieve precision effects with their available forces and systems.

Intelligence Sharing

D-53. Due to classification issues, sharing intelligence with coalition partners may be challenging; the United States may have close intelligence ties with some countries within the coalition and few or none with others. In many cases intelligence personnel from other countries have unique skills (and cultural perspectives) that can significantly contribute to the friendly intelligence effort. Reports from some coalition members may fill intelligence gaps for Army forces and the coalition as a whole. On the other hand, few countries have the sophisticated intelligence collection assets available to Army forces and information that Army units may provide could be critical to a coalition partner's force protection and overall mission success.

Information Management

D-54. National capabilities to collect, process, store, display, and disseminate information and establish a common operational picture vary considerably between coalition partners. As the capabilities of Army information systems increase, those of our future partners may not keep pace. Even when capabilities are similar, equipment may not be compatible. The complex nature of the urban environment and coalition language differences (and linguist shortages) will only serve to exacerbate incompatibilities and limitations. Therefore commanders must understand capabilities and limitations to ensure that orders, control measures, and information can be appropriately disseminated, displayed, and understood. Significantly, they must understand the coalition's ability to accurately locate and track friendly forces as well as identify friend from foe. Commanders must take the appropriate steps to ensure that essential intelligence and relevant information is provided to coalition partners in a form (and language) that they can accept and use. This will be critical to establishing the common operational picture necessary to facilitate rapid decision making and effective command and control across the coalition.

Logistic Requirements

D-55. Logistically, agreement among the multinational forces should include support consolidation whenever possible. Creating a multinational logistics office or coordination center works best when accomplished early. This office can coordinate local contracts as well as already agreed upon host-nation support. This coordination among participating nations will reduce the competition for local assets that could otherwise have detrimental effects on one or more participating nations. Additionally, many nations providing forces for a coalition effort may not be able to support them logistically and will rely almost entirely on the United States or another coalition army for logistic support. Finally, pooled resources such as transportation assets and maintenance personnel could potentially reduce deployment requirements for one or more coalition partners. As the coalition matures, the role of this office or center may be expanded to include command activities.

This page intentionally left blank.

Source Notes

These are sources quoted or paraphrased in this publication. They are listed by page number. Quotations are identified by the first few word of the quote. Where a quote is embedded within a paragraph, the paragraph number is listed. Boldface indicates the title of historical vignettes.

Chapter 1–The Urban Outlook

1-1 "Today's security environment demands more...": FM 1, *The Army*, 14 June 2005: 1-19.

1-4 **Rome: A Microcosm of Urban Warfare**: R. Ernest Dupuy and Trevor N. Dupuy, *The Encyclopedia of Military History: From 3500 B.C. to the Present* (San Francisco: Harper & Row, 1977).

1-6 **The Three Block War**: as quoted in Russell W. Glenn, Steven Hartman, and Scott Gerwehr, *Urban Combat Service Support Operations: The Shoulders of Atlas* (Santa Monica, CA: RAND, 2003): 26.

1-8 **Winning the Peace in Iraq: The Requirement for Full Spectrum Operations**: Peter W. Chiarelli and Patrick R. Michaelis, "Winning the Peace: The Requirement for Full-Spectrum Operations," *Military Review* (July/August 2005): 4.

Chapter 2–Understanding the Urban Environment

2-1 "From a planning perspective, commanders view cities....": FM 3-0, *Operations*, 14 June 2001: 6-19.

2-16 "Me and Somalia Against the World": cited in Andrew S. Natsios, "Food Through Force: Humanitarian Intervention and U.S. Policy," *The Washington Quarterly* (winter 1994): 136.

2-18 **Understanding the Effects of Unit and Soldier Actions in Iraq**: *Initial Impressions Report: Operation Iraqi Freedom—Stability Operations-Support Operations,* Center for Army Lessons Learned (CALL) Newsletter No. 04-13 (Fort Leavenworth, KS: Center for Army Lessons Learned, U.S. Training and Doctrine Command [TRADOC], 2004): ii.

2-21 "Whoever coined the phrase... ": General Sir Rupert Smith, "Wars in Our Time–A Survey of Recent and Continuing Conflicts," *World Defence Systems*, volume 3:2 (London: Royal United Services Institute for Defence Studies, 2001): 121.

2-25 **Understanding the Urban Environment: Paris – 1944**: Russell W. Glenn, "Urban Combat is Complex," *Proceedings* (February 2002): 65. Russell W. Glenn, Steven Hartman, and Scott Gerwehr, *Urban Combat Service Support Operations: The Shoulders of Atlas* (Santa Monica, CA: RAND, 2003): 10-11.

Chapter 3–Understanding the Urban Threat

3-1 "...[T]he United States could be forced to intervene...": *Quadrennial Defense Review Report*, 30 September 2001: 6.

3-3 "...Iraq made no direct effort...": Gregory Fontenot, E. J. Degen, and David Tohn, *On Point: US Army in Operation IRAQI FREEDOM* (Fort Leavenworth, KS: Combat Studies Institute Press, 2004): 388.

3-4 **Tempo**: 26th Infantry Regimental Association, *Aachen. Military Operations in Urban Terrain* (Lititz, PA: 26th Infantry Regimental Association, 1999); Irving Werstein, *The Battle of Aachen* (New York: Thomas Y. Crowell Company, 1962); Charles Whiting, *Bloody Aachen* (New York: Stein and Day, 1976).

3-6 "Chechen fighters sometimes disguised...": Olga Oliker, *Russia's Chechen Wars 1994–2000: Lessons from Urban Combat* (Santa Monica, CA: RAND, 2001): 21.

3-7 **Identifying Threats from Noncombatants**: Olga Oliker, *Russia's Chechen Wars 1994–2000: Lessons from Urban Combat* (Santa Monica, CA: RAND, 2001).

3-9 **Information and the Media**: Eric Hammel, *Fire in the Streets: The Battle for Hue, Tet 1968* (New York: Dell Publishing, 1991).

3-12 **Cultural and Religious Instability**: Sid Heal, "Crowds, Mobs and Nonlethal Weapons," *Military Review* (March/April 2000): 45–50.

3-13 **Food and Water Shortages**: Leif R. Rosenberger, "The Strategic Importance of the World Food Supply," *Parameters* (spring 1997): 84–105.

3-15 **Insurgencies and the Urban Society**: Roger Trinquier, *Modern Warfare: A French View of Counterinsurgency* (New York: Praeger Publisher, 1964): 16-17.

3-16 **Crime and Criminal Organizations**: Eugene Linden, "The Exploding Cities of the Developing World," *Foreign Affairs* (January/February 1996): 52–65; Andrew S. Natsios, "Commander's Guidance: A Challenge of Complex Humanitarian Emergencies," *Parameters* (summer 1996): 50–66.

Chapter 4–Understanding the Urban Environment's Effects on Warfighting Functions and Tactics

4-1 "War is, above all things, an art...": Francis V. Greene, "The Important Improvements in the Art of War During the Past Twenty Years and Their Probable Effect on Future Military Operations," *Journal of the Military Service Institution of the United States* 4, no. 13 (1883): 41.

4-13 "Fighting in a city is much...": Vasili I. Chuikov, *The Battle for Stalingrad* (New York: Holt, Rinehart, and Winston, 1964): as cited by JP 3-06, *Doctrine for Urban Operations*, 16 September 2002: 1-8.

4-14 "I heard small-arms fire and...": Mark A. B. Hollis, "Platoon Under Fire," *Infantry* (January/April 1998): 29–30.

4-17 **Example of Simple Communications Innovation: Israel's Six-Day War – 1967**: Trevor N. Dupuy and Paul Martell, *Flawed Victory: The Arab-Israeli Conflict and the 1982 War in Lebanon* (Fairfax, VA: Hero Books, 1986); Richard A. Gabriel, *Operation Peace for Galilee: The Israeli-PLO War in Lebanon* (New York: Hill and Wang, 1984).

Chapter 5–Contemplating Urban Operations

5-1 "We based all our further calculations...": S. M. Shtemenko, *The Soviet General Staff at War, 1941-1945: Book One* (Moscow: Progress Publishers, 1981): 317–318.

5-19 "Four hostile newspapers...": Napoleon Bonaparte, Justin Wintle (ed.), *The Dictionary of War Quotations* (New York: The Free Press, 1989): 73.

5-24 **Conventional and Special Forces Integration**: Gregory Fontenot, E. J. Degen, and David Tohn, *On Point: US Army in Operation IRAQI FREEDOM* (Fort Leavenworth, KS: Combat Studies Institute Press, 2004): 403.

Chapter 6–Foundations for Urban Operations

6-1 "Utilities such as electricity…": Richard Connaughton, John Pimlott, and Duncan Anderson, *The Battle for Manila: The Most Devastating Untold Story of World War II* (Novata, CA: Presidio Press, 1995): 103.

6-7 **Applying the Urban Operational Framework: Panama – 1989**: Ronald H. Cole, *Operation Just Cause* (Washington DC: Joint History Office, 1995); Thomas Donnelly, Margaret Roth, and Caleb Baker, *Operation Just Cause: The Storming of Panama* (New York: Lexington Books, 1991); John T. Fishel, *The Fog of Peace: Planning and Executing the Restoration of Panama* (Carlisle, PA: Strategic Studies Institute, 1992); Malcolm McConnell, *Just Cause: The Real Story of America's High-Tech Invasion of Panama* (New York: St. Martin's Press, 1991); John Embry Parkerson, Jr., "United States Compliance with Humanitarian Law Respecting Civilians During Operation Just Cause," *Military Law Review* (1991): 31–140; Jennifer Morrison Taw, *Operation Just Cause: Lessons for Operations Other Than War* (Santa Monica, CA: RAND, 1996).

6-12 "I'm talking about attacking...": William S. Wallace quoted in Michael R. Gordon, "Baghdad Targets Picked if Hussein Holes Up There," *New York Times* (7 March 2003): A11.

6-13 "Do not try to..." T. E. Lawrence, "Twenty-Seven Articles," *The Arab Bulletin* (20 August 1917): as cited in David H. Petraeus, "Learning Counterinsurgency: Observations from Soldiering in Iraq," *Military Review* (January/February 2006): 4–5.

Chapter 7–Urban Offensive Operations

7-1 "…Capture Suez City...": Abraham Adan, *On the Banks of the Suez* (Novato, CA: Presidio Press, 1980): 409.

7-2 **Operational Context of Urban Operations: Brittany Ports – 1944**: Kent H. Butts, "The Strategic Importance of Water," *Parameters* (spring 1997): 65–83; Paul Carell, *Stalingrad: The Defeat of the German 6th Army* (Atglen, PA: Schieffer Publishing, 1993); R. Ernest Dupuy and Trevor N. Dupuy, *The Encyclopedia of Military History: From 3500 B.C. to the Present* (San Francisco: Harper & Row, 1977); Ralph Peters, "Our Soldiers, Their Cities," *Parameters* (spring 1996): 43–50; John J. Peterson, *Into the Cauldron* (Clinton, MD: Clavier House, 1973).

7-10 **Forms of Attack in the Urban Offense: Metz – 1944**: John Colby, *War From the Ground Up: The 90th Division in WWII* (Austin, TX: Nortex Press, 1991); Anthony Kemp, *The Unknown Battle: Metz, 1944* (New York: Stein and Day, 1981).

7-17 **Isolating an Urban Area: Hue, Vietnam – 1968**: James R. Arnold, *Tet Offensive 1968: Turning Point in Vietnam* (London: Osprey, 1990); Eric Hammel, *Fire in the Streets: The Battle for Hue, Tet 1968* (New York: Dell Publishing, 1991).

7-21 "Often what seems to be...": Russell W. Glenn, *An Attack on Duffer's Downtown* (Santa Monica, CA: RAND, 2001): 28.

7-22 **Creative Task Organization: Using Artillery in the Direct Fire Role**: 26th Infantry Regimental Association, *Aachen. Military Operations in Urban Terrain* (Lititz, PA: 26th Infantry Regimental Association, 1999); Ronald H. Cole, *Operation Just Cause* (Washington DC: Joint History Office, 1995); Thomas Donnelly, Margaret Roth, and Caleb Baker, *Operation Just Cause: The Storming of Panama* (New York: Lexington Books, 1991); David Eshel, *Mid-East Wars: The Lebanon War 1982* (Hod Hasharon, Israel: Eshel-Dramit, 1983); Richard A. Gabriel, *Operation Peace for Galilee: The Israeli-PLO War in Lebanon* (New York: Hill and Wang, 1984); Malcolm McConnell, *Just Cause: The Real Story of America's High-Tech Invasion of Panama* (New York: St. Martin's Press, 1991); Irving Werstein, *The Battle of Aachen* (New York: Thomas Y. Crowell Company, 1962); Charles Whiting, *Bloody Aachen* (New York: Stein and Day, 1976); Bruce Allen Watson, *Sieges: A Comparative Study* (Westport, CT: Praeger, 1993).

7-26 **Bold Operational Maneuver to Seize an Urban Area: Inchon and Seoul, Korea – 1950**: Robert Debs Heinl, Jr., *Victory at High Tide: The Inchon-Seoul Campaign* (Annapolis, MD: Nautical & Aviation Publishing Company of America, 1979); Douglas MacArthur, *Reminiscences* (New York: McGraw-Hill, 1964); Shelby L. Stanton, *Ten Corps in Korea, 1950* (Novato, CA: Presidio Press, 1989).

Chapter 8–Urban Defensive Operations

8-1 "Generally, a modern city magnifies...": Christopher R. Gabel, "Military Operations on Urbanized Terrain: The 2d Battalion, 26th Infantry, at Aachen, October 1944," *Urban Combat Operations: tactics, Techniques and Procedure.* Center for Army Lessons Learned (CALL) Newsletter No. 99-16 (Fort Leavenworth, KS: Center for Army Lessons Learned, U.S. Training and Doctrine Command [TRADOC], November 1999): 1-9.

8-5 **Urban Defense in a Major Operation: Stalingrad –1942 to 1943**: Anthony Beevor, *Stalingrad* (New York: Penguin Books, 1998); Vasili I. Chuikov, *The Battle for Stalingrad* (New York: Holt, Rinehart, and Winston, 1964); William Craig, *Enemy at the Gates: The Battle for Stalingrad* (New York: Reader's Digest Press, 1973).

8-8 "What is the position...": Michael Dewar, *War in the Streets: The Story of Urban Combat from Calais to Khafji* (New York: Sterling Publishers, 1992): 8.

8-10 **Defensive Combat Power: Suez – 1973**: Abraham Adan, *On the Banks of the Suez* (Novato, CA: Presidio Press, 1980).

Chapter 9–Urban Stability and Civil Support Operations

9-1 "The Rangers were bound...": Mark Bowden, *Black Hawk Down: A Story of Modern War* (New York: Atlantic Monthly Press, 1999): 18.

9-5 "In wars of intervention...": Antoine-Henri Jomini, "The Art of War," in *Roots of Strategy: Book 2* (Harrisburg, PA: Stackpole Books, 1987): 441.

9-8 **Assessment of Security and Force Protection: Belfast, Northern Ireland**; Michael Dewar, *British Army in Northern Ireland* (New York: Sterling Publishers, 1985); Desmond Hamill, *Pig in the Middle: The Army in Northern Ireland, 1969-1984* (London: Methuen, 1985).

9-11 "If there is any lesson...": Frederick Funston, "How the Army Worked to Save San Francisco," *Cosmopolitan Magazine* (July 1906). Available at http: //www.sfmuseum.org/1906/cosmo.html.

9-12 **Civil Support and Coordination with Civilian Authorities: Los Angeles – 1992**: James D. Delk, *Fires & Furies: The LA Riots* (Palm Springs, CA: Etc. Publications, 1995); William W. Mendel, *Combat in Cities: The LA Riots and Operation Rio* (Fort Leavenworth, KS: Foreign Military Studies Office, 1996).

9-14 **Restraint: An Najaf, Iraq – 2003**: Gregory Fontenot, E. J. Degen, and David Tohn, *On Point: US Army in Operation IRAQI FREEDOM* (Fort Leavenworth, KS: Combat Studies Institute Press, 2004).

9-16 "I met immediately with the...": Jeffrey B. Van Sickle, "Stability Operations in Northern Iraq: Task Force Altun Kupri," *Infantry* (January/February 2005): 26.

Chapter 10–Urban Sustainment

10-1 "Even supply is different…" Ralph Peters, "Our Soldiers, Their Cities," *Parameters* (spring 1996): 48.

10-2 **Base Security: Tan Son Nhut, Vietnam – 1968**: James R. Arnold, *Tet Offensive 1968: Turning Point in Vietnam* (London: Osprey, 1990); Keith William Nolan, *The Battle for Saigon: Tet 1968* (New York: Pocket Books, 1996).

10-4 "[Sustainment] planning accounts for...": FM 3-0, *Operations*, 14 June 2001: 6-19.

10-8 "Greater friendly force density would appear...": Russell W. Glenn, Steven Hartman, and Scott Gerwehr, *Urban Combat Service Support Operations: The Shoulders of Atlas* (Santa Monica, CA: RAND, 2003): 4.

10-13 **Host Nation Repair Operations: OPERATION IRAQI FREEDOM**: Steven Leonard, personal experience written specifically for FM 3-06.

10-18 **Combat Stress: Chechnya – 1994 to 1996**: Timothy L. Thomas and Charles P. O'Hara, "Combat Stress in Chechnya: 'The Equal Opportunity Disorder'," *Army Medical Department Journal* (January/March 2000): 46–53.

10-22 "…[I]n an endeavor like that in Iraq…": David H. Petraeus, "Learning Counterinsurgency: Observations from Soldiering in Iraq," *Military Review* (January/February 2006): 4–5.

Appendix A–Siege of Beirut: An Illustration of the Fundamentals of Urban Operations

A-1 **Siege of Beirut: An Illustration of the Fundamentals of Urban Operations**: Trevor N. Dupuy and Paul Martell, *Flawed Victory: The Arab-Israeli Conflict and the 1982 War in Lebanon* (Fairfax, VA: Hero Books, 1986); David Eshel, *Mid-East Wars: The Lebanon War 1982* (Hod Hasharon, Israel: Eshel-Dramit, 1983); Richard A. Gabriel, *Operation Peace for Galilee: The Israeli-PLO War in Lebanon* (New York: Hill and Wang, 1984).

A-1 "The [Israeli Defense Forces] IDF had...": Richard A. Gabriel, *Operation Peace for Galilee: The Israeli-PLO War in Lebanon* (New York: Hill and Wang, 1984): 128.

Appendix B–An Urban Focus to the Intelligence Preparation of the Battlefield

B-1 "Maneuvers that are possible...": *Infantry in Battle, 2d ed.* (Richmond, VA: Garret & Massie, 1939; Reprint, Fort Leavenworth, KS: U.S. Army Command and General Staff College, 1981): 69.

B-5 **Analysis of an Urban Area's Underlying Terrain: Mitrovica, Kosovo – 1999**: Jamison Jo Medby and Russell W. Glenn, *Street Smart: Intelligence Preparation of the Battlefield for Urban Operations* (Santa Monica, CA: RAND, 2002).

B-10 "…[T]he adversaries of freedom...": John F. Kennedy, Message to Congress, 25 May 1961. Available at <http://www.jfklibrary.org/Historical+Resources/Archives>.

B-12 **Shifting Civilian Interests and Intent**: James D. Delk, *Fires & Furies: The LA Riots* (Palm Springs, CA: Etc. Publications, 1995); Jamison Jo Medby and Russell W. Glenn, *Street Smart: Intelligence Preparation of the Battlefield for Urban Operations* (Santa Monica, CA: RAND, 2002).

Appendix C–Operations in Somalia: Applying the Urban Operational Framework to Stability Operations

C-1 **Operations in Somalia: Applying the Urban Operational Framework to Stability and Reconstruction Operations**: Kenneth Allard, *Somalia Operations: Lessons Learned* (Washington, DC: National Defense University Press, 1995).

C-1 "It's impossible for an American mother...": Thomas M. Montgomery, interview, "Ambush in Mogasdishu," *Frontline*, WGBH Boston, 29 September 1998, <http://www.pbs.org/wgbh/pages/frontline/programs/1998.html>.

C-2 C-5 …" to use all necessary…": *United Nations Security Council Resolution 794* (1992) (New York: United Nations, 1992), <http://daccessdds.un.org/doc/UNDOC/GEN/N92/772/11/PDF>.

C-2 C-5 "When directed by the…": CENTCOM mission statement as cited in Kenneth Allard, *Somalia Operations: Lessons Learned* (Washington, DC: National Defense University Press, 1995): 13-14.

Appendix D–Joint and Multinational Urban Operations

D-1 "[Joint force commanders] synchronize the actions…" JP 3-0, *Doctrine for Joint Operations*, 10 September 2001: II-4.

This page intentionally left blank.

Glossary

SECTION I – ACRONYMS AND ABBREVIATIONS

AI	air interdiction
AIDS	acquired immune deficiency syndrome
AMC	Air Mobility Command
AO	area of operations
ARVN	Army of the Republic of Viet Nam
ASCOPE	areas, structures, capabilities, organizations, people, and events
BCT	brigade combat team
BDAR	battle damage assessment and repair
BENELUX	Belgium, Netherlands, and Luxembourg
C2	command and control
CA	civil affairs
CAA	combat aviation advisory
CASEVAC	casualty evacuation
CBRN	chemical, biological, radiological, nuclear
CBRNE	chemical, biological, radiological, nuclear, high-yield explosive
CCIR	commander's critical information requirement
CI	counterintelligence
CEA	captured enemy ammunition
CERP	commander's emergency response program
CLS	combat lifesaver
CMO	civil-military operations
CMOC	civil-military operations center
CNA	computer network attack
CND	computer network defense
CNE	computer network exploitation
CNO	computer network operations
COA	course of action
COG	center of gravity
CONUS	continental United States
COP	common operational picture
DA	direct action
DNBI	disease and nonbattle injury
DOTMLPF	doctrine, organization, training, materiel, leadership and education, personnel, and facilities
EAC	echelons above corps
EEFI	essential elements of friendly information

EOD	explosive ordnance disposal
EW	electronic warfare
FHP	force health protection
FID	foreign internal defense
FM	field manual; frequency modulation
FMI	field manual, interim
FSB	forward support battalion
G-9	assistant chief of staff, civil affairs
GIS	geographic information system
GTL	gun-target line
HRS	human resources support
HUMINT	human intelligence
IDF	Israeli Defense Forces
IED	improvised explosive device
IMINT	imagery intelligence
INFOSYS	information system
IO	information operations
IPB	intelligence preparation of the battlefield
ISR	intelligence, surveillance, and reconnaissance
JDAM	Joint Direct Attack Munition
JFC	joint force commander
JOA	joint operations area
JP	joint publication
JTF	joint task force
JUO	joint urban operations
LAN	local area network
LNO	liaison officer
LOC	line of communications
LOGCAP	logistics civilian augmentation program
LOS	linc of sight
LPT	logistics preparation of the theater
MAGTF	Marine air-ground task force
MANPADS	man-portable air defense system
MEDEVAC	medical evacuation
METT-TC	mission, enemy, terrain and weather, troops and support available, time available, civil considerations
MEU	Marine expeditionary unit
MEU(SOC)	Marine expeditionary unit (special operations capable)
MHE	materials handling equipment
MLRS	Multiple Launch Rocket System
MP	military police

MSC	Military Sealift Command
MST	maintenance support team
NATO	North Atlantic Treaty Organization
NCW	naval coastal warfare
NGA	National Geospatial-Intelligence Agency
NGO	nongovernmental organization
NKPA	North Korean People's Army
NVA	North Vietnamese Army
OAKOC	observation and fields of fire, avenues of approach, key terrain, obstacles, and cover and concealment
ODA	operational detachment alpha
OIF	Operation Iraqi Freedom
OPSEC	operations security
PA	public affairs
PDF	Panamanian Defense Force
PGM	precision-guided munitions
PLO	Palestinian Liberation Organization
PMCS	preventive maintenance checks and services
POW	prisoner of war
PSYOP	psychological operations
QRF	quick reaction force
ROE	rules of engagement
RPG	rocket-propelled grenade
SBU	special boat unit
SDDC	Surface Deployment and Distribution Command
SEAL	sea-air-land (team)
SF	special forces
SIGINT	signals intelligence
SJA	Staff Judge Advocate
SO	special operations
SOF	special operations forces
SOFA	status-of-forces agreement
SOP	standing operating procedure
SOWT	special operations weather team
SPOD	seaport of debarkation
SPE	seaport of embarkation
SR	special reconnaissance
STT	special tactics team
SWET	sewer, water, electricity, trash
SWEAT	sewer, water, electricity, academics trash
SWEAT-MS	sewer, water, electricity, academics trash-medical, security

TACON	tactical control
TC	training circular
TCC	transportation component command
TF	task force
TIM	toxic industrial material
TTP	tactics, techniques, and procedures
UAS	unmanned aircraft system
UN	United Nations
UNITAF	Unified Task Force
UNOSOM	United Nations Operations in Somalia
UNSCR	UN Security Council resolution
UO	urban operations
USACE	United States Army Corps of Engineers
USAID	United States Agency for International Development
USCENTCOM	United States Central Command
USFORSOM	United States Forces, Somalia
USCG	United States Coast Guard
USTRANSCOM	United States Transportation Command
UXO	unexploded ordnance
UW	unconventional warfare
VC	Viet Cong
WMD	weapons of mass destruction

SECTION II – Terms and Definitions

agility – The ability to move and adjust quickly and easily. (FM 3-0)

air interdiction – Air operations conducted to destroy, neutralize, or delay the enemy's military potential before it can be brought to bear effectively against friendly forces at such distance from friendly forces that detailed integration of each air mission with the fire and movement of friendly forces is not required. (JP 1-02)

area defense – A type of defensive operation that concentrates on denying enemy forces access to designated terrain for a specific time rather than destroying the enemy outright. (FM 3-0)

area of interest – That area of concern to the commander, including the area of influence, areas adjacent thereto, and extending into enemy territory to the objectives of current or planned operations. This area also includes areas occupied by enemy forces who could jeopardize the accomplishment of the mission. (JP 1-02)

area of operations – An operational area defined by the joint force commander for land and naval forces. Areas of operations do not typically encompass the entire operational area of the joint force commander, but should be large enough for component commanders to accomplish their missions and protect their forces. (JP 1-02)

ARFOR – The senior Army headquarters and all Army forces assigned or attached to a combatant command, subordinate joint force command, joint functional command, or multinational command. (FM 3-0)

assessment – (Army) The continuous monitoring of the current situation and progress of an operation.(FMI 5-0.1)

asymmetry – Dissimilarities in organization, equipment, doctrine, and values between other armed forces (formally organized or not) and US forces. Engagements are symmetric if forces, technologies, and weapons are similar; they are asymmetric if forces, technologies, and weapons are different, or if a resort to terrorism and rejection of more conventional rules of engagement are the norm. (FM 3-0)

attack – An offensive operation that destroys or defeats enemy forces, seizes and secures terrain, or both. (FM 3-0)

attainability - One of the eight characteristics of combat service support: generating the minimum essential supplies and services necessary to begin operations. (FM 4-0)

attrition - The reduction of the effectiveness of a force caused by loss of personnel and materiel. (JP 1-02)

avenue of approach – An air or ground route of an attacking force of a given size leading to its objective or to key terrain in its path. (JP 1-02)

axis of advance – (Army) The general area through which the bulk of a unit's combat power must move. (FM 3-90)

battle – A set of related engagements that lasts longer and involves larger forces than an engagement. (FM 3-0)

battlefield organization – The allocation of forces in the area of operations by purpose. It consists of three all-encompassing categories of operations: decisive, shaping, and sustaining. (FM 3-0)

battle rhythm – The sequencing of command and control activities within a headquarters and throughout the force to facilitate effective command and control. (FMI 5-0.1)

branch – (Army) A contingency plan or course of action (an option built into the basic plan or course of action) for changing the mission, disposition, orientation, or direction of movement of the force to aid success of the current operation, based on anticipated events, opportunities, or disruptions caused by enemy actions. Army forces prepare branches to exploit success and opportunities, or to counter disruptions caused by enemy actions. (FM 3-0)

breakout – An operation conducted by an encircled force to regain freedom of movement or contact with friendly units. It differs from other attacks only in that a simultaneous defense in other areas of the perimeter must be maintained. (FM 3-90)

buffer zone – A defined area controlled by a peace operations force from which disputing or belligerent forces have been excluded. A buffer zone is formed to create an area of separation between disputing or belligerent forces and reduce the risk of renewed conflict. Also called **area of separation** in some United Nations operations. (JP 1-02)

bypass – A tactical mission task in which the commander directs his unit to maneuver around an obstacle, position, or enemy force to maintain the momentum of the operation while deliberately avoiding combat with an enemy force. (FM 3-90)

campaign – A series of related military operations aimed at accomplishing a strategic or operational objective within a given time and space. (JP 1-02)

canalize – (Army) A tactical mission task in which the commander restricts enemy movement to a narrow zone by exploiting terrain coupled with the use of obstacles, fires, or friendly maneuver. (FM 3-90)

capability - The ability to execute a specified course of action (a capability may or may not be accompanied by an intention) (JP 1-02)

casualty evacuation – (Army) A term used by nonmedical units to refer to the movement of casualties aboard nonmedical vehicles or aircraft. (FM 8-10-6)

centers of gravity – Those characteristics, capabilities, or sources of power from which a military force derives its freedom of action, physical strength, or will to fight. (JP 1-02)

civil affairs – Designated Active and Reserve component forces and units organized, trained, and equipped specifically to conduct civil affairs activities and to support civil-military operations. (JP 1-02)

civil considerations – The influence of manmade infrastructure, civilian institutions, and attitudes and activities of the civilian leaders, populations, and organizations within an area of operations on the conduct of military operations. (FM 6-0)

civil disturbance – Group acts of violence and disorder prejudicial to public law and order. (JP 1-02)

civil-military operations – The activities of a commander that establish, maintain, influence, or exploit relations between military forces, governmental and nongovernmental civilian organizations and authorities, and the civilian populace in a friendly, neutral, or hostile operational area in order to facilitate military operations, to consolidate and achieve US objectives. Civil-military operations may include performance by military forces of activities and functions normally the responsibility of the local, regional, or national government. These activities may occur prior to, during, or subsequent to other military actions. They may also occur, if directed, in the absence of other military operations. Civil-military operations may be performed by designated civil affairs, by other military forces, or by a combination of civil affairs and other forces. (JP 1-02)

civil-military operations center – An ad hoc organization, normally established by the geographic combatant commander or subordinate joint force commander, to assist in the coordination of activities of engaged military forces, and other United States Government agencies, nongovernmental organizations, and regional and international organizations. There is no established structure, and its size and composition are situation dependent. (JP 1-02)

civil support operations – Domestic operations that address the consequences of man-made or natural accidents and incidents beyond the capabilities of civilian authorities. (FM 1)

clear – (Army) 1. A tactical mission task that requires the commander to remove all enemy forces and eliminate organized resistance within an assigned area. (FM 3-90) 2. The total elimination or neutralization of an obstacle that is usually performed by follow-on engineers and is not done under fire. (FM 3-34.2)

close air support – Air action by fixed- and rotary-wing aircraft against hostile targets that are in close proximity to friendly forces and that require detailed integration of each air mission with the fire and movement of those forces. (JP 1-02)

close combat – Combat carried out with direct fire weapons, supported by indirect fire, air-delivered fires, and nonlethal engagement means. Close combat defeats or destroys enemy forces or seizes and retains ground. (FM 3-0)

coalition – An ad hoc arrangement between two or more nations for common action. (JP 1-02)

collateral damage – Unintentional or incidental injury or damage to persons or objects that would not be lawful military targets in the circumstances ruling at the time. Such damage is not unlawful so long as it is not excessive in light of the overall military advantage anticipated from the attack. (JP 1-02)

combatant command – A unified or specified command with a broad continuing mission under a single commander established and so designated by the President, through the Secretary of Defense and with the advice and assistance of the Chairman of the Joint Chiefs of Staff. Combatant commands typically have geographic or functional responsibilities. (JP 1-02)

combat power – The total means of destructive and/or disruptive force which a military unit/formation can apply against the opponent at a given time. (JP 1-02)

combined arms – The synchronized or simultaneous application of several arms—such as infantry, armor, field artillery, engineers, air defense, and aviation—to achieve an effect on the enemy that is greater than if each arm were used against the enemy separately or in sequence. (FM 3-0)

combined arms team – (Army) Two or more arms mutually supporting one another, usually consisting of infantry, armor, cavalry, aviation, field artillery, air defense artillery, and engineers. (FM 3-90).

command and control system – (Army) The arrangement of personnel, information management, procedures, and equipment and facilities essential for the commander to conduct operations. (FM 6-0)

commander's critical information requirements – (Army) – Elements of information required by commanders that directly affect decisionmaking and dictate the successful execution of military operations. (FM 3-0)

commander's intent – (Army) A clear, concise statement of what the force must do and the conditions the force must meet to succeed with respect to the enemy, terrain, and civil considerations that represent the operation's desired end state. (FMI 5-0.1)

commander's visualization – The mental process of developing situational understanding, determining a desired end state, and envisioning how the force will achieve that end state. (FMI 5-0.1)

common operational picture – (Army) An operational picture tailored to the user's requirements, based on common data and information shared by more than one command. (FM 3-0)

computer network attack – Actions taken through the use of computer networks to disrupt, deny, degrade, or destroy information resident in computers or computer networks, or the computers and networks themselves. (JP 1-02)

computer network defense – Actions taken through computer networks to protect, monitor, analyze, detect and respond to unauthorized activity within Department of Defense information systems and computer networks. (JP 1-02)

concept of operations – (Army) How commanders see the actions of subordinate units fitting together to accomplish the mission. As a minimum, the description includes the scheme of maneuver and concept of fires. The concept of operations expands the commander's selected course of action and expresses how each element of the force will cooperate to accomplish the mission. (FM 3-0)

constraint – A restriction placed on the command by a higher command. A constraint dictates an action or inaction, thus restricting the freedom of action a subordinate commander has for planning. (FM 5-0)

control – (Army) 1. The regulation of forces and warfighting functions to accomplish the mission in accordance with the commander's intent. (FMI 5-0.1) 2. A tactical mission task that requires the commander to maintain physical influence over a specified area to prevent its use by an enemy. (FM 3-90)

controlled exchange – The removal of serviceable parts, components, or assemblies from unserviceable, economically reparable equipment and their immediate reuse in restoring a like item of equipment to a combat operable or serviceable condition. (FM 4-30.3)

control measure – A means of regulating forces or warfighting functions. (FMI 5-0.1)

counterattack – (Army) A form of attack by part or all of a defending force against an enemy attacking force with the general objective of denying the enemy his goal in attacking. See FM 3-0.

counterdeception – Efforts to negate, neutralize, diminish the effects of, or gain advantage from a foreign deception operation. Counterdeception does not include the intelligence function of identifying foreign deception operations. (JP 1-02)

counterdrug – Those active measures taken to detect, monitor, and counter the production, trafficking, and use of illegal drugs. (JP 1-02)

counterinsurgency – Those military, paramilitary, political, economic, psychological, and civic actions taken by a government to defeat insurgency. (JP 1-02)

counterintelligence – Information gathered and activities conducted to protect against espionage, other intelligence activities, sabotage, or assassinations conducted by or on behalf of foreign governments or elements thereof, foreign organizations, or foreign persons, or international terrorist activities. (JP 1-02)

countermobility operations – The construction of obstacles and emplacement of minefields to delay, disrupt, and destroy the enemy by reinforcement of the terrain. The primary purpose of countermobility operations is to slow or divert the enemy, to increase time for target acquisition, and to increase weapon effectiveness. (JP 1-02)

counterpropaganda – Programs of products and actions designed to nullify propaganda or mitigate its effects. (FM 3-05.30)

counterterrorism – Operations that include the offensive measures taken to prevent, deter, preempt, and respond to terrorism. (JP 1-02)

course of action – 1. Any sequence of activities that an individual or a unit may follow. 2. A possible plan open to an individual or a commander that would accomplish or is related to the accomplishment of a mission. 3. The scheme adopted to accomplish a job or mission. 4. A line of conduct in an engagement. 5. A product of the Joint Operation Planning and Execution System concept development phase. (JP 1-02)

cover – (Army) 1. Protection from the effects of fires. (FM 6-0) 2. A form of security operation whose primary task is to protect the main body by fighting to gain time while also observing and reporting information and preventing enemy ground observation of and direct fire against the main body. Unlike a screening or guard force, the covering force is a self-contained force capable of operating independently of the main body. (FM 3-90)

dead space – 1. An area within the maximum range of a weapon, radar, or observer, which cannot be covered by fire or observation from a particular position because of intervening obstacles, the nature of the ground, the characteristics of the trajectory, or the limitations of the pointing capabilities of the weapon. 2. An area or zone which is within range of a radio transmitter, but in which a signal is not received. 3. The volume of space above and around a gun or guided missile system into which it cannot fire because of mechanical or electronic limitations. (JP 1-02)

debarkation – The unloading of troops, equipment, or supplies from a ship or aircraft. (JP 1-02)

decision point – (Army) An event, area, or point in the battlespace where and when the friendly commander will make a critical decision. (FM 5-0)

decision support matrix – An aid used by the commander and staff to make battlefield decisions. This matrix is a staff product of the wargaming process that lists the decision point, location of the decision point, the criteria to be evaluated at the point of decision, the action or operations to occur at the decision point, and the unit or element that is to act and has responsibility to observe and report the information affecting the criteria for the decision. (FM 5-0)

decision support template – (Army) A staff product initially used in the wargaming process that graphically represents the decision points and projected situations, and indicates when, where, and under what conditions a decision is most likely to be required to initiate a specific activity or event. (FM 5-0)

decisive engagement – In land and naval warfare, an engagement in which a unit is considered fully committed and cannot maneuver or extricate itself. In the absence of outside assistance, the action must be fought to a conclusion and either won or lost with the forces at hand. (JP 1-02)

decisive operation – The operation that directly accomplishes the task assigned by the higher headquarters. Decisive operations conclusively determine the outcome of major operations, battles, and engagements. (FM 3-0)

decisive point – A geographic place, specific key event, critical system or function that allows commanders to gain a marked advantage over an enemy and greatly influence the outcome of an attack. (JP 1-02)

defeat – A tactical mission task that occurs when an enemy force has temporarily or permanently lost the physical means or the will to fight. The defeated force's commander is unwilling or unable to pursue his adopted course of action, thereby yielding to the friendly commander's will, and can no longer interfere to a significant degree with the actions of friendly forces. Defeat can result from the use of force or the threat of its use. (FM 3-90)

defeat in detail – Concentrating overwhelming combat power against separate parts of a force in sequence rather than defeating the entire force at once. (FM 3-90)

defense in depth – The siting of mutually supporting defense positions designed to absorb and progressively weaken attack, prevent initial observations of the whole position by the enemy, and to allow the commander to maneuver the reserve. (JP 1-02)

defensive information operations – The integration and coordination of policies and procedures, operations, personnel, and technology to protect and defend information and information systems. Defensive information operations ensure timely, accurate, and relevant information access while denying adversaries the opportunity to exploit friendly information and information systems for their own purposes. (FM 3-0)

defensive operations – Operations that defeat an enemy attack, buy time, economize forces, or develop conditions favorable for offensive operations. Defensive operations alone normally cannot achieve a decision. Their purpose is to create conditions for a counteroffensive that allows Army forces to regain the initiative. (FM 3-0)

delay – A form of retrograde in which a force under pressure trades space for time by slowing the enemy's momentum and inflicting maximum damage on the enemy without, in principle, becoming decisively engaged. (FM 3-90)

depth – The extension of operations in time, space, and resources. (FM 3-0)

destroy – 1. A tactical mission task that physically renders an enemy force combat-ineffective until it is reconstituted. 2. To damage a combat system so badly that it cannot perform any function or be restored to a usable condition without being entirely rebuilt. (FM 3-90)

direct action – Short duration strikes and other small-scale offensive actions conducted as a special operation in hostile, denied, or other politically sensitive environments and which employ specialized military capabilities to seize, destroy, capture, exploit, recover, or damage designated targets. Direct action differs from conventional offensive actions in the level of physical and political risk, operational techniques, and the degree of discriminate and precise use of force to achieve specific objectives. (JP 1-02)

direct approach – The application of combat power directly against the enemy center of gravity or the enemy's principal strength. (FM 3-90)

direct fire – Fire delivered on a target using the target itself as a point of aim for either the weapon or the director. (JP 1-02)

disinformation – Information disseminated primarily by intelligence organizations or other covert agencies designed to distort information or deceive or influence United States decisionmakers, United States forces, coalition allies, key actors, or individuals via indirect or unconventional means. (FM 3-13)

displaced person – (Army) A civilian who is involuntarily outside the national boundaries of his or her country or as an internally displaced person is a civilian involuntarily outside his area or region within his country. (FM 2-0)

display – (Army) An information management activity: to represent relevant information in a usable, easily understood audio or visual form tailored to the needs of the user that conveys the common operational picture for decisionmaking and exercising command and control functions (FM 6-0)

disposition – 1. Distribution of the elements of a command within an area, usually the exact location of each unit headquarters and the deployment of forces subordinate to it. (JP 1-02)

disrupt – 1. A tactical mission task in which a commander integrates direct and indirect fires, terrain, and obstacles to upset an enemy's formation or tempo, interrupt his timetable, or cause his forces to commit prematurely or attack in piecemeal fashion. (FM 3-90) 2. An engineer obstacle effect that focuses fire planning and obstacle effort to cause the enemy to break up his formation and tempo, interrupt his timetable, commit breaching assets prematurely, and attack in a piecemeal effort. (FM 90-7)

distribution system – That complex of facilities, installations, methods, and procedures designed to receive, store, maintain, distribute, and control the flow of military materiel between the point of receipt into the military system and the point of issue to using activities and units. (JP 1-02)

doctrinal template – A model based on known or postulated adversary doctrine. Doctrinal templates illustrate the disposition and activity of adversary forces and assets conducting a particular operation unconstrained by the effects of the battlespace. They represent the application of adversary doctrine under ideal conditions. Ideally, doctrinal templates depict the threat's normal organization for combat, frontages, depths, boundaries and other control measures, assets available from other commands, objective depths, engagement areas, battle positions, and so forth. Doctrinal templates are usually scaled to allow ready use with geospatial products. (JP 1-02)

doctrine – Fundamental principles by which the military forces or elements thereof guide their actions in support of national objectives. It is authoritative but requires judgment in application. (JP 1-02)

economy – One of the eight characteristics of combat service support: providing the most efficient support at the least cost to accomplish the mission. (FM 4-0)

economy of force – One of the nine principles of war: allocate minimum essential combat power to secondary efforts. (FM 3-0)

effect – (Army) A result, outcome, or consequence of an action. (FMI 5-0.1)

electromagnetic deception – The deliberate radiation, re-radiation, alteration, suppression, absorption, denial, enhancement, or reflection of electromagnetic energy in a manner intended to convey misleading information to an enemy or enemy electromagnetic-dependent weapons, thereby degrading or neutralizing the enemy's combat capability. Among the types of electromagnetic deception are: a. **manipulative electromagnetic deception**—Actions to eliminate revealing, or convey misleading, electromagnetic telltale indicators that may be used by hostile forces; b. **simulative electromagnetic deception**—Actions to simulate friendly, notional, or actual capabilities to mislead hostile forces; c. **imitative electromagnetic deception**—The introduction of electromagnetic energy into enemy systems that imitates enemy emissions. (JP 1-02)

electromagnetic spectrum – The range of frequencies of electromagnetic radiation from zero to infinity. It is divided into 26 alphabetically designated bands. (JP 1-02)

electronic attack—That division of electronic warfare involving the use of electromagnetic energy, directed energy, or antiradiation weapons to attack personnel, facilities, or equipment with the intent of degrading, neutralizing, or destroying enemy combat capability and is considered a form of fires. Electronic attack includes: 1. actions taken to prevent or reduce an enemy's effective use of the electromagnetic spectrum, such as jamming and electromagnetic deception, and 2. employment of weapons that use either electromagnetic or directed energy as their primary destructive mechanism (lasers, radio frequency weapons, particle beams), or antiradiation weapons. (JP 1-02)

electronic warfare – Any military action involving the use of electromagnetic and directed energy to control the electromagnetic spectrum or to attack the enemy. The three major subdivisions within electronic warfare are: electronic attack, electronic protection, and electronic warfare support. (JP 1-02)

electronic warfare support—That division of electronic warfare involving actions tasked by, or under direct control of, an operational commander to search for, intercept, identify, and locate or localize sources of intentional and unintentional radiated electromagnetic energy for the purpose of immediate threat recognition, targeting, planning, and conduct of future operations. Thus, electronic warfare support provides information required for immediate decisions involving electronic warfare operations and other tactical actions such as threat avoidance, targeting, and homing. Electronic warfare support data can be used to produce signals intelligence, provide targeting for electronic or destructive attack, and produce measurement and signature intelligence. (JP 1-02)

embarkation – The process of putting personnel and/or vehicles and their associated stores and equipment into ships and/or aircraft. (JP 1-02)

encirclement – An operation where one force loses its freedom of maneuver because an opposing force is able to isolate it by controlling all ground lines of communications. (FM 3-0)

end state – (Army) At the operational and tactical levels, the conditions that, when achieved, accomplish the mission. At the operational level, these conditions attain the aims set for the campaign or major operation. (FM 3-0)

engagement – A tactical conflict, usually between opposing lower echelon maneuver forces. (JP 1-02)

envelopment – (Army) – A form of maneuver in which an attacking force seeks to avoid the principal enemy defenses by seizing objectives to the enemy rear to destroy the enemy in his current positions. At the tactical level, envelopments focus on seizing terrain, destroying specific enemy forces, and interdicting enemy withdrawal routes. (FM 3-0).

essential elements of friendly information – (Army) The critical aspects of a friendly operation that, if known by the enemy, would subsequently compromise, lead to failure, or limit success of the operation, and, therefore, must be protected from enemy detection. (FM 3-13)

essential task – A specified or implied task that must be executed to accomplish the mission. Essential tasks are always included in the unit's mission statement. (FM 5-0)

exfiltration – The removal of personnel or units from areas under enemy control by stealth, deception, surprise, or clandestine means. (JP 1-02)

exploitation – 1. Taking full advantage of success in military operations, following up initial gains, and making permanent the temporary effects already achieved. 2. An offensive operation that usually follows a successful attack and is designed to disorganize the enemy in depth. (JP 1-02)

explosive ordnance disposal – The detection, identification, on-site evaluation, rendering safe, recovery, and final disposal of unexploded explosive ordnance. It may also include explosive ordnance which has become hazardous by damage or deterioration. (JP 1-02)

firepower – 1. The amount of fire which may be delivered by a position, unit, or weapon system. 2. Ability to deliver fire. (JP 1-02)

fires – The effects of lethal or nonlethal weapons. (JP 1-02)

fire support coordinating measure – A measure employed by land or amphibious commanders to facilitate the rapid engagement of targets and simultaneously provide safeguards for friendly forces. (JP 1-02)

fix – (Army) 1. A tactical mission task where a commander prevents the enemy from moving any part of his force from a specific location for a specific period of time. 2. An engineer obstacle effect that focuses fire planning and obstacle effort to slow an attacker's movement within a specified area, normally an engagement area. (FM 3-90)

flexibility – One of the eight characteristics of combat service support: being able to adapt combat service support structures and procedures to changing situations, missions, and concepts of operations. (FM 4-0)

force projection – The ability to project the military element of national power from the continental United States (CONUS) or another theater in response to requirements for military operations. Force projection operations extend from mobilization and deployment of forces to redeployment to CONUS or home theater. (JP 1-02)

force protection – Actions taken to prevent or mitigate hostile actions against Department of Defense personnel (to include family members), resources, facilities, and critical information. These actions conserve the force's fighting potential so it can be applied at a decisive time and place and incorporates the coordinated and synchronized offensive and defensive measures to enable the effective employment of the joint force while degrading opportunities for the enemy. Force protection does not include actions to defeat the enemy or protect against accidents, weather, or disease. (JP 1-02)

force tailoring – The process of determining the right mix and sequence of units for a mission. (FM 3-0)

foreign humanitarian assistance – Programs conducted to relieve or reduce the results of natural or manmade disasters or other endemic conditions such as human pain, disease, hunger, or privation that might present a serious threat to life or that can result in great damage to or loss of property. Foreign humanitarian assistance (FHA) provided by US forces is limited in scope and duration. The foreign assistance provided is designed to supplement or complement the efforts of the host nation civil authorities or agencies that may have primary responsibility for providing FHA. FHA operations are those conducted outside the United States, its territories, and possessions. (JP 1-02)

foreign internal defense – Participation by civilian and military agencies of a government in any of the action programs taken by another government to free and protect its society from subversion, lawlessness, and insurgency. (JP 1-02)

fratricide – The unintentional killing or wounding of friendly personnel by friendly firepower. (FM 3-0)

frontal attack – (Army) A form of maneuver in which the attacking force seeks to destroy a weaker enemy force or fix a larger enemy force in place over a broad front. (FM 3-0)

full spectrum operations – The range of operations Army forces conduct in war and military operations other than war. (FM 3-0)

graphic control measure – A symbol used on maps and displays to regulate forces and warfighting functions. (FMI 5-0.1)

gun-target line – An imaginary straight line from gun to target. (JP 1-02)

hazardous material – Any substance which has a human health hazard associated with it. Special storage, use, handling, and shipment safety procedures and protocols must be followed to help protect against accidental exposure. Hazardous materials are specifically identified under federal law. (FM 3-100.4)

high angle fire – (Army) Fire delivered to clear an obstacle (such as a hill) that low-angle fire cannot, or fire delivered to attack targets on the reverse side of an obstacle (such as a hill) that cannot be attacked with low-angle or direct fire. (FM 6-30).

host nation – A nation that receives the forces and/or supplies of allied nations, coalition partners, and/or NATO organizations to be located on, to operate in, or to transit through its territory. (JP 1-02)

host-nation support – Civil and/or military assistance rendered by a nation to foreign forces within its territory during peacetime, crises or emergencies, or war based on agreements mutually concluded between nations. (JP 1-02)

human intelligence – A category of intelligence derived from information collected and provided by human sources. (JP 1-02)

imagery intelligence – Intelligence derived from the exploitation of collection by visual photography, infrared sensors, lasers, electro-optics, and radar sensors such as synthetic aperture radar wherein images of objects are reproduced optically or electronically on film, electronic display devices, or other media. (JP 1-02)

implied task – A task that must be performed to accomplish a specified task or mission, but is not stated in the higher headquarters order. (FM 5-0)

indirect fire – Fire delivered on a target that is not itself used as a point of aim for the weapons or the director. (JP 1-02)

infiltration – (Army) 1. A form of maneuver in which an attacking force conducts undetected movement through or into an area occupied by enemy forces to occupy a position of advantage in the enemy rear while exposing only small elements to enemy defensive fires. (FM 3-90)

information – (Army) 1. In the general sense, the meaning humans assign to data. 2. In the context of the cognitive hierarchy, data that have been processed to provide further meaning. (FM 6-0)

information fratricide – The results of employing information operations elements in a way that causes effects in the information environment that impede the conduct of friendly operations or adversely affect friendly forces. (FM 3-13)

information management – The provision of relevant information to the right person at the right time in a usable form to facilitate situational understanding and decisionmaking. It uses procedures and information systems to collect, process, store, display, and disseminate information. (FM 3-0)

information operations – (Army) The employment of the core capabilities of electronic warfare, computer network operations, psychological operations, military deception, and operations security, in concert with specified supporting and related capabilities, to affect and defend information and information systems and to influence decisionmaking. (FM 3-13)

information requirements – (Army) All information elements the commander and staff require to successfully conduct operations, that is, all elements necessary to address the factors of METT-TC. (FM 6-0)

information superiority – The operational advantage derived from the ability to collect, process, and disseminate an uninterrupted flow of information while exploiting or denying an adversary's ability to do the same. (JP 1-02)

information system – (Army) The equipment and facilities that collect, process, store, display, and disseminate information. This includes computers—hardware and software—and communications, as well as policies and procedures for their use. (FM 3-0)

insurgency – An organized movement aimed at the overthrow of a constituted government through the use of subversion and armed conflict. (JP 1-02)

integration – One of the eight characteristics of combat service support: the total inclusion of Army combat service support into the operations process (plan, prepare, execute, assess), as well as into other logistics comnponents of the unified force. (FM 4-0).

intelligence – 1. The product resulting from the collection, processing, integration, analysis, evaluation, and interpretation of available information concerning foreign countries or areas. 2. Information and knowledge about an adversary obtained through observation, investigation, analysis, or understanding. (JP 1-02)

intelligence preparation of the battlefield – The systematic, continuous process of analyzing the threat and environment in a specific geographic area. Intelligence preparation of the battlefield (IPB) is designed to support the staff estimate and military decisionmaking process. Most intelligence requirements are generated as a result of the IPB process and its interrelation with the decisionmaking process. (FM 34-130)

intelligence, surveillance, and reconnaissance – An activity that synchronizes and integrates the planning and operation of sensors, assets, and processing, exploitation, and dissemination systems in direct support of current and future operations. This is an integrated intelligence and operations function. (JP 1-02)

intelligence, surveillance, and reconnaissance plan – An integrated plan for collection of information from all available sources and analysis of that information to produce intelligence to meet requirements. Specifically, a logical plan for transforming priority intelligence requirements into orders or requests to reconnaissance and surveillance assets to collect pertinent information within a required time limit. (FM 34-3)

interagency – Activities or operations conducted by or through coordination with two or more agencies or an agency and one or more Services of the same nation. (FM 3-07)

interdict – A tactical mission task where the commander prevents, disrupts, or delays the enemy's use of an area or route. (FM 3-90)

interior lines – A force operates on interior lines when its operations diverge from a central point. (FM 3-0)

isolate – A tactical mission task that requires a unit to seal off—both physically and psychologically—an enemy from his sources of support, deny an enemy freedom of movement, and prevent an enemy unit from having contact with other enemy forces. (FM 3-90)

joint force – A general term applied to a force composed of significant elements, assigned or attached, of two or more Military Departments, operating under a single joint force commander. (JP 1-02)

joint force commander – A general term applied to a combatant commander, subunified commander, or joint task force commander authorized to exercise combatant command (command authority) or operational control over a joint force. (JP 1-02)

joint operations – A general term to describe military actions conducted by joint forces, or by Service forces in relationships (e.g., support, coordinating authority), which, of themselves, do not create joint forces. (JP 1-02)

joint task force – A joint force that is constituted and so designated by the Secretary of Defense, a combatant commander, a subunified commander, or an existing joint task force commander. (JP 1-02)

key terrain – Any locality, or area, the seizure or retention of which affords a marked advantage to either combatant. (JP 1-02)

landing zone – Any specified zone used for the landing of aircraft. JP 1-02)

law of war – That part of international law that regulates the conduct of armed hostilities. (JP 1-02)

liaison – That contact or intercommunication maintained between elements of military forces or other agencies to ensure mutual understanding and unity of purpose and action. (JP 1-02)

line of communications – A route, either land, water, and/or air, that connects an operating military force with a base of operations and along which supplies and military forces move. (JP 1-02)

lines of operations – (Army) A line that defines the orientation of the force in time and space, or purpose, in relation to an enemy or objective. (FMI 5-0.1).

logistics – The science of planning and carrying out the movement and maintenance of forces. In its most comprehensive sense, those aspects of military operations which deal with: a. design and development, acquisition, storage, movement, distribution, maintenance, evacuation, and disposition of materiel; b. movement, evacuation, and hospitalization of personnel; c. acquisition or construction, maintenance, operation, and disposition of facilities; and d. acquisition and furnishing of services. (JP 1-02)

logistics-over-the-shore operations – The loading and unloading of ships without the benefit of deep draft-capable, fixed port facilities; or as a means of moving forces close to tactical assembly areas dependent on threat force capabilities. (JP 1-02)

logistics preparation of the theater – Actions taken by combat service support personnel to optimize means—force structure, resources, and strategic lift—of supporting the joint force commander's plan. (FM 4-0)

major operation – A series of tactical actions (battles, engagements, strikes) conducted by various combat forces of a single or several Services, coordinated in time and place, to accomplish operational and, sometimes, strategic objectives in an operational area. These actions are conducted simultaneously or sequentially in accordance with a common plan and are controlled by a single commander. (JP 1-02)

maneuver – One of the nine principles of war: place the enemy in a disadvantageous position through the flexible application of combat power. (FM 3-0)

Marine air-ground task force – The Marine Corps principal organization for all missions across the range of military operations, composed of forces task-organized under a single commander capable of responding rapidly to a contingency anywhere in the world. The types of forces in the Marine air-ground task force (MAGTF) are functionally grouped into four core elements: a command element, an aviation combat element, a ground combat element, and a combat service support element. The four core elements are categories of forces, not formal commands. The basic structure of the MAGTF never varies, though the number, size, and type of Marine Corps units comprising each of its four elements will always be mission dependent. The flexibility of the organizational structure allows for one or more subordinate MAGTFs to be assigned. In a joint or multinational environment, other Service or multinational forces may be assigned or attached. (FM 1-02)

Marine expeditionary force – The largest Marine air-ground task force (MAGTF) and the Marine Corps principal warfighting organization, particularly for larger crises or contingencies. It is task-organized around a permanent command element and normally consists of one or more Marine divisions, Marine aircraft wings, and Marine force service support groups. The Marine expeditionary force is capable of missions across the range of military operations, including amphibious assault and sustained operations ashore in any environment. It can operate from a sea base, a land base, or both. In a joint or multinational environment, it may contain other Service or multinational forces assigned or attached to the MAGTF. (FM 1-02)

Marine expeditionary unit – A Marine air-ground task force (MAGTF) that is constructed around an infantry battalion reinforced, a helicopter squadron reinforced, and a task-organized combat service support element. It normally fulfills Marine Corps' forward sea-based deployment requirements. The Marine expeditionary unit provides an immediate reaction capability for crisis response and is capable of limited combat operations. In a joint or multinational environment, it may contain other Service or multinational forces assigned or attached to the MAGTF. (FM 1-02)

mass – One of the nine principles of war: the effects of combat power at the decisive place and time. (FM 3-0)

mass casualty – Any large number of casualties produced in a relatively short period of time, usually as the result of a single incident such as a military aircraft accident, hurricane, flood, earthquake, or armed attack that exceeds logistical support capabilities. (JP 1-02)

measure of effectiveness – (Army) A criterion used to assess changes in system behavior, capability, or operational environment that is tied to measuring the attainment of an end state, achievement of an objective, or creation of an effect. (FMI 5-0.1)

medical evacuation – The timely and efficient movement of the wounded, injured, or ill while providing en route medical care to and between medical treatment facilities. (FM 4-02)

meeting engagement – A combat action that occurs when a moving force engages an enemy at an unexpected time and place. (FM 3-0)

METT-TC – A memory aid used in two contexts: (1) In the context of information management, the major subject categories into which relevant information is grouped for military operations: mission, enemy, terrain and weather, troops and support available, time available, civil considerations. (2) In the context of tactics, the major factors considered during mission analysis. (FM 6-0)

military deception – Actions executed to deliberately mislead adversary military decisionmakers as to friendly military capabilities, intentions, and operations, thereby causing the adversary to take specific actions (or inactions) that will contribute to the accomplishment of the friendly mission. (JP 1-02)

military decisionmaking process – A process that integrates the activities of the commander, staff and subordinate commanders in developing and operation plan or order. It establishes procedures for analyzing a mission; developing, analyzing, and comparing courses of action; selecting the best course of action; and producing an operation plan or order (FMI 5-0.1).

misinformation – Incorrect information from any source that is released for unknown reasons or to solicit a response or interest from a nonpolitical or nonmilitary target. (FM 3-13)

mission – 1. The task, together with the purpose, that clearly indicates the action to be taken and the reason therefor. 2. In common usage, especially when applied to lower military units, a duty assigned to an individual or unit; a task. (JP 1-02)

mission command – The conduct of military operations through decentralized execution based upon mission orders for effective mission accomplishment. Successful mission command results from subordinate leaders at all echelons exercising disciplined initiative within the commander's intent to accomplish missions. It requires an environment of trust and mutual understanding. (FM 6-0)

mission orders – A technique for completing combat orders that allows subordinates maximum freedom of planning and action in accomplishing missions and leaves the "how" of mission accomplishment to the subordinate. (FM 6-0)

mobile defense – (Army) A type of defensive operation that concentrates on the destruction or defeat of the enemy through a decisive attack by a striking force. (FM 3-0)

mobility corridor – Areas where a force will be canalized due to terrain restrictions. They allow military forces to capitalize on the principles of mass and speed and are therefore relatively free of obstacles. (JP 1-02)

mobility operations – Obstacle reduction by maneuver and engineer units to reduce or negate the effects of existing or reinforcing obstacles. The objective is to maintain freedom of movement for maneuver units, weapon systems, and critical supplies. (FM 3-34)

modified combined obstacle overlay – A joint intelligence preparation of the battlespace product used to portray the effects of each battlespace dimension on military operations. It normally depicts militarily significant aspects of the battlespace environment, such as obstacles restricting military movement, key geography, and military objectives. (JP 1-02)

monitoring – (Army) Continuous observation of the current situation to identify opportunities for the force, threats to the force, gaps in information, and progress according to the plan or order. (FMI 5-0.1)

movement to contact – A form of the offensive designed to develop the situation and to establish or regain contact. (JP 1-02)

naval coastal warfare – Coastal sea control, harbor defense, and port security, executed both in coastal areas outside the United States in support of national policy and in the United States as part of this Nation's defense. (JP 1-02)

naval gunfire support – Fire provided by Navy surface gun systems in support of a unit or units tasked with achieving the commander's objectives. A subset of naval surface fire support. (JP 1-02)

naval surface fire support – Fire provided by Navy surface gun and missile systems in support of a unit or units. (JP 1-02)

neutral – (Army) 1. An individual, group of individuals, organization, or nation that is not hostile to or in any way supportive of only one belligerent force in a hostile environment. (FM 3-07)

no-fire area – A land area, designated by the appropriate commander, into which fires or their effects are prohibited. (JP 1-02)

noncombatant – 1. An individual in an area of combat operations who is not armed and is not participating in any activity in support of any of the factions or forces involved in combat. 2. An individual, such as chaplain or medical personnel, whose duties do not involve combat. (FM 3-07)

noncombatant evacuation operations – Operations directed by the Department of State, the Department of Defense, or other appropriate authority whereby noncombatants are evacuated from foreign countries when their lives are endangered by war, civil unrest, or natural disaster to safe havens or to the United States. (JP 1-02)

nongovernmental organization – A private, self-governing, not-for-profit organization dedicated to alleviating human suffering; and/or promoting education, health care, economic development, environmental protection, human rights, and conflict resolution; and/or encouraging the establishment of democratic institutions and civil society. (JP 1-02)

nonlethal fires – Any fires that do not directly seek the physical destruction of the intended target and are designed to impair, disrupt, or delay the performance of enemy operational forces, functions, and facilities. Psychological operations, electronic warfare (jamming), and other command and control countermeasures are all nonlethal fire options. (FM 6-20)

nonlethal weapons – Weapons that are explicitly designed and primarily employed so as to incapacitate personnel or materiel, while minimizing fatalities, permanent injury to personnel and undesired damage to property and the environment. a. Unlike conventional lethal weapons that destroy their targets through blast, penetration, and fragmentation, nonlethal weapons employ means other than gross physical destruction to prevent the target from functioning. b. Nonlethal weapons are intended to have one, or both, of the following characteristics: (1) They have relatively reversible effects on personnel or materiel. (2) They affect objects differently within their area of influence. (JP 1-02)

objective – (Army) 1. One of the nine principles of war: direct every military operation toward a clearly defined, decisive and attainable objective. (FM 3-0) 2. A location on the ground used to orient operations, phase operations, facilitate changes of direction, and provide for unity of effort. (FM 3-90)

offensive – One of the nine principles of war: seize, retain, and exploit the initiative. (FM 3-0)

offensive information operations – The integrated use of assigned and supporting capabilities and activities, mutually supported by intelligence, to affect enemy decisionmakers or to influence others to achieve or promote specific objectives. (FM 3-0)

offensive operations – Operations which aim at destroying or defeating an enemy. Their purpose is to impose US will on the enemy and achieve decisive victory. (FM 3-0)

operation – 1. A military action or the carrying out of a strategic, operational, tactical, service, training, or administrative military mission. 2. The process of carrying on combat, including movement, supply, attack, defense, and maneuvers needed to gain the objectives of any battle or campaign. (JP 1-02)

operational control – Command authority that may be exercised by commanders at any echelon at or below the level of combatant command. Operational control is inherent in combatant command (command authority) and may be delegated within the command. When forces are transferred between combatant commands, the command relationship the gaining commander will exercise (and the losing commander will relinquish) over these forces must be specified by the Secretary of Defense. Operational control is the authority to perform those functions of command over subordinate forces involving organizing and employing commands and forces, assigning tasks, designating objectives, and giving authoritative direction necessary to accomplish the mission. Operational control includes authoritative direction over all aspects of military operations and joint training necessary to accomplish the missions assigned to the command. Operational control should be exercised through the commanders of subordinate organizations. Normally this authority is exercised through subordinate joint force commanders and Service and/or functional component commanders. Operational control normally provides full authority to organize commands and forces and to employ those forces as the commander in operational control considers necessary to accomplish assigned missions; it does not, in and of itself, include authoritative direction for logistics or matters of administration, discipline, internal organization, or unit training. (JP 1-02)

operational framework – The arrangement of friendly forces and resources in time, space, and purpose with respect to each other and the enemy or situation. It consists of the area of operations, battlespace, and battlefield organization. (FM 3-0)

operations process – The major command and control activities performed during operations: planning, preparation, execution with continuous assessment. These activities occur continuously throughout an operation, overlapping and recurring as required. (FMI 5-0.1)

operations security – A process of identifying critical information and subsequently analyzing friendly actions attendant to military operations and other activities to: a. identify those actions that can be observed by adversary intelligence systems; b. determine indicators hostile intelligence systems might obtain that could be interpreted or pieced together to derive critical information in time to be useful to adversaries; and c. select and execute measures that eliminate or reduce to an acceptable level the vulnerabilities of friendly actions to adversary exploitation. (JP 1-02)

paramilitary forces – Forces or groups distinct from the regular armed forces of any country, but resembling them in organization, equipment, training, or mission. (JP 1-02)

patrol – A detachment of ground, sea, or air forces sent out for the purpose of gathering information or carrying out a destructive, harassing, mopping-up, or security mission. (JP 1-02)

peace enforcement – Application of military force, or the threat of its use, normally pursuant to international authorization, to compel compliance with resolutions or sanctions designed to maintain or restore peace and order. (JP 1-02)

peacekeeping – Military operations undertaken with the consent of all major parties to a dispute, designed to monitor and facilitate implementation of an agreement (ceasefire, truce, or other such agreement) and support diplomatic efforts to reach a long-term political settlement. (JP 1-02)

peace operations – A broad term that encompasses peacekeeping operations and peace enforcement operations conducted in support of diplomatic efforts to establish and maintain peace. (JP 1-02)

penetration – (Army) A form of maneuver in which an attacking force seeks to rupture enemy defenses on a narrow front to disrupt the defensive system. (FM 3-90).

perceptions – Mental images the commander wants the deception target to believe are real. (FM 3-13)

phase line – A line utilized for control and coordination of military operations, usually an easily identified feature in the operational area. (JP 1-02)

physical destruction – The application of combat power to destroy or degrade adversary forces, sources of information, command and control systems, and installations. It includes direct and indirect forces from ground, sea, and air forces. Also included are direct actions by special operations forces. (FM 3-13)

physical security – 1. That part of security concerned with physical measures designed to safeguard personnel; to prevent unauthorized access to equipment, installations, material, and documents; and to safeguard them against espionage, sabotage, damage, and theft. 2. In communications security, the component that results from all physical measures necessary to safeguard classified equipment, material, and documents from access thereto or observation thereof by unauthorized persons. (JP 1-02)

pickup zone – A geographic area used to pick up troops or equipment by helicopter. (FM 90-4)

plan – A design for a future or anticipated operation. (FM 5-0)

planning – The process by which commanders (and staffs, if available) translate the commander's visualization into a specific course of action for preparation and execution, focusing on the expected results. (FMI 5-0.1)

port of debarkation – (DOD) The geographic point at which cargo or personnel are discharged. This may be a seaport or aerial port of debarkation; for unit requirements, it may or may not coincide with the destination. (JP 1-02)

port of embarkation – (DOD) The geographic point in a routing scheme from which cargo and personnel depart. This may be a seaport or aerial port from which personnel and equipment flow to a port of debarkation; for unit and nonunit requirements, it may or may not coincide with the origin. (JP 1-02)

principles of war – Principles that provide general guidance for conducting war and military operations other than war at the strategic, operational, and tactical levels. The nine principles of war are: objective, offensive, mass, economy of force, maneuver, unity of command, security, surprise, and simplicity (FM 3-0)

propaganda – Any form of communication in support of national objectives designed to influence the opinions, emotions, attitudes, or behavior of any group in order to benefit the sponsor, either directly or indirectly. (JP 1-02)

psychological operations – Planned operations to convey selected information and indicators to foreign audiences to influence their emotions, motives, objective reasoning, and ultimately the behavior of foreign governments, organizations, groups, and individuals. The purpose of psychological operations is to induce or reinforce foreign attitudes and behavior favorable to the originator's objectives. (JP 1-02)

public affairs – Those public information, command information, and community relations activities directed toward both the external and internal publics with interest in the Department of Defense. (JP 1-02)

pursuit – An offensive operation designed to catch or cut off a hostile force attempting to escape, with the aim of destroying it. (JP 1-02)

push – In logistics, the delivery of a predetermined amount of supplies to a user on a scheduled basis without the user requesting them. (FM 4-0)

railhead – A point on a railway where loads are transferred between trains and other means of transport. (JP 1-02)

reachback – The process of obtaining products, services, and applications, or forces, or equipment, or material from organizations that are not forward deployed. (JP 1-02)

rear area – (Army) For any particular command, the area extending forward from its rear boundary to the rear of the area assigned to the next lower level of command. This area is provided primarily for the performance of support functions, and is where the majority of the echelon's sustaining functions occur. (FM 3-0)

refugee – A person who, by reason of real or imagined danger, has left his home country or country of nationality and is unwilling or unable to return. (JP 1-02)

relevant information – All information of importance to commanders and staffs in the exercise of command and control. (FM 3-0)

relief in place – (Army) An tactical enabling operation in which, by direction of higher authority, all or part of a unit is replaced in an area by the incoming unit. (FM 3-90)

reorganization – Action taken to shift internal resources within a degraded unit to increase its level of combat effectiveness. (FM 100-9)

reserve – Portion of a body of troops which is kept to the rear or withheld from action at the beginning of an engagement, in order to be available for a decisive movement. (JP 1-02)

responsiveness – One of the eight principles of combat service support: providing the right support at the right place at the right time. (FM 4-0)

retrograde – A type of defensive operation that involves organized movement away from the enemy. (FM 3-0)

riot control agent – (Army) A substance that produces temporary irritating or disabling physical effects that disappear within minutes of removal from exposure. There is no significant risk of permanent injury, and medical treatment is rarely required. (FM 3-11.11)

risk – (DOD) Probability and severity of loss linked to hazards. (JP 1-02)

risk management – The process of identifying, assessing, and controlling risk arising from operational factors, and making decisions that balance risk cost with mission benefits. (JP 1-02)

rules of engagement – Directives issued by competent military authority that delineate the circumstances and limitations under which United States forces will initiate and/or continue combat engagement with other forces encountered. (JP 1-02)

running estimate – A staff section's continuous assessment of current and future operations to determine if the current operation is proceeding according to the commander's intent and if future operations are supportable. (FMI 5-0.1)

search and attack – A technique of conducting a movement to contact that shares many of the characteristics of an area security mission. (FM 3-0)

secure – (Army) 1. A tactical mission task that involves preventing a unit, facility, or geographical location from being damaged or destroyed as a result of enemy action. (FM 3-90)

security – One of the nine principles of war: never permit the enemy to acquire an unexpected advantage (FM 3-0).

seize – (Army) A tactical mission task that involves taking possession of a designated area using overwhelming force. (FM 3-90)

sequel – (Army) An operation that follows the current operation. It is a future operation that anticipates possible outcomes – success, failure, or stalemate – of the current operation. (FM 3-0).

shaping operation – An operation at any echelon that creates and preserves conditions for the success of decisive operations. (FM 3-0)

show of force – An operation designed to demonstrate US resolve that involves increased visibility of US deployed forces in an attempt to defuse a specific situation that, if allowed to continue, may be detrimental to US interests or national objectives. (JP 1-02)

signals intelligence – 1. A category of intelligence comprising either individually or in combination all communications intelligence, electronic intelligence, and foreign instrumentation signals intelligence, however transmitted. 2. Intelligence derived from communications, electronics, and foreign instrumentation signals. (JP 1-02)

simplicity – One of the nine principles of war: prepare clear, uncomplicated plans, and clear, concise orders to ensure thorough understanding. (FM 3-0) [Note: "simplicity" is also one of the eight characteristics of combat service support.]

situational understanding – (Army) The product of applying analysis and judgment to the common operational picture to determine the relationship among the factors of METT-TC. (FM 3-0)

space operations – The employment of space system capabilities that provide the means to enhance command and control, facilitate the maneuver of forces, reduce the commander's uncertainty, and improve fire support, air defense, intelligence collection, and combat service support operations which will support strategic, operational, and tactical missions across the operational continuum in the near, mid, and far term. (FM 3-14)

special operations – Operations conducted in hostile, denied, or politically sensitive environments to achieve military, political, economic, and/or informational objectives employing military capabilities for which there is no broad conventional force requirement. These operations often require covert, clandestine, or low visibility capabilities. Special operations are applicable across the rnage of military operations. They can be conducted independently or in conjunction with operations of conventional forces or other government agencies and may include operations through, with, or by indigenous or surrogate forces. Special operations differ from conventional operations in degree of physical and political risk, operational techniques, mode of employment, independence from friendly support, and dependence on detailed operational intelligence and indigenous assets. (JP 1-02)

special operations forces – Those Active and Reserve Component forces of the Military Services designated by the Secretary of Defense and specifically organized, trained, and equipped to conduct and support special operations. (JP 1-02)

special reconnaissance – Reconnaissance and surveillance actions conducted by special operations forces in hostile, denied, or politically sensitive environments to collect or verify information of strategic or operational significance, employing military capabilities not normally found in conventional forces. These forces provide an additive capability for commanders and supplement other conventional reconnaissance and surveillance actions. (JP 1-02)

stability operations – Operations that promote and protect US national interests by influencing the threat, political, and information dimensions of the operational environment through a combination of peacetime developmental, cooperative activities and coercive actions in response to crisis. (FM 3-0)

status-of-forces agreement – An agreement that defines the legal position of a visiting military force deployed in the territory of a friendly state. Agreements delineating the status of visiting military forces may be bilateral or multilateral. Provisions pertaining to the status of visiting forces may be set forth in a separate agreement, or they may form a part of a more comprehensive agreement. These provisions describe how the authorities of a visiting force may control members of that force and the amenability of the force or its members to the local law or to the authority of local officials. To the extent that agreements delineate matters affecting the relations between a military force and civilian authorities and population, they may be considered as civil affairs agreements. (JP 1-02)

strategy – The art and science of developing and employing instruments of national power in a synchronized and integrated fashion to achieve theater, national and/or multinational objectives. (JP 1-02)

striking force – A committed force organized to conduct the decisive attack in a mobile defense. It normally comprises the maximum combat power available to the commander at the time of the attack. (FM 3-0)

strong point – (Army) A heavily fortified battle position tied to a natural or reinforcing obstacle to create an anchor for the defense or to deny the enemy decisive or key terrain. (FM 3-90)

surprise – One of the nine principles of war: strike the enemy at a time or place or in a manner for which he is unprepared. (FM 3-0)

survivability – (Army) 1. Concept which includes all aspects of protecting personnel, weapons, and supplies while simultaneously deceiving the enemy. Survivability tactics include building a good defense; employing frequent movement; using concealment, deception, and camouflage; and constructing fighting and protective positions for both individuals and equipment. Encompasses planning and locating position sites, designing adequate overhead cover, analyzing terrain conditions and construction materials, selecting excavation methods, and countering the effects of direct and indirect fire weapons. (FM 5-103) 2. One of the eight characteristics of combat service support: being able to shield support functions from destruction or degradation. (FM 4-0)

sustainability - One of the eight characteristics of combat service support: the ability to maintain continuous support, rear area and base security, movement control, terrain management, and infrastructure development. (FM 4-0)

sustaining operation – An operation at any echelon that enables shaping and decisive operations by providing combat service support, rear area and base security, movement control, terrain management, and infrastructure development. (FM 3-0)

synchronization – 1. The arrangement of military actions in time, space, and purpose to produce maximum relative combat power at a decisive place and time. See FM 3-0. 2. In the intelligence context, application of intelligence sources and methods in concert with the operation plan. (JP 1-02)

tactical combat force – A combat unit, with appropriate combat support and combat service support assets, that is assigned the mission of defeating Level III threats. (JP 1-02)

tactical control – Command authority over assigned or attached forces or commands, or military capability or forces made available for tasking, that is limited to the detailed direction and control of movements or maneuvers within the operational area necessary to accomplish missions or tasks assigned. Tactical control is inherent in operational control. Tactical control may be delegated to, and exercised at any level at or below the level of combatant command. When forces are transferred between combatant commands, the command relationship the gaining commander will exercise (and the losing commander will relinquish) over these forces must be specified by the Secretary of Defense. Tactical control provides sufficient authority for controlling and directing the application of force or tactical use of combat support assets within the assigned mission or task. (JP 1-02)

tactical level of war –The level of war at which battles and engagements are planned and executed to accomplish military objectives assigned to tactical units or task forces. Activities at this level focus on the ordered arrangement and maneuver of combat elements in relation to each other and to the enemy to achieve combat objectives. (JP 1-02)

target acquisition – The detection, identification, and location of a target in sufficient detail to permit the effective employment of weapons. (JP 1-02)

targeting – The process of selecting and prioritizing targets and matching the appropriate response to them, taking account of operational requirements and capabilities. (JP 1-02)

target reference point – An easily recognizable point on the ground (either natural or man-made) used to initiate, distribute, and control fires. Target reference points (TRPs) can also designate the center of an area where the commander plans to distribute or converge the fires of all his weapons rapidly. They are used by task force and below, and can further delineate sectors of fire within an engagement area. TRPs are designated using the standard target symbol and numbers issued by the fire support officer. Once designated, TRPs also constitute indirect fire targets. (FM 3-90)

task organization – A temporary grouping of forces designed to accomplish a particular mission. (FM 3-0)

task-organizing – (Army) The process of allocating available assets to subordinate commanders and establishing their command and support relationships. (FM 3-0)

tempo – The rate of military action. (FM 3-0)

terminal control – 1. The authority to direct the maneuver of aircraft which are delivering ordnance, pasengers, or cargo to a specific location or target. Terminal control is a type of air control. 2. Any electronic, mechanical, or visual control given to aircraft to facilitate target acquisition and resolution. (JP 1-02)

terrorism – The calculated use of unlawful violence or threat of unlawful violence to inculcate fear; intended to coerce or to intimidate governments or societies in the pursuit of goals that are generally political, religious, or ideological. (JP 1-02)

terrorist – An individual who uses violence, terror, and intimidation to achieve a result. (JP 1-02)

throughput distribution – The bypassing of one or more intermediate supply echelons in the supply system to avoid multiple handling. (FM 4-0)

time-phased force and deployment data – The Joint Operation Planning and Execution System database portion of an operation plan; it contains time-phased force data, non-unit-related cargo and personnel data, and movement data for the operation plan, including the following: a. In-place units; b. Units to be deployed to support the operation plan with a priority indicating the desired sequence for their arrival at the port of debarkation; c. Routing of forces to be deployed; d. Movement data associated with deploying forces; e. Estimates of non-unit related cargo and personnel movements to be conducted concurrently with the deployment of forces; and f. Estimate of transportation requirements that must be fulfilled by common-user lift resources as well as those requirements that can be fulfilled by assigned or attached transportation resources. (JP 1-02)

toxic industrial material – A generic term for toxic radioactive compounds in solid, liquid, aerosolized or gaseous form. These may be used, or stored for use, for industrial, commercial, medical, military, or domestic purposes. Toxic industrial materials may be chemical, biological, or radioactive and described as toxic industrial chemical, toxic industrial biological, or toxic industrial radiological. (FM 3-11.4)

track – 1. To display or record the successive positions of a moving object. 2. To keep a gun properly aimed, or to point continuously a target-locating instrument at a moving target. (JP 1-02)

troop leading procedures – A dynamic process used by small unit leaders to analyze a mission, develop a plan, and prepare for an operation. (FM 5-0)

turning movement – (Army) A form of maneuver in which the attacking force seeks to avoid the enemy's principal defensive positions by seizing objectives to the enemy rear and causing the enemy to move out of his current positions or divert major forces to meet the threat. (FM 3-0)

unconventional warfare – A broad spectrum of military and paramilitary operations, normally of long duration, predominantly conducted through, with, or by indigenous or surrogate forces who are organized, trained, equipped, supported, and directed in varying degrees by an external source. It includes, but is not limited to, guerrilla warfare, subversion, sabotage, intelligence activities, and unconventional assisted recovery. (JP 1-02)

unexploded explosive ordnance – Explosive ordnance which has been primed, fused, armed, or otherwise prepared for action, and which has been fired, dropped, launched, projected, or placed in such a manner as to constitute a hazard to operations, installations, personnel, or material, and remains unexploded either by malfunction or design or for any other cause. (JP 1-02)

unity of command – One of the nine principles of war: for every objective, ensure unity of effort under one responsible commander.

unity of effort – Coordination and cooperation among all military forces and other organizations toward a commonly recognized objective, even if the forces and nonmilitary organizations are not necessarily part of the same command structure. (FM 6-0)

unmanned aircraft system – A powered aerial vehicle that does not carry a human operator, uses aerodynamic forces to provide lift, can fly automomously or is remotely operated, can be expendable or recoverable, and can carry a lethal or nonlethal payload. Ballistic or semi-ballistic vehicles, cruise missiles, and artillery projectiles are not considered unmanned aircraft systems.

urban area – A topographical complex where manmade construction or high population density is the dominant feature. See also **urban environment**; **urban operations**. (FM 3-06)

urban environment – The physical urban area, as well as the complex and dynamic interaction among its key components—the terrain (natural and manmade), the population, and the supporting infrastructure—as an overlapping and interdependent system of systems. (FM 3-06)

urban operations – Offense, defense, stability, and support operations conducted in a topographical complex and adjacent natural terrain where manmade construction and high population density are the dominant features. (FM 3-0)

versatility – The ability of Army forces to meet the global, diverse mission requirements of full spectrum operations. (FM 3-0)

visualize – To create and think in mental images. (FM 6-0)

vulnerability – 1. The susceptibility of a nation or military force to any action by any means through which its war potential or combat effectiveness may be reduced or its will to fight diminished. 2. The characteristics of a system that cause it to suffer a definite degradation (incapability to perform the designated mission) as a result of having been subjected to a certain level of effects in an unnatural (manmade) hostile environment. 3. In information operations, a weakness in information system security design, procedures, implementation, or internal controls that could be exploited to gain unauthorized access to information or an information system. (JP 1-02)

warfighting function – A group of tasks and systems (people, organizations, information, and processes), united by a common purpose, that commanders use to accomplish missions and training objectives. The six warfighting functions are: a. **command and control** – the related tasks and systems that support the commander in exercising authority and direction; b. **fire support** – the related tasks and systems that provide collective and coordinated use of Army indirect fires, joint fires, and offensive information operations; c. **intelligence** – the related tasks and systems that facilitate understanding of the enemy, terrain, weather, and civil considerations; d. **movement and maneuver** – the related tasks and systems that move forces to achieve a position of advantage in relation to the enemy; e. **protection** – the related tasks and systems that preserve the force so the commander can apply maximum combat power; f. **sustainment** – the related tasks and systems that provide support and services to ensure freedom of action, extend operational reach, and prolon endurance. (FMI 5-0.1)

weapons of mass destruction – Weapons that are capable of a high order of destruction and/or of being used in such a manner as to destroy large numbers of people. Weapons of mass destruction can be high explosives or nuclear, biological, chemical, and radiological weapons, but exclude the means of transporting or propelling the weapon where such means is a separable and divisible part of the weapon. (JP 1-02)

This page intentionally left blank.

References

Military publications are listed by title. When a field manual has been published under a new number for the first time, the old number is provided in parenthesis after the new number.

REQUIRED PUBLICATIONS

These documents must be available to intended users of this publication.

JOINT PUBLICATIONS

Joint publications are available at <http://www.dtic.mil/doctrine/jel/>.

JP 1-02. *Department of Defense Dictionary of Military and Associated Terms*, 12 April 2001.

JP 3-06. *Doctrine for Joint Urban Operations*. 16 September 2002.

ARMY PUBLICATIONS

Army doctrinal publications are available at <http://www.apd.army.mil/>.

FM 1-02 (101-5-1). *Operational Terms and Graphics*. 21 September 2004.

FM 34-130. *Intelligence Preparation of the Battlefield*. 8 July 1994. FM 34-130 will be republished as FM 2-01.3

FM 3-0 (100-5). *Operations*. 14 June 2001.

FM 3-06.11 (90-10-1). *Combined Arms Operations in Urban Terrain*. 28 February 2002.

FM 3-07 (100-20). *Stability Operations and Support Operations*. 20 February 2003. When updated, FM 3-07 will be retitled *Stability Operations*.

FM 3-90. *Tactics*. 04 July 2001.

FM 5-19 (100-14). *Composite Risk Management*. 21 August 2006.

FM 7-0. *Training the Force*. 22 October 2002.

FM 7-1. *Battle Focused Training*. 15 September 2003.

FM 5-0 (101-5). *Army Planning and Orders Production*. 20 January 2005.

RELATED PUBLICATIONS

These sources contain relevant supplemental information.

JOINT PUBLICATIONS

Joint publications are available at <http://www.dtic.mil/doctrine/jel/>.

JP 3-0. *Doctrine for Joint Operations*. 10 September 2001.

JP 3-05. *Doctrine for Joint Special Operations*. 17 December 2003.

JP 3-08. *Interagency Coordination During Joint Operations* (2 Volumes). 9 October 1996.

JP 3-16. *Joint Doctrine for Multinational Operations*. 05 April 2000.

JP 4-0. *Doctrine for Logistic Support of Joint Operations*. 06 April 2000.

JP 5-00.2. *Joint Task Force Planning Guidance and Procedures*. 13 January 1999.

Joint Military Operations Historical Collection. 15 July 1997.

ARMY PUBLICATIONS

Army doctrinal publications are available at <http://www.apd.army.mil/>.

Army Regulations

FM 1. *The Army*. 14 June 2005.

FM 2-0 (34-1). *Intelligence*. 17 May 2004.

FM 2-22.3 (34-52). *Human Intelligence Collector Operations*. 6 September 2006.

FM 3-05 (100-25). *Army Special Operations Forces*. 20 September 2006.

FM 3-05.20 (31-20). *Special Forces Operations*. 1 April 2004.

FM 3-05.30 (33-1). *Psychological Operations*. 15 April 2005.

FM 3-05.105. *Multiservice Tactics, Techniques, and Procedures for Special Operations Forces in Nuclear, Biological, and Chemical Environments*. 28 September 2001.

FM 3-05.212 (TC 31-25). *Special Forces Waterborne Operations*. 31 August 2004.

FM 3-05.301 (33-1-1). *Psychological Operations Tactics, Techniques, and Procedures*. 31 December 2003.

FM 3-05.401 (41-10-1). *Civil Affairs Tactics, Techniques, and Procedures*. 23 September 2003.

FM 3-06 (90-10). *Urban Operations*. 01 June 2003.

FM 3-06.1. *Aviation Urban Operations: Multi-service Tactics, Techniques, and Procedures for Aviation Urban Operations*. 09 July 2005.

FM 3-07.2. *Antiterrorism*. (Future Publication).

FM 3-11.3 ( ). *Multiservice Tactics, Techniques, and Procedures for Chemical, Biological, Radiological, and Nuclear Contamination Avoidance*. 2 February 2006.

FM 3-11.9 (3-9). *Potential Military Chemical/Biological Agents and Compounds*. 10 January 2005.

FM 3-11.14 (3-14). *Multiservice Tactics, Techniques, and Procedures for Nuclear, Biological, and Chemical Vulnerability Assessment*. 28 November 2004.

FM 3-11.19 (3-19). *Multiservice Tactics, Techniques, and Procedures for Nuclear, Biological, and Chemical Reconnaissance*. 30 July 2004.

FM 3-11.21 (3-21). *Multiservice Tactics, Techniques, and Procedures for Nuclear, Biological, and Chemical Aspects of Consequence Management*. 12 December 2001.

FM 3-11.22. *Weapons of Mass Destruction Civil Support Team Tactics, Techniques, and Procedures*. 06 June 2003.

FM 3-13 (100-6). *Information Operations: Doctrine, Tactics, Techniques, and Procedures*. 28 November 2003.

FM 3-14 (100-18). *Space Support to Army Operations*. 18 May 2005.

FM 3-19.4 (19-4). *Military Police Leaders' Handbook*. 04 March 2002.

FM 3-19.15 (19-15). *Civil Disturbance Operations*. 18 April 2005.

FM 3-19.17. *Military Working Dogs*. 22 September 2005.

FM 3-19.40 (19-40). *Military Police Internment/Resettlement Operations*. 01 August 2001.

FM 3-19.50. *Police Intelligence Operations*. 21 July 2006.

FM 3-22.40 (90-40). *NLW: Tactical Employment of Nonlethal Weapons*. 15 January 2003.

FM 3-28. *Civil Support Operations*. (Future Publication).

FM 3-31. *Joint Force Land Component Commander Handbook (JFLCC)*. 13 December 2001.

FM 3-31.1 (90-31). *AMCI: Army and Marine Corps Integration in Joint Operations*. 21 November 2001.

FM 3-34 (5-100, 5-114). *Engineer Operations*. 02 January 2004.

FM 3-34.2. *Combined Arms Breaching Operations*. 31 August 2000. FM 3-34.2 will be republished as FM 3-90.11

FM 3-34.230 (5-105). *Topographic Operations*. 03 August 2000.

FM 3-61.1. *Public Affairs Tactics, Techniques and Procedures*. 01 October 2000.

FM 3-90.12 (90-13). *Combined Arms Gap Crossing Operations*. TBP.

FM 3-97.6 (90-6). *Mountain Operations*. 28 November 2000.

FM 3-100.12. *Risk Management: Multiservice Tactics, Techniques, and Procedures for Risk Management*. 15 February 2001.

FM 3-100.38. *UXO: Multi-service Tactics, Techniques and Procedures for Unexploded Ordnance Operations*. 18 August 2005.

FM 4-0 (100-10). *Combat Service Support*. 29 August 2003.

FM 4-01.45. *Tactical Convoy Operations: Multiservice Tactics, Techniques, and Procedures for Tactical Convoy Operations*. 24 March 2005.

FM 4-02.51 (8-51). *Combat and Operational Stress Control*. 6 June 2006.

FM 4-30.16. *EOD: Multiservice Procedures for Explosive Ordnance Disposal in a Joint Environment*. 27 October 2005.

FM 4-30.51. *Unexploded Ordnance (UXO) Procedures*. 13 July 2006.

FM 5-01.12 (90-41). *JTF LNO Integration: Multiservice Tactics, Techniques, and Procedures for Joint Task Force (JTF) Liaison Officer Integration*. 27 January 2003.

FM 5-33. *Terrain Analysis*. 8 September 1992. FM 5-33 will be republished as FM 3 34.330.

FM 5-103. *Survivability*. 10 June 1985. FM 5-103 will be republished as FM 5-34.112.

FM 5-104. *General Engineering*. 12 November 1986. FM 5-104 will be republished as FM 3-34.250.

FM 5-170. *Engineer Reconnaissance*. 05 May 1988. FM 5-170 will be republished as FM 3 34.212.

FM 5-415. *Fire-Fighting Operations*. 09 February 1999. FM 5-415 will be republished as FM 3-34.281.

FM 6-0. *Mission Command: Command and Control of Army Forces*. 11 August 2003.

FM 6-22.5 (22-9). *Combat Stress*. 23 June 2000.

FM 7-15. *The Army Universal Task List (Incl. changes 2)*. 6 July 2006.

FM 20-32. *Mine/Countermine Operations*. 1 April 2005.

FM 22-51. *Leaders' Manual for Combat Stress Control*. 29 September 1994. FM 22-51 will be republished as FM 4-02.22.

FM 22-100. *Army Leadership: Be, Know, Do*. 31 August 1999. FM 22-100 will be republished as FM 6-22.

FM 23-10. *Sniper Training*. 17 August 1994. FM 23-10 will be republished as FM 3-22.10.

FM 27-10. *The Law of Land Warfare*. 18 July 1956. FM 27-10 will be republished as FM 1-04.10.

FM 27-100. *Legal Support to Operations*. 01 March 2000. FM 27-100 will be republished as FM 1-04.

FM 31-20-3. *Foreign Internal Defense Tactics, Techniques, and Procedures for Special Forces*. 20 September 1994. FM 31-20-3 will be republished as FM 3-05.202.

FM 31-50. *Combat in Fortified and Built-Up Areas*. March 1964.

FM 34-3. *Intelligence Analysis*. 15 March 1990. FM 34-3 will be republished as FM 2-33.4.

FM 34-60. *Counterintelligence*. 03 October 1995. FM 34-60 will be republished as FM 2-01.2.

FM 34-81. *Weather Support for Army Tactical Operations*. 31 August 1989. FM 34-81 will be republished as FM 2-33.2.

FM 34-81-1. *Battlefield Weather Effects*. 23 December 1992. FM 34-81-1 will be republished as FM 2-33.201.

FM 41-10. *Civil Affairs Operations*. 14 February 2000. FM 41-10 will be republished as FM 3-05.40.

FM 44-8. *Combined Arms for Air Defense*. 01 June 1999. FM 44-8 will be republished as FM 3-01.8.

FM 46-1. *Public Affairs Operations*. 30 May 1997. FM 46-1 will be republished as FM 3-61.

FM 90-8. *Counterguerrilla Operations*. 29 August 1986. FM 90-8 will be republished as FM 3-07.11.

FM 90-10. *Military Operations on Urbanized Terrain (MOUT) (How to Fight)*. 15 August 1979.

FM 100-5. *Operations*. 1944.

FM 100-8. *The Army in Multinational Operations.* 24 November 1997. FM 100-8 will be republished as FM 3-16.

FM 100-11. *Force Integration.* 15 January 1998. FM 100-11 will be republished as FM 3-100.11.

FMI 2-91.4. *Intelligence Support to Operations in the Urban Environment.* 30 June 2005. FMI 2-91.4 will be republished as FM 2-91.4.

FMI 3-07.22. *Counterinsurgency Operations.* 01 October 2004. FMI 3-07.22 will be republished as FM 3-07.22.

FMI 3-34.119. *Improvised Explosive Device Defeat.* 21 September 2005.

FMI 4-30.50. *Modular Explosive Ordnance Disposal Operations.* 31 July 2006.

FMI 5-0.1. *The Operations Process.* 31 March 2006.

TC 3-15. *Nuclear Accident and Incident Response and Assistance (NAIRA).* 27 December 1988.

TC 5-230. *Army Geospatial Guide for Commanders and Planners.* 28 November 2003.

TC 7-98-1. *Stability and Support Operations Training Support Package.* 05 June 1997.

TC 9-21-01. *Soldiers Improvised Explosive Device (IED) Awareness Guide: Iraq and Afghanistan Theaters of Operation.* 28 May 2004.

TC 90-1. *Training for Urban Operations.* 01 April 2002.

DEPARTMENT OF DEFENSE PUBLICATIONS

Defense Intelligence Agency Manual 58-11. *Defense Human Resources Intelligence Collection Management.* 20 August 1968.

Quadrennial Defense Review Report. 30 September 2001.

U.S. Army Training and Doctrine Command (TRADOC), Deputy Chief of Staff for Intelligence (DCSINT). *TRADOC DCSINT Handbook No. 1: A Military Guide to Terrorism in the Twenty-First Century* (Version2.0). Fort Leavenworth, KS: U.S. Government Printing Office, 2004.

NONMILITARY PUBLICATIONS

26th Infantry Regimental Association. *Aachen: Military Operations in Urban Terrain.* Lititz, PA: 26th Infantry Regimental Association, 1999.

Adan, Abraham. *On the Banks of the Suez.* Novato, CA: Presidio Press, 1980.

Akers, Frank H., Jr. and George B. Singleton. "Task Force Ranger: A Case Study Examining the Application of Advanced Technologies in Modern Urban Warfare." Oak Ridge, TN: National Security Program Office, November 2000.

Allard, Kenneth. *Somalia Operations: Lessons Learned.* Washington, DC: National Defense University Press, 1995.

Arnold, James R. *Tet Offensive 1968: Turning Point in Vietnam.* London: Osprey, 1990.

Bass, Bernard M. "Leading in the Army after Next." *Military Review* (March/April 1998): 46–58.

Beevor, Anthony. *Stalingrad.* New York: Penguin Books, 1998.

Bowden, Mark. *Black Hawk Down: A Story of Modern War.* New York: Atlantic Monthly Press, 1999.

Butts, Kent H. "The Strategic Importance of Water." *Parameters* (spring 1997): 65–83.

Carell, Paul. *Stalingrad: The Defeat of the German 6th Army.* Atglen, PA: Schieffer Publishing, 1993.

Celeski, Joseph D. "Joint Urban Operations: Special Forces in Urban Campaign Planning." *Special Warfare* (fall 2001): 2–13.

Chiarelli, Peter W. and Patrick R. Michaelis. "Winning the Peace: The Requirement for Full-Spectrum Operations." *Military Review* (July/August 2005): 4–17.

Chiarelli, Peter W., Patrick R. Michaelis, and Geoffrey A. Norman. "Armor in Urban Terrain: The Critical Enabler." *Armor* (March/April 2005): 7–12.

Chuikov, Vasili I. *The Battle for Stalingrad.* New York: Holt, Rinehart, and Winston, 1964.

Colby, John. *War From the Ground Up: The 90th Division in WWII.* Austin, TX: Nortex Press, 1991.

Cole, Ronald H. *Operation Just Cause.* Washington, DC: Joint History Office, 1995.

Connaughton, Richard, John Pimlott, and Duncan Anderson. *The Battle for Manila: The Most Devastating Untold Story of World War II.* Novata, CA: Presidio, 1995.

Craig, William. *Enemy at the Gates: The Battle for Stalingrad.* New York: Reader's Digest Press, 1973.

Delk, James D. *Fires & Furies: The LA Riots.* Palm Springs, CA: Etc. Publications, 1995.

Dewar, Michael. *British Army in Northern Ireland.* New York: Sterling Publishers, 1985.

______. *War in the Streets: The Story of Urban Combat from Calais to Khafji.* New York: Sterling Publishers, 1992.

Donnelly, Thomas, Margaret Roth, and Caleb Baker. *Operation Just Cause: The Storming of Panama.* New York: Lexington Books, 1991.

Dupuy, R. Ernest and Trevor N. Dupuy. *The Encyclopedia of Military History: From 3500 B.C. to the Present.* San Francisco: Harper & Row, 1977.

Dupuy, Trevor N. and Paul Martell. *Flawed Victory: The Arab-Israeli Conflict and the 1982 War in Lebanon.* Fairfax, VA: Hero Books, 1986.

Edwards, Sean J. A. *Freeing Mercury's Wings: Improving Tactical Communications in Cities.* Santa Monica, CA: RAND, 2001.

Eshel, David. *Mid-East Wars: The Lebanon War 1982.* Hod Hasharon, Israel: Eshel-Dramit, 1983.

Fishel, John T. *The Fog of Peace: Planning and Executing the Restoration of Panama.* Carlisle, PA: Strategic Studies Institute, 1992.

Fontenot, Gregory, E. J. Degen, and David Tohn. *On Point: US Army in Operation IRAQI FREEDOM.* Fort Leavenworth, KS: Combat Studies Institute Press, 2004.

Funston, Frederick. "How the Army Worked to Save San Francisco." *Cosmopolitan Magazine* (July 1906): <http://www.sfmuseum.org/1906/cosmo.html>.

Gabel, Christopher R. "Military Operations on Urbanized Terrain: The 2d Battalion, 26th Infantry, at Aachen, October 1944." *Urban Combat Operations: tactics, Techniques and Procedure.* Center for Army Lessons Learned (CALL) Newsletter No. 99-16. Fort Leavenworth, KS: Center for Army Lessons Learned, U.S. Training and Doctrine Command [TRADOC], November 1999: 1-9–1-13.

Gabriel, Richard A. *Operation Peace for Galilee: The Israeli-PLO War in Lebanon.* New York: Hill and Wang, 1984.

Glenn, Russell W. *An Attack on Duffer's Downtown.* Santa Monica, CA: RAND, 2001.

______. *Heavy Matter: Urban Operations' Density of Challenges.* Santa Monica, CA: RAND, 2000.

______. "Urban Combat is Complex." *Proceedings* (February 2002): 62–65.

______. "…We Band of Brothers": *The Call for Joint Urban Operations Doctrine.* Santa Monica, CA: RAND, 1999.

Glenn, Russell W., Steven Hartman, and Scott Gerwehr. *Urban Combat Service Support Operations: The Shoulders of Atlas.* Santa Monica, CA: RAND, 2003.

Gordon, Michael R. "Baghdad Targets Picked if Hussein Holes Up There." *New York Times* (7 March 2003): 1A2.

Grau, Lester W. "Something Old, Something New: Guerrillas, Terrorists, and Intelligence Analysis." *Military Review* (July/August 2004): 42–49.

Grau, Lester W. and Timothy Smith. *A 'Crushing' Victory: Fuel-Air Explosives and Grozny 2000.* Fort Leavenworth, KS: Foreign Military Studies Office, 2000.

Hamill, Desmond. *Pig in the Middle: The Army in Northern Ireland, 1969–1984.* London: Methuen, 1985.

Hammel, Eric. *Fire in the Streets: The Battle for Hue, Tet 1968*. New York: Dell Publishing, 1991.

Heal, Sid. "Crowds, Mobs and Nonlethal Weapons." *Military Review* (March/April 2000): 45–50.

Heinl, Robert Debs, Jr. *Victory at High Tide: The Inchon-Seoul Campaign*. Annapolis, MD: Nautical & Aviation Publishing Company of America, 1979.

Herzog, Cham. *The War of Atonement, October, 1973*. Boston: Little, Brown and Company, 1975.

Higham, Robin. *Bayonets in the Streets: The Use of Troops in Civil Disturbances*. Lawrence, KS: University Press of Kansas, 1969.

Hoar, Joseph P. "A CINC's Perspective." *JFQ Forum* (autumn 1993): 56–63.

Hollis, Mark A. B. "Platoon Under Fire." *Infantry* (January/April 1998): 27–34.

Infantry in Battle. 2d ed. Garret & Massie: Richmond, VA, 1939. Reprint, Fort Leavenworth, KS: U.S. Army Command and General Staff College, 1981.

Initial Impressions Report: Operation Iraqi Freedom—Stability Operations-Support Operations Center for Army Lessons Learned (CALL) Newsletter No. 04-13 (Fort Leavenworth, KS: Center for Army Lessons Learned, U.S. Army Training and Doctrine Command [TRADOC], 2004).

Initial Impressions Report: Operation Iraqi Freedom—Stability Operations-Support Operations. Center for Army Lessons Learned (CALL) Newsletter No. 03 (Fort Leavenworth, KS: Center for Army Lessons Learned, U.S. Army Training and Doctrine Command [TRADOC], 2003).

Jomini, Antoine-Henri. "The Art of War" in *Roots of Strategy: Book 2*. Harrisburg, PA: Stackpole Books, 1987.

Kemp, Anthony. *The Unknown Battle: Metz, 1944*. New York: Stein and Day, 1981.

Kennedy, John F. Message to Congress, 25 May 1961. Available at <http://www.jfklibrary.org/Historical+Resources/Archives>.

Linden, Eugene. "The Exploding Cities of the Developing World." *Foreign Affairs* (January/February 1996): 52–65.

Mabry, Robert L. et al. "United States Army Rangers in Somalia: An Analysis of Combat Casualties on an Urban Battlefield." *Journal of Special Operations Medicine* (fall 2001): 24–40.

MacArthur, Douglas. *Reminiscences*. New York: McGraw-Hill, 1964.

Manwaring, Max G. *Street Gangs: The New Urban Insurgency*. Carlisle, PA: Strategic Studies Institute, 2005.

Maxwell, Mark D. "The Law of War and Civilians on the Battlefield: Are We Undermining Civilian Protections?" *Military Review* (September/October 2004): 17–25.

McConnell, Malcolm. *Just Cause: The Real Story of America's High-Tech Invasion of Panama*. New York: St. Martin's Press, 1991.

Medby, Jamison Jo and Russell W. Glenn. *Street Smart: Intelligence Preparation of the Battlefield for Urban Operations*. Santa Monica, CA: RAND, 2002.

Mendel, William W. *Combat in Cities: The LA Riots and Operation Rio*. Fort Leavenworth, KS: Foreign Military Studies Office, 1996.

Monmonier, Mark. *How to Lie with Maps*. Chicago: University of Chicago Press, 1996.

Montgomery, Thomas M. Interview. "Ambush in Mogasdishu." *Frontline*, WGBH Boston. 29 September 1998. Available at <http://www.pbs.org/wgbh/pages/frontline/programs/1998.html>.

Moore, John C. "Sadr City: The Armor Pure Assault in Urban Terrain." *Armor* (November/December 2004): 31–37.

Natsios, Andrew S. "Commander's Guidance: A Challenge of Complex Humanitarian Emergencies." *Parameters* (summer 1996): 50–66.

______. "Food Through Force: Humanitarian Intervention and U.S. Policy." *The Washington Quarterly* (winter 1994): 136.

Nolan, Keith William. *The Battle for Saigon: Tet 1968*. New York: Pocket Books, 1996.

Oakley, Robert B. "An Envoy's Perspective." *JFQ Forum* (autumn 1993): 44–55.

Oliker, Olga. *Russia's Chechen Wars 1994–2000: Lessons from Urban Combat*. Santa Monica, CA: RAND, 2001.

Operation Restore Hope Lessons Learned Report: 3 December 1992 – 4 May 1993, Operations Other than War. Fort Leavenworth, KS: Center for Army Lessons Learned (CALL), U.S. Army Combined Arms Command (CAC), 1993. Reprint, Washington, DC: U.S. Government Printing Office, 1996.

Parkerson, John Embry, Jr. "United States Compliance with Humanitarian Law Respecting Civilians During Operation Just Cause." *Military Law Review* (1991): 31–140.

Peters, Ralph. "Our Soldiers, Their Cities." *Parameters* (spring 1996): 43–50.

Peterson, John J. *Into the Cauldron*. Clinton, MD: Clavier House, 1973.

Petraeus, David H. "Learning Counterinsurgency: Observations from Soldiering in Iraq." *Military Review* (January/February 2006): 2–12.

Ritchie, Elspeth C. and Robert Mott. "Caring for Civilians During Peace Keeping Missions: Priorities and Decisions." *Military Medicine* (Vol. 167, Supplement 3, 2002): 14–16.

Rosenberger, Leif R. "The Strategic Importance of the World Food Supply." *Parameters* (spring 1997): 84–105.

Schadlow, Nadia. "War and the Art of Governance." *Parameters* (autumn 2003): 85–94.

Shelton, H. Hugh and Timothy D. Vane. "Winning the Information War in Haiti." *Military Review* (November/December 1995): 3–9.

Shtemenko, S. M. *The Soviet General Staff at War, 1941-1945: Book One*. Moscow: Progress Publishers, 1981.

Smith, Rupert. "Wars in Our Time: A Survey of Recent and Continuing Conflicts." *World Defence Systems,* volume 3:2. London: Royal United Services Institute for Defence Studies, 2001.

Stanton, Shelby L. *Ten Corps in Korea, 1950*. Novato, CA: Presidio Press, 1989.

Taw, Jennifer Morrison. *Operation Just Cause: Lessons for Operations Other Than* War. Santa Monica, CA: RAND, 1996.

Thomas, Timothy L. "The Battle for Grozny: Deadly Classroom for Urban Combat." *Parameters* (summer 1999): 87–102.

Thomas, Timothy L. and Charles P. O'Hara. "Combat Stress in Chechnya: 'The Equal Opportunity Disorder'." *Army Medical Department Journal* (January/March 2000): 46–53.

Trinquier, Roger. *Modern Warfare: A French View of Counterinsurgency*. New York: Praeger Publisher, 1964.

United Nations Security Council Resolution 794 (1992). New York: United Nations, 1992. Available at <http://daccessdds.un.org/doc/UNDOC/GEN/N92/772/11/PDF>.

U.S. Agency for International Development, Bureau for Humanitarian Response, Office of Foreign Disaster Assistance. *Field Operations Guide for Disaster Assessment and Response, Version 3.0*. Washington, DC: U.S. Government Printing Office, 2003. Available at <http://www.usaid.gov/our_work/humanitarian_assistance/>.

U.S. Department of Transportation, Research and Special Programs Administration, Office of Hazardous Materials Initiative and Training. *2004 Emergency Response Guidebook*. Washington, DC: U.S. Government Printing Office, 2004. Available at <http://hazmat.dot.gov/pubs/erg/erg2004.pdf>.

U.S. Environmental Protection Agency, Office of Emergency and Remedial Response. *HAZMAT Team Planning Guidance*. Washington, DC: U.S. Government Printing Office, 1990.

Van Sickle, Jeffrey B. "Stability Operations in Northern Iraq: Task Force Altun Kupri." *Infantry* (January/February 2005): 25–29.

Watson, Bruce Allen. *Sieges: A Comparative Study*. Westport, CT: Praeger, 1993.

Werstein, Irving. *The Battle of Aachen*. New York: Thomas Y. Crowell Company, 1962.

Whiting, Charles. *Bloody Aachen*. New York: Stein and Day, 1976.

Wintle, Justin, ed. *The Dictionary of War Quotations*. New York: The Free Press, 1989.

Yates, Lawrence A. "Military Stability and Support Operations: Analogies, Patterns and Recurring Themes." *Military Review* (July/August 1997): 51–61.

RECOMMENDED READINGS

The following sources contain relevant supplemental information.

Alexander, John B. *Future War: Non-Lethal Weapons in Twenty-First-Century* Warfare. New York: St. Martin's Press, 1999.

Ashworth, G. J. *The City and War*. London: Routledge, 1991.

Barker, A. J. *Dunkirk: The Great Escape*. New York: David McKay, 1977.

Baumann, Robert F., Lawrence A. Yates, and Versalle F. Washington. "My Clan Against the World": *U.S. and Coalition Forces in Somalia, 1992-1994*. Fort Leavenworth, KS: Combat Studies Institute Press, 2004.

Birtle, Andrew J. *U.S. Army Counterinsurgency and Contingency Operations Doctrine: 1860-1941*. Washington, DC: Center of Military History, 2001.

Blaufarb, Douglas S. *The Counterinsurgency Era: U.S. Doctrine and Performance 1950 to Present*. New York: The Free Press, 1977.

Breuer, William B. *Hitler's Fortress Cherbourg: The Conquest of a Bastion*. New York: Stein and Day, 1984.

Brown, Stephen D. "PSYOP in Operation Uphold Democracy." *Military Review* (September/October 1996): 57–76.

Bulloch, Gavin. "Military Doctrine and Counterinsurgency: A British Perspective." *Parameters* (summer 1996): 4–16.

Urban Combat Operations: Tactics, Techniques and Procedures. Center for Army Lessons Learned (CALL) Newsletter No. 99-16 (Fort Leavenworth, KS: Center for Army Lessons Learned, U.S. Army Training and Doctrine Command [TRADOC], November 1999).

Cordesman, Anthony H. and Abraham R. Wagner. *The Lessons of Modern War—Volume II: The Iran-Iraq War*. Boulder, CO: Westview Press, 1990.

Corum, James S. and Wray R. Johnson. *Airpower in Small Wars: Fighting Insurgents and Terrorists*. Lawrence, KS: University Press of Kansas, 2003.

Desch, Michael C., ed. *Soldiers in Cities: Military Operations on Urban Terrain*. Carlisle Barracks, PA: Strategic Studies Institute, 2001.

Duffy, Christopher. *Fire & Stone: The Science of Fortress Warfare, 1660–1860*. Mechanicsburg, PA: Stackpole Books, 1996.

______. *The Fortress in the Age of Vauban and Frederick the Great, 1660–1789: Siege Warfare Volume I*. London: Routledge & Kegan Paul, 1985.

______. *Siege Warfare: The Fortress in the Early Modern World, 1494–1660*. London: Routledge & Kegan Paul, 1979.

Edwards, Sean J. A. *Mars Unmasked: The Changing Face of Urban Operations*. Santa Monica, CA: RAND, 2000.

Eikenberry, Karl W. "Take No Casualties." *Parameters* (summer 1996): 109–118.

Erickson, John. *The Road to Berlin: Continuing the History of Stalin's War with Germany*. Boulder, CO: Westview Press, 1983.

Gerwehr, Scott and Russell W. Glenn. *The Art of Darkness: Deception and Urban Operations*. Santa Monica, CA: RAND, 2000.

Glenn, Russell W. *Combat in Hell: A Consideration of Constrained Urban Warfare*. Santa Monica, CA: RAND, 1996.

______. *Marching Under Darkening Skies: The American Military and Impending Urban Operations Threat*. Santa Monica, CA: RAND, 1998.

Glenn, Russell W., ed. *Capital Preservation: Preparing for Urban Operations in the Twenty-First Century: Proceedings of the RAND Arroyo-TRADOC-MCWL-OSD Urban Operations Conference, March 22-23, 2000*. Santa Monica, CA: RAND, 2001.

Glenn, Russell W. et al. *Denying the Widow-Maker, Summary of Proceedings, RAND-DDBL Conference on Military Operations on Urbanized Terrain*. Santa Monica, CA: RAND, 1999.

______. *Getting the Musicians of Mars on the Same Sheet of Music: Army, Joint, Multinational, and Interagency C4ISR Interoperability*. Santa Monica, CA: RAND, 2000.

Glenn, Russell W. et al., eds. *The City's Many Faces: Proceedings of the RAND Arroyo-MCWL-J8 UWG Urban Operations Conference, April 13-14, 1999*. Santa Monica, CA: RAND, 2000.

Gouré, Leon. *The Siege of Leningrad*. Stanford, CA: Stanford University Press, 1962.

Grau, Lester W. and Jacob W. Kipp. "Urban Combat: Confronting the Specter." *Military Review* (July/August 1999): 9–17.

Groves, John R. "Operations in Urban Environments." *Military Review* (July/August 1998): 30–39.

Habeck, Mary R. Lecture. "Jihadist Strategies in the War on Terrorism." Washington, DC: The Heritage Foundation, 8 November 2004. Available at http://www.heritage.org/research/nationalsecurity/hl855.cfm.

Hahn, Robert F., II and Bonnie Jezior. "Urban Warfare and the Urban Warfighter of 2025." *Parameters* (summer 1999): 74–86.

Harman, Nicholas. *Dunkirk: The Patriotic Myth*. New York: Simon and Schuster, 1980.

Heckstall-Smith, Anthony. *Tobruk: The Story of a Siege*. New York: W. W. Norton, 1959.

Hibbert, Christopher. *The Battle of Arnhem*. London: Batsford, 1962.

House, John M. "The Enemy After Next." *Military Review* (March/April 1998): 22–28.

Jackson, W. G. F. *The Battle for Italy*. New York: Harper & Row, 1967.

Karslake, Basil. *1940 – The Last Act: The Story of the British Forces in France After Dunkirk*. Hamden, CT: Archon Books, 1979.

Kostof, Spiro. *The City Shaped: Urban Patterns and Meanings Through History*. London: Thames and Hudson, 1999.

______. *The City Assembled: The Elements of Urban Form Throughout History*. Boston: Little Brown and Company, 1999.

Kuehl, Daniel T. "Airpower vs. Electricity: Electric Power as a Target for Strategic Air Operations." *Journal of Strategic Studies* (March 1995): 237–266.

Larson, Eric. *Casualties and Consensus: The Historical Role of Casualties in Domestic Support of U.S. Military Operations*. Santa Monica, CA: RAND, 1996.

Lieven, Anatol. *Chechnya: Tombstone of Russian Power*. New Haven: Yale University Press, 1998.

______. "The World Turned Upside Down: Military Lessons of the Chechen War." *Armed Forces Journal International* (August 1998): 40–43.

McKee, Alexander. *Caen: Anvil of Victory*. New York: St. Martin's Press, 1964.

McLaurin, R. D. "Military Forces in Urban Antiterrorism." Aberdeen Proving Ground, MD: U.S. Army Human Engineering Laboratory, 1989.

McLaurin, R. D. and R. Miller. "Urban Counterinsurgency: Case Studies and Implications for U.S. Military Forces." Aberdeen Proving Ground, MD: U.S. Army Human Engineering Laboratory, 1989.

Mercier, Charles E. "Terrorists, WMD, and the U.S. Army Reserve." *Parameters* (autumn 1997): 98–118.

Milton, T. R. "Urban Operations: Future War." *Military Review* (February 1994): 37–46.

Moskin, J. Robert. *Among Lions: The Battle for Jerusalem, June 5–7, 1967*. New York: Ballantine Books, 1982.

Northam, Ray M. *Urban Geography*. New York: Wiley, 1975.

Peters, Ralph. "The Culture of Future Conflict." *Parameters* (winter 1995 96): 18–27.

______. "The Future of Armed Warfare." *Parameters* (autumn 1997): 50–59.

______. "The Human Terrain of Urban Operations." *Parameters* (spring 2000): 4–12.

Robertson, William G., ed. *Block by Block: The Challenges of Urban Operations*. Fort Leavenworth, KS: U.S. Army Command and Staff College Press, 2003.

Rosenau, William G. "'Every Room is a New Battle': The Lessons of Modern Urban Warfare." *Studies in Conflict and Terrorism* (October–December 1997): 371–394.

Scalard, Douglas. "People of Whom We Know Nothing: When Doctrine Isn't Enough." *Military Review* (July/August 1997): 4–11.

Scales, Robert H. "The Indirect Approach: How U.S. Military Forces Can Avoid the Pitfalls of Future Urban Warfare." *Armed Forces Journal International* (October 1998): 68–74.

Siegel, Adam B. *Eastern Exit: The Noncombatant Evacuation Operation (NEO) from Mogadishu, Somalia, in January 1991*. Alexandria, VA: Center for Naval Analyses, 1992.

Spiller, Roger J. *Sharp Corners: Urban Operations at Century's End*. Fort Leavenworth, KS: Combat Studies Institute, 2001.

Stech, Frank J. "Winning CNN Wars." *Parameters* (autumn 1994): 37–56.

Stevenson, Jonathan. *Losing Mogadishu: Testing U.S. Policy in Somalia*. Annapolis, MD: Naval Institute Press, 1995.

Taw, Jennifer Morrison and Bruce Hoffman. *The Urbanization of Insurgency: The Potential Challenge to U.S. Army Operations*. Santa Monica, CA: RAND, 1994.

Tully, Andrew. *Berlin: Story of a Battle*. Westport, CT: Greenwood Press, 1977.

United Nations Department of Economic and Social Affairs, Population Division. *World Urbanization Prospects: The 2003 Revision*. New York: United Nations, 2003.

Vick, Alan et al. *Aerospace Operations in Urban Environments: Exploring New Concepts*. Santa Monica, CA: RAND, 2000.

Wykes, Alan. *The Siege of Leningrad*. New York: Ballentine Books Inc., 1968.

Index

This page intentionally left blank.

FM 3-06
26 October 2006

By Order of the Secretary of the Army:

PETER J. SCHOOMAKER
General, United States Army
Chief of Staff

Official:

JOYCE E. MORROW
Administrative Assistant to the
Secretary of the Army
0627803

DISTRIBUTION:

Active Army, Army National Guard, and U. S. Army Reserve: To be distributed in accordance with initial distribution number 111233, requirements for FM 3-06.

***FM 3-34**(FM 5-100)

Field Manual
No. 3-34

HEADQUARTERS
DEPARTMENT OF THE ARMY
Washington, DC, 2 January 2004

Engineer Operations

Contents

Distribution Restriction: Approved for public release; distribution is unlimited.

*This publication supersedes FM 5-100, 27 February 1996, and FM 5-114, 13 July 1992.

Page

Page

Page

Preface

Field Manual (FM) 3-34 is the Engineer Regiment's capstone manual for operating in today's operational environment within the framework of the Army transformation and, although focused at the operational level, is applicable for all levels of war. Engineers are a combat arm (a branch comprised of combat, combat support [CS] and combat service support [CSS] components) that enables joint and maneuver commanders to achieve their objectives through strategic movement and tactical maneuver by providing unique combat, geospatial, and general engineering capabilities. It has been the engineer creed to support the maneuver commander since June 16, 1775, when the Continental Congress organized an Army with a chief engineer and two assistants. Engineers contributed to the hardest fought battles in the Revolutionary War, including Bunker Hill, Saratoga, and the final victory at Yorktown. At the end of the Revolutionary War, the engineers were mustered out of service. However, their unique skills were realized and they were called back to active duty in 1794 when Congress organized a Corps of Artillerists and Engineers and later in 1802 as a separate Corps of Engineers. Today's *FM 3-34* includes engineer doctrine that has evolved for over 200 years.

FM 3-34 is the capstone doctrinal manual for engineer operations and is linked to joint and Army doctrine to ensure its usefulness for all joint and Army level commanders and staff. All other engineer FMs *(see Appendix A)* are based on the principles and tenets found in this manual and are synchronized with their respective joint publications. These principles and tenets are founded on the successful employment of engineers, past and present. In today's complex operational environment, the engineers' warfighting focus produces a full-spectrum force that meets the needs of the land component commander (LCC) in war, conflict, and peace.

The primary audience for *FM 3-34* is engineer commanders and staffs down to and including engineer companies, maneuver force commanders, and battalion and task force (TF) organizations. The focus includes Army Service Component Command responsibilities for conducting operations as part of a multinational force. Information contained in this manual will assist multinational forces and other services and branches of the Army to plan and integrate engineer capabilities. This doctrine also will assist Army branch schools in teaching the integration of engineer capabilities into Army operations. Engineer involvement is a given for nearly every military operation. The degree of involvement will include one or more of the roles associated with engineers performing combat, CS, or CSS missions.

While the nature of war remains constant throughout history, the conduct of war is continually changing in response to new concepts, technologies, and requirements. The contemporary threat is continually evolving and adapting to friendly engineer capabilities. No matter how many engineer capabilities are embedded into Army systems, it is the engineer soldier that must recognize shortfalls and develop new concepts and methods to overcome any doctrine, organization, training, materiel, leadership and education, personnel, and facilities (DOTMLPF) shortfalls. Failure to adapt to these changes may result in the engineer branch not being ready to confront the challenges of future threats. Therefore, we must recognize that it is the adaptable and professional engineer soldiers of the Regiment that are most important to our future. *FM 3-34* furnishes the authoritative foundation for subordinate engineer doctrine and terminology, force design, materiel acquisition, professional education, and individual and unit training. This manual introduces several new terms, including assured mobility, geospatial engineering, maneuver support (MANSPT) *(see Appendix B)*, and field force engineering (FFE) *(see Appendix C)*.

FM 3-34 is built directly on the concepts of *FMs 3-0, 3-90, and 3-07* blending key points of *Joint Publications (JPs) 2-03, 3-0 3-15, 3-34, and 4-04* into its approach to ensure that Army elements of a joint force use all engineer assets to their fullest extent. Given the magnitude of doctrinal changes in recent years, you will need to be familiar with these documents to effectively use *FM 3-34*. This manual addresses engineer roles and functions within a multinational operation, under potentially multinational or interagency leadership and within diverse command relationships. Finally, this manual focuses on the key functional planning considerations for employing engineers at the strategic, operational, and tactical levels of war.

The proponent for this publication is HQ TRADOC. Send comments and recommended changes on Department of the Army (DA) Form 2028 directly to Commandant, United States Army Engineer School (USAES), ATTN: ATSE-DOT-DD, Directorate of Training, 320 MANSCEN Loop, Suite 336, Fort Leonard Wood, Missouri 65473-8929.

NOTE: The bibliography lists FMs by the new number, followed by the old number in parentheses.

Unless this publication states otherwise, masculine nouns or pronouns do not refer exclusively to men.

Chapter 1

The Army and the Role of Engineers

Essayons (Let us try!)

--Motto of the Corps of Engineers

The Army organizes, trains, and equips its forces to fight and win the nation's wars and achieve directed national objectives. The Army also protects the nation's sovereign borders and national interests against aggressors. The engineer's role in the Army's mission of national defense is critical. This chapter outlines the role of the engineer and the mission essential tasks doctrinally performed by engineers, defining the engineer battlespace functions across the spectrum of conflict. This chapter also discusses how engineer leaders interact not only with other Army forces but also with joint, interagency, and multinational organizations to perform the engineer battlespace functions.

SECTION I - THE ENGINEER REGIMENT

1-1. Army forces are normally the decisive component of land warfare in joint and multinational operations. The engineers are Army enablers for success in these operations. They organize and fight with the Army's maneuver forces to win the nation's wars and achieve its national objectives. They also support the Army's ability to deter war by maintaining their deployability and war-fighting skills, while simultaneously leading the world in advances in technology, such as standoff detection systems, intelligent munitions, and doctrinal concepts such as assured mobility. The engineer force is always ready to respond as a combat-ready force prepared to deal with the full spectrum of potential operations. Engineer forces can be tailored to support operations in austere environments, with little or no infrastructure, and provide mobility and enhance force protection through countermobility and survivability. They also provide geospatial (formerly topographic engineering) and general engineering support across the spectrum of potential operations. Engineers support light, heavy, and special operations forces (SOF) in all types of terrain and operational environments (OEs). This provides the joint forces' land, sea, and air component commander the greatest flexibility to package a force that can rapidly deploy, assist in deterring adversaries, and preclude our enemies from gaining an operational advantage in an area of operations (AO). Engineers are trained to operate with and support other service requirements, integrate with other service engineers and contractors, and assume command and control (C2) of other services or branches.

1-2. As Army forces fight and win the nation's wars, they also deter them. The Army's war-fighting focus enables a diverse (full-spectrum) force to meet the needs of the joint force commander (JFC) in war, conflict, and peace. In war, Army forces form the nucleus of the land component—imposing their will on enemies and causing them to collapse. In conflict, Army forces deploy quickly into an AO to deter adversaries and potential enemies from establishing their forces and to preclude them from gaining an operational advantage. If deterrence fails, Army forces defeat the enemy, terminate conflict to achieve national objectives, and establish self-sustaining postconflict stability. Early movement of Army forces retains initiative and freedom of action by providing the JFC complementary means to conduct decisive offensive operations at a time and place of the commander's choosing. If theater circumstances require it, Army forces provide the means to block the enemy's offensive and deliver the counteroffensive blow necessary to win as rapidly as possible. In peace, Army forces train for war and provide military support to civil authorities when necessary. They also help shape the international security environment through engagement activities and nationally, they provide support to civilian authorities both at home and abroad in response to homeland security (HLS) for natural or man-made disasters. Regardless of the type of commitment of Army forces, the degree of engineer participation is likely to be high.

1-3. The Engineer Regiment contributes to the Army's war-fighting abilities and focus. It consists of all Active Component (AC) and Reserve Component (RC) engineer organizations (as well as the Department of Defense [DOD] civilians and affiliated contractors and agencies within the civilian community) with a diverse range of capabilities. The Chief of Engineers leads the Engineer Regiment and is triple-hatted as the chief of the engineer branch, the staff officer advising the Chief of Staff of the Army (CSA) on engineering matters, and the Commander of theUnited States Army Corps of Engineers (USACE). The AC of the Engineer Regiment consists of USACE and AC military engineer units within the combatant commands (COCOMs) and major Army commands (MACOMs). The RC consists of the Reserve and National Guard and provides support to the theater engineer commands (ENCOMs). The RC engineer force consists of more than three fourths of all military engineer forces and includes a wide range of specialized capabilities. Additionally, certain types of units are found only in these two components. For example, facility engineer detachments (FEDs) are only found in the RC. The Regiment is joint in its integration capabilities and supports the planning, preparing, and executing of joint operations. The Regiment is experienced at interagency support and leveraging nonmilitary and nongovernmental engineer assets to support mission accomplishment. At the operational/strategic level, the Regiment is represented as shown in *Figure 1-1*. The Regiment is represented by the various engineer organizations and capabilities reflected in *Figure 1-2, page 1-4* at the tactical/operational level. *Appendix D* provides a more in-depth view of the organizations depicted in *Figure 1-2*.

THE THREEFOLD BRANCH

1-4. The main component of the Engineer Regiment is the Engineer Branch. Engineer officers and engineer enlisted soldiers with combat engineer military occupational specialties (MOSs) are combat arms soldiers. The Engineer Branch reinforces and complements the effects of the other branches

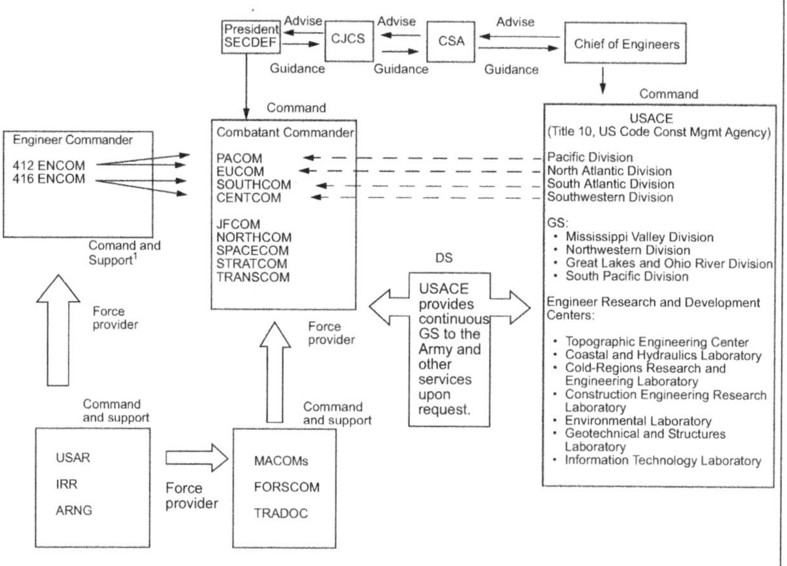

[1]**Command and support to the land component commander for the combatant commander.**

Figure 1-1. The Engineer Regiment at the Strategic/Operational Level

in the Army and the other services. While most branches are identified as being either combat, CS, or CSS, engineers are identified in all of these categories and have significant overlap within the branch and the roles, missions, and functions that they perform. The only other branch that shares this distinction and challenge is the Aviation Branch.

COMBAT ARMS

1-5. Combat arms are those units and soldiers who close with and destroy enemy forces or provide firepower and destructive capabilities on the battlefield. Many engineer units perform combat arms roles. The commander task-organizes combat engineer units with maneuver units and integrates them into a combined arms formation. The engineer units provide demolition and reduction capabilities to the combined arms team. Additionally, engineer units can fire and maneuver to employ direct-fire weapons systems to aid in employing obstacles and breaching obstacles. Regardless of the mission, armored engineer vehicles are combat vehicles and provide a significant contribution to the combat power of the entire formation.

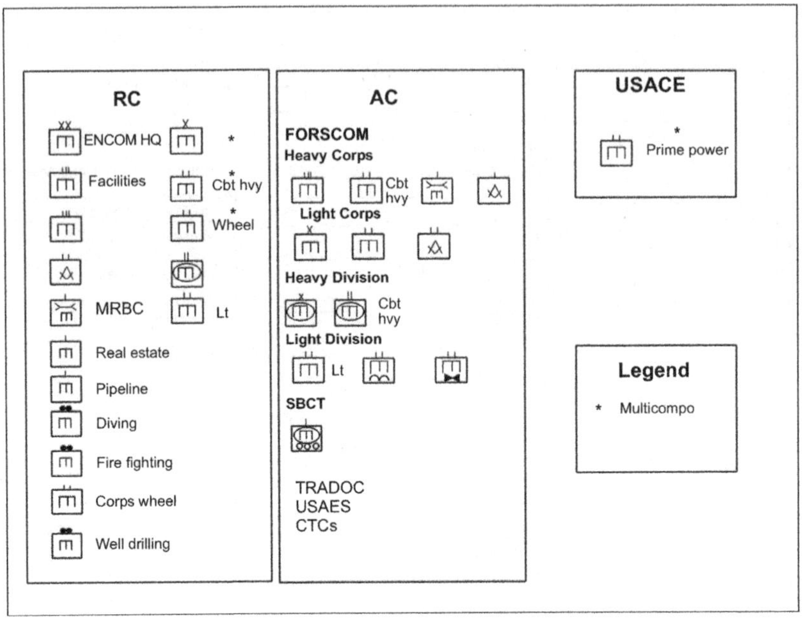

Figure 1-2. The Engineer Regiment (by type) at the Tactical/Operational Level

1-6. Combat engineer units have a secondary mission to fight as infantry. Reorganizing and employing engineers as infantry requires serious consideration. Engineers provide far more combat power in their primary mission than when reconfigured as infantry. Stopping engineer work may reduce the combat power of the commander's entire force. Because of the long-term impact, reorganizing engineers to fight as infantry is an operational issue and careful consideration should be given to what level of command is authorized to order reorganization (normally corps commander or higher). Reorganization implies withdrawing engineer units to a rear area to issue infantry equipment (such as antitank [AT] systems) and providing them with other combined arms maneuver battalion assets (such as a fire support team and enhanced combat medical support). It also implies a period of training with these newly acquired assets.

COMBAT SUPPORT

1-7. CS engineer elements provide the maneuver commander with significant amounts of additional combat power and capabilities not typically organic to the maneuver division. These elements support the maneuver commander using established command and support (C&S) relationships. CS engineers include combat heavy battalions, corps combat battalions, and bridge companies. CS encompasses critical combat functions provided by units and soldiers, in conjunction with combat arms units and soldiers, to secure victory. Engineers operate as integral members of the combined arms team to provide full combat, geospatial, and general engineering capabilities. CS engineers may also advise the maneuver commander on the effective use of terrain; construction efforts; improvement and maintenance of routes, bridges, and airfields; and reorganization to fight as infantry, when required.

1-8. In offensive operations CS engineers concentrate their efforts to support the maneuver by reducing and crossing obstacles and gaps, assisting in the assault of fortified positions, and emplacing obstacles to protect the flanks of friendly attacking forces. In defensive operations, CS engineers reinforce the terrain to anchor the defense in critical areas, maximize the effects of defenders' fires, provide maximum protection to friendly fighting positions, and facilitate the movement of counterattack forces. They may also upgrade routes and construct life support areas.

1-9. Geospatial capabilities are typically a CS element represented by topographic battalions, companies, and terrain teams that manage the flow of geospatial information from the unit level to the level of national service organizations, such as the National Imagery and Mapping Agency (NIMA). Topographic units develop geospatial products (to include detailed terrain analysis products, maps, and digital terrain data) that assist commanders from company to theater levels. Geospatial information and services assist the commander with building a common operational picture (COP) by identifying avenues and routes, covered and concealed areas, obstacle locations, key terrain features, engagement areas (EAs), unit positions, and possible target areas.

COMBAT SERVICE SUPPORT

1-10. CSS engineers sustain the momentum of the force they support. The primary role of tactical CSS units is to sustain Army forces. The CSS effort is successful only when it concentrates and supports forces by focusing on sustaining and reconstituting tactical units. General engineering is the primary CSS engineer function. It involves a variety of roles—civil engineering, fire fighting, maintaining lines of communications (LOCs), employing force protection, assessing area damage control (ADC), managing inland waterways, generating power, and providing environmental support. Much, but not all, of the Regiment's reach capabilities tend to be in the category of CSS.

1-11. CSS engineers at the operational level include theater construction engineering; real estate support; contract services; facilities development; well drilling; base camp development and management; and pipeline, rail, and port construction. USACE elements actively participate in support of engineer CSS functions.

MISSION-ESSENTIAL TASKS

1-12. The Army's mission-essential tasks derive from statutory requirements, operational experience, strategies for the employment of the nation's military forces, and operational requirements of the combatant commanders. They are the operational expression of the Army's core competencies contained in *FM 1 (100-1)*. Army forces develop their battle-focused mission-essential task lists (METLs) in concert with *FM 7-15*. To perform these tasks, the Army continuously integrates DOTMLPF in accordance with Chairman of the Joint Chiefs of Staff Instruction (CJCSI) (S) 3010.02A.

1-13. The six Army mission-essential tasks are—

- Shape the security environment.
- Respond promptly to crisis.
- Mobilize the Army.
- Conduct forcible entry operations.
- Dominate land operations.
- Provide support to civil authorities.

1-14. Engineers are critical to meeting these Army mission-essential tasks. These tasks use engineers in unified action with the other services and in conjunction with multinational and interagency organizations. Engineers are primary players as the Army responds to missions across the spectrum of operations and in line with their mission-essential tasks. Army forces expand the JFC's range of military options in full-spectrum operations, the range of operations Army forces conduct in war and military operations other than war (MOOTW).

1-15. The Engineer Regiment's mission-essential tasks derived from the Army's METL are—

- Shape the security environment.
- Respond promptly to crisis.
- Mobilize engineer forces.
- Support forcible entry operations.
- Support assured mobility to dominate land operations.
- Provide support to civil authorities.
- Provide quality, responsive engineering services to the nation.

SHAPE THE SECURITY ENVIRONMENT

1-16. Engineers shape the security environment by supporting the combatant commander's vision. They engage other nations by working with allies and potential coalition partners in engineer institutions, by participating in international exercises, and by supporting stability operations in foreign countries. Engineers can often lead combatant commander engagement initiatives through military-to-military contact, instruction, or construction projects.

RESPOND PROMPTLY TO CRISIS

1-17. Engineers support the LCC to organize operations in time and space to present the enemy with simultaneous and multidimensional threats via land, air, sea, and space. Some of these forces will come from elements stationed in Germany, Korea, and other outside the continental United States (OCONUS) locations. Engineer forces are ready to deploy by land, sea, air, and rail quickly and with all their capabilities. As combatant commanders develop plans to enter an AO rapidly, engineers provide a myriad of capabilities, such as joint rapid airfield construction and repair, port construction, diving, bridge and road construction/repair, and prime power support. These capabilities are provided by modular packages that meet the specific needs of the mission. Engineers also contribute to strategic responsiveness by providing geospatial terrain products for the commander to visualize the land dimension early as a foundation for base planning.

MOBILIZE ENGINEER FORCES

1-18. Engineers maintain the preponderance of their military force in the United States Army Reserve (USAR) and Army National Guard (ARNG). These reserve forces provide unique skills that in some cases are only, or primarily, in the RC (well drilling, fire fighting). RC engineers are ready to deploy and fulfill the combatant commander's needs or requirements for war or national emergencies. Engineer RC forces provide the Secretary of Defense (SECDEF) a means to expand the Army to confront unforeseen threats to national security. RC forces can also provide peacetime mobilization of engineer units to support HLS and other combatant commander engagement policies.

SUPPORT FORCIBLE ENTRY OPERATIONS

1-19. Engineers provide the Army combat (mobility, countermobility and survivability [M/CM/S]), geospatial, and general engineering capabilities to strike contested areas from the air, land, and sea. Engineer forces support establishing lodgments to project follow-on forces and sustain the maneuver force. During the Cold War, forces fell in on predeployed equipment and supplies in mature theaters with host nation support (HNS). Since then, deployments have become increasingly expeditionary, which requires a greater engineering effort to develop secure ports and forward operating bases (FOBs) in austere or underdeveloped regions with largely unskilled indigenous labor. Engineer forces with airborne and air assault forces support assured mobility by restoring seized airfields and developing storage sites. Engineers can also support Army forces involved with amphibious operations with general engineering support for logistics-over-the-shore (LOTS) operations.

SUPPORT ASSURED MOBILITY TO DOMINATE LAND OPERATIONS

1-20. Army forces are the dominant land force in the world. Engineers support decisive, shaping, and sustaining operations to facilitate success. They support assured mobility to maneuver forces to close with and destroy the enemy to achieve campaign objectives in support of national objectives. Geospatial engineering provides the combatant commander with detailed terrain and other geospatial products that facilitate visualization of the AO. Combat engineering provides freedom of movement to maneuver forces throughout the AO by providing bridging, clearing, detecting, and neutralizing mines; and removing natural and man-made obstacles. Engineers provide unimpeded movement through terrain and interdict threat mobility from interfering with Army maneuvers. They enhance force protection for vulnerable assets by providing survivability support. They provide general engineering support from the joint level down to the most forward tactical units.

PROVIDE SUPPORT TO CIVIL AUTHORITIES

1-21. The Army can adapt and tailor its war-fighting capabilities to complement and support civil authorities and agencies at home (under the guidance of the President and SECDEF) and abroad (according to the Status of Forces Agreement [SOFA] and other formal agreements between the United States [US] and other host nations [HNs]). Engineers provide a full range of engineering capabilities and expertise to reinforce or fill critical requirements beyond the immediate capabilities of civil authorities and agencies. Army engineers are uniquely postured to provide this support by accessing the analytical and construction expertise of USACE, ENCOMs, USAES, and other engineer organizations through TeleEngineering and other consultation. This reach consultation, in combination with unique engineer unit capabilities, can enable commanders to restore basic support infrastructure (transportation networks, utility systems, and facilities), alleviate refugee suffering, and reinstate local governmental authority. Engineers in the AC and RC forces have robust capabilities to support disaster relief, crisis resolution, and HLS. Capabilities include, but are not limited to, prime power, bridging, fire fighting, well drilling, and the broad spectrum of FFE. Engineers continue sustained support until civil authorities no longer require the Army's assistance. USACE supports this requirement with specialized personnel. Engineers, while supporting full-spectrum operations, may have to support civil authorities while simultaneously conducting combat operations. Engineers at the joint level are linked to interagency elements and facilitate a seamless transition between operations.

PROVIDE QUALITY, RESPONSIVE ENGINEERING SERVICES TO THE NATION

1-22. Engineers support the nation by planning, designing, building, and operating water resources and other civil-works projects (navigation, flood control, environmental protection, and disaster response) to facilitate national requirements and prevent disasters. USACE, as the Title 10 Construction Management Agent, also designs and manages the construction of military facilities for the Army and the Air Force. Engineers also provide design and construction management support for other DOD and federal agencies. This is a critical responsibility for interagency and international services.

1-23. Engineers support the Regiment's METL through operations characterized by the engineer battlespace functions at the operational level addressed in *JP 3-34*. Combat engineering enables the commander to maneuver forces freely (mobility), attack the enemy's ability to maneuver (countermobility), and support force protection (survivability). Geospatial engineering provides key geospatial information and services, including analysis products of the terrain within the operational area to enhance the picture within and throughout the theater. General engineering provides for the facilities and infrastructure that play a critical role in shaping the AO and are essential to force projection for decisive operations. Army engineer units maintain critical capabilities within each of these functions that independently may support, or be combined or leveraged with, other service engineers to support joint operations.

1-24. For a closer look at how the engineers perform these functions across the battlefield operating systems (BOSs), *see FM 7-15.* While geospatial engineering is not specifically identified in *FM 7-15* under a particular BOS, it is generally aligned with and supports the intelligence BOS. All of these relationships are further discussed in *Chapter 4* of this manual. Remember that the Army Universal Task List (AUTL) is focused on the tactical level. For linkage to joint operations, refer to the *Universal Joint Task List (UJTL), CJCSM 3500.04C.* The relationship of geospatial engineering to intelligence is better documented in this manual, *CJCSM 3500.04C.*

FULL-SPECTRUM OPERATIONS

1-25. Army doctrine includes the range of full-spectrum operations (*see Figure 1-3, page 1-10*). The 2001 Quadrennial Defense Review (QDR) describes a capabilities-based approach to defense planning that provides broader military options across the operational spectrum, from pre- to postconflict operations. The force-sizing construct–1-4-2-1–takes into account the number, scope, and simultaneity of tasks assigned to the military. It sizes the force for the defense of the US homeland by (1) providing forward deterrence in four critical regions, (4) conducting simultaneous war-fighting missions in two regions, (2) preserving the President's option to call for decisive victory in one of those conflicts, (1) and participating in multiple, smaller contingency operations. Army commanders at all echelons may combine different types of operations simultaneously and/or sequentially to accomplish missions in war and MOOTW. For each mission, the JFC and Army component commanders determine the emphasis Army forces place on each type of operation. Offensive and defensive operations normally dominate military operations in war and some smaller-scale contingencies (SSCs). Stability and support operations will tend to dominate in operations that include certain SSCs and peacetime military engagements (PMEs) and in MOOTW.

1-26. Full-spectrum operations include offensive, defensive, stability and support operations (*see Figure 1-4, page 1-11*). Missions in any contemporary environment require Army forces to conduct or be prepared to conduct any combination of the following primary operations.

OFFENSIVE OPERATIONS

1-27. Offensive operations aim at destroying or defeating an enemy. Their purpose is to impose US will on the enemy for decisive victory. The primary focus for engineers enabling operational and tactical maneuver through combat engineering support to lines of communication and combined arms and breaching river crossing operations.

DEFENSIVE OPERATIONS

1-28. Defensive operations defeat enemy attacks, buy time, economize forces, and develop conditions favorable for offensive operations. Defensive operations alone normally cannot achieve a decisive victory. Their purpose is

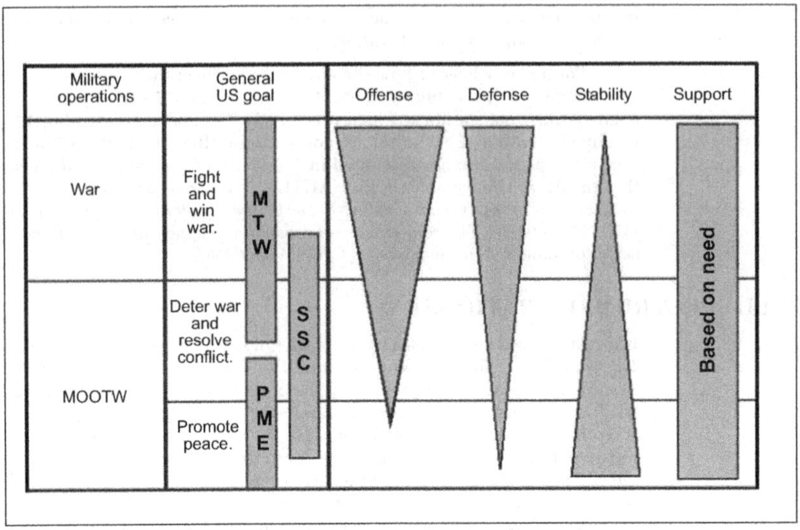

Figure 1-3. Range of Army Operations

to create conditions for a counteroffensive that will regain the initiative. The primary focus for engineers is on combat engineering to enable combined arms obstacle integration and ensure mobility to friendly repositioning or counterattacking forces.

STABILITY OPERATIONS

1-29. Stability operations promote and protect US national interests by influencing diplomatic, civil, and military environments. Engineers conducting missions provide stability, security, and resources to assist in disaster or theater response. Rapid and effectively emplaced sustainment operations reduce human injuries/fatalities and harden infrastructure. Regional security is supported by a balanced approach that simultaneously enhances regional stability and economic prosperity. Army force presence promotes a stable environment. Engineers are focused on assisting in stabilizing a region by improving the infrastructure and integrating with and supporting maneuver forces. While engineers may be focused on one or more of the engineer functions, all functions operate simultaneously in support of stability operations.

SUPPORT OPERATIONS

1-30. Support operations employ Army forces to assist civil authorities, foreign or domestic, as they prepare for or respond to crises and relieve suffering. Domestically, Army forces respond only when directed by the SECDEF. ARNG forces may be activated by order of the state governor.

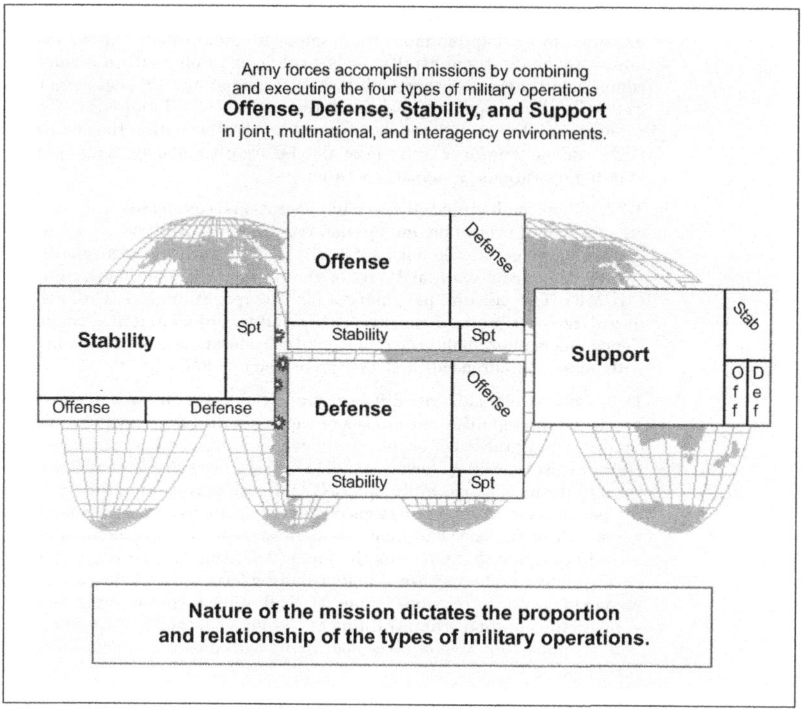

Figure 1-4. Full-Spectrum Operations

Engineers conducting missions provide stability, security, and resources to assist in disaster or theater response. Rapid and effectively emplaced support operations reduce human injuries and fatalities and harden infrastructure. Army forces operate under the lead federal agency (LFA) and comply with provisions of US law, to include the Posse Comitatus Act and the Stafford Act. Typically, the focus for engineers will be on general and geospatial engineering, although all of the engineer functions may be required.

CONDUCTING FULL-SPECTRUM OPERATIONS

1-31. When conducting full-spectrum operations, commanders combine and sequence the four types of operations, as appropriate, to accomplish the mission. As missions change from promoting peace to deterring war, and from resolving conflict to engaging in war, the combinations of and transitions between these operations require skillful planning, preparation, and execution. Engineers may also be required to support civil authorities conducting HLS. Missions can range from providing disaster relief support and humanitarian assistance to urban search and rescue. Operations designed to accomplish more than one strategic purpose may be executed simultaneously, sequentially, or both. For example, within a combatant commander's area of responsibility (AOR), combat and CS engineers may be executing large-scale offensive operations while CS and CSS engineers conduct stability operations. It is very possible that within the combat zone (CZ), various engineer units may also be simultaneously participating in stability operations or support operations.

1-32. When conducting full-spectrum operations, commanders must balance environmental protection and mission requirements. Military environmental-protection principles do not necessarily override other operational factors; rather they are a standard part of the military decision-making process (MDMP). The mission parameters for the operational area identify and quantify the time and resources devoted to environmental protection. Commanders must analyze environmental considerations and impacts in concert with mission requirements and force protection (*see FM 3-100.4*).

1-33. Commanders allocate different proportions of their force to each type of operation during different phases of the mission. Large units are likely to conduct simultaneous offensive, defensive, stability, and support operations. Units at progressively lower echelons receive missions that require fewer combinations. For example, an ENCOM supporting the joint force land component may allocate two corps engineer battalions to the attack (offense), while a third engineer battalion assists in securing and upgrading a port and airfield complex that sustains the force (defense). Around the airfield and port, engineer units may assist in distributing food and providing facilities for medical support to refugees (support). Still other corps engineer units may support the equipping and training of HN forces (stability). Engineer units will be supporting each of these operations with one or more of the engineer functions.

1-34. In the same manner that commanders combine and sequence the four types of military operations, engineers will contribute varying portions of each of the three engineer battlespace functions (combat, general, and geospatial) to each of the four types of military operations. The type, variety, and relative percentages will vary from mission to mission across the spectrum of conflict. *Figure 1-5* highlights the generic relationships between and among the engineer functions and the four types of military operations.

1-35. Everyday, the Army trains soldiers and units while developing leaders. Battle-focused training (BFT) for combat, CS, and CSS tasks prepares soldiers, units, and leaders to fight and win the nation's wars. Training is the cornerstone of success. It is a full-time job for commanders in peacetime, and it continues while deployed in full-spectrum operations as well. Training to high standards is essential for a versatile force. Every commander, soldier, and unit in a strategically responsive Army must be trained and ready to deploy, fight, and win. Upon alert, engineer forces deploy immediately, conduct operations, and complete any mission-specific training in country, if necessary. Commanders conduct abbreviated mission rehearsal exercises (MREs)... if necessary, based on available time and resources. The expectation is that engineer units train, alert, deploy, and sustain.

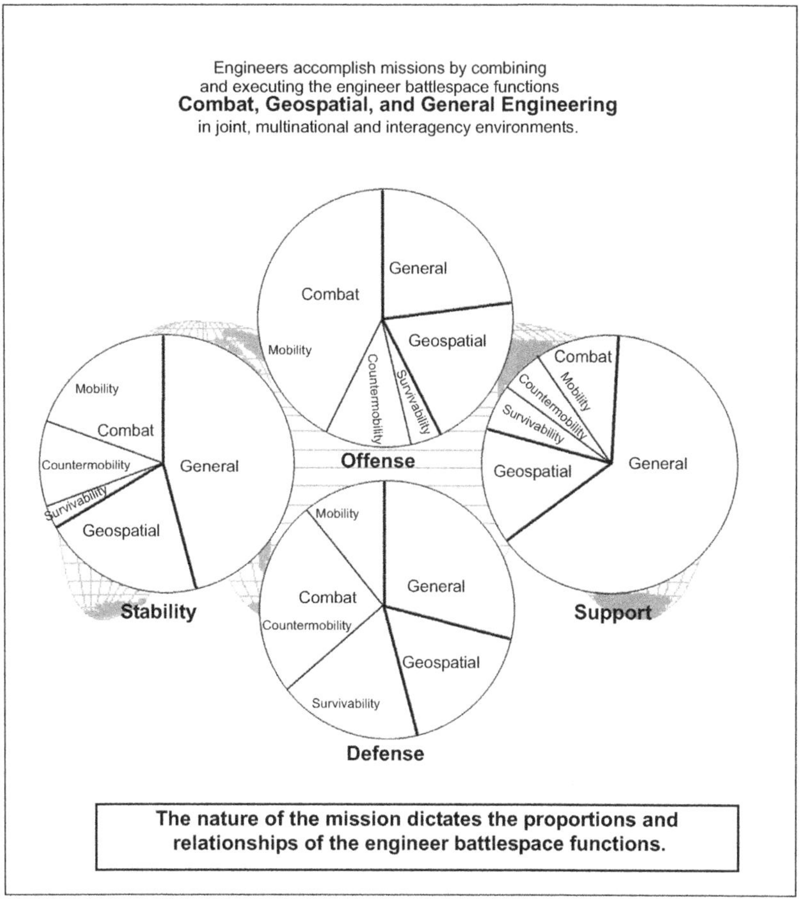

Figure 1-5. Engineer Contributions Across the Spectrum of Conflict

1-36. The METL development process outlined in *FM 7-0 (25-100)* and *FM 7-1 (25-101)* describes the link between mission and training. Commanders focus their METL, training time, and resources on combat tasks, unless directed otherwise. Because Army forces face diverse threats and mission requirements for full-spectrum operations, commanders may need to adjust their METL temporarily from battle-focused tasks and manage their training focus and resources to attain a readiness posture for anticipated missions. MACOM and numbered Army and corps commanders determine the battle focus, resources, and METL that maintain the required readiness posture for anticipated offense, defense, stability, and support (ODSS) operations in war or MOOTW. MACOM commanders decide for operational-level units, and corps commanders decide for corps units. Commanders at lower levels conduct BFT unless otherwise directed. Geospatial and general engineering tasks performed tend to be very similar across the spectrum of conflict for offensive, defensive, stability, or support missions. The primary variable is the condition, with some adjustment to the standard.

1-37. The full-spectrum operations training process is all a part of the cycle—train, alert, (train), deploy, and sustain. After alert, additional specialized training may be required. An example is the training to meet the unique or specialized skills associated with a mission to support another government agency in performing a firefighting mission.

BATTLEFIELD ORGANIZATION

1-38. Within the discussion of the operational framework *(see FM 3-0)* is the concept of the battlefield organization. The battlefield organization *(see FMs 3-0 and 3-07)* is allocating forces in the AO by purpose. Battlefield operations consist of three all-encompassing categories of operations: decisive, shaping and sustaining. The purpose unifies all elements of the battlefield organization by providing common focus for all actions. Commanders organize forces according to purpose by determining whether each unit's operation will be decisive, shaping, or sustaining. This decision is the basis of the concept of operations. When circumstances require a spatial reference, commanders describe the AO in terms of deep, close, and rear areas. These spatial categories are especially useful in operations that are generally contiguous and linear and feature a clearly defined enemy force. Combat engineering primarily tends to support the categories of decisive and shaping operations. Geospatial engineering supports all of the categories with a primary focus on the categories of decisive and shaping operations. General engineering may support all three, and while it is not a separate engineer function, sustainment engineering often describes those engineer missions that support the category of sustaining operations.

DECISIVE OPERATIONS

1-39. Decisive operations are those that directly accomplish the task assigned by the higher headquarters (HQ). They conclusively determine the outcome of major operations, battles, and engagements. In stability operations, decisive operations are usually those that achieve and maintain stability, protect lives and property, or promote peace. Unlike decisive operations in offense or defense, decisive operations in stability operations do not always have immediate impacts; sometimes it takes years to achieve results. In support operations, decisive operations normally prevent or mitigate the effects of natural or man-made disasters. They relieve or reduce conditions such as disease, hunger, or privation. Decisive operations could include stabilizing areas by providing security for personnel, facilities, or capabilities; rendering certain services to populations; or reestablishing critical infrastructure. In certain missions, the collective engineer missions may constitute the decisive operation. In all decisive operations, commanders must analyze environmental considerations and impacts in concert with mission requirements and force protection. As with other considerations, the importance of environmental considerations should be clearly articulated in the higher commander's guidance.

SHAPING OPERATIONS

1-40. Shaping operations, at any echelon, create and preserve conditions for the success of decisive operations. Shaping operations include lethal and nonlethal activities conducted throughout the AO. They support the decisive operation by affecting enemy capabilities and forces or by influencing enemy decisions. Shaping operations use all elements of combat power to neutralize or reduce enemy capabilities. They may occur before, concurrently with, or after the start of the decisive operation. They may involve any combination of forces and occur throughout the AO. In stability operations, shaping operations often convert temporary gains into long-term political successes. The capabilities required to exploit stability often differ from those needed to achieve stability. This may require such measures as rotating different types of units in and out of the AO as the operation progresses through different stages. In support operations, shaping operations may include influencing perceptions, ideas, and information, as well as maintaining legitimacy. Shaping operations will entail transferring tasks to civilian agencies or the local government. For military engineers, this may also include transferring responsibilities to USACE, civilian contract engineers, or HN organizations.

SUSTAINING OPERATIONS

1-41. Sustaining operations are operations at any echelon that enable shaping and decisive operations by providing CSS, rear area and base security, movement control, terrain management, and infrastructure development. Engineering performed in support of sustainment operations is often referred to as sustainment engineering. The vast majority of this support will be general engineering and include at least some geospatial engineering.

SECTION II - ENGINEER BATTLESPACE FUNCTIONS

1-42. The five traditional engineer functions are organized at the operational level into three engineer battlespace functions: combat (M/CM/S), geospatial, and general engineering. They are critical in understanding how the Regiment supports the Army and the foundation of engineer doctrine. The current trend of military operations is an ever increasing US involvement in global responses. Operations Restore Hope and Continued Hope in Somalia, Operation Restore Democracy in Haiti, and Operation Joint Endeavor in Bosnia are all examples of the expanding role of the Engineer Regiment in support of the Army and joint operations. Couple this with the global trend toward quantitatively smaller but technologically and qualitatively better military forces and the need for relevant doctrine becomes apparent. Engineer integration into staff planning requires increased emphasis, since synchronizing the operation or battle is increasingly complex. Engineer C2 must occur rapidly to be responsive to the dynamics of an ongoing operation. Integrating environmental considerations must begin early in the MDMP to ensure that appropriate emphasis is given to these factors. Terrain analysis and its products assist in faster planning and provide a unique perspective of the AO. Requirements for fortifications and protective shelters and the Army's standards for the quality of life (QOL) for its soldiers have gained importance. Obstacle systems retain their importance, while aggressive doctrine within *FM 3-0* places greater emphasis on operational mobility, obstacle breaching, and rapid gap crossing. It is the engineer battlespace functions that will allow the Regiment to address these needs.

COMBAT ENGINEERING

1-43. Combat engineering (M/CM/S) is focused on the support of combined arms maneuver. Combat engineering enhances operational movement, maneuver, and force protection by facilitating M/CM/S operations. At the tactical level, these three subordinate functions are the basis for the engineer tasks included in the M/CM/S BOS, as articulated in *FM 7-15*. Combat engineering, as well as general engineering tasks, provide operational mobility tasks and countermobility to the force at large. The distinction is made by what level of effects are gained from the engineer tasks accomplished. At the operational level, engineer mission analysis must be closely linked with higher-level logistics and movement control planning to ensure that the force sustainment is not encumbered from a lack of operational mobility. The principles of engineer operations still apply, regardless of the level at which they are conducted.

MOBILITY

1-44. Mobility is the first component of combat engineering and includes operations ranging from engineer reconnaissance to countermine/ counterobstacle operations and from gap crossing to the construction of combat roads and trails. It enables the force commander to maneuver tactical units into advantageous positions over the enemy. In the attack, engineers aggressively execute drills to reduce enemy obstacles and assault and destroy enemy fortifications. The commander designates routes for ground forces well in advance of their intended use so that engineer units can upgrade the routes, as necessary, and keep them open or repaired. Engineers may be involved with port construction and improvement of the reception and staging areas for deploying forces. Engineers also prepare field sites that the Army, Air Force, and Marine Corps aviation assets use to support operations. *(See FMs 3-34.2 [90-13-1]* and *90-13* for more information on techniques and procedures for mobility.) Mobility applies at strategic, operational, and tactical levels of war.

1-45. Mobility enables the force commander to maneuver units into advantageous positions. The Army commander relies on mobility to achieve surprise and mass at the critical time and to maintain momentum. Operation Desert Storm provides an excellent example of operational mobility (typically included under general engineering). As it became apparent that Iraqi forces were content to occupy Kuwait and brace for the coming attack, COCOM planners formulated the offensive plans for the sweep north that included two Army corps. The turning movement around Kuwait required displacing the XVIII Airborne Corps from its defensive positions in central Saudi Arabia to the Northwest on the Iraqi border. Marshaling areas to upload tracks and road enhancements allowed for the rapid displacement of the corps into tactical assembly areas (AAs) without providing the Iraqi force with a hint of the coalition's intentions. Army nondivisional engineers aided in breaching the elaborate Iraqi defensive system, thereby allowing divisional engineers to remain integrated with the maneuver force. In a similar vein, the bridge across the Sava River into Bosnia (and the accompanying crossing site support areas) displayed the criticality of operational mobility during Operation Joint Endeavor.

1-46. Mobility includes countermine/counterobstacle, gap crossing, combat roads/trails, forward aviation combat engineering (FACE), and engineer reconnaissance. Obstacle reduction and lane marking equipment are used to increase synergy and survivability on the future battlefield.

COUNTERMOBILITY

1-47. Countermobility is the second component of combat engineering. It augments natural terrain with obstacle systems according to the commander's concept, adding depth to the battle in space and time by attacking the enemy's ability to maneuver its forces. With its movement disrupted, turned, fixed, or blocked, the enemy is vulnerable to friendly forces. Countermobility includes mine systems, obstacle development, and emplaced obstacle planning and control, which will be enhanced through digital communications links in the future. Intelligent minefields with turn on/turn off and sensor capabilities will provide real time intelligence along with increased situational awareness to the combined arms team in the future as well. Engineers advise the force commander on the best means to reinforce the terrain and emplace obstacles that support the force commander's plan. *FM 90-7* is the primary reference for countermobility planning; however, for more information on tactics and techniques for countermobility, *see FM 20-32* and *JP 3-15.*

1-48. Engineers ensure obstacle integration through the proper exercise of obstacle C2, focusing on obstacle emplacement authority, the authority that a unit commander has to emplace reinforcing obstacles, and obstacle control. The LCC or ARFOR commander usually has the authority to emplace obstacles. Generally, the ARFOR commander delegates the authority to corps commanders who may further delegate it to division commanders. Obstacle control ensures that obstacles support current and future operations. The ARFOR commander's control mechanisms to ensure that subordinate commanders do not emplace obstacles which will interfere with future operations are known as obstacle zones and obstacle-free restrictive areas. The nature of obstacle integration from the ARFOR level to company and/or team level leads to echelonment in obstacle planning. At each lower level, engineers conduct more detailed planning. Operational planning consists of developing obstacle restrictions and granting obstacle emplacement authority to subordinate elements. Obstacle planning requires engineers at each level to provide subordinate units with the right combination of positive control and flexibility. The engineer is also an important advisor/partner in deep-targeting discussions and the coordination focal point concerning obstacle barriers and mines for joint managers and coalition forces. Timely, accurate reporting of obstacles from the emplacement unit all the way to the ARFOR HQ—

- Enhances fratricide avoidance—the fourth component of protection *(see FM 3-0)*.
- Allows for dissemination as boundaries change or units pass through areas occupied by others.
- Provides critical information in planning the forward passage of lines (FPOL) and rearward passage of lines (RPOL).
- Enhances demining operations at the conclusion of contingency operations.

SURVIVABILITY

1-49. Survivability is the third component of combat engineering. It provides cover and mitigates the effects of enemy weapons on personnel, equipment, and supplies while simultaneously deceiving the enemy regarding the intentions of the force. Survivability operations range from employing camouflage, concealment and deception to the hardening of facilities, C2 nodes, and critical infrastructure. Engineers may be called on to mass their skills and equipment to augment combat units in developing defensive positions into fortifications or strong points and improve defensive positions. More often, however, engineers participate in and provide staff advice on camouflage, concealment, and deception (CCD) measures and the hardening of facilities to resist the destruction of C2 facilities (as part of integrated plans), air-defense weapons systems, and support structures within the communications zone (COMMZ). Within a missile threat environment, engineers provide field fortification support to harden key assets against missile attacks. ADC and force protection includes survivability engineering applications to HN facilities and US-operated facilities as protective measures against terrorist or extremist groups that threaten US forces or national interests. *See FM 3-34.112 (5-103)* for more information on techniques and procedures for survivability and force protection. Survivability also includes providing concealment and protective shelter from the effects of enemy weapons. Engineers—

- Have the technical knowledge, skills, and equipment to assist other units in developing defensive positions into fortifications and improve defensive positions.
- Provide technical advice on camouflage.
- Prepare fighting positions beyond the combat units' organic capabilities.
- Harden facilities to resist destruction by the enemy.
- Provide the equipment support necessary to establish nuclear, biological, and chemical (NBC) decontamination points and assist in route and area decontamination.
- Use digital position navigation systems to mark survivability positions positively on the battlefield.

GEOSPATIAL ENGINEERING

1-50. Geospatial engineering support is provided based on requirements that are determined and validated by the geospatial information and services (GI&S) officer at the joint level and by the engineer filling that role at tactical levels. That support encompasses those tasks that provide GI&S to commanders and staffs across the full range of military operations. Geospatial engineering provides terrain analysis, digitized terrain products, nonstandard tailored map (NTM) products, map production, precision survey, terrain data management, and baseline survey data—all of which significantly contribute to providing a COP within the joint force. This provides products that assist in identifying key and decisive terrain as well as feasible courses of action (COAs) from a terrain perspective. NIMA produces digital terrain and feature data. The Defense Logistics Agency (DLA) distributes this data to units. The geospatial engineer can request imagery (both NTM and commercial) sources through the Assistant Chief of Staff, G2, (intelligence) (G2) or directly, which can be used for spatial and temporal reasoning or multispectral analysis (CCD or inframetric) products that are customized to meet particular operational requirements. Imagery enhances three-dimensional (3-D) and fly-through perspectives. This support is developed and provided during all phases of an operation throughout the theater of operation (TO), in accordance with *JP 2-03*.

1-51. All engineer officers are charged to be terrain experts and advise commanders on how to conceptualize the battlespace effectively. They must be supported by terrain analysts to fully assist others to use the terrain more effectively. As the proponent for mobility, engineers identify and/or recommend avenues and routes, restricted terrain, danger areas, and potential obstacle locations, environmental hazards, EAs, and unit battle positions. During times of crisis or operations, source data may come directly from NIMA or coalition partners. Terrain teams use digital terrain data to develop a detailed terrain analysis. *See FM 3-34.230 (5-105)* for more information on geospatial techniques and procedures.

1-52. Geospatial engineering provides commanders with information about the terrain and assists them in conceptualizing the battlespace more accurately to make knowledgeable decisions. At the ARFOR or LCC level, the topographic battalion and the planning and control detachment integrate echelons above division (EAD) support to Army forces as well as joint and multinational forces. The staff of the Stryker brigade combat team (SBCT) contains an organic terrain team that provides products to support the brigade commander. The heavy division staff's terrain teams are distributed down to brigade level and provide products for division- and brigade-level planning. Creative and productive use of digital terrain data aids the ARFOR or LCC's visualization of the battlefield, by determining factors such as, but not limited to—

- Avenues and routes for friendly/enemy forces.
- Terrain limitations to enemy capabilities.
- Obstacle zone locations.
- Environmentally significant areas (water resources, hazards).
- Major EAs.

- Unit positions.
- Deep operation targets and their impact on future operations.
- Rescue operation parameters.
- Flood prediction models.
- Mission planning and rehearsal data.

1-53. *See FM 3-34.230* and *JP 2-03* for more information on geospatial support to Army and joint systems. GI&S support requirements, products, and capabilities are included in the GI&S annex (joint) or appendix (Army) to each contingency plan (CONPLAN), operation plan (OPLAN) or operation order (OPORD).

GENERAL ENGINEERING

1-54. General engineering encompasses the construction and repair of LOCs, main supply routes (MSRs), airfields, utility systems, and logistic facilities to support joint and Army military operations. It may be performed in direct support (DS) of combat operations such as battle damage repair (BDR). Facilities are fundamental to the success of force projection and the conduct of military operations and play a critical role in shaping the AO and infrastructure to support the force. When facilities and real estate can be obtained through HNS and commercial leases, or through international agreements (SOFAs), facilities acquisition and real estate management become an important component of general engineering. For further information, *see FM 5-104* and *JP 4-04.*

1-55. Typically, general engineering helps establish and maintain the infrastructure necessary for sustaining military operations in theater. General engineering tasks—

- May include construction or repair of existing logistics-support facilities, supply and line of communication (LOC) routes, airfields, ports, water wells, power plants, pipelines, and base camps/force bed down.
- May be performed by a combination of joint engineer units, civilian contractors and HN forces.
- Usually require large amounts of construction materials, which must be planned and provided for in a timely manner.
- May include the production of construction materials.

1-56. *See FM 5-104, FM 3-100.4;* and *FM 7-15* for additional information on general engineering. General engineering tasks include—

- Constructing and/or repairing the following:
  - Existing logistics support facilities.
  - Supply routes, airfields, and heliports.
  - Railroads.
  - Ports.
  - Water wells.
  - Utilities (electric, heat, and water) and sanitation (sewage, hazardous waste, and solid waste).

  - Power plants.
  - Pipelines.
  - Base camps/force bed down.
- Providing electrical distribution expertise.
- Providing expertise on environmental considerations and protection.
- Supporting ADC.

1-57. General engineering support begins with the supporting requirements for the initial reception of the force projection Army (receiving equipment and soldiers).

- Is maintained throughout the operation, providing the infrastructure for the logisticians to sustain the force.
- Provides the support structure to redeploy the force.
- Ends with environmental restoration and the return of the facilities that were used by the deployed forces to HN control.

1-58. As the force advances forward on the battlefield, the corps's rear boundary will be drawn forward continually. Therefore, general engineering also invokes force protection through operations such as demining after the tactical breach and assisting explosive ordnance teams in clearing unexploded ordnances (UXOs) within the expanded COMMZ to the extent necessary to conduct military operations safely.

1-59. At times, the military strategy may be to extend general engineering support to restore facilities, power, and life-support systems within the infrastructure of the combatant countries. This effort aids in the recovery and the transition to preconflict conditions. Central to planning and executing these tasks is construction standards. The challenge is in establishing measures of success and conditions for transitioning to civil support. Within the modern framework of operations, all these efforts are likely to be performed by a combination of joint engineer units, civilian contractors, and HN forces. Once again, these efforts require that large amounts of construction materials and specialized resources be planned and provided for in a timely manner.

1-60. Army leadership can achieve and enforce environmental protection standards more easily when environmental considerations are integrated into the Army's decision making process and activities. Planning for all Army operations and strategies should include efforts to minimize the release of hazardous substances into the environment, protect cultural and natural resources, and prevent pollution. By maintaining environmental stewardship, relations with the HNs are enhanced and cleanup efforts in postconflict periods can be minimized. Guidelines on requirements for overseas operations are outlined in the SOFA with HNs or the DOD overseas environmental baseline guidance document (OEBGD). *See FM 3-100.4* for more information.

SECTION III - UNIFIED ACTION: JOINT/INTERAGENCY/MULTINATIONAL ENVIRONMENT

DESCRIPTION

1-61. In full-spectrum operations, Army engineers operate as part of a joint force and often within a multinational and interagency environment. The term *Unified action* describes the wide scope of actions (including the synchronizing of activities with governmental and nongovernmental agencies) taking place within unified commands, subordinate unified commands, or joint task forces (JTFs) under the overall direction of the commanders of those commands. Public law charges combatant commanders with employ military forces through unified action. Under unified action, commanders integrate joint, single-service, special, and supporting operations with interagency, nongovernmental, and multinational forces (to include United Nations [UN]) operations *(see JP 0-2).*

1-62. Unified action transfers subordinates to the combatant commander under COCOM authority. Multinational, interagency, and nonmilitary forces work with the combatant commander through cooperation and coordination. Regardless of the task or the nature of the threat, combatant commanders employ air, land, sea, space, and SOFs and coordinate with multinational and interagency partners to achieve strategic and operational objectives. They formulate theater strategies and campaigns, organize joint forces, designate operational areas, and provide strategic guidance and operational focus to subordinate commanders. The aim is to achieve unity of effort among many diverse agencies in a complex environment. Subordinate JFCs synchronize joint operations in time and space, direct the action of other military forces (multinational operations), and coordinate with governmental and nongovernmental organizations (NGOs) (interagency coordination) to achieve the same goal. The conduct of operations by ENCOMs is almost always in a joint environment.

THE LEVELS OF WAR

1-63. The three levels of war—strategic, operational, and tactical—are doctrinal perspectives that clarify the links between strategic objectives and tactical actions, although there are no finite limits or boundaries between them. Understanding the interdependent relationship of all three levels helps commanders visualize a logical flow of operations, allocate resources, and assign tasks. Actions within the three levels are not associated with a particular command level, unit size, equipment type or force, or component type. Instead, actions can be defined based on their effect on or contribution to achieving strategic, operational, or tactical objectives *(see Figure 1-6).* Engineers support all three levels of war. The senior engineer at a given echelon typically supports the commander at that level, but in some cases, the engineer may be the commander. We will use the framework of the levels of war in our discussion of engineer operations.

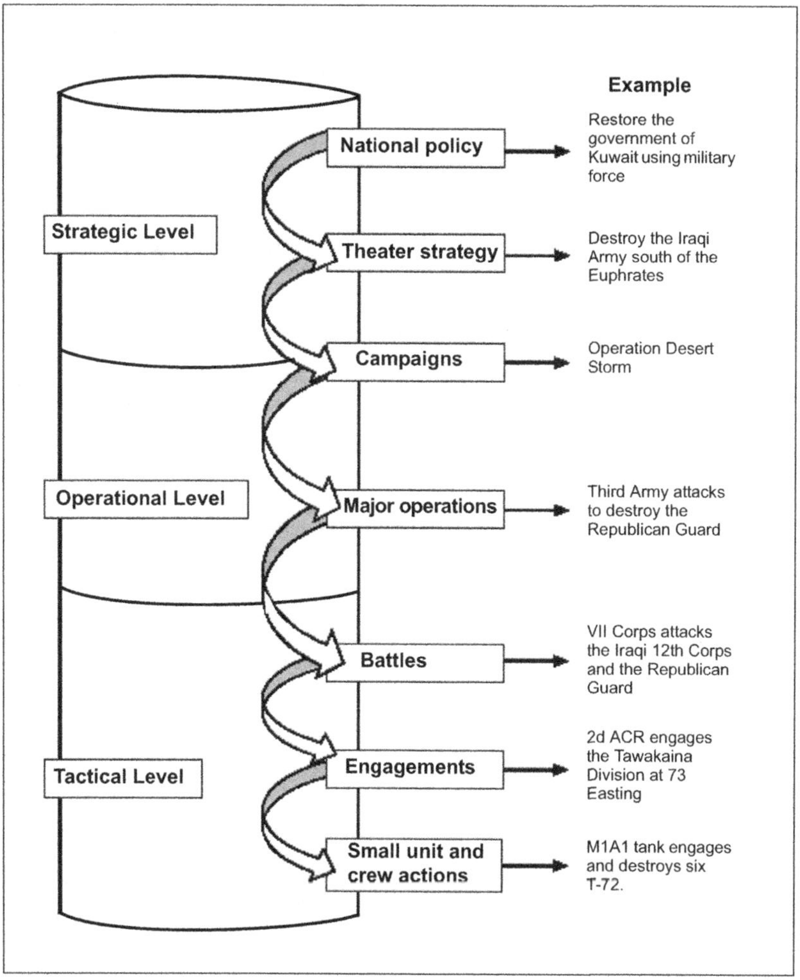

Figure 1-6. Levels of War

THE STRATEGIC LEVEL

1-64. The strategic level is the level of war in which a nation, often as a member of a group of nations, determines national and multinational security objectives and guidance and develops and uses national resources to accomplish these objectives. The term *strategy* is the art and science of developing and employing armed forces and other instruments of national power in a synchronized fashion to secure national or multinational objectives. The Office of the Secretary of Defense translates policy into national strategic military objectives. These military objectives facilitate theater strategic planning. Military strategy, derived from policy, is the basis for all operations *(see FM 3-0* and *JP 3-0).* The engineer organizations that operate at this level are typically the ENCOMs and USACE.

1-65. Engineer activities include force planning, engineer policy and doctrine, and planning and execution of operations and exercises. During operations, strategic engineer activity focuses primarily on the means and capabilities to generate, mount, sustain, and recover forces. Additionally, infrastructure development is a critical aspect of enabling and sustaining force deployments and places a heavy demand on engineer requirements. Engineers advise on terrain, infrastructure, force generation, priorities of engineer support, engineer considerations affecting joint targeting, humanitarian assistance, environmental protection, engineer interoperability, input to the rules of engagement (ROE), and force protection.

THE OPERATIONAL LEVEL

1-66. The operational level of war is the level at which campaigns and major operations are planned, conducted, and sustained to accomplish strategic objectives within theaters or AOs. It links the tactical employment of forces to strategic objectives. The focus at this level is on *operational art*—the use of military forces to achieve strategic goals through the design, organization, integration, and conduct of theater strategies, campaigns, major operations, and battles. Operational art determines when, where, and for what purpose major forces will be employed to influence the enemy disposition before combat. It governs the deployment of those forces, their commitment to or withdrawal from battle, and the arrangement of battles and major operations to achieve operational and strategic objectives. The engineer organization operating at this level and functioning as the senior engineer is typically the ENCOM or the corps engineer brigade.

1-67. Operational art is translated into an OPLAN through operational design. A well-designed plan and successfully executed operation shape the situation for tactical actions. Executed skillfully, a good plan increases the chances of tactical success by creating advantages for friendly forces and disadvantages for the enemy. A flexible plan gives tactical commanders freedom to seize opportunities or react effectively to unforeseen enemy actions and capabilities. Flexible execution maintains the operational initiative and maximizes tactical opportunities *(see FM 3-0).*

1-68. Joint operational engineers assist in preparing and shaping the joint operational area. They develop geospatial products for the commander and staff and develop recommendations to the commander on joint fires, preserving and protecting the force, and sustaining the infrastructure.

THE TACTICAL LEVEL

1-69. Tactics is the employment of units in combat. It includes the ordered arrangement and maneuver of units in relation to each other, the terrain and the enemy, to translate potential combat power into victorious engagements and battles. An engagement is a small tactical conflict between opposing maneuver forces, usually conducted at brigade level and below. Engagements are usually short, lasting minutes, hours, or a day. A battle consists of a set of related engagements that last longer and involve larger forces than an engagement. Battles can affect the course of a campaign or major operation *(see FM 3-90* for more information). The tactical echelon is where many engineer commanders and units, especially combat and CS engineers, will operate.

1-70. Tactical engineers focus on providing geospatial products for the commander to conceptualize the battlespace and make decisions. They provide combat engineering support to tactical maneuver forces. Tactical-level engineers are not solely an activity in the land component. They may support amphibious operations by developing beachheads and air operations by constructing and repairing airfields.

JOINT/INTERAGENCY/MULTINATIONAL OPERATIONS

1-71. In unified action, Army forces synchronize their actions with other participants to achieve unity of effort and accomplish the combatant commander's objectives. The capabilities of joint, multinational, and interagency partners expand strengths, compensate for limitations, and provide operational and tactical depth to Army forces. Given their unique capabilities and the flexible nature of the engineer branch to perform combat, CS, and CSS functions, Army engineers are frequently the unit of choice committed to support these operations. The Engineer Regiment's METL supports these operations as well.

JOINT OPERATIONS

1-72. An operation that includes forces of two or more services under a single commander is a joint operation. Land operations and joint operations are mutually enabling—land operations are inherently joint operations. Joint integration allows the JFC to attack an opponent throughout the depth of a JFC's AO to seize the initiative, maintain momentum, and exploit success. Effective joint integration does not require joint commands at all echelons but does require an understanding of joint synergy at all levels of command. Joint synergy extends the concept of combined arms synergy familiar to land forces. The strengths of the different services combine to overcome each service's limitations and reinforce each other's effects. The combination of multiple and diverse joint force capabilities creates military power more potent than the sum of its parts. The focus of joint operational engineers is to provide the combatant commander operational maneuver options through repair maintenance and construction of aerial pods of debarkation (APODs) and sea port of debarkation (SPODs), enabling river crossing operations and LOC repair or construction. Army engineers provide unique capabilities and strengths to the joint arena. The primary focus of joint engineer operations is to achieve the commander's intent by coordinating engineer support throughout the joint AO.

THE OTHER ARMED FORCES

1-73. US armed forces defend US national interests worldwide. The Constitution commits the federal government and, by extension, the armed forces to protecting the US and US territories. Through Title 10, United States Code (USC) and DOD Directive (DODD) 5100.1, Congress organized the national defense and defined the function of each branch of service. All US armed forces—Army, Air Force, Navy, Marine Corps, Coast Guard, and SOFs—are required to provide globally responsive assets to support combatant commanders' theater strategies and the national security strategy. The capabilities of the other armed forces complement those of Army forces. During joint operations, they provide support consistent with the JFC's directed missions.

EMPLOYING ARMY ENGINEERS IN JOINT OPERATIONS

1-74. Joint doctrine describes the employment of all US military forces in joint operations. Army force commanders are always subordinate to a JFC or are, themselves, designated as the JFC with elements from the other armed forces subordinate to them. The C2 relationships between those forces are the key to effective joint military operations.

Army Engineers in Unified Commands

1-75. Except for forces exempted by the SECDEF, military departments assign all forces—to include nonfederalized ARNG and unmobilized USAR forces—under COCOM of combatant commanders. The Joint Strategic Capabilities Plan apportions major Army forces, by type, to combatant commanders for deliberate planning. In addition to forces assigned in peacetime, Army forces are allocated to combatant commanders in response to crises. The SECDEF, through the Chairman, Joint Chiefs of Staff (CJCS), directs other combatant commanders to reinforce the supported combatant commander with augmentation forces.

Chain of Command

1-76. The SECDEF exercises authority and control of the armed forces through a single chain of command with two branches. One branch goes from the SECDEF to combatant commanders to the various service component commands and subordinate joint commands for the conduct of operations and support. The other branch goes from the SECDEF to the military departments to their respective major service commands. An administrative control (ADCON) relationship exists between the secretary of the military department to the respective service component commands to carry out their Title 10 responsibilities of recruiting, manning, equipping, training, and providing service forces to the combatant commanders. Although the service branch of the chain of command is separate and distinct from the operating branch, the Army service component commander (ASCC) and the ARFOR operate within the combatant commander's chain of command in the theater.

Command Relationships

1-77. At the theater level, when Army forces operate outside the US, they are assigned under a JFC *(see JP 0-2; JP 3-0;* and *FM 100-7).* A JFC is a combatant commander, subunified commander, or JTF commander. At the theater level, the combatant commander provides strategic direction and operational focus to forces by developing strategy, planning the theater campaign, organizing the theater, and establishing command relationships for effective unified action. The JFC plans, conducts, and supports the campaign in the theater of war, subordinate theater campaigns, major operations, and battles. The four joint command relationships are COCOM, OPCON, tactical control (TACON), and support with only a combatant commander authorized to exercise COCOM over a joint force. *Chapter 7* contains a more in-depth discussion of C&S relationships and their application at all echelons for engineers.

The Army Service Component Commander

1-78. An ASCC is responsible for Army Title 10 requirements in support of a combatant commander. This includes recruiting, organizing, supplying, equipping, training, servicing, mobilizing, demobilizing, administering, and maintaining the construction, outfitting, and repair of military equipment; the construction, maintenance, and repair of buildings, structures, and utilities; and the acquisition of real property. The ASCC may also be responsible for significant DOD and combatant commander-designated common-user logistics (CUL) functions. The ASCC provides administrative and logistic services to assigned Army forces and the ARFORs of subordinate JFCs. When appropriate, the ASCC may delegate authority for support tasks to a single theater support command (TSC) or another subordinate Army HQ, such as the ENCOM or the medical command (MEDCOM), when the focus of support suggests this as the best solution. USACE is also often involved with supporting the ASCC, and will generally operate through the ENCOM, if one is present.

The Army Force

1-79. Designated ASCCs, other numbered armies, and corps HQ can serve as the base for ARFORs, joint force land components, and JTFs. Division HQ with augmentation may serve as the ARFOR to a JTF in SSCs. Army force HQ can perform in a variety of capacities and rapidly reorganize to meet diverse mission requirements. This versatility is a prerequisite for full-spectrum operations. However, all Army organizations require augmentation to function as a joint HQ. The optimum ARFOR organization and C2 structure is based on the mission, the OE, and the JFC's C2 requirements. It is entirely possible that in some specialized support missions, the senior engineer may also be the ARFOR commander. The ARFOR staff engineer at echelons above corps (EAC) is not dual-hatted as the staff engineer and commander. There is no organic engineer commander at these echelons. Therefore, the senior engineer commander task-organized at that level will be responsible for the

C2 of all subordinate engineer units, and the staff engineer will continue to synchronize and plan engineer operations.

INTERAGENCY COORDINATION

1-80. The intricate linkages among the instruments of national power demand that commanders consider all capabilities and agencies to help achieve the common end state. Interagency coordination forges a vital link between military operations and activities conducted by such organizations as US government agencies; agencies of partner nations; NGOs; and regional, international, and UN organizations, as well as agencies of an HN. Theater strategies routinely employ the capabilities of the entire US interagency network. The National Security Act of 1947 establishes an interagency process for national security-related issues. The National Security Council provides national oversight of the interagency process *(see JP 3-08)*. Because of their unique capabilities, combat, CS, and CSS engineers will frequently be involved with interagency coordination involving all of the engineer functions.

MULTINATIONAL OPERATIONS

1-81. Although the US sometimes acts unilaterally, it pursues its national interests through alliances and coalitions when possible. In Operation Desert Shield and Operation Desert Storm, more than 800,000 military personnel from 36 nations combined their wills, forces, and resources to oppose the Iraqi armed forces. The coalition increased the size of the overall force, shared the cost of waging the war, and enhanced the legitimacy of the strategic aims. These multinational operations demonstrated the advantage of successful multinational warfare over the unilateral efforts of a single state. It is the responsibility of the senior engineer commander to coordinate engineer operations and ensure interoperability issues. Multinational operations are conducted within the structure of an alliance or a coalition *(see JP 3-16* and *FM 100-8)*.

Chapter 2

The Operational Environment

A general in all his projects should not think so much about what he wishes to do as much as what his enemy will do; that he should never underestimate this enemy, but he should put himself in his place to appreciate difficulties and hindrance the enemy could interpose; that his plans will be deranged at the slightest event if he has not foreseen everything and if he has not devised means with which to surmount the obstacles.

Frederick the Great: Instructions to his Generals, iii, 1747

To win the next war, we have to think differently.

President George W. Bush 2001

An OE, as defined in *JP 1-02,* is a composite of all conditions, circumstances, and influences which affect the employment of military forces and bear on the decisions of the unit commander. An OE consists of both conventional and unconventional threats. These threat strategies dictate the key variables affecting operations, define the characteristics of the AO, and link to their strategic aims and will use a broad range of ways and means to enable their strategies. Operational and tactical designs are the manifestation of the application of those ways and means. *JP 1-02* also states that the contemporary operational environment (COE) is the overall operational environment that exists today and in the near future (out to the year 2020). It is imperative that engineers understand the OE to best inform the commander at all levels how to effectively employ engineer capabilities to support the Engineer Regimental METL.

GENERAL DESCRIPTION

2-1. The general current OE that exists in today's world is expected to exist until a peer competitor (similar to the Soviet Union during the Cold War) arises. There are multiple threats to US interests, ranging from regional powers to transnational groups interested in terrorism, illicit drug trading, or other illegal activities. They are a mix of radicalism and advanced technological capabilities. This broad range of activities threatens US interests and citizens at home and abroad.

2-2. Our current force of corps and divisions is resourced, designed, organized, and equipped to deploy to a mature theater, with HNS. Before the publication of *FM 3-0*, the Army was trained to meet an echeloned, doctrinally based enemy who shaped a linear battlefield with complex obstacles. Now we are challenged to deploy to austere, underdeveloped theaters to face either conventional forces or nonlinear, asymmetric terrorist adversaries across the spectrum of conflict.

CRITICAL VARIABLES

2-3. The Army and its engineers will encounter a variety of conflicts in a number of different OEs. The OE is a function of a number of variables. There are eleven critical factors or variables that define the nature of the OEs in which those conflicts or other US military activities may occur. These factors are variables, because the exact nature of the conditions, circumstances, and influences that make up the OE will vary according to the particular situation. The variables include—

- Physical environment.
- Nature and stability of the state.
- Military capabilities.
- Technology.
- Information.
- External organizations.
- Sociological demographics.
- Regional and global relationships.
- National will.
- Time.
- Economics.

2-4. These variables are interrelated and sometimes overlap. Different variables will be more or less important in different situations, but they are all common to any OE. Nevertheless, the collective content of these variables will define any OE the Army and its engineers could face, whether they are involved in PME, SSCs, or major theater war (MTW). Each OE is different because the content of the variables is different. The most difficult aspect of analyzing the OE is that the variables will evolve over time. They will not remain fixed but will morph and change. Therefore we can expect the physical environment we are operating in to change over time.

ASYMETRIC, ADAPTIVE APPROACHES

2-5. The concept of asymmetric warfare is critical to understanding the COE. Asymmetry is a condition of ideological, cultural, technological, or military imbalance that exists when there is a disparity in comparative strengths and weaknesses. In the context of the COE, asymmetry means an adaptive approach to avoid or counter US strengths without attempting to oppose them directly, while seeking to exploit weaknesses. The asymmetric approach is not a new phenomenon, but given the position and capabilities of the US as opposed to its potential enemies, it is more likely to be used against the US and its allies by other nations and nonstate adversaries. Potential opponents will seek to avoid US strengths while exploiting perceived US weaknesses.

2-6. Various countries and nonstate entities have studied how the US fights and have begun to devise ways to fight a technologically superior force, if necessary, and win. Most nations of the world and other actors of consequence have devoted considerable effort in studying the US and how we fight. Since it is difficult to predict whom we might have to fight, we don't always have the luxury of having studied these nations or other actors in detail. We must be prepared to refocus quickly, learn fast, and apply what we have learned rapidly. Flexibility and initiative are key to being able to adapt our DOTMLPF solutions to meet the challenges of a given adversary and the asymmetric approaches that he may apply against us.

THREATS AND OTHER INFLUENCES

2-7. Engineers must be prepared to go into any OE and perform its full range of missions while dealing with a wide range of threats and other influences. Some threats come in the form of nation-states; this may be a country or a coalition of countries. Threats can also come from entities that are not states; these can include insurgent, terrorist, drug trafficking, and other criminal organizations. The nonstate actors may use force of arms to further their own interests and threaten the interests of the US or other nation-states. Nonstate threats may exist in isolation or in combination with other nonstate or nation-state threats.

2-8. No single nation-state actor is expected to present a peer or near-peer threat to the US until 2020 or beyond. However, this does not mean that the US and its armed forces will not face serious challenges in the next two decades. The sum total of all the possible conflicts and the level of difficulty of those conflicts could present a challenge equivalent to that of having a near-peer competitor. The net effect of several OEs when considered in its totality from a global, strategic perspective stretches our combat power in ways we have never before encountered. Remember that combat power is composed of maneuver, firepower, leadership, protection, and information. *See FM 3-0* for an in-depth discussion of combat power and its subordinate elements.

2-9. We must be ready to counter all possible threats and, at the same time, be prepared to deal with various third-party actors, such as international humanitarian relief agencies, news media, refugees, and civilians on the battlefield. These groups may be working in cooperation with us or allied efforts or may be operating independently. It is essential that commanders define our relationships with these organizations and include them in the ROE, especially regarding their status within our force protection structure. The engineer effort should be attentive to Title X limitations and how they affect potential cooperation between our organizations. When not properly integrated, their presence can change or constrain the nature of the conflict and can influence the outcome.

2-10. As depicted in *Figure 2-1, page 2-4*, recurring consideration needs to be given as to where a particular geopolitical actor is on the continuum of relative interests and what the affect of that is on our forces. Adversaries share a common characteristic—the capacity and intent to conduct violence against Army forces to negatively influence mission accomplishment. All potential adversaries need to be evaluated in this light. Using the continuum of relative interests, we may find a neutral population becomes an obstacle or adversary over time. An aide such as an NGO could also become an obstacle over time. Therefore, it is imperative that we maintain a current knowledge of all parties in an AO and discern where they are in relation to the continuum and how that might affect engineer operations.

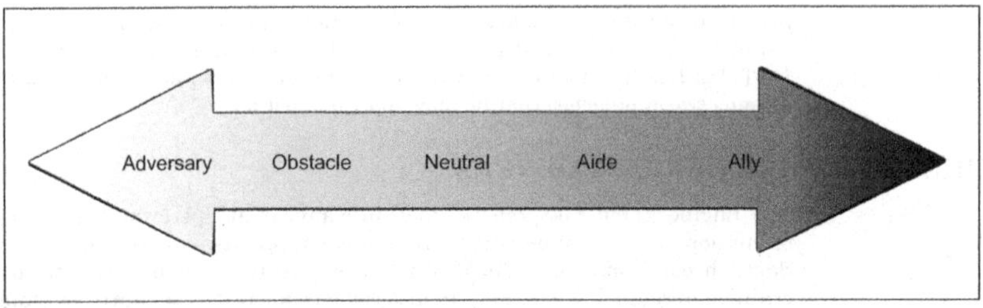

Figure 2-1. Continuum of Relative Interests

WARFIGHTING IN THE CONTEMPORARY OPERATIONAL ENVIRONMENT

2-11. Most participants in conflicts around the world will have no intent to fight the US—they look for ways to keep us out of conflicts or from staying involved. They will not fight us in the same manner as their peers or lesser forces in their region. We can expect potential adversaries to adapt their methods of fighting, using a combination of principles, to include—

- Controlling access into the region.
- Changing the nature of the conflict.
- Employing operational shielding.
- Controlling the tempo.
- Neutralizing technological overmatch.
- Causing politically unacceptable casualties.
- Allowing no sanctuary.
- Entering into urban areas.

2-12. These principles attempt to exploit perceived weaknesses of vulnerabilities in the US force's activities, force structure, or ROE.

IMPLICATIONS FOR EACH OF THE LEVELS OF WAR

STRATEGIC IMPLICATIONS

2-13. Based on threats and the National Security Strategy, the combatant commander should continue to promote their engagement strategies. Engineers will support the commanders with planning support, including geospatial studies; advice on the integrating environmental considerations; infrastructure construction; training; and assistance to international disasters. In times of war, engineers provide a rapid strategic response. Engineers from the AC and RC are prepared to deploy on short notice with their units or just provide tailored capabilities. RC personnel mobilize for specific periods of time to provide their unique skills. Engineers improve interoperability by standardizing with allies in the North Atlantic Treaty Organization (NATO); American, British, Canadian, and Australian (ABCA) coalition; and other international organizations to simplify and facilitate interoperability between international and service engineers and maneuver units. By potentially deploying to austere environments, engineers with limited haul capabilities or manpower can access theater engineers or USACE to contract for the Logistics Contract Augmentation Program (LOGCAP), HN, and local support (and contractors). They can use FFE and the associated reach capabilities to access analyses or expertise on other general-engineering-related challenges. They can also reach engineer centers of expertise for unique solutions to unforeseen and one-of-a-kind challenges or reach laterally to other engineer organizations and assets for expertise and support.

OPERATIONAL IMPLICATIONS

2-14. Operational implications. Many threats or threat forces (rogue organizations) can only fight the US and its allies using asymmetric techniques to nullify technological advantages. However, they may also possess or gain access to weapons of mass destruction (WMD) that cause significant implications for operations, to include tremendous environmental considerations. Threat forces may initially fight force on force and then integrate their forces into urban areas and use civilians to complicate operations. Engineers must conduct their battlefield assessment and work closely with intelligence officers to analyze the threat during the intelligence preparation of the battlefield (IPB). In today's OE, engineers have difficulty predicting an enemy course of action (COA) based on battlefield geometry alone. Engineers may describe an enemy force more in terms of functions (fixing, assaulting, exploiting, shielding) of its subelements and not in terms of where it might be found on a linear battlefield (2d echelon, main defense belt). US engineers will have to develop methods to discern threat engineer patterns of behavior. Engineers have to match training and material capabilities to specific requirements. Operational engineer commanders have proponency for training soldiers on common soldier skills such as mine awareness, detection, avoidance, and extraction and environmental considerations before, during, and after combat operations.

TACTICAL IMPLICATIONS

2-15. Tactical implications. Tactical missions are more complex than ever when preparing to fight both symmetric and asymmetric threats over a noncontiguous AO. Terrain analysis and an understanding of threat capabilities are more important than ever as engineers template potential asymmetric capabilities. Since engineers do not have one tool to accomplish all tasks or counter every threat impediment, they need to gather very specific threat information. They must be able to discern and identify patterns. Engineers work to develop specific detection strategies based on the threat reinforcing obstacle type. The obstacle intelligence (OBSTINTEL) collection plan can then be integrated into the intelligence, surveillance, and reconnaissance (ISR) collection plan and resourced appropriately. The proliferation of mines around the world requires engineers to continuously develop new DOTLMPF procedures to address evolving countermine requirements. Training and Doctrine Command (TRADOC) engineers work closely with United States Army Forces Command (FORSCOM) engineers and other service engineers to address lessons learned and new DOTMLPF procedures. Engineers will have to operate mine information coordination cells (MICCs) and establish databases to track existing mines and other obstacles such as the tactical minefields. Engineers will establish close relationships with explosive ordnance disposal (EOD) units to facilitate a COP of mine/UXO/improvised explosive device (IED) threats.

IMPACT OF THREAT ENGINEERS ON OUR DOCTRINE

2-16. As stated, threat forces may oppose the Army symmetrically or asymmetrically, as seen in Bosnia and Afghanistan. Engineers can anticipate dealing with increased mines, booby traps, IEDs and UXOs throughout linear or nonlinear AOs. The threat may no longer come from mines on the ground but may now include side and top attack (offensive) mines that are command-detonated, remotely or via hard wire. The threat's countermobility will be dispersed throughout the AO to impede friendly maneuver ability and attack our will to fight. The possibility of an opposing threat using the urban environment to diminish US capabilities is increasing. *See FM 3-06(90-10)* and *FM 3-06.11(90-10-1)* for more information. Recent urban combat experiences in Northern Ireland, Grozny, Jenin, Afghanistan, and Iraq reveal an accelerated adaptation to countermeasures developed by friendly forces in urban areas. Experience shows the threat adapting and using IEDs to complicate our detection and neutralization efforts.

2-17. Examples of recent IEDs found around the world include remote detonating devices using electronic, radio control, or cell phones to initiate the systems. These methods are all shared between terrorist organizations via the Internet. What British Forces encounter in Northern Ireland may soon confront our soldiers in another urban AO. Some IED techniques include—

- Coupling. Coupling links one mine or explosive to another, usually with detonating cord. When the first device is detonated, it detonates the linked explosive. This technique is often used to defeat countermine equipment such as mine rollers. When the linked devices are directional fragmentation mines, they can create a large, lethal EA.
- Boosting. Boosting involves stacking buried low-metal mines on top of one another and fuzing the farthest from the surface. This reduces the probability of detection and increases the force of the blast.
- Sensitizing. Sensitizing uses AT mines. On some nonmetallic AT mines, the pressure plate can be cracked and the spring removed to reduce the pressure required to initiate the mine. Similarly, the pressure plate can be removed from metallic AT mines to obtain the same effect. Alternatively, a pressure-fuzed antipersonnel (AP) mine can be placed on top of an AT mine, creating a very large AP mine.
- Daisy-chaining. Daisy-chaining uses command-detonated AP mines. Enemy forces link the mines with trip wires or detonating cord. When the initial mine is detonated, the other mines will detonate. This creates a large, lethal EA.

2-18. Threat forces may also use environmental damage as a weapon. Enemies, using our sense of environmental stewardship and our desire to avoid/prevent environmental disaster, may deliberately release environmental hazards or hide them in structures that, when attacked, may release unacceptable environmental damage. A good example is the deliberate burning of oil wells in Iraq during Desert Storm. Threat forces could use industrial chemicals not banned by international law but which could cause widespread contamination if released. Rebels used these chemicals against the Russians in Grozny.

2-19. Another threat operational capability includes area inundation. This is the idea of releasing large bodies of water from dam systems to temporarily flood areas as operational countermobility. Threat regions with a large number of dam and dike systems are a planning consideration for potential engineer bridging requirements.

2-20. The threat commander understands the importance of Army engineers and will focus assets to eliminate their capabilities, especially when they operate independently and are most vulnerable. He will develop new techniques to counter friendly mobility and countermobility operations. Army engineers must coevolve with their adversaries and adapt new procedures concurrently in the AO. Using internal training or mobile training teams (MTTs) in the field will facilitate this necessary use of new techniques and equipment.

2-21. Engineers will be challenged to understand the OE they face and apply their knowledge and capabilities to support the force. They should understand the variables that constitute the OE to best understand how to leverage the engineer battlespace functions to support the Regimental METL. Future adversaries have developed a wide range of asymmetric capabilities that will challenge the engineer functions. It is imperative that engineers understand how to analyze the threat differently than in the past and develop solutions that consider all the variables associated with the OE.

Chapter 3

Engineer Battlespace Functions and Regimental Capabilities

The European operations were frequently referred to as an Engineer war. The fact that Engineers so ably performed all tasks which fell to their lot indicates the basic soundness of Engineer doctrine, Engineer training, and European troop organization.

Final Report of the Chief of European Theater of Operations 1942-1945

The three primary engineer battlespace functions are combat (M/CM/S), geospatial, and general engineering. These functions, when integrated within the operational framework in support of the elements of combat power, assist the commander in achieving successful decisive, shaping, and sustaining operations. At the strategic level, *JP 3-34* combines the combat engineering functions of M/CM/S, while *JP 4-04* explains civil engineering (general engineering in the Army). *JP 2-03* describes engineer functions and the contributions of geospatial engineering. As these functions are used at the operational and tactical levels, they tend to blur and overlap. Therefore, they are addressed as five separate functions to differentiate the capabilities within each function. For instance, a MSR upgrade for tactical units in their zone or sector is a mobility function but a LOC upgrade at the operational and strategic levels is a general engineering function. The engineer commander has the responsibility to integrate safety and environmental considerations when implementing these functions. The staff engineer is also the primary integrator of environmental considerations the planning process and writing of the environmental appendix to the OPLAN/OPORD. Regimental capabilities exist to support each of the echelons of command and across the spectrum of conflict. Which of those capabilities support an echelon of command depends upon the requirements and challenges associated with mission, enemy, terrain and weather, troops and support available, time available, and civil considerations (METT-TC).

SECTION I - ENGINEER BATTLESPACE FUNCTIONS

3-1. The engineer battlespace functions are the baseline functions that contribute to supporting the METL of the Engineer Regiment. The functions are each generally aligned with a specific BOS, although they have impact in and across the other BOSs. Combat engineering is aligned primarily with the M/CM/S BOS, geospatial engineering with the intelligence BOS, and general engineering with the CSS BOS *(see Figure 3-1)*. Each BOS is also related to and a part of assured mobility.

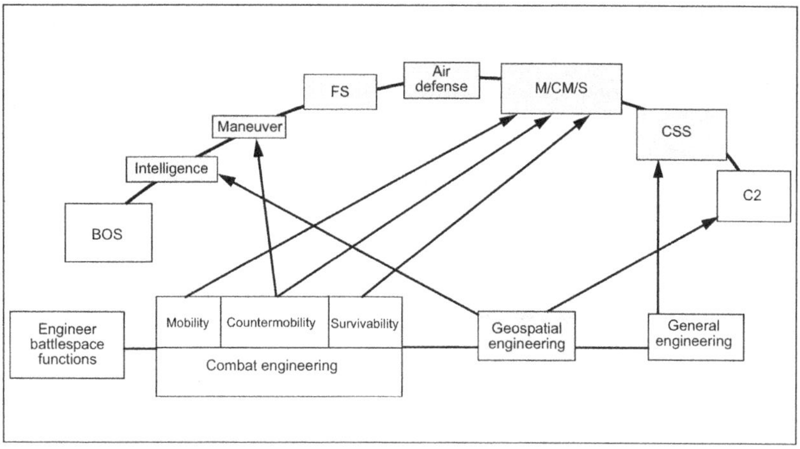

Figure 3-1. Engineer Primary Relationships to the Battlefield Operating System

COMBAT ENGINEERING

3-2. Combat engineering is an integral part of a combined arms unit's ability to maneuver. It is focused on the support of close combat forces. Combat engineers enhance the force momentum by physically shaping the AO to make the most efficient use of space and time to generate mass and speed while denying the enemy maneuver. Combat engineers accelerate the concentration of combat power, increasing the velocity and tempo of the force necessary to exploit critical enemy vulnerabilities. By reinforcing the natural restrictions of the AO, combat engineers limit the enemy's ability to generate tempo and velocity. These limitations increase the enemy reaction time and physically and psychologically degrade his will to fight. Combat engineering includes the subordinate functions of M/CM/S.

KEY FUNDAMENTALS OF COMBAT ENGINEERING

3-3. The following key fundamentals of combat engineering improve the probability of a successful operation.

Evaluate the Terrain

3-4. Engineers should evaluate how the threat may employ obstacles to shape the battlefield or create casualties given the terrain in the AO. They should also consider how the threat will use the terrain to shape his scheme of maneuver. Engineers should then integrate how the terrain will impact on friendly operations and how they can leverage their efforts to achieve the commander's intent.

Understand How the Threat Fights

3-5. Engineers must be familiar with the history of the AO and the conflict. Knowledge of threat doctrine and engineer methods and functions is critical. Engineers should consider patterns of obstacle employment—are they in compliance with mine employment? Not all armies or belligerents mark their minefields and many non-first-world countries rely on IEDs to mark minefield locations and hazard areas. Many local factions, militia, or military units lay mines and mark them with readily available materials rather than formal marking methods adhering to a doctrine. These markings are generally used to warn their own troops and local civilians of the presence of mines. Friendly units operating in these threat environments must know and understand these markings. This is very important because the Army has entered many areas of ongoing or previous conflicts where opposing forces exchanged terrain on multiple occasions leaving behind altered or unrecorded mines and booby traps. The zone of separation (ZOS) region in Bosnia-Herzegovina and areas fought over by the Taliban and the Northern Alliance are recent examples.

Use all Available Engineer Assets

3-6. Engineers seek to predict where obstacles are emplaced, detect or avoid them, and reduce or clear obstacle areas. Engineers use all of the friendly assets available to focus resources to find and avoid obstacle areas or to develop intelligence and fidelity about obstacle areas that must be traversed. At the operational level, engineers should seek OBSTINTEL collected by intelligence specialists, SOF, psychological operations (PSYOP), and civil affairs, and distribute this information to military planners and patrols to complete the COP. All OBSTINTEL should be reported and distributed quickly to enhance force mobility and protect the force.

MOBILITY

3-7. There is a great deal of confusion about the doctrinal definitions and use of the term mobility. DOD defines mobility as quality or capability of military forces which permits them to move from place to place while retaining the ability to fulfill their primary mission. This definition applies to all services and is very broad. The Army defines mobility as "those activities that enable a force to move personnel and equipment on the battlefield without delays due to terrain or obstacles." This definition applies to all branches as they seek to move throughout the AOs. Mobility refines the definition as facilitating the momentum and freedom of movement for the maneuver of forces by reducing or negating the effects of existing or reinforcing obstacles. It includes activities such as reconnaissance, countermine/counterobstacle operations, gap crossing operations, and combat roads and trails. *FM 3-34.2* is the primary reference for mobility planning. Mobility, as an element of the BOS, is broader than the engineer mobility function and includes reducing obstacles, providing bridge and raft support, increasing battlefield circulation, improving or building roads, and identifying routes and contaminated areas. It is the engineer function of mobility that we are addressing when we speak of mobility operations—obstacle reduction by maneuver and engineer units to reduce or negate the effects of existing or reinforcing obstacles. The objective is to maintain freedom of movement for maneuver units, weapon systems, and critical supplies *(see FM 101-5-1)*.

3-8. Mobility is a quality or capability of military forces that permits them to move in time and space while retaining their ability to fulfill their primary mission. A commander must be able to mass forces quickly, at a chosen place and time, to accomplish the assigned mission. The commander must be able to achieve superior tempo through a quick observe, orient, decide, and act (OODA) loop than the enemy. Mobility is critical to achieving this situation and maintaining it for extended periods of time over great distances. The combat-engineer mobility focus must support the requirements of the close combat force.

3-9. Sustained ground combat operations require coordinated movement and effective concentration of combat power against the enemy, in spite of enemy interdiction efforts. Air defense, air and ground transportation, reconnaissance and security, service support, and traffic control operations are among the chief concerns as these large movements occur. Ground combat operations have the best chance of success when they are synchronized with air superiority and air interdiction operations. The LCC or ARFOR commander directs the movement of subordinate forces, ensuring that by the end of a distinct phase of a major operation, decisive and shaping forces are positioned in a way that enables rapid transition to subsequent phases. Additionally, the ARFOR command capability to conduct sustaining operations may set the tempo of the overall joint operation. Within the framework of decisive, shaping, and sustaining operations, the ARFOR will designate and shift the main effort.

3-10. Operational mobility is linked closely to the concept of movement and maneuver operations, which includes the functions of providing mobility for operational forces and countering the mobility of enemy operational forces. Operational-level engineer units provide support for mobility and countermobility operations. Facilitating the maneuver of major formations without delays includes countering the effects of operationally significant obstacles, as well as enhancing operational movement, by preparing and improving facilities and routes critical to major operations. Operational countermobility delays hinder the movement of enemy operational formations, to include selecting and emplacing systems of obstacles for operational effect. Operational-level engineer units provide support for mobility and countermobility.

3-11. Terrain, natural and man-made, significantly influence operational mobility. It often dictates the capacity of LOCs, which can limit the size and composition of supported forces. In war, the operational-level commander considers the effect of terrain features on ground movement and the ability of air power to influence that movement by detecting ground forces and subsequently delaying, disrupting, and destroying these forces. In peacetime, the Army commander may consider how these features affect accomplishing missions that support peacekeeping or humanitarian operations.

ASPECTS OF MOBILITY

3-12. Mobility operations are intended to maintain freedom of tactical maneuver and operational movement through the following five functional areas:

- Countermine activities detect, neutralize (through a combined arms breach or bypass), mark, and proof mined areas.
- Countering employs tactics and equipment to breach or bypass and ultimately reduce obstacles other than mines.
- Gap crossing fills gaps in the terrain to allow personnel and equipment to pass.
- Constructing combat roads and trails expediently prepares or repairs routes of travel for personnel and equipment. This includes temporary bypasses for damaged roads and bridges.
- FACE prepares or repairs expedient landing zones (LZs), forward arming and refueling points (FARPs), landing strips, or other aviation support sites in the forward combat area.

Countermine Activities

3-13. Countermine operations are all efforts taken to counter an enemy's mine effort. Countermine operations are difficult because detection systems are imperfect and mine neutralization systems are only partially effective. Normally, countermine operations using explosive systems are conducted under enemy observation and fire. Countermine operations include mine detection, enemy minefield reconnaissance, combined arms breaching, and enemy mine operations prevention.

3-14. **Mine Detection**. The detection of mines is linked to proper reconnaissance techniques to include geospatial engineering assistance. A proper analysis of enemy techniques and devices lays the groundwork for effective reconnaissance. *FM 20-32,* Mine/Countermine Operations, is the primary reference for a discussion of mine detection.

3-15. **Enemy Minefield Reconnaissance**. Engineer reconnaissance and the specific considerations of minefield reconnaissance are discussed in *FM 5-170.* Proper reconnaissance creates the conditions necessary for successful obstacle breaching.

3-16. **Combined Arms Breaching**. As engineers plan for mobility operations, they may realize that they will have to conduct breaching operations. Enemy obstacles that disrupt, fix, turn, or block the force can affect the timing and force of the operation. Most obstacles can and will be observed by the enemy and will be protected with fires. They should be bypassed if possible. For those that must be breached, constant coordination and integration of all elements of the TF are vital for success. Combat engineers are key to the orchestration of the operation and are responsible for employing the tactics and techniques necessary to penetrate obstacles in the path of the force. Combined arms breaching operations are some of the most complex of modern warfare but are not an end in themselves. They exist only as a part of the maneuver forces' operation focused on the objective. The goal of breaching operations is the continued, uninterrupted momentum of ground forces to the objective; therefore, these operations should be planned and executed in support of the ground forces' needs to ensure that actions at the objective are supported by actions at the breach. Combined arms breaching is explained in depth in *FM 3-34.2*.

3-17. Successful breaching operations are characterized by applying breaching tenets. These tenets should be applied whenever an obstacle is encountered in the AO, whether during an attack or a route clearance operation. These breaching tenets include—

- Intelligence.
- Breaching fundamentals.
- Breaching organization.
- Mass.
- Synchronization.

3-18. Combined arms breaching operations require the constant application of the factors of METT-TC and the concentrated use of supporting arms. Fundamentals of combined arms breaching operations have evolved in concert with the fundamentals of ground combat and provide a logical and time-proven set of rules. These fundamentals include suppress, obscure, secure, reduce, and assault (SOSRA).

3-19. **Enemy Mine Operations Prevention**. The most effective means of countering a mine threat is to prevent mine laying. Proactive countermine operations destroy enemy mine manufacturing and storage facilities or mine-laying capabilities before the mines are laid. Planners must consider enemy storage and mine production facilities and assets for inclusion on the target lists. In addition to destroying mine manufacturing and storage facilities, units must consider targeting enemy engineers and equipment capable of laying mines. During the IPB process, engineers work with the Intelligence Officer (U.S. Army) (S2)/G2 to create an initial enemy obstacle template and variations of the template to show what the enemy is capable of creating over a period of time. Use of this information may place enemy engineers and equipment on the high-payoff target list (HPTL) with collection assets focused to find them.

Countering an Obstacle

3-20. Many issues encountered in counterobstacle operations apply to nonmine obstacles. Engineer reconnaissance should focus on collection efforts to detect the presence of enemy obstacles, determine their types, and provide the necessary information to plan appropriate combined arms breaching or bypass operations to negate the impact on the friendly scheme of maneuver. Reconnaissance also allows friendly forces to anticipate when and where the enemy may employ obstacles that could impede operations. It is prudent to incorporate plans whenever possible to deny the enemy the opportunity to establish effective obstacles. Achieving this goal can be accomplished by—

- Occupying the area before the enemy can exploit it.
- Preplanning artillery and close air support to deny or harass enemy units attempting to establish obstacles.
- Looking for or creating alternative routes for friendly force units.

- Using engineering knowledge of obstacles to create contingency plans for combined arms breaching or bypassing for quick neutralization of enemy obstacles.

Gap Crossing

3-21. Engineers focus on projecting combat power over gaps by tailoring the appropriate resources for the specific mission set. Engineer planners task-organize the appropriate bridge units to support gap crossing operations. Combat engineers can assist with an assault gap crossing using organic armored vehicle-launched bridges (AVLBs) and Wolverines or their heavy equipment to modify the existing gap or by using expedient bridging (rope bridges, small nonstandard bridging using local materials). Operational-level engineers resource subordinate units that do not possess the necessary organic standard bridging equipment. River crossing is a unique gap crossing mission that requires specific and dedicated assets from the BOSs. For a discussion of river crossing, refer to *FM 90-13.*

Combat Roads and Trails

3-22. The ability to move personnel and equipment is essential to maneuver warfare. This ability provides the commander with the means to increase tempo and speed, as well as concentrate mass at crucial times and places. The construction and maintenance of trails and roads are normally considered general engineering tasks and are, therefore, performed by engineering support units. However, locations near the forward line of own troops (FLOT) or time restrictions may require forward combat engineer units to perform these functions in an expedient manner or for short durations until support engineers are available. The two most likely scenarios that would involve this requirement would be bypass operations or support FACE operations. It is important for the engineer commander and staff to perform this function only in support of the maneuver plan. They should not allow engineering assets to be dissipated, rendering them unable to perform their primary role of supporting the commander's operational scheme of maneuver. Engineers should always strive to take full advantage of existing infrastructure and natural terrain features when constructing combat trails and roads.

Forward Aviation Combat Engineering

3-23. With the advent of airpower and its associated support requirements, engineers have acquired a mission to support aviation assets. This frontline support will normally take the form of creating LZs for helicopters and vertical and/or short take-off and landing aircraft or parachute drop zones for personnel, equipment or supplies. Engineers should always strive to take full advantage of existing infrastructure and natural terrain features when constructing expeditionary landing and/or drop zones. Airpower and helicopters are important to support maneuver warfare. The use of forward landing and/or drop zones can increase the speed and tempo of operations. For example, a closer proximity decreases turnaround time for aircraft and helicopters (for example, FARP sites); decreases personnel, equipment, and supply travel times from rear areas to the forward combat area; and decreases the response times of close air and helicopter support missions.

COUNTERMOBILITY

3-24. Countermobility is the augmentation of existing obstacles with reinforcing obstacles integrated with direct- and/or indirect-fire systems to disrupt, fix, turn, or block the enemy. The maneuver commander destroys enemy combat capabilities making use of the increased time for target acquisition. Once again, the countermobility function should not be confused with the countermobility BOS. The countermobility BOS is expanded to include not only obstacle construction but also smoke generation. The engineer commander focuses on proper obstacle integration with the maneuver plan, adherence to obstacle emplacement authority, and rigid obstacle control. *FM 90-7* is the primary reference for countermobility planning.

3-25. Countermobility is the physical shaping of the AO to alter the scheme of maneuver of the enemy. Countermobility operations block, fix, turn, or disrupt the enemy's ability to maneuver, giving the commander opportunities to exploit enemy vulnerabilities or react effectively to enemy actions. When planning countermobility obstacles, it is important to understand the commander's intent, timetable, and scheme of maneuver. These factors, along with available manpower, equipment, and materials, ultimately determine what is feasible to support the OPLAN. Two key actions in obstacle plans include—

- Avoiding obstacle plans that require so much material and the manpower they cannot be emplaced in a timely manner to provide useful support to the maneuver plan—for example, the enemy maneuver elements bypass the obstacle field before it is completed.
- Assuming that friendly forces are not impeded with friendly obstacles later in the operation.

3-26. Nonengineer units may need to augment engineer units with security and personnel to execute countermobility operations. The operational commander and various unit commanders must be aware of this possible requirement during planning.

SURVIVABILITY

3-27. Survivability encompasses developing and constructing protective positions such as earth berms, dug in positions, and overhead protection as a means to mitigate the effectiveness of enemy weapon systems. The survivability function is differentiated from survivability in the M/CM/S BOS since the BOS includes deception, camouflage, operations security (OPSEC), and NBC defense measures. Survivability is also often confused with force protection. *FM 3-0* states that force protection is one of four components of protection (force protection, field discipline, safety, and fratricide avoidance). Survivability operations are just one means to enhance force protection. *FM 5-103* is the primary reference for survivability planning.

3-28. Survivability is the ability of personnel, equipment, and facilities to continue to operate within a wide range of conditions faced in a hostile environment. It includes all aspects of protecting personnel, weapons, and supplies. Units must be able to reduce the exposure to threat acquisition, targeting, and engagement. Engineer-supported tasks such as the construction of field fortifications; hardening of command, communication, and combat training locations; improvements to weapons systems firing positions and infantry fighting positions, are critical to this effort.

Field Fortifications

3-29. DOD/NATO defines a field fortification as an emplacement or shelter of a temporary nature which can be constructed with reasonable facility by units requiring no more than minor engineer supervisory and equipment participation. Engineers construct fighting positions for combat vehicles, direct-fire weapons systems, artillery, and air defense artillery. Field fortifications provide a degree of protection from the effects of enemy weapons systems and a more stable weapons platform from which to sustain accurate volumes of fire. They sustain confidence in a soldier's ability to fight effectively where he otherwise could not survive.

Strong Points

3-30. DOD/NATO defines a strong point as a key point in a defensive position, usually strongly fortified and heavily armed with automatic weapons, around which other positions are grouped for its protection. Strong points are heavily fortified battle positions that cannot be overrun quickly or bypassed easily by enemy forces. They consist of an integrated series of well-protected fighting positions connected by covered routes and reinforced with extensive protective obstacles. They are designed to withstand artillery fire, air strikes, and mounted and dismounted assaults. The enemy can reduce them only by expending significant time, personnel, and equipment assets in overwhelming force.

3-31. It takes significant engineer effort to create a strong point. As a rule of thumb, it will require an engineer company to create a company-sized strong point. Altering a portion of an urban area to turn it into a strong point is generally easier than creating, but it still requires significant effort. Strong points smaller than company size are not usually discussed, except in urban areas.

GEOSPATIAL ENGINEERING

3-32. Geospatial engineering is collecting, developing, disseminating, and analyzing positionally accurate terrain information that is tied to some earth reference. These actions provide mission-tailored data, tactical decision aids, and products that define the character of the zone for the maneuver commander. Key aspects of the topographic mission are geospatial databases, analysis, positional control, and printed maps. These aspects provide the commander a common view of the terrain, which leads to a COP that he uses for C2. Engineer officers at theater, corps, division, brigade, and battalion levels are responsible for geospatial engineering. They provide commanders a clear understanding of the physical environment by enabling visualization of the terrain and explaining its impact on friendly and enemy operations. They identify terrain aspects that the commander can exploit to gain advantage over the enemy, as well as those that the enemy will exploit.

3-33. Geospatial engineering is a subjective evaluation of the terrain's physical attributes of the terrain and the physical capabilities of the vehicles, equipment, and people that must cross or occupy the terrain. Engineer officers as geospatial engineers also coordinate requirements for collecting and generating higher-resolution data and write geospatial annexes/appendices to OPLANs/OPORDs for their supported echelon. Geospatial engineering depends on topographic support for managing, developing, disseminating, and analyzing positionally referenced terrain information. The geospatial engineer must—

- Continually provide current and accurate digital information databases to maintain the commanders' vision of the AO.
- Analyze the effects of operations on the ground and weather factors contributing to maneuverability.
- Perform the critical task of maintaining a common operational background throughout all echelons.
- Provide a means for the commander to visualize the AO.

3-34. The geospatial function primarily supports the C2 and intelligence BOSs. *FM 3-34.230* and *JP 2-03* are the primary references for geospatial engineering.

GENERAL ENGINEERING

3-35. General engineering encompasses those engineer tasks that establish and maintain the infrastructure required to conduct and sustain military operations. Such tasks include the construction and repair of LOCs, MSRs, airfields, utilities, and logistical facilities. *FM 5-104* is the primary reference for general engineering planning. It is directly related to *JP 4-04*.

3-36. General engineering tasks are typically performed to the rear of the division boundaries but can also be performed in forward areas. This function is usually performed by engineer units above division level and consists of repair and construction tasks. Well-developed theaters have a preponderance of repair tasks and lesser developed theaters require more construction tasks. The principles of general engineering in an AO are speed, economy, flexibility, decentralization of authority and establishment of priorities.

3-37. A major requirement for general engineering is a suitable and sufficient supply of construction materials. Engineer commanders have the burden of locating or manufacturing many of these resources. All engineer units must be capable of exploiting available materials. Outsourcing may become an option to access materials as well. Quality control is imperative to ensure that materials meet the critical construction tolerances/standards.

SECTION II - INTEGRATING THE ENGINEER FUNCTIONS

3-38. For the current force and Stryker organizations, the engineer battlespace functions must be integrated to achieve synergy with the other BOSs. The engineer coordinator (ENCOORD) at each echelon is responsible for integration and synchronization of the engineer effort. The engineer battlespace functions are integrated across doctrine through the application of the two primary integration constructs of assured mobility and FFE.

ASSURED MOBILITY

3-39. Linking to the operational framework established in *FM 3-0*, the engineer battlespace functions support the maneuver commander's use of the elements of combat power for decisive, shaping, and sustaining operations within the full spectrum of operations. How these engineer functions are integrated across the elements of combat power is the question. Assured mobility as depicted in *Figure 3-2* integrates all of the engineer battlespace functions, however, assured mobility should not to be confused with the mobility function.

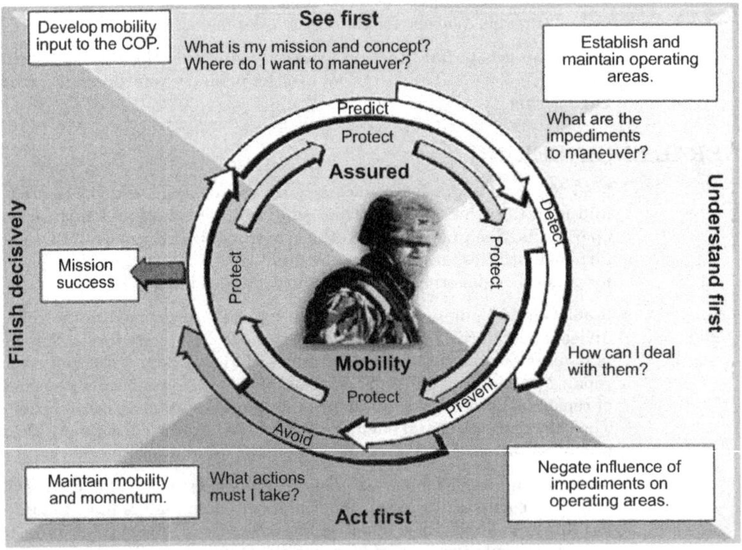

Figure 3-2. Assured Mobility and the Engineer Fundamentals

3-40. Assured mobility encompasses those actions that give the force commander the ability to deploy, move, and maneuver where and when he desires, without interruption or delay, to achieve the mission. A relatively new doctrinal framework, the imperatives and fundamentals of assured mobility are what enable friendly forces to exploit superior situational understanding (SU) and, therefore, gain unsurpassed freedom of movement. Put simply, this framework describes the processes that enable the commander to see first, understand first, act first, and finish decisively.

3-41. Assured mobility supports the maneuver commander's use of the elements of combat power to achieve decisive, shaping, and sustaining operations across the full spectrum of operations and conflict. The framework of assured mobility entails four imperatives that are linked to the elements of combat power *(see Figure 3-3)*. These imperatives are proactive, not reactive, and assure mobility only if integrated into the MDMP.

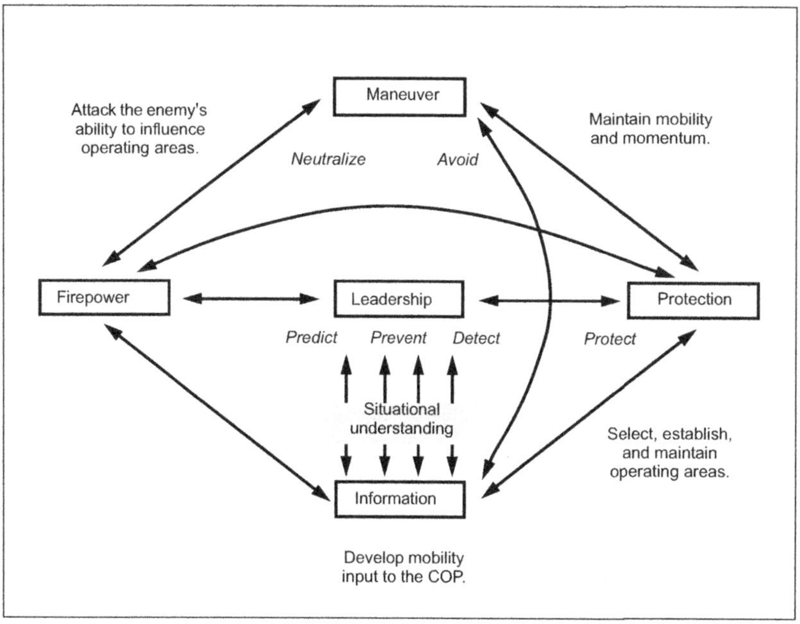

Figure 3-3. Imperatives of Assured Mobility and the Elements of Combat Power

3-42. The first imperative, develop mobility input to the COP, is the collection and integration of geospatial, cultural, and enemy information (aided by automated mobility planning tools) to establish mobility input to the COP for the entire AO. This information allows quick development of the initial and follow-on, real-time modified combined-obstacle overlay (MCOO) that enables the maneuver commander to select the focused operating areas within the AO that best provide positions of advantage. The operating areas are smaller areas designated within the AO that allow the commander to focus collection assets and efforts. The MCOO is defined by the desired endstate and is updated with new information to reflect real-time mobility aspects.

3-43. Knowledge of existing obstacles and monitoring of existing traffic patterns are two examples that allow us to see the battlefield in near-real time. Where obstacles are is as important as where obstacles are not. This information allows the maneuver commander to determine where he may maneuver, what resources will be required to get there, and how the enemy forces may attempt to influence this maneuver plan. The mobility input to the COP enables the maneuver commander to identify the operating areas in the AO and the associated mobility challenges. It is linked to ISR operations and continuously updates the commander and leaders with real-time mobility information. This imperative links the information element to the leadership element to provide SU.

3-44. The second imperative, select, establish, and maintain operating areas, means identifying the threat and restrictive terrain and location of a countermobility effort. This imperative is the linkage between seeing first and understanding first. With the aid of automated tools, critical mobility choke points within operating areas will be identified and a shaping plan. This plan includes predicting enemy actions and intents and the required sensor coverage to fill any information voids within the operating areas. Battlefield terrain reasoning and awareness (BTRA) allows us to template potential obstacles and locations of where the enemy might place obstacles. Sensors dedicated within the collection plan will be employed, or focused from other assets, on the critical areas to fill the information voids or improve our SU and maintain mobility input to the COP. The system allows the commander to track existing obstacles and catalog threat patterns to facilitate a predictive analysis of what the threat may do to terrain to impede friendly maneuver.

3-45. To solidify the COP, sensors are employed at critical areas to protect them from enemy influence. In coordination with these sensors, conducting standoff attacks on selected enemy capabilities to perform mine or countermine activities (or deceiving the enemy to focus his attention on other areas) will fix the current mobility SU. Sensors and forces may be emplaced at choke points, such as bridges, to counter enemy attempts to disrupt operations. Being able to control and monitor critical mobility areas is coordinating a mobility plan in conjunction with the scheme of maneuver. Simply put, "sensor staring" will enable the force to "own" the operating areas. By identifying the locations that may impede movement, the leadership element can resource where he wants to maneuver to avoid or neutralize these impediments while protecting the force and minimizing risks. This imperative supports the elements of protection and maneuver.

3-46. The third imperative, attack the enemy's ability to influence operating areas, is accomplished by interdicting the enemy's countermobility efforts and providing the maneuver commander with multiple avenue options. This imperative is a proactive means of employing standoff detection and obstacle neutralization systems to maintain mobility within the operating areas and assist with ensuring freedom of maneuver. The focus is on shaping the noncontiguous operating areas for decisive operations.

3-47. Standoff detection of obstacles, standoff neutralization, and attacking these obstacles in coordination with the maneuver plan, is critical to preserve resources. This is a proactive attack of the enemy's ability to employ obstacles. Attacking the enemy's ability to shape the AO against us is a key enabler for mobility. Current sensor systems and future intelligent munitions and antipersonnel (AP) land mine alternatives (APLA) will be used to proactively attack the threat.

3-48. The fourth imperative, maintain mobility and momentum, is accomplished by using each of the engineer battlespace functions to support firepower and maneuver. The first three imperatives are command, control, computers, communications, intelligence, surveillance, and reconnaissance (C4ISR) intensive. This imperative considers the fact that the threat is thinking and will adapt to our operations. The purpose of this imperative is to allow maneuver forces to neutralize the effects of obstacles in the context of their maneuver. Their goal is freedom of maneuver to seize objectives without delay along multiple and parallel routes. Detecting obstacles (including side-attack and wide-area mines) and neutralizing their effects without allowing them to adversely affect the maneuver plan or the momentum of the maneuver force are critical. As a last resort, an organic breaching capability exists to overcome enemy obstacles that may be encountered when it is necessary for the force to cross through the obstacles. Marking systems that provide visual, virtual and active identification of obstacles and cleared or safe areas are required.

3-49. In addition to supporting the assaulting force, maintaining mobility and momentum is an important role to ensure that the force can be supported. This includes maintaining and operating routes for pulse logistics as well as forward landing areas for air resupply. Engineers must be able to neutralize the effects of obstacles without disrupting our maneuver momentum in one of four ways: detect and avoid, detect and destroy from standoff, detect and reduce, or withstand the effects. FFE is used as an enabler for the engineer battlespace functions to support assured mobility and the elements of combat power *(see Figure 3-3, page 3-12)*, and the discussion of FFE.

3-50. The application of assured mobility is performed by all branches of the Army although the engineer branch has primary staff responsibility for it. As engineers support the full spectrum of operations, assured mobility is overlaid and integrated with those fundamentals of full-spectrum operations, as identified in *FM 3-0*. *Figure 3-4* visually represents the linkages and relationships between those fundamentals, assured mobility, the BOS, the engineer battlespace functions, and FFE. Each is related to and imbedded in those basic fundamentals.

3-51. Integrating new capabilities within the imperatives, engineers provide a combined arms proactive approach to mobility. To do this, engineers use information to maximize avoidance without committing into the threat's massed effects. Engineers support maneuver on multiple parallel routes and provide computing power used to identify numerous routes to the objective. They focus on attacking or neutralizing individual mines (just the ones that limit our mobility) instead of reducing lanes in minefields.

3-52. The fundamentals of assured mobility are predict, detect, prevent, avoid, neutralize, and protect. They are part of the full spectrum of operations that follow a continuous cycle of planning, preparing, and executing engineer operations that support decisive, shaping, and sustaining operations *(see Figure 3-4)*. Achieving assured mobility rests on applying the six

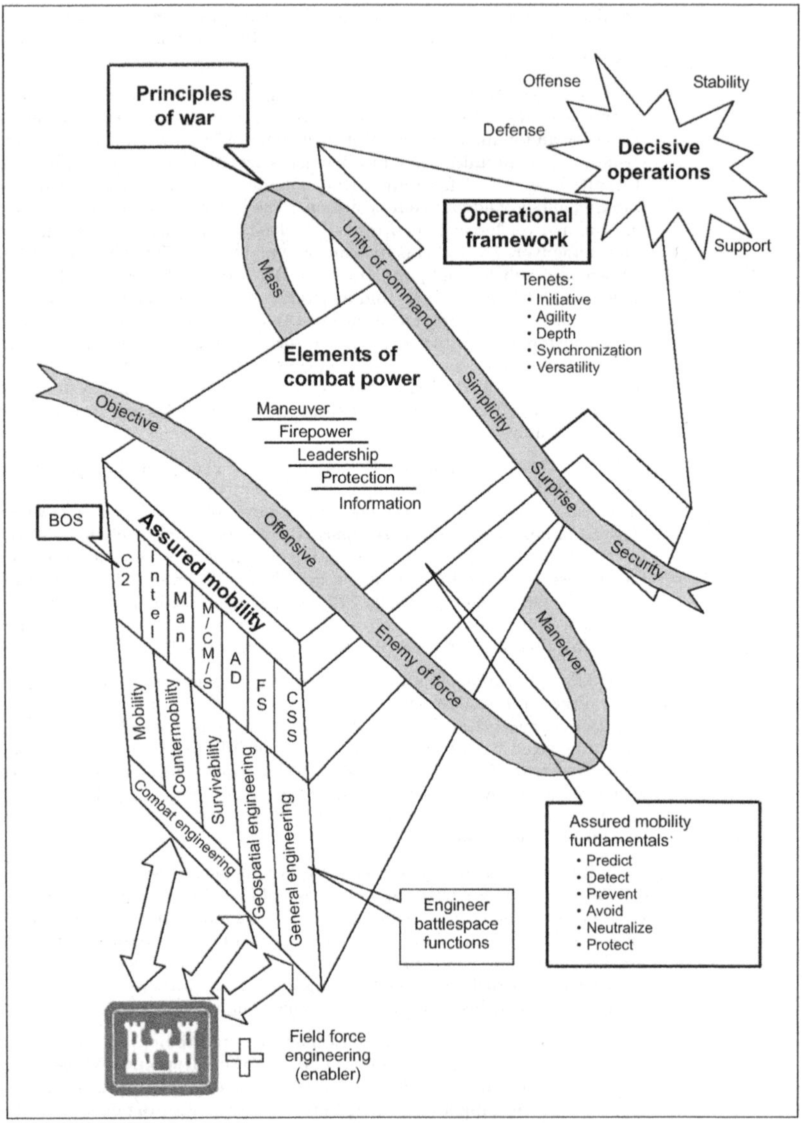

Figure 3-4. Fundamentals of Full Spectrum Operations and Engineer Battlespace Functions

fundamentals listed below. In essence, these fundamentals describe actions that sustain friendly maneuver ability and preclude enemy maneuver ability. They depend on superior SU, shared knowledge, and decisive execution.

- Predict. Predict actions and circumstances that could affect the ability of the force to maintain momentum.
- Detect. Detect early indicators of impediments to battlefield mobility and identify solutions through the use of ISR assets.
- Prevent. Prevent potential impediments to maneuver from affecting the battlefield mobility of the force by acting early. Political considerations and ROE may hinder our ability to apply the fundamental early in a contingency.
- Avoid. Avoid detected impediments to battlefield mobility of the force, if prevention fails.
- Neutralize. Neutralize, reduce, or overcome (breach) impediments to battlefield mobility that cannot be prevented or avoided. The breaching tenets and fundamentals apply when forced to neutralize an obstacle.
- Protect. Protect against enemy countermobility effects.

3-53. The most critical aspect of this framework is the linkage between the fundamentals. For instance, the linkage between predict and prevent, between detect and prevent, between detect and avoid, and between detect and neutralize are essential for success. A failure of any of these linkages will diminish a commander's ability to achieve decisive results.

3-54. The fundamentals of assured mobility are applicable from the strategic to the tactical level. *Table 3-1* provides examples of what resources might be used at various command levels, using the fundamentals against potential impediments. The list is not all-inclusive but intended as a tool to capture the distinctions of assured mobility at various levels.

Table 3-1. Potential Assured Mobility Resources

	Mobility Impediments	Predict	Detect	Prevent	Avoid	Neutralize	Protect
Strategic	Ground Routes	National assets	National assets	Blockades	Planning	Dive teams	Coalition support
Combatant Cdr	APOD	SOF	NGOs	Embargoes	Mine database	Construction units	HLS
	SPOD	FBI/Local police	HUMINT	Interdiction targeting of production facilities		TeleEngineering technical data	Key node force protection
	APOE	Request NIMA Data	UN demining data	PSYOPS	Alternate deployment routes	FBI/local police	Threat condition status
	SPOE	Schedule rapid terrain data acquisition	FBI/local police	FBI/local police	Research and Development Programs	USACE construction	FBI/local police
	Rail	Develop doctrinal template	Research and development programs	Research and development programs		Research and development programs	Research and development programs
		Research and development programs					
Operational	Theater LOCs	DTSS—terrain analysis	ACR	Interdiction targeting	Mine database	EOD	Danger areas isolation
JTF/Corps	ISB's	Terrain data acquisition—terrain IR	UAV	PSYOPS	Movement control	Airfield repair teams	Security operations
	Theater log bases	Refine/develop doctrinal template	SOF	CA/CI activities	Campaign areas of attack determination	Fuel-air bombs	
	Theater airfields	Develop SITEMP	NGOs			TeleEngineering technical data	
			HUMINT			Construction units	
			SIGINT				

Table 3-1. Potential Assured Mobility Resources (Continued)

	Mobility Impediments	Predict	Detect	Prevent	Avoid	Neutralize	Protect
Tactical	Routes	DTSS—tactical terrain products	RSTA units	Preparatory fires	Mine database	Area clearance	Armored vehicles
Corps/ Division/ Brigade	MSR/ASRs	UAV	TUAV	Route security patrols	Movement control	RTE clearance	Security patrols
	FOB	Refine / Develop SITEMP	STAMIDS	Raids	Alt corridors	CA breaching	Local security
	Axis		PSS-12	Interdiction targeting		EOD	
	Operations bases		HUMINT	PSYOPS			
	Road network		SIGINT			River Crossing	
	Rail network						
	Obstacles						
	Gaps						
	Rivers						
	Defiles						
	Urban areas						

3-55. With the development of the COP, commanders can use the fundamentals of assured mobility to ensure freedom of maneuver. During planning, engineers predict areas that the enemy may use to impede movement, using geospatial terrain products and ISR capabilities. During preparation, engineers, in conjunction with the intelligence community, employ sensors to detect impediments in the AO. ISR provides SU in the AO and populate selected corridors with collection capabilities. Engineers apply the prevent fundamental by denying the enemy the ability to influence operating areas. We do so by employing detection capabilities, countermobility systems, and obstacle neutralization systems to attack enemy assets that attempt to influence corridors. During execution, the commander uses the COP to maneuver his force and avoid restrictions. If a force cannot avoid restrictions, a commander neutralizes the impediment using the breaching tenets (intelligence, breach organization, breach fundamentals, mass, and synchronization). Protection is the enforcement of force protection measures to deny the enemy the ability to harm the force as the force maneuvers and performs its engineer functions during all phases of planning, preparing, and executing a mission.

3-56. Assured mobility is built from relationships with the functions and the BOS. The maneuver support (MANSPT) cell is responsible for integrating the BOS contributions to the fundamentals of assured mobility. MANSPT representative ensures the maneuver force integrates the imperatives for execution.

FIELD FORCE ENGINEERING

3-57. Even though the Army is reducing in size and is becoming more expeditionary in nature, engineers are still required to perform all of the broad engineer functions in support of the force commander. No one engineer unit METL is broad enough to allow units to train in every functional area, *(see FM 7-0)*. Engineers must balance multifunctional capabilities with a limited internal degree of expertise. FFE is one method to access the specialized aspects of the engineer functions via TeleEngineering. FFE applies Engineer Regiment capabilities across the range of engineer functions (although primarily general engineering intensive) and in all phases and types of operations ODSS through reach and forward presence. (*See Appendix C* for more detail.) FFE recognizes the critical need for early integrated engineer participation in planning and optimizing engineer capabilities for mission development and execution. This virtual collaboration with the FFE integrates the strengths of Army engineers in table of organization and equipment (TOE)/table of distribution and allowance (TDA) units (AC/RC), USACE (services and technical engineer expertise), and USAES (tactical expertise with geographically dispersed units in the field). This capability allows engineers to develop feasible solutions in a timely manner with early elimination of ad hoc solutions. Based on the current operational environment with more complex requirements and operating conditions, FFE addresses concerns about engineer availability and synchronization, given the general limited technical engineering and environmental capability in most TOE engineer units. FFE allows engineers to reduce their footprint, maximize their capabilities by deploying only the necessary engineer skills (specialized

USACE teams and ENCOM modules) on the ground (reaching to access other technical expertise on a situational/as needed basis), and reduce the associated vulnerabilities and other negative aspects of deployment into the AO.

SECTION III - THE ENGINEER REGIMENT

3-58. The Engineer Regiment contains a wide range of organizations that contribute performing and supporting capabilities to engineer functions at every level. Determining the type and numbers of Engineer Regiment assets and capabilities to support any given operation will be performed by the combatant commander and/or the ARFOR, in conjunction with his engineer staff.

UNITED STATES ARMY CORPS OF ENGINEERS

3-59. The Chief of Engineers has aligned USACE divisions with combatant commanders as they reinforce and extend the capabilities of the Regiment *(see Figure 1-1, page 1-3)*. This relationship with the combatant commander allows direct access to USACE resources to support engagement strategies and wartime operations. The USACE mission covers the full spectrum of operations with five major functions:

- War fighting operations provide full-spectrum engineering and contingency support.
- Disaster relief operations respond to local, national and global disasters.
- Infrastructure builds and sustains the critical facilities for military installations and the public.
- Environmental operations restore, manage, and enhance ecosystems, local and regional.
- Water resource development creates synergy between water resources development and the environment.

3-60. USACE capabilities include access to the expertise of the seven Engineer Research and Development Centers (ERDCs) and all of the resources within the divisions, districts, and other sources. USACE is the primary proponent of FFE, which enables the engineer battlespace functions. FFE is provided through deployed tactical engineer units and USACE personnel (deployed and at their home station). The engineer commander maintains his flexibility and determines the mix of capabilities (troop, USACE civilian, and contractor) based on the tactical situation, time-phased requirements, required/available capabilities, funding, and force caps. The USACE division commander task-organizes division capabilities to meet the varying time-phased requirements. The division's capabilities depend heavily on reach through systems such as TeleEngineering. The FFE concept is applicable in joint and combined operations to provide a better engineer solution that can be implemented faster and with a smaller footprint. The Air Force and Navy have similar capabilities. The Air Force's FFE is called geo-reach and the Navy's FFE is called engineer recon teams (not to be confused with tactical Army engineer reconnaissance teams [ERTs]). The USACE objective of FFE is to more effectively execute its roles (engineering expertise, contract construction, real estate acquisition and disposal, and environmental engineering) in all operations. Additionally, the USACE FFE objective is to maximize use of reach to provide technical assistance and enable the engineer battlespace functions for the combatant commander. USACE accomplishes this by training, equipping, and maintaining specialized, deployable forward engineer support teams (FESTs).

ECHELONS ABOVE CORPS

3-61. At EAC there are engineer C2 HQ called ENCOMs and/or theater Army (TA) brigades and engineer groups that are assigned the responsibility of commanding and controlling how forces execute engineer functions in a theater AO. As the combatant commander's LCC establishes the theater backbone structure, the ENCOMs focus on operational issues such as enabling theater entry and supporting sustained forces. The ENCOM commander establishes a relationship with the Army or land component staff engineer to synchronize the engineer effort. CS assets task-organized to the component commander cannot be task organized to the staff engineer because he does not have command authority. This allows corps and division engineers (DIVENGs) to remain focused on the conduct of the tactical operations. The actual structure of the theater-level engineer force is not firm. The organization is tailored to the anticipated mission. Its capabilities are diverse and specialized. The theater force engineer uses the building block approach to determine the engineer structure based on the varying operational requirements. The theater force engineer has TDA and modified table of organization and equipment (MTOE) units from AC and RC organizations to tailor its capabilities. These unit capabilities include prime power, facility engineering, real estate, topography (and other geospatial capabilities), combat heavy, port construction, utilities, fire fighting, quarry, pipeline, diving, well drilling, and bridging. *Figure 3-5, page 3-22*, shows the existing operational engineer force structure for EAC and EAD.

3-62. The theater force engineer support spans more than just theater sustainment. He can support the maneuver force engineers by accomplishing missions within the combat zone. Control measures such as engineer work lines (EWL) are used to delineate engineer responsibilities between EAC and corps and below engineer units. ENCOMs and TA brigades are prepared to deploy capability modules to provide tailored capabilities to the ARFOR/ combatant commander. These modules are based on unique competencies that include C2, theater planning, facilities engineering, design and survey, and geospatial engineering.

CORPS

3-63. The corps engineer brigade commands and controls all engineer support to the corps and is assigned all engineer units that are not organic to divisions, separate maneuver brigades, and cavalry regiments. The brigade provides combat, geospatial, and general engineering support to the corps based on METT-TC. Corps geospatial engineering support normally is

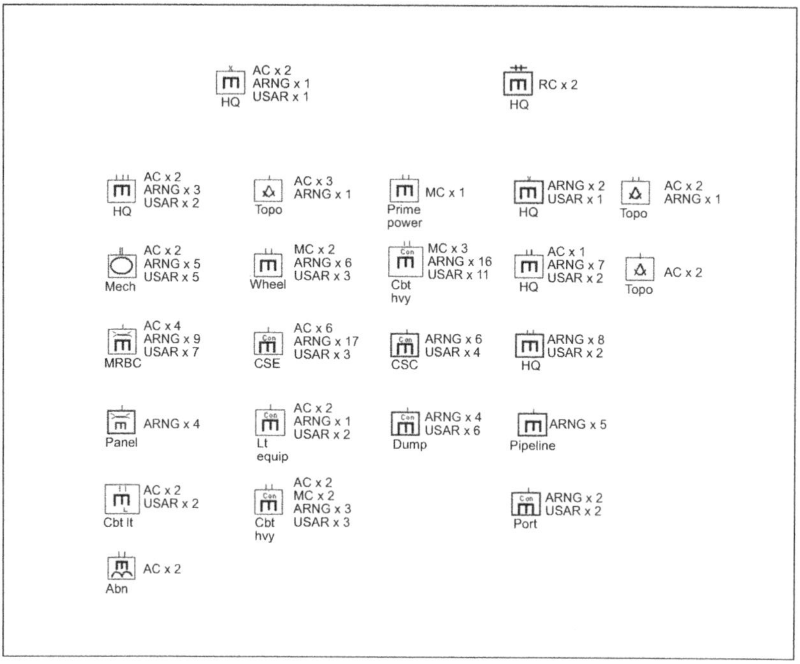

Figure 3-5. Engineer Units at Echelons Above Division and Echelons Above Corps

provided by a topographic engineer company placed in DS of the corps from the TA topographic engineer battalion. The corps engineer brigade augments engineers organic to divisions, separate maneuver brigades, and cavalry regiments. The corps engineer brigade may contain various numbers of engineer groups, corps engineer battalions (mechanized, wheeled, airborne, and light), combat-heavy engineer battalions, and separate engineer companies (fixed and assault float bridge, combat support equipment [CSE], light equipment [LE], and topographic). Other specialized engineer organizations will augment the corps engineer effort as the mission dictates. The brigade provides oversight and support for theater engineer forces from EAC units operating in the corps area. These units include prime power, construction support, pipeline construction, dump truck, and port construction, utilities, well-drilling, fire fighting, and other special teams and detachments.

3-64. The corps engineer brigade commander also serves as the corps engineer special staff officer. The corps staff engineer section (SES) provides engineer functional-area expertise to all corps staff elements. The SES provides recommendations to the corps staff on the use of engineer assets and ensures that the engineer battlefield functions are fully planned, integrated, synchronized, and executed to support the corps commander's intent and scheme of maneuver. The corps engineer tasks and prioritizes the work effort of the DS corps topographic company. In force projection theaters where no forward-based theater engineer structure exists, the corps engineer brigade could initially function as the theater engineer HQ and regional-contingency engineering manager (RCEM). To do this, the brigade needs special augmentation from an ENCOM and/or the USACE in the areas of construction management, real estate acquisition, and construction contracting support. The brigade would execute this function until an ENCOM, TA engineer brigade, or engineer construction group arrives in theater. In the absence of follow-on deployment of an ENCOM, TA engineer brigade, or engineer construction group, the corps engineer brigade (with the special augmentation listed above) may have to act as the theater engineer HQ indefinitely.

DIVISION AND BELOW

3-65. DIVENGs in an analog division serve two critical roles for the division. First, they provide engineer expertise at every echelon of command from the division to the company or team. Second, they provide the structure necessary to command engineer units at these echelons. Both of these roles involve all of the engineer battlefield functions: combat (M/CM/S), geospatial, and general engineering. As a combat multiplier, engineer units focus on maintaining division freedom to maneuver and attacking the threat's freedom to maneuver on the battlefield. As part of the division staff, the DIVENG focuses on integrating and synchronizing engineer missions to support the division commander's intent and scheme of maneuver.

3-66. In the Force XXI (FXXI) division, the engineer brigade HQ is replaced by an engineer cell in the division tactical (TAC) command post (CP), Assistant Chief of Staff, G3, Operations and Plans (G3) division main, and DIVENG section in the division main. This staff is tasked by the DIVENG to form an organic division staff engineer section (DSES). The DIVENG has no units task-organized directly to him. He provides the division commander with planning, integration, and synchronization of engineer operations in the deep, decisive, and sustaining fight. Integrating ISR requirements is critical to this end. The DIVENG's knowledge on maneuver and force integration allows for effective force integration of EAD engineer forces as mission enhancements. This includes acquiring and managing information, maintaining SU, and estimating the necessary direction of subordinate forces.

3-67. The M/CM/S element of the division TAC is comprised of chemical and engineer personnel. Their function is to enhance division mobility and degrade enemy mobility. This element is also mission with the survivability by avoiding or at least minimizing the effects of WMD. The M/CM/S element:

- Monitors, directs, and coordinates the activities of chemical and engineer units.
- Has an embedded engineer team that conducts reconnaissance to continuously update the division knowledge base and ensure that the

COP accurately reflects all obstacles (both friendly and enemy) and trafficability information concerning the AO.

- Has engineer and chemical elements that work closely with the G2 and FSE to track and plot known obstacles.

3-68. The G3 M/CM/S element in the division main consists of three distinct sections—engineer, chemical, and MP. The cell is focused on mobility (both friendly and enemy), countermobility (friendly and enemy) and survivability (protection) of the force. The engineer staff will also integrate the geospatial and general engineering requirements. The engineer section—

- Prioritizes, makes allocation recommendations, and task-organizes divisional engineer assets and other engineer units received from EAD.
- Assists the plans team as it develops OPLANs and OPORDs. Makes recommendations for the appropriate use of engineers and provides detailed knowledge of their capabilities and limitations.
- Assists the G2 section of the IPB cell and the G2 plans officer with the terrain analysis portion of the IPB.
- Maintains an accurate status and location of all friendly and known enemy obstacles, to include impediments to movement caused by the weather or terrain.
- Advises and coordinates with the DIVENG for planning and executing rear-area missions/tasks in support of sustainment operations across all five of the engineer functions. This includes providing terrain products through the use of the Digital Topographic Support System (DTSS) and providing terrain products used while developing or analyzing COAs, thus enhancing the division commander's ability to make informed battlefield decisions. CS assets task-organized to the FXXI division cannot be task-organized to the division staff engineer because he does not have command authority.

3-69. In the FXXI division, the engineer sections can electronically move and share information with subordinate and higher units. This allows for rapid synchronization of operations and dissemination of orders. Each process is complemented by the use of the Maneuver Control System-Engineer (MCS-ENG), FXXI Battle Command Brigade and Below (FBCB2), All-Source Analysis System-Remote Workstation (ASAS-RWS), DTSS and C2 personal computer (C2PC). As part of the division staff, the DIVENG section focuses on integrating and synchronizing all engineer missions to support the commander's intent and scheme of maneuver. Engineers of the FXXI and the analog divisions provide engineer expertise at each echelon of command and planning and synchronization for engineer operations in support of the maneuver commander's intent and the scheme of maneuver.

3-70. Each armored-cavalry regiment (ACR) has a regimental staff engineer and an organic engineer company. The regimental engineer provides engineer planning expertise to regimental and squadron commanders. Although the regimental engineer has no command authority, he remains the senior engineer advisor to the regimental commander. CS engineer units task-organized to the regiment work in conjunction with the regimental engineer to conduct parallel planning and battle tracking. The regimental engineer remains the primary engineer staff officer in the regimental staff even if the task-organized engineer commander is senior. CS units may be abruptly task-organized away; the regimental engineer provides continuity. The regimental engineer role involves all of the engineer battlefield functions: combat (M/CM/S), geospatial, and general engineering. The engineer company commander is a combat multiplier, focused on maintaining the Regiment's freedom of maneuver and mission accomplishment. As part of the regimental staff, the regimental engineer focuses on integrating and synchronizing engineer missions to support the regimental commander's intent and scheme of maneuver.

3-71. The SBCT has a brigade engineer and a single engineer company that is designed and organized principally for operating in SSCs, stability actions, support actions, and MTWs. The company provides combat engineering and limited general engineering (primarily sustainment), and embedded geospatial engineering to support the force in preventing, containing, stabilizing, or terminating a crisis. The brigade engineer is the primary advisor to the brigade commander and may also serve as the MANSPT coordinator. As the MANSPT coordinator, integrates and synchronizes all MANSPT assets into the planning process and conducts battle tracking. The brigade engineer has no command authority, but remains the primary advisor on engineer matters, even if a senior engineer commander is task-organized to support the SBCT.

SECTION IV - INTEGRATION INTO THE COMBINED ARMS TEAM

3-72. Each maneuver force echelon from corps down to battalion/TF level has an ENCOORD to integrate engineers into the combined arms fight. For some higher echelons, the ENCOORD is solely a staff officer; however, the ENCOORD is usually an engineer unit commander/leader with an associated staff officer. At the non-FXXI division level, for example, they are the DIVENG and assistant division engineer (ADE), respectively. In either case, the engineer is a special staff officer who is a member of the echelon battle staff and plays an integral part in developing plans and orders. Engineers work with all members of the battle staff and must understand their capabilities to integrate and synchronize the M/CM/S BOS and contribute to integrating the intelligence and CSS BOS.

3-73. The primary contribution of the staff engineer at the operational level is a solid engineer battlefield assessment (EBA). Without a good EBA, the staff engineer cannot properly contribute to developing a scheme of engineer operations (SOEO) and a task organization to support the plan. A good EBA distributed to subordinate engineers also facilitates the parallel planning process and integration of a BOS.

3-74. At the strategic and operational planning stages, engineer input makes an essential contribution to setting the conditions for operational and tactical success. It enhances the effectiveness of joint targeting by advising on the selection of targets within the enemy's national and military infrastructure. This is done to optimize the required effect on his forces and population while minimizing unwanted effects on nonmilitary targets and joint forces in subsequent operations. This allows attacks to be focused, reducing the effort required to achieve the desired strategic or operational effect and improving the joint capability to attack the enemy's war effort without causing disproportionate collateral damage to civilians and the environment. In DS to imminent operations, engineers must be included in targeting planning at all levels to prevent destruction of key infrastructure essential to future operations and to provide guidance on efficient achievement of the required countermobility effects. The engineer staff should also be involved in battle damage assessment (BDA).

3-75. Engineer support allows the maneuver commander to reinforce cohesion and HN cooperation by preserving HN infrastructure and relieving the effects of conflict on its environment and its population, including refugees. There are a wide variety of activities or tasks that could be conducted by engineers to support building routes and camps for displaced persons and refugees and mine awareness training for civilians. Engineers directly support preserving and protecting the force by hardening the C2 and other facilities, providing collective protection, and removing threats to troops and morale (UXO, mines, scatterable munitions, and IEDs). Engineers provide support for cover, concealment, and camouflage deception efforts, beyond the supported formation capabilities, to counter enemy intelligence operations and protect the force from the effects of enemy fires.

3-76. Engineers sustain the joint force directly through activities such as construction of fuel supply networks and production and supply of water, electricity, and accommodations. They are responsible for construction, maintenance, and post conflict restoration of services and facilities essential to operations. Engineers also provide essential support to logistic activities. In addition to assessing and recommending the most efficient use of existing infrastructure, they develop, construct, repair, and maintain a wide range of logistic facilities.

SECTION V - OPERATIONAL CONSIDERATIONS

FIGHTING AS ENGINEERS

3-77. Combat and CS engineers are the vanguard of the Army because they fight beside combat arms units. When conducting combat operations in close battle, they must be prepared to fight and employ their combat skills, using fire and maneuver to accomplish their engineer mission. On today's battlefield, the enemy can detect and engage engineers quickly, regardless of their location. Consequently, all combat engineers/sappers are organized, trained, and equipped to fight and destroy the enemy in addition to their primary responsibilities within combat engineering. This section addresses aspects of engineers in close combat, organized to fight as engineers.

3-78. Combat and CS engineers are organized, trained, and equipped to engage in close combat to accomplish their engineer missions and—

- Support a movement to contact (MTC) or attack as part of a maneuver formation in the movement to accomplish the formation mission.

- Fight as the reduction force during combined arms breaching operations.
- Assist the supported organization to defeat an unexpected attack.
- Protect a reserve obstacle that must remain passable until friendly forces are able to withdraw.
- Maintain security at a work site.
- Protect themselves in an AA or on the march.

3-79. The enemy will attempt to kill combat engineers as well as dismounted infantry or mechanized forces. It is imperative that engineers are trained to be physically aggressive and tactically competent. The combat engineers' secondary mission is to reorganize into infantry units and fight as infantry.

ENGINEER COMBAT ORGANIZATION

3-80. The 21B combat engineer is trained to accomplish the same basic tasks as the 11B infantryman, in addition to specialized engineer-unique tasks. Engineer squads/platoons/companies are trained to move rapidly and fight violently, either by themselves or in maneuver battalions as a part of a combined arms formation. The additional skills of the 21B make him a potent and versatile member of the combined arms team.

3-81. Mechanized (and the motorized SBCT) combat-engineer squads are organized around their armored vehicle and are armed with an array of rifles, squad automatic weapons, grenade launchers, light and heavy machine guns, and AT weapons. The squads also carry an array of demolition materials, configured into satchel and combat demolition charges, and are able to attack rapidly and violently with demolitions as well as fire. In the platoon, they carry a basic load of conventional and scatterable mines (SCATMINEs) sufficient to emplace a minefield quickly and defend it, if necessary.

3-82. Wheeled combat engineers are organized and equipped much the same as mechanized combat engineers. The major difference is the squad carrier—a 5-ton dump truck or family of medium tactical vehicles (FMTV). When dismounted, a squad and platoons are trained to function much like a dismounted infantry organization.

3-83. Light engineers move on foot, carrying critical tools and equipment as well as demolition materials. As squads or platoons, light engineers move as part of the light infantry formation. Capable of using fire and movement techniques, they also contribute employing demolition and fire to a close-combat fight.

3-84. Combat (heavy), topographic engineer units and specialized engineer companies/detachments are armed primarily with rifles and have a limited number of crew-served weapons. They are not organized to move within combined arms formations or apply fire and maneuver. They are capable of engaging in close combat with fire and movement primarily in a defensive role.

ENGINEER COMBAT CAPABILITIES

3-85. During offensive operations, combat engineer units are task-organized with maneuver units and are integrated into the combined arms formation.

The engineer unit is designed to provide demolition and reduction capabilities to the combined arms team. The engineer unit can also employ direct-fire weapons systems to aid in employing demolitions and reduction assets. Regardless of the mission, armored engineer vehicles are combat vehicles that provide a significant contribution to the combat power of the entire formation. To accomplish the mission, engineers fire and move under the direction of the formation commander, as necessary, using demolition skills when appropriate. Fire and movement techniques are based on rifle, automatic rifle, and grenadier covering fire, allowing the placement of demolition charges within striking range.

3-86. When involved in an assault, engineers fight dismounted on the objective. However, they focus on reducing the close-in protective obstacles as well as demolition tasks against positions and dug-in vehicles. Demolition charges produce significant shock and concussion effects on defenders, as well as destroy critical positions, munitions, and combat vehicles.

3-87. Combat engineers employed on reserve obstacles in the defense execute the technical procedures necessary to ensure target destruction. However, the engineer demolition party responds to enemy contact. They assist the demolition guard in securing the target by holding it open or gaining time to ensure for destruction. The engineer force may assist in target defense by installing obstacles to support the defensive scheme.

3-88. Engineer units engaged in emplacing obstacle systems provide their own local security. Within their capability, they employ close combat techniques against attackers to ensure that the obstacle system is completed. Construction, topographic, and other specialized engineer organizations also provide their own local security. In rear operations, they participate in base cluster defense. They install local protective obstacles and fight from defensive positions on the perimeter. They also form reaction forces that can expel or destroy enemy forces that penetrate a base cluster.

REORGANIZING TO FIGHT AS INFANTRY

3-89. Throughout history, engineer organizations have been required to fill the role of infantry as a secondary mission. The combat engineer/sapper is part of a well-armed and well-equipped organization capable of executing infantry tasks in conjunction with other combat units. Organizational deficiencies include the lack of organic fire control personnel, communications equipment, and medical personnel. If an engineer battalion has been designated to fight as infantry (a maneuver unit), then it requires the same support and integration of other maneuver elements (such as armor, AT, fire support, or air defense artillery [ADA] elements) into its task organization to accomplish its mission. While engineers always fight as engineers, reorganizing them and employing them as infantry requires operational considerations.

EMPLOYMENT CONSIDERATIONS

3-90. Reorganizing engineers as infantry is an operational-level decision that requires corps commander or higher authorization due to the nature and considerations involved with reorganization and the impact (resourcing and training) on the engineer unit and the higher organization. However, a commander must carefully weigh the gain in infantry strength against the loss of engineer support. Engineers provide far more combat power in their primary mission than when configured as infantry. Stopping the engineer work may reduce the combat power of a commander's entire force. Because of the long-term impact, a commander must notify higher HQ when employing engineers as infantry. A commander must carefully analyze infantry and engineer demands before deciding to employ an engineer unit as infantry. Employing engineers merely implies the higher commander using the engineers for a short period of time. On the other hand, reorganization is an action taken to shift internal resources within a degraded unit to increase its level of combat effectiveness. This requires resources, time, and training and is for an extended period of time.

3-91. An immediate requirement for the infantry may not require reorganization; engineers are simply committed to the fight. Reorganization occurs when time allows moving unneeded engineer elements and equipment from the battle area and augmenting the engineer structure with additional capabilities. A commander normally considers reorganizing when forecasting a shortage of infantry before a future operation or phase of an operation. The commander makes a decision after weighing METT-TC factors and determining an acceptable risk level. Reorganizing as infantry takes time and other assets.

Corps And Above

3-92. Corps combat engineer battalions working in either the division or the corps rear may be employed more easily as a separate infantry force. These units frequently work under the control of their battalion HQ and are not dispersed and integrated into other formations. They are also located to move forward and join the force in contact, form a reserve, or prepare and occupy blocking positions. The commander directing this employment should provide early warning to allow the unit time to assemble, reorganize, and prepare for the mission before commitment. The engineer unit must provide immediate liaison to the gaining maneuver command to facilitate planning and integration. This generally requires about 24 hours, unless the unit has previously prepared for a similar mission.

3-93. When an engineer unit is employed as infantry, one major consideration for the commander is to store engineer equipment, such as bulldozers, bucket loaders and road graders in tactical assembly areas. Equipment not used in the infantry role may be attached to other units for C2 purposes or to accomplish other engineer tasks. This decision is METT-TC driven and generally based on the overall concept of the operation.

3-94. The commander directing the employment should augment the engineer unit with additional medical personnel and vehicles, air defense teams, and fire support teams. The unit should also be augmented with armor, AT weapons teams, mortar teams, and other combined arms capabilities, as required, to perform the expected mission.

Division and Below

3-95. Generally, DIVENG battalions are task-organized throughout the division's area and are closely integrated with other maneuver arms. Engineers fight and conduct their operations in this configuration. Engineers in a combat vehicle or dismounted formation (with satchel charges or rifle fire) fight, as required, under the formation commander. Engineers who prepare defensive positions fight from those positions alongside the defenders, if attacked. DIVENGs use their close combat skills as infantry in an emergency while performing their engineer mission.

CONSIDERATIONS

3-96. Engineer units employed as infantry do not have the same capabilities as conventional infantry units. Because they normally operate the same as infantry organizations and have the same basic weapons, this is generally not a problem at squad and platoon levels. At higher levels, this is a concern.

3-97. The engineer company can effectively control other arms as a company/team because it normally works closely with them. The engineer company routinely maneuvers alone and is well suited to protect TF flanks during MTC operations or as the TF breach force during deliberate attacks.

3-98. To be fully effective at the battalion level, engineer units need to be augmented with heavy AT weapons and mortars, as well as the normal CS provided to any infantry unit. Engineer battalions frequently maneuver as battalions, so their training makes them effective in offensive operations and in a defensive role when employed as infantry.

3-99. Employing engineers as infantry is likely to occur when the reserve force has been committed and no other reserve force is possible. The engineer reserve force can be used in two ways: as a reinforcing force for units in contact with or as a blocking force against an attack or counterattack. It can accomplish this mission by building and occupying a strong point. Other uses of an engineer reserve force include—

- Augmenting an armor battalion with infantry to build a TF.
- Augmenting an infantry battalion with an additional company.
- Operating as the breach force commander for a brigade breaching operation.
- Operating separately in an economy of force role or as part of a brigade defense.
- Providing air-assault forces for seizing critical terrain.
- Providing a specialized assault force for urban operations (UO) or the attack of an enemy strong point.

Chapter 4

Planning Engineer Operations

A military engineer must have broad engineering experience between wars.

Major General Julian L. Schley
(Chief of Engineer, 1937-1941)

The Engineer Regiment is a large and diverse organization that provides support to commanders at every level. The lines of responsibility between USACE, MTOE, and TDA units have blurred with our involvement in full-spectrum operations and new threats. As echelons of responsibilities have blurred, we realize that every engineer unit is capable of supporting a maneuver unit at any level of war. For example, a combat heavy battalion may deploy a subordinate unit or capability to support a JTF at the strategic level or provide support for an Army corps at the operational level or even a mechanized battalion at the tactical level. Engineer planning is conducted at the strategic, operational, and tactical levels. The engineer staff should participate in all aspects of planning for operations, particularly during the early stages, to ensure the successful engineer preparation of the operational area. Engineer planners should consider a wide range of requirements.

SECTION I - INTRODUCTION TO PLANNING

4-1. Engineer doctrine addresses the range of full-spectrum operations, which include offensive, defensive, stability, and support operations. Engineers support these operations, or a combination of these operations in any environment. Versatility is a key principle for engineers participating in executing and transitioning between these operations. Essentially, any engineer organization is capable of operating and supporting the commander through the entire range of operations. This may be accomplished by tapping into the expertise, skills, and products available outside of their organization. Expertise may come from international and/or other service engineers. Internally, the Regiment has become a network-centric organization that allows every organization to access each other's unique skills via reach (backward and laterally), including TeleEngineering capabilities embodied in FFE. An MTOE engineer unit brings its core strengths and is reinforced by USACE, ENCOMs, USAES, or other assets within the Regiment for the specific operation in which it participates.

ARMY TRANSFORMATION AND THE ROLE OF ENGINEERS

4-2. The Army is currently engaged in a transformation to become a more strategically responsive force across the full spectrum of operations. The Army and its engineers are moving along three major paths- the Future Force, the current force, and the Stryker force. This FM supports the doctrinal requirements for current engineer operations, and it lays out the framework for operations during the Army's transformation. It addresses engineer operations in support of modern and digitally enhanced forces and Stryker force. By addressing these forces, we provide engineer doctrine that will pull us toward the future force and fulfill the role of being an engineer operational capstone manual.

4-3. The critical transformation path leads to the Future Force. It will be more strategically responsive, deployable, sustainable, versatile, agile, lethal, and survivable than the current force structure. The engineers enabling the Future Force will have these characteristics in order to provide proactive support and imbed combat (M/CM/S), geospatial, and general engineering capabilities. Engineers are a primary enabler of assured mobility. Today, the science and technology community is working hard to achieve advanced capabilities such as stand-off mine detection and neutralization for the close fight. Engineers will develop the systems and techniques necessary to stay ahead of the potential asymmetric capabilities of our adversaries.

4-4. At the same time, the Army will retain those forces (the current force) as we know them today. These forces are being recapitalized through the continuation of existing modernization programs, such as digital technologies. This will be part of the force used for war should they be required anytime in the next 10 years or so. Engineers will continue to field digital capabilities, and provide geospatial support, and in some cases, Bradley engineer fighting vehicles (BEFVs), Wolverines, and other necessary systems to provide modern capabilities to the current force.

4-5. The Stryker force is bridging the gap in capabilities between today and the Future Force. The Stryker force will validate an organizational and operational model for the Future Force. This organization is based on a medium-weight combat vehicle, the Stryker, that allows a more responsive and deployable unit. The engineer squad vehicle (ESV) Stryker variant requires adjustments in the tactics, techniques, and procedures (TTP) involved with SBCT breaching operations. The organic terrain team in the SBCT provides improved geospatial capabilities to the brigade. ISR capabilities and sensors are distributed to the lowest-level units and contribute to the collection plan and OBSTINTEL. Engineers in this structure will adhere to their base doctrine in *FM 3-34.221*.

ENGINEER PLANNING FOCUS

4-6. The planning process and considerations for the engineer are very similar, regardless of whether involved at the joint or the Army level. The omission of engineer planning in any phase of an operation adversely affects the entire plan. In many cases, the engineer commanders may be required to educate supported commanders on engineer capabilities, methods, and depth of the Engineer Regiment.

4-7. Engineer operations require engineer planning at all levels across the range of military operations. Engineer planners must determine the mobilization, deployment, employment, sustainment, and redeployment requirements of the combatant commander or senior Army commander's concept of operations. Operational planning merges the operation plan of the joint force, specific engineer mission, and available engineer forces. Tactical planning occurs primarily at the level of the unit specified to accomplish the tactical task or mission. The other service components can also accomplish tactical planning. When planning joint/Army operations, engineer planners should consider the wide range of diverse requirements for engineer support operations. Successful engineer support to the joint force requires the early involvement of engineers in the planning process and in all phases of joint operations. Understanding terrain and how engineers support the joint force and enhance air, land, and sea operations provides the essential background for planning engineer operations. Early development of a comprehensive plan for engineer operations ensures the availability of engineer forces, equipment, and materiel in support of joint/Army operations.

4-8. The challenges of planning successful engineer operations in support of joint operations within diverse theaters are vast and varied *(see Figure 4-1, page 4-4)*. The engineer staff must be involved in planning from the initial stage of the process by providing geospatial products to the commander and staff. Understanding how engineers affect air, land, and sea operations equips the planner with the background to form a comprehensive plan of engineer actions. This universal application of engineers is crucial at all levels of war.

4-9. Early in the planning process, the engineer staff produces an EBA. During operations, engineer planners should consider such things in their EBA as—

- Geospatial terrain analysis.
- Intelligence requirements.
- Topographic support.
- Construction support (including construction safety requirements and safety criteria).
- Countermine operations.
- Force protection.
- HN forces.
- HN infrastructure.
- Multinational operations.
- Interagency operations.
- Contractor support.
- Materiel acquisition.
- Operational phases.
- Environmental considerations.
- Funding requirements.
- Resource management.
- Other critical considerations, including threat engineer capabilities, terrain, and weather conditions having an impact on particular AO.

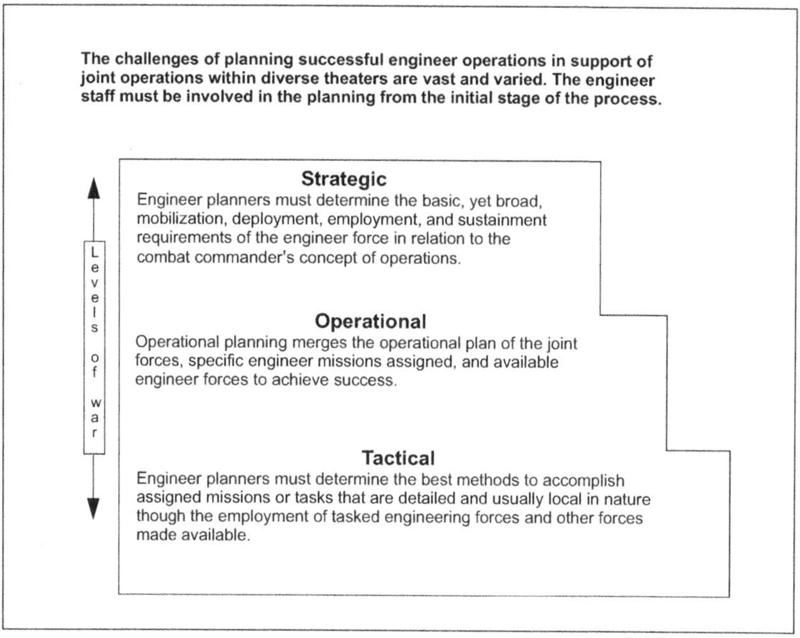

Figure 4-1. Engineer Planning

4-10. *Planning* is the means by which the commander envisions a desired outcome, lays out effective ways of achieving it, and communicates to his subordinates his vision, intent, and decisions, focusing on the results he expects to achieve. Planning is the first step of the operations process that leads to preparation and execution. Throughout the planning process, commanders focus on the results they expect to achieve. This is true at all echelons and levels of war.

4-11. Planning is a dynamic and continuous process. It applies the elements of operational design to a specific situation to accomplish a mission. Commanders assess enemy and friendly centers of gravity, decisive points, and the associated objectives that allow them to attack or defend. The engineer commander's vision, as summarized within the intent, expresses the military conditions that achieve the end state. The concept of operations describes how the various elements of the command will operate together and how they are nested in purpose, space, and time to support the higher commander's or HQ's intent.

4-12. When engineers plan for operations (threat manifested as an opponent or a natural disaster), they must—

- Understand the higher commander's or HQ's intent and be an integral member of the planning staff.
- View the battlefield/AO; have extensive knowledge of the terrain and geospatial products available, OBSTINTEL, and threat capabilities; and anticipate enemy COAs (ECOAs). ISR assets are integrated into finding and answering combat, geospatial, and general engineering intelligence requirements to predict, detect, and prevent threat capabilities. Engineers should visualize the end state and all engineer tasks and purposes from start to finish and transition to future operations.
- Use engineer systems to the best advantage. There is no single solution to engineer problems. Engineer commanders know their equipment and configure/package it in the best combination to address impediments to the friendly maneuver plan. They should resource equipment and organizations to accomplish the identified tasks within the time allotted, identify reduced engineer capabilities, and identify methods to mitigate the risks and leverage all capabilities in the Regiment.
- Concentrate overwhelming engineer capabilities. Engineers mass capabilities to accomplish all missions to facilitate maneuver operations and impede threat movement. During defensive operations, engineers leverage all capabilities to isolate the threat and protect the force.
- Shock and overwhelm the threat. Engineer units move quickly and synchronize to facilitate maneuver operations and counter all threat impediments. They should deny the threat time to prepare and execute their plan. During stability and support operations, engineers may look to assist the population recover from the shock and damage of a threat, such as a natural disaster.
- Provide continuous engineering support. Engineers cannot just look at the tip of the spear but must consider the depth and ensure engineer capabilities are available for freedom of movement from the rear boundary forward. This includes the integration of environmental considerations.
- Man, arm, fuel, fix, move, and sustain the sappers. Engineers ensure that all logistical requirements are thought through to the end state and resourced to accomplish the mission and facilitate future operations.

4-13. Typically, the engineer commander is dual-hatted commander and senior engineer staff advisor or ENCOORD. At the TA level and above and currently in FXXI divisions, the senior staff engineer may no longer be the commander. However, it is still critical in either case that the commander or senior staff engineer conduct parallel planning using the MDMP with task-organized or subordinate engineers to facilitate a realistic assessment. The engineer commander's intent and planning guidance direct the activities of the staff and subordinate commanders. The staff assists the commander with the coordination and detailed analysis necessary to convert the planning guidance and commander's intent into a plan. It becomes a common point of reference for operations.

4-14. Operational and tactical planning complement one another but have different aims. Operational planning prepares the way for tactical activity on the most favorable terms (proper resourcing) and continually seeks to foster and exploit tactical success. Major operations depend on the creative use of tactical action to accomplish a strategic or operational purpose within a specific situational context against an adaptive opponent. Tactical planning emphasizes flexibility and options. Comprehensive planning may be feasible only for the first event or phase in an operation; succeeding actions depend on the enemy response and circumstances and require sound branches and sequels.

4-15. Scope, complexity, and length of planning horizons differ between operational and tactical planning. Campaign planning coordinates major actions across significant time periods and distances to achieve operational objectives. Planners mesh service capabilities with joint and multinational formations as well as interagency and NGOs. Tactical planning has the same clarity of purpose as operational planning but has a shorter planning horizon. The plan guides subordinates as they progress through each phase of operations. Comprehensive, continuous, and adaptive planning characterizes successful operations at the operational and tactical levels of war.

4-16. The joint combatant commander or senior Army commander's engineer planning concepts focus on the relationship of geography and force projection infrastructure to the concept of operations. Engineer planners must determine the basic, yet broad, mobilization, deployment, employment, and sustainment requirements of the combatant commander's concept of operations. At all levels of planning, the senior engineer commander and/or the ENCOORD at each echelon must support the supported commander's OPLAN/OPORD and the internal OPLAN/OPORD for the engineer organization. In selected circumstances, the senior engineer commander may also be the supported commander.

ENGINEERS AND THE LEVELS OF WAR

STRATEGIC LEVEL

4-17. In strategic level of war, art and science are used to develop and employ armed forces as part of the President's instruments of power (diplomatic, information, military, and economic). The instruments of power are synchronized to secure national and multinational objectives. Combat, CS, and CSS engineer capabilities available to the JFC are described in *JP 2-03, JP 3-34,* and *JP 4-04.* Engineers provide unique capabilities derived through USACE and AC and RC engineer forces to allow the JFC the greatest flexibility when developing a strategy to achieve the LCC's objectives. At the strategic level, the availability of ports, roads, airfields, and other infrastructure affects the sequencing of units and the tempo of entry operations. Engineer support to force projection is essential to operational mobility. The USACE and the Army's ENCOMs are the critical strategic resources that support these force projection missions.

4-18. The ENCOMs have established relationships in each major theater of war. The Army's two ENCOMs (412th and 416th) are aligned to the LCC for specific combatant commanders and have forward cells co-located with the supported LCC. Additionally, subordinate organizations of the USACE are aligned with the combatant commanders and provide a bridge of support and expertise to every level, including tactical engineers. The Regiment aligned four USACE divisions with the four overseas geographic combatant commands at the direction of the CSA. Since the Transatlantic Programs Center was aligned to support Central Command (CENTCOM)/United States Army Forces Central Command (Third US Army) (ARCENT) but is neither a division nor a general officer command, the Chief of Engineers aligned the Southwestern Division to CENTCOM/ARCENT, with the Transatlantic Programs Center in support. These strategic support relationships are shown in *Table 4-1* and further articulated in a CSA letter to the Chief of Engineers dated 28 July 2000. Note that strategic engineer support to Northern Command (NORTHCOM) is to be determined.

Table 4-1. Strategic Engineer Support Relationships

COCOM	ASCC	ENCOM	USACE
EUCOM	USAREUR	412th	North Atlantic Division
CENTCOM	Third United States Army/ Army CENTCOM	416th	Transatlantic Programs Center
PACOM	USARPAC and Eighth Army (Korea)	412th	Pacific Ocean Division
SOUTHCOM	USARSO	416th	South Atlantic Division
NORTHCOM	TBD	TBD	HQUSACE

4-19. Engineers strategically respond to the combatant commander's requirements to deploy both AC and RC capabilities, ranging from modularly configured capabilities to entire organizations with complete C2. In the past, the RC was mobilized, en masse, to support the large military that opposed the Soviet threat. Today, the Army has become more expeditionary and the AC and RC have had to become more responsive in order to provide specific required capabilities not available locally/organically to a combatant commander. Often, individual RC units with specific skill sets are activated for short periods to augment the AC forces. In an effort to minimize the force footprint and reduce sustainment requirements, engineers deploy the minimal force necessary as outlined in strategic restrictions. Military engineers have many efforts underway to modularize their expertise and skill sets into deployable C2, ASCC augmented, facility engineering, troop construction, and geospatial teams. Reach is another method engineers use to reduce their AO footprint. Reach is a form of FFE *(see Appendix C)*. In peacetime, engineers participate in improving HN infrastructure to support the combatant commander's engagement strategy. Engineers also provide assistance to HLS and international disasters. Engineers are assigned, apportioned, allocated, and used to augment the Army and component commander's operations. The civilians and engineers of the USACE provide support to the nation at the strategic level by managing the civil works and military construction projects to improve the facilities used by our armed forces.

OPERATIONAL LEVEL

4-20. Operational planning merges the OPLAN/OPORD of the joint force, specific engineer missions assigned, and available engineer forces to achieve success. At the operational level, engineers contribute to battles by prioritizing limited assets and mitigating risks. Engineers seek ways to contribute to decisive, shaping, and sustaining operations by setting the conditions for success and facilitating the component or operational commander's objectives. Engineer commanders anticipate requirements and resource them with engineer units from their command the request engineer support from outside of their organizations. Engineer commanders and staffs also leverage resources available to the ENCOMs, USAES, and USACE ERDC through FFE teams and contractors. Engineers seek ways to leverage other service engineer and interagency capabilities. Engineer organizations may also act as the controlling HQ for certain focused missions.

4-21. Many times operational engineering is general engineering, but even tactical-level combat engineering tasks achieve operational results. For example, clearing an airfield of mines and other hazards is a tactical engineering mission that provides new operational-level capabilities. This tactical task facilitates the rapid flow of light forces and supplies into a forward area by air instead of over unsecured land routes. Operational-level engineering and tasks are not limited to aerial port of debarkation (APOD) and seaport of debarkation (SPOD) and COMMZ areas.

4-22. The defense planning guidance (DPG) directs forces to develop strategies to overcome antiaccess or area denial efforts by a threat that prevents forces from performing entry operations and gaining access to areas where they can operate from a tactical- or operational-level distance. This requires an operational-level engineer analysis that is linked closely to higher- level logistics and movement planning. Operational-engineer considerations include—

- Coordinating with logisticians to leverage existing infrastructure and identify restrictions that can be widened by engineer effort (such as increasing the maximum on the ground (MOG) for airfields, improving marshalling/staging areas.
- Examining methods to facilitate rapid intra-theater repositioning for forces/assets (such as converting stretches of highways into airfields).
- Using reverse planning from the objective to the entry point to identify how best to support operational maneuver (such as reconstruction of major bridges).
- Planning for success—post-hostility stability operations demand infrastructure requirements beyond military needs (contracting for rebuilding, moving civilians, and performing reconstruction).
- Coordinating early and synchronizing with other interagency organizations (such as prioritizing efforts and defining roles and responsibilities between the UN, governmental organizations, NGOs, USACE, other military organizations).

- Emphasizing that operational and strategic-level engineer and logistician sustainment planning efforts merge.

TACTICAL LEVEL

4-23. The commanders or planners of the unit or units specified to accomplish the tactical task or mission normally perform tactical planning; however, the other service components can also accomplish tactical planning. Engineer planners determine the best methods to accomplish assigned missions or tasks that are detailed and usually local in nature by employing tasked engineering forces and other forces that are made available. The JFC should ensure that engineer forces are placed properly in this arrangement and employed to influence the joint force AO.

4-24. The tactical level employs units within the full spectrum of operations. For combat operations, engineers are employed as part of a maneuver force to achieve tactical objectives. Engineers are resourced by the operational commander to accomplish M/CM/S, geospatial, and general engineering tasks as part of an engagement or battle. Engineers are task-organized to provide combat engineer capabilities, sustain combat power with their CSS, and deal with uncertainty. Engineers may also be called upon based on the METT-TC, to function as a maneuver TF or company team commander in special situations. In stability or support operations, engineers may become the priority of effort to provide general engineer support such as constructing life-support areas (LSAs) and improving force protection measures.

SECTION II - ENGINEER RESPONSIBILITIES AND FORCE PROJECTION

FORCE PROJECTION

4-25. Force projection is the ability to project the military element of national power from the continental United States (CONUS) or another theater, in response to requirements for military operations. Force projection operations extend from the mobilization and deployment of forces to the redeployment to CONUS or home theater *(see JP 3-35).*

4-26. The most important characteristic of force projection is synchronizing assets at all levels of war and projecting forces rapidly in response to a crisis or other military requirement. Force projection operations usually begin as contingency operations, involving imminent or actual involvement during war, or as conflict on a regional scale or stability or support mission. Force projection also applies to rapidly deploying forces to respond to a HLS requirement or national emergency or disaster. In combat operations, theater aims may be achieved faster by committing a smaller forward presence force rather than waiting for a larger, but less timely, response option. In this case, US forces could be opposed; however, force projection may occur unopposed. Unopposed operations could afford forces time to continue to build combat power, train, and acclimate after they arrive in theater. The engineer will conduct force projection as part of the overall joint and, possibly, multinational force operation. Engineer support efforts require close coordination with joint and coalition military engineer forces and other agencies to meet force projection requirements. Operational requirements for force projection enablers may require creating or upgrading an intermediate staging base (ISB), a rapid port enhancement (RPE), or similar support. These missions would require extensive use of engineer support in the earliest stages of force projection.

4-27. Force projection operations are undertaken in response to a crisis. That crisis can occur in isolation, as is the expected case in stability operation. However, a crisis also can occur during a major operation. Force projection operations can be received as a series of processes that can occur in an overlapping or sequential manner. When the contingency occurs during a major operation, these processes assist in both resolving the crisis and in returning the contingency forces back into the ongoing operation as rapidly as possible.

4-28. Force projection encompasses five processes—mobilization, deployment, employment, sustainment and redeployment—that normally occur in a continuous, overlapping and iterative sequence for the duration of the mission *(see Figure 4-2 and FM 100-7).*

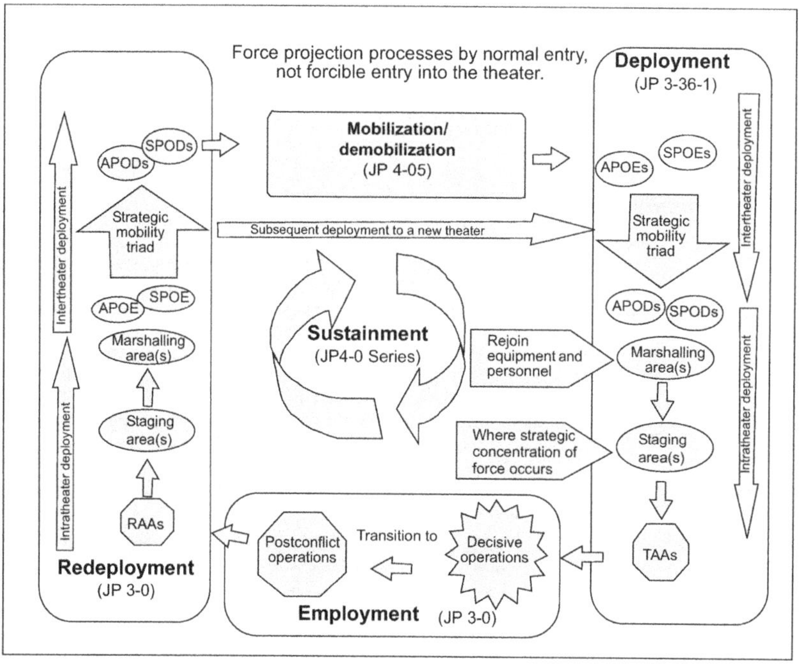

Figure 4-2. Force Projection Process

4-29. Deployment activities, for example, might be so closely followed by decisive operations that they are indistinct from one another. Decisive operations may begin well before the force has completely deployed. At a minimum, commanders and staffs must consider the:

- Coordination of sequencing and phasing of forces (maneuver, operational support, operational protection, SOF, and C2/intelligence and surveillance [IS]/ISR).
- Requirement and time frame to establish and build up the theater infrastructure.
- Environmental issues and the need to reduce the risk of environmental impacts on force health protection.
- Protection of forces, to include rear area security operations (tactical combat force [TCF]) and ADC.
- Preparation time for deployment and operational readiness—types of units and their readiness.
- Combatant commander's critical items list in the time-phased force and deployment data (TPFDD) flow.
- Requirement and level of in-theater stocks.
- HN capability and availability.

4-30. Any particular force projection operation may not include all of the processes. For example, a force projection operation may be the first phase of an evolving major operation. Redeployment of all forces may not begin until the end of the subsequent phase, of which the force projection was a single phase.

FORCE PROJECTION CONSIDERATIONS

4-31. Force projection operations challenge all leaders. Early critical decisions made under uncertain circumstances will be required at all levels. These decisions can greatly affect future conditions for successful mission accomplishment. During the initial planning stages, consideration should be given to the entire deployment cycle, including postconflict operations. It is imperative for engineer leaders to understand the anticipated end state to conduct reverse planning and properly plan the ordered flow of units and the resources to properly take them through reception, staging, onward movement, and integration (RSOI). Unit mobilization and deployment can occur at the same time, or sequentially, and are based on force requirements and strategic aims. When an engineer unit deploys, it will do all that is necessary to meet the demands of the overall mission. *Figure 4-2, page 4-10* shows the deployment cycle as stated in *FM 100-7.*

4-32. Operational design and unit execution must be capable of overcoming any unforeseen obstacles. In most force projection scenarios, combat engineers will be unable to cope with the requirements for general engineering, real estate support, and related technical services required to develop and maintain the operational support base. However, they can provide some of the initial C2 and planning until additional engineer support becomes available. Typically, ENCOMs, with a combination of CS and CSS engineers, respond to these operational-level engineering requirements—with a mix of military and contractor capabilities, integrating tailored organizations, and elements from USACE. *FM 3-0* describes several key considerations that apply to force projection operations. The following are engineer mission capabilities that may arise during the force projection process:

LETHALITY FOR THE DEPLOYING FORCE

4-33. An important strategic/operational consideration for planning contingency operations is to introduce credible lethal forces early. The early entry force must possess the required lethality to accomplish the mission and protect the force the moment it arrives in theater. The lethality of combat forces securing operational objectives can be enhanced through a combination of the engineer battlespace functions. Engineers may contribute to the lethality of the early combat force by emplacing minefields and other obstacles. They protect lodgments by constructing secure C2 nodes, logistics bases, and other necessary fortifications and survivability positions.

ANTICIPATION AND INTELLIGENCE

4-34. Force projection anticipation is the expectation of being alerted and deployed. To properly anticipate mission requirements, engineer commanders should make every effort to ensure that equipment and personnel shortages are identified before being alerted. If faced with a unique support mission (non METL related), anticipation implies the possibility of fielding new equipment. Rapidly introducing US forces to an area requires accurate, detailed, timely, and continuous intelligence. If units have been assigned a region of focus in peacetime, planning can occur before an alert and deployment. Commanders/operators need direct access to engineer assessments of the theater infrastructure for developing COAs and evaluating the infrastructure. Engineers provide geospatial products of likely contingency areas to support the IPB process for all current and potential operations. When assigned as the ENCOORD, engineers advise the G2/S2 of significant environmental factors and ensure that these impacts are integrated into the IPB process. They also assess the available infrastructure for possible general engineering requirements, which include airfields, MSRs, ports, utilities, and logistics facilities. Engineers must anticipate requirements for construction, construction standards, and critical resources to support deployed forces. Other intelligence and preparation might include learning the enemy engineers' capabilities and preparing for them accordingly.

FORCE TAILORING AND TEAMWORK

4-35. Force tailoring is the process of determining the right mix and sequence of the following:

- AC/RC component engineers.
- Combat engineers.
- CS engineers.
- CSS engineers.
- Contractors and contracted engineers.

4-36. Rapid deployment forces may have little opportunity to tailor forces. Follow-on forces can be tailored to meet the specific concerns of the long-term mission. Proper planning should give the operational commander the resources and dispositions to deal with anything that might jeopardize the mission of protecting the force. Commanders consider METT-TC, strategic lift, pre-positioned assets, civilian-contractor support, and HNS when tailoring forces. Deploying units must be very flexible and versatile, to perform missions beyond the unit METL but within the broad scope of the engineer functions.

BATTLE COMMAND

4-37. During force projection, engineer commanders must develop an appreciation for the types of roles/missions that the organization may support across the spectrum of operations. They must quickly visualize the battlefield environment through to the end state and articulate clear guidance that meets the intent of the supported maneuver commander. The commander's intent and other guidance provide the impetus for staff planning and set the conditions for anticipating engineer requirements.

4-38. Commanders must deal with deployment, entry, and potential combat, concurrently, while adjusting to the evolving conditions of each. Deployment can cause the physical separation of units in space and time and the separation of the unit from the next higher HQ. Additionally, units may be placed under unfamiliar organizations. Therefore, simplicity and the ability to adapt and adjust are key considerations. Engineers must support—

- Deployment of themselves and other forces.
- Lodgments and base camps with construction and leasing.
- Maneuver operations with combat engineering.
- Federal, state, and local agencies.

4-39. Engineers will accomplish these tasks at the small-unit level. Engineers at all echelons are challenged to ensure that they understand the mission due to separation in time and space. Engineer commanders at all levels should describe and direct their intent to influence the situation. Moreover, commanders must demonstrate confidence in their subordinates and stimulate a level of initiative and motivation that accomplishes the mission with very little supervision.

LOGISTICS

4-40. Like the initial entry forces in TOs, logistics must be tailored and flexible. The availability of ports, airfields, roads (infrastructure), and other assets will affect the sequencing of units and the tempo of entry operations, as well as the overall logistics planning. ENCOMs classically support force projection logistics operations by leasing or constructing forward support bases, ISBs, and lodgments; improving APODs and SPODs; and providing engineer support for joint logistics over-the-shore (JLOTS) operations. Engineers must work with logistics planners to synchronize the flow of engineer logistics with the flow of engineer units into the TO, including working closely with HN and contracted logistics support.

MEDIA IMPACT

4-41. Emerging information communication technologies, the evolving global information environment, and the media's ability to provide live coverage from anywhere in the world to everywhere throughout the world bridges the gap between the tactical, operational, and strategic levels. Media coverage of Army operations can influence public opinion, political decisions; and the direction, range, and duration of operations. In the emerging information environment, working effectively with the media is a critical element of mission success.

4-42. Engineer operations, especially those executed in support of contingency operations, are likely to attract significant media coverage. Engineers at all levels—

- Must be prepared to operate in a media-intense environment.
- Need to understand that the Army has a vital interest in facilitating media coverage and communicating the Army's perspective.
- Must be prepared to support open and independent reporting and access to units as early as possible.
- Should be trained to interact with media representatives and confidently provide complete, accurate, and timely information.
- Must know that public affairs support is available.

POST CONFLICT ACTIVITIES

4-43. Issues related to the strategic end state, postconflict activities, and transition to peace are considered throughout the force projection operational planning and execution. This is also applicable to post crisis or disaster operations. Engineers tend to play a significant role in conducting postconflict activities due to the long-term military presence required for force protection and general engineering requirements.

4-44. Engineer planners must plan for success in postconflict planning if they are to be prepared to support postconflict operations. The major consideration is the renewal of traffic and the demands on the civilian support infrastructure to keep the AO stabilized after combat operations. Title X and the Army's minimum military requirement focus for posthostilities are often at odds, as engineers consider the transition to stability or support operations. Engineer planners must consider in advance how C2 and resources and military engineer effort and materials will be used in the transition vis-à-vis any international or other US agency effort. This is most significant during the transition from combat to stability operations where critical infrastructure support is destroyed or damaged from the war but other agencies are not yet available or in position to assume the mission. Military engineer operational planning for this transition period is particularly important to the stability of the postconflict theater.

SECTION III - THE MILITARY DECISION-MAKING PROCESS

4-45. There are two Army doctrinal planning procedures defined in *FM 101-5*. In units with a formally organized staff, the military decision-making process

MDMP helps commanders and staffs develop estimates, plans, and orders. It provides a logical sequence of decision and interaction between a commander and his staff. The MDMP provides a common framework that supports the maximum use of parallel planning for all staffs. However, at the lowest tactical echelons, commanders do not have a staff. Consequently, they (and their subordinate leaders) follow troop leading procedures (TLP). Both procedures hinge on the commander's ability to visualize and describe the mission/operation. Both are means to an end, and their value lies in the result, not in the process. Each process can be performed in detail, if time permits, or in an abbreviated fashion in a time-constrained environment. TLP will not be addressed in this manual, but rather in the respective engineer unit manuals (such as *FM 3-34.221*) focused on the specific TLP associated with that unit and echelon of command.

4-46. The different methods of planning are described in *FM 5-0*. For joint planning methods refer to *JP 5-0* and *JP 5-00.2*. The particular procedures used in joint planning depend on the time available to accomplish them. However, engineering considerations are similar for both deliberate and crisis action planning. Engineering considerations in the deliberate planning steps encompass the same engineering considerations as in the crisis action planning steps. This method applies to the full spectrum of operations. A correlation of these steps is provided, and considerations are outlined in the phases of the deliberate planning process.

4-47. The different methods of planning are described in *FM 101-5*. For joint planning methods refer to *JP 5-0* and *JP 5-00.2*.

4-48. Engineer commanders or the senior ENCOORD should develop a parallel planning process with their task-organized engineer units to facilitate realistic assessments. This near parallel process feeds into the force commander's MDMP *(see Figure 4-3, page 4-16)* and provides input for an engineer OPORD or annex to be published, nearly simultaneously, maximizing the time available for execution. This helps to ensure that the engineer capabilities (and included specialized areas such as environmental considerations) are integrated into the MDMP.

RECEIPT OF MISSION

4-49. During this phase, the Joint/Army force engineer assembles the resources required to support the mission analysis and COA development tasks for the concept development phase. Engineers must immediately begin development of mission specific geospatial products after the receipt of the mission in order for them to be available during mission analysis. As a general rule, commander allocates a minimum of two-thirds of the available time for subordinate units to conduct their planning and preparation, leaving one-third of the time for the commander and staff planning. Issuance of a warning order (WO) facilitates parallel planning, allowing subordinates the maximum time to conduct their own planning.

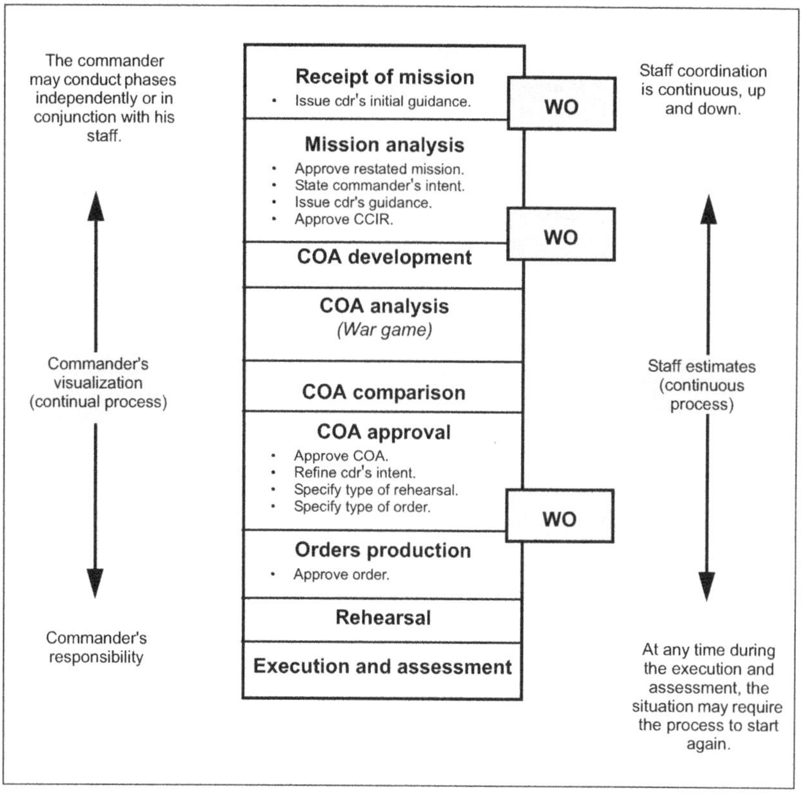

Figure 4-3. The Military Decision Making Process

MISSION ANALYSIS

4-50. A mission analysis has seventeen subordinate steps that provide the framework for success in the MDMP. The engineer has a role to play in each of these steps, although some are more critical than others. *See FM 101-5.*

4-51. The end product of the mission analysis process is a restated engineer mission statement and the development, preparation, and issuance of planning guidance to the staff and subordinate commands. Developing an engineer end state and a clear understanding throughout the chain of command contributes to the mission accomplishment and achievement the desired objectives. During the EBA, the ENCOORD identifies the specified and implied engineer tasks (may be more than M/CM/S) and their associated purposes. From these tasks, the ENCOORD will recommend essential mobility survivability tasks (EMSTs) to the higher commander. Engineer considerations of the EBA during this step of planning include identifying the threat, the terrain, and the friendly forces, to include—

- Threat engineer capabilities, order of battle, tactics, and TTPs.
- Threat engineer templates eventually linked to the collection plan.
- Existing mines, UXOs, and so forth, including access to the databases.
- A terrain analysis in support of battlefield visualization.
- An infrastructure and facilities assessment (bulk fuel storage, ports, airfields).
- Environmental considerations, to include environmentally sensitive areas and associated impacts.
- The availability of HNS.
- Digital mapping requirements.
- The capabilities of assigned engineer forces.
- The capabilities of multinational engineers.
- All engineer capabilities.
- The identification of EMSTs.
- The engineer end state.
- The recommended information requirements (IR).

PLANNING GUIDANCE DEVELOPMENT

4-52. During this phase, the ENCOORD receives the maneuver commander's guidance and priorities for engineer operations. The ENCOORD also seeks to gain the maneuver commander's approval of his proposed EMSTs. The engineer commander assesses all available information derived from the mission-analysis process to provide the commander and staff with the input required to develop and support the initial COAs. The maneuver commander communicates the desired engineer tasks to achieve the desired purpose and effects. The joint/Army force engineer uses this combined assessment to identify the resources required to support each COA, make recommendations based upon available time and resources, and recommend force tailoring to best support the combatant commander's intent. This assessment is the linchpin of successful engineer integration into operations.

COURSE OF ACTION DEVELOPMENT

4-53. COA development is the critical phase of the planning process. In this phase, engineers fully integrate and synchronize combined arms operations as part of the maneuver plan. The ENCOORD synchronizes the SOEO and confirms the EMSTs required to achieve the desired purpose and effect. He then assigns resources and recommend priorities to accomplish the SOEO.

4-54. During COA development, the engineer may consider the following:

- All identified engineer tasks and resources required.
- The detailed reverse planning process.
- River crossing planning.
- Situational obstacle planning.

- The options for joint force operational movement, maneuver, and protection.
- The C2 options to best employ the engineer capabilities of the joint force.
- Recommendations for ISBs, FOBs, forward logistic support sites, and avenues of approach (AAs).
- MSRs and available LOC facilities.
- The development of EMSTs.

COURSE OF ACTION ANALYSIS (WAR GAME)

4-55. COA analysis allows the staff to synchronize the BOS for each COA and identify which COA best accomplishes the mission. Through war gaming, the staff attempts to foresee the action-reaction-counteraction dynamics of the battlefield that would result from each COA. Engineers should develop and recommend the criteria for evaluation of EMSTs during this phase to subjectively assess the war game. Engineers should focus on the impact of the threat engineer functions and the friendly engineer functions with respect to the impact of the terrain on the operation. The ENCOORD synchronizes the methods to achieve the desired effects and ensures that they are accomplished. The process of war gaming allows the staff to understand the advantages and disadvantages of each COA and forms the basis for the commander's course of action comparison and decision. COA war gaming may identify branches and potential sequels that require additional planning by the engineer and the staff.

COURSE OF ACTION COMPARISON

4-56. Based on the identified criteria for evaluating the war gamed COAs, the engineers highlight the advantages and disadvantages from a SOEO perspective for each COA. The actual comparison may follow any technique that results in a recommendation. An effective technique for comparing COAs is a comparison matrix. Using specific criteria, each COA is compared to the others. While comparing COAs, engineers determine which SOEO best supports the mission.

COURSE OF ACTION APPROVAL

4-57. The staff recommends the best COA to the maneuver commander based on an outline of the COA and its advantages and disadvantages. The force commander considers the staff recommendation presented by the Operations Directorate (J3), G3, or Operations and Training Officer (U.S. Army) (S3); approves the SOEO to include EMSTs; and announces a decision and concept/intent for the approved COA. At this point, the engineer commander can issue another WARNO using the force commander's updated information. This facilitates the planning for engineer subordinate units. The engineer staff officer makes a recommendation to the commander during the decision brief. The type and amount of detail that the engineer briefs depends on the needs and preferences of the individual commander. In general, it covers the—

- Concept of engineer support.
- Engineer mission priorities.
- Critical engineer events/actions.
- Task organization and command/support relationships.
- Obstacle overlay (including SCATMINE employment authority and concept for use by the system type).
- Survivability estimate and priority.
- Critical tasks directed to subordinate units.
- Engineers' work/construction time line.

4-58. Other members of the battle staff brief information that the engineer provides input for during the estimate process. This is particularly true of the intelligence portion of the decision brief and the comparison of COAs. Once the commander makes a decision, the estimate provides the bulk of the information needed to prepare the maneuver force's OPLAN or OPORD.

ORDERS PRODUCTION

4-59. During the planning development phase, an approved COA is expanded into a complete OPLAN, OPLAN in concept format (with timed-phased force and deployment data [TPFDD]), or functional plan. The process is the same for all plan types. To support this phase, the engineer provides input for the appropriate annexes and appendices of the plan to include the SOEO and the EMSTs, as found in *FM 101-5* and *Chairman of the Joint Chiefs of Staff Manual (CJCSM) 3122.03A*. Engineers prepare Annex F and its associated appendices for Army plans and Annex D (Appendix 6) and Annex L for joint plans. Besides the civil engineering support plan (CESP) and Annex L, the joint/Army engineer staff usually provides input, depending on the mission and combatant commander's intent. Refer to *JP 4-04* for further information on the CESP and Annex L.

JOINT ORDERS/PLANS

4-60. Although not included in *CJCSM 3122.03A*, engineers should seriously consider preparing an engineer appendix to Annex C. Major operational engineering considerations may affect the success of a joint operation. The emphasis should focus on the overall priority of the engineer effort by phase, as well as by any unique C2 considerations and support relationships between combat engineers and supporting theater-level engineers.

4-61. Engineers should review the entire order/plan for adequacy, feasibility, acceptability, and consistency with joint and Army doctrine for all combat, geospatial, and general engineering considerations. Special attention should be given to the following areas:

Task Organization, Annex A

4-62. Refer to Annex A to ensure sufficient capability to meet identified requirements. Additionally, planners should provide input to the flow of the engineer force, as detailed on the TPFDD.

Operation, Annex C

4-63. Engineers review and participate in the writing of Annex C. Their participation may be required in many of the subordinate appendices. Although not listed in *CJCSM 3122.03A* as a designated appendix, engineers may also find it appropriate to prepare a supporting engineer appendix to the appendix to *Annex C*. The format for this appendix would be similar to Annex F in *FM 101-5*.

Air Base Operability, Appendix 8, Annex C

4-64. Engineer considerations may factor heavily into the overall air base operability (ABO) concept of Appendix 8, Annex C, including the concepts of the five basic functions to—

- Defend (installations).
- Survive (provide expedient protection).
- Recover (assess the damage and effect repairs).
- Generate (alternatives for damaged systems).
- Support (the recovery effort).

Counterattack, Appendix 12, Annex C

4-65. The engineer will be a critical contributor to Appendix 12, Annex C. Any counterattack will be directly affected and, in some cases, defined by the availability and capabilities of combat, topographic, and general engineering support.

Explosive Ordnance Disposal, Appendix 13, Annex C

4-66. Appendix 13, Annex C should provide an outline of expected EOD operations in support of the plan. EOD personnel assist in identifying and removing UXO to support mission requirements and/or return facilities to a usable status. At the joint level, EOD is considered an engineer responsibility rather than a logistic one.

Force Protection, Appendix 15, Annex C

4-67. The combatant commander may require the joint force engineer to provide a list of the forces available to support the protection plan for Appendix 15, Annex C. Based on the commander's intent, activities of the enemy, and the predominant threat, the joint force engineer may provide appropriate assistance, including—

- Facility hardening.
- Revetments.
- Berms.
- ADC.
- Installation security improvements (barriers, perimeter fencing, monitors, and cameras).

Critical Infrastructure Protection, Appendix 16, Annex C

4-68. The combatant commander may require the joint force engineer to provide engineer expertise to support the critical infrastructure protection plan. The joint force engineer may provide assistance in the writing and implementation of the plan.

Mobility and Transportation, Appendix 5, Annex D.

4-69. Engineers provide input to this appendix to support the given OPLAN/ CONPLAN. Most of the support at this level will be general or topographic engineering related. All three tabs will likely require engineer input.

Civil Engineering Support Plan, Appendix 6, Annex D

4-70. Engineers provide Appendix 6, Annex D as an action plan for the six broad categories of facilities necessary to support the given OPLAN/ CONPLAN. The appendix defines the general character and magnitude of the civil engineering support (synonymous with general engineering for this appendix).

Civil Affairs, Annex G

4-71. Engineers should focus on areas affecting dislocated civilians, humanitarian demining operations (HDOs) efforts, government stability, and the destruction or degradation of civilian infrastructure. Engineers play a key role in providing shelter (tent cities) and performing essential repairs to facilities and utilities (water, sanitation, and power production). Engineering personnel can also help identify local resources to assist in civil-military operations (CMOs).

Joint Operation Planning and Execution System Operation Plan/operation Order/ contingency Plan, Annex L

4-72. Annex L, which contains three appendices, is the equivalent of Appendix 2, Annex F (environmental considerations). It is the parallel document for a joint staff and is generally written by the joint force engineer.

Geospatial Information and Services, Annex M

4-73. Annex M, which contains three appendices, identifies the geospatial forces assigned or attached, their manner of employment, and the required geospatial products and services. Engineers assist the combatant commander by identifying available geospatial assets to support the plan and making recommendations for additional support. In a Joint Operation Planning and Execution System (JOPES) OPLAN/OPORD/CONPLAN, this is the equivalent of Appendix 3, Annex F (geospatial information and services). It is the parallel document for a joint staff.

ARMY ORDERS/PLANS

4-74. Engineers should review Annex A to ensure sufficient capability to meet the identified requirements. The engineer staff officer lists the engineer units under the proper control HQ with the correct command or support relationships. Additionally, planners provide input to the flow of the engineer force as detailed on the TPFDD.

4-75. Engineers review Annex C and ensure the inclusion of obstacle effects or other graphics to assist in conveying the commander's intent on the overlay.

4-76. In Annex D engineers work with the fire support officer/staff to integrate obstacles with fire. Of particular interest are SCATMINEs and confirming that all obstacles are covered by fire.

4-77. Annex F is the principal means through which the engineer defines engineer support to the maneuver commander's intent, the SOEO; EMSTs, and coordinating instructions to subordinate commanders. It is not intended to function as the internal order for an engineer organization, where the engineer commander will articulate intent; the concept of operations; and coordinating instructions to subordinate, supporting, and supported commanders.

4-78. An integral component of the deliberate decision making process is the production of Annex F. The preparation of the annex seeks to clarify the engineer support to the OPLAN/OPORD and includes the—

- Priorities of work to shape the theater/AO (not in a tactical level engineer annex).
- Operational project planning, preparation and execution responsibilities (not in a tactical level engineer annex).
- Engineer organization for combat.
- The SOEO to include EMSTs.
- EMSTs for subordinate units.

NOTE: Guidance to maneuver units on obstacle responsibilities should be listed in the body of the basic order under paragraph 3a, Tasks to maneuver units, not here.)

- Allocations of Class IV and Class V (obstacle).

4-79. For Appendix 1, Engineer Overlay, Annex F, the engineer/ENCOORD may produce this overlay in conjunction with Annex C to highlight obstacle information (to include water obstacles) or breaching operations. A deliberate river crossing operation may require a separate annex as part of an order.

4-80. The engineer/ENCOORD performs as the staff integrator and advisor to the commander for environmental considerations. Appendix 2, Annex F, is a parallel document to Annex L in a joint OPLAN/OPORD/CONPLAN. *See FM 3-100.4* for an example of this appendix. When specific command procedures dictate, other staff officers include some environmental considerations in logistics and medical annexes. Unit planning at the regiment or brigade level and below will normally include only those elements required by the higher HQ orders or plans that are not already included in a unit standing operating procedure (SOP). If this appendix is not written, appropriate material will be placed in the coordinating instructions (paragraph 3, Execution) of the basic order.

4-81. The engineer/ENCOORD performs as the staff integrator and advisor to the commander for geospatial information and services. Appendix 3, Geospatial Information and Services, Annex F, is a parallel document to Annex M in a joint OPLAN/OPORD/CONPLAN. *See FM 3-34.230* for an example of this appendix.

SECTION IV - OPERATIONAL CONSIDERATIONS

4-82. The MDMP is valid at the operational level. At the operational level, the supported OPORD/OPLAN may be either joint or Army-specific, depending on the forces involved. *See CJCSM 3122.03A* for the formats that go with the JOPES process.

4-83. The engineer support planning process at the operational level focuses on the following interrelated activities:

- Engineer facilities study. The study is derived from the Joint Engineer Planning and Execution System (JEPES) computer model, which analyzes data. The study is used to develop a CESP and becomes part of Appendix 6, Annex D of the joint OPORD/OPLAN.
- Engineer annex. Annex F to the OPLAN is prepared using the results from the engineer facilities study and the CESP. The annex provides instructions for executing the engineer part of the OPLAN.

ENGINEER FACILITIES STUDY

4-84. According to *CJCSM 3122.03A*, the combatant commander's engineer planners use the JEPES computer model to prepare estimates of theater-level wartime engineer requirements for the following items in support of an OPLAN:

- Facilities.
- Engineer man-hours.
- CESP.
- Engineer annex.
- Environmental annex/appendix.

4-85. The primary purpose of JEPES is to assist the combatant commander and service-component engineer planners in determining whether the OPLAN provides the correct amount of engineer capability at the right place and is timed correctly to support deploying forces.

4-86. The JEPES is one of several tools the commander has to assess the validity and the accuracy of an engineer plan. The JEPES data, along with the engineer analysis and command guidance, provides a commander with another means to check the supportability of the engineer plan for a specific OPLAN.

4-87. In deliberate planning, the combatant commander includes a CESP within the logistics annex of the OPLAN. Independent of the combatant commander's plan, the Army's service-component engineers routinely develop their service plan as a means of detailed, deliberate planning.

4-88. The TPFDD is the primary driver of the JEPES model. This model extracts information such as the unit type, the destination location, the arrival time, and the population from the TPFDD. Given this input, the JEPES estimates construction man-hour and facility-type requirements to support the bed down of forces deploying into a theater. The JEPES also computes the engineer assets (man-hours) that are available to meet the estimated requirements. The JEPES provides Class IV output in the form of long tons and short tons. The results from the analyzed JEPES data are gross estimates that are used in the deliberate planning process for analyzing COAs for engineer support to the OPLAN. Because of the integral relationship between the JEPES model, the OPLAN, and the TPFDD for a theater, the JEPES does not readily lend itself to crisis planning in theaters where an OPLAN and TPFDD have not been prepared.

4-89. The JEPES model algorithms are based primarily on the support facilities necessary for the RSOI of all inbound forces. The model calculates facility requirements for a unit's final destination on the TPFDD but does not compute other engineer missions and support requirements within the theater. Aspects of the estimate that are not automatically calculated by the JEPES include—

- The construction of forward logistics bases and enemy prisoner of war (EPW) and displaced persons camps.The construction, maintenance, and repair of MSRs.
- The survivability of command, control, and communications (C3) nodes.
- The construction, expansion, or maintenance of port activities.
- Support for JLOTS operations.
- The construction of attack aviation strips, theater ammunition storage points, and fuel pipelines.
- Support for tactical elements in combat engineering (M/CM/S).

4-90. The JEPES has a capability for manually inputting specific requirements such as EPW camps; petroleum, oils, and lubricants (POL) pipelines; MSR construction; and other requirements, specified or implied in the mission analysis and planning guidance. This data is entered into a user's input file, which the JEPES combines with the other TPFDD requirements.

4-91. The output from the JEPES is a gross estimate, reflecting engineer capabilities, as depicted by the TPFDD. The accuracy and reliability of the information generated by the JEPES model is directly affected by the—

- Accuracy of the unit data on the TPFDD.
- Level of accuracy of the assets in the joint operations area (JOA).
- Level of specificity on the TPFDD.
- Assumptions for HN-provided facilities.

4-92. The product of the JEPES output analysis is the engineer facilities study. Engineer planners at the combatant commander and component levels use this study to prepare their CESP and the engineer annex to the OPLAN. The study also becomes Tab C of the engineer annex in joint orders.

CIVIL ENGINEERING SUPPORT PLAN

4-93. The CESP is used in preparing the engineer annex to the OPLAN. It is the primary planning document in which the engineer staff considers the minimum essential facilities and the construction capabilities required to support the commitment of military forces. When developing the CESP, the engineer staff should consider the—

- Engineer facilities study.
- Engineer intelligence of the theater.
- HN capabilities and HNS agreements.
- Construction contract capabilities.
- Mission of other US and allied forces.
- War damage estimates.
- Facility engineering responsibilities.
- Logistics support plans.
- Subordinate units.
- Any other aspect of the operation that impacts general engineering support.

4-94. In deliberate planning, the combatant commander includes the CESP within the logistics annex of the OPLAN. Independent of the combatant commander's plan, the Army's service-component engineers routinely develop their service plan as a means of detailed, deliberate planning. The CESP differs from the engineer facility study by considering other planning aspects that the JEPES was not designed to estimate. The CESP usually addresses the following engineer aspects in addition to the JEPES data:

- Information on available resources, facilities, and characteristics within the region relevant to the construction mission and construction capabilities.
- Restrictions imposed on the use of bases and installations.
- Major construction resources and their allocations.
- Future construction standards to be used as the theater matures.
- Responsibility for construction management among components.
- Responsibility for determining the facility's use in light of competing requirements from the components.
- Priorities at different phases during the conflict.
- Provisions for withdrawal, such as base denial and the movement of residual assets and stored critical Class IV supplies.
- Contract construction agent (CCA) missions and responsibilities and their relationship to engineer assets.
- Engineer support guidance and agreements for support commands and the area support group (ASG).
- Class IV construction materials availability.

4-95. Supported by the JEPES data analysis or other studies, the CESP states the priorities, programs, and general policy when seeking general engineering support. The format for the CESP is governed by *CJCSM 3122.03A* and included as Appendix 6 of Annex D in joint orders.

ENGINEER ANNEX

4-96. In JOPES, the engineer annex is primarily covered by the engineer responsibility to write Annex L, Annex M, and Appendix 6 to Annex D, although the engineer has significant participation in other portions of the OPORD and OPLAN as identified in *paragraphs 4-59 to 4-69* above. A separate engineer appendix to Annex C may be necessary or desirable in certain situations. In Army orders, Annex F will include Appendix 1, Appendix 2, and Appendix 3.

Chapter 5

Operational Principles of Engineer Operations and Organization

I would desire to have companies of Sappers formed—they should be instructed in every thing that relates to the construction of Field works—how to dispose of the Earth—to cut the Slopes—face with turf or sods—make fascines—arrange them properly—cut and fix Palisades, etc.

Louis Duportail
Chief of Engineers, Continental Army
January 18, 1778

Engineers must understand and be knowledgeable of basic Army tasks first, but this responsibility is closely followed by having a firm grounding in the fundamental Army missions, roles, and functions that are supported by the engineer battlespace functions. They must not only have an appreciation for Army engineer organizations, but also have a working knowledge of the organizations and capabilities of each of the other elements that make up the Engineer Regiment. This knowledge must then be applied to the tactical, geophysical, and political environments in which they are operating. The engineer branch contains all three of the force structure categories: combat, CS, and CSS. As such, it is a threefold branch with diverse capabilities that interact across all areas of the Army. When you add the other elements of the Engineer Regiment, the Army engineer must be multifunctional and versatile.

ENGINEER OPERATIONAL PRINCIPLES

5-1. Engineer operational principles are extracted chiefly from the principles of war and the Army tenets. There are a number of principles that engineers adhere to in order to optimize the engineer battlespace functions and to achieve success in the context of today's technologies and organization. The principles apply to all types of engineer units and to the full spectrum of operations. The eight operational principles associated with engineer operations are—

- Mass.
- Versatility.
- Security.
- Economy of force.
- Momentum.
- Simplicity.

- Synchronization.
- Initiative.

MASS

5-2. Engineers mass the effects of their combat power in time, space, and purpose to overwhelm the enemy or gain control of a situation. Engineer assets may be massed in conjunction with maneuver forces or independently. Mass does not necessarily equate to more engineer assets, it refers to the massing of engineer capabilities and effects. Effects can be massed by requesting special capabilities and assets (to include, but not limited to divers, pipeline construction, bridging organizations, FFE, and geospatial engineering capabilities) from higher HQ for special or unique requirements. Massing of effects is facilitated by accessing expertise through reach and synchronizing them to achieve the desired effect(s) at the times and places required to facilitate the maneuver commander's success. The most critical massing of engineer capability may be that of massing engineer expertise and "brain power" from a variety of locations other than the mission site. Note that the main effort of the maneuver force may not be where the engineer effort is massed. Engineer resources may be massed in a sustainment role supporting disaster relief in a foreign country or performing specialized combat missions within the maneuver plan, such as a river crossing.

5-3. Engineers must begin their work early to complete it on time. To effectively mass engineer effects, they have to anticipate future missions and reposition their unit, if necessary, while accomplishing current missions. Engineers must share the commander's vision and be proactive in anticipating mission requirements. As the operation transitions through the subsequent phases of the operation, engineers must be able to synchronize their actions to meet mission requirements and achieve operational and tactical success. Engineers focus on the success of mission rather than on habitual support to particular organizations. This does not mean that habitual combat engineer associations are not vital or retained whenever possible. Improved SU allows engineer commanders to anticipate requirements and mass M/CM/S, geospatial, and general engineering effects over simultaneous targets in the AO. Assured mobility and the selected massing of engineer assets depend on the ability to draw information from the COE and develop them into a shared SU.

5-4. Engineers organic to maneuver forces do not stay with those forces when they are held out of the fight. They remain out of action only long enough to refit or reconstitute after a major action before they are recommitted. Plans must ensure timely return to the parent maneuver force before their commitment to combat to ensure adequate integration and preparation time. Combat engineer forces can be positioned to support the combined arms reserve or counterattack forces and are considered to be committed, not held in reserve. There are time and distance considerations that affect how to plan and organize the massing of tactical engineer effects. Unlike fire effects, which can be reprioritized and massed quickly, tactical engineer effects are often ground-based and subject to time-distance factor limitations.

VERSATILITY

5-5. Engineers are capable of operating in multiple environments and performing a myriad of tasks that include M/CM/S, geospatial, and general engineering. These missions can occur during offensive, defensive, stability, and/or support operations under the C&S of a combined arms, joint, coalition, or interagency organization. It is the engineer's ability to transition quickly between these operations that reveals their versatility. (As an example, both combat and combat heavy engineers may be required to transition between conducting combat operations and repairing utilities or facilities while simultaneously providing support to civilians during UOs.) It is not unusual for engineers to have a battle-focused METL and be alerted to support a fire fighting mission. Engineers may also perform additional missions with augmentation and the approval of the higher commander. Engineers can organize and perform C2 operations during defensive attack, or stability and support operations. In these roles, engineers typically require other BOS elements to be task-organized with them to achieve the proper C2 structure.

SECURITY

5-6. Engineers never permit the enemy to acquire an advantage while performing their engineer battlespace functions. Army engineers are well integrated into the ISR plan and attempt to have SU at all levels. They embrace all measures available to protect themselves and conserve their assets. They integrate themselves into maneuver formations when possible for maximum security. When operating independently in remote parts of the AO, engineers provide their assets to maintain local security at all times. The same tactical security requirements apply in a support operation. Equipment and personnel are vulnerable to crime and hazardous materials (HAZMATs).

ECONOMY OF FORCE

5-7. Understanding the principle of mass, engineers know that they cannot provide troops or assets to every identified task in the AO. Assets and/or effort are task-organized to mass the effects and yet reduce the risks for the commander throughout the AO. Engineers recognize that risk may be required in military operations and that resources are based on priorities. They may reduce the risks by improved SU and task-organizing limited engineer units or assets to accomplish the mission as an economy of force. The higher commander should always understand the risks associated with engineer missions. Economy of force implies that the senior engineer commander must prioritize engineer effort. In a support mission, combat heavy engineers may have to mass effort in one region and minimize construction efforts in an economy of force with specific priorities, providing the minimum essential needs of a populated area.

MOMENTUM

5-8. Engineers ensure that the force maintains offensive momentum regardless of the enemy, the terrain, or any other impediment within the fundamentals of assured mobility and in line with the commander's intent. They are proactive, not reactive, according to the commander's intent and scheme of maneuver. Momentum implies that engineers work through the problems they encounter using resources (such as FFE) at their disposal or improvising other means to prevent the mission from stalling or failing. Engineer operations enable the momentum of the force throughout the AO in support of decisive, shaping, and sustaining operations.

SIMPLICITY

5-9. Engineers achieve simplicity by providing clear, uncomplicated plans and unambiguous, concise orders to ensure thorough understanding. Whenever possible, engineers resource subordinate units with substantial assets to prevent complicated movement of formations/assets/materials across the battlefield/AO and to simplify the plan. Orders are simplified and direct to alleviate confusion. Rehearsals are used to identify confusing areas and simplify them for subordinates. Combat engineer elements may, at times, remain with a given maneuver organization rather than be shifted to another organization or mission in the interest of simplicity. CS and CSS engineers use SOPs to provide simplicity when task-organizing between units. Combat heavy engineers maintain simplicity for construction projects using primarily basic theater construction standards.

SYNCHRONIZATION

5-10. Engineer commanders and his staff members serve as the senior member of the M/CM/S BOS and the senior coordinator of MANSPT to ensure that they synchronize MANSPT functions. The commander's staff synchronizes the plan and the commander synchronizes the execution. The staff promotes synchronization through SU and a COP of all friendly and enemy engineer effort in the AO. Additionally, engineers integrate engineer efforts across BOSs operating as combat, CS, or CSS forces.

5-11. Engineers are closely integrated with maneuver and fires. The scheme of maneuver governs the engineer plan. Fire, maneuver, and mobility and countermobility form a triad. Neither fire nor maneuver is truly effective if the combat formation cannot move at will and deny battlefield maneuver to the enemy. Engineers operate well forward to integrate mobility and countermobility into the triad.

INITIATIVE

5-12. Engineer leaders and soldiers at every level will operate with the commander's intent and common sense to use initiative in the absence of orders. Leaders know their missions, including reconnaissance, breaching, obstacle emplacement, and airfield and port construction and repair. They set the terms on their own, according to the commander's intent, when they are in the absence of specific orders. Initiative is enabled when engineers are able to develop and sustain a good SU.

ENGINEER ORGANIZATIONAL PRINCIPLES

5-13. Engineers are organized and equipped to support linear and nonlinear operations *(see Chapter 6)*. Engineers are a portion of a larger BOS and, as such, are required to task-organize to accomplish specific missions. To leverage their capabilities for the maneuver commander, engineer units are organized for success to perform all of the battlespace functions to support full-spectrum capabilities. The following principles apply to combat, CS, and CSS engineers:

- Deployability.
- Modularity.
- Multifunctionality.
- Span of C2.
- Sustainability.
- Commercial integration.

DEPLOYABILITY

5-14. To perform their missions, engineer units must be organized and prepared to rapidly deploy to support the unit they are task-organized to support. AC and RC engineer units are well-rehearsed and prepared to deploy with configured basic loads for a variety of potential missions ranging from combat operations to HLS support. Soldiers are qualified according to Army regulations (ARs), FMs, and unit guidance on their weapons and individual and collective tasks. Units that require contractor support to operate have a plan for autonomous operations or a plan to maintain contract support in the AO. Concepts for future engineer equipment is primarily designed for rapid deployment and ease of loading and unloading onto railcars, aircraft, and sealift. These concepts are balanced with design for M/CM/S, geospatial, and/or general engineering capabilities. Deployability is directly enhanced by the modularity of units.

MODULARITY

5-15. Engineer units have a baseline standard suborganization with the ability to quickly deploy and be task-organized. Multifunctional soldiers provide the flexibility required in full-spectrum operations. Flexibility must exist at all levels of the organization and within each module. This enables rapid force tailoring before deployment and during employment. Modular design provides the versatility and agility to configure into the smallest unit building blocks to deploy the minimum capabilities for assignment to higher C2 HQ (either engineer or maneuver forces). The mission accomplishment should not be delayed or adversely affected by an unnecessary footprint of larger engineer units.

5-16. Mission requirements drive the size and composition of the engineer force. A mix of different units and organizations is often necessary to achieve the proper balance of capabilities. Modular engineer support capabilities may be provided from various sources other than Army engineer units. Engineers from other services, USACE, other military branches, and outside agencies can all contribute to ensuring the right mix of engineer support. This mix may change as the operation progresses through its various phases. These modular capabilities must be organized to maintain responsiveness, flexibility, and effectiveness. The use of reach capabilities (to include TeleEngineering) within the Engineer Regiment makes it possible to minimize the footprint of engineers in theater while still supplying the necessary engineer support to committed forces.

MULTIFUNCTIONALITY

5-17. Engineer units must be capable of performing multiple engineer functions and should anticipate these requirements. Units are organized to perform specific engineer functions but are prepared to conduct other engineer tasks outside of their unit METL, but within the broad range of engineer battlespace functions, when missions are anticipated. As units support full-spectrum operations, they may have to quickly transition from combat operations to support operations requiring a completely different set of tasks. Units operating in noncontiguous areas are capable of performing limited engineer functions outside their area of proficiency when empowered and enabled with reach capabilities and expertise.

SPAN OF C2

5-18. Engineer operations are often characterized by supporting a large geographic area and having many diverse organizations under its control. Therefore, engineer C2 capabilities strive to be interoperable with all Army, joint, and multinational organizations. The C2 is capable of scaling larger or smaller in size and capability, as required by the subordinate or geographic requirements. Engineers are prepared to provide C2 to other nonengineer units when METT-TC requires stability or support operations. Timely and accurate information is power; therefore, C2 facilities must function vertically and horizontally, monitoring and tracking combat operations and sending accurate reports to the right person at the right time. Monitoring and reporting the status of engineer missions are critical to C2 decision making as is adjusting engineer forces and assets to meet planned or unanticipated mission requirements. Mission control and effective communication down to the lowest organizational level of engineers is critical to achieving this goal.

SUSTAINABILITY

5-19. Engineer units are organized with the appropriate logistical support to move, maintain, and fuel, or they coordinate the appropriate service, HN, or contractor for support. Engineer resources are always limited. Materiel, transportation assets, and time restrictions limit the engineer's ability to execute missions. Engineer unit sustainment and the supporting logistical structure must be planned in detail. Logistical limitations may restrict the mission accomplishment by the engineer force. Engineer operations are often resource-intensive. Engineer commanders should exhaust unit capabilities before tapping into external logistical support from the HN or LOGCAP.

COMMERCIAL INTEGRATION

5-20. Engineer units are capable of operating commercial engineer equipment and using available resources to expand their versatility. This allows for smaller deployment requirements while improving efficiency and, in some cases, the overall effectiveness of the engineer unit. Engineer deployments require coordination with contractors, ENCOMs and the USACE to facilitate access to commercial equipment, assets, and real estate. Engineer resources (labor, services, equipment, and materials) belonging to the HN, other services, and multinational forces are present in every theater. At all levels, but especially at the tactical level, improvising may be necessary to convert on-site materials and equipment for military use. HN resources may be well-suited for general engineering support, especially in the rear areas. Local resources may augment available engineers, releasing more engineer units for other missions. Through the processes of communication technology, engineers also have the capability to reach from nearly any location worldwide to obtain products, specialized expertise, and other necessary information. Reach capability may help to minimize the footprint of engineer forces in a given forward AO.

Chapter 6

Battlefield Framework

The battlefield is the epitome of war. All else in war, when war is perfectly conducted, exists but to serve the forces of the battlefield and assure success on the field.

BG S.L.A Marshall, 1947

The Army's warfighting focus enables a diverse (full-spectrum) force to meet the needs of the JFC in war, conflict, and peace. In war, Army forces form the nucleus of the land component—imposing their will on enemies and causing their collapse. In conflict, Army forces deploy quickly into an AO to deter adversaries and potential enemies from establishing their forces and to preclude them from gaining an operational advantage. If deterrence fails, Army forces defeat the enemy, terminate conflict to achieve national objectives, and establish self-sustaining postconflict stability. Early movement of Army forces retains initiative and freedom of action by providing the JFC complementary means to conduct decisive offensive operations at a time and place of the commander's choosing. If theater circumstances require it, Army forces provide the means to block the enemy's offensive and deliver the counteroffensive blow necessary to win as rapidly as possible. In peace, Army forces train for war. They also help shape the international security environment through engagement activities and support civilian authorities both at home and abroad in response to natural or man-made disasters. Regardless of the type of commitment of Army forces, the degree of engineer participation is likely to be high.

BATTLEFIELD ORGANIZATION

6-1. Within the discussion of the operational framework *(see FM 3-0)* is the concept of battlefield organization. Battlefield organization is the allocation of forces in the AO by purpose *(see FMs 3-0* and *3-07 [100-20])*. It consists of three all-encompassing categories of operations: decisive, shaping, and sustaining. The purpose unifies all elements of battlefield organization by providing common focus for all actions. Commanders organize forces according to purpose by determining each unit's operation will be decisive, shaping, or sustaining. These decisions are the basis of the concept of operations. When circumstances require a spatial reference, commanders describe the AO in terms of deep, close, and rear areas. These spatial categories are especially useful in operations that are generally contiguous and linear and feature a clearly defined enemy force. Combat engineering tends to support the categories of decisive and shaping operations. Geospatial engineering supports all categories, with a primary focus on decisive and shaping operations. General engineering may also support categories. While it is not a separate engineer battlespace function, sustainment engineering is the term often used to describe those general engineer missions that support the category of sustaining operations.

LINEAR OPERATIONS

6-2. Linear operations are how the Army has traditionally fought, with ground forces sharing boundaries and oriented against a similarly organized enemy force. Linear operations employ deep, close, and rear areas within the battlefield framework of decisive, shaping, and sustaining operations.

6-3. Operations in deep areas are conducted in a designated area forward of the close area, with the purpose of shaping enemy forces before their arrival into the close area. The deep area is defined in terms of time and space. Engineers in the offense use reconnaissance assets to predict and detect the enemy to prevent and avoid obstacles that may impede the friendly force. When required to neutralize, engineers may coordinate for fires where they anticipate the enemy is overwatching obstacles and to set the SOSRA conditions for the eventual close fight; breaching, or gap crossing operations. In the defense, engineers may emplace scatterable munitions or conventional obstacles covered by direct or indirect systems to disrupt the enemy's maneuver in the deep fight.

6-4. Operations in close areas occur where the commander envisions close combat as imminent. The maneuver commander masses his combat capabilities to achieve a decisive action. In the offense, engineers balance their limited assets between the requirement to mass with redundancy and the risk with an economy of force. Engineer effort is focused on avoiding impediments then on neutralizing. In the defense, engineers enhance force protection providing survivability positions. Engineers emplace obstacle groups to reinforce the maneuver commander's plan and increase his target acquisition time in the EA.

6-5. Sustaining operations are conducted behind the commander's rear boundary. They assure freedom of action and continuity of operations, sustainment, and C2. Forces in this area are often prone to attack and take enhanced force protection measures to protect forces and facilities. Engineers provide support to enhance force protection by providing survivability to protect vulnerable assets. Engineers are integrated to provide rear commanders with assured mobility support to assist in predicting, detecting, preventing, avoiding, neutralizing, and protecting in the rear area.

NONLINEAR OPERATIONS

6-6. Nonlinear operations are how the Army fights with ground forces in noncontiguous and simultaneous operations while allocating forces in the AO by purpose. Engineers contribute to decisive, shaping, and sustaining operations by establishing the imperatives of assured mobility. Developing a COP leverages information by using geospatial tools to combine terrain data and an integrated ISR collection plan to attain a level of understanding within the elements of combat power.

Decisive Operations

6-7. Decisive operations are those that directly accomplish the task assigned by the higher HQ. Decisive operations conclusively determine the outcome of major operations, battles, and engagements. There is only one decisive operation for any major operation, battle, or engagement for any given echelon. The decisive operation may include multiple actions conducted simultaneously throughout the AO. Commanders weight the decisive operation by economizing on combat power allocated to shaping operations. In the offense and defense, engineers normally focus on freedom of maneuver through assured mobility. Decisive operations integrate all aspects of the assured mobility imperatives, with emphasis on maintaining mobility and momentum. This requires the senior engineer to synchronize the M/CM/S BOS capabilities to protect and sustain the ability to maneuver when and where the maneuver commander desires, enabling him to maintain pressure and lethality upon the enemy. It also requires synchronizing those engineer tasks that are embedded in the other BOSs.

6-8. In stability operations, decisive operations are usually those that achieve and maintain stability, protect lives and property, or promote peace. Unlike decisive operations in offense or defense, decisive operations in stability operations do not always have immediate impacts; sometimes results take years to achieve. In support operations, decisive operations normally prevent or mitigate the effects of natural of man-made disasters. They relieve or reduce conditions such as disease, hunger, or privation. Decisive operations could be stabilizing areas by providing security for personnel, facilities, or capabilities; rendering certain services to populations; or reestablishing critical infrastructure. Engineers play a significant role in both stability and support operations, and engineer actions may contribute to or actually be the decisive operation, especially during support operations.

Shaping Operations

6-9. Shaping operations at any echelon create and preserve conditions for the success of decisive operations. Shaping operations include lethal and nonlethal activities conducted throughout the AO. They support the decisive operation by affecting enemy capabilities and forces, or by influencing an adversary's decisions. Shaping operations use all elements of combat power to neutralize or reduce enemy capabilities. They may occur before, concurrently with, or after the start of the decisive operation. They may involve any combination of forces and occur throughout the AO. Typically prior to decisive operations, engineers contribute to shaping operations by establishing and maintaining operating areas. They gain control of friendly and enemy EAs, areas of interest (AOIs), operating areas, and mobility corridors that connect these areas. Engineers may then attack the enemy's ability to influence friendly operating areas. This proactive shaping allocates combat power and ISR capabilities to secure friendly maneuver areas and/or attack threat operating areas to prevent the threat ability to influence friendly maneuver operations. In stability operations, shaping operations often convert temporary gains into long-term political successes. The capabilities required to exploit stability often differ from those needed to achieve stability. This may require such measures as rotating different types of units in and out of the AO as the operation progresses through different stages. In support operations, shaping operations may include influencing perceptions, ideas, and information, as well as maintaining legitimacy. Shaping operations entail transferring tasks to civilian agencies or the local government. For military engineers, this may also include transferring responsibilities to USACE, civilian contract engineers, or HN organizations. Regardless of the type of operation, commanders may designate a successful shaping operation as a decisive operation. In that case, commanders weight the new decisive operation with combat power from other shaping operations. The concept of operations clearly describes how shaping operations support the decisive operation.

Sustaining Operations

6-10. Sustaining operations generate and maintain combat power. They are operations at any echelon that enable shaping and decisive operations by providing CSS, rear-area and base security, movement control, terrain management, and infrastructure development. Engineers contribute to sustaining operations by providing general engineering support to establish the theater backbone structure. Sustaining operations include rear-area and base security measures taken by military units, activities, and installations to protect themselves from acts designed to impair their effectiveness. Sustaining operations has four components: intelligence, base and base cluster self-defense, response force operations, and combined arms TCF operations. Engineers provide geospatial products that are distributed throughout the force to develop the COP from the FLOT to the rear most boundaries. FESTs from USACE and theater level ENCOMs provide design and labor capabilities to construct staging bases, logistic bases, FOBs, and LSAs. Engineer support to infrastructure development applies to all fixed and permanent installations, fabrications, or facilities that support and control military forces. Infrastructure development focuses on facility security modifications and includes ADC and repairs. Engineer operations provide mobility support throughout all LOCs for unimpeded sustainment support to forward forces. Engineering performed in support of sustainment operations may be referred to as sustainment engineering. (This is not another category or addition to the engineer battlespace functions.) The vast majority of this support will be general engineering and include at least some geospatial engineering.

FRAMEWORK CONSIDERATIONS

THEATER ECHELON

6-11. The ARFOR or ASCC engineer tailors the engineer structure to meet the requirements the JFCs plans or orders. While focused on operations in the CZ, these engineers also support the theater with general engineering support. The ARFOR or ASCC engineer must be closely tied into current and future operational planning and have his own C2 structure to ensure the timely and proper execution of the intent and scheme of maneuver. He must also anticipate and plan for future mission requirements. Engineer mission responsibilities at the operational level include constructing, maintaining and rehabilitating the theater support base, and providing support to other services, agencies, and other multinational forces in the TO. This may include support of the initial force projection requirements to create or improve the force projection enablers. The ability of CSS units to conduct sustainment operations, as well as moving and sheltering combat and CS forces, depends on adequate, responsive engineer support.

6-12. The number and type of operational-level engineer organizations depends on the operational environment, mission requirements, HN infrastructure, and availability of existing engineer support to the TO. Operational-level engineer units provide—

- Geospatial information and services support to the theater.
- Construction management for the ARFOR or ASCC.
- Troop construction and repair to all US elements in the COMMZ.
- Contract construction support.
- LOC development and sustainment support.
- Infrastructure enhancement and sustainment support.
- Real estate and facility acquisition, management, and maintenance support.
- General, geospatial, and combat engineering support to tactical-level organizations on an area or mission basis.

6-13. Establishing EWLs is a standard engineer C2 method used by the ARFOR or ASCC engineer, in conjunction with the task organization, to designate AORs for subordinate engineer organizations. EWLs are generally established from the corps rear to the rear of the COMMZ in a TO. EWLs may also be established to designate engineer support boundaries in operations other than war. These lines generally match ASG and HN military and political boundaries. EWLs will be adjusted as an operation transitions. Because engineers focus on mission requirements rather than area support, EWLs may be independent of other control measures (for example, to permit operational-level engineers to concentrate forward in the corps area to free up corps engineer assets or to conduct specific missions on a task basis forward of the EWL.)

6-14. Engineers are nearly always required in a TO. This force must conduct a carefully tailored, well-planned, and well-rehearsed mission. Contingency operations require a greater proportion of engineers to support the force.

6-15. Requests for engineer support (whether Army, other service, or HN) in the COMMZ are received and passed from an ASG to the supporting engineer group or from a TSC to a supporting engineer brigade or the ARFOR or ASCC engineer. The ENCOORD assigns missions based on ASCC priorities, drawn from the combatant commander or JFC priorities. Engineer units in the COMMZ execute the missions under control of the ASCC task organization. Unfulfilled requests for COMMZ engineer support are passed by dual routes to the higher ENCOM and the ASCC engineer, as well as through ASG channels to the TSC for resolution at the ASCC and joint levels.

6-16. Theater construction management often spans multiservice requirements. The combatant commander may establish a theater regional contingency engineering management (TCEM) or RCEM cell to prioritize engineer construction activities and augment the joint force staff. Army representation in this cell is normally drawn from the ARFOR/ASCC engineer staff. Other service component engineer personnel are also part of this cell.

6-17. The combatant commander may also establish joint boards to address engineer issues outside of normal operations and ensure coordination among the services. Such boards include the Joint Facilities Utilization Board (JFUB) and the Joint Environmental Management Board (JEMB). Each is discussed in *JP 3-34* and *JP 4-04*. Again, Army representation is normally drawn from the ARFOR or ASCC engineer staff and includes other service component engineers.

6-18. The TCEM or RCEM cell and joint boards are used to coordinate other military and non military resources. They synchronize and integrate the total engineer effort. Other military resources available to the combatant commander and the JFC include USACE teams, Navy construction engineers (Seabee teams), Rapid Engineers Deployable Heavy Operations Repair Squadron, Engineer (RED HORSE) units, or multinational engineer units. HN support, including local contractors and other engineer and logistical contracting options such as LOGCAP, provides a vast variety of engineer capabilities.

Combat Engineering

6-19. Theater and operational-level engineer units typically only perform combat engineer (M/CM/S) missions when operating in the division area as reinforcement to divisional engineers. Operational mobility and countermobility tasks are included under the general engineering function. Combat engineering tasks may also be performed in the theater rear area but tend to be associated with support of the TCF in its rear-area mission. *See FM 3-90* for a discussion of this mission.

Geospatial Engineering

6-20. The theater topographic battalion HQ is typically located in the COMMZ and will generally fall under the ENCOM, if an ENCOM is in theater. With the topographic planning and control team at the theater HQ, the battalion coordinates with the Topographic Engineering Center (TEC), HN, NIMA, and the National Ground Intelligence Center to fulfill theater mapping and satellite imagery products. They may also integrate USACE products to provide a more complete product in some cases. The general support (GS) topographic company supports noncorps units at the theater level and assists the DS companies in corps areas. They are normally collocated with the battalion HQ and may send elements to other locations, to include an EAC intelligence center.

General Engineering

6-21. Theater construction management often spans multiservice requirements. The theater commander may require the establishment of a TCEM or RCEM to prioritize and control engineer activities, with particular emphasis on construction activities. This management responsibility may be performed by the senior engineer staff and leadership (usually an ENCOM) on behalf of the ASCC. The engineer brigades and groups plan, coordinate, and supervise their assigned work in their respective AORs.

6-22. New construction, maintenance, and repair work in the COMMZ support and sustain combat operations forward. This includes roads, railways, pipelines, bridges, airfields, ports, buildings, utilities, EPW camps and installations, displaced persons camps, and any other requirements to support or take the load off combat operations.

6-23. Combat heavy engineer battalions have a variety of equipment, tools, and skills to do all types of construction. Some capabilities, such as asphalt and paving, require pairing with separate engineer companies and teams to accomplish the mission. Construction support companies operate and maintain equipment to augment combat engineer battalions, heavy, and other units as well.

6-24. Dump truck companies provide additional haul assets. Pipeline construction support companies provide technical personnel and specialized equipment to support pipeline construction and related facilities. Port construction companies provide technical personnel and specialized equipment for constructing and restoring ports, to include JLOTS facilities, inland waterway facilities, and POL marine terminals.

6-25. It is not uncommon for the ASCC to draw on other military and nonmilitary resources to integrate into the total engineer effort. Other military sources include USACE teams and support, Seabee teams, RED HORSE units, or other similar multinational engineer units. HNS, civilian contractors, and civil works agencies/businesses from outside the AO also provide a vast variety of potential augmentation capabilities.

6-26. While not exclusively a general engineering concern, environmental considerations *(see Appendix K)* come under general engineering for AUTL purposes. Engineers proactively advise the commander on environmental issues which promote awareness in the unit. Commanders protect the natural environment in which US military forces operate, to the greatest extent possible, consistent with operational requirements. Commanders must ensure compliance, to the greatest extent possible and as far as practicable within the confines of mission accomplishment, with all applicable environmental laws and agreements, including those of the HN. For more information on environmental regulations, laws, and treaties, *see FM 3-100.4.* The goal of compliance is to minimize the potential adverse impacts on human health and the environment while maximizing readiness and operational effectiveness. *FM 3-100.4,* and *JPs 3-0, 3-34, and 4-04* each highlight the critical integration of environmental considerations into all operations and training.

6-27. The ASCC has overall responsibility for real-property maintenance activities (RPMA). The TSC, through its ASGs, normally provides the needed RPMA support. Principal RPMA functions in a TO include operating, repairing, and maintaining facilities and utilities; fire prevention and protection; and refuse collection and disposal. The TSC provides technical RPMA guidance to the ASGs. ASGs provide RPMA support to all Army facilities in their AOs, to include leased facilities, unless HNS is or has available leased facilities. RPMA may also be provided for other service component facilities within the AO.

6-28. The theater's organizational RPMA requirements that exceed the TSC and ASG capabilities are forwarded to the supporting engineer group or brigade for execution according to the theater priorities. Configuring engineer units to support the ASGs is largely based on the expected RPMA workload. They are tailored to the specific ASG that they will be supporting to accomplish RPMA missions.

6-29. RPMA may be administered on a centralized or decentralized basis. If decentralized, the TSC, through its ASGs, uses supporting engineer teams, facility engineer detachments (FEDs), USACE, other CCA contracted RPMA services', or HNS to accomplish the mission. ASGs have small cells to receive, prioritize, and coordinate local engineer support requirements. They assess workloads, sequence critical requirements, and request assistance for the engineer group in GS.

6-30. In operations where a robust TSC/ASG structure doesn't exist, but base camps, logistic bases, and other operating bases have been established, the FEDs have taken on the larger task of conducting Directorate of Public Works (DPW) bare-base facilities management services. Leveraging their internal capabilities with other engineer resources in the theater, the FEDs become the focal point of administering RPMA to their individually assigned bases.

6-31. Many specialized engineer teams can be tailored to meet the needs of a particular ASG and support it under the C2 of an engineer HQ team. With the exception of the specialized team assigned to the ASG, the engineer group typically controls all of the engineer units in its assigned area. The ENCOM may also administer RPMA on a centralized basis. *FM 5-116* contains additional details on engineer operations in support of RPMA.

6-32. Contingency real estate support teams (CRESTs) are called upon early in a contingency operation to support acquiring real estate and facilities in support of reception, staging, RSOI bed down, and unit operations. Generally, requirements are centralized and front loaded to meet the anticipated theater level operational needs.

CORPS ECHELON

6-33. The corps generally fights at the tactical level of war; however, it may fight at the operational level during SSC operations. Although the corps has an engineer brigade assigned to it, the number and type of engineer units assigned to the brigade depend primarily on the mission of the corps, the number and type of their divisions and separate brigades, and the operational environment. Normally, each committed division of the corps will include elements of an engineer group from the corps engineer brigade. *See FM 5-100-15* for more information on corps engineer operations. *See Figure 6-1* for a notional corps engineer brigade organization.

6-34. Engineers perform battlespace functions (combat, geospatial, and general engineering) and have missions in all parts of the corps area. The general engineer mission in the corps AO is to construct and maintain LOCs and tactical march routes. This mission is continuous due to the effects of enemy actions, heavy traffic, and weather.

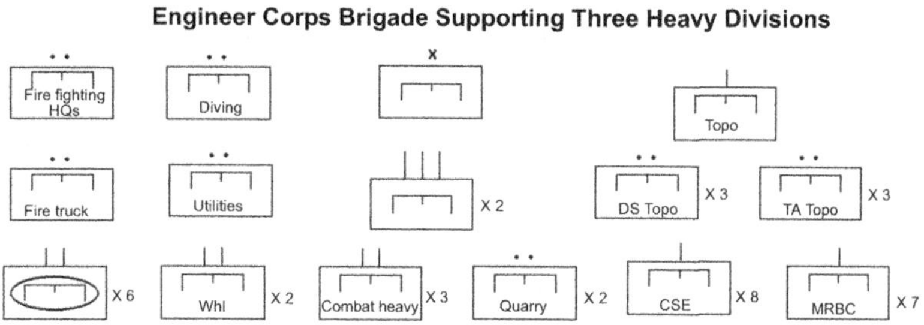

Engineer brigade competencies—

- Control corps engineer units not organic to divisions, separate maneuver brigades and ACR.
- Provide combat (M/CM/S), geospatial, and general engineering support to the corps.
- Augment engineer support to the divisions and other subordinate units.
- Require a major commitment of corps units to keep the LOCs and tactical march routes open to sustain the committed forces, shift other forces, and implement corps-directed attacks. Corps combat battalions perform those tasks with CSEs companies augmenting other units with horizontal equipment to maintain roads and airstrips.
- Provide wet and dry bridging multirole bridge company (MRBC) for use in LOC and other routes. A diving team performs scuba diving in forward combat areas (such as in river and far-shore reconnaissance), clears underwater obstacles, and supports assault float-bridging operations.
- Provide construction assets (combat heavy battalions and quarry teams) for operational construction requirements such as forward log bases, heliports, airfields, bed downs, and MSRs.
- Provide geospatial engineering for the corps; terrain teams provide geospatial (topographic) support for each assigned division.
- Provide fire fighting assets to support aviation units and petroleum units.
- Require utilities teams to support the corps support command (COSCOM) with RPMA support in the corps area.

Figure 6-1. Notional Corps Engineer Brigade Laydown

6-35. Corps engineer organizations provide guidance on protective shelters and camouflage measures to units and support corps units with geospatial information and services, as directed by the mission. Corps engineer units operating in the forward area reinforce divisional engineers in combat engineering roles. Many of the tasks performed by corps battalions in the corps rear area are also done in the forward area while performing missions in support of division operations.

Combat Engineering

6-36. Corps engineer units are typically combat support engineers. They usually perform (M/CM/S) missions when operating in the division area as reinforcement to divisional engineers. These tasks may also be performed in the corps rear area but tend to be associated with support of the TCF in its rear-area mission. *See FM 3-90* for a discussion of this mission.

6-37. Separate corps brigades and ACR have an organic engineer company, which is usually not sufficient to handle all required engineer tasks when these units are committed. These companies are designed and focused for supporting the mobility and countermobility portions of combat engineering. The corps engineer brigade reinforces these organizations with additional combat battalions and separate companies based on the mission and the situation.

Geospatial Engineering

6-38. A DS topographic company of the theater topographic battalion typically provides all forms of geospatial support for a corps. It is positioned in the corps rear area, providing a terrain analysis team to the corps CP.

General Engineering

6-39. General engineering activities occur throughout the theater and are not limited to the COMMZ. Within the corps area, operational construction requirements, such as forward logistics bases, heliports, and MSRs are needed to ensure the continuity of support for combat operations. These activities are in addition to the corps-generated construction requirements (bed down, logistics bases, and rehearsal ranges) that keep corps construction assets fully engaged. The nature of corps operations generally limits construction to the austere, essential facilities needed to sustain the current fight or support near-term major operations in offensive, defensive, and selected stability operations.

6-40. Corps combat engineer battalions perform general engineering tasks in the division area. CS equipment companies augment the combat battalions with equipment to move earth and maintain or create horizontal surfaces such as roads and airstrips. Combat battalions may also assemble tactical bridges provided by panel-bridge companies or allocated from theater stocks for use on LOCs and other routes. MRBC erect their own tactical bridges (fixed and floating) to support river crossing operations.

6-41. Light corps combat battalions and light equipment companies reinforce light DIVENGs, particularly during their initial deployment. The airborne corps battalion can parachute into an operational area and construct an assault airstrip. The light equipment company augments light forces with additional air-transportable earth-moving equipment.

6-42. Utility teams support the COSCOM and provide RPMA and base operations (BASOPS) support throughout the corps area. Additional support may be available through HNS or corps engineer brigade assets. The ENCOM provides additional support and technical guidance, as necessary. Frequently, in stability or support operations, USACE contractors (or other contracting agents) provide RPMA engineer services under the general oversight of Army engineers (the FED, which provides DPW bare-base facilities management).

DIVISION ECHELON AND BELOW

6-43. Divisions perform major tactical missions and can conduct sustained battles and engagements. The engineer forces organic to each division are tailored specifically to support that type of division. The corps engineer brigade provides additional engineer units based on the division's specific mission and tactical situation. The engineer infrastructure at the higher echelons makes it possible to commit and sustain divisions in combat.

Combat Engineering

6-44. The division combat engineer battalion is designed to perform the combat engineering function (M/CM/S) and to participate in the close fight. Corps combat battalions also participate in the division close fight. A corps commander will usually place at least one corps combat battalion in a command relationship to a division or engineer group supporting a division to reinforce the close fight. Corps and DIVENG elements often mix and cross attach elements to enhance the relative capabilities and strengths of each. The HQ of the habitually associated engineer battalion generally commands all engineers supporting the major effort of a brigade. A notable exception to this may be in a light division where only an engineer company is habitually associated to a maneuver brigade. There is also only a single organic engineer company in the SBCT. Each committed legacy heavy maneuver brigade normally needs the equivalent of an engineer battalion or one company per battalion TF. These levels of engineer support are adjusted based on the METT-TC analysis. Additional engineer capabilities may be task-organized from within the division or from EAD and EAC engineer units.

6-45. Additional corps engineer battalions operate in the division on an area or mission task basis. Separate engineer companies, especially bridge companies, operate in support of the division as required. When a division has the priority and need for a large number of corps engineers, it will likely also have an engineer group in support to control the activities of these engineers. When allocated to a division, the group HQ controls all engineer operations within the division's rear area. It also allocates and controls engineer organizations and resources to perform specific or selected tasks within maneuver brigade areas. This group HQ, or another group HQ may also function as the planning and control HQ for engineer tasks affecting the division as a whole (as in divisional assault river crossing of a breach by one or more maneuver brigades of the division).

Geospatial Engineering

6-46. The DS topographic company in the corps area also provides a terrain analysis team to the division. The team typically locates at the DIVENG HQ, with DTSS teams allocated to the assistant brigade engineers (ABEs). The DIVENG team works with the division's G2. TerraBase™ and other newly-developed programs are valuable tools to assist engineers at brigade and TF levels. The engineer cell of the SBCT contains an organic terrain team.

General Engineering

6-47. Combat engineer units rarely perform general engineering tasks within the division since their focus is on combat engineering tasks. General engineering tasks within the division are typically performed by corps level or higher CS engineer units. Tasks performed by combat engineers that obstruct and channel enemy maneuver (barriers and obstacles), protect friendly forces (defilade positions and shelters), and enhance forward mobility (combat trails and roads) are considered to be combat engineering functions. These missions are incorporated into the OPLAN and OPORD of the division and support the maneuver brigade/TF in accomplishing their respective missions and the commander's intent.

6-48. RPMA and BASOPS needs are generally very limited at the division echelon and below, although certain contingency operations will heighten their importance. Support is provided by the corps engineer brigade and higher echelon ENCOMs.

INTEGRATING OTHER ENGINEER ORGANIZATIONS AND CAPABILITIES

6-49. Integrating the variety and special capabilities of engineer organizations requires an understanding of the various capabilities and limitations of the engineer assets available for any given mission. Besides Army engineers, there are a variety of other organizations that may be available to support the overall effort.

6-50. The first of these are the other members of the Engineer Regiment, primarily USACE. They may have assets directly integrated into the military C2 structure and linked to the ENCOMs or be already operating under contract in theater. Whether providing construction contract and design support in the AO or outside of the contingency area, USACE can obtain the necessary data, research, and specialized expertise not present in the theater through TeleEngineering and other reach capabilities. *See Appendices I* and *J* for additional USACE capabilities.

6-51. Other US and allied military service engineers may also have engineer units and capabilities assigned to a given mission or operation. *See Appendices D, E, F, G, H,* and *I* for more information on other US and multinational military service engineer capabilities. HN engineer capabilities may be available if an adequate infrastructure exists. Potentially, this could include a wide array of civil and public works organizations. *See Appendix H* for a discussion of potential HNS.

6-52. It is increasingly common to contract a wide range of engineer services with local or third party national organizations and civilian contractors. These assets are typically used to free up military assets, minimizing the military footprint in a theater, when requirements exceed military capabilities or when the engineer operations and requirements are to be conducted in areas that are relatively safe from active combat.

Chapter 7

Battle Command for Engineers

In the profoundest sense, battles are lost and won in the mind of the commander.
Captain Sir Basil Liddell Hart, 1934

Battle command applies the leadership portion of the elements of combat power. It is based on the commander having a profound understanding of the conduct of full-spectrum operations and being able to visualize, describe, and direct the actions of subordinates. An engineer commander has a difficult responsibility in performing dual roles as the organic engineer commander and the senior engineer advisor to the maneuver commander. In today's operating environment, the engineer commander's leadership is essential to ensuring mission success. Understanding *FM 6-0 (101-5)* and how battle command relates to C2 operations is essential for all commanders.

BATTLE COMMAND AND COMMAND AND CONTROL OPERATIONS

7-1. The JFC and ARFOR organize assigned forces to meet mission requirements and use the resources available to them in the most effective manner. Engineer forces are extremely adaptable and can be tailored to best meet mission requirements. Each contingency operation will be unique, incorporating any of a range of different combinations of engineer entities and supporting relationships. Additionally, the structure and requirements that developed for engineer support will almost certainly vary as the operation transitions. Engineers will generally support the mission effort. However, they may also be called on to lead or command the operation, particularly when the mission is focused on disaster relief or nation building.

7-2. For most missions, the senior engineer commander in theater will C&S engineer forces as a specialized command with an independent command structure, maintaining technical and operational ties to strategic-level agencies and establishing working relationships with their ARFOR staff counterparts. Command authority is delegated by the ASCC or ARFOR commander, as is any command relationship with the TSC.

7-3. C2 operations are the exercise of authority and direction by a designated commander over assigned and attached forces to accomplish the mission *(FM 6-0)*. The engineer commander must leverage available systems to most effectively achieve his intent. These systems include FBCB2 and MCS-ENG. As stated in *FM 3-0*, the goal of C2 is mission accomplishment. The engineer commander seeks to accomplish this by understanding the end state and ensuring that his assets are postured for transitions and future operations. He typically exercises authority and direction over engineer forces made available to him through establishing command or support relationships. However, he must also work within specific command or support relationships as well. Within these constraints, he is responsible for organizing resources and directing their operations. *See FM 6-0* for a more in depth discussion of battle command and all of the other primary aspects of C2.

7-4. *Figure 7-1* represents C2 visually and depicts the organizing principle of Army C2 doctrine. The focus of C2 is the commander. He determines what to do and directs actions. C2 is a continuous process; the staff must diagnose what is happening, even if unexpected, and develop appropriate solutions for the commander's decision.

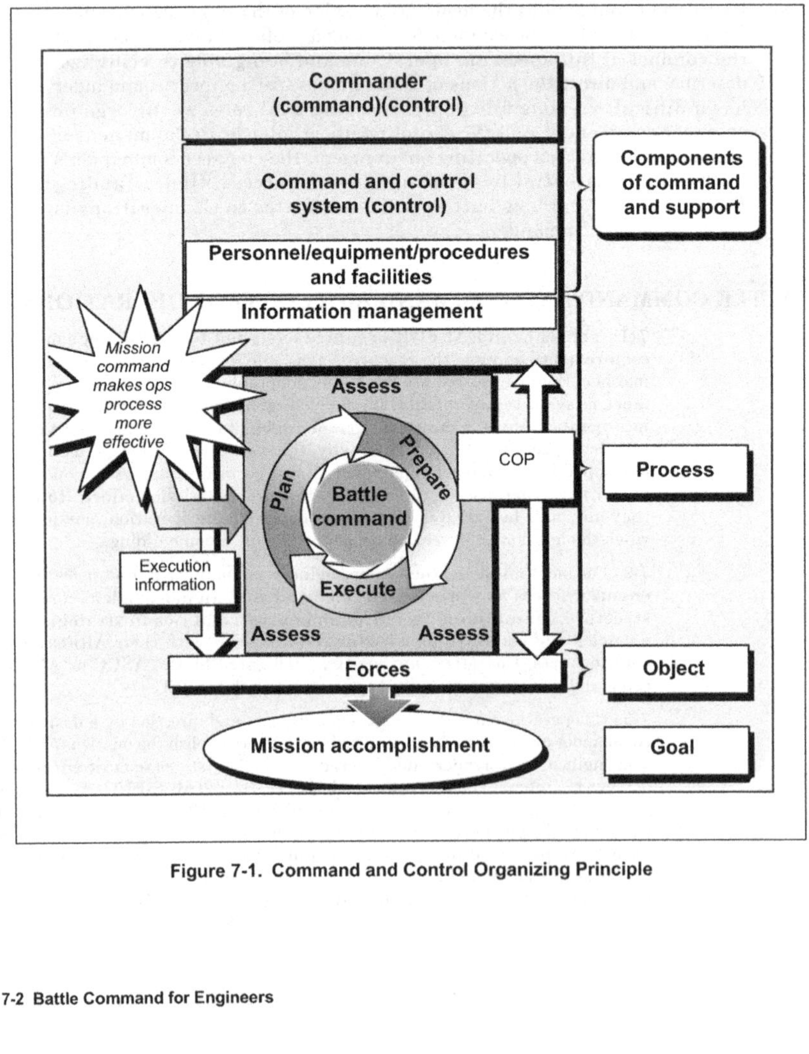

Figure 7-1. Command and Control Organizing Principle

LEADERSHIP ASPECTS—VISUALIZE, DESCRIBE, DIRECT

7-5. Commanders at all levels provide skilled judgment based on study, training, experience, and creative thinking. *FM 3-0* describes the critical aspects of leadership—visualize, describe, and direct *(see Figure 7-2, page 7-4).*

7-6. Upon receipt of the order, the engineer commander begins to formulate (visualize) a plan. He assesses his capabilities weighed against enemy capabilities on the terrain or support mission requirements. The commander conducts an informal engineer assessment after he reads the order, with input from his staff estimate. His staff prepares an EBA as part of the staff estimate. Using METT-TC, Army tenets, principles of war, engineer principles, and operational framework the commander visualizes the potential engineer tasks and the kind of information he needs to finalize a plan. The force commander may request products from geospatial engineering assets and FFE for technical support to improve his ability to visualize the combat, geospatial, and general engineering requirements for the operation. Engineer commanders may use sketches of the AO and locations of all the potential engineer tasks with purpose to visualize how the operation will occur.

7-7. The commander identifies (describes) what each subordinate commander, two levels down, must understand in the absence of orders. The commander uses the framework of the battlefield as a common reference to describe his vision to subordinates. Descriptions are provided of how he envisions supporting the operation regarding time, space, resources, purpose, and action for each type of operation—decisive, shaping, sustaining. He concludes with the commander's intent for engineer operations and planning guidance. The engineer commander's intent expresses the conditions for success and the desired end state of engineer forces on the objective.

7-8. The commander uses his focused intent (direct) as the foundation for this staff to develop the SOEO and supporting engineer concept of operations. This effort is synchronized by the senior engineer, chemical, and MP commanders within the M/CM/S BOS and through the MANSPT cell coordinator in the SBCT. The engineer commander directs his efforts by resourcing engineer mission requirements with personnel, time, and equipment.

ROLES OF ENGINEER COMMANDERS AND STAFFS

7-9. Whether performing as a commander or a supporting engineer staff element (referred to as the ENCOORD), engineers at each echelon monitor current operational activities, execute and control engineer missions, and plan future engineer operations in conjunction with the supported commander and forces. The depth and breadth of engineers involvement depends on the role the engineer plays within the military organization and the nature of the overall operation. His responsibilities may range from planning combat engineer operations to planning and coordinating environmental protection, critical areas, and protection levels. Success as an ENCOORD is critical to the success of the engineer unit. Commanders are challenged to generate a force that is tailored to current and anticipated missions, flexible in changing circumstances, and able to meet the constraints of lift capabilities. The factors of METT-TC drive the initial tailoring of the force. Prioritizing and leveraging available assets for acceptable degrees of risk guide the commander in

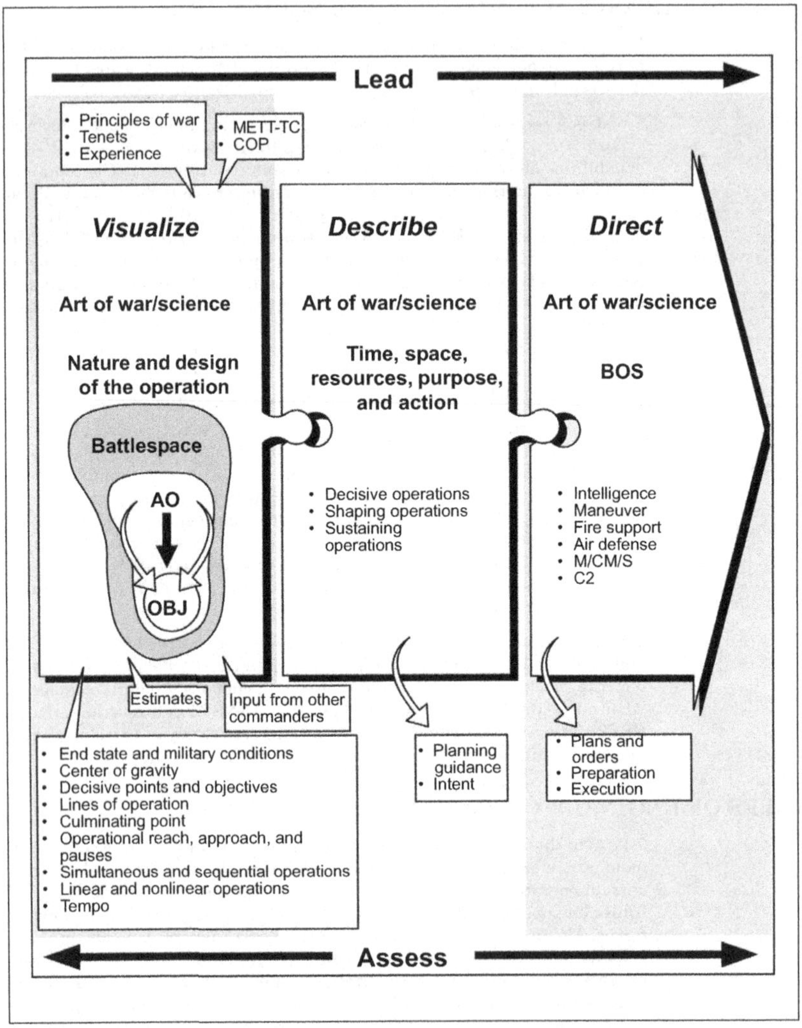

Figure 7-2. Leadership Aspects

determining the timing, number, and type of units to employ. Engineer units that make up the architecture of Army engineer forces are diverse and possess highly specialized capabilities. *(See Appendix D.)* Because the demands of any given operation vary, there is no set engineer structure; however, certain maneuver organizations do have a specific engineer element organic to them, and a building block approach based on need is used to determine the ultimate force refinement.

7-10. *Figure 7-3* is a useful depiction of the multiple engineer organizations that may comprise the theater engineer task organization. This organization lays the foundation to meet mission requirements. This mature structure develops incrementally, over time during a contingency response. *Appendix D* catalogs the deployable MTOE units within this structure as well as TDA organizations associated with the Engineer Branch. A discussion of other service engineer organizations is included in *Appendices E* through *G*. The discussion of potential contract engineering and other potential engineering support is included in *Appendices H, I,* and *J*.

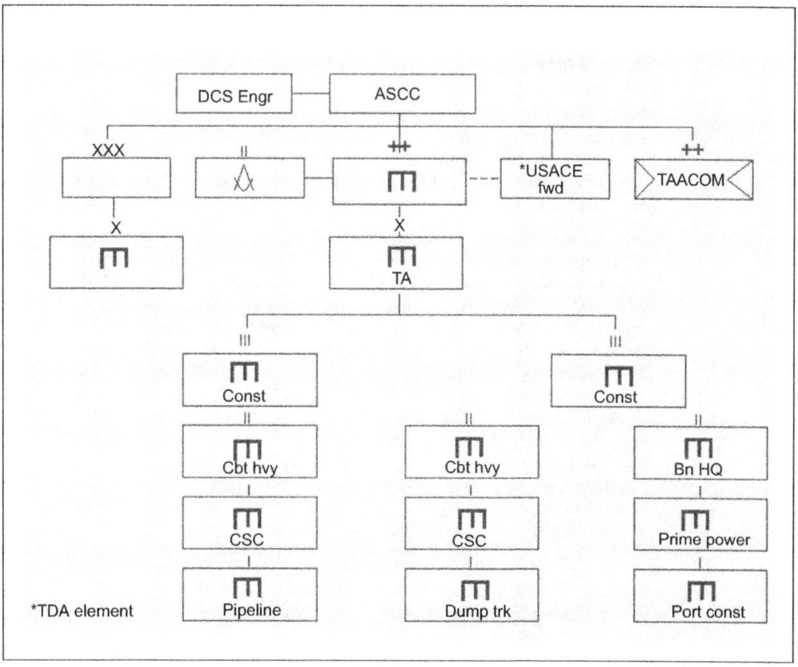

Figure 7-3. Notional Army Service Component Commander Theater Engineer Laydown

7-11. Each military operation is unique in itself, possessing varying degrees of the four basic operational components—ODSS. Engineers have a key role in each of these components, and they execute combat (M/CM/S), geospatial, and general engineering missions. The nature of the missions executed, both sequentially and simultaneously, determines the engineer's role.

7-12. As operations change and move across the spectrum of conflict, many of the products that engineers deliver at the unit level do not change. They continue to improve roads, upgrade infrastructure, lay pipeline, dig, and build. Force protection support is an engineer mission at every point on the spectrum. C2 structures and hierarchies are similarly organized, although at the lower end of the spectrum, engineers tend to play a proportionately greater role and may even be the controlling senior HQ in some instances. The prominence of the engineer role will be particularly high in austere environments. Additionally, it is in these environments and situations where USACE plays a larger role outside of its traditional boundaries and extends the participation of the entire Engineer Regiment.

THE ENGINEER COORDINATOR

7-13. The ENCOORD is the senior engineer officer within the force. However, in the SBCT, FXXI organized DIVENG staffs, ACRs, and Army level staffs, the ENCOORD is also the senior engineer commander in the force. He is responsible for advising the force commander on the best use of available engineer support resources, developing the SOEO, issuing necessary orders in the name of the commander, and implementing the approved SOEO. The ENCOORD is critical to the targeting process as well. His input is essential to the success of assured mobility and logic predict, detect, and prevent.

7-14. Typically, the commander of the organic engineer unit supporting the command or a senior staff officer is the ENCOORD. When an organization has a commander as the ENCOORD, the assistant or deputy ENCOORD is a permanent staff officer, representing the ENCOORD in his absence. An ENCOORD is normally located at every level of command within the theater, from the combatant commander or theater level to the maneuver battalion. Depending on the composition and missions of the engineer forces and the levels of operation, the ENCOORD can take on numerous forms at respective levels, each with its own scope and span of control and associated responsibilities. Whatever the Army echelon or level, the ENCOORD becomes the commander's means for accessing support from the various elements of the Engineer Regiment. When joint functions apply, the ARFOR ENCOORD may be designated as the joint ENCOORD. The ENCOORD's specific responsibilities include the following:

- Planning and controlling the engineer battlespace functions of combat (M/CM/S), geospatial, and general engineering.
- Recommending engineer organization for operations, to include combat.
- Advising on and obtaining engineer services and support through the reach capabilities of the Engineer Regiment.
- Maintaining the engineer COP of current operations.
- Advising on and developing policies for real estate acquisition, management, and disposal.
- Planning and coordinating with the S3, G3, J3, and fire support coordinator (FSCOORD) on integrating obstacles and fires.

- Planning and coordinating with the FSCOORD using artillery-delivered SCATMINEs.
- Advising the commander on the use of organic and nonorganic engineer assets.
- Advising the commander on employing and reducing obstacles.
- Advising the commander on environmental issues (to include the command environmental program), coordinating with other staff officers to determine the impact of operations on the environment, and helping the commander integrate environmental considerations into the decision-making process.
- Planning and coordinating environmental protection, critical areas, and protection levels.
- Providing a terrain mission folder to determine its effect on friendly and enemy operations.
- Managing the digital terrain data storage device (coordinating with the S2, G2, Intelligence Directorate [J2] for planning and distribution).
- Planning and supervising the construction, the maintenance, and the repair of camps and facilities for friendly forces, EPWs, and civilian internees.
- Assisting the S2, G2, and J2 in the IPB, including preparing the EBA portion of the engineer estimate.
- Participating in the targeting meeting.
- Participating in the JFUB, generally as the chair.
- Participating in the Joint Civil-Military Engineering Board (JCMEB), generally as the chair.
- Participating in the JEMB, generally as the chair.
- Providing information on the status of the engineer assets on hand.
- Making recommendations based on technical information to the appropriate operations and logistics staffs for the route of the MSRs and the logistics areas.
- Planning the reorganization of engineers to fight as infantry combat units when the appropriate commander deems their emergency employment necessary.
- Coordinating with interagency department engineers, such as the Federal Bureau of Investigation (FBI).

ASCC ENGINEER

7-15. The ASCC staff normally includes the ASCC engineer as a member of the commander's special staff. He integrates engineers into the ASCC's plan to sustain Army forces and support other services or allied forces. The senior engineer commander may be dual-hatted as the ASCC engineer or his deputy may assume this position as a primary role.

7-16. In a developed contingency operation that includes a COMMZ or rear-area support structure and boundaries, the senior engineer commander would most likely be an ENCOM commander with C2 of numerous engineer units, capabilities (topographic, design, and RPMA and assets (contract construction and HN). As such, the ENCOM commander executes the ASCC's goals of operational sustainment and mobility in synchronization with the roles already identified above. This is done through operational, planning, and technical staffs that are integrated within the ASCC as well as those organic to staff within the ENCOM. The ENCOM typically functions as a separate and specialized command under the ASCC. A theater engineer brigade or engineer group could also provide this support.

CORPS ENGINEER

7-17. The commander of the corps engineer brigade is the corps engineer or ENCOORD at the corps level. He plans and executes engineer operations in the corps area. As the engineer brigade commander, the ENCOORD commands all nondivisional engineer units in the corps. As the engineer special staff officer, he is responsible to the corps commander for all engineer-related matters in the corps AO.

7-18. The engineer brigade commander has a staff element located in the corps CPs. This staff is under the direction of the assistant corps engineer (ACE), who integrates engineers into the corps planning process. The ACE conducts planning with the corps staff and provides advanced warning of future corps operations through engineer channels to the corps engineer brigade, DIVENG brigade, separate brigades, engineer groups, and (ACR) engineer companies. He receives reports from these levels to keep the corps staff informed on current engineer operations.

DIVISION ENGINEER

7-19. The commander of the division (armor/mechanized and light) organic engineer brigade or engineer battalion is dual-hatted as the commander and the special staff DIVENG or ENCOORD. The DIVENG or ENCOORD plans and executes engineer operations in the division area. The division order often gives control of the elements of the engineer brigade to the maneuver commanders, yet the DIVENG commander retains command responsibilities and employs those elements left under his direct control. As the engineer special staff officer, the DIVENG is responsible (to the division commander) for all engineer-related matters in the division AO. He remains responsible, regardless of the rank of the senior corps or EAC engineer unit commander in the division AO. Nondivisional engineer units in the division AO provide liaison to the DIVENG.

7-20. To assist the commander in the role as a special staff officer, the DIVENG has a staff element located in the division CPs. It is under the direction of the ADE, who integrates engineer operations during the division's planning process. Through engineer channels, the ADE provides advance warning of future division operations to the division and supporting corps and EAC engineer battalions and maneuver brigades. He also receives reports from these levels to keep the division staff and the higher HQ ENCOORD informed on current engineer operations in the division. With staff assistance, the DIVENG controls corps and EAC units in the division area. He should task EAD engineer units in his role as the engineer brigade (heavy)/battalion (light) commander.

7-21. The senior engineers within FXXI divisions and ACRs are not commanders. They are not resourced with a full staff and CP and are not capable of independent C2 operations. In FXXI and ACRs, the senior engineers are special staff officers imbedded within the maneuver CP. EAD engineer units may not be task-organized directly to the special staff officer.

BRIGADE ENGINEER

7-22. In an armored or mechanized division, the organic engineer brigade generally provides an engineer battalion to each armor/mechanized maneuver brigade. In a light, air assault, or airborne division, the organic engineer battalion provides an engineer company to each maneuver brigade. The respective engineer battalion commander and company commander assume the brigade engineer (or ENCOORD) responsibility. In the SBCT, the brigade engineer is a special staff officer, not a commander, and functions as both the ENCOORD and BOS integrator for the MANSPT cell. The brigade engineer integrates engineers into the brigade planning process and coordinates current engineer operations in the brigade AO. As the brigade engineer, he commands all engineer units in the brigade. As the brigade engineer special staff officer, he is responsible to the brigade commander for all engineer-related matters in the brigade AO. Separate maneuver brigades and ACRs have a staff engineer organic to the brigade or regiment.

BATTALION AND TASK FORCE ENGINEER

7-23. The battalion or TF engineer often establishes a normal association between an engineer company and a heavy maneuver battalion or a light engineer platoon with a light maneuver battalion. Maintaining that association is one factor to consider in the tactical planning process, since there are advantages to the company commander or platoon leader functioning as the TF engineer. The company commander or platoon leader normally associated with a battalion/TF is also the staff engineer and advisor to the TF commander. The company commander has a small HQ section to assist the TF commander in integrating engineers into the TF planning process and in executing the engineer portion of the operation. The light engineer platoon leader will not have this luxury. The battalion/TF engineer makes operational reports through the TF S3 and provides other required reports through engineer channels, as necessary.

7-24. Since the SBCT has only one organic engineer company, engineer platoons will not be habitually task-organized to maneuver battalions/TFs. As a result, METT-TC will determine the level of engineer support task-organized to each maneuver battalion/TF, if any. In most cases, engineers will be organized to weight the main effort. In that instance, the company commander will serve as the TF engineer. When allocated engineers, the other maneuver battalions/TFs will usually have a platoon leader as the TF engineer.

7-25. When additional engineers operate with the TF, the associated company commander or platoon leader normally remains the TF engineer. However, the reinforcing engineer unit commander/platoon leader and staff task-organized to the maneuver unit or organic engineer will assist with the detailed planning.

COMMAND AND SUPPORT RELATIONSHIPS

7-26. Engineers have a number of specialty units with unique capabilities. Commanders task-organize these units to subordinate maneuver commanders to provide these capabilities. The commander can task-organize them in a variety of command support relationships. Based on the hierarchy, a subordinate commander may not impose a more restrictive command/support relationship if the unit is task-organized further down.

JOINT RELATIONSHIPS

7-27. At the theater level, when Army forces operate outside the US, they are assigned under a JFC *(see JP 0-2, JP 3-0,* and *FM 100-7).* A JFC is a combatant commander, subunified commander, or JTF commander, authorized to exercise COCOM or OPCON over a joint force. At the theater level, the combatant commander provides strategic direction and operational focus to forces by developing strategy, planning the theater campaign, organizing the theater, and establishing command relationships for effective unified action. The JFC plans, conducts, and supports the campaign in the theater of war, and during subordinate theater campaigns, major operations, and battles. The four joint command relationships are COCOM, OPCON, TACON, and support *(see Table 7-1).*

Table 7-1. Joint Command Relationships and Inherent Responsibilities

Inherent responsibilities are:	If relationship is—		
	COCOM	**OPCON**	**TACON**
Has command relationship with—	Gaining combatant commander, gaining service component commander	Gaining command	Gaining command
May be task organized by—	Gaining combatant commander, gaining service component commander	Gaining command	Parent unit
Receives logistic support from—	Gaining service component commander	Service component command; parent unit	Parent unit
Assigned positions or AO by—	As required by gaining component commander	As required by gaining command	As required by gaining command
Provides liaison to—	As required by gaining component commander	As required by gaining command	As required by gaining command
Establishes and maintains communications with—	As required by gaining component commander	As required by gaining command	As required by gaining command and parent unit
Has priorities established by—	Gaining component commander	Gaining command	Gaining command
Gaining unit can impose further command relationship/ authority of—	OPCON, TACON, DS, mutual support, GS, close support	OPCON, TACON, DS, mutual support, GS, close support	DS, mutual support, GS, close support

7-28. COCOM is a nontransferable command authority, exercised only by unified or specified combatant commanders, unless otherwise directed by the SECDEF. Combatant commanders exercise authority over assigned forces. COCOM provides full authority to organize and employ commands and forces the combatant commander considers necessary to accomplish missions. The combatant commander exercises COCOM through subordinate commands, to include subunified commands, service component commands, functional commands, and JTFs.

7-29. OPCON is inherent in COCOM. It is the authority to perform those functions of command over subordinate forces involving organizing and employing commands and forces, assigning tasks, designing objectives, and giving authoritative direction necessary to accomplish the mission. OPCON may be exercised at any echelon at or below the level of the combatant commander and can be delegated or transferred. Army commanders use it routinely to task-organize forces. The transfer of OPCON of units between two combatant commanders must be approved by the SECDEF. ENCOMs typically operate in an OPCON status.

7-30. TACON is the authority normally limited to the detailed and specified local direction of movement and maneuver of the tactical forces to accomplish an assigned task. It allows commanders below COCOM level to apply force and direct the tactical use of logistic assets but does not provide authority to change organizational structure or direct administrative or logistic support. The combatant commander uses TACON to limit the authority to direct the tactical use of combat forces. TACON is often the command relationship established between forces of different nations in a multinational force. It may be appropriate when tactical-level Army units are placed under another service HQ. TACON does not provide organizational authority or authoritative direction for administrative and logistic support. The commander of the parent unit continues to exercise those responsibilities unless otherwise specified in the established directive. Army commanders make one Army force TACON to another when they want to withhold authority to change the subordinate force organizational structure and leave responsibility for administrative or logistic support with the parent unit of the subordinate force.

7-31. Joint doctrine establishes support as a command authority. Commanders establish it between subordinate commanders when one organization must aid, protect, or sustain another. Under joint doctrine, the four categories of support are general, mutual, direct, and close *(see JP 0-2* and *JP 3-0)*. GS and DS describe the supporting command's focus. Mutual and close support are forms of activity based on proximity and combat actions. Army doctrine establishes four support relationships: direct, reinforcing, general, and GS reinforcing *(see FM 3-0)*.

ARMY RELATIONSHIPS

7-32. Commanders build combined arms organizations using C&S relationships. Command relationships define command responsibility and authority. Support relationships define the purpose, scope, and effect desired when one capability supports another. *Table 7-2* summarizes Army C&S relationships.

Table 7-2. Army Command and Support Relationships with Inherent Responsibilities

	If relationship is:	Has command relationship with—	May be task organized by—	Receives CSS from—	Assigned position or AO by—	Provides Liaison to—	Establishes/ maintains communications with—	Has priorities established by—	Gaining unit can impose further command or support relationship /authority of—
COMMAND	Attached	Gaining unit	Gaining unit	Gaining unit	Gaining unit	As required by gaining unit	Unit to which attached	Gaining unit	Attached, OPCON, TACON, GS,GSR, R, DS
COMMAND	OPCON	Gaining unit	Parent unit and gaining unit, gaining unit may pass OPCON to lower HQ. *See note 1.*	Parent unit	Gaining unit	As required by gaining unit	As required by gaining unit and parent unit	Gaining unit	OPCON, TACON, GS, R, DS
COMMAND	TACON	Gaining unit	Parent unit	Parent unit	Gaining unit	As required by gaining unit	As required by gaining unit and parent unit	Gaining unit	GS, GSR, R, DS
COMMAND	Assigned	Parent unit	Parent unit	Parent unit	Gaining unit	As required by parent unit	As required by parent unit	Parent unit	Not applicable
SUPPORT	DS	Parent unit	Parent unit	Parent unit	Supported unit	Supported unit	Parent unit, supported unit	Supported unit	*See note 2*
SUPPORT	Reinforcing (R)	Parent unit	Parent unit	Parent unit	Parent unit	Reinforced unit	Parent unit reinforced unit	Reinforced unit then parent unit	Not applicable
SUPPORT	General support reinforcing (GSR)	Parent unit	Parent unit	Parent unit	Parent unit	Reinforced unit and as required by parent unit	Reinforced unit and as required by parent unit	Parent unit then reinforced unit	Not applicable
SUPPORT	GS	Parent unit	Parent unit	Parent unit	Parent unit	As required by parent unit	As required by parent unit	Parent unit	Not applicable

NOTE 1: In Nato, the gaining unit may not task organize a multinational unit (see TACON).
NOTE 2: Commander of units in DS may further assign support relationships between their subordinate units and elements of the supported unit after coordination with the supported commander.

7-33. The following are C&S relationships:

- Assigned.
- Attached.
- OPCON.
- DS.
- GS.
- TACON.

7-34. ADCON is the direction or exercise of administration and support authority over subordinate or other organizations, including the organization of service forces, resources and equipment control, personnel management, unit logistics, individual and unit training, readiness, mobilization, demobilization, discipline, and other matters not included in operational missions of the subordinate or other organizations. ADCON is synonymous with administration and support responsibilities identified in Title 10 USC. Unless otherwise specified by the SECDEF, ADCON of Army forces remains within the Army chain of command, from the lowest levels to the ASCC and back to the Secretary of the Army. ADCON is always subject to the command authority of the combatant commander.

COMMAND RESPONSIBILITIES

7-35. The engineer commander leads his engineers with the authority placed in his position. The commander balances his effort as a commander and senior advisor and integrator of the M/CM/S BOS for the combatant or force commander. In this unique role as a dual-hatted commander and staff integrator and coordinator, the engineer commander is first and foremost responsible for integrating and synchronizing the M/CM/S BOS and geospatial and general engineering into the concept of operations. It is this approach to support the bigger picture that allows engineers and their units to achieve success. Planning is part of the broader field of C2 and is a fundamental responsibility of commanders. While the staff completes much of the detailed analysis and preparations of plans and orders, the commander plays a central role in planning through the commander's intent and planning guidance. These guide the activities of the staff and subordinate commanders.

STAFF RESPONSIBILITIES

7-36. The senior engineer commander or staff engineer is first and foremost responsible for integrating and synchronizing all combat (M/CM/S), geospatial, and general engineering functions. This senior staff coordinator assists the maneuver commander with the coordination and detailed analysis necessary to convert the commander's intent and planning guidance into a plan. When another engineer organization arrives with a commander of equal or junior grade to the ENCOORD or the organic senior staff engineer, the organic ENCOORD will remain the senior advisor to the maneuver commander.

Chapter 8

Full-Spectrum Operations

Every age has its own kind of war, its own limiting conditions and its own peculiar preconceptions.

MG Carl von Clausewitz, 1832

The Engineer Regiment has a very challenging task to support its METL for a broad range of military operations ranging from decisively winning wars to keeping the peace and responding to other requirements within HLS. The Regiment, as a whole, is organized and equipped to respond to the broad range of full-spectrum operations. However, it is the challenge of the commander to plan, prepare, and execute thes ODSS missions. It is the operations process of planning, preparing, and executing with continuous assessment that the commander uses as a means to the end.

PLANNING, PREPARING, AND EXECUTING

8-1. The operations process of planning, preparing, and executing has proven successful to all commanders. It is a means for the commander or ENCOORD to break down responsibilities to subordinates and the staff. The engineer will face many challenges as he thinks through the potential for engaging in the full spectrum of operations. He needs to consider the complex tasks associated with transitions between these types of operations as well. This chapter provides some planning, preparing, and executing considerations for the spectrum of operations the engineer will face. It is the commander's ability to continuously reassess through the use of the operations process that allows him to dictate the terms of the operation and support the force commander or LFA in an HLS operation.

8-2. Planning for operations leads to an engineer commander making decisions before and during execution. At the core of decision making is knowing if, when, and what to decide. It includes understanding the consequences of decisions. Decisions are the means by which the commander translates the vision of the end state of the operation. Decision making is both science and art. At the higher echelons, art tends to govern over science and vice versa at the lower levels. Many aspects of engineer operations are quantifiable, such as the number of mines available or the materials required to construct a life support area. This is the science of war. Other aspects such as leadership, understanding the complexity of the operation, and anticipating the best place to breach a threat's defense in spite of the uncertainty of the enemy's actions are part of the art of war.

8-3. Preparing for operations includes activities conducted by engineers before executing to improve their ability to conduct an operation. At a minimum, these activities include plan refinement, rehearsals, reconnaissance and surveillance, coordination, inspections, and movement. Reconnaissance and rehearsals are of critical importance to engineers. Engineers do not have just one solution for all the tasks they may encounter. Therefore, it is essential to conduct reconnaissance rehearsals to identify the means and methods to accomplish known engineer missions and build flexibility into the plan for possible contingencies.

8-4. Executing is putting a plan into action by applying engineer capabilities to accomplish the mission. Inherent to executing operations is deciding when to execute the plan and when to adjust the plan based on the changing situation. Commanders use judgment and initiative to assess the situation and make decisions based on the model of assess, decide, and direct.

8-5. Because of the complexity of full-spectrum operations, engineers are critical in the transition between operations. As operations transition from defense to offense to stability or support, engineers have to show their versatility by rapidly resourcing and moving assets and capabilities to maintain momentum. Engineer capabilities are massed for periods of time and then redistributed to support the new operation, while continuing to support existing operations.

OFFENSIVE OPERATIONS

8-6. Surprise, concentration, tempo, and audacity all characterize offensive operations. Assured mobility provides a framework for the commander to examine using the operations process. Effective offensive operations capitalize on accurate intelligence and relevant information regarding enemy forces, weather, and terrain. The commander maneuvers his force to advantageous positions before contact. Security operations and defensive information operations keep or inhibit the enemy from acquiring accurate information about friendly forces. Contact with enemy forces before a decisive operation is deliberate, designed to shape an optimal situation for a decisive operation. A decisive operation is a sudden, shattering action that capitalizes on subordinate initiative and a COP to expand throughout the AO. The commander executes violently without hesitation to break the enemy forces' will or destroy them. *FM 3-0* discusses the four characteristics of offensive operations.

8-7. Army forces conduct offensive operations simultaneously with defensive, stability, and support operations while performing joint, multinational, and interagency operations. Similarly, while engineers may be focused on one or more of the engineer functions, all five functions operate simultaneously in support of offensive operations. *Figure 8-1* shows that relationship.

8-8. For offensive operations, prediction is a critical engineer contribution. By gathering the products available from geospatial engineering sources, engineers contribute to the commander's ability to see the terrain and frame the COP.

8-9. Engineers then describe the threat in terms of his subelements' functions on the terrain with weather effects. The threat is identified in terms of—

- How and where the threat is able to conduct engineer related functions such as assaulting and fixing?

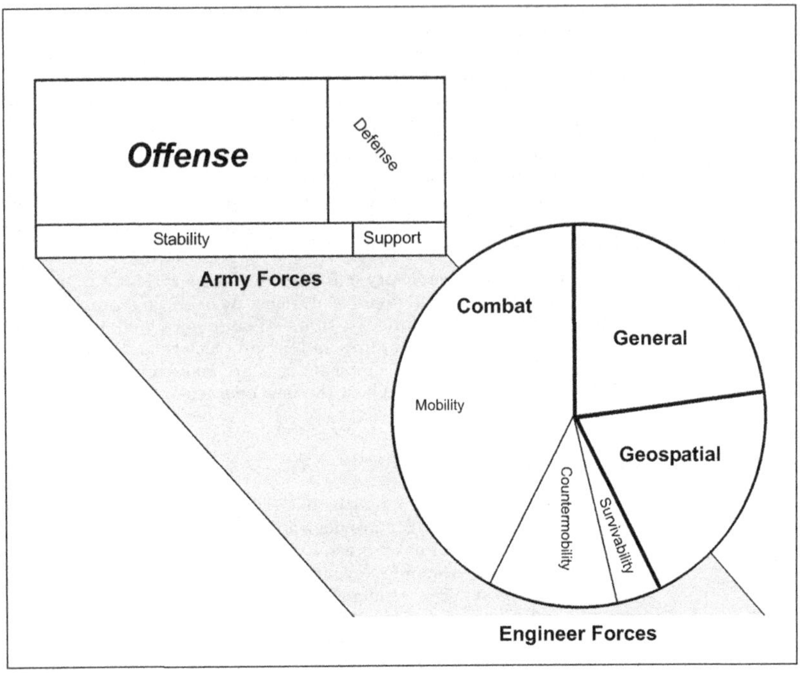

Figure 8-1. Generic Offensive Operations Relationships

- What assets are available to him?
- Is the threat in range to transport mines with assets available?
- How many mines can he carry?
- What terrain optimizes their use and impedes friendly maneuver?

8-10. Engineers should use sensors to observe these areas in the intelligence collection plan. It is during the planning phase that the commander identifies the links that predict must have with prevent and detect.

8-11. During the preparation phase, engineer assets are postured with their task-organized C2 HQ. The focus is on preventing impediments from impacting the friendly forces scheme of maneuver. This is accomplished by identifying viable routes around them or proactively planning for the removal of the impediments if an alternate route is not possible or desirable by the maneuver commander. There must be established links from the collection plan that are focused on preventing the threat from interdicting the friendly scheme of maneuver. Engineers also prevent enemy impact by participating in combined arms rehearsals to facilitate rapid execution to avoid or neutralize an obstacle. The threat actions that cannot be prevented must then be detected. Prediction efforts are linked to identifying and detecting impediments to the maneuver forces' freedom of movement. The engineer commander and his staff coordinate with the maneuver commander for ISR assets to detect potentially hazardous areas. Engineers also prevent enemy impact by conducting combined arms rehearsals to facilitate rapid-capability execution to avoid or neutralize an obstacle.

8-12. When executing offensive operations, the maneuver force uses its COP to link its detection to avoid obstacles along the route and provide assured mobility. The maneuver force can actively avoid an obstacle by interdicting threat countermobility before emplacement or passively avoid by identifying, marking, and bypassing an obstacle. If the friendly force commander is compelled to neutralize the obstacle to his front, the force will employ the breach tenets of intelligence, breach organization, breach fundamentals, mass, and synchronization. While supporting the maneuver commander, the engineer contributes to protecting the force by interdicting threat ability to influence friendly maneuver, constructing life support areas, or by digging in assets such as radars.

FORMS OF MANEUVER

8-13. Assured mobility supports all the forms of maneuver identified in *FM 3-90*. Unimpeded maneuverability is essential to the commander to achieve decisive operations. The five forms of maneuverability are specified by the higher commander. However, his guidance and intent, along with the mission that includes the implied tasks, may impose constraints such as time, security, and direction of attack that narrow the forms of offensive maneuver to one alternative. The fundamentals of assured mobility allow the commander to focus his assets and effort to maneuver on the battlefield within these constraints. Engineers synchronize their capabilities across the BOS to support assured mobility for each of the forms of offensive maneuver.

Envelopment

8-14. An envelopment is a form of maneuver in which an attacking force seeks to avoid the principal enemy defenses by seizing objectives to the enemy's rear to destroy him in his current position. It avoids the enemy's strengths and obstacles. At the tactical level, envelopments focus on seizing terrain, destroying specific enemy forces, and interdicting enemy withdrawal routes. A force seeking to execute an envelopment must have a substantial mobility advantage over the enemy.

Turning Movement

8-15. A turning movement is a form of maneuver in which the attacking force seeks to avoid the enemy's principle defensive positions by seizing objectives to the enemy's rear and causing him to move out of his current positions, or divert major forces to face the threat.

Penetration

8-16. A penetration is a form of maneuver in which an attacking force seeks to rupture enemy defenses on a narrow front to disrupt the defensive system.

Destroying the continuity of that defense allows the enemy's subsequent isolation and defeat, in detail, by exploiting friendly forces. The penetration extends from the enemy's security area through his main defensive positions and into his rear. A commander employs a penetration when there is no assailable flank, enemy forces are overextended and weak spots are detected in his defenses, or time pressures do not permit an envelopment.

Infiltration

8-17. An infiltration is a form of maneuver in which an attacking force conducts undetected movement through, or into, an area occupied by enemy forces to occupy an advantageous position in the enemy's rear while exposing only small elements to enemy defensive fires.

Frontal Attack

8-18. A frontal attack is a form of maneuver in which an attacking force seeks to destroy a weaker enemy force or fix a larger enemy force in place over a broad front.

ENGINEER PLANNING CONSIDERATIONS

8-19. To support planning, the ENCOORD develops an engineer estimate. The four key elements of the engineer estimate include developing an SOEO, refining an SOEO, identifying EMSTs, and the EBA. Planning for each form of offensive maneuver at the operational level requires the engineer commander to use the assured-mobility framework. Although the forms of offensive maneuver have different intentions, the planning phase must always begin with predicting the adversary's intent through a thorough understanding of the threat, his engineer capabilities, and how the terrain will effect operations. Geospatial products and information become the foundation and common reference for planning. Of all the forms of maneuver, knowledge of the threat's disposition is especially critical and required for an infiltration or penetration due to the requirements for stealth and surprise. A greater degree of planning is required for a penetration from the breach to the ultimate control of the decisive objective.

8-20. Planning includes how to best organize available engineer assets. The reverse planning procedure—identifying the combat engineering task required from the objective back to the start point—is essential for achieving the assured mobility fundamentals. As the ENCOORD resources each task, he identifies the shortages and develops a plan to ask for more resources, develops decision points to shift engineer assets, or receives a decision on where to accept risk. A frontal attack is generally a shaping operation that requires a broad front to fix a threat. Engineers may identify areas where no engineers are task-organized and elements of the maneuver force can provide necessary resources in critical areas. All of these plans require planning to protect the flanks of the maneuver from interdiction. Countermobility is the focus of flank protection while protection of the force focuses on the survivability effort and protection of priority assets.

8-21. Planning is the basis for developing a collection plan, which is the foundation for the fundamentals of prevention and detection. A thorough understanding of how the enemy fights allows the staff to develop a focused collection plan that resources ISR assets to prevent and detect potential impediments to force movement. During offensive operations, engineer units tend to have command relationships with maneuver commanders. OPCON is the most common command relationship for engineers during offensive operations because it allows engineers to be responsive and provides the maneuver commander the greatest flexibility.

8-22. Effective preparation for offensive operations requires engineer units to establish early linkups with the maneuver units and commanders they will support. As engineer units prepare for offensive operations, they should focus on inspections and combined arms rehearsals that address the avoidance of restrictive areas by creating combat trails or bypass routes. If the plan is to neutralize or breach through obstacles, the rehearsals should cover all the breaching tenets—intelligence, breach organization, breach fundamentals, mass, and synchronization. The rehearsal should also include the participants from each element of the breach organization. A current assessment of the ISR collection plan is required for final adjustments to the plan, and this may affect the fundamentals of assured mobility.

8-23. Executing the forms of maneuver translates into four types of offensive operations at the tactical level—MTC, attack, exploitation, and pursuit *(see FM 3-90)*. Engineers may measure the successful execution of offensive operations when assured mobility facilitates orienting and massing combat power on the enemy. Successful execution includes integrating all of the BOS and other branches as they contribute toward their pieces of assured mobility. Engineers continue to update their running or continuous staff estimate and ensure it still meets the needs of the current situation. Changes that affect the functions of prevention and detection must be distributed to all forces, rapidly. As new impediments are detected, engineers should be organized to remain flexible and maintain momentum for the force so they can support avoiding or neutralizing threats to maneuver forces.

DEFENSIVE OPERATIONS

8-24. Defensive operations defeat an enemy attack, buy time, economize forces, or develop conditions favorable for offensive operations. Defensive operations alone normally cannot achieve a decision. Their purpose is to create conditions for a counteroffensive that allow for a regaining of the initiative. Other reasons for conducting defensive operations include retaining decisive terrain or denying a vital area to the enemy, attriting or fixing the enemy as a prelude to offensive operations, countering surprise action by the enemy, and increasing the enemy's vulnerability by forcing him to concentrate his force. While the offense is the most decisive type of combat operation, the defense is the stronger type. The inherent strengths of the defense include the defenders' ability to occupy his positions before the attacker and use the available time to prepare his defenses. Preparations end only when the defender retrogrades or begins to fight. The defender can study the ground and select defensive positions that mass the effects of his fires on likely approaches. He combines existing and reinforcing fires to disrupt, fix, turn, or block.

8-25. As a part of full-spectrum operations, the purpose of defensive operations is to impose US will on the enemy for decisive victory. Engineers play a significant role in defensive operations as a combat multiplier with a focus on combat engineering (M/CM/S), although geospatial engineering and general engineering are also major contributors. As such, engineers provide support throughout the depth of the battlefield, through continuous support to decisive, shaping, and sustaining operations.

8-26. Army forces conduct defensive operations simultaneously with offensive, stability, and support operations while performing joint, multinational, and interagency operations. Similarly, while engineers may be focused on one or more of the engineer functions, all five functions operate simultaneously in support of defensive operations. *Figure 8-2* shows that relationship.

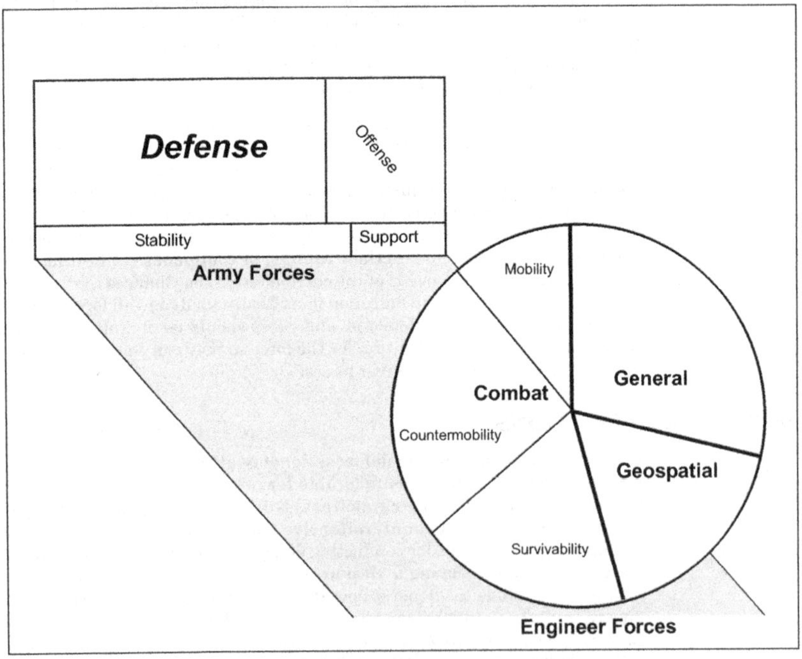

Figure 8-2. Generic Defensive Operations Relationships

8-27. Planning for defensive operations is inextricably linked to offensive operations and, for planning purposes, must consider the transition from offensive operations, and the follow-on offensive operations. During defensive operations, engineers use terrain products to best position the units within the defense. Engineers then work with intelligence staff to describe the threat functions to predict where the threat is likely to attack friendly forces. Engineers work in conjunction with intelligence personnel to determine which sensor capabilities to leverage and best predict and prevent the threat from maneuvering freely into the defended area. The consideration of counterattack planning or support for the mobile strike force is the same as in the offense. The ENCOORD works with other BOS's and uses the fundamentals of assured mobility to ensure that the counterattack force can mass its effects on the enemy for decisive operations. The type of defensive operation will define the amount of engineer effort required. An area defense will typically require a greater amount of effort. A mobile defense's effort will be to a lesser degree because it has greater flexibility and takes advantage of the terrain in depth. During defensive operations, engineer units tend to have support relationships to the maneuver commander.

8-28. During the preparation phase, engineer assets are postured with their task-organized C2 HQ and initiate engineer work effort. The focus is on resourcing the materials for creating obstacles and integrating them with friendly fire effects. Engineers are designated to support mobility operations and selected countermobility operations in support of the reserve or mobile strike force. The engineer commander and his staff coordinate for ISR assets to detect enemy engineer and breaching capabilities. Engineers provide protection from the threat systems through survivability operations to key support systems.

8-29. There are three types of defensive operations—area defense, mobile defense, and retrograde *(see FM 3-90)*. When executing defensive operations, engineers employ SCATMINEs and conventional mines and obstacles to develop EAs. They should also provide mobility to assist the reserve or mobile strike force with enemy destruction. Within the defense, engineers contribute to protecting the force by interdicting the threat's ability to maneuver, constructing life support areas, or digging in assets such as radars.

STABILITY OPERATIONS

8-30. Stability operations are part of full-spectrum operations intended to promote and protect US interests. This is done by influencing the threat, political, and information dimensions of the operational environment through a combination of peacetime developmental, cooperative activities and coercive actions. Stability operations are not necessarily sequential. They may occur independently, concurrently, or at the conclusion of combat operations. The focus is on sustaining the outcome achieved from combat operations to prevent the threat or the conditions for a threat to return and realize strategic results. Stability operations, unlike combat operations, may have an unambiguous threat. As discussed in *Chapter 2*, the continuum-of-relative interests model shows that groups may become more or less adversarial as the stability of the area changes. The engineer must be aware of the impact this will have on supporting stability operations.

8-31. Army forces may conduct stability operations simultaneously or in combination with offensive, defensive, and support operations while performing joint, multinational, and interagency operations. Similarly, while engineers may be focused on one or more of the engineer functions, all five functions operate simultaneously in support of stability operations. *Figure 8-3* shows that relationship.

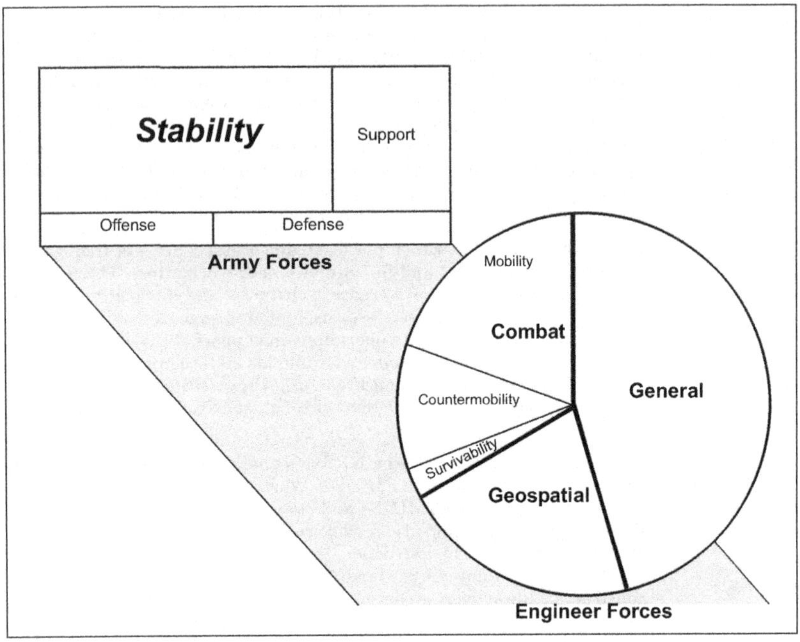

Figure 8-3. Generic Stability Operations Relationships

8-32. When planning for stability operations, engineers should consider a broad range of potential requirements. The EBA, although similar in format for combat operations, looks at different aspects of the terrain and friendly and threat capabilities. Terrain products continue to have a great deal of importance, but political and cultural considerations may be more important than strictly a combat terrain analysis. Terrain analysts work with the intelligence staff to develop a usable product for the commander to reflect this information if available. When analyzing the troops available, the ENCOORD should consider if there are HN, third-party NGOs, or other multinational forces involved. Interaction with these other parties requires engineers to address interoperability, common standards, and mutual agreements. Engineers should also plan for engineer units operating amongst civilians or in conjunction with NGOs and other international organizations.

8-33. The ENCOORD typically receives priorities from the joint staff on general engineering tasks. He then coordinates to secure the assets to perform these missions. As stability operations evolve, there may be periods of threat activity. If this occurs, the adversaries may use IEDs to target soldiers and affect the nation's will to continue the stability operation. Engineers are a major player, particularly in urban areas, as they plan general engineering tasks. Engineers are critical to predicting threat patterns in conjunction with the intelligence staff *(see FM 3-07)*. It is important for engineers to understand the goals and objectives of the higher HQ and address these issues with the maneuver commander to best assist him in preparing for expanding missions and considering the unique equipment and personnel capabilities.

8-34. Preparing for stability operations is more difficult than preparing for combat operations because of the broad range of potential missions engineers are expected to participate in during stability operations. An early on-the-ground assessment by engineers is absolutely critical to tailor the engineer force properly and to support the follow-on engineer contingency operations force logistically. Results of this assessment are quickly passed to deployment planners to ensure that an adequate engineer support force arrives in the AO in a timely manner. The failure to provide these engineers may cause inadequate troop bed down, sanitation, and force protection to the deployed force. This early on-the-ground engineer assessment identifies the—

- Status of the infrastructure in the AO, to include airfields, roads, ports, logistics bases, and troop bed-down facilities; real estate acquisition; environmental standards, conditions, and considerations; construction material supply; construction management; and line-haul requirements.
- Theater and situation-specific force protection requirements.
- Existing geospatial product availability and requirements for new terrain products.
- Specialized engineer requirements such as prime power, well drilling, and fire fighting support.
- Engineer C2 requirements, including HQ staffing, communications, and information systems support.
- Engineer liaison requirements, including linguists and civil affairs personnel.
- Requirements for officers with contracting officer's representative (COR) or USACE experience.
- Potential requirements for LOGCAP, contractor responsibilities, contract-construction procedures, and initial work areas.
- Potential requirement to establish a MICC to coordinate for HDOs.

8-35. There are ten types of stability operations and each is unique in its execution *(see FM 3-07)*. The ten stability operations include peace operations (PO), foreign internal defense (FID), security assistance (SA), humanitarian and civic assistance (HCA), support to insurgencies, support to counterdrug operations, combating terrorism (CBT), noncombatant evacuation operations (NEOs), arms control, and show of force. Engineers may or may not play a role in each type of stability operations. However, these areas are addressed for potential engineer considerations.

PEACE OPERATIONS

8-36. The broadest type of stability operation is POs, either as part of a UN, NATO, or multinational force or performed unilaterally. Peace operations inside peacekeeping operations (PKOs), peace enforcement operations (PEOs), and support to diplomatic efforts as its three basic forms.

Peacekeeping Operations

8-37. PKOs are military operations undertaken with the consent of all major parties in a dispute. They are designed to monitor and facilitate the implementation of an agreement (cease-fire, truce, or other such agreement) and support diplomatic efforts to reach a long-term political settlement *(see JP 1-02)*. Before PKOs begin, a credible truce or cease-fire must be in effect, and the parties to the dispute must consent to the operation. PKOs take place following diplomatic negotiation and agreement among the parties to a dispute, the sponsoring organization, and the potential force-contributing nations. Engineers participate as part of a combined arms force and may construct and maintain roads, airfields, LZs, ports, pipelines, and other associated missions such as land mine detection and removal. Removal of mines by engineers during PKOs is based on tactical necessity. Humanitarian demining organizations provide the preponderance of mine removal.

Peace Enforcement Operations

8-38. A PEO is the application of military force or the threat of its use, normally pursuant to international authorization, to compel compliance with resolutions or sanctions designed to maintain or restore peace and order *(see JP 1-02)*. PEOs are, by definition, coercive in nature and rely on the threat or use of force. However, the impartiality with which the peace force treats all parties and the nature of its objectives separates PEOs from war. The purpose of a PEO is not to destroy or defeat an enemy but to use force or threat of force to establish a safe and secure environment so that peace building can succeed. Engineers may participate in disarming to include seizing ammunition, collecting, and destroying weapons and supplies, closing weapons and ammunition factories, and preventing resupply.

8-39. HDOs are a part of the disarmament program *(see CJCSI 3207.1)*. Because of the threat posed by land mines to peace and safety, HDOs have become a significant activity in POs. In many demining missions, military ordnance specialists and civilian organizations participate in partnership. Demining is ultimately an HN responsibility. US Army participation in HDOs focuses on mine-awareness education and, most importantly, training of HN personnel in the surveying, marking, and clearing or lifting of mines. US Army personnel do not remove mines (they are prohibited by DOD); however, they may assist and train others in demining techniques and procedures. Specialists, such as EOD personnel *(see Appendix L)*, may be called on to destroy UXO, including munitions shells, fuses, grenades, and rockets. Special Forces, civil affairs, and PSYOP personnel are currently the primary means to execute the geographic combatant commanders' programs of HDOs. EOD and engineer personnel are also included in these programs and integrated into operational training missions.

Operations in Support Of Diplomatic Efforts

8-40. Military support of diplomatic efforts improves the chances for success in the peace process by lending credibility to diplomatic actions and demonstrating resolve to achieve viable political settlements. In addition to or as an integral part of PO, Army forces may conduct operations in support of diplomatic efforts to establish order before, during, and after a conflict. While these activities are primarily the responsibility of civilian agencies, the military can support these efforts within its capabilities. Army forces may support diplomatic initiatives such as preventive diplomacy, peacemaking, and peace building. Engineers may support executing PO with geospatial engineering support, lodgment, and theater infrastructure development, including the construct and repair of protective facilities, roads, airfields, parts, and troop life-support facilities.

FOREIGN INTERNAL DEFENSE

8-41. FID is a form of stability operation but is specifically a program to support friendly nations operating in or threatened with potential hostilities. FID promotes regional stability by supporting an HN program of internal defense and development (IDAD). These national programs free and protect a nation from lawlessness, subversion, and insurgency by emphasizing the building of viable institutions that respond to the needs of society. FID can include training; material, technical, and organizational assistance; advice; infrastructure development; and tactical operations. Military assistance is often necessary to provide a secure environment for these efforts to become effective *(see JP 3-07.1)*. Army engineer units and individuals can be tasked to provide this military assistance. Typically, engineers will be subject matter experts (SMEs) when tasked to support this program.

SECURITY ASSISTANCE

8-42. SA is the means through which the US provides defense articles, military training, and other defense-related services to eligible foreign governments or international organizations to further US national policies and objectives. These programs include foreign military sales, international military education and training, HCA, humanitarian assistance, humanitarian demining programs, international peacekeeping, the Warsaw Initiatives Program, and the Partnership for Peace Program. SA is a group of programs, not a mission assigned to Army units specifically. However, Army units and soldiers participate in SA programs through peacetime engagement activities and by training, advising, and assisting allied and friendly armed forces.

8-43. There are four primary methods of training:

- Mobile training teams are used when an HN element requires on-site training or needs surveys and assessments of training requirements. Engineers may deploy as part of a single-service, joint, or conventional force team. Each team is tailored to the training the HN requires.
- Extended training service specialists teams are employed on a permanent change of station to assist the HN in attaining readiness

on weapons or other equipment. These teams train the HN instructor cadre so the HN can assume responsibility for the training.

- Technical-assistance field teams are also deployed on a permanent change of station basis and train HN personnel in equipment-specific military skills.
- International military education and training and mobile education teams provide HN personnel with training opportunities in the continental US and in the host country. This training not only meets the immediate HN requirement of increased training, but also has a long-term impact of improving US and HN relations.

HUMANITARIAN AND CIVIC ASSISTANCE

8-44. HCA is a stability operation that provides assistance to the local populace by predominantly US forces with military operations and exercises. Such assistance must fulfill unit training requirements that incidentally create a humanitarian benefit to the local populace. The assistance that engineers may provide under HCA is limited to—

- Constructing rudimentary surface transportation systems.
- Well drilling and constructing basic sanitation facilities.
- Rudimentary construction and repair of public facilities.
- Detecting and clearing landmines, including activities relating to furnishing education, training, and technical assistance regarding detecting and clearing.

8-45. US forces, to include engineers, may be tasked to provide the C2 support necessary to plan and execute the ground portion of any humanitarian assistance operation. Engineers may also be tasked to provide the logistics support necessary to relieve human suffering or the forces to secure an area to allow the humanitarian relief efforts of other agencies to proceed. Engineer assistance may also include—

- Constructing and repairing rudimentary surface transportation systems, basic sanitation facilities, and rudimentary public facilities and utilities.
- Drilling water wells.
- Constructing feeding centers.
- Disposing of human and hazardous wastes.

SUPPORT TO INSURGENCIES

8-46. Engineers may provide limited support to insurgencies. On order, Army forces support insurgencies that oppose regimes that threaten US interests or regional stability. While any Army force can be tasked to support an insurgency, Army Special Operations Forces (ARSOF) usually receive these missions. Engineer support to insurgency forces is limited to providing geospatial products and construction of SOF operating bases located outside the AO. Engineer missions for counterinsurgency operations are similar to those for humanitarian and nation assistance water supply and sanitation improvements; road, airfield, and port construction; and multinational training.

SUPPORT TO COUNTERDRUG OPERATIONS

8-47. Two principles guide Army support to counterdrug operations. The first principle is using military capabilities to benefit the supported agency and to train our soldiers and units. The second is to ensure that the military members do not become directly involved in law enforcement activities. Engineers supporting domestic counterdrug operations perform missions focused on supporting local law enforcement agencies. Engineers are sensitive to the legal aspects of support to civilian authorities and abide by the Posse Comitatus Act. They are also aware of the capabilities of the threat, primarily heavily armed narcotics traffickers. Typical support tasks include—

- Constructing or rehabilitating law enforcement target ranges; helipads; and fuel-storage, billet, CP, and maintenance facilities.
- Producing geospatial products of likely counterdrug operations areas.
- Constructing or upgrading access roads for drug interdiction patrols.
- Clearing observation fields for counterdrug teams.

COMBATTING TERRORISM

8-48. After the events of September 11, 2001, the Army can expect to be involved with CBT. Terrorism is the calculated use of unlawful violence or threat of unlawful violence, intended to coerce or to intimidate governments or societies in the pursuit of goals that are generally political, religious, or ideological *(JP 1-02)*. CBT involves opposing terrorist actions across the threat spectrum. These actions include offensive and defensive components. The offensive form of action is counterterrorism (CT). The defensive form of action is antiterrorism. Army commanders at all echelons must protect their soldiers, equipment, and installations. Army personnel and units conduct antiterrorism anywhere in the world. Engineers may become targets for terrorists because of how and where they perform their missions, especially construction projects and other wide-area missions. Equipment parks and supply yards are large and difficult to defend. Soldiers operating equipment or hauling materials are vulnerable to ambush by fire, mines, and booby traps. In support of antiterrorism, engineer leaders develop force protection measures whenever they conduct engineer missions.

NONCOMBATANT EVACUATION OPERATIONS

8-49. NEOs are conducted to support evacuating noncombatants and nonessential military personnel from a foreign nation to an appropriate safe haven. These operations evacuate US citizens whose lives are endangered by war, civil unrest, or natural disaster. Such operations also may include evacuating selected HN citizens or third-country nationals. NEOs usually involve a swift insertion of a force, temporary occupation of an objective, and planned withdrawal upon completion of the mission *(see JP 3-07.5)*. Engineers that support a NEO generally operate as part of a joint force and may conduct a wide variety of tasks, such as—

- Constructing temporary facilities and protective structures in country or in another country for either US forces or evacuees.
- Providing needed geospatial products and data for the operation.

- Conducting route reconnaissance and mobility operations for land evacuation.
- Repairing airfields and clearing helicopter LZs for use in air-evacuation operations.

ARMS CONTROL

8-50. Another stability operation that requires limited engineer support is arms control. The overarching goal of arms control is to prevent or deter war, promote stability, reduce the potential damage of a conflict, and reduce defense expenditure. Arms control focuses on promoting strategic military stability. It encompasses any plan, arrangement, or process controlling the numbers, types, and performance characteristics of weapons, C2, logistics-support, and intelligence-gathering systems. Engineers may support executing arms control operations by providing geospatial products used to verify treaty compliance and by constructing logistics support facilities involved with the arms control process.

SHOW OF FORCE

8-51. A show of force is an operation designed to demonstrate US resolve by deploying or using military forces. The US conducts shows of force for three principal reasons. To bolster and reassure allies, to deter potential aggressors, and to gain or increase influence. Army units are not usually assigned the mission to conduct a show of force but rather to conduct other operations, such as those listed below, for the purpose of showing force. A show of force is normally executed as—

- A permanent forward deployment of military forces.
- Combined training exercises.
- The introduction or build-up of military forces in a region or area.
- An increase in the readiness status and level of activity of designated forces.

8-52. Engineer support to demonstrations and shows of force normally a joint and multinational effort. Engineer tasks are very similar to those described in POs. Overt use of engineers during shows of force may aid in the political intent of the operation.

SUPPORT OPERATIONS

8-53. Support operations provide essential services, assets, or specialized resources to help civil authorities deal with situations beyond their capabilities. Support operations usually involve actions that help civil authorities or NGOs provide the required DS to the affected population. Army forces may provide relief or assistance directly, when necessary; however, they normally support the overall effort controlled by another agency. When visualizing a support operation, commanders recognize that they will have to use a different definition of the enemy. In support operations, the adversary is often disease, hunger, or the consequences of disaster. These types of support operations are domestic support operations (DSO) and foreign humanitarian assistance (FHA) operations. They share four forms: relief operations; support to chemical, biological, radiological, nuclear, and high-yield explosive (CBRNE) consequence management; support to civil law enforcement; and community assistance. The types of support operations occur in varying degrees in DSO and FHA operations. Since domestic emergencies can require Army forces to respond with multiple capabilities and services, these operations may be conducted simultaneously during a given operation. *Figure 8-4*, shows the relationship of the five engineer functions to support operations. General engineering tasks provide the preponderance of support.

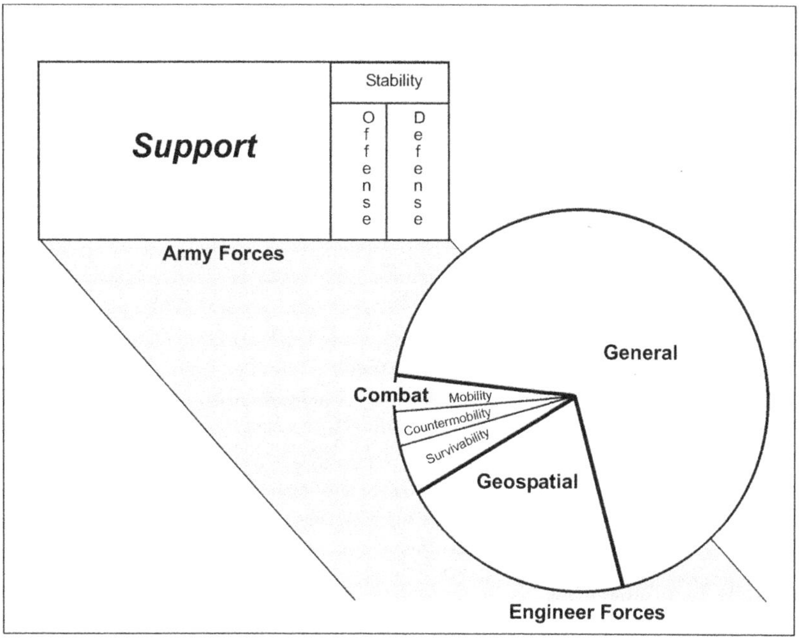

Figure 8-4. Generic Support Operations Relationships

8-54. Planning for support operations is significantly different from offense, defense, or stability operations—assessment of the threat is the reason. The threat may be a natural disaster with unpredictable behavior. Additionally, engineers must be aware of the number of statutes and regulations that restrict the Army's interaction with other government agencies and civilians for support operations. Different federal agencies have responsibility for the emergency support functions (ESFs) identified by Congress. *Table 8-1* contains a list of the ESFs and the federal and DOD proponents.

8-55. Army commanders will assume a support role to a designated agency. It is important for engineer commanders and staff to know who the proponent is

Table 8-1. ESFs and Proponents

ESF		
ESF	**Primary FED Agent**	**DOD POC**
Transportation	DOT	USTRANSCOM
Communications	NCS	OASD (C3I)
Public Works	DOD	USACE
Firefighting	USDA	USACOM
Info & Planning	FEMA	DOMS
Mass Care	ARC	DLA
Resource Support	GSA	DLA
Health/Medical Services	DHHS	OTSG (ARMY)
Urban Search & Rescue	FEMA	DOMS
Hazmat	EPA	DON
Food	USDA	DLA
Energy	DOE	USACE

and its responsibilities. Initial coordination and training with these agencies will facilitate the planning process.

8-56. There is usually little time for preparation of support operations. Relief operations and support to CBRNE consequence management require an immediate response. It is essential for commanders to have contingency plans developed and units identified so that they may prepare for such contingencies. Support to civilian law enforcement and community assistance allows greater leeway to plan and prepare.

8-57. During the execution of support operations, engineers may provide support to all four types. Relief operations respond to and mitigate the effects of natural or man made disasters (including CBRNE incidents). They maintain or restore essential services and activities to mitigate damage, loss, hardship, or suffering. Relief is primarily a state, local, or HN responsibility. To support the efforts of local authorities or the lead agency, the President and SECDEF can employ Army forces before, during, or after an event to save lives; protect property, public health, and safety; or to lessen or avert the threat of catastrophe.

8-58. Engineers can expect to be involved in planning for support of relief operations with geospatial products and analysis of potential locations to establish LSAs. Engineers may be called on to provide manpower support or general engineering support to units with unique capabilities such as water purification operations, temporary shelter construction, power generation operations, and firefighting operations.

8-59. Support to domestic and foreign CBRNE consequence management is another major support operation that requires engineer support. It has, by far, the most extensive support requirements for military personnel. Other US government agencies have primary responsibility for responding to domestic terrorist incidents. Local authorities are the first to respond to a CBRNE incident. However, Army forces have a key supporting role and can quickly respond when authorized. In a permissive overseas environment, the President and SECDEF may make Army assets available to assist a foreign government after a CBRNE incident. Such assistance may be linked to concurrent relief operations. CBRNE incidents are deliberate or unintentional events that produce catastrophic loss of life or property.

8-60. Due to the requirement to respond to consequence management and to control a situation before it can get worse, engineers can expect to respond with mobility assets. Engineer equipment is best suited to remove rubble and debris associated with consequence management. Other contributors are public works and engineering support that includes technical advice and assessments, engineering services, construction management and inspection, emergency contracting, emergency repair of wastewater and solid waste facilities, and real estate support. Other engineering considerations include—

- The need for heavy equipment for camp construction and power generation.
- The emergency clearance of debris to allow for reconnaissance missions and the passage of emergency personnel.
- The temporary construction or repair of emergency access routes.
- The emergency restoration of critical public services and facilities.
- Technical assistance and damage assessment.
- Emergency demolition operations.
- An effective maintenance program which is vital to ensure that required support is provided.
- The consideration of all classes of supply, particularly Class IV.

8-61. Support to civil law enforcement includes support to CT and counterdrug operations, civil disturbances, and GS. The Army assists civil law enforcement by providing personnel, equipment, training, and expert advice within the limits of applicable laws. Units in state status (Title 32 USC) provide the primary source of military assistance to state and local law enforcement agencies. They may assist civil authorities in instances where federal units are precluded due to the restrictions of the Posse Comitatus Act. Engineers may have a role in supporting law enforcement. There support may include geospatial products, map support, and manpower support. There may be some general engineering requirements, particularly in the enforcement of counterdrug operations.

8-62. Community assistance applies the skills, capabilities, and resources of the Army to the needs and interests of the US and local communities. Supporting and participating in events and activities that benefit the Army and the civilian community build on a long tradition of America's Army helping American communities. Community assistance can have a large impact because AC, ARNC, and USAR units are located in thousands of towns and cities across the nation. What a command does, or fails to do, for the community will affect the attitudes of the American people on whom the Army depends for its support and existence. Every commander should identify opportunities to conduct initiatives that meet specific needs; have specific start points and end states; enhance readiness; and advance the interests of the nation, the Army, and local communities. Engineers may be called on to provide everything from hometown recruiters to support of holiday events. Installations may have memorandum of agreements with neighboring towns to provide emergency snow removal or other capabilities not readily available.

TRANSITIONS BETWEEN OFFENSIVE, DEFENSIVE, STABILITY OPERATIONS, AND SUPPORT OPERATIONS

8-63. The transition between full-spectrum operations is one of the most difficult tasks for engineer planners to prepare for. As forces transition between operations, there will be gaps or voids between combat forces and other support forces, resources, and/or agencies. The reason is because operations may transition more rapidly than expected or other forces or agencies are not readily available. As a result, engineers and other forces may expect to perform tasks not on their METL. *Figure 8-5, page 8-20* represents the broad scope of tasks that occur in each operation and the wide array of tasks that must occur to transition between them.

8-64. Engineer planners and commanders will be faced with the difficult challenge of planning for success and identifying what tasks engineers may be asked to perform during the void and having available resources, forces, or agencies to perform their requirements. Some tasks that engineers may expect to participate in while awaiting legitimate civilian authorities or agencies include—

- Coordinating with the civil-military operations center (CMOC) to identify the tasks NGOs will perform.
- Coordinating with the J4 to identify JCMEB requirements and priorities.
- Closing LSAs and other areas that will be handed over to the HN, NGOs, or other agencies.
- Reestablishing utilities, sanitation, infrastructure, and governmental responsibilities.
- Establishing a MICC to coordinate with the Humanitarian Operations Center and related demining operations.

8-65. In addition to these considerations, engineers will be affected by the decisions and priorities of higher HQ. These decisions and priorities are established at convened boards. Examples of joint boards that will affect transition operations include the—

- JCMEB. The JCMEB establishes policies, procedures, priorities, and overall direction for civil military construction and engineering requirements in the theater. A primary concern of the board is to deconflict requirements between the military and civilian portions of a joint operation. The JCMEB also arbitrates issues referred to it.
- JFUB. The JFUB is used to manage facilities. The JFUB evaluates and reconciles component requests for real estate, use of existing facilities, interservice support, and construction to ensure compliance with priorities established by the JCMEB.

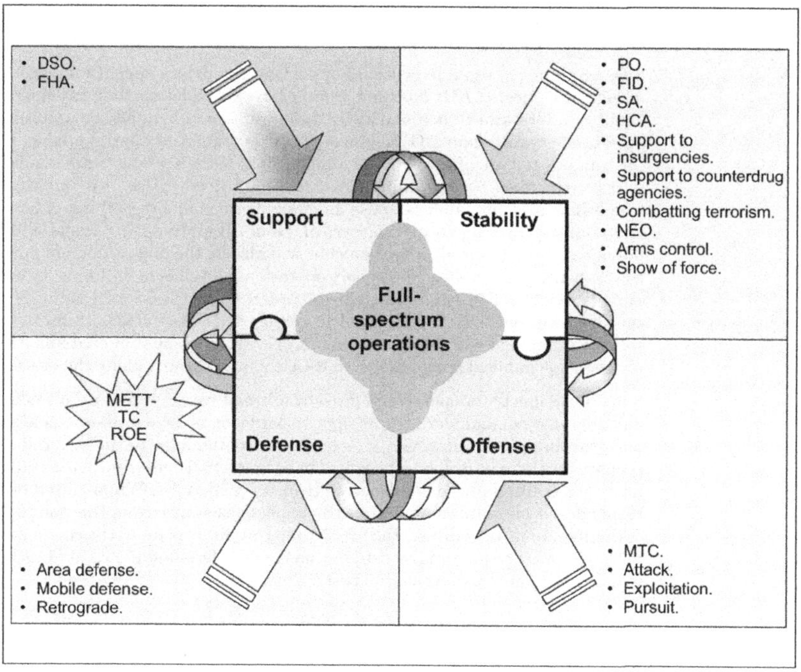

Figure 8-5. Full-Spectrum Operations

- JEMB. The JEMB board establishes policies, procedures, priorities, and the overall direction for environmental management requirements in the operational area.

8-66. The senior ENCOORDs must be involved with the J4 to coordinate priorities and synchronize the matching of capability requirements well in advance to ensure successful execution of the transition periods.

SPECIAL CONSIDERATIONS: URBAN OPERATIONS

8-67. Army commanders will likely determine that operations in the urban environment will be essential to mission accomplishment. They need to assess the relevance and impact of one or more urban areas as part of their mission. They will also need to determine whether full-spectrum UO may be the commander's sole focus or only one of several tasks nested in an even larger operation. Although UO can be conducted as a single battle, engagement, or operation, they will more often be conducted as a major ODSS operation requiring joint resources. *FM 3-06* provides a new framework—assess, shape, dominate, and transition—for UO. These are not phases or sequential operations but rather a means to visualize the fight.

8-68. Engineers provide critical support to any urban operation, whether foreign or domestic. *FMs 3-06* and *3-06.11* have more details, but commanders should understand that historically large numbers of engineers units have been task-organized for UO. Engineers provide unique geospatial products for the complex terrain of cities. Three-dimensional terrain products are available and continue to be developed. Assured mobility will be an important framework for commanders to use as maneuver commanders think of how to shape and dominate in urban terrain. General engineering tasks will be prevalent throughout all operations but will also be the major function during the transition to stability or support operations. Engineers will have to work closely with all branches of the M/CM/S BOS. They will have close coordination with EOD personnel to reduce IEDs and UXOs to minimize collateral damage with MP personnel to provide movement of civilians along routes, and chemical personnel to address chemical agents along the route.

8-69. Full-spectrum operations present a broad range of potential tasks to any engineer commander. It may appear daunting as he considers his METL and training plans. However, it is up to the commander to understand and prepare for the missions he can expect to encounter. There is no substitute for having a trained and disciplined unit in its core tasks. When called on to respond to a mission, commanders can expect assistance from the remainder of the Regiment to facilitate the unit's preparation. It is up to the commander to be aware of the potential considerations and understand the right questions to ask and explore to develop the best plans.

Chapter 9

Engineer Combat Service Support and Logistics Considerations

The troop engineer must have labor and transport with which to do his work, and be provided with a horse to inspect and speed its progress. He must serve many and accomplish much with little, yet, since he cannot do everything for all, he is beloved by none.

King Charles I (Articles of War 1625)

I don't know what the hell this 'logistics' is that Marshall is always talking about, but I want some of it!

Fleet Admiral E.J. King (to a staff officer in 1942)

Field Marshal Rommel once remarked that battles are won or lost on the back of logistics before the first rounds are fired. Admiral King's statement and Field Marshal Rommel's belief place a significance on logistics that underscores the relevance of logistic matters in doctrinal discussions. *JP 1* emphasizes that logistics sets the campaign's operational limits. The lead time needed to arrange logistics support and resolve concerns require continuous integration of logistic considerations into the operational planning process. This chapter will discuss special engineer logistic considerations and the internal implementation of CSS for engineer organizations.

FORCE SUSTAINMENT

9-1. Logistics is the science of planning and executing the movement and maintenance of forces. A force projection Army depends on making the right logistics decisions before the onset of operations. There is normally little time for last-minute logistics fixes when the decision to employ forces is made. Success is largely enabled through mental agility. Although the J4/G4/S4 is the lead for many aspects of support, engineers provide significant support to the logistician's overall plan. The logistics officer and staff planner should consider the—

- Force capabilities.
- Resources available in the AO.
- Construction standards.
- Environmental concerns.
- Contracting procedures.
- Force protection concerns.

9-2. Engineer force sustainment is critical to maintaining and multiplying combat power. Logistics operations must accurately anticipate engineer CSS needs. The ENCOORD staff assists the logistician in anticipating those unique engineer-related logistics. A mature, preplanned operational environment and extensive HN support cannot be assumed. The HN may not want or be able to deliver the necessary support in terms of quality, quantity, or timelines. Many engineer needs are unique requirements that demand the logistician to improvise, oftentimes causing the logistics system to become strained. Special engineer equipment is of low density, requiring intensive management to ensure that it is available for mission use. Engineer mission materials are normally bulky, heavy, and hard to transport. They must be requisitioned, transported, stockpiled, and issued in a streamlined manner. Engineer materials do strain the logistical system, but engineers balance the system with their contributions to theater logistics operations by constructing and upgrading logistics bases, troop bed-down facilities, airfields, ports, and MSRs. This chapter focuses on sustaining engineer units and engineer support to logistics operations. For more information, *see FMs 3-100-7, 3-0* and *4-0 (100-10).*

THE UNDERPINNINGS OF LOGISTICS

9-3. The objective of logistics is to ensure that operations succeed and facilitate the commander's ability to generate and mass combat power at a decisive time and place. Strategic logistics support wars, campaigns, and major operations. Operational logistics support encompasses those activities required to sustain campaigns and major operations and enable success at the tactical level of operations. Engineers closely support operational logistics operations in areas such as—

- Constructing base camps and bed-down facilities for force reception, onward movement, and sustainment.
- Opening and maintaining ports and airfields to support deployment, sustainment, and redeployment.
- Assisting in distribution management and movement of material and personnel by constructing MSRs and other logistics support facilities.
- Providing real estate management.
- Providing facilities operations and maintenance.
- Assisting in analyzing and integrating environmental considerations.
- Generating and distributing power.
- Providing enablers to facilitate JLOTS operations.

9-4. Operational and tactical logistics encompass all the CSS and engineer activities required to sustain the commander's ability to fight operations, battles, and engagements. Successful tactical logistics provide the right support at the right time and place to units. Engineers receive tactical logistics support from CSS agencies located at the maneuver brigade through the TSC level in the areas of manning, arming, fueling, fixing, moving, and sustaining operations. Each operational and tactical task is related or supportive of a specific BOS. It is important that engineer planners synchronize the implications of support for all types of forces and functions. Engineers support operational and tactical logistics operations in the following areas:

- Constructing FARPs.
- Digging ammunition supply points (ASPs) and corps logistics C2 nodes.
- Erecting fixed bridging along forward supply routes.
- Constructing the infrastructure for an ISB.

9-5. Regardless of the level of military operations, the engineer CSS structure and resource requirements depend on the METT-TC. The engineer logistics support structure fully supports the commander's intent and is integrated into the commander's concept of operation. Trade-offs between combat and general engineering capabilities directly affects this CSS capability.

COMBAT SERVICE SUPPORT CHARACTERISTICS

9-6. Successful CSS is balanced between being effective and efficient. CSS operations are characterized by being able to anticipate requirements, integrate joint and multinational CSS, improvise solutions, and be responsive and continuous. The characteristics of CSS (responsiveness, simplicity, flexibility, attainability, sustainability, survivability, economy and integration) are defined in *FM 4-0.* These characteristics facilitate effective, efficient CSS and enable operational success. They apply in war and contingency operations. These imperatives act as a guide for planners and operators to synchronize logistics on the battlefield. The engineer unit commander and staff understand and use these while planning engineer operations. The following paragraphs describe these eight characteristics and the engineer considerations for each.

9-7. Force commanders visualize and describe the concept of logistics support in concert with their CSS commanders. Commanders view CSS characteristics through an operational perspective to assist them in describing the considerations required to plan and execute successful operations. CSS characteristics are integrated throughout the operational framework, guide prudent planning, and assist the staff in developing the support plan. *FM 4-0* discusses the relationship between joint logistics and joint personnel principles and the US Army CSS characteristics). CSS characteristics seldom exert equal influence, and their importance varies by the situation. The commander identifies CSS characteristics having priority that, in turn, become the foundation for preparing the concept of logistics support.

RESPONSIVENESS

9-8. Responsiveness is the key characteristic of CSS. It means providing the right support in the right place at the right time. Responsiveness includes the ability to anticipate operational requirements and is the keystone of all logistics principles. Engineers ensure they identify all logistical requirements in advance, taking into consideration the support relationships of subordinate units. This information must be passed into CSS channels and tracked through delivery. It involves identifying, accumulating, and maintaining the minimum assets, capabilities, and information necessary to meet the support requirements. On the other hand, the force that accumulates enough material and personnel reserves to address every possible contingency usually cedes the initiative to the enemy.

9-9. Versatile CSS systems—

- Enhance the engineer unit's responsiveness.
- Adapt engineer change requirements without interrupting the flow of support.

9-10. In this respect, responsiveness is closely tied with improvisation. Theater logistics planners structure the logistics force to be versatile enough to complement engineer plans and operations yet be robust enough to ensure that engineer services are not interrupted. The structure is responsive enough to allow the engineer commander to seize and maintain the initiative.

9-11. Engineers plan to meet the changing requirements of the operation on short notice. The engineer sustainment system should be versatile enough to keep pace with rapid decision cycles and mission execution and react rapidly to crises or opportunities. Engineer planners are sensitive to engineer task-organization changes. Engineer units can normally respond to a change in task organization much quicker than theater CSS packages. Because of this, contingency engineer sustainment plans are normally developed.

9-12. The planner who anticipates is proactive—not reactive—while conducting before, during, and after operations. The ability of the force to seize and maintain the initiative, synchronize activities along the entire depth of the battlefield, and exploit success depends on the commanders', logisticians', and engineers' abilities to anticipate requirements. Engineers consider joint, multinational, contract civilian, and interagency assets when planning support for engineer operations. They—

- Use all available resources to the fullest, especially HN assets.
- Prioritize critical engineer activities based on the concept of operations.
- Anticipate engineer requirements based on wargaming by incorporating experience and historical knowledge.
- Do not think linearly or sequentially, but organize and resource for simultaneous and noncontiguous operations.
- Participate in and evaluate the engineer significance of each phase of the operation during the entire command estimate process, to include the mission analysis and COA development, analysis, and wargaming recommendation and execution.

SIMPLICITY

9-13. Simplicity is avoiding complexity and often fosters efficiency in planning and executing logistics operations. Mission-type orders and standardized procedures contribute to simplicity. Engineer commanders establish priorities and allocate classes of supply and services to simplify logistics support operations. Engineers use preconfigured loads of specialized classes of supply to simplify transport.

ECONOMY

9-14. Economy is the provision of support at the least cost. At some level and to some degree, resources are always limited. When prioritizing and allocating resources, the engineer commander and staff may not be able to provide a robust support package. Priority of effort will be established while balancing mitigation of risk to the operation. Engineer commanders may have to improvise to meet the higher intent and mitigate the risks. Commanders consider economy in prioritizing and allocating resources. Economy reflects the reality of resource shortfalls, while recognizing the inevitable friction and uncertainty of military operations.

FLEXIBILITY

9-15. Flexibility is the ability to adapt logistical availability based on changing situations, missions, and concepts of operation. Due to the inherently changing world environment, engineer missions will change. Engineers require flexibility and coordination with higher staffs to provide logistical support. As ENCOORDs work future branch or sequel plans, they ensure that logistical support is planned and resourced.

9-16. CSS plans and operations must be flexible enough to achieve responsiveness and economy. Flexibility may include improvisation, which is the ability to make, invent, or arrange what is needed from what is on hand. Improvised methods and support sources can maintain CSS continuity when the preferred method is undefined or not usable to complete the mission.

9-17. Extraordinary methods may be necessary to ensure success on the battlefield. Logistics planners attempt to push support to forward engineer units to ensure smooth combat operations. Sometimes this is not feasible or supportable. In such cases, engineers improvise by making, inventing, devising, or fabricating what is needed from what is on hand. One example is creating a demolition cratering charge using common fertilizer and diesel fuel.

9-18. Commanders must be aware of the environmental impacts of their actions. They must weigh the implications of holding out for logistical support against the environmental collateral damage that they will cause. They must ensure that a proper environmental risk assessment is done before beginning any action.

9-19. Specific damage assessment and repair procedures have been developed based on the need to improvise during the operation. Improvisation is not a substitute for good planning; requirements must be anticipated. However, improvisation can be a great strength; engineer personnel must recognize it as an advantage during emergencies.

ATTAINABILITY

9-20. Attainability is generating the minimum essential supplies and services necessary to begin operations. Commanders determine that minimum levels of support are acceptable to initiate operations. The engineer, in conjunction with the logistician, completes the logistic estimate and initiates resource identification based on the supported commander's requirements and priorities. An operation should not begin until minimum essential levels of support are on hand.

9-21. For engineers, attainability is at the very core of decisions that are made. Trade-offs may be necessary to attain a given goal or level and quality of product. While attainable, the cost may make other things unattainable. Since engineer materials must meet specific technical requirements, engineers work closely with the logistics staff to help them understand these requirements and obtain acceptable and suitable alternatives when trade-off decisions are required.

SUSTAINABILITY

9-22. Sustainability is a measure of the ability to maintain logistic support to engineers throughout the AO for the duration of the operation. Sustainability focuses on the engineer commander's attention for long-term objectives and capabilities of the engineer forces. Long-term support is a challenge for the engineer staff, which must not only attain the minimum essential materiel levels to initiate operations but sustain those operations through the end state. The ENCOORD must ensure that logistical requirements are known and are flowing based on the available transportation assets.

9-23. The engineer commander needs continuous logistics capability to gain and maintain the initiative. Pauses for rebuilding power impede momentum and rob the command of the initiative. Engineer planners synchronize all CSS assets to ensure that the support operation is not an impediment to the engineer commander. Continuity of operations is critical to success.

9-24. Engineers are either committed to the current operation or preparing for the next. The tempo of the battlefield requires a constant vigilance by the logistician and engineer commander to ensure a constant flow of support. Supplies are pushed (through unit distribution) forward whenever logistically feasible. This is especially crucial to engineer units because they do not usually have lulls in their operations that would allow them to resupply using the supply point method.

9-25. General engineering and contract support also support sustaining operations. General engineering involves constructing, repairing, operating, and maintaining infrastructure and facilities to enhance the provision of sustainment and services *(see JP 4-04 and FM 5-104).* Contracting support provides supplies, services, and construction labor and materiel. Contracting often provides a responsive option or enhancement to support the force *(see FM 100-10-2).*

SURVIVABILITY

9-26. The CSS characteristic of survivability is related but not exactly the same as the BOS of the same name. It is based on being able to protect support functions from destruction or degradation. Engineers contribute to ensuring logistics are survivable by constructing logistic bases and clearing LOCs.

INTEGRATION

9-27. Integration consists of synchronizing CSS operations with all aspects of Army, joint, interagency, and multinational operations. The concept of operations achieves integration through a thorough understanding of the commander's intent and synchronization of the CSS plan. Integration includes coordination with and mutual support among Army, joint, multinational, and interagency CSS organizations.

9-28. Operational and tactical plans integrate all logistics support so that it creates a synergy with the concept of operation. Engineer planners participate in and evaluate the logistics significance of each phase of the operation during the entire command estimate process. They create a clear and concise concept of support that integrates the commander's intent and concept of operation. This includes analyzing the mission; developing, analyzing, wargaming, and recommending a COA; and executing the plan.

9-29. Engineers will support joint and multinational operations. The theater commander integrates operations in their AOR, which often includes engineers from other services or countries and possibly civilian engineering contractors. FFE is one example of the integration of military and civilian engineers *(see Appendix H)*.

LOGISTICS CONSIDERATIONS FOR ENGINEERS

PREDEPLOYMENT AND DEPLOYMENT PHASES

9-30. As soon as the commander begins the operational planning and development of COAs, the engineer planners should assess the potential support infrastructure and develop a CESP. The engineer should then apprise the logistics and operations planners of the infrastructure capabilities and options for engineer support. Since all military operations depend on a theater infrastructure system, base development should be placed high on the force projection theater commander's priority list. A foreign country's infrastructure cannot be developed overnight to support force projection operations. Normally, a country's infrastructure (to include roads, bridges, inland waterways, seaports, and airports) is built only to sustain indigenous population and industry, with minimal additional capacity available to support US and coalition forces.

9-31. The theater needs to have the capacity to move large numbers of units through its air and seaports. The theater also needs facilities to manage such functions as implementing C2, storing and transferring ammunition, maintaining equipment, storing and moving bulk petroleum, generating and distributing power, and staging and billeting rear area troops, which are necessary for a synchronized flow of support to occur.

9-32. Moving follow-on forces and supplies is critical to success. Engineer planners work closely with the logistics staff to develop a suitable transportation infrastructure (roads, bridges, and airfields). Anticipating engineer requirements is crucial to ensure that adequate time is available to complete a robust infrastructure. These facilities can also be improved with the foresight of using engineer assets before the operation and during contingency operations. HN or civilian contractor personnel can perform much of this work. Alternate plans should be developed for if HN support is not available, such as underdeveloped or uncooperative theaters.

9-33. Base development does not end once the operation begins. Base development increases, depending on the size of the force involved in the operation. Each time the force expands or contracts, planners review facilities and LOC requirements to ensure that they are adequate to accomplish the mission.

ENTRY AND OPERATIONS PHASES

9-34. The nature of engineer entry and operations phases places an extraordinary burden on the logistics structure. Rates of fuel consumption, repair parts, construction and obstacle materials, mines, and explosives dictate the commitment of a large amount of maintenance and transportation assets in support of engineers. Engineer construction of EPW and/or displaced civilian camps remains logistically intensive during this period. Engineer operations, by their nature, are dangerous. Engineer planners, therefore, anticipate and provide for replacing engineer losses.

POSTCONFLICT OPERATIONS

9-35. Upon the cessation of hostilities, engineer forces may be tasked to restore areas destroyed or damaged during combat operations or as a result of other factors. Depending on the political and social factors of an operation, devastation may require forces to do some restoration. This requires attention to detail in logistics anticipation planning and, most likely, in rotating follow-on engineer units working in concert with HN and civilian construction contractors. Access to ENCOM and USACE capabilities and expertise is vital as assessments are submitted to the JFCs staff for prioritization. Engineers can expect to support the requirement to enable the return to normalcy for the civilian population, including movement, commerce, and construction. Engineers must plan for success and be prepared to quickly transition to stability- or support-type missions.

ENGINEER AND THEATER COMBAT SERVICE SUPPORT RELATIONSHIPS

9-36. The relationship between the ENCOM and subordinate engineer elements and the senior CSS elements in the theater depends on both the echelon at which the elements are operating and the operational environment in which they are operating. The basic relationships are shown in *Figure 9-1.* The ENCOM and TSC (if it exists) command relationship with and between specialized commands is determined by the ASCC or ARFOR commander. The ENCOM routinely provides a liaison officer to the TSC support operations cell, which lacks organic engineer planning capability.

9-37. Specialized engineer detachments and teams (such as CRESTs or FEST–elements, MTOE utility detachments, or facilities engineering [FE] teams) are task-organized within engineer brigades and groups to provide area support services. These specialized detachments and teams may be temporarily under the ADCOM of supported ASGs, but typically provide support through DS or GS relationships.

LOGISTICS REQUIREMENTS

9-38. Engineer force sustainment maintains and multiplies combat power. Engineer commanders must understand their internal organizational CSS system and know where to obtain logistics support. Many engineer needs are

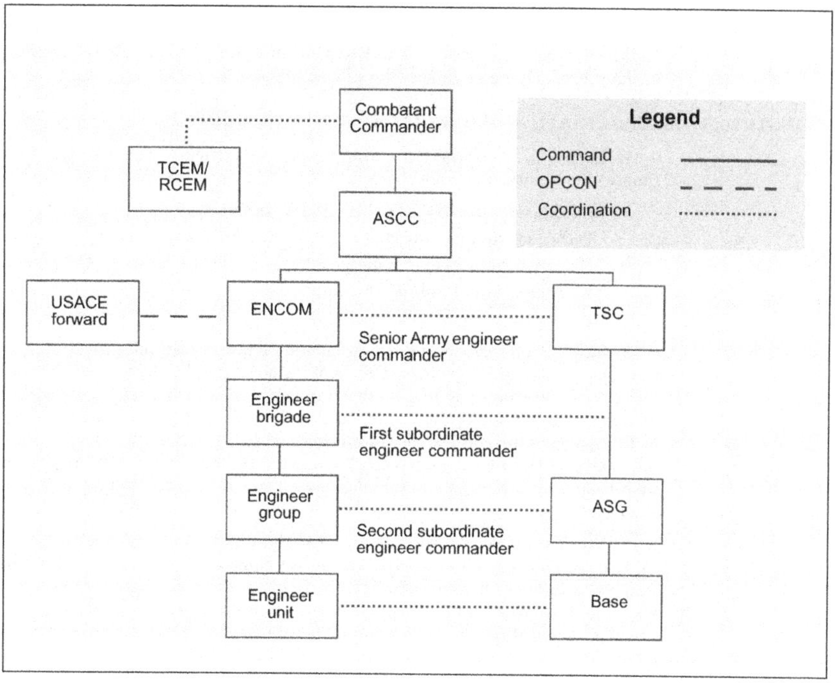

Figure 9-1. Engineer and Combat Service Support Relationships in Theater

unique requirements that stress the logistics system, such as massive requirements for Class IV barrier and construction materials. The key classes of supply are addressed in the following paragraphs:

9-39. Class III bulk and POL products are critical to the sustained operation of engineer equipment. Inadequate POL availability results in lost time that may affect the force commander's plan or result in a loss of momentum. While engineers are in a support relationship, the force commander should consider task-organizing fuel assets to the engineer force. Engineer equipment will operate for extended hours and may be out of synchronization with normal logistic resupply periods. Therefore, dedicated assets should be planned for and allocated. Engineer commanders will reflect on environmental considerations and take appropriate actions to manage risk when dealing with large amounts of Class III supplies.

9-40. Adequate Class IV supplies are central to the ability of engineer units to construct and maintain facilities in support of the ASCC's operations and the supporting sustainment base. For this reason, the senior Army engineer HQ (typically an ENCOM) plays a key role in establishing distribution protocols, stockage levels, and construction material allocation in theater. The joint force commanders, through TCEM/RCEM cells, often control critical Class IV supplies across the theater. The ENCOM may send a liaison to support the packaging of critical engineer Class IV supplies.

CONSTRUCTION MATERIALS AND SPECIAL CLASSES OF SUPPLY

9-41. In the TO, construction materials may be difficult to obtain. Required construction materials might be—

- Supplied by the CONUS through the supply system.
- Procured from neighboring countries within the region of operations.
- Obtained from local suppliers.
- Extracted from local natural sources.
- Produced by engineer units.

9-42. Ideally, construction materials should be procured from sources close to the construction site; however, this is often not possible. Because of the time and cost involved in moving construction materials, facility designs are often adapted to make maximum use of locally available materials. Quality control at the manufacture or delivery site is critical to ensuring that the materials meet any critical construction tolerances and standards. The quality of materials from HN suppliers may not meet these tolerances and standards or the necessary associated safety and environmental considerations.

9-43. Vertical materials, such as plywood and dimensional lumber, are Class IV items. These materials are universal commodities that are used by logisticians for overhead protection, flooring in high traffic areas, and shelving. However, equally important to the engineer are the materials that are instrumental in the success of horizontal (earthwork) construction.

9-44. Aggregate (crushed rock or river run and sand) are locally purchased or provided by engineer units. Other items, such as geotextiles or sand grid used to bolster poor soil conditions and improve trafficability, must be procured and shipped from CONUS. Soil stabilization and dust abatement require admixtures to the soil, such as lime, portland cement, asphalt emulsions, or resins. These products are bulky and may or may not be available within the region of current operations. Poor drainage can hinder operations. Therefore, ample supplies of culvert materials are a frequent concern of engineers.

9-45. Because of the multiple applications of many Class IV items, it is important that engineers remain involved in distribution decisions. This is particularly true regarding critical materials to ensure that they are not squandered on less critical missions. Engineers are not equipped to run Class IV yards, nor do they have the transportation assets to distribute Class IV supplies. However, the absence of engineer insight and critical assets management could lead to devastating theater shortages and mission impairment. Priority for transportation support is essential to supply operations, as Class IV items and engineer equipment are large and bulky and may require convoy credits or special haul assets.

9-46. Class V materials include ammunition, mines, and explosives. These materials are critical to the combat and CS operations that engineers conduct during combat operations. Additionally, these types of materials are usually not available for local procurement. Therefore, it is paramount for the ENCOORD to have an understanding of unit basic loads, anticipated missions, and availability of Class V resources. For instance, Class V Volcano mines may have a controlled supply rate (CSR). If this occurs, the ENCOORD should identify other types of mines available to provide a similar capability. Particular care must also be dedicated to planning for the transport of these HAZMATs with the transportation officer.

9-47. Class IX repair parts (less medical-peculiar repair parts)—all repair parts and components to include kits, assemblies, and subassemblies (repairable and nonrepairable)—required for maintenance support of all equipment. Engineers have a unique position in that their equipment tends to be low-density or civilian-leased equipment. Due to the low-density requirements for parts, the military logistic system may not be as responsive as local procurement of the commercially common parts. Engineers should be aware of all local vendors in the AO during the EBA.

9-48. Class X are materials to support nonmilitary programs, such as agricultural economic development, not included in Classes I through IX. These materials may be required by engineers during stability or support operations to assist local nationals and assist in economic engagement programs.

CONTRACTORS ON THE BATTLEFIELD

9-49. Contractors on the battlefield are addressed in *AR 715-9, FM 100-10-2,* and *FM 3-100.21.* Experience shows that there are never sufficient engineers to accomplish all missions for extended periods of time within the AO. The shortage of engineers may be partially overcome through the use of support contracts.

9-50. Engineers may be involved with a variety of contractors due to the large amount of general engineering requirements. Theater support contractors support deployed operational forces under prearranged contracts or contracts awarded from the mission area by contracting officers serving under the direct contracting authority of the theater principal assistant for contracting (PARC). They provide goods, services, and minor construction, usually through local vendors, to meet the immediate needs of operational commanders. Examples are organizations such as Brown and Root that work with engineer forces in many theaters.

9-51. External support contractors provide support to deployed operational forces that are separate and distinct from either theater support or support provided by system contractors. To support the mission, they may perform according to prearranged contracts or contracts awarded during the contingency itself. Engineers may receive logistical support from external support contractors while deployed on engineer specific deployments. Contracting officers who award and administer external support derive their contracting authority from PARC's outside the theater.

9-52. System contractors support deployed operational forces under prearranged contracts awarded by program executive officers, program managers, and the Army Materiel Command (AMC) to provide specific support to materiel systems throughout their life cycle, during both peacetime and contingency operations. These systems include, but are not limited to, vehicles, weapon systems, aircraft, C2 infrastructure, and communications equipment. Engineers have an increased likelihood of working with system contractors to assist with unique engineer equipment. When contractors support engineer operations, they must be managed, deployed, protected, and sustained.

9-53. The LOGCAP is a DA capstone program that employs contractor support to augment the Army's organic planning and CS and CSS capabilities. This program applies to CONUS and OCONUS. Before implementing LOGCAP, the combatant commander considers the use of AC and RC, other services, contract resources, and HN support. During a contingency operation, the combatant commander normally establishes an acquisition review board to determine the best means of fulfilling requirements. Board considerations include criticality, timeliness, quality, administration, effort, and cost. LOGCAP is used when contractor support is determined to be the most effective, expeditious, or cost effective method.

9-54. LOGCAP applies primarily in areas where multilateral or bilateral agreements or treaties exist. However, it can be employed in areas with formal HN agreements, where contractors are involved or peacetime support contracts exist. LOGCAP can also be used during mobilization to assist the CONUS support base and help units prepare for war or other contingencies. LOGCAP resolves shortfalls; it does not replace force structure. It includes all preplanned logistics-, engineering- and construction-oriented contingency contracts already awarded and peacetime contracts with contingency clauses. Preplanned weapon system sustainment contracts, ASCC contingency contracts, and the AMC support contracts are examples of contracts that fall under this program.

PLANNING CONSIDERATIONS

9-55. One of the primary responsibilities of the senior Army engineer is to forecast the types and the quantities of engineer materials required for the theater. *Chapter 3* described the CESP, which is generally used to establish the initial requirements during preconflict planning. *See CJCSM 3122.03A* for an example and further guidance for the CESP. Planning during the conflict requires good intelligence as to the damage caused by enemy action and good forecasting of the additional facilities that are needed. The Theater Construction Management System (TCMS) can be used as a guide in determining material requirements for necessary facilities. In many cases, existing facilities can be modified to meet military requirements, thus conserving Class IV materials and expediting construction completion.

9-56. For horizontal construction and nonstandard projects, material requirements must be estimated manually because TCMS provides standard Army designs. The senior Army engineer must also aid the logisticians in determining the adequacy and the most expeditious source of materials, either from local sources, regional sources, or a CONUS projection base. Materials from within the AOR may be from local manufacturers, commercial stockpiles, or HN government assets. Materials not locally available must either be procured out of theater (directly or through a service contract such as LOGCAP) or produced in theater by engineer units. Materials that fall in the latter category include aggregate, concrete, construction water, asphalt, and lumber. A local procurement system must be established to expedite procuring local materials. Such procurement may be restricted or centrally controlled in some theaters to avoid—

- Inflating the cost of construction materials in the HN.
- Creating a bidding war between all participants vying for materials or services within a specific area.
- Depleting specific resources.
- Overspending of operational funds.

DESIGN CONSIDERATIONS

9-57. When designing projects for the TO, designers consider the—

- Availability of construction materials.
- Local materials and property and their suitability.
- Construction practices to accommodate HN maintenance and repairs after US forces depart.
- HN environmental requirements.

9-58. A recent study by the Construction Engineering Research Laboratory (CERL) to update Class IV supply planning factors found that the requirement for local material, especially sand and gravel, could be substantial. Even in the most developed countries, problems exist in finding and hauling locally procured construction materials. Engineer planners develop realistic estimates of the available quantities of local materials as early as possible and assess the feasibility of the plans they have developed.

9-59. Many designs may not be practical because of logistics considerations. For example, although the TCMS's designs are adjusted for various climates (desert, tropic, and arctic), they may be difficult to construct because required construction materials are unavailable in the region. Suitable materials could be brought from the CONUS; however, the level or the length of the US commitment may not support this action. The engineers have developed theater-specific design books that consider regional requirements and standards. In European Command (EUCOM), the Red Book, and in CENTCOM, the Sand Book, are the theater standards.

9-60. Military designers must be knowledgeable of local construction standards and materials commonly used in the particular region. Designs must include the use of local materials or provide flexibility within the design for use of substitute materials. The construction standard for the TO is generally one of the following:

- Initial standard (up to six months of expected use).
- Temporary standard (up to twenty-four months of expected use).

9-61. Since the design life is short, only essential utilities, such as heating and cooling, are provided. This also reduces engineer material requirements. A heavy reliance on occupying existing facilities, either provided by the HN or leased, also minimizes construction material requirements. There are design requirements associated with existing facilities because modifications to structures are probable to meet the needs of the US force mission. As a contingency is drawn out and soldiers are deployed for greater lengths of time, the engineer should expect to upgrade facilities to enhance the QOL as the mission, time, and materials allow.

CONSTRUCTION AND OTHER SPECIAL CONSIDERATIONS

9-62. In the material estimation process, ten percent is usually added to the estimated quantity as an overage factor. The overage in the TCMS's bills of material (BOMs) provides for material loss, damage, and waste and minor field modifications. Although this overage may seem minor, the combined effect of material overage has a significant impact on the supply system. On the other hand, engineers should not succumb to pressures to eliminate this overage factor when ordering materials under the assumption that waste material can be eliminated through maximizing efficiency. Refuse material is unavoidable. However, all who are involved with actual construction can limit construction material waste by—

- Ordering and using optimum lengths.
- Providing clear designs (understandable to the construction unit).
- Inspecting the vendor's stocks to ensure satisfactory quality.
- Providing for proper storage, security, and handling of construction materials.
- Providing proper worker training to limit waste.
- Reusing materials.
- Constructing the facility right the first time.

9-63. Another construction consideration more fundamental than ensuring adequate quantities are ordered is securing transportation to haul materials to the construction site. For some materials, transportation is available within engineer resources. Aggregate, for example, can be hauled by engineer dump truck companies. Small quantities of dimensional lumber and plywood for small jobs can be hauled using organic assets within the combat heavy battalions. Larger quantities of Class IV supplies require coordination with the theater-level movement control center (MCC) and the Theater Army Area Command (TAACOM) to—

- Schedule convoys on MSRs.
- Allocate lift assets to move the material from the stockage points to the area of construction.
- Coordinate material drop-off requirements at engineer supply points (ESPs).

9-64. Beyond lumber and aggregate, other engineer items of bulk requiring coordinated lift assets include inland pipeline distribution systems (IPDSs), mines, and other special engineer Class IV barrier and Class V materials.

9-65. Unlike other supply classes, Class IV construction materials and Class II geospatial items are not provided based on documented consumption rates; there are no anticipated, preprogrammed surge rates for these items. With technological advances in material management, logistics planners can reduce the time required to get mission material to engineers. The management practices, however, cannot change the physical constants in the execution—

engineer mission materials are normally bulky, heavy, and require dedicated transportation. Prepositioning Class IV stocks reduces lift requirements during the initial stages of force projection.

9-66. Construction material is expensive and susceptible to weather damage; therefore, only limited quantities are on hand in the areas of major war plans. For this reason, the ASCC validates needs and initiates requisitions in advance of deployment or operations. Initial material forecasts are submitted by the senior Army engineer staff, using data from the TCMS and the JEPES and the base development plans in the theater-specific, mission-oriented CESP.

9-67. Successful execution of the theater's construction program depends on sufficient materials, logistics units to process the materials, and construction capability. Typically, during the early stages of a conflict, war damage repair and the construction of mission-essential facilities dominate engineer construction activities. The RSOI mission places heavy demands on engineers, as well as logisticians, to receive and launch the combatant forces into forward operating areas. As the theater matures, more substantial facilities are required and more construction forces are available. The ENCOM (or senior Army engineer) forecasts adequate construction materials that are flexible enough to meet the varied construction requirements throughout all operational phases. Working closely with the TSC's support operations cell, engineers ensure the control of Class IV materials. The ENCOM (or senior engineer commander) is a member of the RCEM and the TCEM teams. Given an understanding of both the combatant commander's and the ASCC's work priorities, the ENCOM (or the senior engineer commander) prioritizes the allocation of critical Class IV for the TSC Class IV manager, who implements the actual distribution of these materials.

9-68. Special logistics support considerations for topographic units are equally important. Equipment such as the multispectral imagery processor (MISP) and the Combat Terrain Information System (CTIS) require specialized critical low-density supplies. For example, during the Dayton Peace Accord negotiations, November 1995, the Defense Mapping Agency (DMA) (currently known as NIMA) distributed some 30,000 paper copies of maps covering the revised interentity boundary (IEB) on Bosnia-Herzegovina. As troops deployed to monitor the authoritative ceasefire line (the IEB and their respective buffer zones), thousands more copies of the 1:50,000 scale tactical line maps were needed. Engineer planners anticipated for demand surges on geospatial products before operations and recognized that topographic units were not equipped to run map warehouses, and did not have the transportation assets to distribute geospatial products.

9-69. Engineer bridging beyond the organic equipment available in the AO will either have to be purchased or transferred from another area of operations or from strategic and operational reserves stocks. It is the ENCOORD's responsibility to identify these materials and coordinate for their arrival.

ENGINEER ROLE IN PLANNING AND COORDINATION

9-70. The engineer's efforts to plan and coordinate engineer logistics efforts are essential to full integration of engineer units into the theater sustainment structure. All engineers and logisticians work closely to synchronize the logistics planning and coordination processes. They facilitate sound and timely plans or orders and necessary sustainment for engineer units.

9-71. After receiving a WO for a mission, the staff engineer immediately initiates an engineer logistics estimate as part of the EBA. This estimate is specifically focused on sustaining all subordinate engineer units. Class I, III, IV, and V supplies and personnel losses are the essential elements in the logistics estimate process. Close integration with the logistics staff can simplify and speed the estimate process by using their automatic data processing (ADP) systems. During continuous operations, the estimate process may need to be abbreviated due to time constraints. The staff engineer aggressively maintains an accurate logistics and combat status of all engineer units, which is critical to shortening the engineer logistics estimate process.

9-72. After developing an estimate to determine unit sustainment and mission supply requirements, the staff engineer compares the requirements with the reported status of subordinate units to determine the specific amount of supplies needed to support the operation. These requirements are then coordinated with the logistics staff to ensure that necessary supplies are identified and resourced. At the same time, the SES develops a required supply rate (RSR) to support engineer mission requirements. Based on how much of the required materiel is on hand and how much is needed, the staff engineer, in coordination with the logistics staff, assesses the availability of these supplies in stocks. They also analyze the capability to transport mission supplies to engineer units.

9-73. After identifying the requirements for unit sustainment and mission supplies and their availability, the staff engineer develops a projected engineer status, based on the current engineer sustainment status. The staff engineer then analyzes the requirements to support the plan and translates them into specific plans that are used to determine the supportability of COAs. After determining a COA, the specific engineer logistics input into the basic OPORD and paragraph 4 of the engineer annex are developed and incorporated. Current engineer sustainment operations may require redirection based on the new plan. If so, the OPORD and engineer annex is sent to subordinate engineer units for coordination and execution.

9-74. Engineer units provide unit and mission logistics status to the staff engineer so that the engineer can do a similar logistics staff planning process. Accurate and timely status reporting assists the staff engineer in providing accurate engineer unit status to the commander. It also energizes the staff engineer support to intercede in critical sustainment problems when necessary. The staff engineer ensures that mission-required supplies needed by engineer units to execute missions are integrated into logistics plans.

COMBAT SERVICE SUPPORT FOR ENGINEER ORGANIZATIONS

9-75. Engineer organizations are not organized with adequate internal and organic CSS to sustain the organization. They receive additional CSS from logistics elements at each echelon through their higher HQ support organization or, if in a command relationship, from the unit they are supporting.

9-76. Logistics elements at each echelon are organized to provide engineers with supply, maintenance, transportation, and field services. How the theater's logistics system supports a particular engineer unit depends on—

- That unit's organization.
- Its location in the TO.
- The command or support relationship under which it is operating.
- The maturity of the theater.

ARMY SERVICE COMPONENT COMMANDER

9-77. Operational-level engineer units are supported by subordinate ASGs of the ASCC's TSC. Other modular commands that may support engineers include personnel, transportation, finance, and medical commands. The ENCOM monitors logistics support to operational-level engineer forces.

JOINT TASK FORCE, JOINT FORCE LOGISTIC SUPPORT COMMAND, AND ARMY FORCES

9-78. Engineer units assigned to a JTF, Joint Force Logistic Support Command (JFLSC), or ARFOR are normally supported by an austere logistics support structure. A JFLSC may be established to support all forces in theater until an ASCC TSC structure is required. The senior JTF, JFLSC, or ARFOR engineer HQ monitors logistics support to deployed engineer forces.

CORPS

9-79. The COSCOM provides corps-level logistics support and health services support to corps engineer units and operational-level engineer units attached to the corps. The division support command (DISCOM) supports corps units attached to the divisions. All other corps units operating in division areas receive logistics support from COSCOM units operating in nearby areas. The division's medical support structure provides health services support for corps units operating in division areas on an area basis. Operational-level engineer units working in the corps's area generally receive support for common classes of supply and maintenance from the COSCOM. The corps engineer coordinates for required logistics support for all engineer units working in the area.

DIVISION

9-80. The DISCOM usually needs augmentation from the COSCOM to support corps engineer units attached to the division. In the heavy division, the forward support battalion (FSB) normally supports engineer equipment maintenance and Class IV and engineer Class V requirements. In light divisions, the main support battalion (MSB) supports engineer requirements.

Nondivisional engineer units working in the division area generally receive support for common classes of supply and common maintenance from the DISCOM. The DIVENG coordinates required logistics support for all engineer units working in the division area.

BRIGADE

9-81. The DISCOM forward support elements located in the BSA support the DIVENGs operating in the brigade's area. Nondivisional engineer units working in the brigade's area generally receive support for common classes of supply and common maintenance. The brigade engineer coordinates the required logistics support for all engineer units working in the brigade area.

MAINTENANCE SUPPORT

9-82. An effective maintenance program ensures that engineer units are capable of providing continuous, responsive support to combatant forces and the sustainment base. An effective supply system for repair parts and lubricants is also necessary. Engineer units consider the risks involved in mobility versus ample coverage for prescribed load list (PLL) repair parts. Logisticians anticipate and provide forward stockage of engineer-specific maintenance supply support requirements. The materiel management center (MMC) coordinates with tailored engineer forces for stockage selection and execution to maintain the dominant tempo of operations.

9-83. Military engineer units have the potential to use foreign commercial construction equipment. The source of this equipment is varied; for example, equipment may be—

- Provided by the HN.
- Leased from commercial venders.
- Obtained through a grant or loan.
- Bought by benefactor countries supporting the contingency operation.
- Purchased or leased through the LOGCAP.

9-84. Such was the case for the 34th Engineer Battalion (Combat) (Heavy) during the Gulf War. They deployed without their organic construction equipment and used the commercial equipment that the Government of Japan provided. Other battalions augmented their construction capabilities using additional leased equipment. These alternatives to increase construction capabilities pose challenges in maintenance to be considered, such as the—

- Lack of manuals and English labels on controls and switches.
- Lack of on-vehicle equipment (OVE), particularly safety equipment.
- Lack of service records or even a service schedule.
- Incompatibility or unavailability of repair parts within the Army's maintenance system.

9-85. The lack of manuals tends to make preventive maintenance haphazard. Nonstandard repair parts require mechanics to remove the part, take it to a vender to match up with a suitable replacement, and establish a PLL from scratch with no maintenance history. These challenges are not insurmountable; however, prior knowledge of the equipment condition and preventative maintenance status helps the commander establish a reputable maintenance program when assuming responsibility for foreign commercial construction equipment.

OTHER SUPPORT

9-86. The ENCOM and its operational units assist in managing critical Class IV and engineer Class V supplies, primarily in the establishment of theater policy or advisory capacity to the ASCC's logistics planner or the TSC. For other classes of supply and services, engineer units are consumers within the theater.

9-87. Engineer units obtain personnel service support (PSS), including finance services, from EAC PSS assets. Engineers need finance support to pay contractors and other local providers. The TSC's finance support center provides finance support. The nearest medical facility provides medical support, including medical supplies. The TSC's EOD detachments provide EOD support.

9-88. The ASG's supply and services battalion provides Class I supplies (rations). Rations are obtained based on the unit strength reports that the battalion Adjutant (U.S. Army) (S1) prepares. When working on an air base, engineer units obtain Class I support from the Air Force.

9-89. The property book officer (PBO), at the battalion level or in separate companies, requisitions Class II supplies from the ASG's supply company. Requisitions for regulated or command-controlled items are processed through the command channels. Engineer units request Class VII items through the material readiness officer at the MMC.

9-90. The ASG's supply and services battalion provides Class III supplies (POL). This is true for both bulk and packaged products. Requisitions from the PBO are required to obtain the needed supplies. When working on an air base, engineer units obtain Class III support from the Air Force.

9-91. Class V supplies (ammunition) are obtained from the nearest ASP that the TSC's ammunition battalion operates. Requisitions must be processed by the PBO. The TSC requires input on recommended stockage levels for engineer-specific Class V supplies, such as mines and demolition items, based on plans and anticipated usage rates.

9-92. Transportation and material transfer support are important when planning for Class IV and engineer Class V material. Supplies are often shipped by class. Transportation priorities for Class V supplies are often higher than those for Class IV supplies. Synchronizing mission Class IV and engineer Class V transportation and material transfer support is desirable to use mission loads effectively.

9-93. Logistics is an integral part of the engineer planning process. Engineer units that are not organic to a division are tailored for specific operations. Therefore, the staff engineer must articulate his CSS requirements from deployment planning—through the engineer estimate for mission loads—to the actual delivery of services and material. The coordination of units and resources facilitates the anticipation of engineer unit sustainment requirements and the execution of CSS. The staff engineer ensures that CSS planners understand engineer unit sustainment requirements and have a plan to meet those requirements.

ENGINEER COMBAT SERVICE SUPPORT COMMAND AND CONTROL

9-94. Each engineer CP has specific responsibilities in identifying unit and mission logistics requirements, estimating resources, integrating the logistics requirements into the operational planning cycle, and monitoring the execution of engineer missions supporting logistics operations.

REAR COMMAND POST ENGINEER SECTION

9-95. The rear CP engineer section—

- Is the engineer's primary integrator into rear CPs for executing logistics support for subordinate engineer units.
- Coordinates sustainment for current engineer operations and plans and prepares for implementing future operations.
- Maintains updated logistics status of engineer units.
- Provides the main CP engineer section with detailed logistics estimates to help formulate plans and orders.
- Ensures that engineer sustainment plans are synchronized with the logistics staff.

MAIN COMMAND POST ENGINEER SECTION

9-96. The main CP engineer section—

- Develops engineer sustainment plans.
- Writes the engineer logistics portions of the basic OPLAN or OPORD and paragraph 4 of the engineer annex.
- Integrates engineer sustainment through coordination with the main CP logistics cells.
- Ensures that the immediate engineer sustainment requests received from the tactical command post (TACCP) or assault CP are forwarded to the rear CP.

TACTICAL OR ASSAULT COMMAND POST ENGINEER SECTION

9-97. The TACCP or assault CP engineer section has limited capability to impact engineer logistics support from this location. Its primary logistics duties are receiving and forwarding reports and influencing the redirection of sustainment priorities for forward operating engineer units.

ENGINEER BRIGADE AND GROUP COMBAT SERVICE SUPPORT CELLS

9-98. The engineer brigade and group CSS cells—

- Support the brigade and group S1 and S4 officers in developing the engineer sustainment plans and writing paragraph 4 for brigade and group OPORDs and OPLANs.
- Monitor current engineer logistics status through periodic personnel and logistics status reports from subordinate units.

- Recommend logistics priorities to the brigade or group commander.
- Identify critical personnel and supply shortages, along with maintenance or transportation problems, that affect engineer unit and mission sustainment.
- Redirect logistics support.

9-99. The engineer brigade CSS cell maintains constant communication with subordinate engineer unit logistics sections, the rear CP engineer section, the COSCOM CP, and the G4. The engineer group CSS cell maintains constant communication with the brigade CSS cell, the subordinate engineer units, and the supporting maneuver unit logistics support units, if required.

ENGINEER BRIGADE AND GROUP HEADQUARTERS AND HEADQUARTERS COMPANY COMMAND POSTS

9-100. The engineer brigade and group headquarters and headquarters company (HHC) establish CPs at or near the brigade or group TOC. The HHC CP is responsible for sustaining the brigade and group CP. This includes establishing accounts with designated logistics support units, setting up life support areas for CP personnel, and locating vehicle maintenance areas. The HHC CP may also be designated as a Base Defense Operations Center (BDOC) or Base Cluster Operations Center (BCOC).

Appendix A

Engineer Field Manuals and Related Joint Publications

This appendix shows the comprehensive hierarchy of Army and joint engineer field manuals and the relationships between them.

A-1. *Table A-1, page A-2,* is a hierarchy chart of the Engineer Branch proponent manuals and the related joint engineer manuals. It is current as of the publication of this manual, but is subject to change as manuals are updated, renumbered, and published. Always refer to the latest official publications.

A-2. Army and engineer proponent manuals are in the process of being renumbered to align them with the joint manuals numbering system. *Table A-1* lists manuals by the current number, followed by the old number in parenthesis. In the new numbering system—

- The prefix 3 represents alignment with the category of joint publications operations manuals.
- The suffix 34 represents an engineer publication.
- The extensions designate the mission content (such as combined arms or engineer operations) or a reference focus for the manual.

A-3. All Army manuals can be downloaded and viewed through the Reimer Digital Library (RDL). Manuals that require passwords can be accessed by applying for a password on the web site. The RDL is located at http:// 155.217.58.58/atdls.htm.

The JPs can be located at http://www.dtic.mil/doctrine. Those of specific interest to engineers include—

- JP 2-03.
- JP 3-15.
- JP 3-34.
- JP 4-04.

Table A-1. Engineer FMs and Related Joint Publications

JP 2-03 JTTP for Geospatial Information and Services Support to Joint Operations

JP 3-15 Joint Doctrine for Barriers, Obstacles, and Mine Warfare

JP 3-34 Engineer Doctrine for Joint Operations

JP 4-04 Joint Doctrine for Civil Engineering Support

FM 3-34 (5-100) Engineer Operations

Combat Engineering

Mobility

- FM 3-34.2 (90-13-1) Combined-Arms Breaching Operations
- FM 90-13 River-Crossing Operations
- FM 5-170 Engineer Reconnaissance
- TC 5-210 Military Float Bridging Equipment
- FM 5-177 M2 Bailey Bridge
- FM 5-212 Medium Girder Bridge
- FM 3-34.343 (5-446) Military Non-standard Fixed Bridging

Countermobility

- FM 90-7 Combined Arms Obstacle Integration
- FM 5-250 Explosives and Demolitions
- FM 5-34 Engineer Field Data

Survivability

- FM 5-103 Survivability
- FM 20-3 Camouflage, Concealment and Decoys

Geospatial Engineering

- FM 3-34.230 (5-105) Topographic Operations
- FM 5-33 Terrain Analysis
- FM 21-31 Topographic Symbols
- FM 3-34.331 (5-532) Topographic Surveying
- TM 5-235 Special Surveys

General Engineering

- FM 20-11 Military Diving
- FM 5-490 Engineer Diving Operations
- FM 5-104 General Engineering
- FM 5-125 Rigging Techniques Procedures & Applications
- FM 5-146 Military Nonstandard Fixed Bridging
- FM 5-134 Pile Construction
- FM 5-412 Project Management
- FM 5-553 / TMs 5-581B & 5-704 General Drafting
- FM 5-233 Construction Surveying
- FM 5-410 Military Soils Engineering
- FM 5-472 Materials Testing
- TM 5-545 Geology
- FM 5-430-00-1 P&D of R/A/H in to Road Design
- FM 5-430-00-2 P&D of R/A/H in to Airfields & Heliport Design
- TC 5-340 Air Base Damage (Pavement Repair)
- FM 5-434 Earthmoving Operations FM3-34.465 (TM 5-432 /FM 5-331C) Pits And Quarries
- FM 5-436 / TM 5-337 Paving & Surfacing Operations
- FM 5-499 Hydraulics
- FM 5-420 / 5-163 Plumbing, Pipefitting And Sewerage
- FM 5-424 Theater Of Operations Electrical Systems
- FM 5-426 Carpentry
- FM 5-428 Concrete And Masonry
- TM 5-323 MSH Generator Sets, Electrical Measurement
- FM 5-484 Multiservice Procedures For Well Drilling Opns
- FM 5-422 Engineer Prime Operations
- FM 5-480 Port Construction and Repair
- FM 5-482 Military Petroleum Pipeline Systems
- TM 5-349 Arctic Construction

Operations

- FM 20-32 Mine/ Countermine Operations
- FM 5-116, Engineer Operations: EAC
- FM 5-100-15 Corps Engineer Combat Operations
- FM 5-71-100 Division Engineer Combat Operations
- FM 5-71-2 Armored TF ENGR CBT OPNS
- FM 5-2 Engineer Operations -- SBCT
- FM 5-7-30 BDE ENG & ENG CO CBT OPS (ABN, AASLT, LT)
- FM 5-71-3 BDE ENGR CBT OPNS (Armored)
- FM 5-10 Combat Engineer PLT
- FM 5-415 Firefighting Operations
- TC 20-401 The Soldier and The Environment
- FM 3-100.4 Environmental Considerations
- TC 5-150 Engineer Qualification Tables

Appendix B

Maneuver Support

Maneuver Support (MANSPT) and the grouping of those aspects of CS activities and elements are not recent ideas, although the operational reality of this is only recently being worked with in the SBCT as the Army looks at other ways to functionally organize staffs to enhance performance. The creation of a MANSPT cell, just like the creation of a fire effects cell and others, is an initial step toward what staff structures will look like in the future. As such, this appendix will look at MANSPT, its definition, today's realities, and its likely future. The primary premise for a MANSPT cell is built around the M/CM/S BOS. At the core of MANSPT are the engineer, chemical, MP, and other branches such as the EOD portion of the Ordnance Branch. By establishing the Maneuver Support Battle Lab and consolidating the chemical, engineer, and MP schools at Fort Leonard Wood, Missouri, the Army is furthering the concept of MANSPT and developing relationships between the three primary branches. This appendix will discuss current and future opportunities for this evolving concept.

DEFINITION

B-1. MANSPT is defined as staff integration of the M/CM/S BOS, with the remaining BOS focused on enabling assured mobility for the friendly force. Staff integration is focused on enhancing tactical freedom of maneuver and force protection using the assured mobility imperatives and fundamentals as the framework.

TACTICAL LEVEL: MANEUVER SUPPORT IN THE STRYKER BRIGADE COMBAT TEAM

B-2. The SBCT MANSPT cell is organic to the SBCT HQ and is responsible for planning, integrating, and synchronizing MANSPT for all SBCT operations. It is composed of an engineer and an MP section. Although not organic to the MANSPT cell, significant coordination is required with the chemical section to integrate MANSPT considerations into the cell. While no organic EOD capability exists in the SBCT, this too is an area of MANSPT that must be integrated into this cell. If the brigade engineer is also the senior officer with the MANSPT cell, he may also function as the MANSPT coordinator. In this case, he must rely heavily on the MP and chemical sections for their expertise to ensure integrated and synchronized MANSPT to the SBCT. For more information related to the specific duties of the MANSPT coordinator, *see FM 3-21.31*.

OPERATIONAL-LEVEL CONSIDERATIONS

B-3. The key aspects or imperatives of MANSPT include the following:

- Understand the environment—consider the space, air, water, land, subterranean, weather, light, infrastructure, hazards, and populations.
- Enable theater access—access installations through air, land, and seaports of embarkation (SPOEs) and move between them; mitigate the adverse effects on the environment, and protect and facilitate multiple entry points.
- Provide assured mobility—guarantee the force commander the ability to deploy, move, and maneuver at all times without interruption.
- Deny enemy freedom of action—leverage the environment to isolate, deny, impede, or canalize the enemy.
- Enable force protection and security—provide cueing, early warning information, protected movement of forces between operating areas, and layered protection.
- Engage and control populations—control EPWs, refugees, detainees, internees, mass transiting civilian populations, and disaster evacuees.
- Neutralize hazards and restore the environment—reduce or eliminate hazards through avoidance, mitigation, neutralization, and restoration of the environment.

THE ROAD AHEAD

B-4. MANSPT as a future concept and framework for the future force is much broader than the SBCT and encompasses the means to enable, enhance, and protect the strategic, operational, and tactical freedom of action of the force, as well as the denial of comparable freedom of action to the adversary. MANSPT accomplishes this by shaping, leveraging, or mitigating the effects of the operational environment. For an offensively oriented, maneuver-based force, the focus is on enabling force movement and maneuver at all levels of war and echelons and across the full spectrum of military operations. This is critical with the increased emphasis on operational and strategic maneuver of ground forces for the future force.

B-5. MANSPT concentrates on two interrelated components: freedom of maneuver and force protection. For a maneuver-based force, there is significant overlap and synergy between these two components, with MANSPT assets, systems, and soldiers able to readily support both components simultaneously. This translates into an effective and efficient application of forces. MANSPT takes on added dimensions and significance due to the increasing emphasis on battlefield frameworks that encompass extended, dispersed, distributed, noncontiguous, and 3-D battlefield and nonlinear operations, to include forcible and early entry operations and operational maneuverability. Additionally, the Army's vision of the operational environment recognizes that adversaries will apply a wide array of conventional and unconventional or asymmetric means to deny the access and movement of friendly forces, starting with their strategic deployment platforms and continuing within the TO and in individual tactical AO.

Finally, the physical environment, to include terrain, infrastructure, weather, hazards, and the presence of the local populace, will affect any ground force, sometimes more than enemy actions will.

B-6. For further thoughts on the future of MANSPT and its application as a mission area see the soon to be published *TRADOC Pamphlet 525-3-25.*

Appendix C

Field Force Engineering

FFE is a focused area for the Engineer Regiment that specifically targets improving general (and to a lesser degree geospatial) engineering support to contingency operations. The focused area reduces the ad hoc solutions previously applied to meet contingency requirements. These requirements have grown, and the Engineer Regiment has responded by doctrinally accepting the best ad hoc solutions, using lessons learned and actively seeking to add technological advances as they impact the engineer mission. Another reason for the development of FFE as a focus area is to provide specialized engineer support to help to quickly establish and maintain the necessary infrastructure to support forces in contingency AOs. The nature of contingencies has made it necessary to provide installation-type engineer support to tactically deployed combined arms forces through the creation and use of base camps. In Bosnia alone, over 2,500 engineer soldiers from a host of countries established and maintained base camps. During the Bosnia and Kosovo operations, a modular expert personnel package called the Base Camp Coordination Agency (BCCA) was developed and deployed to assist these efforts. The original package, partially staffed by members of the ENCOMs, consisted of a cell with a tailored capability of engineer skill sets. After 1997, the package consisted largely of civilian and military personnel from elements within the USACE, using military individual mobilization augmentation personnel, mobilized RC personnel, civilian personnel from elements within USACE, and selected individuals within CONUS (largely pickup teams, not units). These teams quickly develop and implement techniques, tactics and procedures to accomplish their mission. Typical missions included base camp design and closure, force protection analysis, ASP upgrade, MSR maintenance and basic DPW missions in support of base camps. A book of standards, commonly referred to as the "Red Book," was developed by United States Army, Europe (USAREUR), Deputy Chief of Staff, Engineers (DCSENGR), and provided a blueprint for planning and requirements relating to base camp construction and maintenance. A similar book of standards know as the "Sand Book" was developed for the CENTCOM AOR. Additionally, a critical piece of FFE and the reach capability includes the associated technology of the TeleEngineering communication suite.

THE CHALLENGE

C-1. The challenge for engineers is to be prepared to support contingency operations and all missions within the full spectrum of conflict. *Figure C-1, page C-2* graphically demonstrates the changing requirements for engineers over time in any given contingency operation. Different portions of the Engineer Regiment will play varying roles in supporting these requirements at different times. Added to this are the factors of METT-TC that may limit or require the use of one or more portions of the Engineer Regiment. At a particular part of a contingency operation, it may be desirable to limit either the footprint of engineers in the AO, or at least the number of green suit engineers. In this case, extensive use of reach capability and/or the use of USACE civilians may be preferable to military engineers out of the ENCOMs. In another contingency operation, the threat level may be so high that the decision is made to keep as many civilians as possible out of the AOR, leaving a pure military engineer solution. In any case, FFE allows for a doctrinal solution to meet each challenge in an organized fashion.

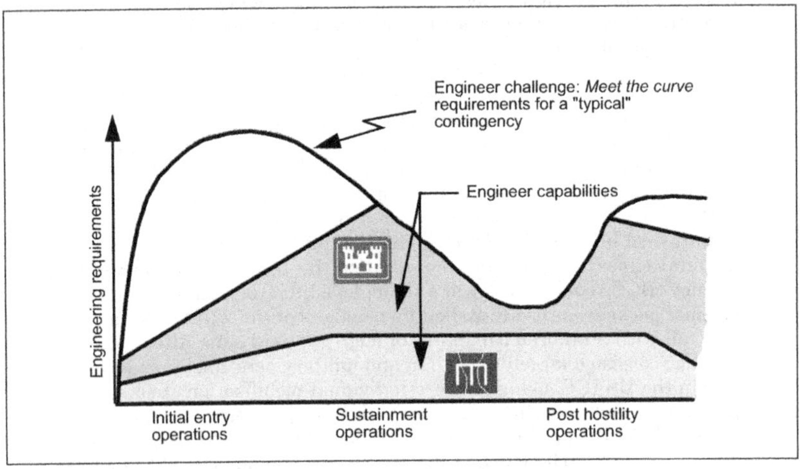

Figure C-1. Engineer Support Shortfall

OVERVIEW

C-2. The application of FFE is Army-wide, and, due to the joint nature of operations, applicable to the support of not only Army forces, but other Services and governmental agencies outside of DOD. *Figure C-2* highlights the majority of capabilities the Engineer Regiment offers to support contingency operations. FFE works to provide seamless general (and associated geospatial) engineering support for the war fighter by fusing the capabilities resident in USACE, USAES, ENCOMs, public works, and civilian contractors. Modular teams are available to meet the needs of C2, augmentation of ASCC, or other headquarters for engineering staff, theater planning, facilities engineering, troop construction, topographic engineering, contract construction, technical engineering, real estate acquisition, and environmental engineering. The mission of FFE is to support the senior engineer in theater by providing focused planning, coordination, and execution of nontactical engineer missions, including the construction, maintenance, technical engineer support, and final disposition of infrastructure.

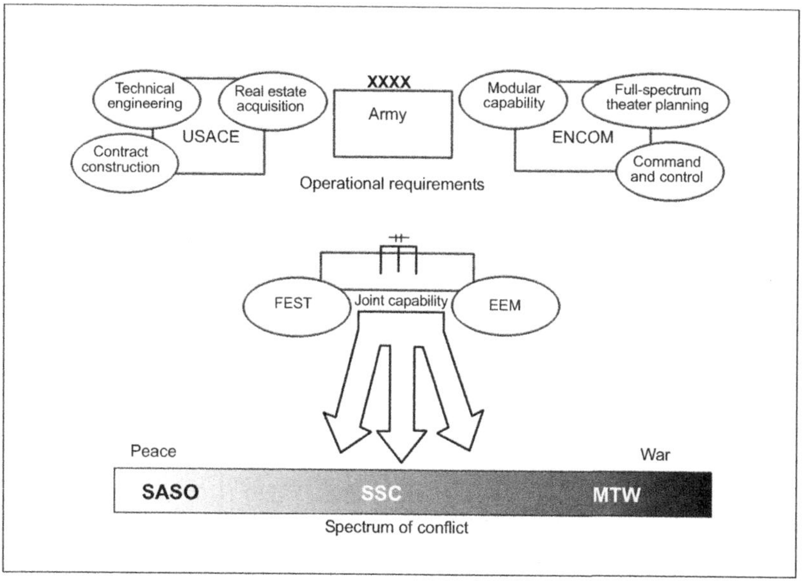

Figure C-2. Field Force Engineering

PLANNING STRUCTURE

C-3. The USACE and the ENCOMs provide the primary elements to create the integrated planning structure for FFE support. Operational missions addressed include: LOC integrity, AOR infrastructure stability, construction management, contractor oversight, and oversight of facilities maintenance missions. The forward elements are responsible for daily interaction with the MACOMs and joint staff as required. They provide the senior engineer with a level of expertise and linkage with other engineer capabilities that will allow comprehensive construction and facilities maintenance planning. *Figure C-3, page C-4* demonstrates the links of the USACE and the ENCOMs with the various combatant commanders, JFCs and ASCC commanders in both peace and war. The current established relationship with particular combatant commanders and other HQs is identified in Appendix I under the USACE and later in this appendix. The ENCOMs currently have permanent forward planning cells with all of the ASCCs except the United States Army Forces Southern Command (USARSO). The ultimate goal of these linkages is to provide proactive and appropriate support for any type of operation. It establishes a more unified approach to using all assets of the Engineer

Regiment across the spectrum of conflict and acknowledges the reality of needed tailored engineer support assets.

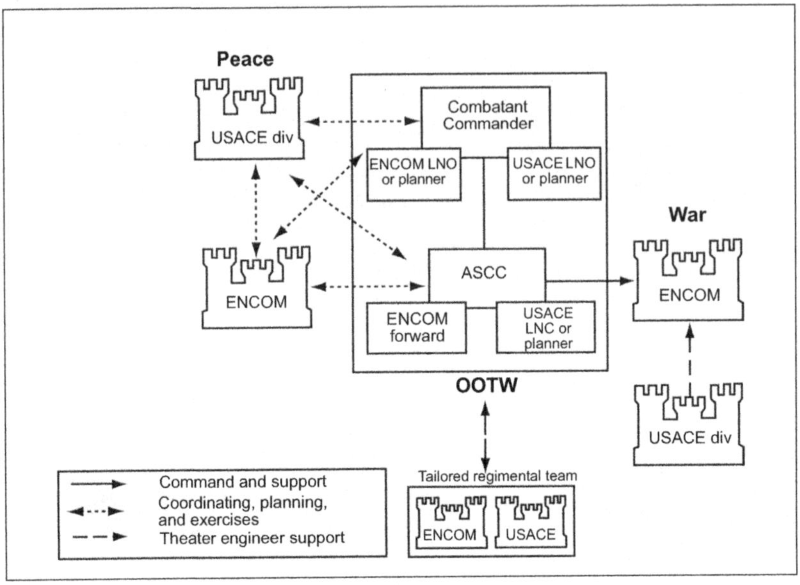

Figure C-3. Field Force Engineering Planning Structure

MODULAR DEVELOPMENT

C-4. The Army's first major challenge in a contingency operation is mobilizing and deploying forces into theater. The space availability on military strategic sealifts and airlifts will always constrain the size and content of the deploying force. Early deployment of one or more engineer modular packages provides specific and necessary expertise from the beginning of operations. Modular deployment minimizes the footprint of units and personnel. As the theater matures and sealifts and airlifts become available, more robust and complete capabilities may be brought into theater if necessary and desirable. There is recognition that most of the focused expertise in the FFE is nontactical in nature and may be supported by civilian contractors. Reach may be used to obtain much of the specialized support required. One method for this is the use of the USACE TeleEngineering Emergency Operations Center (TEOC) and a forward deployed TeleEngineering communication suite. Examples of modular capabilities are listed below:

C-5. USACE personnel are organized for employment as FESTs. The FEST planning cell provides direct augmentation to the combatant commander, JFC, and ASCC staff engineers; or the ENCOM if they are the senior engineer element. The FEST execution team provides C2 for missions for which the USACE has the lead.

- A forward engineer support team-main (FEST-M) is a deployable USACE organization that executes the USACE mission in the AO. It is usually subordinate to the senior engineer commander in the AO.
- A forward engineer support team-augmentation (FEST-A) is a deployable planning augmentation cell that augments the engineer staff of other organizations including the combatant commander, JFC, ASCC, and ENCOMs. If a FEST-M is also deployed, the FEST-A will come under the command of the FEST-M commander.
- A CREST is a deployable 5- or 6-person team and typically a component of a FEST. A CREST can quickly execute real property (land and facilities) leases at forward locations.
- An environmental support team (ENVST) is a deployable 4-person environmental team, typically a component of a FEST. An ENVST provides baseline and other environmental studies at forward locations. It is staffed with environmental specialists that can provide technical expertise to staff engineers and ultimately the commander of the combined arms force.
- A base development team (BDT) is a nondeployable team that can quickly provide base development engineering and planning and facilities design for staging bases, base camps, forward operating bases, displaced persons camps, and any similar requirement.
- An infrastructure assessment team (IAT) is a nondeployable team that provides engineering infrastructure assessments for military deployments and civil military operations in forward areas. Focus areas for the IAT are infrastructure related to USACE missions and aspects of the AO impacting contract construction, including roads, utilities, water resources, and HNS.

C-6. ENCOMs typically support a combatant commander through the ASCC and plans for flexible modular entry into theater via early entry modules (EEMs). EEMs may range from several soldiers to a functional ENCOM C2 element. They have the ability to put soldiers into theater within days of mobilization through the use of three primary means. This includes the use of AC and Title 10 active USARNG and USAR soldiers, hasty mobilization teams (a new process recently endorsed by the Chief of USAR) and soldiers on temporary tours of active duty and active duty for special work. For situations requiring military organizations, the ENCOMs contain the following types of organizations *(refer to Appendix D as well):*

- A facility engineer group (FEG) is a 405-person TDA or TOE organization responsible for design and technical engineering support, including facility and environmental management, BASOPS, force protection, and other areas of expertise. Only one FEG exists in the force structure, and it is under the C2 of the 416th ENCOM during peacetime operations. An FEG is comprised of six 15-person TO&E FED and thirty-person TDA facility engineer teams (FETs) aligned under four geographic facility engineer centers (FECs) that provide additional peacetime command, control, and administrative support.

Each organizational element of the FEG is assigned a unique Unit Identification Code (UIC) for rapid and efficient deployment.

- Three FEDs are war-traced to each of the two major contingency operations (MCOs) ENCOMs and are further aligned as one per ASG. During peacetime, the detachments provide environmental and facility assessments, engineer design, and other support to the USAR engineer, with the other half of their effort dedicated to exercise participation, engineering planning, and design review. The FEDs are deployable elements structured to provide the full range of DPW facilities management for troop concentrations in a theater of operations. Their composition includes all the necessary disciplines to manage, operate, design, and assess bare base facilities and their component systems.
- FETs deploy during war to conduct DPW, BASOPS, environmental, and facility assessment missions. When deployed, they come under the C&S of the senior engineer in their particular AO. During peacetime, they provide environmental and facility assessments, engineer design, and support to the USAR engineer. FETs are deployable organizations that can support contingency operations.

C-7. A partial list of potential missions performed by deployed FFE units and organizations includes—

- Project management, to include supervising work done by military engineers from all branch services, LOGCAP contractors, and local contractors.
- Expertise in force protection engineering and engineering services, including areas such as wastewater treatment, nontactical power generation, water production, snow and ice removal, and entomology.
- Expertise in theater operational (such as in the IPDS) and infrastructure pipeline construction.
- Expertise in mining and demining operations.
- The development and maintenance of base camp standards according to ASCC guidance.
- The development of prioritized facilities requirements with supporting construction engineers.
- The management of the installation and regional and AOR master plan development process, including the—
  - Development of base camp master plan(s).
  - Review and approval of base camp master plans developed by others.
  - Development of regional master plans in support of the operational support structure.
- Base camp maintenance, including—
  - Structures.
  - Utilities.
  - Roads and airfields.
  - Provisions for water.
  - Wastewater treatment.

    - Solid waste disposal.
    - Training ranges.
- Management of base camp facilities use, including—
    - Environmental compliance.
    - Maintenance of real property records.
    - Plans for future expansions or contractions of base camps.
    - Quality control for construction projects and facilities upgrades.
    - The establishment of a fire prevention program and management of fire fighting teams.
    - Facility engineering supplies management.

LEVERAGING TECHNOLOGY

C-8. Reach and TeleEngineering are terms used to describe the ability to conduct engineering analysis and support using expertise (professional engineers, scientists, and technicians), as well as databases that are not resident in the theater or the AO. The source of this expertise and data is real time via secure voice and video information technology to the forward mission site. This results in faster response times and the application of greater levels of expertise examining and solving problems and providing solution options. An additional set of benefits includes a reduced deployment footprint, increased associated support and force protection requirements, and a general reduction in cost.

Appendix D

Army Engineer Organizations and Capabilities

Army engineers operate as an integral member of the joint force to provide a complete range of engineering capabilities to the JFC across the full spectrum of military operations. Whether performing combat engineering missions, providing general engineering support throughout the operational area, or providing geospatial engineering support to commanders, engineers do not fight alone. They are an important part of the combined arms team, increasing the team's combat effectiveness.

The engineer force structure has been developed to support various missions from front to rear, with more survivable and mobile forces in the forward CZ. Any engineer unit may be pushed forward, if the mission dictates.

A wide variety of engineer units provide specialized technical capabilities required to accomplish essential tasks throughout the theater. The engineer architecture depicts an organization that is responsive to commanders at all echelons. This appendix addresses all Army engineer organizations and select architectures. For information on USACE organizations and capabilities, *see Appendix I.*

The following information identifies the types of engineer units available by their base TOE or TDA. For further information, consult the USAES, Directorate of Combat Developments.

SECTION I – OPERATIONAL-LEVEL UNITS

HEADQUARTERS, ENGINEER COMMAND

D-1. The ASCC commander tailors the engineer structure to the theater requirements based on staff advice from the DCSENGR. The operational-level engineer commander provides C2 operations and a central organizational framework for the engineer effort. Engineer forces outside the corps focus on reinforcing and augmenting corps engineer efforts, developing the theater support base, and maintaining an infrastructure for sustainment. This focus involves—

- Planning.
- Ensuring combat, geospatial, and general engineering support at the operational level.
- Coordinating all theater engineer assets.

- Providing construction guidance, RPMAs, LOC sustainment, engineer logistics management, and base development.

D-2. Engineers must be closely tied into current and future operations because they provide versatility to the operational commander. All engineer units focus on operations in the CZ. Additionally, they support the theater by providing general engineering support at the operational level. The engineer's operational-level topography unit and a variety of specialized engineer teams and detachments support or augment engineer forces throughout the theater. Combat heavy engineers weight the main effort and provide sea, air, and land operational and strategic mobility. A typical operational-level engineer organization is shown in *Figure D-1*. *FM 5-116* discusses the operational-level engineer function.

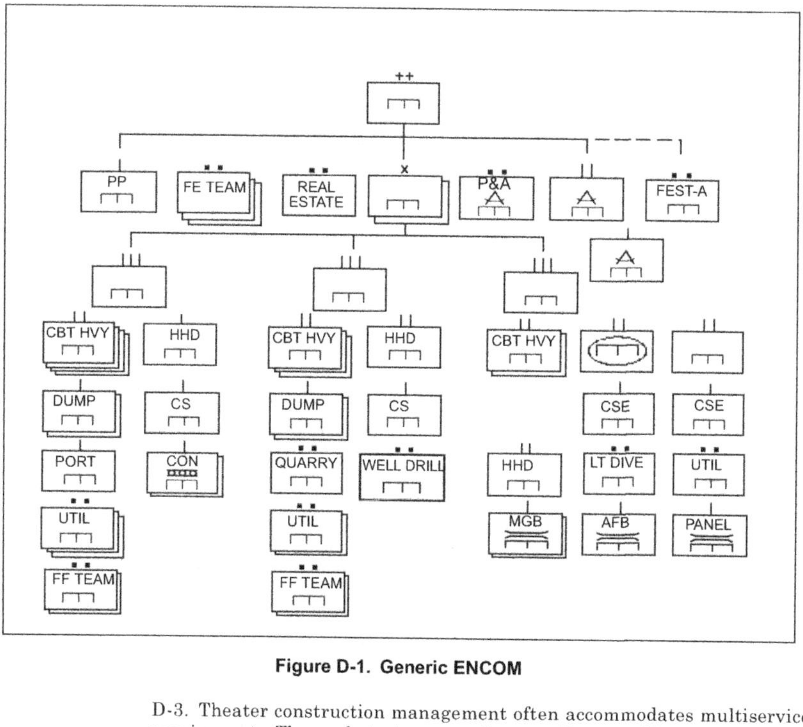

Figure D-1. Generic ENCOM

D-3. Theater construction management often accommodates multiservice requirements. The combatant commander may direct establishing a RCEM to control all theater level engineering. The Army ENCOM commander can perform this role if the combatant commander designates the ASCC commander as the RCEM and the ASCC designates the ENCOM commander as the agent. To support force projection requirements, an engineer element may deploy in sections to meet highly variable workloads and situations.

D-4. Theater infrastructure development is a primary concern of the ENCOM. The ENCOM is responsible for developing plans, policies, procedures, and programs for engineer support for the ASCC. This support includes requirements determination, M/CM/S (at the operational level), other general engineering (ADC, military construction, engineering design, construction materiel and maintenance, and repair of real property), and geospatial engineering (production of military geographic intelligence). Engineer units are responsible for infrastructure planning, development, construction, and maintenance *(see Figure D-2)*.

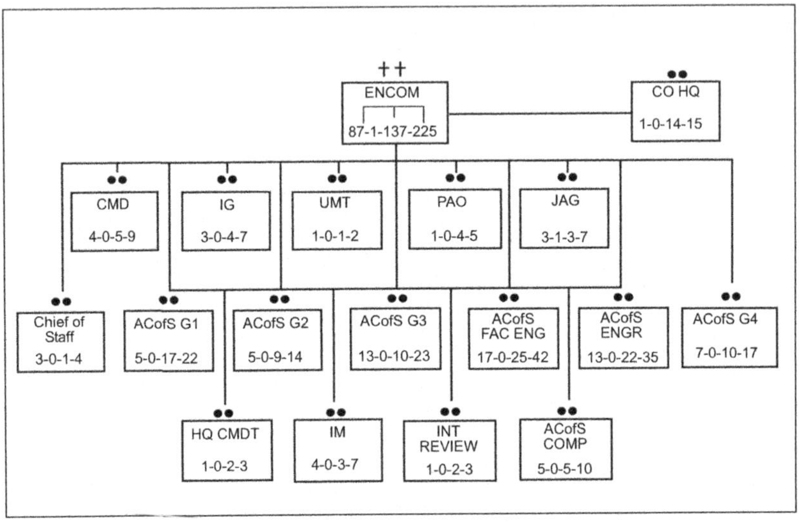

Figure D-2. Headquarters and Headquarters Company, Engineer Commands

D-5. Typical operational-level engineer missions include the following:

- Planning; designing; supervising, coordinating and providing guidance and technical assistance; constructing or rehabilitating general troops; and providing contractual support and support to the Army, other services, and allies throughout the theater to include the construction of—
  - Airfields, ports, pipelines, bridges, roads, railroads, and inland waterways.
  - Hospitals, base camps, EPW and civilian internee compounds, bulk petroleum storage and distribution systems, dry cargo and ammunition storage areas, and equipment maintenance facilities.

   - Missile sites, air defense emplacements, protective shelters, and local security measures.
- Performing emergency runway repairs that exceed Air Force capabilities.
- Conducting rear operations missions, to include ADC.
- Augmenting ASG RPMA capabilities.
- Providing and coordinating topographic, geospatial, and military geographic intelligence support.
- Maintaining critical LOCs.
- Managing HN and contractual construction and labor engineering efforts, to include US, indigenous, and third-country personnel.
- Providing fire fighting support, as specified by the field commander.
- Providing environmental support.

D-6. The HQ element of the ENCOM provides staff supervision over all engineer operations in the COMMZ and directs engineer support to all Army, joint, and multinational commands and other theater elements, as directed by the JFC and ASCC commander. It provides policy and technical guidance to all Army engineer units in the theater. The HQ element maintains a technical relationship with the ASCC staff engineer to help establish engineer policy for the theater and maintains required coordination links with JFC and multinational command engineering staffs.

D-7. Several critical specific functions of the ENCOM HQ include—

- Monitoring engineer support for real-property management and coordinating reports from engineer organizations and additional support elements supporting TSC operations, HNS, and contracting directorates.
- Monitoring requirements and priorities for Class IV and engineer Class V materials and coordinating the flow of these materials with the TSC supply and maintenance directorate, the distribution management center (DMC), and MMC.
- Performing the EBA and coordinating with the ASCC engineer, G2, and TSC support operations section to communicate critical enemy capabilities, affecting support facilities and operations.

D-8. *FM 5-116* provides more details on engineer operations in a theater.

HEADQUARTERS, ENGINEER BRIGADE, THEATER ARMY

D-9. The HQ, Engineer Brigade *(see Figure D-3)* is normally assigned as an ASCC and may be attached to an ENCOM. This HQ—

- Commands, controls, conducts planning for, and supervises attached engineer units that are engaged in CS operations and the construction and rehabilitation of facilities in support of a TO.
- Commands, controls, plans, and supervises the attached units' activities.
- Allocates resources in support of engineer operations.

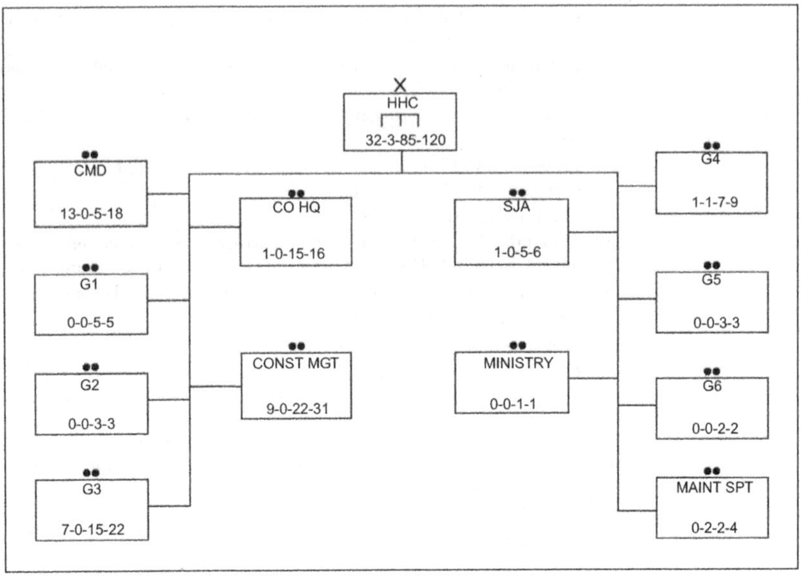

Figure D-3. Headquarters and Headquarters Company, Engineer Brigade (Theater Army)

- Supervises engineer units that are constructing and rehabilitating roads, structures, air-landing facilities, and petroleum storage and distribution facilities.
- Supervises HN personnel and administers contractual construction and labor.

D-10. Utility teams (TOE 05530LH00) normally augment the capabilities of this unit when facilities engineering support is required.

HEADQUARTERS, ENGINEER GROUP, FACILITY

D-11. The United States Army Facility Engineer Group (USAFEG) is a composite TDA and TOE organization headquartered in Chicago, Illinois. This USAR unit contains the Army's highest concentration of 21Ds and facility engineer personnel. The USAFEG is the Army's only FEG. (There are no AC or ARNG equivalent units.) The USAFEG consists of four FECs and a mobilization planning support cell (MPSC).

D-12. The FECs are O6 directorates and are located throughout the country in the northeast, southeast, northwest, and southwest. Each of the three FECs consists of 10 seven-person FETs that are dispersed geographically from their FEC HQ. The fourth FEC is comprised of six 15-person FEDs, which are TOE organizations. These FEDs are war-traced with three detachments going to each of the two Army ENCOMs.

D-13. The MPSC provides mobilization and training support to the FECs and their teams. They also support FORSCOM during activation in times of need.

D-14. The USAFEG has a diverse membership that, with civilian skills, include licensed professional engineers, environmental engineers, electrical engineers, mechanical engineers, civil engineers, architects, construction contractors, facility managers, and project managers. The civilian experience of these engineers uniquely qualifies them to perform their peacetime and wartime missions.

D-15. In peacetime, the USAFEG supports the United States Army Reserve Command (USARC). Teams are responsible for assessing over 1,200 USAR facilities comprising 25 million square feet of workspace. Assessments focus on facility maintenance requirements, environmental compliance, port maintenance, energy conservation, and safety. As force protection becomes increasingly relevant to the safety of soldiers, the teams could also assess and recommend security improvements to meet these new requirements.

D-16. The USAFEG wartime mission is to provide facility engineer services, including base camp design and limited master planning. In times of national emergency or contingency operations to Army components worldwide, this mission is accomplished using the small footprint TDA component— a highly modular skill structure engineer unit—the FE team. In these situations, a seven-person team is mobilized and deployed into the theater of operations for each base camp that is or will be operational. Properly constructed and maintained operations and logistics facilities are instrumental in sustaining military operations in theater.

D-17. In contingency operations, the FET is normally assigned to the ASG and has a coordination relationship with the ENCOM. The missions of these teams range from the construction design and management of building base camps, as soldiers are initially deployed into an area, to managing the maintenance and repair of established bases. In all cases, the FET operates as a DPW. As a TDA unit, the FET has no organic communication, transportation, or defensive equipment and must rely on the ASG for this support.

D-18. During major combat operations, the FEDs aligned with the appropriate theater can be activated to provide those same services. These 15-man units bring their civilian engineering experience and knowledge to assist in all facility work, including base camp planning, design, and layout. These soldiers utilize their civilian skills as master planners; civil, mechanical, structural, and electrical engineers; contracting specialists; and project managers to provide a force multiplier in these essential task areas. These units are aligned with the ENCOM responsible for the theater.

D-19. As the DPW for contingency base camps, the FET and/or FED, in coordination with the base camp commander,—

- Controls, supervises, and coordinates all engineer and construction assets, including troop labor, contracted HN construction assets, and LOGCAP construction.
- Plans and designs projects for infrastructure and operations and living facilities.

- Allocates engineer troops, materials, and equipment for construction projects.

HEADQUARTERS, ENGINEER GROUP, CONSTRUCTION

D-20. The HQ of this engineer group is assigned to a divisional engineer brigade, ASCC, or corps engineer brigade. This engineer group commands and controls assigned or attached units to include engineer combat heavy battalions, corps combat battalions, and separate engineer companies and detachments *(see Figure D-4).* It possesses a design and management section for specific engineering tasks at EAC and can command, control, plan, and supervise engineer units with construction capabilities throughout the depth of the AO. Normally, the HQ commands and controls three to seven engineer battalions. When task-organized to a division, the group may assume a command or support responsibility for divisional engineer assets. This HQ—

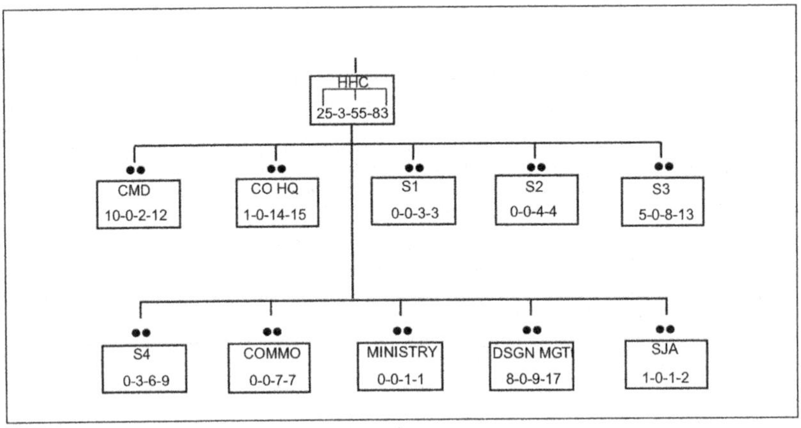

Figure D-4. Headquarters and Headquarters Company, Engineer Group, Construction

- Plans, supervises, and coordinates the activities of the assigned and attached engineer units engaged in survivability, general engineering, and limited mobility and countermobility operations at the operational level.
- Plans and supervises units performing general engineering and selected combat engineering tasks such as constructing and maintaining combat routes, MSRs, bridges, fords and river crossing sites, landing strips, heliports, port facilities, railroads, and battle positions. This includes the engineer technical reconnaissance associated with these missions.
- Plans and supervises the construction of base camps, logistical bases, EPW and displaced civilian camps, pipelines, railroads, and so forth.
- Plans and supervises ADC operations.

ENGINEER BATTALION, PRIME POWER

D-21. The engineer battalion, prime power is assigned to an ASCC and attached to either an ENCOM or a TA engineer brigade. This battalion consists of one HHC and two line companies *(see Figure D-5)*. Each line company contains an HQ section and six prime power platoons. This battalion—

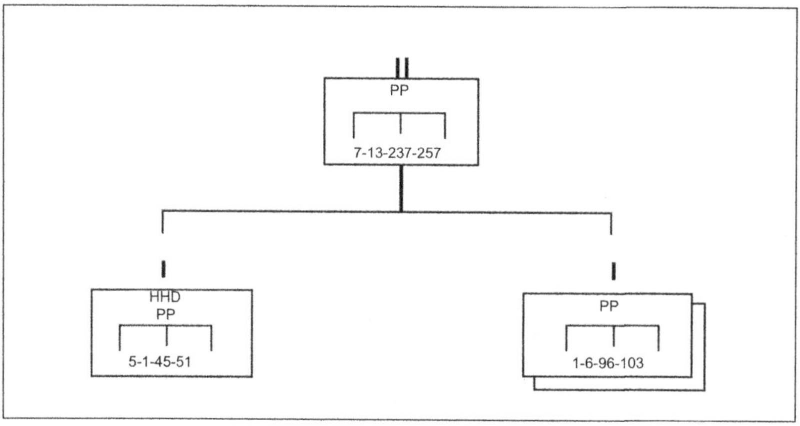

Figure D-5. Engineer Battalion (Prime Power)

- Generates electrical power and provides advice and technical assistance on all aspects of electrical power and distribution systems in support of military operations.
- Produces electrical power, up to 36 megawatts, in support of C2 sites, hospitals, weapons systems, and logistics support areas and provides relief for tactical generators at fixed sites and critical facilities.
- Repairs and maintains organic power production and distribution equipment, distributing power produced with organic systems.
- Operates, maintains, and performs minor repairs to other electrical power production equipment, to include HN fixed plants.
- Provides electrical engineering support, such as limited design and analysis capabilities, to commanders and senior engineers for all aspects of electrical power systems.
- Performs electrical surveys and electrical-related COR assistance.
- Manages and coordinates worldwide prime power requirements.
- Supports RPMA and power reliability enhancement programs.
- Designs and constructs power plants and distribution systems to the panel box for base camps.

- Provides quality control (QC) for contractor designs and the construction of electrical power plants and power distribution systems.

D-22. *FM 5-442* provides more details on engineer prime power operations.

HEADQUARTERS AND HEADQUARTERS DETACHMENT, ENGINEER BATTALION

D-23. Assigned to an engineer brigade, an ASCC, or an engineer group, the number of battalion headquarters and headquarters detachments (HHDs) authorized is based on the number of separate engineer companies and teams. This unit or detachment provides—

- Command, control, and administrative support for separate engineer companies and teams, to include construction support companies, pipeline companies, port construction companies, dump truck companies, and utilities teams.
- C2 for three to seven companies and teams performing engineer-related tasks.

HEADQUARTERS AND HEADQUARTERS COMPANY, ENGINEER TOPOGRAPHIC BATTALION

D-24. The headquarters and headquarters company, engineer topographic battalion is normally assigned to an ENCOM. It—

- Provides C2 for operating, planning, and supervising topographic units in a TO.
- Provides DS and GS maintenance for topographic, reproduction, air-conditioning, and power generation equipment to support engineer topographic units.
- Provides technical supplies for and performs maintenance on organic equipment, as well as unit maintenance on communications and electronics equipment.
- Maintains liaison with allied topographic units, NIMA, and appropriate support unit staff elements.

SECTION II – SEPARATE OPERATIONAL-LEVEL COMPANIES

ENGINEER COMPANY ECHELONS ABOVE CORPS TOPOGRAPHIC BATTALION

D-25. Normally assigned to a TA, this engineer company (EAC) topographic battalion—

- Provides geospatial engineering support to EAC.
- Compiles controlled, semicontrolled, and uncontrolled image maps and mosaics.

- Revises existing maps and other geospatial data within its capabilities.
- Drafts special maps, overprints, overlays, and other geospatial products.
- Reproduces monochrome and multicolor maps, map substitutes, overlays, overprints, and other topographic products by offset lithography and photocopy.
- Provides terrain intelligence and terrain analysis products.
- Performs topographic surveys and provides survey information to the EAC.
- Interprets and measures remote-sensed imagery.
- Extends horizontal and vertical controls into corps and division areas.
- Stores and distributes special topographic products that the company produces.
- Provides a survey information system and maintains digital point-positioning databases (DPPDBs).

ENGINEER COMPANY (CONSTRUCTION SUPPORT)

D-26. Normally assigned to an engineer battalion (heavy) to provide additional horizontal construction capability at the operational level, the engineer company (construction support)—

- Provides construction support, equipment, and personnel for rock crushing, bituminous mixing, paving and storage facility construction, and airfield operations.
- Provides support to one engineer brigade or group engaged in construction projects that—
  - Require up to 150 to 225 tons per hour (tph) of crushed rock and sand from rock quarries and gravel pits for a two-shift operation.
  - Require up to 75 tph of washed and sized precrushed rock for a two-shift operation.
  - Require up to 150 to 225 tph of bituminous mixes and blends for paving projects for a one-shift operation.
  - Require equipment and operators to support selected construction for a two-shift operation and personnel with the capability of supervising contractual labor and indigenous personnel and assisting in the supervising contractual construction.

ENGINEER COMPANY (DUMP TRUCK)

D-27. Normally assigned to an engineer brigade or group, the engineer company (dump truck) is often linked to an engineer construction company. It—

- Operates dump trucks for moving bulk materials in support of other engineer units.
- Provides a haul capacity of up to 600 tons of bulk material (such as gravel, earth fill, and crushed stone) per cycle.

ENGINEER COMPANY (PIPELINE CONSTRUCTION SUPPORT)

D-28. The engineer company (pipeline construction support) is normally assigned to an engineer brigade, ASCC, or engineer group. It typically augments a combat heavy battalion to give it the capability to perform pipeline operations. The engineer company (pipeline construction support)—

- Provides technical personnel and specialized equipment to assist combat heavy battalions or other construction units in constructing, rehabilitating, and maintaining pipeline systems.
- Provides advisory personnel to support up to three engineer companies engaged in pipeline construction, pipe coupling, storage tank erection, and pump station and dispensing facility construction.
- Provides specialized tools, equipment, and personnel to operate on a two-shift basis.
- Is capable of transporting 21,000 linear feet of 6-inch pipe, 16,200 linear feet of 8-inch pipe, or 9,000 linear feet of 12-inch pipe over unimproved roads in two lifts.

ENGINEER COMPANY (PORT CONSTRUCTION)

D-29. The engineer company (port construction) is normally assigned to an engineer brigade, ASCC, or engineer group. It typically augments a combat heavy battalion with specialized equipment required for port construction and repair operations. The engineer company (port construction)—

- Provides specialized engineer support in developing, rehabilitating, and maintaining port facilities, including LOTS operations.
- Constructs, rehabilitates, and maintains offshore facilities, including mooring systems, jetties, breakwaters, and other structures required to provide safe anchorage for ocean-going vessels.
- Constructs, rehabilitates, and maintains piers, wharves, ramps, and related structures required to load and off-load cargo.
- Constructs facilities for roll-on/roll-off (RO/RO), break-bulk, and containerized cargo handling.
- Maintains tanker discharge facilities, including repairing or replacing existing POL jetties and submarine pipelines.
- Provides limited dredging and removes underwater obstructions.
- Installs the offshore petroleum discharge system (OPDS) in support of Army LOTS operations when no naval units are assigned.
- Provides operators for selected items of equipment for a two-shift operation.
- Depends on the pipeline construction engineer company for the radiographic inspection of pipeline welds, when required, and the lightweight diving team for underwater construction, survey reconnaissance, and recovery.

SECTION III – SEPARATE ENGINEER TEAMS (OPERATIONAL-LEVEL)

ENGINEER TEAM, FIRE FIGHTING HEADQUARTERS

D-30. The engineer team, fire fighting HQ provides C2 and administrative support. One HQ team (AA) can control three to seven fire fighting teams (AB). A team commander serves as the fire marshal of an installation/facility or within the AOR. An AA team—

- Plans for fire defense on an installation.
- Conducts fire-prevention inspections.
- Conducts fire investigations.
- Establishes a fire department communications network between the HQ, the MP, the airfield, and the fire fighting teams.
- Commands the fire fighting teams.
- Maintains and refills fire extinguishers.
- Makes minor repairs to fire hoses.
- Inspects and maintains fixed fire-protection systems on an installation/in an AO.
- Coordinates for the resupply of fire fighting assets, agents, self-contained breathing apparatus (SCBA); air, and fuel.
- Coordinates mutual aid with other services and HN fire-protection assets.

D-31. *FM 5-415*, provides more details on engineer fire fighting operations.

ENGINEER TEAM, FIRE TRUCK (AB)

D-32. The team provides fire protection, administers first aid, provides an initial response to HAZMAT incidents, and implements a fire prevention program. A commander's primary task list determines the team's assignments. An AB team—

- Provides crash and rescue support for medical evacuation (MEDEVAC) and normal flight or maintenance standbys.
- Conducts fire prevention inspections on an installation or airfield.
- Provides C2 of the nonfire fighting assets used to support natural cover fire fighting operations (such as heavy equipment and personnel).
- Conducts fire fighting operations (such as structural, crash/rescue, and natural cover) on an installation and in an AO.
- Provides emergency medical assistance to victims.
- Conducts an initial response to HAZMAT incidents.
- Conducts the training of unit level fire brigades.
- Assists with medical resources during mass casualty incidents.
- Assists in HNS, as a commander requires.

ENGINEER TEAM, QUARRY

D-33. A quarry team is normally assigned or attached to an engineer battalion (heavy) and does the following:

- Performs rock-crushing operations, which increases the capabilities of the construction group in major horizontal construction projects such as roads, storage facilities, and airfields.
- Provides personnel and equipment for a 24-hour period to operate the 75 (68.25 metric tons) or 150-tph (136.5 metric tons) crushing plant and issue its product to users.
- Provides personnel and equipment for drilling and blasting operations that are required to produce raw stone for operating the 75 (68.25 metric tons) or 150-tph (136.5 metric tons) crushing plant.
- Is capable of hauling 30 tons (27.3 metric tons) of rock per trip from the quarry to the processing plant.

ENGINEER TEAM, WELL-DRILLING HEADQUARTERS

D-34. Assigned or attached to an engineer brigade or group, this team can be deployed to support units from divisional to operational level. For more information *see FM 5-484*. The team HQ—

- Provides personnel and equipment for drilling and developing water wells.
- Is capable of drilling and casting two complete water well holes of 5 7/8 inches (14.92 centimeters) in diameter to a depth of 2,500 feet (758.33 meters).
- Installs casings, screens, and pumps and develops the well to provide water at the wellhead.
- Can sustain two-shift operations.

ENGINEER TEAM, WELL-DRILLING

D-35. The engineer team, well-drilling, is assigned or attached to an engineer brigade or group, and can be deployed to support other units. This team—

- Provides personnel and equipment for drilling and developing water wells.
- Is capable of drilling and casting two complete water well holes of 5 7/8 inches (14.92 centimeters) in diameter to a depth of 2,500 feet (758.33 meters).
- Installs casings, screens, and pumps and develops the well to provide water at the wellhead.
- Can sustain two-shift operations.

ENGINEER HEAVY DIVING TEAM

D-36. The engineer heavy diving team is normally assigned to a TA and attached to an ENCOM to support commanders in ports, harbors, and coastal zones. This team—

- Assists in constructing port facilities, LOTS structures, and floating barriers.
- Repairs damaged piers, docks, wharves, seawalls, and breakwaters.
- Clears underwater obstructions and marks navigational waterways.
- Reduces and emplaces underwater obstacles and mines.
- Reduces structures with underwater demolitions.
- Recovers sunken materiel and vessels.
- Installs and maintains vessel moorings.
- Repairs Army lighters and vessels.
- Installs and maintains the underwater portion of off-shore petroleum and water distribution systems.
- Protects land forces, vessels, and underwater structures from underwater threats, reducing the probability of underwater structural damage.

ENGINEER LIGHT DIVING TEAM

D-37. An engineer light diving team is assigned to a port construction company, an ASCC, or a corps HQ. This team—

- Performs scuba, lightweight, or deep-sea surface diving to a maximum depth of 190 feet (57 meters). Diving is done to support light-salvage, harbor clearance, underwater pipeline, fixed bridge, and port construction repair and rehabilitation.
- Performs ship husbandry, LOTS, underwater demolition, cutting, welding, and multiple diving operations.
- Depends on the control and support team for specialized supplies, DS and GS maintenance of life-support systems, and augmentation of personnel and equipment for deep sea and heavy salvage operations.
- Performs scuba diving in forward combat areas to perform river crossing site reconnaissance (near shore and far-shore); assists in preparing landing sites, riverbank, and exit routes for crossing forces; clears underwater obstacles as part of combat operations; and supports assault float bridge operations.
- Repairs damaged bridges, locks, dams, pipelines, canals, and levees.
- Constructs underwater bridge structures, obstacles, and floating barriers.
- Recovers submerged weapons systems.
- Protects land forces, river crossing equipment, and underwater structures from underwater threats and deceives enemy forces of friendly underwater and waterborne intentions.

ENGINEER TEAM, REAL ESTATE

D-38. An engineer team, real estate, is assigned to an ASCC. This team performs functions related to acquiring, using and disposing of real property for military purposes. A real estate team—

- Manages real estate.

- Investigates and processes real estate claims.
- Records, documents, and prepares reports on the real estate in any area that the Army uses, occupies, or holds.
- Coordinates with other agencies or friendly HNs to execute joint real estate functions.

ENGINEER TEAM, UTILITIES

D-39. An engineer team, utilities, is assigned to an ASCC, TAACOM, ASG, or a corps HQ. This unit provides limited facility engineering support in the areas of carpentry, masonry, electrical, plumbing, and road maintenance and repair.

ENGINEER TEAM, TOPOGRAPHIC PLANNING AND CONTROL

D-40. An engineer team, topographical planning and control, is assigned to an HHC, engineer topographic battalion. This team—

- Performs geospatial operational planning for units and agencies engaged in producing and supplying military geographic information and topographic products.
- Determines the requirements and provides programs for and coordination of engineer topographic units assigned or attached to the theater.
- Coordinates with the NIMA, the HN or allied nations' topographic support activities, and the higher HQ to accomplish the mission.

SECTION IV – CORPS ENGINEER UNITS

HEADQUARTERS, ENGINEER BRIGADE, CORPS

D-41. Refer to *Figure D-6, page D-16* for a diagram of the HQ engineer brigade corps. This brigade—

- Provides C2 operations to assigned and attached units and coordinates engineer activities.
- Plans and coordinates the operations of the engineer units that are engaged in CS and constructing and rehabilitating facilities in support of the corps or the airborne corps.
- Plans and supervises the activities relating to river crossing, barrier placement, and counterobstacle and countermine operations.
- Supervises the engineer units that are constructing, rehabilitating, and maintaining trails, structures, air landing facilities, and petroleum storage facilities.
- Supervises indigenous and HN personnel and administers contractual construction and labor.
- Provides an engineer staff element to the corps or the airborne corps HQ.
- Plans and supervises terrain intelligence and topographic operations.

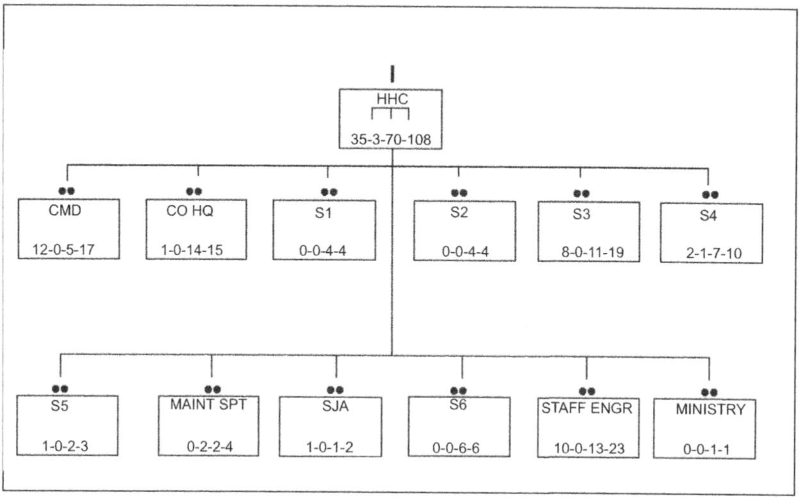

Figure D-6. Headquarters and Headquarters Company, Engineer Brigade, Corps

ENGINEER COMBAT BATTALION, CORPS, MECHANIZED

D-42. Engineer battalion, corps mechanized, is normally assigned to a combat group HQ. This battalion consists of one HHC and three line companies *(see Figure D-7)*. The number of allocated mechanized battalions is based on the number of heavy divisions allocated to the corps. This battalion—

- Increases the combat effectiveness of supported units by accomplishing combat (M/CM/S) and limited general engineering tasks.
- Reinforces heavy divisional engineer units, heavy brigades, and ADC engineer units.
- Reorganizes and fights as infantry, when required.
- Participates in joint military operations, when required.
- Constructs tactical obstacles, defensive positions, and fixed and floating bridges.
- Supports combined arms breaching and river crossing operations.
- Emplaces and maintains assault bridges to span twelve 60-foot gaps, simultaneously.
- Constructs, repairs, and maintains landing strips, heliports, CPs, LOCs, and tactical routes, culverts, fords, and other general engineering horizontal-related tasks.
- Provides technical advice, assistance, and training in mine warfare, field fortification, and camouflage, demolition, and engineer reconnaissance techniques.

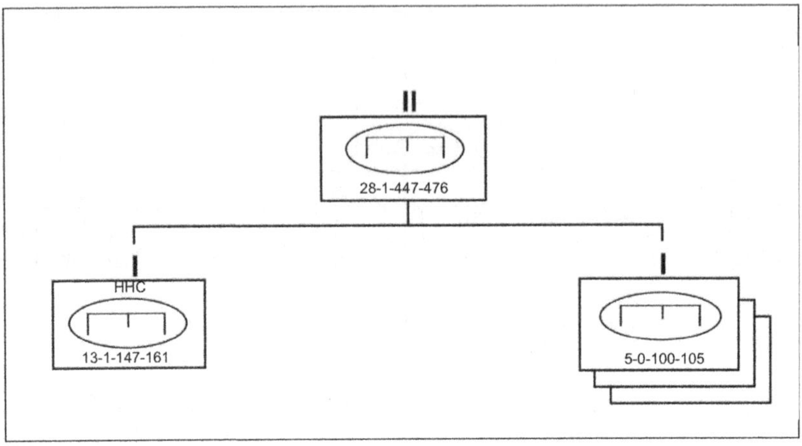

Figure D-7. Engineer Battalion, Corps Mechanized

- Emplaces conventional and SCATMINEs.

ENGINEER COMBAT BATTALION, CORPS, WHEELED

D-43. The engineer combat battalion, corps, wheeled, is normally assigned to a combat group HQ. The number of wheeled battalions allocated is based on the number and type of divisions in a corps. An engineer combat battalion consists of one HHC and three line companies *(see Figure D-8, page D-18).* This battalion—

- Increases the combat effectiveness of the corps by accomplishing combat (M/CM/S) and general engineering tasks.
- Reinforces divisional engineer units, when required.
- Reorganizes and fights as infantry, when required.
- Participates in joint military operations, when required.
- Commands, controls, and supervises assigned and attached units.
- Provides engineer support in constructing obstacles and defensive positions.
- Emplaces conventional and SCATMINEs.
- Employs fixed and floating bridges.
- Performs general engineering tasks (such as constructing, repairing, and maintaining landing strips, heliports, CPs, LOC, tactical routes, culverts, fords, supply installations, buildings, and structures) and other horizontal related construction tasks, as required.
- Supports limited combined arms breaching operations.
- Provides engineer support in river crossing operations.

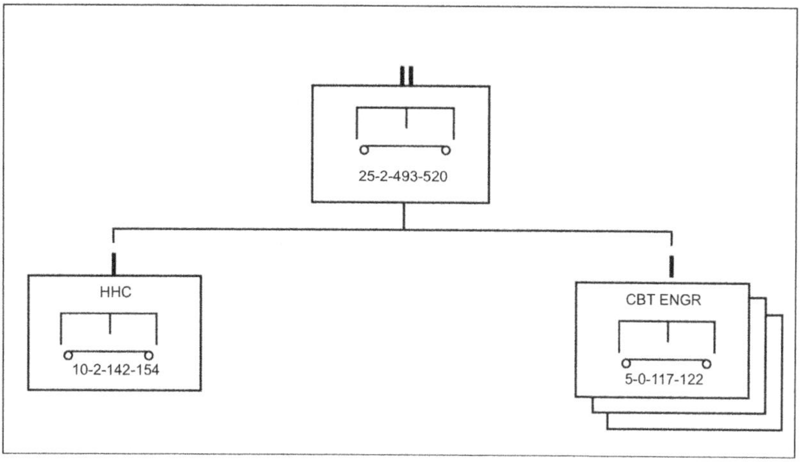

Figure D-8. Engineer Battalion, Corps Wheeled

- Provides technical advice, assistance, and training in mine warfare, field fortification, and camouflage, demolition, and engineer reconnaissance techniques.
- Supports ADC missions.

ENGINEER COMBAT BATTALION, CORPS, AIRBORNE

D-44. The engineer combat battalion, corps airborne, is normally assigned to a combat group HQ supporting an airborne corps. This battalion consists of one HHC and three line companies *(see Figure D-9)*. The battalion—

- Increases the combat effectiveness of supported units by accomplishing combat (M/CM/S) and limited general engineering tasks.
- Augments engineer operations and capabilities with engineer equipment to support light force operations.
- Can parachute or be air-delivered to work sites.
- Reinforces divisional engineer units, when required.
- Reorganizes and fights as infantry, when required.
- Participates in joint military operations, when required.
- Provides C2 operations and supervises assigned and attached units.
- Provides engineer support in constructing obstacles and defensive positions.
- Employs fixed and floating bridges.
- Provides engineer support in river crossing operations.
- Performs general engineering tasks (such as constructing, repairing, and maintaining landing strips, heliports, CPs, LOCs, tactical routes,

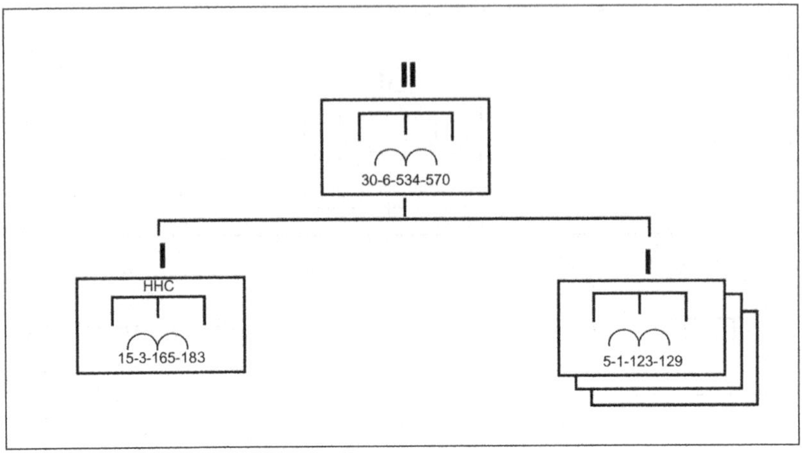

Figure D-9. Engineer Combat Battalion, Corps, Airborne

culverts, fords, supply installations, buildings, and structures) and other horizontal related construction tasks, as required.
- Conducts limited breaching operations and assists in assaulting fortified positions.
- Provides technical advice, assistance and training in mine warfare, field fortification, and camouflage, demolition, and engineer reconnaissance techniques.

ENGINEER COMBAT BATTALION, CORPS, LIGHT

D-45. The engineer combat battalion, corps, light, is normally assigned to a combat group HQ. This battalion consists of one HHC and three line companies and normally supports a light division *(see Figure D-10, page D-20)*. This battalion—

- Increases the combat effectiveness of supported units by accomplishing combat (M/CM/S) and limited general engineering tasks.
- Augments engineer operations and capabilities with engineer equipment to support light force operations.
- Reinforces divisional engineer units, when required.
- Reorganizes and fights as infantry, when required.
- Participates in joint military operations, when required.
- Commands, controls, and supervises assigned and attached units.
- Provides engineer support when constructing obstacles and defensive positions.
- Provides engineer support in river crossing operations.

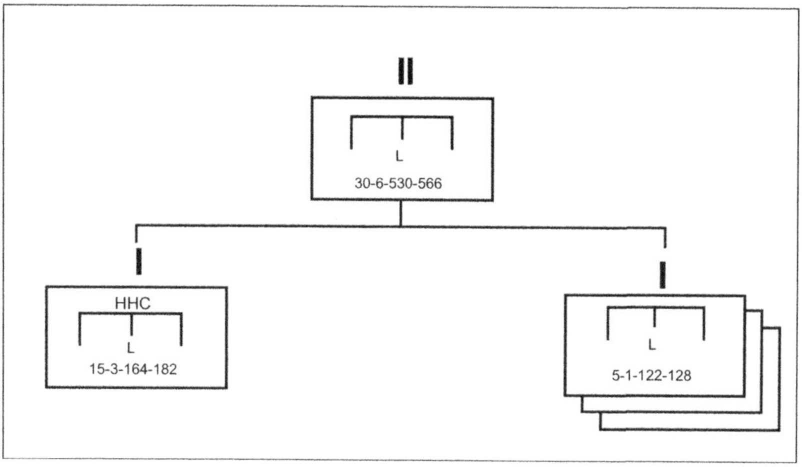

Figure D-10. Engineer Combat Battalion, Corps, Light

- Performs general engineering tasks (such as constructing, repairing, and maintaining landing strips, heliports, CPs, LOCs, tactical routes, culverts, fords, supply installations, buildings, and structures) and other horizontal related construction tasks, as required.
- Conducts limited breaching operations and assists in assaulting fortified positions.
- Provides technical advice, assistance, and training in mine warfare, field fortification, and camouflage, demolition, and engineer reconnaissance techniques.

ENGINEER COMBAT BATTALION, HEAVY

D-46. Engineer combat battalion, heavy is assigned to a corps, based on one battalion per division, and at the operational level, based on the number of battalions per the workload driven requirements. This battalion consists of one headquarters, headquarters and headquarters company, and three line companies *(see Figure D-11)*. This battalion—

- Increases the combat effectiveness of the division, corps, and TAs forces by accomplishing combat (M/CM/S) and limited general engineering tasks.
- Constructs, rehabilitates, repairs, maintains, and modifies landing strips, airfields, CPs, MSRs, supply installations, building structures, bridges, and other related structures as required, normally to the rear of the division.
- Repairs, and on a limited basis, reconstructs railroads and sewage and water facilities.

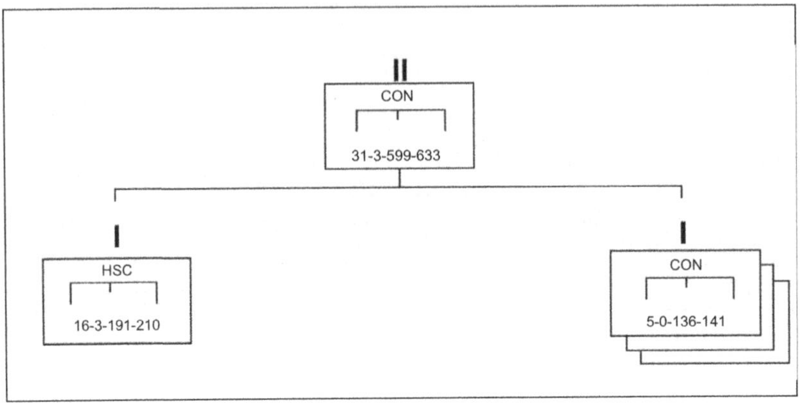

Figure D-11. Engineer Combat Battalion, Heavy

- Supervises contractual construction, skilled construction labor, and unskilled indigenous personnel.
- Constructs protective obstacles to degrade enemy mobility in rear areas.
- Clears obstacles as part of an area clearance operation, not as part of an assault-breaching operation.
- Provides bituminous paving operations and quarrying and crushing operations, rehabilitates ports, constructs petroleum pipelines and storage facilities, distributes power, and restores and constructs major airfields (when specialized personnel and equipment are attached).
- Performs rear area security operations, when required.
- Provides field engineering assistance and support to divisional engineers preparing protective positions.
- Performs rear area operations, including infantry combat missions, within the limitations of organic weapons and equipment.
- Conducts area damage clearance and restoration operations.

SECTION V – SEPARATE CORPS COMPANIES

ENGINEER COMPANY, COMBAT SUPPORT EQUIPMENT

D-47. This engineer company is normally assigned to a corps and attached to an engineer brigade or group *(see Figure D-12, page D-22)*. The number of companies allocated is based on the number of combat wheeled and mechanized battalions assigned to a corps AO. This company—

- Supports engineer combat operations within corps and division areas by conducting general engineering tasks.

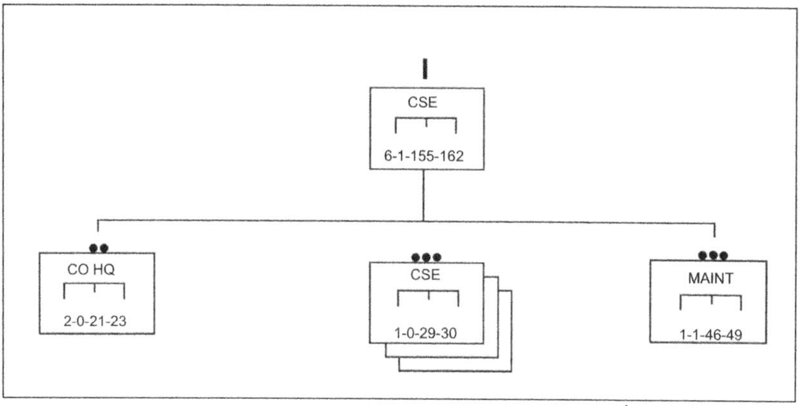

Figure D-12. Combat Support Equipment Company

- Performs survivability and countermobility tasks, general engineering along MSRs and combat trails in other corps close operation areas, and countermobility operations in a corps area.
- Provides manned engineer construction equipment to construct, rehabilitate, repair, maintain, and modify landing strips, airfields, base camps, CPs, MSRs, and LOCs.

ENGINEER COMPANY, LIGHT EQUIPMENT, AIRBORNE

D-48. This engineer company, light equipment, airborne, is normally assigned to a corps and is usually aligned with the engineer combat battalion, corps, air assault/light. This company—

- Augments engineer operations with the capabilities to support airborne operations with engineer equipment that is downsized and rapidly deployable. Can parachute or be air-delivered to work sites.
- Provides earthmoving equipment to support survivability and general engineering missions.
- Provides a cross-country dump truck capability of about 50 cubic yards or 45 tons (40.95 metric tons) per lift.
- Works with airborne battalions in early deployment with forces to establish forward logistics bases until heavier corps and theater engineer assets arrive.

ENGINEER COMPANY, LIGHT EQUIPMENT, AIR ASSAULT

D-49. This engineer company, light equipment, air assault, is normally assigned to a corps and is usually aligned with the engineer combat battalion, corps, air assault and light. This company—

- Augments engineer operations with the capabilities to support air assault and light operations with engineer equipment that is downsized and rapidly deployable.
- Provides earthmoving equipment to support survivability and general engineering missions.
- Provides a cross-country dump truck capability of about 50 cubic yards or 45 tons (40.95 metric tons) per lift.
- Works with air assault, light battalions in early deployment with force protection forces to establish forward logistics bases until heavier corps and theater engineer assets arrive.

ENGINEER COMPANY, MULTIROLE BRIDGE

D-50. The engineer MRBC, is normally task-organized to a corps or a divisional engineer battalion or a combat engineer group to support bridging operations *(see Figure D-13).* This company—.

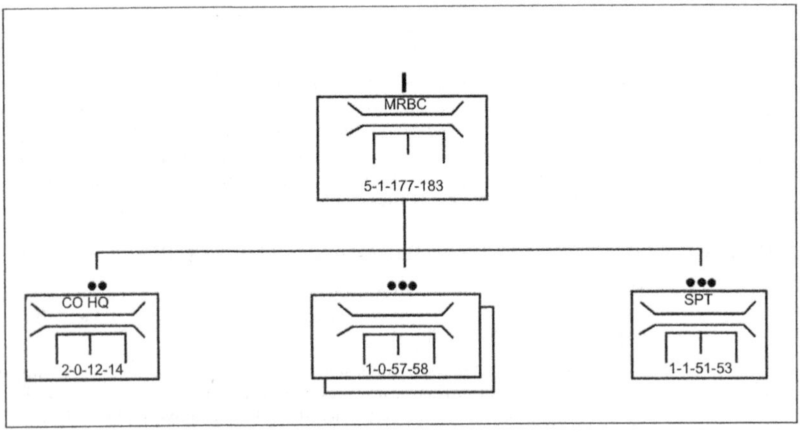

Figure D-13. Engineer Company, Multirole Bridge

- Provides personnel and equipment to transport, assemble, disassemble, retrieve, and maintain all standard US Army bridging systems:
  - Performs fixed bridging with the medium girder bridge (MGB). Maintains four MGB sets with sufficient components for the assembly of bridges with various spans and classes. Under normal conditions, the sets provide four 31.1 meters (102 feet) of Class 16 bridges (reinforceable to Class 70 with reinforcement kits) or two 49.4 meters (162 feet) of Class 60 bridge (reinforceable to Class 70 with reinforcement kits).

  - Performs float bridging using the ribbon bridge. Maintains about 213 meter (705 feet) of bridge with load classes up to Class 140 depending on current speed, or maintains four rafts with load classes up to class 140 depending on current speed.
  - Performs float bridging using the ribbon bridge. Maintains about 213 meters (705 feet) of Class 75 (tank) and Class 96 (wheel) float bridge or 6 rafts of Class 75 (tank) and Class 96 (wheel) bridge, based on a 0 to 1 meter per second (0 to 3 feet per second (fps)) water velocity.
- Provides transportation for palletized load system (PLS) configured cargo.
- Provides one panel bridge set (Bailey bridge) with sufficient components and a cable reinforcement set for erecting bridges of various spans and load classes. The set includes two 24.4-meter (80 feet) double-truss, single-story (Class 50 wheeled/Class 60 tracked) bridges without a cable reinforcement set and one 58.5-meter (180 feet) triple-truss, single-story (Class 50 wheeled/Class 60 tracked) bridge with a cable reinforcement set. Additional bridge components for spans over 100 feet are available.
- Provides technical supervision to assist other engineer units in bridge construction.
- Constructs bridges using organic personnel (in emergencies).

D-51. *FM 90-13* discusses the use of the Bailey bridge

ENGINEER COMPANY (CORPS), TOPOGRAPHIC

D-52. This engineer company (corps), topographic is normally assigned to a TA engineer battalion (topographic). This company—

- Provides geospatial engineer support to the corps.
- Compiles controlled and uncontrolled photomaps and mosaics.
- Revises existing maps and other topographic data within its capabilities.
- Drafts special maps, overprints, overlays, and other topographic products.
- Reproduces monochrome and multicolor maps, map substitutes, overlays, overprints, and other topographic products by offset lithography and photocopy.
- Provides terrain intelligence and analysis products.
- Performs topographic surveys and provides survey information.
- Interprets and measures remotely sensed imagery.
- Extends horizontal and vertical controls into corps and division areas.
- Stores and distributes special topographic products.
- Provides a survey information system and maintains the DPPDB.

SECTION VI – DIVISION ENGINEER UNITS

HEADQUARTERS, HEADQUARTERS DETACHMENT, ENGINEER BRIGADE, HEAVY DIVISION

D-53. Headquarters, headquarters detachment, engineer brigade, heavy division—

- Provides staff planning for C2 operations and supervision of engineer units that are assigned, attached, or supporting units engaged in combat (M/CM/S), geospatial, and general engineering tasks.
- Advises the division commander and staff on engineer operations and their impact on division operations.
- Plans and coordinates engineer operations for the units constructing tactical obstacles, defensive positions, and fixed and floating bridges; breaching or clearing obstacles; and conducting river crossing operations.

ENGINEER COMBAT BATTALION, HEAVY DIVISION

D-54. Three engineer combat battalions are organic to an engineer brigade, heavy division. Each battalion consists of one HHC and three line companies *(see Figure D-14)*. The engineer combat battalion, heavy division, increases the combat effectiveness of a division by accomplishing combat (M/CM/S) and limited general engineering tasks; it may fight as infantry, when required. Special capabilities of this battalion include, but are not limited to—

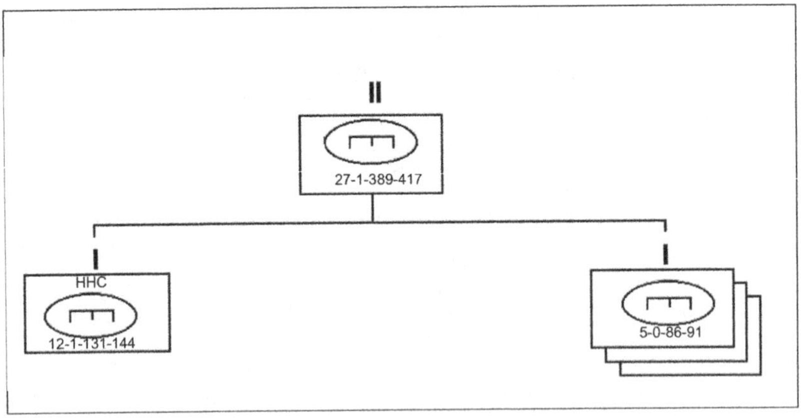

Figure D-14. Engineer Combat Battalion, Heavy Division

- Emplacing and maintaining assault bridges to span twelve 20 meter (60-foot) gaps simultaneously.

- Supporting combined arms breaching operations.
- Constructing tactical obstacles and defensive positions.
- Performing expedient repair of essential combat trails, bridges, fords, and roads in the brigade area.
- Performing engineer technical reconnaissance in support of maneuver units and other elements of the division.

ENGINEER COMBAT BATTALION, FORCE XXI DIVISION

D-55. There are three engineer combat battalions per FXXI division—one per maneuver brigade. Each battalion consists of one HHC and three line companies. The engineer combat battalion, FXXI division, increases the combat effectiveness of a division by accomplishing combat (M/CM/S) and limited general engineering tasks; it may fight as infantry, when required. Special capabilities of this battalion include, but are not limited to—

- Emplacing and maintaining assault bridges to span twelve 20 meter (60-foot) gaps simultaneously.
- Supporting combined arms breaching operations.
- Constructing tactical obstacles and defensive positions.
- Performing expedient repair of essential combat trails, bridges, fords, and roads in the brigade area.
- Performing engineer technical reconnaissance in support of maneuver units and other elements of the division.

ENGINEER COMBAT BATTALION, ENHANCED HEAVY SEPARATE BRIGADE

D-56. One engineer combat battalion is organic to a maneuver-enhanced (armor/mechanized), heavy separate brigade. The engineer combat battalion consists of one HHC and three line companies *(see Figure D-15)*. This battalion increases the combat effectiveness of an enhanced heavy separate brigade by accomplishing combat (M/CM/S) and limited general engineering tasks; it may fight as infantry, when required. Special capabilities of this battalion include, but are not limited to—

- Emplacing and maintaining assault bridges to span twelve 20 meter (60-foot) gaps simultaneously.
- Supporting combined arms breaching operations.
- Constructing tactical obstacles and defensive positions.
- Performing expedient repair of essential combat trails, bridges, fords, and roads in the brigade area.

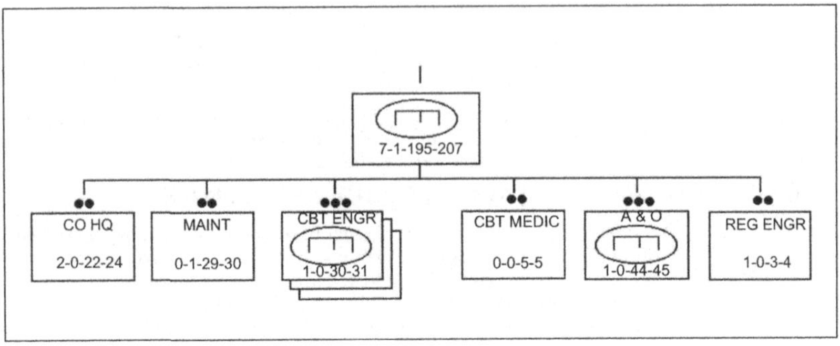

Figure D-15. Engineer Combat Battalion, Enhanced Separate Heavy Brigade

- Performing engineer technical reconnaissance in support of maneuver units and other elements of the brigade.

SECTION VII – LIGHT DIVISION ENGINEER UNITS

ENGINEER BATTALION, AIRBORNE DIVISION

D-57. Organic to an airborne division, an engineer battalion, airborne division, consists of one HHC and three line companies *(see Figure D-16, page D-28* It increases the combat effectiveness of an airborne division by accomplishing combat (M/CM/S) and limited general engineering tasks; it may fight as infantry, when required. This battalion—

- Prepares and maintains expedient combat routes in the forward battle area, to include ingressing and egressing to block positions and river crossing sites and conducting expedient repair of essential bridges, fords, and culverts.
- Assists in assaulting fortified positions.
- Constructs tactical obstacles.
- Supports combined arms breaching operations.
- Performs engineer technical reconnaissance in support of maneuver units and other elements of the division.

D-58. Organic to an airborne division, an engineer battalion, airborne division, consists of one HHC and three line companies *(see Figure D-16).* It increases the combat effectiveness of an airborne division by accomplishing combat (M/CM/S) and limited general engineering tasks; it may fight as infantry, when required. This battalion—

- Prepares and maintains expedient combat routes in the forward battle area, to include ingressing and egressing to block positions and river crossing sites and conducting expedient repair of essential bridges, fords, and culverts.
- Assists in assaulting fortified positions.

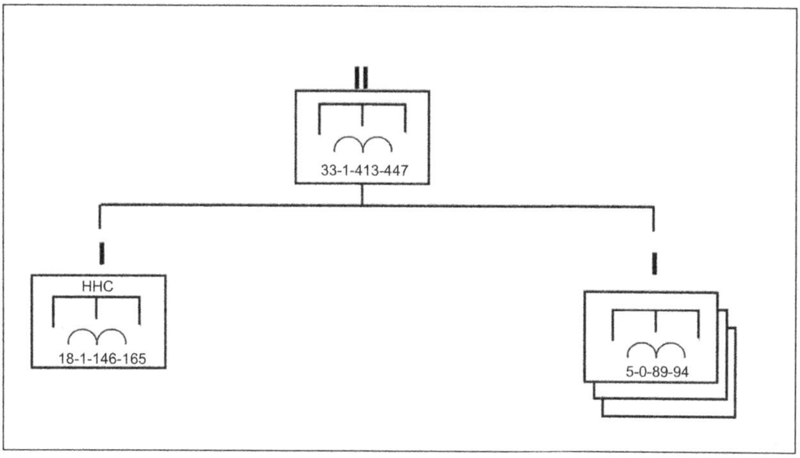

Figure D-16. Engineer Battalion, Airborne Division

- Constructs tactical obstacles.
- Supports combined arms breaching operations.
- Performs engineer technical reconnaissance in support of maneuver units and other elements of the division.

ENGINEER BATTALION, AIR ASSAULT DIVISION

D-59. Organic to an air assault division, this battalion consists of one HHC and three line companies *(see Figure D-17)*. It increases the combat effectiveness of an air assault division by accomplishing combat (M/CM/S) and limited general engineering tasks; it may fight as infantry when required. This battalion—

- Prepares and maintains expedient combat routes in the forward battle area, to include ingressing and egressing to block positions and river crossing sites and conducting expedient repair of essential bridges, fords and culverts.
- Assists in assaulting fortified positions.
- Constructs tactical obstacles.
- Supports combined arms breaching operations.
- Performs engineer technical reconnaissance in support of maneuver units and other elements of the division.

ENGINEER BATTALION, LIGHT INFANTRY DIVISION

D-60. Organic to a light infantry division, this battalion consists of one HHC and three line companies *(see Figure D-18)*. It increases the combat effectiveness of a light infantry division by accomplishing combat (M/CM/S)

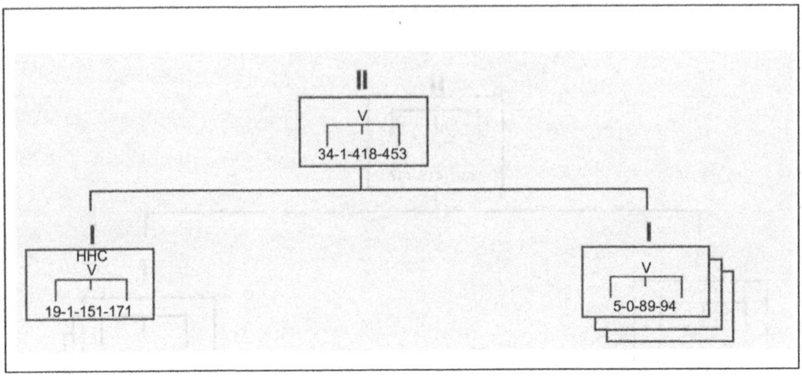

Figure D-17. Engineer Battalion, Air Assault Division

and limited general engineering tasks; it may fight as infantry, when required. This battalion—

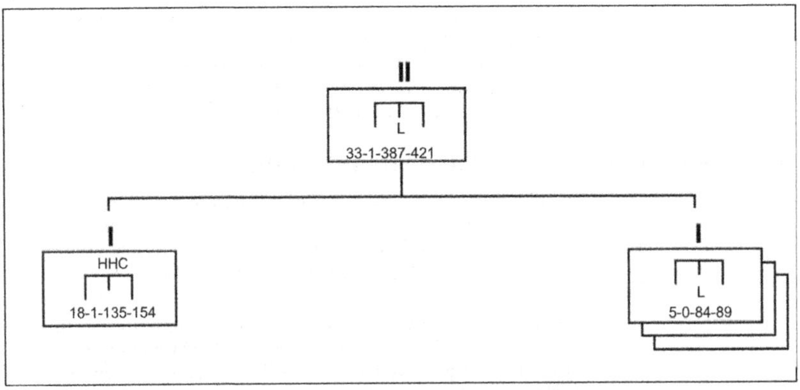

Figure D-18. Engineer Battalion, Light Infantry Division

- Prepares and maintains expedient combat routes in the forward battle area, to include ingressing and egressing to block positions and river crossing sites and conducting expedient repair of essential bridges, fords, and culverts.
- Assists in assaulting fortified positions.
- Assists maneuver forces in the assault breach of obstacles and minefields.
- Constructs tactical obstacles.
- Supports combined arms breaching operations.

- Performs engineer technical reconnaissance in support of maneuver nits and other elements of the division.

SECTION VIII – SEPARATE COMPANIES

ENGINEER COMPANY, ARMORED CAVALRY REGIMENT

D-61. Organic to an ACR, this company *(see Figure D-19)* increases the combat effectiveness of an ACR by accomplishing combat (M/CM/S) and limited general engineering tasks; it may fight as infantry, when required. This engineer company—

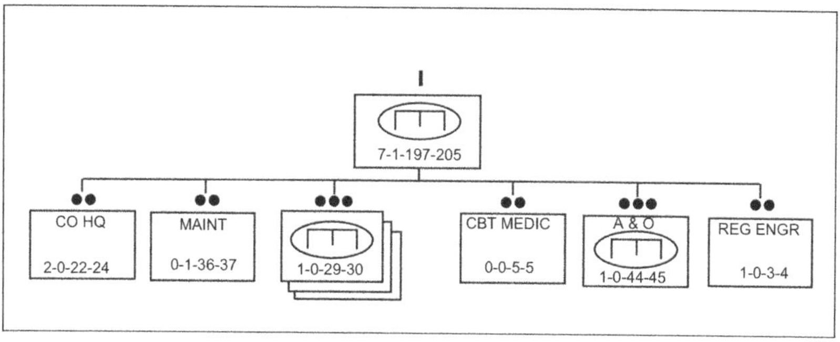

Figure D-19. Engineer Company (Armored-Cavalry Regiment)

- Advises the maneuver commander on engineer unit capabilities and the impact on the maneuver plan.
- Prepares and maintains essential combat trails in forward areas.
- Conducts expedient repair of essential bridges, fords, and roads.
- Provides, emplaces, and maintains assault bridges to span six 20 meter (60-foot) gaps simultaneously.
- Constructs tactical obstacles, defensive positions, and fixed and floating bridges.
- Supports combined arms breaching operations and assault river crossing operations.
- Constructs, repairs, and maintains CPs, combat trails, site damage, chemical decontamination sites, and logistics field sites.
- Performs engineer technical reconnaissance in support of maneuver units and other elements of the regiment.

ENGINEER COMPANY, LIGHT ARMORED CAVALRY REGIMENT

D-62. Organic to a light ACR, this company increases the combat effectiveness of a light ACR by accomplishing combat (M/CM/S) and limited general engineering tasks; it may fight as infantry, when required. This company—

- Provides limited combat (M/CM/S) engineering capability to support one cavalry squadron.
- Advises the maneuver commander on engineer unit capabilities and the impact on the maneuver plan.
- Prepares and maintains combat trails in forward areas; repairs essential bridges, fords, and roads expediently; conducts breaching operations; and constructs tactical obstacles and defensive positions.
- Performs engineer technical reconnaissance in support of maneuver units and other elements of the regiment.

ENGINEER COMPANY, HEAVY SEPARATE BRIGADE

D-63. Organic to a heavy separate brigade, this company increases the combat effectiveness of a heavy separate brigade by accomplishing combat (M/CM/S) and limited general engineering tasks; it may fight as infantry, when required. This company—

- Advises the maneuver commander on engineer unit capabilities and the impact on the maneuver plan.
- Prepares and maintains essential combat trails in forward areas.
- Conducts expedient repair of essential bridges, fords, and roads.
- Provides, emplaces, and maintains assault bridges to span six 20 meter (60-foot) gaps simultaneously.
- Constructs tactical obstacles, defensive positions and fixed and floating bridges.
- Supports combined arms breaching operations and assault river crossing operations.
- Constructs, repairs, and maintains CPs, combat trails, damaged sites, chemical decontamination sites, and logistics field sites.
- Performs engineer technical reconnaissance in support of maneuver units and other elements of the brigade.

ENGINEER COMPANY, STRYKER BRIGADE COMBAT TEAM

D-64. Organic to a SBCT, this company *(see Figure D-20, page D-32)* increases the combat effectiveness of an SBCT by accomplishing combat (M/CM/S) and limited general engineering tasks; it may fight as infantry, when required. This company—

- Advises the brigade commander, the staff, and other maneuver commanders on allocating engineer resources and capabilities available to assist in accomplishing the unit's missions.
- Prepares and maintains essential combat routes in the brigade area, to include ingressing and egressing to block positions and river crossing sites and conducting expedient repair of essential bridges, fords, and culverts.
- Assists in assaulting fortified positions.

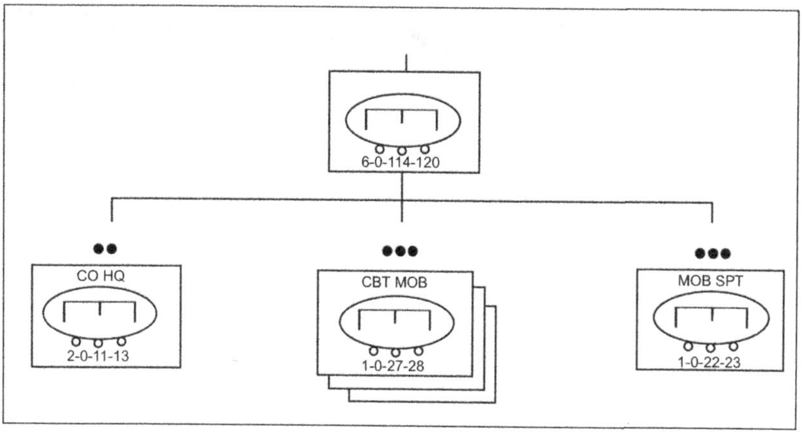

Figure D-20. Engineer Company (Stryker Brigade Combat Team)

- Supports combined arms breaching operations.
- Performs engineer technical reconnaissance in support of maneuver units and other elements of the brigade.

ENGINEER COMPANY, SEPARATE INFANTRY BRIGADE

D-65. Organic to a separate infantry brigade, this company increases the combat effectiveness of a separate infantry brigade by accomplishing combat (M/CM/S) and general engineering tasks; it may fight as infantry, when required. This company—

- Advises the brigade commander, the staff, and other maneuver commanders on allocating engineer resources and capabilities available to assist in accomplishing the unit missions.
- Prepares and maintains essential combat routes in the brigade area to include ingressing and egressing to block positions and river crossing sites and conducting expedient repair of essential bridges, fords, and culverts.
- Assists in assaulting fortified positions.
- Supports combined arms breaching operations.
- Performs engineer technical reconnaissance in support of maneuver units and other elements of the brigade

SECTION IX – SEPARATE TEAMS

TOPOGRAPHIC TERRAIN DIRECT SUPPORT TEAM, HEAVY DIVISION

D-66. Organic to a heavy division, the topographic terrain DS team is normally collocated with the G3/G2 planning and operations staff. It provides staff advice and assistance to a supported division and control of terrain analysis teams (heavy) in DS of geospatial missions.

TOPOGRAPHIC TERRAIN ANALYSIS TEAM, HEAVY DIVISION

D-67. Organic to a heavy division, the topographic terrain analysis team—

- Produces terrain intelligence for a heavy division.
- Provides qualified personnel who collect, evaluate, and disseminate terrain data.
- Analyzes the effects of terrain on military operations.
- Advises the supported commander on all terrain-related matters.

TOPOGRAPHIC DETACHMENT, FORCE XXI DIVISION

D-68. Organic to a FXXI division, the topographic detachment—

- Produces terrain intelligence for a FXXI division.
- Provides qualified personnel who collect, evaluate, and disseminate terrain data.
- Analyzes the effects of terrain on military operations.
- Advises the supported commander on all terrain-related matters.

TOPOGRAPHIC TERRAIN ANALYSIS TEAM, LIGHT DIVISION

D-69. Organic to a light division, topographic terrain analysis team—

- Produces terrain intelligence for a light division.
- Provides qualified personnel who collect, evaluate, and disseminate terrain data.
- Analyzes the effects of terrain on military operations.
- Advises the supported commander on all terrain-related matters.

FACILITY ENGINEER TEAM

D-70. The FE team is a TOE (SRC 05530L100) and a reserve TDA unit with specific installation responsibilities. TOE teams are generally assigned to an ENCOM and/or a TSC or ASG. In times of national emergencies and contingency operations, it may be mobilized and attached to an ASCC, TAACOM, ASG, corps HQ or a staff element of a TF HQ. When three or more FE teams are operating in a theater, an engineer support group is usually formed. It provides an additional depth of FE capability and has C2 over its assigned teams. The engineer support group has the capability to perform the same missions as the FE teams. The group—

- Manages engineer resources in support of facility and civil engineering; performs master planning and allocates resources, including inspecting facilities and identifying, prioritizing, and conducting work; conducts planning boards; and develops facility and civil engineering projects.
- Manages real property and controls real estate engineer teams and coordinates their activities.
- Manages housing and space use.
- Manages the theater environmental compliance and prevention programs, to include environmental compliance assessments, recommendations for corrective actions, and proper reporting.
- Performs limited design capability, including preparing drawings, specifications, and cost estimates in support of facility and civil engineering projects.
- Reviews the designs of contract architects and engineers to ensure that they conform to the user requirements, missions, and codes.
- Manages utilities services, and maintenance and repair efforts and can control engineer utility teams and coordinate their activities.
- Inspects and ensures that the quality standards of construction projects by contract or troop labor are met.
- Manages BASOPS, including sanitation and landfill operations and can control and coordinate the activities of fire fighting and utility teams.
- Performs limited supervision of troop labor and indigenous personnel.
- Manages facility engineering supplies through assigned unit.

Appendix E

Navy Engineer Organizations and Capabilities

Naval engineering units are composed of the subordinate elements of the naval construction force (NCF), better known as Seabees. They are deployable naval engineering units with the primary mission of providing responsive contingency construction support for US military forces in the operational area. The common thread that is woven throughout all SEABEE units is responsiveness and flexibility.

NAVY CAPABILITIES

NAVY ENGINEER UNITS

E-1. The NCF performs generalized and specialized construction missions in support of the joint force. Naval mobile construction battalions (NMCBs) possess robust vertical construction, bridging, and heavy earthmoving capabilities. NMCBs construct roads for supply routes, extend or construct airfields, construct ammunition supply points, install expeditionary airfields, and provide all types of force bed-down and logistic facilities. NCF units are necessary to reinforce and augment the Marine Corps's limited general engineering capability. The normal employment of the NCF is as a major subordinate element within the Marine air-ground task force (MAGTF) to maximize engineering capabilities available to the MAGTF commander.

SPECIALIZED CAPABILITIES

E-2. Navy engineer units have specialized capabilities for performing engineering work at the water and shore interface in support of amphibious operations or other beach operations.

NAVAL CONSTRUCTION FORCE

E-3. NCF SEABEE units are organized under two naval construction brigades that include both AC and RC units. They are under the administrative command (ADCOM) of the naval component commander of their home port geographic combatant commander. When deployed during peacetime, their OPCON chain normally falls under the theater naval component commander. During contingencies when NCF SEABEE units deploy to support Marine forces, their OPCON chain falls under the theater Marine forces component commander.

E-4. Amphibious construction battalions (ACBs) are organized under two naval beach groups (NBGs) that fall under the Atlantic/Pacific surface forces. The NBGs are OPCON and ADCON to the Atlantic and Pacific fleets. Like the NCF SEABEEs, OPCON is normally chopped to the theater combatant commander and delegated to the naval component command in which the units are deployed.

E-5. The NCF is organized, trained, and equipped to construct, maintain, operate, and repair advanced bases and their associated logistic pipelines. The force also provides disaster control and relief assistance and performs civic action projects to complement military or other national programs. NCF units are organized for quick and effective response, as required by planning assumptions and guidance. When directed, SEABEE assets are transferred to the COCOM (command authority) of the combatant commander.

E-6. In executing assigned projects, NCF units maintain a significant self-defense capability for their construction sites and can be employed as part of a perimeter defense force. All SEABEE units are equipped with small arms. The larger units (NMCBs) have organic, indirect-fire weapons systems (mortars). Their weapons are identical to those in the United States Marine Corps (USMC) inventory. A USMC advisor is resident to the NMCB staff, and NMCB personnel receive semiannual training on military skills and tactics.

E-7. SEABEE resources are continually forward deployed to provide a quick response to any location where a contingency may occur. The forward-deployed presence of SEABEE resources minimizes lift requirements and ensures prompt engineering support to the commanders. The following SEABEE units make up the NCF:

- Naval construction brigade (NCB).
- Naval construction regiment (NCR).
- NMCB.
- Construction battalion maintenance unit (CBMU).
- Construction battalion unit (CBU).
- Underwater construction team (UCT).
- Naval construction force support unit (NCFSU).
- ACB.

NAVAL CONSTRUCTION BRIGADE

E-8. The NCBs provide forces to fulfill the operational requirements of a combatant commander. There are two active NCBs. The NCB commander normally exercises C2 over two or more construction regiments. The NCBs are deployable but rely on subordinate units for tactical and logistic support. When deployed, NCB personnel can perform limited work in an NBC environment, possessing individual personal protective gear.

NAVAL CONSTRUCTION REGIMENT

E-9. The NCRs, both AC and RC, act as C2 elements for subordinate NCF units and report to the NCBs. There are two line NCRs and two home port training and logistical support NCRs. The regiment primarily functions as a planning, training, and oversight group and is subordinate to the command exercising OPCON of NCF units. The home port NCR performs administrative, training, project-tasking, and logistical support functions for

NCF units. The logistical support provided by the home port NCR includes planning for and movement of personnel.

NAVAL MOBILE CONSTRUCTION BATTALION

E-10. The NMCBs' primary mission is the expeditionary construction of advanced base facilities in support of the Navy, USMC, other services, and allied and coalition partners engaged in military operations. NMCBs construct base and troop bed down facilities; install, repackage, and redeploy Bailey bridges and MGBs; and conduct defensive combat operations, as required by the operational commander. They may also conduct rapid runaway repair (RRR) and BDR for airfields. Additional functions include the repair, the maintenance, and the construction of shore facilities and LOC during contingency, emergency, or disaster recovery operations.

E-11. NMCBs are operationally self-sustainable for up to 60 days, requiring only replenishment of consumables (Classes I, II, and V limited to 5, 3, and 15 days, respectively). They are capable of limited operations in an NBC environment, possessing individual personal protective gear, limited chemical and radiological detection equipment, and decontamination apparatus capable of decontaminating facilities and equipment organic to the unit. Each battalion is organized into one HQ and four line companies. Other configurations of the battalion's personnel and equipment, such as those pre-positioned on maritime pre-positioning force ships (MPSs), have been tailored to meet various missions. An NMCB can task-organize and deploy away from its main body a number of detachments, details, and teams, depending on the operational tasking and theater requirements. Units may be reconfigured into modular packages. An NMCB can organize and deploy the following standard organic detachments and teams:

AIR DETACHMENT

E-12. The air detachment is a readily deployable (within 48 hours), self-sustainable, company-sized detachment, completely air-transportable via a C-130 lift, with a minimum of 89 personnel and 300 short tons of equipment. The unit is capable of sustained operations in a contingency or wartime environment for 30 days without resupply (except for consumables). The air detachment's primary mission is to repair war damage and construct urgent projects required by OPLANs and/or OPORDs during the early stages of a contingency operation.

REINFORCED AIR DETACHMENT

E-13. The reinforced air detachment includes two additional platoons and is capable of providing increased construction and engineer support tailored to the operational scenario.

OPERATIONS DETACHMENT

E-14. The operations detachment is a reinforced, company-sized detail deployed from the main body of the NMCB, to perform light to medium construction and engineer support.

REINFORCED OPERATIONS DETACHMENT

E-15. The reinforced operations detachment is expanded to two companies in strength and is capable of providing increased construction and engineering support tailored to the operational scenario.

LIMITED OPERATIONS DETACHMENT

E-16. The limited-operations detachment is smaller, typically platoon-sized, deployed from the NMCB, and able to perform light construction and engineering support.

ENGAGEMENT TEAM

E-17. The engagement team is a squad-sized detachment deployed from the NMCB for task-specific light construction projects that are accomplished within 30 to 120 days.

TRAINING TEAM

E-18. The training team is a fire team-sized detachment deployed from the NMCB to provide training and oversight of task-specific construction and engineering projects performed by others.

MAIN BODY

E-19. The main body is the core of the deployed NMCB. The size of the main body ranges from 812 personnel at full wartime end strength with no detachments to as few as 250 personnel with all other NMCB personnel deployed with detachments.

MARITIME PREPOSITIONING

E-20. Three NMCB tables of allowances (TOAs) (P25M) are configured for maritime pre-positioning within the three existing strategically deployed maritime pre-positioning force squadrons (MPSPRON), in support of the Marines. Each TOA P25M is configured and spread-loaded on three ships within a squadron in support modules that can be deployed in various sizes. The modules are configured as: core, basic, and heavy. The core module contains enough tools and equipment to sustain 250 SEABEEs ashore in support of a Marine expeditionary unit (MEU). The basic module contains additional vertical capability and can be added to the core module, should the mission require it. The heavy module contains additional horizontal capability and can be added to any of the other core modules, with or without a basic module as the situational tasking dictates. All of the modules together, plus the fly in echelon, which contains NMBC personnel, weapons and communications equipment not on the MPSRONs, constitute a full NMCB. The MPS survey, liaison, and reconnaissance party (SLRP); the MPF advance party; and the MPF offload preparation party (OPP) are not organic elements of the NMCB; they are temporary, subordinate elements of the MPF. Each element includes a commander, an MPF, naval coastal warfare, Navy support element (NSE), Navy cargo handling and port group, an MPF MAGTF, and NCF and fleet hospital personnel. The SLRP, advance party and OPP only exist temporarily to facilitate establishing the MPF MAGTF ashore.

- The MPF SLRP is a detachment of personnel that deploys as the lead element in support of an MPF operation. The size of the SLRP is situation-dependent and typically consists of three to eight NCF personnel.
- The MPF advance party is the advance element of an NMCB which assists with offload operations and receives NMCB equipment and supplies in support of the MAGTF in the MPF.
- The MPF OPP deploys to the MPF squadron four days before offload operations to prepare the equipment for debarkation.

CONSTRUCTION BATTALION MAINTENANCE UNIT

E-21. The CBMU provides follow-on public works (PW) operations to maintain and repair existing advanced base shore facilities or facilities constructed by NMCBs during contingency operations. As a secondary mission, a CBMU can backfill a PW organization at existing bases during emergency operations or due to increased operational requirements. The unit is capable of equipping, manning, and maintaining water production, as well as steam and electrical power generation and distribution systems for advanced base facilities of up to 5,000 personnel. They perform war damage repairs to central base camps, power, sewage, POL, and water systems. A CBMU also operates and maintains automotive and construction equipment, including materials handling equipment (MHE). Possessing personal protective gear, limited chemical detection equipment, and decontamination apparatus capable of limited decontamination of equipment organic to the unit, the unit is capable of limited operations in an NBC environment. CBMUs are operationally self-sustainable for up to 60 days, requiring only replenishment of consumables (Classes I, II, and V limited to 5, 3, and 15 days, respectively).

CONSTRUCTION BATTALION UNIT

E-22. CBUs support Navy fleet hospitals. Normally, two CBUs deploy to provide the engineering, constructing, maintaining and repairing services necessary to support a fleet hospital. Following the erection of fleet hospital facilities, the CBU provides operation, maintenance and repairs and augments security for the facility. The unit is capable of deploying within 48 hours and erecting a CZ fleet hospital within 10 days. The unit can perform limited operations in an NBC environment when attached to the fleet hospital, which provides CBU personnel with individual protective gear and limited chemical-detection equipment. CBUs have a secondary mission to conduct disaster recovery.

UNDERWATER CONSTRUCTION TEAM

E-23. UCTs are specially trained and equipped units that provide underwater engineering, construction, repair, and inspection capabilities to meet the Navy and Marine Corps or joint force operational requirements.

E-24. UCTs perform complex inshore and deep ocean underwater construction tasks in any climate, including the arctic. They provide ocean bottom surveys for the appropriate site selection of underwater facilities. They are capable of diving and working to a depth of 60 meters (190 feet) and rely on scuba gear and/or surface-supplied air. Typical projects include measurement conversions, underwater repair of wharves, piers, pipelines, moorings, boat ramps, and underwater cable systems. The unit can also support offshore petroleum discharge equipment by installing single-anchor leg moorings. An UCT possesses individual protective gear and limited chemical detection capabilities and can perform limited operations in an NBC environment. UCTs have a secondary mission to conduct disaster recovery, humanitarian assistance, and civil-action operations.

E-25. A UCT is divided into three identical squad-sized air detachments and a platoon-sized sea echelon. The unit can deploy as one unit or separately. Each air detachment carries a surface decompression chamber as part of its TOA. The sea echelon accompanies larger unit equipment for sustained operations. The shore component of a UCT is used for equipment and follow-on logistics coordination.

NAVAL CONSTRUCTION FORCE SUPPORT UNIT

E-26. NCFSUs provide logistics-oriented construction and engineering support for the NCR as well as other NCF units, including specialized civilian engineer support equipment (CESE), materials, repair parts, and technical expertise. NCFSU echelons are attached to other units to manage, maintain, and inventory material, transportation, and construction equipment to augment the NMCB TOA. NCFSUs have secondary missions to provide limited construction and defensive capabilities during a contingency. NCFSUs are operationally self-sustainable for up to 60 days, requiring only replenishment of consumables (Classes I, II, and V limited to 5, 3, and 15 days, respectively). The unit provides support in the following areas:

- The operation, maintenance, and repair of local- and long-haul transportation equipment.
- The operation and support of quarry and rock-crushing operations, asphalt and concrete production and placement, and soil analysis and stabilization equipment.
- The production and storage of potable water using nine reverse osmosis purification units 600 gallons (2500 liters per hour) and nine 3,000 gallon (12000 liter) water bladders.
- The construction material management, including requisitioning, expediting, receiving, controlling, storing, issuing, and delivering material.
- The advanced facility planning, designing, estimating, and engineer support (as required) to execute construction projects.
- The maintenance, custody, inventory control, and issuing of special SEABEE support equipment.

AMPHIBIOUS CONSTRUCTION BATTALION

E-27. There are two ACBs in the NCF. The ACBs provide over-the-shore logistics movement and construction support to amphibious forces during the initial assault and early phases of an operation. ACBs are part of the NSE and report to the NBG, which is responsible for in-stream off loading of maritime ships in support of amphibious military operations. They are not placed OPCON or ADCON to an NCB. The primary tasks of the ACBs are to—

- Provide, assemble, and operate causeway barge ferries.
- Provide side-loadable warping tugs and powered causeway sections for installing piers, elevated causeways systems (ELCAS), ship-to-shore bulk fuel, and water systems and the OPDS.
- Provide, assemble, maintain, and operate the ELCAS and amphibious assault bulk fuel and bulk water systems.
- Install the OPDS, with support from two UCT air detachments.
- Establish and operate a support camp and provide limited construction support for the NBG.

E-28. The unit can operate in an NBC environment and has limited detection and decontamination capability.

ARMY-NAVY ENGINEER CONSIDERATIONS

E-29. During force projection operations, the initial US naval engineering capability available in theater is most likely NMCB air detachments and MAGTF amphibious forces. NMCBs are also deployed quickly to force projection theater locations to construct necessary naval facilities. The Army engineer staff must consider the following when coordinating joint engineer plans and operations with the Navy:

- Requesting the latest engineer intelligence data from deployed or deploying NMCB air detachment elements to assist in identifying force projection theater Army engineer requirements and enemy engineer capabilities. (These requirements include soils data, the availability of construction materials, and HN construction support.)
- Establishing engineer staff links between the Navy forces (NAVFOR) and ARFOR engineer staff sections through the JTF or theater engineer staff and HQ.
- Providing necessary Army engineer LNO support.
- Developing the joint task-organization relationships that enhance NCR capabilities, following the deployment of Army engineer units.
- Assessing the need for NMCB support following the arrival of Army combat and construction units in theater.
- Determining if NMCB units need augmentation from Army combat and construction units.
- Developing procedures for Army engineer units to be able to acquire Class IV construction materials from NCFSUs.

E-30. For additional information on NCF units, *refer to Operational Naval Instruction (OPNAVINST) 5450.46K*. For additional information on ACB units, *refer to OPNAVINST 3501.93D*.

Appendix F

Air Force Engineer Organizations and Capabilities

Air Force mobility engineering forces, organized as Prime base engineer emergency force (BEEF) or RED HORSE units, provide support ranging from expeditionary civil to general engineering services across the range of military operations. Both units have limited combat engineering capabilities focused on the defense of deployed forces and base denial. Capabilities include airfield construction, maintenance, repair, and sustainment operations, as well as other specialized functions, such as EOD, fire, and limited NBC detection and response capability. Air Force engineering units can deploy either as a part of an air expeditionary force (AEF) or as detached units operating in support of specific missions and operational taskings.

AIR FORCE CAPABILITIES

AIR FORCE ENGINEER UNITS

F-1. Prime BEEF and RED HORSE units are deployed in unit type code (UTC) sets. Depending on the mission and engineering requirements, UTCs can be added or subtracted as needed. Air Force engineering units can deploy either as part of an AEF or as detached units operating in support of specific missions and operational taskings. Prime BEEF lead UTC teams to deploy with the necessary command, control, and communications (C3) to conduct independent engineer; fire, crash, and rescue, NBC defense; and EOD operations to establish, sustain, defend and recover a bed-down location. Follow teams combine with lead teams to satisfy support at larger locations. Prime BEEF teams deploy with their individual protective clothing, hand tools, and selective team equipment; however, most construction equipment and supplies and all vehicles (except EOD) must be provided at the deployment location. Prime BEEF forces can conduct construction; maintenance; repair; fire protection, NBC, including disaster preparedness planning, training, detection, and decontamination within service limitations; force protection; and EOD operations. RED HORSE squadrons and their associated UTC configurations provide highly mobile, largely self-sufficient, rapidly deployable echelons to support major force bed-down requirements and repair heavy war damage. RED HORSE units are stand-alone squadrons, not tied to peacetime base support. They provide a dedicated, flexible airfield and base heavy construction and repair capability that allows JFCs to move and support missions as requirements dictate. RED HORSE squadrons are organized and deployed for austere, independent operation to execute heavy horizontal and vertical construction projects. The projects include site development; construction and repair of runways, taxiways and ramps, roads, and revetments; heavy earthwork; and construction and repair of facilities and infrastructure. General capabilities shared by Prime BEEF and RED HORSE include site surveys, bare base construction using mobility assets (Harvest Falcon and Harvest Eagle), concrete and asphalt paving, and utility system installation (water, waste and electrical). RED HORSE expands on this capability by providing automatic building machine (K-span) support, large-scale concrete and paving operations, heavy horizontal and vertical construction, water well drilling, water purification, and quarry operations.

SPECIALIZED CAPABILITIES

F-2. Air Force engineer units have specialized capabilities to support all aspects of airfield operations, including RRR, the installation of aircraft arresting systems, war damage repair, and force bed down. By deploying civil engineering teams, Air Force engineer units also perform other engineer functions, such as fire protection (structural, aircraft crash fire and rescue, and HAZMAT response), EOD (munitions support, antiterrorism, and civil assistance), and NBC support (including disaster preparedness planning, training, detection, and decontamination within service limitations). Specialized support includes installing and operating emergency airfield lighting systems, mobile aircraft arresting systems, reverse osmosis water purification, munitions storage site construction, and HAZMAT response.

CIVIL ENGINEERING TECHNICAL AND CONTRACT SUPPORT

F-3. A variety of other supporting organizations separate from Prime BEEF and RED HORSE units provide technical engineering, environmental, and contract support for the joint force. The Air Force Civil Engineering Support Agency (AFCESA) provides technical support and training and administers the Air Force Contract Augmentation Program (AFCAP). Additionally, AFCESA provides specialized civil engineer maintenance, inspection, and repair teams to assess pavement and runway conditions, as well as teams to conduct field maintenance and repair of specialized equipment. The Air Force Center for Environmental Excellence also provides a wide range of environmental program support and contract services.

PRIME BASE ENGINEER EMERGENCY FORCE

F-4. Prime BEEF is the primary organizational structure for supporting mobility and in-place contingency requirements. The principal objective of deploying Prime BEEF teams is to bed down and support an AEF. Force bed down is generally divided into three categories: aircraft, personnel, and infrastructure support. Aircraft support provides the maintenance shops, hangars, squadron operations centers, munitions storage, fuel storage, and other facilities directly supporting the flying mission. Personnel support provides the housing, dining facilities, latrines, showers, administrative offices, and other indirect-support facilities. Infrastructure support provides the utility systems, solid and hazardous waste disposal, roads, and communications that serve the bed-down site. Bed-down locations range from main operating bases with adequate existing facilities to bare bases (BBs) with no facilities other than runways, taxiways, and aircraft parking aprons with a source of water.

F-5. Prime BEEF tasks executed during the deployment include the following:

- Airfield support operations—pavements, lighting, and mobile aircraft arresting systems.
- Fire protection—aircraft crash, fire, and/or rescue; structural fire; and HAZMAT response operations.
- Fuel systems set up and support.
- EOD operations and planning.
- NBC training, reconnaissance, and operations.
- Force protection and base defense.
- RRR.
- Expedient facility repair and rapid utility repair.

F-6. Prime BEEF UTCs include—

- 4F9EA—Prime BEEF AEF team A.
- 4F9EB—Prime BEEF AEF team B.
- 4F9EP—Prime BEEF AEF team C.
- 4F9SA—staff augmentation team.
- 4F9FL—engineer sustainment team.
- 4F9AP—power production team.
- 4F9DA—NBC full-spectrum threat response heavy team.
- 4F9DB—NBC full-spectrum threat response light team.
- 4F9DC—NBC full-spectrum threat response augmentation team.
- 4F9DD—NBC full-spectrum threat response theater and/or joint task force planning and support staff.
- 4F9DE—NBC full-spectrum threat response contamination control team.
- 4F9DF—lightweight NBC reconnaissance system.
- 4F9DG—NBC collective protection system.
- 4F9DH—open air contamination control area set.
- 4F9DJ—full-spectrum threat response personnel augmentation team.
- 4F9DK—lightweight NBC reconnaissance team.
- 4F9DL—NBC collective protective equipment set.
- 4F9FJ—fire protection incident command team.
- 4F9FN—fire protection management augmentation.
- 4F9FP—fire protection operations team.
- 4F9XA—EOD C2.
- 4F9XB—EOD en route.
- 4F9X1—EOD lead team.
- 4F9X2—EOD follow team.
- 4F9X3—EOD base support team.
- 4F9X4—EOD augmentation.
- Special teams include—
  - 4F9AC—civil engineering maintenance, inspection, and repair team—aircraft arresting systems, power systems, heating,

ventilation, and air conditioning (HVAC) systems and bare base systems.

- 4F9AC—pavement evaluation team.
- Prime BEEF self-sustainability package.
- Snow removal team.
- En route airlift support team.
- EOD contingency team.
- Disaster preparedness high-threat teams.
- Armored base recovery vehicle team.
- Armored munitions clearance vehicle team.

RAPID ENGINEERS DEPLOYABLE HEAVY OPERATIONS REPAIR SQUADRON, ENGINEER

F-7. RED HORSE squadrons accomplish major construction (both vertical and horizontal) in forward locations, often in advance of the main deploying force. With their organic capabilities and stand-alone operations, RED HORSE squadrons are essential assets for early entry operations and joint reception, staging, onward movement, and integration (JRSOI) preparation. In addition to providing heavy horizontal (earthmoving and pavements) and vertical (facility and utility skills including POL, structural, electrical, mechanical, and power generation) capabilities, RED HORSE units also have the following special capabilities:

- Concrete and asphalt batch plant and paving operations.
- Mobile aircraft arresting system and navigational aid (NAVAID) installation.
- Field dispensary and air transportable clinic operations.
- Automated building machine (K-span) operations.
- Material testing.
- Quarry and rock-crushing operations.
- Revetment erection and facility hardening.
- Water well drilling and sanitation system construction.
- Road, airfield, and ramp construction.
- Expedient site and airfield repair.
- Base denial operations.

F-8. RED HORSE mobility UTCs deploy with self-sustainment capabilities not found in Prime BEEF units, including organic logistics, services, medical, transportation and financial management, and contracting capabilities. Team equipment can be airlifted, shipped via sealift, drawn from prepositioned stockpiles, or obtained from HN sources (contracting). Each RED HORSE squadron has the resources to conduct site development and heavy repair work at four different locations. The mobility echelons include a—

- 4F9R-1 advanced team, which is deployable in 16 hours (12 hours with prepositioned assets). Priority tasks include advanced airfield surveys, site layout, and planning for the establishment and future development of an operational location during contingencies. The team is self-sustaining at the deployed location, when water and fuel is available, for five days.

- 4F9R-2 team, which forms the hub of the construction capability. It is largely self-sufficient at the deployed location for 30 days. This team includes the C2 function of the entire squadron and contains personnel, skills, vehicles, and equipment to execute heavy horizontal and vertical construction. It can deploy its full mission capability (MISCAP) in 96 hours and deploy personnel in only 24 hours.
- 4F9R-3 UTC, which contains the specialized skills necessary to complete site development, construct and repair runways, taxiways and ramps, and complete other horizontal construction. This UTC provides heavy horizontal and light vertical construction capabilities in austere environments. It can deploy its full MISCAP in six days and deploy personnel only in twenty-four hours.
- 4F9R-4 UTC, which contains the specialized skills necessary to construct and repair facilities and infrastructure. It has limited capability to do earthwork, road, and airfield operations. This UTC provides the heavy vertical and light horizontal construction capabilities in austere environments. It can deploy its full MISCAP in eight days and deploy personnel in only 24 hours.
- 4F9H1, heavy equipment package, designed to supplement 4F9R-2 and 4F9R-3 teams and provide full heavy earthwork operations. The package should come by sealift, but many items could come by C-5 or C-17 aircraft if necessary. This package includes T-9 dozers, belly scrapers, 4-cubic yard front-end loaders, size 7 graders, and a sheepsfoot roller.
- 4F9H2, specialized building construction, including K-span, a large crane, forms for footings and stem walls, and concrete placement tools.
- 4F9H3, well drilling, including a well drilling rig, tow vehicle, trencher, casing, and drill bits.
- 4F9H4, asphalt batch plant, including asphalt batch plant and conveyor belts.
- 4F9H5, concrete batch plant, including a concrete batch plant, and a concrete transit, mixer, and conveyor.
- 4F9H6, quarry operations, including rock drills, a crusher, and rock dumps.

PRIME READINESS IN BASE SUPPORT

F-9. Prime Readiness in Base Support (RIBS) units are worldwide combat morale, welfare, recreation, and services (MWRS) forces, organized and trained for wartime support. The Prime RIBS program organizes forces that can deploy with 22 to 28 hours notice to support global or major regional conflict operations on military operating bases (MOBs), collocated operating bases (COBs), forward operating locations (FOLs), APODs, aerial ports of embarkation (APOEs), and BBs or to support essential MWRS missions at critical CONUS bases. Prime RIBS units can—

- Provide initial food service, billeting, recreation programs, and mortuary operations support for up to 1,200 people.

- Support an independent or dependent combat aviation squadron of 16 to 24 fighter aircraft or a significant aviation deployment less than squadron size in a major deterrent force posture.
- Support, when augmented, organizational field laundry operations, personnel fitness programs, and tactical field exchange resale operations.

ARMY AND AIR FORCE ENGINEER CONSIDERATIONS

F-10. During force projection operations, the initial available Air Force engineering capabilities in theater will probably be RED HORSE elements that establish APODs. Prime BEEF and RIBS units are also deployed quickly to force projection theater locations to operate at major air bases. The Army staff engineer should consider the following when coordinating joint engineer plans and operations with the Air Force:

- Requesting the latest engineer intelligence data from deployed or deploying RED HORSE elements to assist in identifying force projection theater Army engineer requirements and capabilities. These requirements include soils data and the availability of construction materials and HN construction support.
- Establishing engineer staff links between the Air Force forces (AFFOR) and ARFOR engineer staff sections through the JTF or theater engineer staff and HQ.
- Providing the necessary Army engineer LNO support.
- Developing the joint task organization relationships that enhance RED HORSE and Prime BEEF capabilities, following the deployment of Army engineer units.
- Assessing the need for RED HORSE airfield maintenance and repair support following the arrival of Army construction units in theater.
- Determining if Prime BEEF units need augmentation from Army construction units, especially in the area of RRR.

F-11. For additional information on Air Force engineers and their doctrine and associated TTPs, refer to Air Force Doctrine Document (AFDD) 2-4, and AFDD 2-4.4.

Appendix G

Marine Corps Engineer Organizations and Capabilities

The Marine Corps is organized into three Marine expeditionary forces (MEFs). Each MEF contains a command, ground combat, aviation combat, and CSS element. Each element, except the command element, contains organic engineer support. The MEF can form smaller task-organized MAGTFs to conduct expeditionary operations in littoral environments or other places as directed. A MAGTF ranges in size from a full MEF to a Marine expeditionary brigade (MEB), down to a MEU or smaller unit. MAGTFs normally contain task-organized elements from each of the primary engineer units.

MARINE CORPS CAPABILITIES

EXPEDITIONARY FORCE

G-1. The engineer force is organized to accomplish specific tasks of limited duration. Marine Corps civil engineering is temporary in nature due to its organization and mission, which focus on providing expeditionary engineering support to the MAGTF. Civil (general) engineering capabilities include but are not limited to construction of expeditionary airfields, FOBs, LZs, roads for supply routes, and small-scale construction operations. The Navy provides civil engineering support to the Marine Corps through NCF units under the OPCON of a MAGTF commander. NCF units are necessary to reinforce and augment the Marine Corps's civil engineering capability.

SPECIALIZED CAPABILITIES

G-2. The specialized capabilities of Marine engineers include erecting standard bridging and rafting, constructing nonstandard bridging, providing mobile electric power, producing and storing potable water supplies, and providing EOD. Marine engineers have the capability to construct and maintain expeditionary airfields up to 600 meters (2,000 feet) in length with arresting gear and perform RRR. Marine engineers also maintain and operate bulk fuel systems capable of storing up to 600,000 gallons. Marine engineers are integrated with their combat elements under the MAGTF concept of operations.

G-3. Geospatial intelligence is not a combat engineer function in the USMC. A topographic platoon resides in the production and analysis (P&A) company of the intelligence battalion within the MEF. It provides tailored terrain, littoral and infrastructure studies, and factor overlays. It is a MEF-level asset. It disseminates products such as graphic decision aids to support IPB and COA development. These topographic platoons use the same ArcView software as Army topographic units.

COMMAND STRUCTURES

COMBAT ENGINEER BATTALION

G-4. The Combat Engineer Battalion (CEB) *(Figure G-1)* enhances the M/CM/S of the Marine division and provides close combat engineer support and limited general engineer support to the Marine division. It enhances the momentum and tempo of maneuver units and helps shape the battlespace for the Marine division. The CEB contains a headquarters and service (H&S) company, engineer support company, and four combat engineer companies. An engineer support company contains heavy engineer equipment (bulldozers, cranes, and forklifts), motor transport equipment (including dump trucks), and utilities equipment (shower units, mobile electric power generators, water purification equipment).

G-5. Of note, the CEB does not have AVLBs. Their AVLBs are organic to their tank battalions. Also the MK-1545 mine clearing line charges ([MICLICs], triple shot) are organic to the assault amphibian vehicle (AAV) battalion. Normally, a combat engineer company, reinforced with detachments of equipment and personnel from the support company and H&S company, provides the close combat engineer support to an infantry regiment. A combat engineer platoon, reinforced with detachments of equipment and personnel from a support company and an H&S company, provides close combat engineer support to an infantry battalion. A combat engineer squad provides close combat engineer support to an infantry company. While these are typical relationships, the actual level of combat engineer support depends on the situation and the mission analysis. The capabilities of a CEB include field fortifications, planning, and installation.

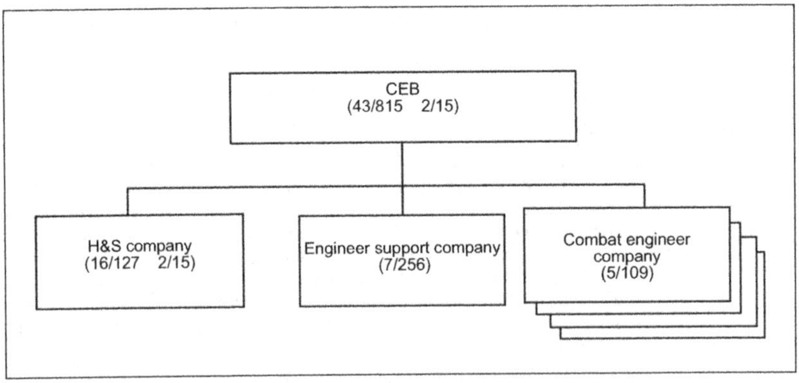

Figure G-1. Combat Engineer Battalion

G-6. The CEB provides the following capabilities and support:

- Conducts engineer reconnaissance to support the division's intelligence collection plan.

- Plans, organizes, and conducts breaching operations.
- Provides barrier planning and installation.
- Conducts demolition operations beyond the ability of other division units.
- Employs assault bridging and other standard bridging systems when augmented with equipment and trained personnel.
- Provides expedient short-span bridges from local material.
- Constructs and maintains limited combat roads and trails in support of division operations (construction and maintenance requirements are limited to those that can be performed with organic equipment and personnel).
- Provide mission-essential, temporary vertical and horizontal construction.
- Provides provisional infantry.

G-7. The battalion requires motor transport support to move the battalion as a unit. However, there are adequate ground transportation assets (medical and logistic support vehicles, including heavy trucks) organic to the battalion to accomplish its primary mission. Most of the battalion's engineer assets require surface transportation; however, some engineer equipment is helicopter transportable.

G-8. CSS capabilities provide:

- Organic supply support.
- Organizational (1st and 2d echelon) maintenance on organic equipment. Intermediate (3d and 4th echelon) maintenance is provided by the maintenance battalion, force service support group (FSSG).
- Organic transportation support required to accomplish its mission.
- Limited general engineering capability.
- Routine and limited emergency medical support to the battalion.
- Administrative, postal, and chaplain support to the battalion.

ENGINEER SUPPORT BATTALION

G-9. An engineer support battalion (ESB) *(Figure G-2, page G-4)* increases the combat effectiveness of the MAGTF by accomplishing general engineering missions of a more deliberate nature. An ESB is organic to the FSSG but supports the entire MAGTF. An ESB contains an H&S company, engineer support company, three combat engineer companies, a bridge company, and a bulk fuel company. An ESB contains EOD capabilities in the H&S company. The engineer support company contains medium bulldozers, road graders, scrapers, compactors, light and medium cranes, light and medium forklifts, mobile electric power generators, water purification units, welders, and limited organic motor transportation—including dump trucks. The ESB typically performs general engineering tasks such as pioneer road construction, expeditionary airfield construction, cantonment or camp planning and construction, and limited vertical and horizontal construction of a temporary nature. It is also capable of EOD, obstacle reduction, and limited power support.

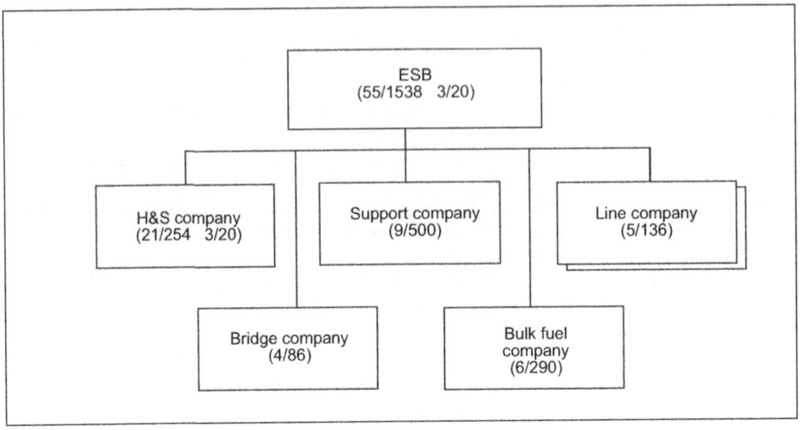

Figure G-2. ESB

G-10. Adequate ground transportation to move essential command and operational elements is organic to the battalion. External motor transport support is required to move all assets (Class III and III [A] bulk fuel), including bulk water supplies, bridging, and heavy engineer equipment.

G-11. CSS capabilities provide—

- Organic supply support.
- Organizational (1st and 2d echelon) maintenance on organic equipment. Intermediate (3d and 4th echelon) maintenance is provided by the maintenance battalion, FSSG.
- Organic transportation support required to accomplish its mission.
- Limited general engineering support.
- Routine and limited emergency medical support to the battalion.
- Administrative, postal, and chaplain support to the battalion.

MARINE WING SUPPORT SQUADRON

G-12. A Marine wing support squadron (MWSS) *(Figure G-3)* provides all essential aviation ground support (AGS) requirements to the Marine aircraft group. MWSSs provide all essential AGS and engineer support requirements to aid components of a Marine Corps FOB. One MWSS supports one Marine aircraft group or the aviation combat element (ACE) of a MEB. A task-organized detachment from the MWSS provides AGS to the ACE of a MEU. The engineers of an MWSS are located in the engineer operations division. MWSS capabilities include providing expeditionary airfield construction; RRR; mobile electric power; water purification and storage; refrigeration; light and medium MHE; light and medium bulldozing; welding, drafting, and survey; and limited vertical construction. The MWSS has a bulk fuel section, under the control of airfield operations, responsible for storing and dispensing aircraft fuel. Limited EOD capabilities provide support for the disposal of UXO, airfield damage assessment, and aircraft emergency landings. Engineer tasks and functions performed by the MWSS are primarily general engineering oriented.

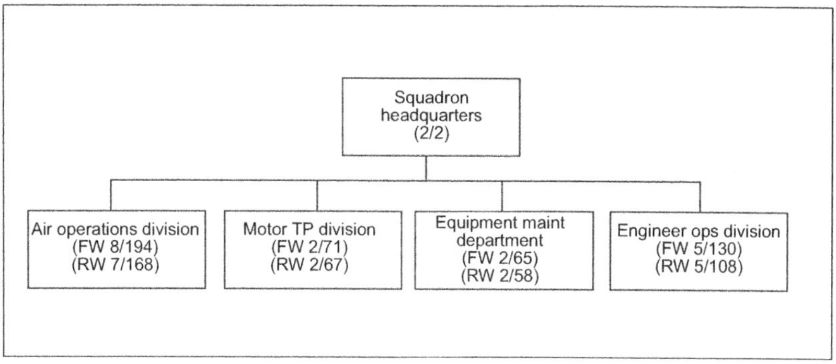

Figure G-3. MWSS

G-13. The engineer operations division of the MWSS organizes to provide limited combat and general engineering support to designated components of the ACE. Engineers located in the airfield operations division receive, store, and dispense aviation and ground fuels from various types of expeditionary fuel systems. The engineer operations division is capable of—

- Providing engineer reconnaissance and survey.
- Repairing, improving, and maintaining existing road networks for the ACE.
- Constructing and maintaining expedient roads.
- Constructing, maintaining, and improving expeditionary airfields, vertical or short takeoff locations, and landing sites.
- Constructing and maintaining mission-essential base camp requirements (temporary bunkers, aircraft revetments, and strongbacks).
- Providing technical and equipment assistance for erecting preengineered buildings.
- Providing utilities support (mobile electric power, water, potable water production, bath and laundry facilities, and refrigeration services).
- Developing, improving, and maintaining drainage systems.
- Providing technical assistance to support camouflage requirements.
- Assessing bomb damage and providing RRR.
- Providing materials and equipment handling services.
- Providing for EOD.

G-14. Organizational vehicles of the squadron provide sufficient transportation for command, control, and routine support activities. The squadron requires external transportation support to displace as a unit.

G-15. CSS capabilities of the squadron provide—

- Organic ground supply support (except for Navy-funded equipment that requires support from a designated aviation supply element).
- Organizational (1st and 2d echelon) maintenance on organic equipment; intermediate (3d and 4th echelon) maintenance by the maintenance battalion, and FSSG.
- Sufficient motor transport equipment to accomplish its mission.
- Limited general engineering capability, as required to support the squadron's mission.
- Service company, H&S battalion, and FSSG service support.

NAVAL CONSTRUCTION FORCE

G-16. The NCF may be assigned as a major subordinate element within the MAGTF to sustain MAGTF operations and maximize naval civil engineering capabilities. Elements of the NCF construct and maintain base facilities; repair battle-damaged facilities; conduct limited defensive operations, as needed; and accomplish disaster control and recovery efforts when required. NCF units contain highly skilled specialists capable of projects of a more permanent nature than those normally accomplished by Marine Corps engineers. *See Appendix E* for a more in-depth discussion of these forces.

MARINE AIR-GROUND TASK FORCE ENGINEERS

G-17. Engineers are normally provided to each corresponding MAGTF as follows:

- **MEF.** A CEB, an ESB, a Marine wing support group, and an NCR.
- **MEB.** A CEB, an ESB, an MWSS, and an NMCB.
- **MEU.** A reinforced platoon from a CEB, an engineer detachment from a MWSS (if a task-organized unit), a detachment from an ESB, and civil engineer support from an NMCB, if required.

G-18. While these are typical support relationships, the actual level of combat engineer support depends on the situation and mission analysis.

ARMY AND MARINE CORPS ENGINEER CONSIDERATIONS

G-19. Marine division CEBs, as part of MAGTF operations, are usually the initial USMC engineering capabilities available in theater during force projection operations. ESBs are also deployed quickly to force projection operations to construct necessary Marine facilities. The Army engineer staff should consider the following when coordinating joint engineer plans and operations with the Marine Corps:

- Requesting the latest engineer intelligence data from deployed or deploying Marine CEB and ESB elements to assist in identifying force projection theater Army engineer requirements and enemy engineer capabilities. These requirements include threat mine and obstacle data, soils data, the availability of construction materials, and HN construction support.

- Establishing engineer staff links between the MAGTF, MARFOR, and ARFOR engineer staff and HQ.
- Providing necessary Army engineer LNO support.
- Developing the joint task organization relationships that enhance Marine engineer capabilities following the deployment of Army engineer units.
- Assessing the need for CEB and ESB support following the arrival of Army combat and construction units in theater.
- Determining if ESB units need augmentation from Army combat and construction units.
- Developing procedures for Army engineer units to be able to acquire Class IV construction materials from ESBs.

G-20. For additional information on Marine Corps engineer units and their doctrine and supporting TTPs, *refer to Marine Corps Warfare Publication (MCWP) 3-17* and *FMs 13-7 and 13-8.*

Appendix H

Multinational, Interagency, Non-Governmental Organizations, and Host Nation Considerations

Military engineers may need to coordinate their activities with multinational, US government, NGO, UN, and HN agencies according to the operational mandate or military objective. In all cases, authority must exist for direct coordination. Military engineers must establish interagency relationships through negotiation. The specific agency will vary, depending on who has federal or state proponency for the situation (for example, disaster relief versus a fire fighting mission). Agreements should be written as memorandums of understanding or terms of reference to ensure understanding and avoid confusion. Most agreements are made at the unified command or JTF level and normally place serious legal restrictions on using military personnel and equipment. These agencies and organizations may have unique engineer capabilities that could be used as part of the overall operational effort. However, these agencies and organizations often request extensive engineer support of their activities and programs. It is critical that an effective engineer liaison is established with the force HQ CMOC to coordinate and execute any engineer support exchange with these agencies.

UNITS AND ORGANIZATIONS

SECTION I - MULTINATIONAL CONSIDERATIONS

H-1. When military operations are considered, the US seeks to develop coalitions rather than conduct unilateral operations. The operation in the Dominican Republic in 1965, for example, was under the aegis of the Organization of American States (OAS). More recently, Operation Provide Comfort (Northern Iraq) and Operation Joint Endeavor (Bosnia) included a coalition of forces from other nations, as well as the US, under the sponsorship of the UN. The US may participate in a US-led coalition such as Operation Restore Hope (Somalia), or a non-US-led coalition, such as Operation Able Sentry (former Yugoslav Republic of Macedonia). The agencies involved in each of these operations are both consumers and possible resources of engineer activity. Army engineer units may be subordinate to, collocated and working alongside, or directing engineer activities and providing oversight or support for the missions assigned to these organizations. The engineer forces' effectiveness to operate within the varied framework surrounding a collective international enterprise can be greatly enhanced by respecting the multinational partners; their construction and engineering techniques; and their ideas, culture, religion, and customs. Equally important and parallel to operating within a US-only joint environment is understanding multinational unit or organization capabilities and training. This understanding ensures the assignment of appropriate missions and avoids the risk of offending national honor or prestige by allocating unsuitable tasks to partners in the multinational endeavor.

MULTINATIONAL ENGINEERS

H-2. The engineer organizations available from deployed national armies are generally a mix of combat and/or construction engineers in company- and battalion-sized units. The training and experience levels and equipment fielding varies among these units. National engineers from Britain, Canada, and Australia have been involved in numerous missions outside their territorial boundaries. The political impact of these missions is important to understand. When German engineers deployed into Somalia in 1992, it took a national legislative amendment to their constitution to allow them to participate in operations off German soil. This was their first experience in multinational efforts outside of NATO. Smaller countries have more regional restrictions on their involvement, and their experience is correspondingly narrow. However, they are also more likely to be attuned to the special circumstances that are relevant to the AO.

MULTINATIONAL ENGINEER CAPABILITIES

H-3. NATO and ABCA engineer capabilities are well known, and data about them is readily available. Standardization agreements (STANAGs) between national armies facilitate engineer interoperability and cooperation. The capabilities of other nation's engineers are normally available through intelligence channels or formal links with the nations concerned. Several nations have engineers that are experts in specific combat engineering tasks, such as mine detection and removal. Other national engineers are focused on specific missions, such as disaster relief. Engineers must have an appreciation for the engineer capabilities and limitations of other nations. Allied Joint Publication (AJP) 3.12 provides the keystone NATO engineer doctrine for working in this allied environment. Engineers must ensure they understand multinational doctrine, terms, and symbols in order to effectively operate in this environment. For example, AJP 3.12 defines two terms that are similar to US Engineer doctrine but the differences must be understood to effectively operate in the multinational environment:

- (NATO)—An Allied term that encompasses those military engineer tasks associated with direct support to current or imminent operations. It is conducted by the military engineers of any service or component to support land, air, maritime, and information operations. With the emphasis on speed of execution, it frequently relies on prefabricated equipment solutions, may involve a degree of improvisation and is likely to fulfill a short-term tactical requirement. It is closely aligned with the Engineer Battlespace function of Combat Engineering.(AJP 3.12)
- (NATO)—An Allied term that describes the deliberate, longer-term preparation for, and indirect support to ongoing or future operations

as well as those military engineering tasks associated with sustaining the joint force throughout all stages of an operation. It will be the predominant engineer focus for precombat and post combat and for rear operations during conflict. It is an area which may involve a greater degree of cross-compartment support and the engineer tasks will usually be more enduring, relying more on purpose designed and built solutions. It is likely to fulfill a longer term, operational requirement. It is closely aligned with the engineer Battlespace function of General Engineering. (AJP 3.12)

MULTINATIONAL ENGINEER COMMAND AND CONTROL

H-4. Depending on the multinational force arrangement in theater, Army engineers may control or work closely with engineers from other nations. Multinational engineer C2 relationships are established to foster cooperation and share information. Critical to this process is providing adequate US engineer LO support, including linguist support, communications equipment, and transportation.

MULTINATIONAL ENGINEER CONSIDERATIONS

H-5. During force projection operations, the initial engineers in theater most likely provide the HN engineering capabilities. As Army engineers deploy into a theater, they may be joined by allied and coalition engineers. The Army engineer staff should consider including the following when coordinating multinational engineer plans and operations:

- Requesting the latest intelligence information concerning the HN, allied and coalition engineer structures and logistics requirements.
- Requesting the latest engineer intelligence data from the HN or deploying allied and coalition engineer elements to help identify force projection theater Army engineer requirements and enemy engineer capabilities. The requirements include threat mine and obstacle data, soils data, construction materials availability, and HN construction support.
- Establishing multinational engineer staff links between the Army, HN, allied, and coalition engineer force staff sections through the JTF or theater engineer staff and HQ.
- Executing NATO multinational C2 using the NATO OPORD format and the NATO decision-making process.
- Providing necessary Army engineer LO support.
- Developing multinational task organization relationships that enhance HN, allied, and coalition engineer capabilities following the deployment of Army engineers.
- Assessing the need for HN, allied, and coalition engineer support following the arrival of Army combat and construction units in theater.
- Determining if multinational engineer units need augmentation from Army combat and construction units.

- Developing procedures for Army engineer units to support multinational engineers with additional Class IV construction materials and engineer equipment.

UNITED NATIONS OPERATIONS

H-6. UN-sponsored operations normally employ a force under a single commander. The force commander is appointed by the Secretary General of the UN, with the consent of the UN Security Council. The force commander reports to a special representative of the secretary general or the chief administrative officer (CAO) or directly to the secretary general. While the force commander conducts day-to-day operations with a fairly wide discretionary authority, the commander refers all policy matters to the CAO or the secretary general for resolution. The CAO not only establishes policy but controls the resources and funding expenditures within a given operation. The US commander, designated by national policy, retains command over all assigned US forces. The US chain of command runs from the President through the combatant commander to the lowest commander in the field. Other agencies may negotiate and exercise OPCON over US military forces subject to prior approval by the President and SECDEF. The degree of OPCON exercised over US forces must be coordinated and agreed upon by the multinational force commander and the combatant commander according to the President's criteria. The US commander is, therefore, responsible for mission success to the UN force commander as well as the theater combatant commander.

NON-UNITED NATIONS ORGANIZATIONS

H-7. Although the US is more likely to undertake involvement in multinational operations with the UN, a number of regional organizations may perform this leadership function, including—

- NATO.
- The Organization of African Unity (OAU).
- The OAS.
- The Organization for Security and Cooperation in Europe (OSCE).

SECTION II - INTERAGENCY AND NONGOVERNMENTAL ORGANIZATION CONSIDERATIONS

INTERAGENCY OPERATIONS

H-8. Interagency operations greatly expand the scope and capabilities of any given response team because of the wide variety of expertise and funding resources that can be tapped to perform functions during a crisis response. This is true whether the response is international or within the territorial confines of the US and its protectorates and territories. Interagency operations increase the resources engaged in any given operation, but they also significantly increase and complicate the coordination necessary to achieve victory. Additionally, they generate mechanisms that reduce efficiency, as organizations work at cross-purposes. The SOPs, report formats, information requirements, and intermediate goals and perceptions of each of these organizations vary greatly. Therefore, coordination and a clear understanding of the commander's intent are absolutely crucial when arranging operational efforts involving multiple interagency organizations. The following are some of the interagency organizations that could be involved:

- Federal Emergency Management Agency (FEMA).
- Environmental Protection Agency (EPA).
- Drug Enforcement Administration (DEA).
- National Oceanic and Atmospheric Administration (NOAA).
- United States Geological Survey (USGS).
- Public Health Service.
- Civil Air Patrol.
- United States Department of Agriculture (USDA).
- Department of State (DOS).
- Office of Foreign Disaster Assistance (OFDA).
- United States Department of the Interior, Fish and Wildlife Agency.
- General Accounting Office (GAO).

H-9. *See JPs 3-07* and *3-08* for an in-depth discussion of interagency coordination during joint operations.

NONGOVERNMENTAL ORGANIZATIONS

H-10. Relationships with international and domestic NGOs must be established through negotiation. Most agreements are made at the strategic level (unified command); however, the operational commander may have some latitude delegated to him. All agreements normally have serious legal restrictions on using military personnel and equipment. Some of these agencies may have unique and significant engineer capabilities and intelligence that could be used as a part of the overall operational concept. These capabilities may be a useful source of Class IV material, not only for the agency's own projects but also as a negotiated barter for services rendered in support of its mission. More often than not, however, these agencies and organizations may request extensive engineer support for their activities and programs. As these organizations play an important part in the combatant commander's achievement of strategic objectives, their demands must be coordinated. Therefore, it is critical that an effective engineer liaison be established and maintained with the force HQ CMOC.

H-11. The UN may designate a regional organization with a greater vested interest and appreciation for the forces at work in a given region as its operational agent to exercise control. Each of these organizations has different operational concepts and organizational procedures; US forces are familiar with some of these concepts and procedures, such as those of NATO. However, there are others with which they are not familiar.

SECTION III - HOST NATION CONSIDERATIONS

HN INTERFACE

H-12. In a forward deployed theater, the ASCC, in conjunction with the other component commanders, the combatant commander, the allies, and the HNs, identifies wartime facility and construction requirements for the Army as part of the deliberate war planning effort. Doctrinal construction requirements for the ASCC may be identified using the planning module in the TCMS. Subsequent analyses further refine construction requirements and provide a basis for—

- Force structuring.
- Procurement.
- Leasing provisions and establishing HN agreements.

H-13. The product of these analyses is the CESP. The goal is to reach HNS agreements in peacetime to provide as many of the facilities as possible that are needed within the theater. Advanced planning and the commitment of resources by HNs reduce the early lift requirements needed to support the RSOI and focus force projection assets C2 and engagement systems. Engineering support from the HN usually involves providing—

- Land.
- Facilities.
- Construction support.
- Manpower.
- Equipment.
- Materials.
- Services.
- Hazardous-waste disposal.

H-14. It is highly desirable to secure written agreements with the HNs regarding support items to foster an understanding of the assistance levels and increase the likelihood of fulfillment upon execution.

REAL ESTATE CONSIDERATIONS

H-15. Real estate operations involves acquiring, managing, and disposing of land and facilities to support joint operations. The JFC should establish real estate acquisition policies and programs in support of contingency operations. The JFC determines what real estate is needed to satisfy operational requirements. Acquiring land and facilities not owned by the United States Government (USG) is accomplished through assignment, international agreements such as SOFAs, and memorandums of agreement (MOAs), leases from the HN, or direct leases from the private sector. Within DOD, the secretaries of the military departments are authorized to lease structures and real property relating to structures that are needed for urgent military purposes in foreign countries (*see Title 10 USC,* initiated as contingency plans and developed to identify land and facility requirements for joint operations).

Real estate acquisition requires special contracting procedures be performed by the USACE, the Naval Facilities Engineering Command (NAVFACENGCOM), or a designated executive agent. Early deployment of real estate personnel is essential to acquire land and facilities in a timely manner.

H-16. Real estate authorities throughout the world have been assigned to components along similar lines corresponding to the designation of DOD construction agents (DODD 4270.5). Within regions designated to the Army, the USACE establishes policies for the acquisition, maintenance, and disposal of real estate, to include leased and rent-free facilities. Real estate teams may be assigned to each ASG or centrally controlled at the senior engineer HQ. These teams coordinate with HN agencies and private owners to acquire and dispose of real estate and establish the terms of lease agreements. Real estate planning and surveys must be initiated as campaign plans are developed to provide timely and adequate facilities to sustain the combat force. Local HN officials can help identify available facilities or land that meets military requirements. Thorough documentation of lease agreements, property conditions at the time of the lease, to include environmental baseline survey (EBS) data and expectations of property conditions at the termination of the lease are crucial to expedite a fair and amiable conclusion of lease activities. Civil affairs and real estate personnel may be required to work through HN governments to settle agreements with property owners. Real estate acquisition is more difficult in contingency operations due to the lack of preparation to identify probable sources of and confirm legal ownership. Real estate is required for—

- Air bases.
- Base camps.
- Medical and logistics complexes.
- Training sites.
- Quarry and borrow sites.
- Trailer transfer points.
- Traffic control points.

H-17. Property is generally acquired by requisition, with all transactions documented thoroughly under the provisions of the combatant commander's directives. Procedures are used that provide the property required for missions while protecting the property owner's legal rights. Using rent-free facilities provided by the host government or a host agency require the same legal responsibilities as using facilities leased from private owners. Real estate policies and procedures are discussed in more detail in *FM 5-104.*

HOST NATION SUPPORT

H-18. Wartime HNS agreements in forward-presence theaters (Europe and Korea) have been negotiated to provide HN construction support, such as facility modifications, LOC maintenance and repair, and utility services. In Southwest Asia, the agreements are less formal and lack the practiced application that accompanies the full-time presence of US forces in Europe and Korea. However, these agreements are no less critical to mission success in the event of an MTW in this region. Such HNS is used whenever possible to free US engineer units for critical missions where HNS alternatives are not viable. Support agreements are negotiated in peacetime on an asset basis. Assets may be facilities, contracts, or equipment. Again, this support is particularly critical during the initial stages of a contingency when RSOI requirements are high and engineer assets are limited.

H-19. Prepositioning equipment within the region reduces the US response time into a particular theater by allowing military forces to deploy by air and fall in on war stocks within the region. These prepositioning locations are a critical element of US force projection national strategy and represent a significant contribution of HNS. The HN's commitment for space, facilities, services, and utility support for these complexes demonstrate the HN's interface with US forces and the partnership of the US and UN in the defense and stability within the region. Beyond direct HNS, allied nations directly or indirectly involved in the crisis may provide other support. Other nations sympathetic to the cause may be limited in their direct participation because of constitutional restrictions or political sensitivities. However, these nations may provide nonlethal equipment or funding much like that provided by the Japanese during the Gulf War.

H-20. During a conflict, the HN may provide construction organizations to repair or construct facilities, usually within the rear area. Construction materials such as cement, asphalt, aggregate, timber, and steel, as well as contract labor, may also be available. HN assets may also be available for local security and the transportation of construction materials and equipment. Third country nationals (TCNs) may also be available, by request through the HN or through direct contact with nationals, to support engineer activities within the rear areas. Engineer reconnaissance and assessment teams engaged in deliberate planning during peacetime or dispatched early in contingency operations are the key to identifying and accessing available HN assets.

Appendix I

Contract Construction Agents

Use of construction contracting and engineering support can play an important role in the support of joint operations. Civilian construction contractors and HN engineering support provide the JFC with a significant engineering capability that becomes a force multiplier when combined with joint force military engineering units. CCAs provide the ability to harness and direct this means of support.

CONSTRUCTION CONTRACTING AND ENGINEERING SUPPORT

I-1. The DOD CCAs are the USACE, the NAVFACENGCOM, or other approved DOD activity *(see DODD 4270.5)*. These organizations and their contractors are a powerful force multiplier, allowing military engineers to concentrate on engineering missions in high-threat areas. The USACE and the NAVFACENGCOM also provide the JFC with a significant engineering capability to be leveraged in joint operations. They are DOD's principal organizations to plan, design, construct, and acquire (lease or buy) facilities and real estate. Inherent in their mission support capabilities is a planning and engineering capability for theater advanced base and infrastructure development. These organizations also maintain in-depth engineering expertise in their operating field organizations and laboratories.

I-2. The responsibilities of the DOD CCAs include designing, awarding, and managing the construction contracts for projects associated with the peacetime military construction program. Overseas, the USACE, the NAVFACENGCOM, and the Air Force are assigned specific geographical areas under DODD 4270.5 *(see Figure I-1, page I-2)*. Related to these responsibilities is the leasing of real estate.

I-3. The combatant commander may also use the USACE and the NAVFACENGCOM as contingency CCAs for designing, awarding, and managing construction contracts in support of military operations. For geographical areas where there is no designated DOD construction agent, the combatant commander usually designates a CCA for support during a contingency. The USACE and the NAVFACENGCOM also provide base development master planning, contract administration, and technical engineering support to JFCs (advanced base master planning, geospatial [topographic] engineering, infrastructure security and force protection engineering, and cold-weather mobility assessments). The Air Force also maintains a limited capability in contract construction in contingencies and facilities and real estate acquisition in England, Turkey, Spain, and Israel.

Designated Geographical Areas of Department of Defense Construction Agents

United States Army Corps of Engineers

Panama Canal Zone, Canada (excluding Newfoundland), Egypt, Europe (excluding Spain, Portugal, Italy, Greece), Greenland, Israel, Japan (including the Ryukyu Islands (Okinawa)), Korea, Taiwan, Marshall Islands, Middle East (including the Saudi Arabian Peninsula), Southern Asia (from Iran to Burma), and Turkey

Naval Facilities Engineering Command

Atlantic Ocean area, Australia, New Zealand, Caribbean Sea area, Greece, Iceland, Indian Ocean area, Italy, Newfoundland, North Africa (including Somalia and Kenya but excluding Egypt) Pacific Ocean area (excluding the Marshall Islands), Portugal (including the Azores), Republic of the Philippines, Southeast Asia, and Spain

United States Air Force

British Isles

Figure I-1. Designated Geographical Areas of Department of Defense Construction Agents

UNITED STATES ARMY CORPS OF ENGINEERS

The word 'miracle' has often been used when reporting the exploits of the American Corps of Engineers during the last war.... In fact there has been no miracl.... there has been only the logical result of a peacetime engineer organization unique in the world, which participates actively... in the development and execution of great public works of National interest.

General Robert J.L. Pinson, 1948, Chief of Engineers, French Army

I-4. USACE is the Army's major command assigned responsibility to execute the following Army and DOD mission areas:

- Engineering and design for the Army and Air Force military construction programs.
- Contract construction.
- Real estate acquisition.
- Technical assistance.
- Geospatial (topographic) engineering support.
- Army's civil works program.
- Public works and engineering for the Federal Response Plan (FRP).

I-5. The USACE's subordinate commands are organized geographically and functionally. It's C2 contains three major organizational structures, including—

- Divisions. The division is the major subordinate C2 organization for the USACE. The division commander provides executive direction to and management of the subordinate district commands. The division's orientation is regional and provides broad interface with regional interests and management of division-wide programs. Division commanders are responsible for contingency support in their directed areas of responsibility.
- Districts. The district command is the operating arm of the division. All CONUS USACE districts in the US have civil works responsibilities. In the US, their boundaries are delineated along major watershed basins, and their work lines are set on state boundaries *(see Figure I-2, page I-4)*. Additionally, some of the districts have military execution responsibilities. The districts maintain in-house capabilities in planning, engineering, construction, operations, project management, and contract administration. USACE provides support through subordinate commands to the following geographic combatant commands and subunified commands:
  - United States Forces Korea by the Pacific Ocean Division through the Far East District—Seoul, South Korea.
  - United States Forces Japan by the Pacific Division through the Japan District—Camp Zama, Japan.
  - Pacific Command (PACOM) by the Pacific Ocean Division through the Honolulu District—Fort Shafter, Hawaii.
  - Southern Command (SOUTHCOM) by the South Atlantic Division through the Mobile District—Mobile, Alabama.
  - EUCOM by the North Atlantic Division through the European District—Wiesbaden, Germany.
  - CENTCOM by Transatlantic Programs Center—Winchester, Virginia. For contingencies, the Transatlantic Programs Center is in support of the Southwestern Division for support to CENTCOM.
- USACE laboratories. The ERDC is the USACE's distributed research and development command. Headquartered in Vicksburg, Mississippi, the ERDC consists of seven unique laboratories that conduct research and development in support of the Army and the other services. By virtue of their engineering expertise, lab personnel provide operational support to the rest of the USACE and the DOD through the aforementioned subordinate commands. Additionally, the USACE maintains several specialized centers of expertise at its districts that provide additional technical engineering services (infrastructure security and force protection, airfield design, electronic systems security, physical security, base camp design, water flow data, and electronic systems security).
  - The Coastal and Hydraulics Laboratory, Vicksburg, Mississippi—provides services in vehicle mobility and trafficability; coastal engineering; structural hardening against weapons effects;

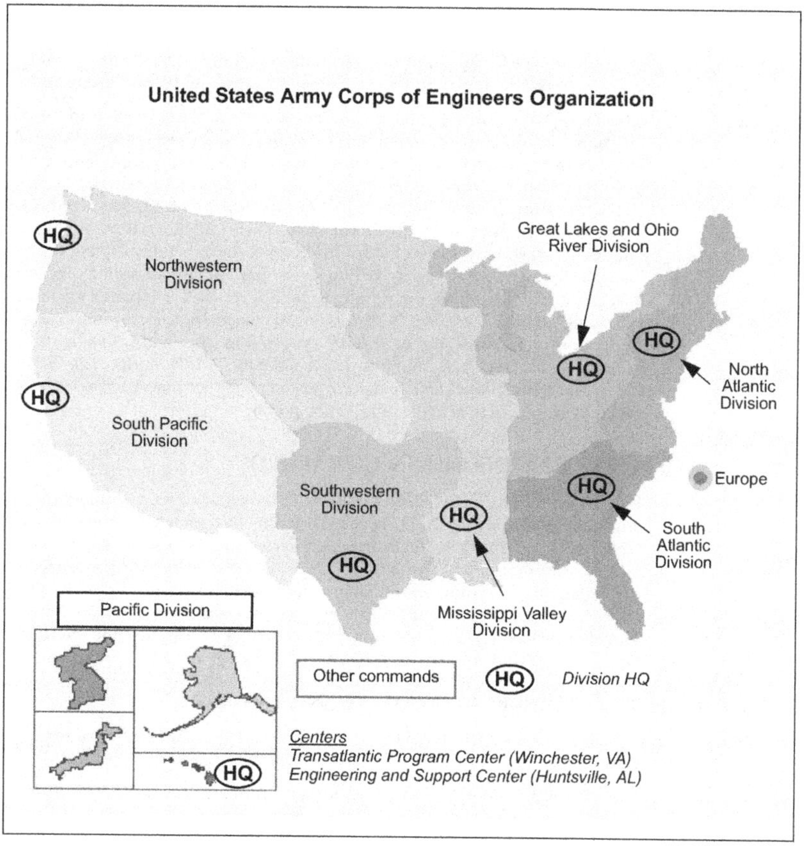

Figure I-2. United States Army Corps of Engineers

facilities camouflage, concealment, and deception technology; geotechnical expertise; and pavement and surface treatment technologies.

- The Cold-Regions Research and Engineering Laboratory (CRREL), Hanover, New Hampshire—provides services in all aspects of engineering research with respect to the conditions unique to cold weather.
- CERL, Champaign, Illinois—provides services in construction management technology and facilities engineering.
- The Environmental Laboratory, Vicksburg, Mississippi.

- The Geotechnical and Structures Laboratory, Vicksburg, Mississippi.
- The Information Technology Laboratory, Vicksburg, Mississippi.
- The TEC, Alexandria, Virginia—specializes in engineer mapping products, analysis of remote sensing data, and terrain analysis.

I-6. The USACE designs and constructs military facilities and supports military installations worldwide. The military engineering expertise of the USACE is focused on the engineering required to plan, design, and construct military facilities and the environmental engineering necessary to execute DOD installation environmental restoration projects. USACE maintains specialized expertise in its laboratories and centers for cold-weather engineering, remote sensing and imagery, force protection design, airfield design, weapons effects (for example, support for operational targeting—assessing the target, recommending the appropriate weapon systems, and providing an attack profile), terrain analysis for mobility and countermobility, topographic engineering, security systems engineering, environmental management, and environmental engineering. Additionally, the USACE's 249th Engineer Battalion (Prime Power) can conduct power assessments and install generators to provide emergency power.

NAVAL FACILITIES ENGINEERING COMMAND

I-7. The NAVFACENGCOM and its field divisions directly support the Navy and Marine Corps and DOD shore establishment throughout the world with a wide variety of military and contract construction, real estate acquisition, and public works support. The NAVFACENGCOM's engineering support units provide project management, planning, design engineering, construction, operations, and maintenance, as well as disposal functions for Navy and Marine Corps shore facilities and real estate. They also provide engineering, logistics, doctrine, and policy support and guidance for NCF units.

NAVAL FACILITIES ENGINEERING COMMAND AND CONTROL

I-8. The NAVFACENGCOM and subordinate commands are organized geographically and functionally *(see figure I-3, page I-6).*

I-9. The NAVFACENGCOM headquarters is located in Washington, DC. Subordinate commands include four engineering field divisions (EFDs), six engineering field activities (EFAs), officers in charge of construction (OICCs), and the Naval Facilities Engineering Service Center (NFESC). Additionally, the NAVFACENGCOM provides technical support to ten regional Navy public works centers, including—

- EFDs and EFAs. The EFD commanders provide executive management and engineering support for all Navy and Marine Corps facilities in designated geographical areas. EFAs provide similar support for smaller CONUS and overseas areas within the greater geographic area of the EFDs. Both EFDs and EFAs have design, engineering, and architecture capabilities and act as technical consultants for all facilities of supported commands. EFDs play a significant role in contingency operations through forward-deployed resident OICCs that can provide construction contracting and real

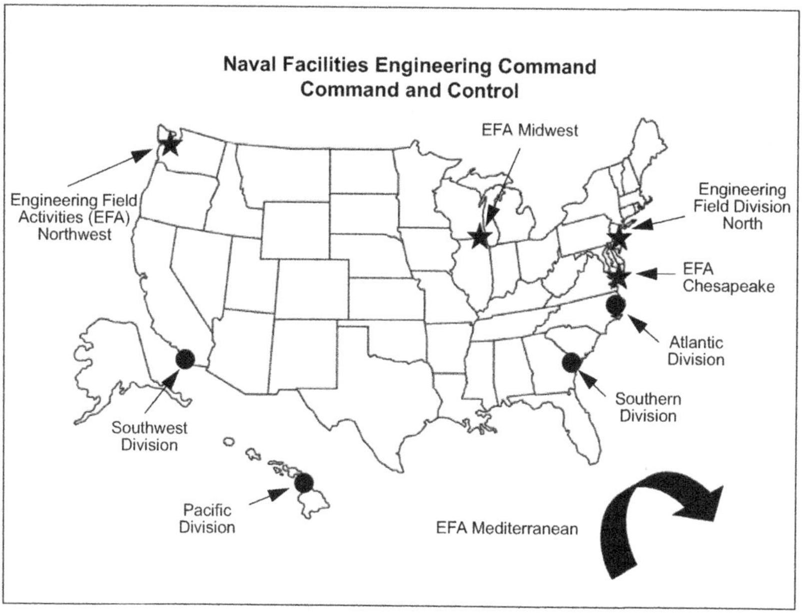

Figure I-3. Naval Facilities Engineering Command, Command and Control

estate support and manage Navy construction capabilities contracts (CONCAPs).

- OICC. The EFD commanders designate OICCs as contracting officers in certain geographical locations where the complexity and contracting tempo warrant. The EFD commander can also assign a resident officer in charge of construction (ROICC) in any geographical location, as required. The ROICC is trained in facilities support, construction, engineering, environmental design, and real estate contract acquisition and management. The ROICC office is especially useful for contingency operations in support of the joint force. The ROICC may also provide construction management support to the Army, Air Force, and DOD, as well as other federal agencies.
- NFESC. The NFESC provides engineering support and expertise to the Navy in undersea and amphibious operations, conventional ammunition storage, mobile utilities and environmental support equipment, and research and consulting engineering services. The NFESC is located in Port Hueneme, California, with a detachment in Washington, DC *(see Figure I-4).*

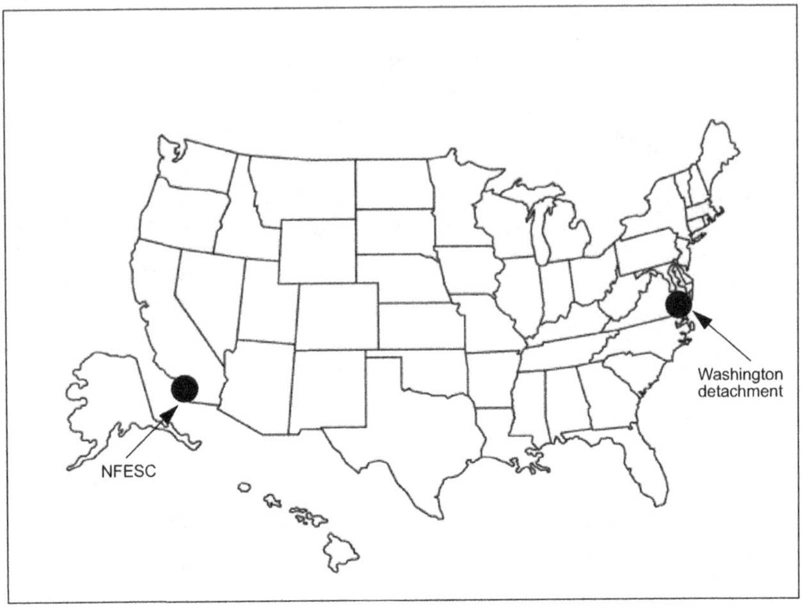

Figure I-4. Naval Facilities Engineering Command and Naval Facilities Engineering Center

NAVAL FACILITIES ENGINEERING COMMAND SUPPORT FOR MILITARY OPERATIONS

I-10. The NAVFACENGCOM supports Navy and Marine Corps operations, DOD missions, and joint force operations around the world, to include—

- Force projection missions. The Navy accomplishes force projection worldwide by using superior naval, air, and amphibious forces, equipment, tactics, and doctrine. This force projection requires shore logistic platforms, such as advanced logistic support sites and ports to receive supplies, equipment, and personnel to be deployed to ships and advanced bases. The NAVFACENGCOM directly supports these activities by providing engineering, construction forces, contract construction, and facilities management, including the disposition of real estate for these ports and bases.
- Theater missions. The NAVFACENGCOM supports the theater through real estate acquisition, management, and final disposition, as well as through the construction of facilities to support Navy, Marine Corps, combatant commands, subordinate joint forces, and other DOD elements. The NAVFACENGCOM provides this support through the subordinate EFD and EFA and provides technical engineering and construction support to other services and DOD agencies through the

NCF and as a designated CCA. The NAVFACENGCOM commander may provide a forward element dedicated to support the geographic combatant commander or subordinate JFC for military operations in contingencies. The forward element may call on the full capability of the entire NAVFACENGCOM organization to assist in providing this support.

- CONUS contingency missions. The NAVFACENGCOM has responsibility for disaster recovery and other contingency operations at Navy and Marine Corps stations in CONUS. With NCF personnel and equipment, in addition to its contracting capability, the NAVFACENGCOM can quickly mobilize and provide significant assets for disaster recovery.

CONTRACTED SUPPORT

I-11. The US military can contract civilian engineering support, as required, based on the threat situation and available resources. These contracts relieve the workload on US military engineer units in such areas as logistics base construction, real estate and facilities acquisition, RPMA, and demining operations. As discussed earlier, overseas construction and other contracting services are available through the USACE, or the NAVFACENGCOM, or through limited Air Force capability during contract construction in contingencies, as well as facility and real estate acquisition (in England, Turkey, Spain, and Israel).

CIVIL AUGMENTATION PROGRAMS

I-12. Civil augmentation programs such as the LOGCAP, the CONCAP, and the AFCAP can provide worldwide logistics and construction support. Civil augmentation programs can play a significant role in mission accomplishment by providing the JFC and joint force engineer with additional options and flexibility in achieving timely civil engineering and logistic support. Civil augmentation programs are managed by a contract agent and are structured with one contractor responsible for providing support that effectively integrates construction, facility maintenance, and logistic support to the joint force. For example, within an operational area, subcontractors, materiel, and personnel may come from many countries within the region. A single contractor prevents multiple agencies and their contractors from bidding against one another for services and materiel in the operational area. Use of civil augmentation programs requires planning and operational oversight and quality control and assurance to ensure costs are effectively controlled, while providing support consistent with the JFC's concept of operations.

ARMY

I-13. The LOGCAP is an Army program funded in peacetime as a component of Army readiness. The LOGCAP is a broad logistic and engineering contingency support contract that encompasses all Army preplanned contingency contracts and contingency components of contracts. Currently, the LOGCAP is a cost plus award fee contract, managed by the United States Army Materiel Command (USAMC). In peacetime, the LOGCAP contractor maintains an on-call, preplanned, ready capability. The contractor demonstrates readiness through the development of a worldwide plan, supporting plans to OPLANs, specific regional plans, and participation in exercises.

- The USAMC support contract provides both engineering construction and general logistic services. USAMC is supported by the USACE for engineering and construction contract administration and by the Defense Contract Management Agency (DCMA), a component of DLA for logistic services contract administration.
- In the operational area, the USAMC logistics support element, subordinate to the senior Army logistics commander in the operational area, manages the LOGCAP support contract. The USACE provides program management, coordinates the LOGCAP requirements with supported MACOMs, and administers the LOGCAP contract. The MACOM G3, G4, engineer, and comptroller are key players in developing the LOGCAP requirements and ensuring the appropriate mix of contractor and troop support. The contract can support all DOD components and missions and is a cost plus award fee service contract that provides support to the joint force in four major areas:
  - Contractor planning expertise to assist contingency planners.
  - Facilities repair and construction.
  - Base operations and maintenance.
  - Logistics services.

I-14. The LOGCAP is especially suited to support reception, onward movement, and sustainment facilities. The LOGCAP can augment engineer units by operating Class IV supply yards, supplying construction equipment, providing facility engineering support, and supporting theater construction.

NAVY

I-15. The Navy's CONCAP program is a cost-reimbursable contract administered by the Atlantic Division, NAVFACENGCOM. The contractor is usually a large construction firm or joint venture firm, with international capability. The contract offers responsive engineering and construction capabilities for a wide range of construction missions. The CONCAP was used for engineering fieldwork in Bosnia and for disaster response and recovery for Hurricane Bertha at Camp Lejeune, North Carolina. This construction-oriented contract may be used worldwide, including in CONUS. Major capabilities include the—

- Engineering, design, and construction of the following:
  - Airfield and port facilities—piers and dredging.
  - Roads, bridges, ordnance facilities, and landfills.
  - Power plants and utility systems.
  - Communication and supply systems.
  - Medical and EPW facilities.
- Operation of the following:

- Facilities and utilities.
- Billeting, food services, and recreation.
- Waste management—refuse and sanitation.

AIR FORCE

I-16. The AFCAP is a cost plus award fee contract that is centrally administered by the AFCESA at Tyndall Air Force Base, Florida, (through the Air Force's major commands' civil engineers). The contract was designed to augment or relieve specified civil engineering support functions participating in MOOTW. The AFCAP capabilities focus on temporary contingency skills to sustain military forces and to support a 10,000-person force in up to eight different locations for periods of 180 or more days. When AFCAP contract support is initiated for a contingency operation, a cadre of contractor personnel responds on-site, as soon as possible, to begin construction tasks (typically within 30 calendar days). Each task order specifies the required response time and the period of performance.

CONTRACTED CIVILIAN ENGINEER CONSIDERATIONS

I-17. The challenge for engineer planners and executers is to achieve the optimal mix of contractor and military engineer unit capabilities. Construction contractors are best suited for the longer duration and for heavy construction work in stable environments. In turn, contractors leverage local resources (labor and materiel) to minimize costs and impacts on intratheater lift and port facilities. The contractor's presence contributes significantly to local area political and economical stabilization and thereby reduces the need for the presence of US security forces. In turn, the US commander in theater must recognize the need for US military oversight of contract and contractor activities in the areas of project management, financial management, quality assurance, and audit.

I-18. During force projection operations, extensive contracted civilian engineer capabilities are probably available only after 30 days (D+30), due to mobilization and deployment time lines. Civilian engineer contracting may be available sooner when deliberately and properly planned during permissive entry conditions. As Army engineers deploy into the theater, they may be joined by contracted civilian engineers. The Army engineer staff should consider the following when coordinating engineer plans and operations with contracted civilian engineers:

- Requesting the latest engineer intelligence data from any contractors working in the theater to help identify force projection theater Army engineer requirements and enemy engineer capabilities. These requirements should include the availability of real estate, construction materials, and facilities; data on threat mines, obstacles, and soils; and construction support from the HN.
- Establishing engineer staff links between the Army and contracted civilian engineer staff through the JTF, the USACE or, NAVFAC and the theater engineer staff and HQ.
- Providing necessary Army engineer LO support.

- Developing time lines that quickly phase in contracted civilian engineer capabilities to relieve deployed Army engineer units of some responsibilities.
- Assessing the need for additional contracted civilian engineer support following the arrival of Army combat and construction units in theater.
- Determining if contracted civilian engineers need augmentation from Army combat and construction units.
- Developing procedures for Army engineer units to draw on contracted Class IV construction materials and engineer equipment.

Appendix J

Contingency Authorities and Funding

It is especially important that engineers understand contingency authorities and the associated funding. These are the tools that set the conditions for success during contingency operations and provide the basis for legal spending to take care of DOD personnel and activities in support of contingency operations. Contingency operations comprise a very large portion of the operations conducted by Army and engineer forces.

LEGAL PERSONNEL

J-1. Legal personnel can provide invaluable advice and guidance on authorities and sources of funding for civil engineering activities in a variety of situations. From the earliest stages of planning, execution, and redeployment, legal professionals play a vital role in preparing the AO by identifying and assisting in the resolution of legal and political constraints, and provide relevant and responsive readiness programs to the individual civil engineering members.

TYPES OF AUTHORIZATIONS AND SOURCES OF FUNDING

J-2. Services are authorized to use annual operation and maintenance (O&M) funds for construction projects costing less than $500,000 ($1 million to correct a life threatening condition or for new construction and $2 million for maintenance and repair of existing facilities). This is a peacetime provision, applicable during contingencies and emergencies; however, life threatening is generally considered a safety issue vice an emergency in the context of contingency operations. During combat or designated contingency operations, O&M may be used to fund construction projects exceeding these thresholds. The JFC must consult with the Staff Judge Advocate (SJA) before making a determination to use O&M in such a case.

J-3. Several broad authorities have been established under Title 10 USC that enable the JFC to carry out contingency construction, including procuring materials for construction by military forces and funding civilian contracts. *Figures J-1 and J-2, pages J-2 and J-3,* depict decision trees for the contingency construction funding options.

J-4. Section 2803, Emergency Construction, authorizes each service to use $30 million per year of unobligated military construction (MILCON) funds for projects that cannot wait for the normal MILCON submission procedures. Projects must comply with a 21-day congressional notice and wait period before proceeding. Generally, a previously congressionally approved project must be canceled to free the $30 million.

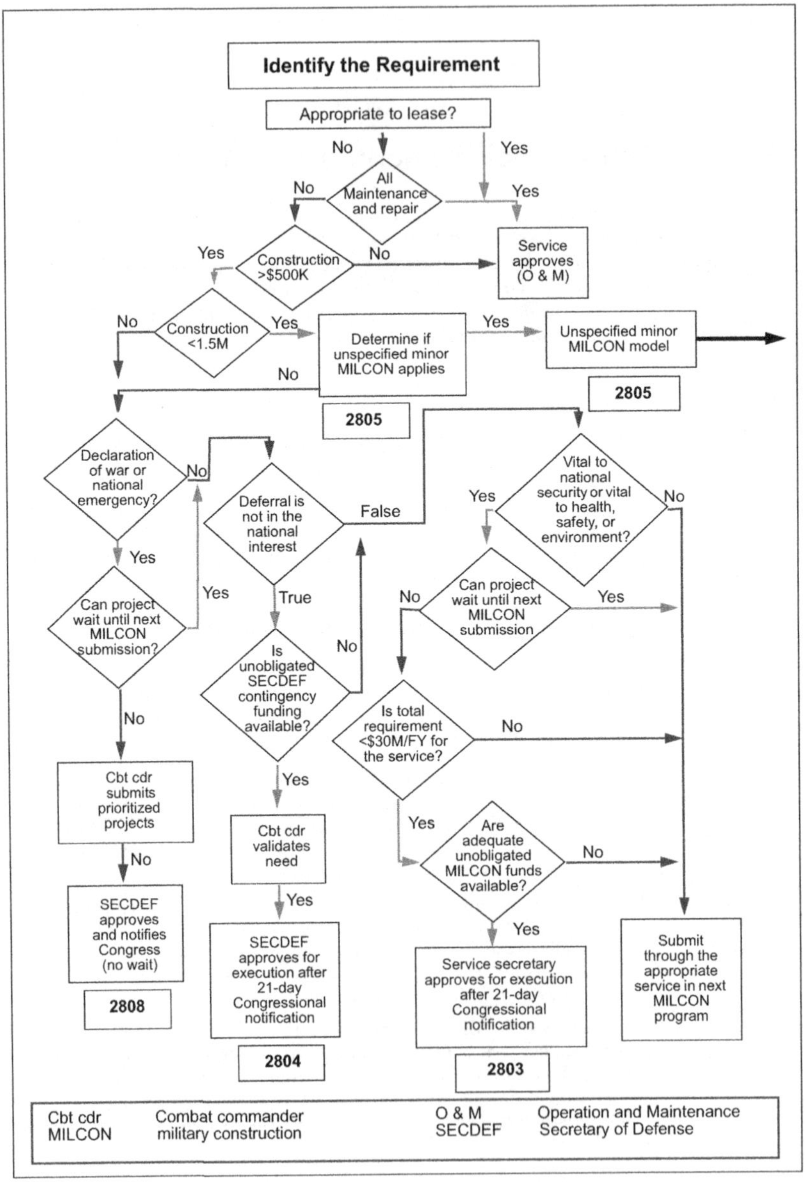

Figure J-1. Contingency Construction Funding Model

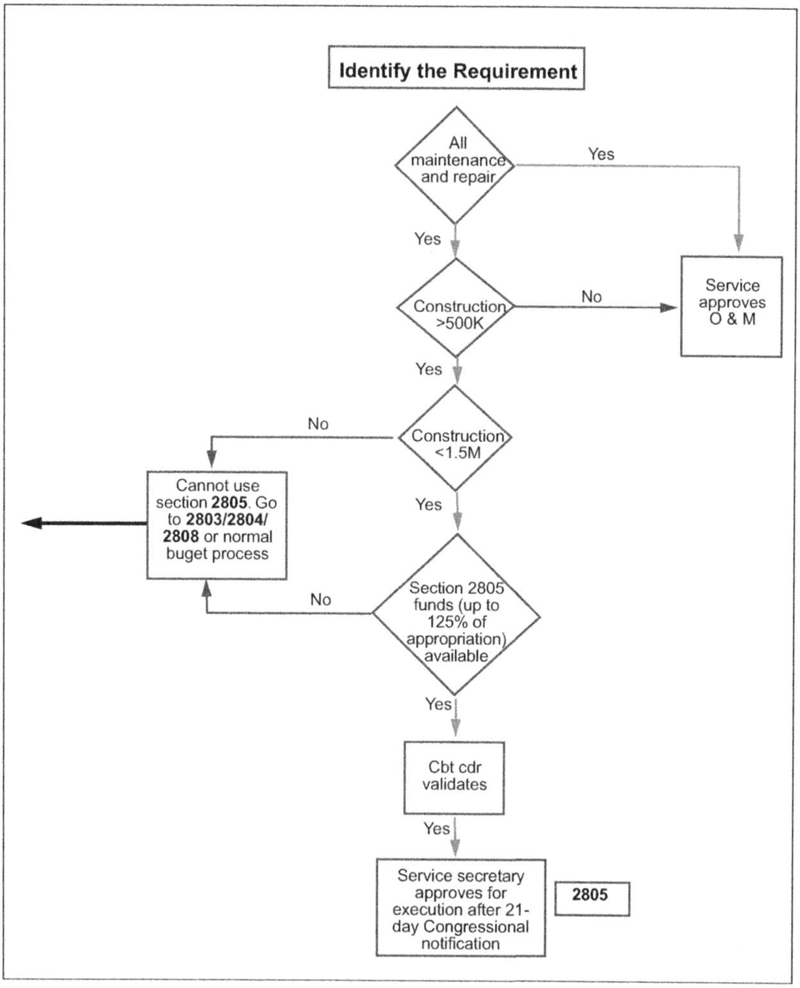

Figure J-2. Contingency Construction Funding Model (Unspecified Minor, $500,000-$1,500,000)

J-5. Section 2804, Contingency Construction, authorizes the SECDEF a specific military construction line item amount for contingency construction projects that cannot wait for the normal military construction program submission process. A project must comply with a 21-day congressional notice and wait period before proceeding. Generally, funding for this section is limited to less than $10 million per year.

J-6. Section 2805, Unspecified Minor Construction, authorizes each service a specific MILCON line item amount that varies annually for unspecified minor construction. Projects must be less than $1.5 million each ($3 million to correct a life threatening safety condition). Projects greater than $500,000 require a 21-day congressional notice and wait period before proceeding.

J-7. Section 2808, Construction Authority requires a Presidential declaration of war or national emergency and authorizes the SECDEF to carry out any military construction project for the war or national emergency within the total amount of unobligated MILCON funds available. Congress must be notified of each project, but there is no wait requirement before the project may begin.

J-8. Combatant commanders do not need specific authority to request projects under Sections 2803 and 2804. To gain approval for a project under either authority, it is necessary to provide the appropriate service secretary or SECDEF with a justification of need, estimated costs, and source of funding.

OTHER AUTHORITIES AND SOURCES OF FUNDING

BURDEN SHARING (TITLE 10 USC 2350.J)

J-9. This regulation provides guidance enabling DOD to accept funds from HN or foreign governments to share the financial burden of construction, supplies, and services.

- Countries must be designated countries which allow DOD to accept cash contributions for these purposes. A listing is held with the Office of the SECDEF (Comptroller).
- If a country is not previously designated, the SECDEF must formally consult with Secretary of State (SECSTATE) for designation of a particular country. The Joint Chiefs of Staff (JCS) would submit a formal request to the SECDEF to initiate this process.
- Contributions can be accepted only for certain costs, including—
  - Compensation for foreign local national employees of DOD.
  - Military construction projects of DOD.
  - Supplies and services for DOD.
- Reporting procedures for military construction projects state that a project is initiated when a report is submitted to congressional committees. There is a 21-day wait period for approval.

SECTION 607A OF THE FOREIGN ASSISTANCE ACT OF 1961 (PL 87-195, AS AMENDED)

J-10. This act provides restoration of HN civil infrastructure. This provision of law allows any USG agency to provide goods and services to friendly countries and NGO agencies on an advance-of-funds or reimbursable basis.

ARMS EXPORT CONTROL ACT OF 1976 (PL 90-629, AS AMENDED)

J-11. HN military facilities may be restored under the foreign military sales provisions of this authority.

ECONOMY IN GOVERNMENT ACT (TITLE 31 USC 1535)

J-12. This act allows USG agencies to support each other provided that the supported agency has the funds and authority to do the work requested.

HUMANITARIAN AND CIVIC ASSISTANCE PROJECTS (TITLE 10 USC 401)

J-13. In HCA facilities projects, the JFC and joint force engineers may work with HN government agencies to repair or improve infrastructure and public facilities. These authorized and funded projects are designed to provide assistance to the HN populace in conjunction with a military operation or exercise. They are usually planned well in advance and are not usually planned in response to disasters, although HCA activities have been executed following disasters. Specific engineer activities for which HCA funds can be used include the construction of rudimentary surface transportation systems, well drilling, construction of basic sanitation facilities, and rudimentary construction and repair of facilities *(refer to Chapter 8).*

FOREIGN HUMANITARIAN ASSISTANCE (TITLE 10 USC 2551)

J-14. In disaster operations, the UN and the DOS, OFDA may generate funded requirements for DOD assistance. FHA programs focus on the use of DOD excess property, emergency transportation support, disaster relief, or other support, as necessary, to alleviate urgent needs caused by some type of disaster or catastrophe in a host country. While other elements of the joint force are focused on immediate humanitarian assistance, civil engineering planning may focus on projects that provide immediate shelter for dislocated civilians. The joint force engineers must work in a close relationship with the representatives of the HN and US country team *(refer to Chapter 8).*

DRAW DOWN OF DEPARTMENT OF DEFENSE ARTICLES AND SERVICES (TITLE 10 USC 506(A)(1), 506(A)(2), AND 552(C)(2)

J-15. Draw-down authority is a means to respond to unforeseen military emergencies or humanitarian relief situations. These recurring authorities have placed annual limitations on the value of articles and services that may be drawn down in any fiscal year.

- Section 506(a)(1). Draw down of DOD stocks and military assistance to countries and NGOs in support of emergencies worldwide. The total is limited to $100 million per year, of which no more than $75 million can come from DOD.
- Section 506(a)(2). Draw down of stocks of any USG agency for counternarcotics, disaster relief, migration, and refugee assistance worldwide. The total is limited to $150 million per year, of which no more than $75 million can come from DOD.

- Section 552(c)(2). Draw down of stocks from any USG agency to countries and international organizations in support of PKOs. The total is limited to $25 million per year.

DEPARTMENT OF DEFENSE DIRECTIVE 5100.46, FOREIGN DISASTER RELIEF

J-16. Normally, DOD components may participate in foreign disaster relief operations only after a determination is made by the DOS. This directive allows the military commander at the scene of a disaster to undertake disaster relief operations without prior approval of the ambassador and/or chief of mission when the emergency is so acute that immediate action is required to save life and property.

Appendix K

Environmental Considerations

"The American people will continue to expect us to win any engagement, but they also expect us to be efficient in protecting lives and resources while accomplishing the mission successfully. Commanders will be expected to reduce the cost and adverse effects of military operations, from environmental disruptions in training to collateral damage in combat." "

Joint Vision 2010

Environmental considerations are the spectrum of environmental media, resources, or programs that may impact on, or are affected by, the planning and executing of military operations. Factors may include, but are not limited to, environmental compliance, pollution prevention, conservation, protection of historical and cultural sites, and protection of flora and fauna (JP 3-34). Military environmental protection is the application and integration of all aspects of environmental considerations as they apply to the conduct of military operations. The purpose of this chapter is to outline the environmental considerations as part of a commander's planning process and specific responsibilities of the engineer/ ENCOORD on the staff.

COMMANDERS AND ENVIRONMENTAL CONSIDERATIONS

K-1. Commanders must be able to identify ways to protect the natural environment while executing their missions by doing the following:

- Considering the natural and the physical environment in planning and decision making in conjunction with other essential considerations of national policy.
- Protecting both the natural and physical environment of the home station and training areas as a means of retaining resources for mission purposes.
- Complying with all federal, state, local, and HN environmental laws and regulations.
- Using environmental risk-assessment and environmental management principles to integrate environmental considerations into mission performance.
- Instilling an environmental ethic in subordinates.

K-2. These actions identify environmental protection as an important undertaking considered in mission planning and execution. The commander will use the staff engineer as the key advisor on all environmental issues. Working with other staff officers, the engineer determines the impact of the operation on the environment and integrates environmental considerations into the decision-making process.

K-3. Commanders, engineers, and other staff officers should refer to *FM 3-100.4* for specific guidance in applying appropriate environmental consideration procedures during military operations.

ENVIRONMENTAL ROLES AND RESPONSIBILITIES

COMBATANT COMMANDER AND SUBORDINATE JOINT FORCE COMMANDER

K-4. The combatant commander and subordinate JFC are responsible for ensuring environmental compliance, within the confines of mission accomplishment, with all domestic environmental laws and international agreements that apply within the joint operations area. JFCs should demonstrate proactive environmental leadership during all phases of joint operations, instill an environmental ethic in subordinate commands, and promote environmental awareness throughout the joint force. JFCs should ensure that environmental considerations are an integral part in the planning and decision-making processes. They should identify specific organizational responsibilities and joint force environmental requirements. These responsibilities should have clearly defined goals, strategies, and measures of success.

COMBATANT COMMAND AND SUBORDINATE JOINT FORCE ENGINEER

K-5. The COCOM and subordinate joint force engineer are responsible for providing guidance to the combatant commander and/or subordinate JFC on environmental considerations in planning and executing joint operations. The COCOM and subordinate joint force engineer and staff develop and assist with implementing the policies, procedures, and practices of the Environmental Considerations annex to an OPLAN and/or OPORD. *See CJCSI 3122.03A* for additional information.

COMBATANT COMMAND AND SUBORDINATE JOINT FORCE STAFF JUDGE ADVOCATE

K-6. The COCOM and subordinate joint force SJA advise the commander on compliance with environmental laws, regulations, treaties, conventions, OEBGDs, and SOFAs and their potential impact on operations. Specifically, the COCOM and subordinate joint force SJA are responsible for legal support in the development of the environmental considerations annex to an OPLAN and/or OPORD to ensure that legal requirements related to environmental considerations are incorporated, as appropriate. The joint force SJA—

- Assists other members of the joint force staff (J-4) and defense agencies in negotiating transit agreements in advance of the actual deployment, to permit the transit of regulated (hazardous) waste and ensure they are disposed of in an environmentally sound manner.
- Helps determine baseline environmental survey requirements and processes civilian claims resulting from environmental damage.

- Participates in the development of baseline environmental survey exemptions.

COMBATANT COMMAND AND SUBORDINATE JOINT FORCE SURGEON

K-7. The COCOM and subordinate joint force surgeon are responsible for health services support (preventive medicine and occupational health) to the joint force. Priorities are on water and wastewater, including water vulnerability assessment support, sanitation, waste disposal (hazardous and infectious waste); health risk assessment (base-camp site selection); environmental health sampling; and surveillance and vector control to protect human health and welfare.

JOINT FORCE PUBLIC AFFAIRS OFFICER

K-8. The joint force public affairs officer (PAO) coordinates with the appropriate staff and commanders to plan and accomplish public relations efforts in support of mission objectives. Special attention should be given to potentially sensitive environmental issues associated with a joint operation. In the US and its territories and possessions, various environmental laws require public involvement. The joint force PAO will be a significant participant in public outreach efforts. He should participate in developing assigned responsibilities in environmental contingency plans.

JOINT FORCE LOGISTICS DIRECTORATE

K-9. The joint force J4 and staff, when augmented by engineer personnel, are responsible for ensuring that waste and effluents from operations materiel and service functions are appropriately controlled. The joint force J4 and staff—

- Are responsible for all aspects of hazardous materials and regulated (hazardous) waste management, to include minimizing the use, storage, transportation, disposition, and redeployment of excess materials.
- Should coordinate closely with the joint force SJA in negotiating transit agreements and establishing procedures for the turn-in of regulated (hazardous) waste for proper treatment and disposal.

JOINT ENVIRONMENTAL MANAGEMENT BOARD

K-10. A JEMB may be established by the combatant commander or subordinate JFC for a joint operation to integrate the environmental-protection efforts of all participating components under a single authority, ensuring unity of effort for environmental protection activities. The JEMB should be chaired by the COCOM or subordinate joint force J4 or joint force engineer and include representatives from each service component and joint force staff representative, as necessary (legal, occupational health, preventive medicine, safety, comptroller, planning, operations, and logistics). The JEMB assists the JFC in establishing the joint force environmental policies, practices, procedures, and priorities and in providing oversight of environmental protection standards and compliance. Establishing a dedicated and appropriately staffed environmental engineering staff, supported by experts from other joint force staff members (legal and medical), may obviate the need for a JEMB in smaller operations.

OTHER FEDERAL AGENCIES AND NGOS

K-11. During operations, such as those involving responses to disasters or support to civilian governmental agencies (cleanup of major oil or hazardous substance spills), the JFC may have to work with other federal agencies or NGOs to ensure successful completion of the operation. Where appropriate, these representatives should be a part of the environmental planning process. The JFC may also consider their participation as ad hoc members of the JEMB.

ARMY FORCES COMMANDER AND STAFF

K-12. Similar to joint-level commanders and staffs, the ARFOR commander and staff have inherent responsibilities within their areas of expertise that require environmental actions. Whether developing the staff estimate, protection levels, or EBS, environmental protection requires active participation from each staff member. Environmental factors may affect or influence a wide range of activities or require a significant expenditure of resources. The unit staff integrates environmental considerations into the planning and execution processes. Common staff duties provide the basis for some environmental responsibilities, and *FM 101-5* provides a basis for others. While some of these responsibilities may depend on the command or commander, all staffs undertake many of them. The commander will use the staff engineer as an integrator of environmental considerations, which is similar to the role of the G2/S2 in the IPB process.

SUBORDINATE COMMANDERS AND STAFFS

K-13. Unit commanders are responsible for complying with the applicable environmental requirements established by the commander in the environmental considerations appendix of the OPLAN and/or OPORD. Subordinate commanders should keep the higher command staff informed of conditions that may result in noncompliance or the potential for noncompliance with this appendix. Unit SOPs at battalion and company levels must incorporate specific responsibilities. Unit commanders should appoint primary and alternate environmental compliance officers (ECOs) to act as liaisons to higher command engineering and environmental staff and communicate applicable issues to the commander and unit personnel in both training and real-world operations.

ENGINEER COORDINATOR

K-14. The ENCOORD is the special staff officer for coordinating engineer assets and operations for the command. As the senior engineer officer in the force, he advises the commander on environmental issues, to include the command environmental program. Working with other staff officers, the ENCOORD determines the impact of operations on the environment and integrates environmental considerations into the decision-making process. He works with the G4 in performing site assessments for installations and facilities. With the SJA, the ENCOORD advises the commander on the necessity for environmental assessments to meet HN or Executive Order (EO) 12114 requirements. He is also responsible for advising the G2 and S2 of significant environmental factors and ensuring these impacts are integrated into the IPB process. *See FM 34-130* for more information. Additionally, *FM 101-5* directs that OPLANs, OPORDs, and CONPLANs contain an appendix to address environmental considerations. Through coordination with the unit staff, the ENCOORD prepares Appendix 2, Environmental Considerations to Annex F, Engineer, for a joint order. *See FM 3-100.4* for the specific appendix format and content.

ENVIRONMENTAL REQUIREMENTS

GENERAL

K-15. Environmental requirements can be divided into overseas requirements and requirements applicable in the US and its territories and possessions, although some US environmental requirements may have extraterritorial application. For example, EO 12114 establishes requirements for the conduct of environmental studies for activities conducted overseas, somewhat similar to the environmental analysis requirements mandated by the National Environmental Policy Act (NEPA) regarding operations conducted within the US. The SJA should be consulted to determine extraterritorial applicability. The activation and incorporation of environmental management systems (EMSs) is critical for all DOD organizations *(see EO 13148)* regardless of whether they operated in CONUS or OCONUS. *See FM 3-100.4* for a more focused discussion on environmental requirements affecting Army operations.

REQUIREMENTS APPLICABLE WITHIN THE UNITED STATES

K-16. All military operations within the US and its territories and possessions are conducted in compliance with applicable federal, state, or local environmental laws and regulations. NEPA requires environmental planning for major federal actions that have the potential for a significant environmental impact. It requires that commanders make environmental considerations an integral part of the mission-planning and decision-making processes. NEPA mandates procedures for environmental planning but does not mandate decisions. Other federal statutes, such as the Clean Water Act (CWA); Clean Air Act (CAA); Resource Conservation and Recovery Act (RCRA); and the Comprehensive Environmental Response, Compensation, and Liability Act (CERCLA), establish environmental requirements that may impact joint operations. During MOOTW, all legal requirements imposed by various jurisdictions during peacetime may still apply. In many cases, US environmental laws allow for national security exemptions for specific activities, but only upon action by the President.

REQUIREMENTS APPLICABLE BEYOND UNITED STATES TERRITORY

K-17. All operations beyond US territory are conducted according to applicable international treaties, conventions, OEBGDs, SOFAs, final governing standards (FGS), and other international agreements. In the absence of definitive environmental guidance within applicable international agreements, geographic combatant commanders and subordinate commanders should establish guidance in the OPLAN and/or OPORD that protect force health, limit adverse public health impacts, consider US liability, and remain consistent with mission goals.

ENVIRONMENTAL PLANNING

THE NEED FOR ENVIRONMENTAL PLANNING

K-18. By considering environmental issues early during the planning process, the commander may continue to achieve operational objectives while minimizing the impact on human health and the environment. Failure to consider the environmental impact of all activities may adversely affect the operation. Potential effects include endangering personnel, delaying operation commencement, limiting future use of exercise or HN areas, and creating an adverse public opinion. Through early assessment of environmental considerations, commanders may become aware of the potential environmental effects or impacts of mission accomplishment while alternatives to address mitigating actions still exist. By planning early, the commander and staff must be aware of the environmental requirements and able to plan more effectively and act to incorporate environmental considerations. Furthermore, careful and visible attention to environmental considerations in military operations can assist in shaping a positive image both internationally and domestically.

K-19. Integrating environmental considerations into planning is very similar to integrating safety and force protection issues. *FM 3-100.4* discusses environmental planning and focuses on how and where the Army integrates environmental considerations into the MDMP. As part of the MDMP, risk management is an effective process to minimize actions that may negatively impact the environment and take appropriate steps to prevent or mitigate damage.

ELEMENTS OF ENVIRONMENTAL PLANNING

K-20. The staff should plan the operation to achieve the mission objectives while minimizing the environmental effects and observing environmental requirements. Although not all elements *(See Figure K-1)* are applicable to all operations (some, such as the identification of alternatives to obtaining objectives, are not required for OCONUS operations), they may prove helpful during planning operations to—

- Identify the operational objectives and the activities that are proposed to obtain these objectives, including the logistics requirements and the identification of HAZMATs that may be used.
- Identify potential alternative means of obtaining the operational objectives. These alternatives may include the use of new technologies to minimize the impact on the environment.
- Identify the environmental requirements applicable to the operational area.
- Identify adverse environmental health and environmental impacts that may result from the operation.

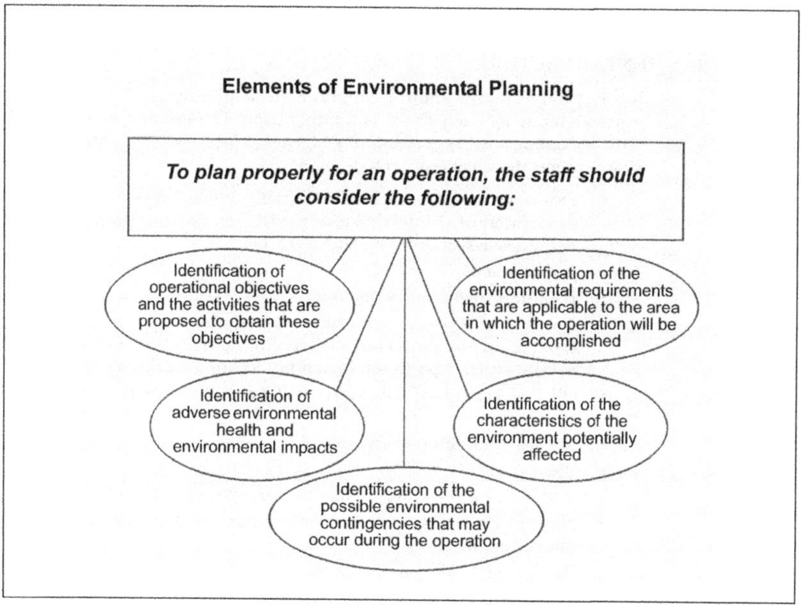

Figure K-1. Elements of Environmental Planning

- Establish formal relationships and coordinate with other disciplines that have roles in environmental planning and operations (medical and legal).
- Identify the environmental characteristics of the potentially affected area.
- Identify possible environmental contingencies that may occur during the operation, such as accidental spills.
- Determine how the environmental contingency would affect the environment in the operational area and how it could be prevented or mitigated should it occur.
- Determine the environmental and operational risk associated with the operation. If risks are unacceptable, identify alternatives that mitigate associated risks.
- Establish standardized identification (signage or markings) of off-limits or high-risk areas.
- Negotiate applicable agreements to allow for the unimpeded transit of HAZMAT or waste by military and contracted assets for environmentally sound treatment or disposal.
- Determine contractor status, to include privileges and immunities in support of the operation.

- Identify environmental resources and reach capabilities.

KEY ENVIRONMENTAL FACTORS

K-21. Commanders should consider environmental and force health protection during each phase of an operation. In planning and conducting military operations, regardless of geographic location, commanders should give appropriate consideration to the following:

- Preexisting environmental conditions impacting site selection and environmental health vulnerabilities, as well as potential US liabilities associated with the operation.
- Air emissions.
- HAZMAT, including pesticides.
- Hazardous waste. Appropriate disposition could include recovery, treatment, or disposal within the operational area or, where necessary, transportation to another location for these purposes.
- Oil and hazardous substance spill prevention, control and response training.
- Medical and infectious waste.
- Solid waste.
- Water and wastewater, to include sanitary wastewater.
- Natural resources, to include endangered or threatened species and marine mammals.
- Historic and cultural resources.
- Noise abatement.
- Resource and energy conservation through pollution prevention practices.
- Camp closure and site cleanup before redeployment.
- Incident reporting and documenting of cleanup actions.
- The transportation of excess material and equipment from the tactical area in an environmentally sound manner.
- Ensuring that contractors and their vehicles have unhindered transit across international borders.

RISK MANAGEMENT

K-22. *FM 101-5* describes risk management as the process of detecting, assessing and controlling risk arising from operational factors and balancing risk with mission benefits. Risk management is an integral part of the MDMP. *FM 100.14* outlines the risk management process and provides the framework for making risk management a routine part of planning, preparing, and executing operational missions and everyday tasks. *FM 3-100.4* further clarifies the elements of risk when focused on environmental considerations. Assessing environmental-related risks is part of the total risk management process. Knowledge of environmental factors is key to planning and decision-making. With this knowledge, leaders quantify risks, detect problem areas, reduce risk of injury or death, reduce property damage, and ensure compliance with environmental laws and regulations. Unit leaders should conduct risk assessments before conducting any training, operations, or logistical activities.

Appendix L

Explosive Ordnance Disposal Organizations and Functions

The mission of EOD is to eliminate or reduce chemical, biological, radiological, nuclear, and CBRNE hazards and protect the commander's combat power. UXO limits battlefield mobility, denies the use of critical assets, and threatens to injure or kill soldiers at levels unprecedented in past wars. UXOs, to include improved conventional munitions (ICM) and IED, have greater emphasis now and in the future because of the potential of significantly reducing the commander's combat power. The continuing development of foreign and US smart and brilliant munitions that disperse hundreds of submunitions and area denial ordnance has lead to the proliferation of UXOs on the battlefield. These munitions are available for a range of weapon systems, including artillery, ballistic and cruise missiles, rockets, and bombs.

INTRODUCTION TO EXPLOSIVE ORDNANCE DISPOSAL

L-1. The EOD force is prepared to deal with the increased quantity, quality, and lethality of UXOs. Sophisticated fuzing and sensors systems developed for the 21st century have the capability to detect, identify, and select specific targets using infrared, proximity, magnetic influence, acoustic, and seismic technologies. Attempts to approach and perform a render-safe procedure (RSP) on these munitions by the traditional 20th century methods will cause detonation of the devices. This is because of their wide kill radius 750 feet (greater than 750 feet [250 meters]) added to the hazards of their antidisturbance, antiremoval, antilift, random-delay, and self-destruct features. UXOs also present access problems from—

- Toxic chemicals from rocket motors and guidance systems.
- Ordnance exposed to fire, especially newer ordnance made of lightweight metals and plastics.
- The location. Destroyed vehicles and aircraft have unique dangers like depleted uranium (DU) and carbon fiber.
- Confined space, bunkers, caves and tunnels.

L-2. EOD missions include—

- Responding to counter terrorism, WMD, and UXO incidents on the sea, in the air, and on land.
- Supporting the United States Secret Service (USSS), DOS, Department of Justice (DOJ), Department of Energy (DOE), and FEMA.
- Advising and assisting civil authorities in the remediation of military ordnance that poses a threat to public safety.

- Providing education on the hazards of UXO.
- Examining, identifying, and reporting new and unusual explosive ordnance for technical intelligence purposes.
- Supporting nuclear and chemical weapons shipments.
- Conducting a range sweep by disposing of UXOs on impact areas.
- Destroying ammunition and routine ammunition stocks and conducting emergency destruction of ammunition to prevent capture by the enemy (including sensitive site exploitation).
- Responding to increased lethality of UXOs, IEDs, and WMD.
- Removing rounds stuck in artillery tubes and other large-caliber weapons.
- Advising and assisting in the instruction of UXOs for humanitarian demining missions.
- Clearing enemy UXOs and booby traps from a captured ASP, airfield, air defense battery, command post, or other key objectives.
- Clearing US dropped UXOs from enemy targets—ASP, airfield, air defense battery, and CP, or other key objectives that have been captured by US forces.
- Responding to an enemy sapper attack clearing satchel charges from targets—aircraft, vehicles, stacks of ammunition, and dead sappers or suicide bombers.
- Responding to terrorist attacks against US facilities with car or truck bombs—American embassies, consulates, and military barracks.
- Retrieving casualties from a minefield and evacuating survivors to medical facilities while assisting graves registration with fatalities.
- Clearing enemy aircraft, armor, artillery, and other materiel identified as having significant intelligence value of booby traps and hazardous ordnance.
- Collecting war souvenirs from departing US forces.

EXPLOSIVE ORDNANCE ORGANIZATION

GROUP HEADQUARTERS

L-3. The EOD group commander serves as the ASCC, G3, and theater EOD special staff officer. The EOD group provides C2 of all Army EOD assets and operations in theater. When directed by the JFC, the EOD group commander becomes the commander of the joint EOD TF and coordinates all EOD assets within theater.

L-4. The EOD group provides C2, mission tasking, theater EOD planning, technical intelligence acquisition and management, and limited administrative and logistics support for two to six EOD battalions. In operations without a fully deployed theater or EOD group a C2 element of an EOD group, will deploy to provide C2 and staff planning for deployed EOD battalions.

BATTALION HEADQUARTERS

L-5. The EOD battalion exercises C2 for three to seven EOD companies in the AO. The commander of the EOD battalion is the EOD officer for the corps. He monitors operations and develops plans to meet the needs of the combatant commander, providing an LO team, as required.

L-6. The EOD battalion provides C2, mission tasking, EOD planning, technical intelligence acquisition and management, and limited administrative and logistics support for up to seven EOD companies.

COMPANY

L-7. The EOD company exercises C2 for two EOD response sections. The primary function of the EOD company is to provide support as directed by the EOD battalion; it does this in a variety of ways. An EOD company provides GS support to assigned AORs and all units within it. Dependent upon METT-TC considerations, this may require the company to perform split-based operations to fully support mission requirements. An EOD company is tasked-organized by the EOD battalion commander and is typically attached to a separate organization for administrative and logistical support. The EOD company commander may further task-organize EOD teams to division or BCT's areas to conduct EOD operations in support of maneuver elements. The commander of an EOD company exercises C2 of the company throughout his assigned AOR, to include split-based operations and fragmented team operations. He must also coordinate and conduct liaison functions with various supported and supporting units, to include civil and HN authorities and other agencies.

L-8. The EOD company provides the ability to eliminate or reduce the hazards of domestic and/or foreign conventional, NBC munitions, and IEDs that threaten personnel, military operations, facilities, and materiel. The EOD company exploits technical intelligence by submitting reports on first-seen ordnance. It provides support to the USSS to protect the President, Vice President, and others, as directed. It also provides support to the FBI and the DOE about CT with emphasis on IEDs.

ENGINEER AND EXPLOSIVE ORDNANCE CONSIDERATIONS

L-9. Army engineers and ordnance (EOD) units have a unique relationship compared to other armies and services around the world. Several allied countries (the United Kingdom, France, Canada, and Australia) teach EOD skills in addition to combat engineer training; we do not. Army engineers and ordnance are two distinct branches and organizations. The role of each function is unique but similar, and this necessitates close coordination when operating in today's operational environment. Engineers are responsible for assuring the mobility of combat forces. The Army's combat engineers approach mine detection and neutralization using three strategies—metal detection, explosive neutralization, and mechanical neutralization. Detection technologies focus on identifying the metal content of mines. Explosive neutralization applies systems such as MICLICs to detonate the mines. The mechanical neutralization applies plows and rollers to push the mines aside or detonate them by pressure. All of these means require the use of intelligence sources and reconnaissance to detect mines and use the appropriate combination of explosive or mechanized means to neutralize and breach conventional minefields. Explosive and mechanized neutralization strategies may not be appropriate in all operations, such as in urban areas. EOD personnel can task-organize directly to a maneuver unit to render safe or neutralize booby traps, UXOs, and IEDs. (Booby traps have increased in sophistication in recent years and may include electronic circuitry—light sensors, motion sensors, and command detonation.) *See FM 3-100.38 (100-38)* and *FM 13-8* for additional multiservice considerations and UXO procedures.

L-10. The engineer commander and/or staff engineer should consider the following when working with EOD units:

- Ensure that you know the EOD LO and work together in the planning process.
- Exchange information on the mines, booby traps, UXOs, and IEDs.
- Work with the S2 and G2 to identify the types of booby traps expected.
- Include the EOD in the planning process and request EOD intelligence of the region.
- Establish and operate a training scenario to teach mine awareness and the most current procedures.
- Use combat engineers to detect, mark, record, and report booby traps.
- Know that EOD is task-organized to evaluate and render safe booby traps, IEDs, and UXOs.
- Know that combat engineers can remove standard structural demolitions that are not booby-trapped.

Glossary

The glossary lists acronyms/abbreviations and terms with Army or joint definitions, and other selected terms. Where Army and joint definitions are different, (Army) follows the term. Terms or acronyms for which *FM 3-34* is the proponent manual (the authority) are marked with an asterisk (*).

1SG	first sergeant
A2C2	Army airspace command and control
A&O	assault and obstacle
AA	assembly area
AA	TOE numbering system variation for a firefighting headquarters
AAP	Allied administrative publication
AAV	assault amphibious vehicle
AB	TOE numbering system variation for a firefighting team
ABCA	American, British, Canadian, and Australia
ABCS	Army Battle Command System
ABE	assistant brigade engineer
ABO	air base operability
ABN	airborne
AC	Active Component
ACB	amphibious construction battalion
ACE	assistant corps engineer
ACofS	assistant chief of staff
ACR	armored cavalry regiment
AD	air defense
ADCOM	administrative command
ACS	assistant chief of staff
ADA	air defense artillery
ADC	area damage control
ADCON	administrative control
ADE	assistant division engineer
ADP	automatic data processing
AECA	Arms Export Control Act
AEF	aerospace expeditionary force
AFB	assault float bridge

AFCAP	Air Force Contract Augmentation Program
AFCESA	Air Force Civil Engineering Support Agency
AFDC	Air Force Doctrine Center
AFDD	Air Force doctrine document
AFFOR	Air Force forces
AFSOC	Air Force Special Operations Command; Air Force special operations component
AGS	aviation ground support
AIGIS	Army imagery and geospatial information systems
AISI	automated integrated survey instrument
AJP	allied joint publication
AL	Alabama
ALSA	Air Land Sea Application (Center)
alt	alternate
ALSS	advanced logistics support site
AMC	Army Materiel Command
AMOPES	Army Mobilization and Operations Planning and Execution System
AO	area of operations
AOI	area of interest
AOR	area of responsibility
AP	antipersonnel
APC	armored personnel carrier
APLA	antipersonnel land mine alternatives
APOD	aerial port of debarkation
APOE	aerial port of embarkation
AR	Army regulation
ARC	American Red Cross
ARCENT	Army Forces Central Command
ARFOR	Army forces
ARNG	Army National Guard
ARSOF	Army Special Operations Forces
ART	Army tactical task
ASAS	All-Source Analysis System
ASAS-RWS	All Source Analysis System—Remote Workstation
ASCC	Army service component commander (FM 101-5-1)
ASG	area support group
ASL	authorized stockage list

ASP ammunition supply (or storage) point

ASR alternate supply route

assault breach A breach tactic used by small unit (company, teams, and platoons) to penetrate an enemy's protective obstacles and seize a foothold within his defense; it is normally a very decentralized operation with SOSRA actions synchronized at the platoon and company level. (FM 3-34.2)

***assured mobility** Actions that give the force commander the ability to maneuver where and when he desires without interruption or delay to achieve the mission. (FM 3-34)

***assured mobility imperatives** There are four imperatives: (1) Develop a mobility COP:—leverage information dominance by using geospatial tools to combine terrain data, and an integrated R&S to gain/attain a level of understanding from friendly and enemy perspectives; (2) select, establish, and maintain operating areas—gain ownership of friendly and identification of enemy; EAs and TAIs, operating areas, and the mobility corridors that are the LOCS connecting those areas; (3) attack the enemy's ability to influence operating areas—proactively allocate combat power and sensors to secure our ability to maneuver) operating areas, LOCs, TAIs and EAs) and or attack the enemy's operating areas and or maneuver to prevent his ability to influence our maneuver; (4) maintain mobility and momentum—synchronize all the BOSs capabilities to protect and sustain our established ability to maneuver when and where we wish, enabling us to maintain pressure and lethality upon the enemy for legacy, interim and future organizations. (FM 3-34)

***assured mobility (fundamentals)** Predict, detect, prevent, avoid, neutralize and protect—these fundamentals support the assured mobility imperatives. (FM 3-34)

ASL authorized stock list

AT antitank

AUTL Army universal task list—a comprehensive listing of Army tactical-level task, missions, and operations; the AUTL complements CJCSM 3500.04B by providing tactical-level Army-specific tasks. (FM 7-15)

AVLB armored vehicle-launched bridge

AWE Army warfighting experiment

bare base (Joint) A base having minimum essential facilities to house, sustain, and support operations to include, if required, a stabilized runway, taxiways, and aircraft parking areas. A bare base must have a source of water that can be made potable. Other requirements to operate under bare base conditions form a necessary part of the force package deployed to the bare base. (JP 3-05.3/JP 1-02)

barrier A coordinated series of obstacles designed or employed to channel, direct, restrict, delay, or stop the movement of an opposing force and to impose additional losses in personnel, time, and equipment on the opposing force; barriers can exist naturally, be man-made, or a combination of both. (JP 3-15/JP 1-02)

barrier, obstacle and mine warfare plan (joint)—A comprehensive, coordinated plan that includes responsibilities; general location of unspecified and specific barriers, obstacles, and minefields; special instructions; limitations; coordination; and completion times; the plan may designate locations of obstacle zones or belts. It is normally prepared as an annex to a campaign plan, OPLAN, or OPORD. (JP 3-15/JP 1-02)

base development (less force beddown) (joint)—Acquisition, development, expansion, improvement, and construction, and/or replacement of the facilities and resources of an area or location to support forces employed in military operations or deployed according to the strategic plans. (JP 4-04/JP 1-02)

baseline environmental survey (joint)—Multidisciplinary site survey conducted before or in the initial stage of a joint operational deployment; the survey documents existing deployment-area environmental conditions determines the potential for present and past site contamination (hazardous substances, petroleum products, and derivatives) and identifies potential vulnerabilities (to include occupational and environmental health risks). Surveys accomplished in conjunction with joint operational deployments that do not involve training or exercises (contingency operations) should be completed to the extent practicable consistent with operational requirements. (JP 4-04/JP 1-02)

BASOPS base operations

battlespace (Joint)—The environment, factors, and conditions that must be understood to successfully apply combat power, protect the force, or complete the mission. Included is the air, land, sea, space, enemy and friendly forces, facilities, weather, terrain, electromagnetic spectrum, and the information environment within the operation areas and AOIs. (JP 1-02)

BB bare base—(joint)—A base having minimum essential facilities to house, sustain, and support operations to include, if required, a stabilized runway, taxiways, and aircraft parking areas. A BB must have a source of water that can be made potable. Other requirements to operate under BB conditions form a necessary part of the force package deployed to the BB. (JP 3-05.3/JP 1-02)

BCAT base camp assistance/assessment team

BCCA Base Camp Coordination Agency

BCOC Base Cluster Operations Center (FM 3-90)

BCT brigade combat team

BDA battle damage assessment

BDOC Base Defense Operations Center

***BDT** base development team—Nondeployable team that can quickly provide base development engineering and planning and facilities design for staging bases, base camps, FOBs, displaced persons camps, and any similar requirement. (FM 3-34)

BDP base development plan

BDR battle damage repair

BEEF base engineer emergency force

BEFV Bradley engineer fighting vehicle

BFT battle-focused training

BFV Bradley fighting vehicle

BOM bill of materials

BOS battlefield operating system—The physical means (soldiers, organizations, and equipment) that tactical commanders use to execute operations and accomplish missions assigned by superior tactical- and operational-level commanders. The seven BOS are (1) intelligence system—the activity to generate knowledge of and products portraying the enemy and the environmental features required by a command planning, preparing, executing, and assessing operations; (2) maneuver system—the movement of forces to achieve a position of advantage with respect to enemy forces. This system includes employing forces in combination with direct fire or fire potential. It also includes the conduct of tactical tasks associated with force protection; (3) fire support system—collective and coordinated use of target-acquisition; data, indirect-fire weapons, fixed-wing aircraft, offensive information operations, and other lethal and nonlethal means against targets located throughout an AO; (4) air defense system—employment of all active measures to nullify or reduce the effectiveness of attack by hostile aircraft and missiles after they are airborne; (5) M/CM/S system—mobility operations preserve the freedom of maneuver of friendly forces, countermobility operations deny mobility to enemy forces, and survivability operations protect friendly forces from the effects of enemy weapon systems; (6) CSS system—the support and services provided to sustain forces during war and MOOTW; (7) C2 system—all collective tasks associated with supporting the exercise of authority and direction by a properly designated commander over assigned and available forces in the accomplishment of the mission. (FM 7-15)

breach A tactical mission task in which the unit employs all available means to break through or secure a passage through an enemy defense, obstacle, minefield, or fortification. (FM 3-90)

breach area Area that is established and fully defined by the higher HQ of the unit conducting breaching operations; it is the area where a breaching operation occurs. (FM 3-34.2)

breach force A combined-arms force that is task-organized with the maneuver and engineer forces necessary to reduce obstacles and create lanes through an obstacle to pass initial assault forces through the lanes. (FM 3-34.2)

breaching fundamentals Suppress, obscure, secure, reduce and assault (SOSRA) are the breaching fundamentals that must be applied when breaching against a defending enemy. These fundamentals will always apply, but they may vary based on the specific battlespace situation. (METT-TC) (FM 3-34.2)

breaching operations Operations conducted to allow maneuver despite the presence of obstacles; breaching is a synchronized combined-arms operation under the control of the maneuver commander. Breaching operations begin when friendly forces detect an obstacle and begin to apply the breaching fundamentals, and they end when battle handover has occurred between follow-on forces and a unit conducting the breaching operation. (FM 3-34.2)

breach tenets Intelligence, breach organization, breach fundamentals, mass, and synchronization. (FM 3-34.2)

BTRA battlefield terrain reasoning and awareness

C2 command and control

C2PC command and control personal computer

C3 command, control, and communications

C4ISR command, control, communications, computers, intelligence, surveillance, and reconnaissance

C&S command and support

CA civil affairs

CAA Clean Air Act

CALL Center for Army Lessons Learned

camouflage (joint)—Use of natural or artificial material on personnel, objects, or tactical positions with the aim of confusing, misleading, or evading the enemy. (JP 1-02/FM 20-3)

CAO chief administrative officer

CAP crisis action plan(ning)

CBMU construction battalion maintenance unit

CBRNE chemical, biological, radiological, nuclear, and high-yield explosive

CBRNE-CM chemical, biological, radiological, nuclear and high-yield explosives consequence management

CBT combating terrorism

cbt combat

CBU construction battalion unit

CCA contract construction agent

CCD camouflage, concealment, and deception

CCIR commander's critical information requirements

CDMP combat decision making process

cdr commander

CEB combat engineer battalion

***CELL** Center for Engineer Lessons Learned

CENTCOM Central Command

CERL Construction Engineering Research Laboratory

CERCLA Comprehensive Environmental Response, Compensation, and Liability Act

CESE civilian engineer support equipment

CESP civil engineering support plan—(joint)—An appendix to the logistics annex or separate annex of an OPLAN that identifies the minimum essential engineering services and construction requirements required to support the commitment of military forces. (JP 4-04/JP 1-02)

CESPG civil engineering support plan group; civil engineering support planning generator

CG commanding general

cgo cargo

civil augmentation program (joint)—Sanding, long-term contracts designed to augment service logistic capabilities with contract support in both preplanned and short notice contingencies; examples include US Army LOGCAP, US Air Force contract augmentation program, and US Navy Construction Capabilities Contract. (JP 4-07)

civil engineering (joint)—Those CS and CSS activities that identify, design, construct, lease, or provide facilities and which operate, maintain, and perform war damage repair and other engineering functions in support of military operations. (JP 4-04/JP 1-02)

CJCS Chairman of the Joint Chief of Staff

CJCSI Chairman of the Joint Chiefs of Staff, Instruction

CJCSM Chairman of the Joint Chiefs of Staff manual

CJTF commander, joint task force

***clear** (1) The total elimination or neutralization of an obstacle that is usually performed by follow-on engineers and is not done under fire (FM 3-34/FM 101-5-1); (2) A tactical mission task that requires the commander to remove all enemy forces and eliminate organized resistance in an assigned area. (FM 3-90/FM 101-5-1)

clearing operation (joint)—Operation designed to clear or neutralize all mines and obstacles from a route or area. (FM 3-34.2/FM 101-5-1/JP 3-15/JP 1-02)

CM consequence management—(joint)—Those measures taken to protect public health and safety, restore essential government services, and provide emergency relief to governments, businesses, and individuals affected by the consequences of an NBC and/or high-yield explosive situation; for domestic consequence management, the primary authority rest with the States to respond and the Federal Government to provide assistance as required. (JP 3-0/JP 1-02)

cmd command

CMO civil-military operations

CMOC civil-military operations center

co company

COA course of action

COB collocated operating base

COCOM combatant command

COE contemporary operational environment—(joint)—Automation services that support the development of the common reusable software modules which enable interoperability across multiple CS application; includes segmentation of common software modules from existing applications, integration of commercial products, development of a common architecture, and development of common tools for application developers. (JP 1-02)

***combat engineering** (joint)—Those engineering task that assist the tactical and/or operational commander to "shape" the battlespace by enhancing mobility, creating the space or time necessary to generate mass and speed while protecting the force, and denying mobility and key terrain to the enemy; these tasks include breaching, bridging and emplacement of obstacles to deny mobility to the enemy. (JP 3-34/JP 1-02); one of the engineer battlespace functions that includes mobility, countermobility and survivability. (FM 3-34)

combat support engineering (NATO)—An Allied term that encompasses those military engineer tasks associated with direct support to current or imminent operations. It is conducted by the military engineers of any service or component to support land, air, maritime, and information operations. With the emphasis on speed of execution, it frequently relies on prefabricated equipment solutions, may involve a degree of improvisation and is likely to fulfill a short-term tactical requirement. It is closely aligned with the Engineer Battlespace function of Combat Engineering. (AJP 3.12)

comdt commandant

commo communications

COMMZ communications zone

COMPT comptroller

COMSEC communications-security

CON construction

CONCAP construction capabilities contract (Navy)

CONPLAN concept or contingency plan

contingency engineering management organization (joint)—An organization that may be formed by the combatant commander, or subordinate joint force commander, to augment the engineering expertise to support both deliberate and crisis action planning and to provide construction management in contingency and wartime operations; the combatant commander may form a theater contingency engineering management cell, and similar organizations may be formed at subordinated levels of command (RCEM cell and/or JTFCEM cell). These organizations should be staffed with expertise in combat engineering, general engineering and topographic (geospatial) engineering. (JP 3-34)

const construction

CONUS continental United States

COP common operational picture—(joint)—An operational picture tailored to the user's requirements, based on common data and information shared by more that one command. (JP 3-0/JP 1-02)

COR contracting officer's representative

COS chief of staff

COSCOM corps support command

countermine (NATO, joint)—To explode the main charge in a mine by the shock of a nearby explosion of another mine or independent explosive charge; the explosion of the main charge may be caused either by sympathetic detonation or through the explosive train and/or firing mechanism of the mine. (JP 1-02); (Army)—the actions taken to detect, bypass, breach, mark, report, record, and eliminate enemy mines or minefields. (FM 20-32)

countermine operations (joint)—In land warfare, an operation to reduce or eliminate the effects of mines or minefields. (JP 1-02) See CJCSI 3207.01.

countermining (joint)—(1) Land mine warfare—tactics and techniques used to detect, avoid, breach, and/or neutralize enemy mines and the use of available resources to deny the enemy the opportunity to employ mines; (2) naval mine warfare—the detonation of mines by nearby explosions, either accidental or deliberate. (JP 1-02)

***countermobility** A component of combat engineering and one of the five engineer battlespace functions. It augments natural terrain with obstacle systems according to the commander's concept. This adds depth to the battle in space and time by attacking the enemy's ability to maneuver his forces. (FM 3-34/FM 101-5-1)

***countermobility operations** (joint)—The construction of obstacles and emplacement of minefields to delay, disrupt, and destroy the enemy by reinforcement of the terrain; the primary purpose of countermobility operations is to slow or divert the enemy, to increase time for target acquisition, and to increase weapons effectiveness. (FM 3-34/FM 101-5-1/JP 3-34/JP 1-02)

covert breach A breach tactic used when the force must reduce lanes through enemy tactical or protective obstacles undetected; in the covert breach, suppression from the support force is a be-prepared task upon detection of the breach force, or an on-order once the breach is completed and the assault is initiated. (FM 3-34.2/FM 101-5-1)

CP command post

CPW Center for Public Works

CRREL Cold Regions Research and Engineering Laboratory

***CREST** contingency real estate support team (FM 3-34)

CS civil support—DOD support to US civil authorities for domestic emergencies and for designated law enforcement and other activities (FM 3-07/JP 3-07.7); combat support.

CSA Chief of Staff, United States Army

CSE combat support equipment

CSS combat service support

CSR controlled supply rate

CSS-BOS The support and service provided to sustain forces during war and MOOTW; contains the subordinate Army Task (ART) of Provide General Engineering Support. (FM 7-15)

CSSE combat service support element—(joint)—Those elements whose primary missions are to provide service support to combat forces and which are a part, or prepared to become a part, of a theater, command, or task force formed for combat operations; the core element of a MAGTF that is task-organized to provide the CSS necessary to accomplish the MAGTF mission. The CSSE varies in size from a small detachment to one or more force service support groups. It provides supply, maintenance, transportation, general engineering, health services, and a variety of other services to the MAGTF. The CSSE itself is not a formal command. (FM 101-5-1/JP 1-02)

CT counterterrorism

CTC Combat Training Center

CTIS Combat Terrain Information System

CUL common-user logistics

cultural resource Monuments, nationally identifiable or distinctive buildings and structures, archives and libraries, ancient artifacts and structures, archaeologically important sites, historically import sites or structures, mosques, cathedrals, temples, other churches or sacred structures, sacred sites or areas, museums, and works of art. (FM 101-5-1)

CWA Clean Water Act

CWD civil works directorate

CZ combat zone

DA Department of the Army

datum (NATO, Joint)—Any numerical or geometrical quantity or set of such quantities which may serve as reference or base for other quantities. Where the concept is geometric, the plural form is "datums" in contrast to the normal plural "data". (JP 1-02)

datum (geodetic) (joint)—(1) A reference surface consisting of five quantities: the latitude and longitude of an initial point, the azimuth of a line from that point, and the parameters of the reference ellipsoid; (2) the mathematical model of the earth used to calculate the coordinates on any map; different nations use different datums for printing coordinates on their maps. The datum is usually referenced in the marginal information of each map. (JP 1-02)

DCD directorate of combat developments

DCMA Defense Contract Management Agency

DCSENG Deputy Chief of Staff for Engineering

DCSINT deputy chief of staff for intelligence

DEA Drug Enforcement Administration

defensive minefield (joint)—In land mine warfare, a minefield laid in accordance with an established plan to prevent a penetration between positions and to strengthen the defense of the positions themselves. (JP 1-02)

delegation of authority (joint)—Action by which a commander assigns part of his authority commensurate with the assigned task to a subordinate commander; while ultimate responsibility cannot be relinquished, delegation of authority carries with it the imposition of a measure of responsibility; the extent of the authority delegated must be clearly stated. (JP 1-02)

deliberate breaching (NATO, joint)—Creation of a lane through a minefield or a clear route through a barrier or fortification, which is systematically planned and carried out. (FM 3-34.2/FM 101-5-1/JP 1-02) (At the tactical level, simply referred to as a breach.) (FM 3-34.2)

deliberate crossing (NATO, joint)—Crossing of an inland water obstacle that requires extensive planning and detailed preparations. (FM 90-13/FM 101-5-1/JP 1-02)

***demining** Activities to remove the hazard of all mines and other unexploded explosive munitions for a defined area. (FM 3-34)

demolition belt (joint)—A selected land area sown with explosive charges, mines, and other available obstacles to deny use of the land to enemy operations, and as a protection to friendly troops; there are two types of demolition belts: (1) primary continuous series of obstacles across the whole front, selected by the division or higher commander. The preparation of such a belt is normally a priority engineer task. (2) subsidiary—a supplement to the primary belt to give depth in front or behind, or to protect the flanks. (JP 1-02)

DHHS Department of Health and Human Services

directed obstacle An obstacle directed by a higher commander as a specified task to a subordinate unit. (FM 3-34.2/FM 101-5-1)

***directed target** A target directed by the responsible commander to be prepared for demolition or destroyed to support his intent. (FM 3-34/FM 101-5-1)

DISCOM Division Support Command

div division

DIVENG division engineer

DLA Defense Logistics Agency

DLEA Drug Law Enforcement Agency

DMA Defense Mapping Agency

DMC distribution management center

doctrine Fundamental principles by which the military forces or elements thereof guide their actions in support of national objectives; authoritative but requires judgment in application. (FM 3-0/FM 101-5-1/JP 1-02)

DOD Department of Defense

DOD construction agent (joint)—USAES, NAVFACENGCOM, or other such approved DOD activity that is assigned design or execution responsibilities associated with military construction programs, facilities support, or civil engineering support to the combatant commanders in contingency operations. (JP 3-34/JP 1-02)

DODD Department of Defense directive

DODI Department of Defense instruction

DOE Department of Energy

DOJ Department of Justice

DOMS Directorate of Military Support

DON Department of the Navy

DOS Department of State

DOT Department of Transportation

DOTMLPF doctrine, organization, training, materiel, leadership and education, personnel, and facilities

DPG defense planning guidance

DPPDB digital point-positioning database

DPW Directorate of public works

DS direct support

DSES division staff engineer section

dsgn design

DSO domestic support operations

DST decision support template

DTED Digital Terrain Elevation Data

DTG date-time group

DTSS digital topographic support system

DTSS-B digital topographic support system-base

DTSS-D digital topographic support system-deployable

DTSS-H digital topographic support system-heavy

DTSS-L digital topographic support system-light

DU depleted uranium

EA engagement area

EAB echelons above brigade

EAC echelons above corps

EAD echelons above division

EBA engineer battlefield assessment

EBS environmental baseline survey

ECO environmental compliance officer

ECOA enemy course of action

ECR Environmental conditions report concise summary of environmental conditions at a base camp site, based on the environmental baseline survey, supported by maps and backup documents, prepared by base camp commanders for each base camp; the ECR documents conditions at the site it claims or other legal challenges arise against the government. (FM 3-100.4)

EEFI essential elements of friendly information

EEM early entry module

EFA engineering field activity

EFD engineering field division

ELCAS elevated causeways system

EMS Environmental Management System

***EMST** essential mobility/survivability task—Specified or implied BOS-specific tasks that are critical to mission success; although ultimately executed by a combined-arms element, the staff (typically MANSPT elements such as engineer, chemical, MP, EOD) identifies the EMSTs. A fully developed EMST has four components including effect (the qualitative or qualitative impact desired by the higher commander on the friendly or adversary caused by the completion of the task); purpose (the desired or intended result of the task stated in terms relating to the purpose of the supported unit); tasks (one or more clearly defined and measurable tasks accomplished by individuals or organizations required to achieve the desired effects); and method (how the task and purpose will be achieved, described by a quantifiable use of assets or capabilities and communicates their priority). (FM 3-34)

ENCOM engineer command

***ENCOORD** engineer coordinator—Special staff officer who coordinates engineer assets and operations for the command, usually the senior engineer officer in the force. (FM 6-0 and FM 3-34)

enabling mine/ countermeasures (joint)—Countermeasures designed to counter mines once they have been laid. This includes both passive and active mine countermeasures. (JP 3-15/JP 1-02)

***engineer battlespace functions** The three battlespace functions include combat (M/CM/S grouped under combat engineering), geospatial and general engineering. (FM 3-34)

***Engineer Regiment** All AC and RC engineer organizations (as well as the DOD civilians and affiliated contractors and agencies within the civilian community) with a diverse range of capabilities that are all focused toward supporting the Army and its warfighting mission. (FM 3-34)

engr engineer

ENSIT enemy situation

environmental area of interest An assessment or study done on an area of interest (a property) to define the environmental state or condition of that property prior to use by US forces; survey is used to determine the environmental impact of property use by US forces and the level of environmental restoration needed before returning the property upon US departure. (FM 3-100.4)

environmental base line survey An assessment or study done on an area of interest (a property) in order to define the environmental state or condition of that property prior to use by US forces. Used to determine the environmental impact of property use by US forces, and the level of environmental restoration needed prior to returning the property upon US departure. (FM 3-100.4)

environmental compliance The unconditional obeying of international, foreign nation, federal, state and local environmental rules, regulation and guidelines that affect current operations. (FM 3-100.4)

environmental considerations (joint)—The spectrum of environmental media, resources or programs that may impact on, or are affected by, the planning and execution of military operations; factors may include, but are not limited to, environmental compliance, pollution prevention, conservation, protection of historical and cultural sites and protection of flora and fauna. (JP 3-34/JP 1-02)

environmental hazard All activities that may pollute, create negative noise related effects, degrade archaeological/cultural resources, or negatively affect threatened or endangered species habitats; also included are environmental health-related hazards. (FM 3-100.4)

environmental protection The application of human ingenuity and resources, through the disciplines of science and engineering, as required by environmental protection laws, regulations, and policies to protect the natural environment. (FM 3-100.4)

environmental protection level The varying level of environmental protection that can reasonably be afforded at any particular time during warfare of battlefield conditions, given the absolute requirement that such a diversion of resources away from the mission at hand does not adversely affect that mission, or any friendly personnel, or indigenous or refugee populations. (FM 3-100.4)

environmental reconnaissance The systematic observation and recording of site or area data collected by visual or physical means, dealing specifically with environmental conditions as they exist, and identifying areas that are environmentally sensitive or of relative environmental concern, for information and decision-making purposes. (FM 3-100.4/FM 101-5-1)

environmental services (joint)—Various combinations of scientific, technical and advisory activities (including modification processes, that is, the influence of man-made and natural factors) required to acquire, produce, and supply information on the past, present, and future states of space, atmospheric, oceanographic, and terrestrial surrounding for use in military planning and decisionmaking processes, or to modify those surroundings to enhance military operations. (JP 1-02)

environmental stewardship (joint)—The integration and application of environmental values into the military mission to sustain readiness, improve quality of life, strengthen civil relations, and preserve valuable natural resources. (JP 1-02)

ENVST environmental support team

EO executive order

EOD Explosive ordnance disposal(Joint)detection, identification, on-site evaluation, rendering safe, recovery, and final disposal of UXO; may also include explosive ordnance which has become hazardous by damage or deterioration. (FM 3-100.38/FM 101-5-1/JP 1-02)

EPA Environmental Protection Agency

EPW enemy prisoner of war

equip equipment

ERC exercise related construction

ERDC Engineer Research and Development Center

***ERP** Engineer regulating point checkpoint to ensure that vehicles do not exceed the capacity of the crossing means and to give drivers final instructions on site-specific procedures and information, such as speed and vehicle interval. (FM 90-13)

ERT engineer reconnaissance team

ESB engineer support battalion

ESF emergency support function

ESP engineer supply points

ESV engineer squad vehicle

EUCOM European Command

EUSA Eighth United States Army

EW electronic warfare

***EWL** Engineer work line coordinated boundary or phase line used to compartmentalize an operational area to indicate where specific engineer units have primary responsibility for the engineer effort; it may be used at division level to discriminate between a sector supported by division engineer assets and a sector supported by DS or GS corps engineer units. (FM 3-34)

executive agent (joint)—Term used to indicate a delegation of authority by the SECDEF to a subordinate to act on the Secretary's behalf; an agreement between equals does not create an executive agent. For example, a service cannot become a DOD executive agent for a particular matter with simply the agreement of the other services; such authority must be delegated by the SECDEF. Designation as executive agent, in and of itself, confers no authority. The exact nature and scope of the authority delegated must be stated in the document designating the executive agent. An executive agent may be limited to providing only administration and support or coordinating common functions, or it may be delegated authority, direction, and control over specified resources for specified purposes. (JP 0-2/JP 1-02)

explosive ordnance (joint)—All munitions containing explosives, nuclear fission or fusion materials, and biological and chemical agents; this includes bombs and warheads; guided and ballistic missiles; artillery, mortar, rocket, and small arms ammunition; all mine, torpedoes, and depth charges; demotion charges; pyrotechnics; clusters and dispensers; cartridge and propellant actuated devices; electro-explosive devices; clandestine and improvised explosive devices; and all similar or related items or components explosive in nature. (JP 1-02)

explosive ordnance disposal unit (joint)—Personnel with special training and equipment who render explosive ordnance safe (such as bombs, mines, projectiles, and booby traps), make intelligence reports on such ordnance, and supervise the safe removal thereof. (JP 1-02)

FAA Foreign Assistance Act

fac facility

FACE forward aviation combat engineering—(joint)—A mobility operation in which engineers perform tasks in support of forward aviation ground facilities; tasks include reconnaissance; construction of low altitude parachute extraction zones, landing strips, and airstrips; and providing berms, revetments, and trenches for forward arming and refueling points. (JP 3-34/JP 1-02)

FAO foreign area officer

FARP forward area refueling point

FASCAM family of scatterable mines—A grouping of munitions that dispenses SCATMINEs by artillery, helicopter, fixed wing, or ground launchers; there are AP and AT mines. (FM 20-37)

fascines Large cylindrical bundles of material, usually wooden poles or plastic or metal pipe loosely bound together, which are dropped into ditches or gaps to create crossings. (FM 3-34.2/FM 101-5-1)

FBCB2 Force XXI Battle Command Brigade and Below

FBI Federal Bureau of Investigation

FE facilities engineer(ing)

FEC facility engineer center

FED facilities engineer detachment

FEG facility engineer group

FEMA Federal Emergency Management Agency

***FEST** Forward engineer support teamdeployable USACE organization that executes the USACE mission in the AO; it is usually subordinate to the senior engineer commander in the AO. (FM 3-34)

***FEST-A** forward engineer support team-advance (FM 3-34)

***FEST-M** forward engineer support team-main (FM 3-34)

FET facility engineer team

FF fire fighting

***FFE** field force engineering—The application of all of the Engineer Regiment's capabilities (to include TeleEngineering) across the range of engineer battlespace functions (although primarily general engineering intensive) and in all phases and types of operations (offense, defense, stability, support) through both reach and forward presence.) (FM 3-34); flame field expedients(Joint)simple, hand-made devices used to produce flame or illumination. (JP 3-15/JP 1-02)

FFIR friendly forces information requirements

FGS final governing standards

FHA foreign humanitarian assistance

FID foreign internal defense

field fortifications (joint)—An emplacement of shelter of a temporary nature which can be constructed with reasonable facility by units requiring no more than minor engineer supervisory and equipment participation. (FM 5-102/FM 101-5-1/JP 1-02)

FLOT forward line of own troops

flt flat

FM field manual; frequency modulation

FMTV family of tactical vehicles

FOB forward operating base

FOL forward operating location

force beddown The provision of expedient facilities for troop support to provide a platform for the projection of force; these facilities may include modular or kit-type facility substitutes. (JP 4-04/JP 1-02)

force support engineering (NATO)—An Allied term that describes the deliberate, longer-term preparation for, and indirect support to ongoing or future operations as well as those military engineering tasks associated with sustaining the joint force throughout all stages of an operation. It will be the predominant engineer focus for precombat and post combat and for rear operations during conflict. It is an area which may involve a greater degree of cross-compartment support and the engineer tasks will usually be more enduring, relying more on purpose designed and built solutions. It is likely to fulfill a longer term, operational requirement. It is closely aligned with the engineer Battlespace function of General Engineering. (AJP 3.12)

FORSCOM United States Army Forces Command

ford A shallow par of a body of water or wet gap that can be crossed without bridging, boats, ferries, or rafts. It is a location in a water barrier where the physical characteristics of current, bottom, and approaches permit the passage of personnel, vehicles, and other equipment where the wheels or track remain in contact with the bottom at all times. (FM 90-13/FM 101-5-1)

FP force protection(joint)—Security program designed to protect service members, civilian employees, family members, facilities, and equipment, in all locations and situations, accomplished through planned and integrated application of combating terrorism, physical security, operations security, personal protective service, and support by intelligence, counterintelligence, and other security programs. (Army)those actions taken to prevent or mitigate hostile actions against DOD personnel (to include family members), resources, facilities, and critical information; these actions conserve the forces's fighting potential so it can be applied at a decisive time and place and incorporates the coordinated and synchronized offensive and defensive measures to enable the effective employment of the joint force while degrading opportunities for the enemy. Force protection does not include actions to defeat the enemy or protect against accidents, weather, or disease. (FM 3-0/FM 101-5-1/JP 3-0/JP 1-02)

FPOL forward passage of lines

fps foot(feet) per second

FRAGO(RD) fragmentary order

FRP Federal Response Plan

FS fire support

FSB forward support base

FSCOORD fire support coordinator

FSSG force service support group (USMC)

***FTEE** forward theater engineer element (FM 3-34)

functions (joint)—The appropriate or assigned duties, responsibilities, missions or tasks of an individual, office or organization; as defined in National Security Act of 1947, as amended, the term function includes functions, powers and duties. (JP 1-02)

fwd forward

FXXI Force XXI

G1 Assistant Chief of Staff, Personnel, General Staff (Division or higher staff)

G2 Assistant Chief of Staff, G2 (intelligence)

G3 Assistant Chief of Staff, G3 (operations and plans)

G4 Assistant Chief of Staff, G4 (logistics)

G5 Assistant Chief of Staff, G5 (Civil Affairs)

G6 Assistant Chief of Staff, G6 (command, control, communications and computer systems)

GAO General Accounting Office

gal gallon

***GE** general engineering—(joint)—Encompasses the construction and repair of LOCs, MSRs, airfields, and logistic facilities to support joint military operations and may be performed in DS of combat operations, such as battle damage repair; these operations include both horizontal and vertical construction, and may include use of both expedient repair methods and more deliberate construction methods characterized by the application of design criteria, advanced planning, and preparation, depending on the mission requirements. (JP 3-34/JP 1-02); one of the engineer battlespace functions. (FM 3-34)

***geospatial engineering** Provides commanders with terrain visualization, operational and tactical terrain analysis, digitized terrain products, nonstandard map products, and baseline survey data; one of the engineer battlespace functions. (FM 3-34)

geospatial information Foundation information upon which all other battlespace information is referenced to form the COP. (FM 5-105/FM 101-5-1)

(GI&S) geospatial information and services—(joint)—The concept for collection; information extraction; storage; and dissemination and exploitation of geodetic, geomagnetic, imagery (both commercial and national source), gravimetric, aeronautical, topographic, hydrographic, littoral, cultural, and toponymic data accurately referenced to a precise location on the earth's surface; these data are used for military planning, training, and operations, including navigation, mission planning, mission rehearsal, modeling, simulation, and precise targeting. Geospatial information provides the basic framework for battlespace visualization. It is information produced by multiple sources to common interoperable data standards. It may be presented in the form of printed maps, charts, and publications; in digital simulation and modeling databases; in photographic form; or in the form of digitized maps and charts or attributed centerline data. Geospatial services include tools that enable users to access and manipulate data, and also includes instruction, training, laboratory support, and guidance for the use of geospatial data. (FM 5-105/FM 101-5-1/JP 2-03/JP 1-02)

GPH gallons per hour

GPR general support reinforcing

GPS global positioning system

GS general support

GSA General Services Administration

Harvest Eagle A US-owned facility substitute of flexible tempered tents with insulation and environmental control units; the structures can be used to house more than 500 people and include a dining facility, showers and latrines.

Harvest Falcon A US-owned facility substitute used to accommodate USAF personnel and required operational facilities when deployed in off-shore environments; this system includes living accommodations, aircraft flight line facilities, air-contingency hospital operations, specialized clinics, and administrative support functions.

hasty breaching (joint)—Rapid creation of a route through a minefield, barrier, or fortification by any expedient method. (JP 1-02)

hasty breaching (land mine warfare) (joint)—Creation of lanes through enemy minefields by expedient methods such as blasting with demolitions, pushing rollers or disabled vehicles through the minefields when the time factor does not permit detailed reconnaissance, deliberate breaching, or bypassing the obstacle. (JP 1-02)

hasty crossing (NATO, joint)—Crossing of an inland water obstacle using the crossing means at hand, or those readily available, and made without pausing for elaborate preparations. (FM 3-34.13/FM 101-5-1/JP 1-02)

hazard (DOD)—Condition with the potential to cause injury, illness, or death of personnel; or damage to or loss of equipment or property; or mission degradation. (FM 3-100.4/FM 100-14)

HAZMAT hazardous material

HCA humanitarian and civic assistance

HD humanitarian demining—(joint)—DOD and DOS program to promote the foreign policy interests of the US by assisting other nations in protecting their populations from landmines and clearing land of the threat posed by landmines remaining after conflict has ended; the HD program includes training of HN deminers, establishment of national demining organizations, provision of demining equipment, mine awareness training, and research development. (JP 3-07/JP 1-02)

HDO humanitarian demining operation

HET heavy equipment transporter

HHC headquarters and headquarters company

HHD headquarters and headquarters detachment

HLC helicopter landing zones

HLD homeland defense—The protection of US territory, sovereignty, domestic population, and critical infrastructure against external threats and aggression. (JP 3-26)

HLS homeland security—The preparation for, prevention of, deterrence of, preemption of, defense against, and response to threats and aggression directed toward US infrastructure; as well as crisis management, consequence management, and other domestic civil support. Homeland Defense and Civil Support are sub elements of HLS. (JP 3-26)

HMMWV high mobility, multipurpose wheeled vehicle

HN host nation

HNS host nation support—(joint)—Civil and/or military assistance rendered by a nation to foreign forces within its territory during peacetime, crisis or emergencies or war, based upon agreements mutually concluded between nations. (FM 3-07/JP 4-0/JP 1-02)

holding line In retrograde river crossing operations, the outer limit of the area established between the enemy and water obstacle to preclude direct and observed indirect fires in the crossing areas. (FM 90-13/FM 101-5-1)

horizontal action mine (joint)—In land mine warfare, a mine designed to produce a destructive effect in a plane approximately parallel to the ground. (JP 1-02)

HPTL high-payoff target list

HQ headquarters

H&S headquarters & service

HSC headquarters support company

humanitarian demining (HD) (joint)—Department of Defense and Department of State program to promote the foreign policy interests of the United States by assisting other nations in protecting their populations from landmines and clearing land of the threat posed by landmines remaining after conflict has ended. The humanitarian demining program includes training of host nation deminers, establishment of national demining organizations, provision of demining equipment, mine awareness training, and research development. Also called HD. (JP 3-07/JP 1-02)

HUMINT human intelligence; human resources intelligence

HVAC heating, ventilation, and air conditioning

HVT high-value target

hvy heavy

***IAT** infrastructure assessment team—nondeployable team that provides engineering infrastructure assessments for military deployments and civil military operations in forward areas; focus areas for the IAT are infrastructure-related to USACE missions and aspects of the AO impacting contract construction, to include roads, utilities, water resources, and HN support. (FM 3-34)

ICM improved conventional munitions

IDAD internal defense and development

IEB interentity boundary

IED improvised explosive device—(joint)—A device placed or fabricated in an improvised manner incorporating destructive, lethal, noxious, pyrotechnic, or incendiary chemicals and designed to destroy , incapacitate, harass, or distract; it may incorporate military stores, but is normally devised from nonmilitary components. (JP 1-02)

IG inspector general

info information

information operations Use of offensive and defensive information means to degrade, destroy, and exploit an adversary's information-based process while protecting one's own. The actions taken affect adversary and influence others' decision making processes, information, and information systems while protecting one's own information and information systems. (FM 3-0)

infrastructure (joint)—All building and permanent installations necessary for the support, redeployment, and military forces operations (for example., barracks, HQ, airfields, communications, facilities, stores, port installations, and maintenance stations). (JP 4-01.8/ JP 1-02)

intel intelligence

IPDS inland petroleum distribution system

intelligence (joint)—Product that results from collecting, processing, integrating, analyzing, evaluating, and interpreting available information concerning foreign countries or areas. (JP 1-02)

intelligence BOS Activity that generates knowledge of and products which portray the enemy and the environmental features required by a command planning, preparing, executing, and assessing operations. (FM 7-15)

intelligence requirement (joint)—(1) Any subject, general or specific, for which there is a need to collect information or produce intelligence; (2) a requirement for intelligence to fill a gap in the command's knowledge or understanding of the battlespace or threat forces. (JP 1-02)

interchangeability (joint)—Condition that exists when two or more items possess such functional and physical characteristics as to be equivalent in performance and durability and are capable of being exchanged one for the other without alteration of the items themselves or of adjoining items, except for adjustment, and without selection for fit and performance. (JP 1-02)

interoperability (joint)—The ability of systems, units, or forces to provide services to and accept services from other systems, units, or forces and to use the services so exchanged to enable them to operate effectively together. (JP 1-02)

IO information operations; international organization

IPB intelligence preparation of the battlefield/battlespace—An analytical methodology employed as part of intelligence planning to reduce uncertainties concerning the enemy, environment, and terrain or all types of operation; IPB is conducted during mission planning to support the commander's decision making and to form the basis for the direction of intelligence operation in support of current and future missions. It utilizes existing databases and identifies gaps in intelligence needed to determine the impact of the enemy, environment, and terrain on operations and presents this in an appropriate form to facilitate operational planning. It forms the basis for situation development. (FM 2-01.3); (Joint) an analytical methodology employed to reduce uncertainties concerning the enemy, environment, and terrain for all types of operations; IPB builds an extensive database for each potential area in which a unit may be required to operate. The database is then analyzed in detail to determine the impact of the enemy, environment, and terrain on operations and presents it in graphic form. IPB is a continuing process. (JP 2-02/JP 1-02/FM 2-0)

IPDS inland petroleum distribution system

IR information requirement

IRR individual ready reserve

IS intelligence and surveillance

ISB intermediate staging base

ISR Intelligence, surveillance, and reconnaissance integrating and synchronizing of all BOS to collect and process information about the enemy and the environment to produce relevant information to facilitate the commander's decision making. (FM101-5-1)

ISR assets Those organizations, systems, sensors, and equipment dedicated to or directed toward the collection of information and the analysis and production of intelligence from the collected information in response to the commander's PIRs. (FM 101-5-1)

ISR plan Integrated plan for collecting information from all available sources and analysis of that information to produce intelligence to meet requirements; specifically, a logical plan for transforming PIRs into orders or requests to reconnaissance and surveillance assets to collect pertinent information within a required time. (FM 101-5-1)

ISYSCON integrated systems control

J2 Intelligence Directorate

J3 Operations Directorate

J4 Logistics Directorate

J7 development directorate, joint staff

JAG Judge Advocate General

JCMEB Joint Civil-Military Engineering Board

JCS Joint Chiefs of Staff

JDWP joint doctrine working party—(joint)—A forum to include representatives of the services, combatant command, and joint staff (represented by the operational plans and joint force development directorate, J7) which meets semiannually to address and vote on project proposals; discuss key joint doctrinal or operational issues; keep up to date on the status of the joint publication projects and emerging publications; and keep abreast of other initiatives of interest to the members. The JDWP meets under the sponsorship of the J7. (JP 1-01/JP 1-02)

JEMB Joint Environmental Management Board

JEPES Joint Engineer Planning and Execution System

JFACC joint force air component commander—(joint)—Derives authority from the JFC, who has the authority to exercise OPCON, assign missions, direct coordination among subordinate commanders, redirect and organize forces to ensure unity of effort in the accomplishment of the overall mission; normally designated by the JFC; responsibilities are assigned by the joint force commander (normally these include, but are not limited to, planning, coordination, allocation, and tasking, based on the JFC's apportionment decision); using the JFC's guidance and authority, and in coordination with other service component commanders and other assigned or supporting commanders, recommends to the JFC apportionment of air sorties to various missions or geographic areas. (JP 3-0/JP 1-02)

JFC joint force commander

JFCOM Joint Forces Command

JFLCC joint force land component commander—(joint)—The commander within a unified command, subordinate unified command, or joint task force who is responsible to the establishing commander for making recommendations on the proper employment of land forces, planning and coordinating land operations, or accomplishing such operational missions as may be assigned; JFLCC is given the authority necessary to accomplish missions and tasks assigned by the establishing commander. JFLCC is normally the commander with the preponderance of land forces and the requisite C2 capabilities. (JP 3-0/JP 1-02)

JFLSC Joint Force Logistic Support Command

JFUB Joint Facilities Utilization Board—(joint)—Joint board that evaluates and reconciles component requests for real estate use of existing facilities, interservice support, and construction to ensure compliance with JCMEB priorities. (JP 4-04/JP 1-02)

JIC joint implementation committee

JIM joint, interagency and multinational

JIMP Joint Vision Implementation Master Plan

JLOTS joint logistics over-the-shore

JIP joint implementation plan

JLP joint logistics plan

JLSP joint logistics support plan

JMC joint movement center

JOA joint operations area

joint doctrine (joint)—Fundamental principles that guide the employment of forces of two or more military departments in coordinated action toward a common objective, it is authoritative; as such. Joint doctrine will be followed except when, in the judgment of the commander, exceptional circumstances dictate otherwise. It will be promulgated by or for the Chairman of the Joint Chiefs of Staff, in coordination with the combatant commands and services. (JP 1-01/1-02)

JOPES Joint Operation Planning and Execution System

JP joint publication

JRSOI joint reception, staging, onward movement, and integration

JTF joint task force

JTFCEM joint task force contingency engineering management

K-span automatic building machine

kW kilowatt

lane A route through an enemy or friendly obstacle that provides a passing force safe passage; the route may be reduced and proofed as part of a breach operation or constructed as part of a friendly obstacle. It is a clear route that moves all the way through an obstacle. (FM 3-34.2)

LCC land component commander

LE light equipment

LFT lead federal agency

LMTV light, medium-tracked vehicle

LNO liaison officer

LO logistics order

LOC line of communications

LOCTDA line of communication tactical decision aid

LOGCAP Logistics Contract Augmentation Program

logistics preparation of the battlefield All actions taken by combat service support to maximize the means of supporting the commander's plans. (FM 4-0)

LOGPAC logistics package

LOS line of sight

LOTS logistics over the shore operations

LSA	life support area
LSE	logistics support element
lt	light
LZ	landing zone
MACA	military assistance to civil authorities
MACOM	major Army command
MAGTF	Marine air-ground task force
maint	maintenance
maintenance engineering	(joint)—Application of technique, engineering skills, and effort, organized to ensure that the design and development of weapon systems and equipment provide adequately for their effective and economical maintenance. (JP 1-02)
MAN	maneuver
***MANSPT**	maneuver support—Includes all of the functions identified in the M/CM/S BOS. (FM 3-34)
MARFOR	Marine Corps forces
MBA	main battle area
MC	multicomponent
MCC	movement control center
MCM	mine countermeasures—(joint)—All methods for preventing or reducing damage or danger from mines. (JP 3-15/JP 1-02)
MCMOPS	mine countermeasures operations
M/CM/S	mobility, countermobility, and survivability (both for the collective components of combat engineering and the M/CM/S BOS).
M/CM/S BOS	mobility operations preserve the freedom of maneuver for friendly forces; These missions include breaching obstacles, increasing battlefield circulation, improving or building roads, providing bridge and raft support, and identifying routes around contaminated areas. Countermobility denies mobility to enemy forces. Survivability operations protect friendly forces from the effects of enemy weapon systems and from natural occurrences. (FM 7-15)
MCO	major contingency operation
MCOO	modified combined-obstacle overlay—(joint)—A joint IPB product used to portray the effects of each battlespace dimension on military operations; it normally depicts militarily significant aspects of the battlespace environment, such as obstacles restricting military movement, key geography, and military objectives. (JP 2-01.3/JP 1-02)
MCS	maneuver control system; modular causeway system
MCS-ENG	maneuver control system-engineer
MCWP	Marine Corps warfare publication

MDMP military decision-making process

MEB Marine expeditionary brigade

MEDEVAC medical evacuation

MEDCOM medical command

MEF Marine expeditionary force

MEP mobile electric power; Military environmental protection application and integration of all aspects of natural environmental considerations, as they apply to the conduct of military operations. (FM 3-34.500)

METL mission-essential task list—Compilation of collective mission-essential tasks an organization must perform successfully to accomplish its wartime mission(s). (FM 7-0)

METT-TC mission, enemy, terrain and weather, troops and support available, time available, and civil considerations (FM 101-5-1)

MEU Marine expeditionary unit

MFO multinational force and observers

MFPF minefield planning folder

MGB medium girder bridge

mgt management

MHE materials handling equipment

MICC mine information coordination cell

MICLIC mine clearing line charge

MILCON military construction—(joint)—Any construction, alteration, development, conversion, or extension of any kind carried out with respect to a military installation. (JP 4-04/JP 1-02)

mine (NATO, joint)—(1) In land mine warfare, an explosive or material, normally encased, designed to destroy or damage ground vehicles, boats, or aircraft or designed to wound, kill, or otherwise incapacitate personnel; it may be detonated by the action of its victim, by the passage of time, or by controlled means; (2) In naval mine warfare, an explosive device laid in the water with the intention of damaging or sinking ships or of deterring shipping from entering an area; the term does not include devices attached to the bottoms of ships or to harbor installations by personnel operating underwater, nor does it include devices which explode immediately on expiration of a predetermined tie after laying. (FM 20-32/FM 101-5-1/JP 3-15/JP 1-02)

minefield (NATO, joint)—In land warfare, an area of ground containing mines emplaced with or without a pattern. (FM 20-32/JP 3-15/JP 1-02)

minefield breaching (NATO, joint)—In land mine warfare, the process of clearing a lane through a minefield under tactical conditions. (FM 20-32/JP 1-02)

minehunting (joint)—Employment of sensor and neutralization systems, whether air, surface, or subsurface, to locate and dispose of individual mines; minehunting is conducted to eliminate mines in a known field when sweeping is not feasible of desirable, or to verify the presence or absence of mines in a given area. (JP 3-15/JP 1-02)

MINEOPS joint minelaying operations

minesweeping (joint)—Technique of clearing mines using either mechanical, explosive, or influence seep equipment; mechanical sweeping removes, disturbs, or otherwise neutralizes the mine. Explosive sweeping causes sympathetic detonation in, damages, or displaces the mine; influence sweeping produces either the acoustic and/or magnetic influence required to detonate the mine. (JP 3-15/JP 1-02)

MISCAP mission capability (Air Force)

MISP multispectral imagery processor

MIW mine warfare—(joint)—Strategic, operational, and tactical use of mines and mine countermeasures; mine warfare is divided into two basic subdivisions: the laying of the mines to degrade the enemy's capabilities to wage land, air, and maritime warfare; and the countering of enemy-laid mines to permit friendly maneuver or use selected land or sea areas. (FM 20-32/JP 3-15/JP 1-02)

MKT mobile kitchen trailer

MLC military load classification—(NATO, joint)—Standard system in which a route, bridge, or raft is assigned class number(s) representing the load it can carry; vehicles are also assigned number(s) indicating the minimum class of route, bridge, or raft they are authorized to use. (FM 101-5-1)

MMC Materiel Management Center

MOA memorandum of agreement

MOB military operating base; mobility

mobility (NATO, joint)—Quality or capability of military forces which permits them to move from place to place while retaining the ability to fulfill their primary mission; (Army)—those activities that enable a force to move personnel and equipment on the battlefield without delays due to terrain or obstacles. (JP 1-02)

mobility analysis (joint)—In-depth examination of all aspects of transportation planning in support of OPLAN and OPORD development. (JP 1-02)

***mobility operations** Obstacle reduction by maneuvering and engineer units to reduce or negate the effects of existing or reinforcing obstacles; the objectives are to maintain freedom of movement for maneuver units, weapon systems and critical supplies. (FM 3-34)

MOMAT mobility matting

MOOTW military operations other than war

MOPMS	modular pack mine system
MOS	military occupational specialty
MOUT	military operations on urbanized terrain (being replaced by UO)
MP	military police
MPF	maritime pre-positioning force
MPS	maritime pre-positioning ship
MPSC	mobilization planning support cell
MPSPRON	maritime pre-positioning force squadrons
MRB	multirole bridge
MRE	mission rehearsal exercise
MSDS	mission specific data sets—(joint)—Further densification of global geospatial foundation data; information created to support specific operations, OPLANs, training or system development. Information conforms to established DOD data specifications. (JP 2-03/JP 1-02)
MSR	main supply route
MS	Mississippi
MSB	main support battalion
MSC	major subordinate command
MRBC	multi-role bridge company
MTC	movement to contact
MTOE	modified table of organization and equipment
MTT	mobile training team
MTW	major theater war
MULTICOMPO	multicomponent
MWRS	morale, welfare, recreation and services
MWSS	Marine wing support squadron
NAI	named area of interest
NATO	North Atlantic Treaty Organization
natural environment	The human ecosystem, including the physical and biological systems that provide resources (clean air, clean water, healthy surroundings, sufficient food) necessary to sustain productive human life; included in the natural environment are man-made structures, such as water and wastewater treatment facilities and natural/cultural resources. (FM 3-100.4)
NAVAID	navigational aid
NAVFAC	naval facilities
NAVFACENGCOM	Naval Facilities Engineering Command
NAVFOR	Navy forces
NAVSTAR	navigation satellite timing and ranging

NBC nuclear, biological, and chemical

NBG naval beach group

NCB naval construction brigade

NCF naval construction force

NCFSU naval construction force support unit

NCR naval construction regiment

NCS National Communications System

NEO noncombatant evacuation operation

NEPA National Environmental Policy Act

neutralization (NATO, joint)—In mine warfare, a mine is said to be neutralized when it has been rendered, by external means, incapable of firing on passage of a target, although it may remain dangerous to handle. (FM 20-32/JP 1-02)

neutralize (joint)—(1) As pertains to military operations, to render ineffective or unusable; (2) a tactical mission task that results in rendering enemy personnel or material incapable of interfering with a particular operation (FM 3-90); (3) to render safe mines, bombs, missiles, and booby traps; (4) to render harmless. (FM 20-32/JP 1-02)

NFESC Naval Facilities Engineering Service Center

NGO nongovernmental organization

NIMA National Imagery and Mapping Agency

NMCB naval mobile construction battalion

NOAA National Oceanic and Atmospheric Administration

NORTHCOM Northern Command

NSE Navy support element

NWP naval warfare publication

OAKOC observation, avenues of approach, key terrain, obstacles and cover and concealment

OAS offensive air support; Organization of American States

OASD Office of the Assistant Secretary of Defense

OAU Organization of African Unity

obstacle (joint)—Any obstruction designed or employed to disrupt, fix, turn, or block the movement of an opposing force and to impose additional losses in personnel, time, and equipment on the opposing force; obstacles can be natural, man-made, or a combination of both. (FM 90-7/FM 101-5-1/JP 3-15/JP 1-02)

obstacle belt (joint)—Brigade-level C2 measure, normally given graphically, to show where within an obstacle zone and ground tactical commander plans to limit friendly obstacle employment and focus the defense; it assigns an intent to the obstacle plan and provides the necessary guidance on the overall effect of obstacles within a belt; (Army)—graphic control measure that corps, division, and brigade echelon commanders use to constrain tactical obstacle employment. (FM 3-90/FM 20-32/JP 3-15/JP 1-02)

obstacle clearing (joint)—The total elimination or neutralization of obstacles. (JP 1-02)

obstacle control measure Specific measures that simplify the granting of obstacle-emplacing authority while providing obstacle control. (FM 3-90/FM 101-5-1/FM 20-32)

obstacle effect block The block effect integrates fire planning and obstacle effort to stop an attacker along a specific AA or prevent him from passing through an EA. (FM 90-7/FM 101-5-1)

obstacle groups One or more individual obstacles grouped to provide a specific obstacle effect. (FM 3-90/FM 101-5-1)

obstacle line A conceptual control measure used at battalion or brigade level to show placement intent without specifying a particular type of linear obstacle. (FM 90-7)

obstacle plan That part of an operation plan or order that is concerned with the use of obstacles to enhance friendly fires or to affect the movement of an enemy. Obstacle plans are used at corps level and below. (FM 90-7)

obstacle restricted areas (joint)—C2 measure used to limit the type or number of obstacles with an area. (FM 90-7/JP 3-15/JP 1-02)

obstacle zone (joint)—Division-level C2 measure, normally done graphically, to designate specific land areas where lower echelons are allowed to employ tactical obstacles; (Army)—a graphic control measure that corps and division commanders use to grant obstacle-emplacement authority to brigades (including armored cavalry regiments and other major subordinate units). FM 3-90/FM 90-7/JP 3-15/JP 1-02)

OBSTINTEL obstacle intelligence—The first breach tenet; an IR prior to conducting combined-arms breaching operations. Obstacle intelligence includes a broad range of information to include location, composition, and orientation of exiting obstacles. (FM 3-34.2)

OCONUS outside the continental United States

ODSS offense, defense, stability, and support

OEBGD overseas environmental baseline guidance document—(joint)—Set of objective criteria and management practices developed by DOD to protect human health and the environment. (JP 4-04/JP 1-02)

OE operational environment

OFDA Office of Foreign Disaster Assistance

OICC officer in charge of construction

O&M operation and maintenance

OODA observe, orient, decide and act

OPCOM (NATO)—operational command

OPCON operational control

OPDS Offshore Petroleum Discharge System

OOTW operations other than war

operational area Subareas defined by the commander and designated within the AO that bounds the specific areas the commander will maneuver through and therefore can focus ISR capabilities to achieve the assured mobility imperative to select, establish, and maintain operating areas. (JP 1-02)

operational environment (joint)—A composite of all conditions, circumstances, and influences which affect the employment of military forces and bear on the decisions of the unit commander. (JP 1-02)

operational countermobility To delay, channel, or stop offensive air, land, and sea movement by enemy operational formations to help create positional advantage for friendly joint and multinational operational forces; operational countermobility exposes enemy decisive points and centers of gravity or high-payoff targets to destruction in conformance with the joint force commander's plans and intent. It includes the execution of sanctions, embargoes, blockades, and no-fly zones. (CJCSM 3500.04B/JP 3-15) (OP 1.4 in CJCSM [UJTL])

operational mobility To facilitate the movement of joint and multinational operational formations in a campaign or major operation without delays due to operationally significant terrain or obstacles. (CJCSM 3500.04B) (OP 1.3 in CJCSM [UJTL])

OPLAN operation plan

OPNAVINST operational naval instruction

opns operations

OPORD operation order

OPP offload preparation party

ops operations

Ops Det operations detachment

OPSEC operations security

OSCE Organization for Security and Cooperation in Europe

OSD Office of the Secretary of Defense

OTSG Office of the Surgeon General

OVE on-vehicle equipment

P&A	production and analysis
PACOM	Pacific Command
PAO	public affairs office(r)
PARC	principal assistant for contracting
PBO	property book officer
PE	peace enforcement
PEO	peace enforcement operations
pers	personnel
PIR	priority intelligence requirements
PKO	peacekeeping operation
PL	phase line; public law
plan	planning
PLL	prescribed load list
PLS	palletized load system
PME	peacetime military engagement
PO	peace operations
POD	port of debarkation
POE	port of embarkation
point obstacle	Any obstruction designed or employed to deny ease of movement on a road, lane, or bridge that is normally part of or completes a larger man-made or natural obstacle that has a designed effect on the attacking enemy force. (FM 90-7/FM 101-5-1)
POL	petroleum, oils, and lubricants
prestock point	Location where mines, explosives, and materials for use in preparation of a planned obstacle are stored. (FM 20-32/FM 101-5-1)
proactive mine countermeasures	(joint)—Measures intended to prevent the enemy from successfully laying mines. (JP 3-15/JP 1-02)
protective minefield	(NATO, joint)—In land mine warfare, a minefield employed to assist a unit in its local, close-in protection. (FM 20-32/FM 101-5-1/JP 1-02)
protective obstacles	Obstacles employed to assist a unit in its local, close-in protection. (FM 90-7)
PSS	personnel service support
PSYOP	psychological operations
pub	public
PW	public works
QC	quality control
QDR	Quadrennial Defense Review
QOL	quality of life
R	reinforcing

RAA Redeployment Assembly Area

RC Reserve Component

RCEM regional-contingency engineering manager

RCRA Resource Conservation and Recovery Act

RDL Reimer Digital Library

***reach** To communicate and access the Engineer Regiment's capabilities via means such as TeleEngineering; reach may occur from forward units to the rear or laterally. (FM 3-34) (The Air Force and Navy have similar capabilities in their services.)

real property (joint)—Lands, buildings, structures, utilities systems, improvements, and appurtenances; includes equipment attached to and made part of buildings and structures (such as heating systems) but not movable equipment (such as plant equipment). (JP 1-02)

rearming (joint)—(1) An operation that replenishes the prescribed stores of ammunition, bombs, and other armament items for an aircraft, naval ship, tank, or armored vehicle (including replacement of defective ordnance equipment) to make it ready for combat service; (2) resetting the fuze on a bomb, or on an artillery, mortar, or rocket projectile so that it will detonate at the desired time. (FM 101-5-1/JP 1-02)

RED HORSE Rapid Engineers Deployable Heavy Operations Repair Squadron, Engineer—(joint)—Air Force units wartime-structured to provide a heavy engineer capability; they have a responsibility across the operational area, are not tied to a specific base, and are not responsible for base operation and maintenance. These units are mobile, rapidly deployable, and largely self-sufficient for limited periods of time. (JP 3-34/JP 1-02)

reduce (1) Tactical mission task that involves the destruction of an encircled or bypassed enemy force (FM 3-90); (2) One of the five breaching fundamentals; the creation of a lane through, over, or around an obstacle. In the case of minefields, refers to destroying, neutralizing, removing, or bypassing mines. (FM 3-34.2/FM 101-5-1)

reduction (joint)—Creation of lanes through a minefield or obstacle to allow passage of the attacking ground force. (JP 3-15/JP 1-02)

reinforcing obstacles (joint)—Those obstacles specifically constructed, emplaced, or detonated through military effort and designed to strengthen existing terrain to disrupt, fix, turn, or block enemy movement. (FM 90-7/JP 3-15/JP 1-02)

reserved demolition target (NATO, joint)—Target for demolition, the destruction of which must be controlled at a specific level of command because it plays a vital part in the tactical or strategic plan, or because of the importance of the structure itself, or because the demolition may be executed in the face of the enemy. (FM 90-7/FM 101-5-1/JP 1-02)

reserved obstacles (joint)—Those demolition obstacles that are deemed critical to the plan for which the authority to detonate is reserved by the designating commander; (Army)—obstacles of any type, for which the commander restricts execution authority. (FM 20-37/JP 3-15/JP 1-02)

RHQ regional headquarters

RIBS readiness base support

river crossing Operation required before ground combat power can be projected and sustained across a water obstacle; it is centrally planned and controlled offensive operation that requires the allocation of external crossing means and a force dedicated to the security of the bridgehead. (FM 90-13/FM 101-5-1)

road block (NATO, joint)—Barrier or obstacle (usually covered by fire) used to block or limit the movement of hostile vehicles along a route. (FM 90-7/FM 101-5-1/JP 1-02)

ROE rules of engagement

ROICC resident officer in charge of construction

RO/RO roll-on/roll-off

***route classification** (NATO, joint)—Classification assigned to a route using factors of minimum width and worst route type, least bridge, raft, or culvert MLC; and obstructions to traffic flow. (FM 3-34/FM 101-5-1)

route reconnaissance Reconnaissance that focuses along a specific LOC, such as a road, railway, or cross-country mobility corridor. (FM 3-90)

RPE rapid port enhancement

RPMA real-property maintenance activities

RPOL rearward passage of lines

RRR rapid runway repair

R&S reconnaissance and surveillance

RSOI reception, staging, onward movement, and integration

RSP render-safe procedure

RSTA reconnaissance, surveillance, and target acquisition

RTE/rte route

RSR required supply rate

rupture Task to create a gap in enemy defensive positions quickly. (FM 3-34.2)

S1 Adjutant (U.S. Army brigade and battalion)

S2 Intelligence Officer (U.S. Army)

S3 Operations and Training Officer (U.S. Army)

S4 Supply Officer (U.S. Army)

S5 Civil Affairs Officer (U.S. Army)

S6 Communications Staff Office (U.S. Army)

SA security assistance

SASO stability and support operation

SBCT Stryker brigade combat team

SCATMINE scatterable mine—(NATO, joint)—In land mine warfare, a mine laid without regard to classical pattern and which is designated to be delivered by aircraft, artillery, missile, ground dispenser, or by hand; once laid, it normally has a limited life. (FM 20-32)

SCBA self-contained breathing apparatus

scuba self-contained underwater breathing apparatus

SD self-destruct

SECDEF Secretary of Defense

SECSTATE Secretary of State

secure (NATO, joint)—In an operational context, to gain possession of a position or terrain feature with or without force, and to make such disposition as will prevent, as far as possible, its destruction of loss by enemy action. [Note: Army classifies this as a tactical mission task]. (FM 3-90); (Army)—one of the five breaching fundamental; those actions which eliminate the enemy's ability to interfere with the reduction and passage of combat power through a lane. Secure may be accomplished by maneuver or by fires. (JP 1-02)

SEE small emplacement excavator

SES staff engineer section

SIC survey information center

SIGINT signal intelligence

situational obstacle An obstacle that a unit plans and possibly prepares before starting an operation, but does not execute unless specific criteria are met; it provides the commander flexibility for emplacing tactical obstacles based on battlefield development. (FM 90-7)

SITEMP situational template

SJA Staff Judge Advocate

SLRP survey, liaison, and reconnaissance party

SMART special medical augmentation response team

SME subject matter expert

SOEO scheme of engineer operations

SOF special operations forces

SOFA Status of Forces Agreement

SOP standing operating procedure

SOSRA suppress, obscure, secure, reduce and assault

SOUTHCOM Southern Command

SPACECOM Space Command

SPCE survey planning and coordination element

SPOD seaport of debarkation

SPOE seaport of embarkation

spt support

SSC smaller-scale contingency

stab stability

STAMIDS Standoff Mine Detection System

STANAG standardization agreement

STRATCOM Strategic Command

strong point (NATO, joint)—Key point in a defensive position, usually strongly fortified and heavily armed with automatic weapons, around which other positions are grouped for its protection (JP 1-02); (Army)—a heavily fortified battle position tied to a natural or reinforcing obstacle to create an anchor for the defense or to deny the enemy decisive or key terrain. (FM 3-90)

SU situational understanding

SUPCOM support command

support force That force in a breaching operation whose mission is to eliminate enemy interference with the breach through suppressive, direct, and indirect fires; the support force is one of the three breach organizations used in breaching operations. (FM 3-34.2)

suppress (1) A tactical mission task that results in temporary degradation of the performance of a force or weapons system below the level needed to accomplish the mission (FM 3-90) (2) One of the five breaching fundamentals; the focus of all fires on enemy personnel, weapons, or equipment to prevent effective fires on friendly forces. The purpose of suppression is to protect forces reducing and maneuvering through the obstacle and to soften the initial foothold. (FM 3-34.2)

***survivability** (joint)—Concept which includes all aspects of protecting personnel, weapons, and supplies while simultaneously deceiving the enemy; survivability tactics include building a good defense; employing frequent movement; using concealment, deception, and camouflage; and constructing fighting and protective positions for both individuals and equipment. (JP 3-34/ JP 1-02) [Note: Army adds, "Encompasses planning and locating position sites, designing adequate overhead cover, analyzing terrain conditions and construction materials, selecting excavation methods, and countering the effects of direct and indirect fire weapons."] (FM 3-34)

survivability operations Development and construction of protective positions, such as earth berms, dug-in positions, overhead protection, and countersurveillance means, to reduce the effectiveness of enemy weapon systems. (FM 5-103/FM 101-5-1)

sustaining operations Operations at any echelon that enable shaping and decisive operations by providing CSS, rear area and base security, movement control, terrain management, and infrastructure development. (FM 3-0)

TA theater Army

TAA tactical assembly areas

TAACOM Theater Army Area Command

TAC Transatlantic Programs Center

TACCP tactical command post

TACON tactical control

tactical minefield (NATO, Joint)—Minefield that is employed to directly attack enemy maneuver as part of a formation obstacle plan and is laid to delay, channel, or break up an enemy advance, giving the defending element a positional advantage over the attacker. (FM 20-32/JP 1-02)

tactical mobility the ability to move rapidly from one part of the battlefield to another, relative to the enemy. (FM 3-90)

tactical obstacles (joint)—Those obstacles employed to disrupt enemy formations, to turn them into a desired area, to fix them in position under direct and indirect fires, and to block enemy penetrations. (FM 90-7/JP 3-15/JP 1-02)

TAI targeted area of interest

TC training circular

TCEM theater contingency engineering management

TCF tactical combat force

TCMS theater construction management system

TCN third country national

TDA table of distribution and allowance

TEC Topographic Engineering Center

***TeleEngineering** Reflects an engineering telepresence to the force under the proponency of USAES; this capability is focused on assisting engineers and the commanders they support in planning and executing their operational and tactical missions. It is a piece of the capabilities inherent in FFE. The overarching concept is the exploitation of the Army's C3 architectures to provide a linkage between engineers and the appropriate nondeployed SMEs for resolution of engineer challenges. (FM 3-34)

tempo the rate of military action (FM 3-0/FM 101-5-1)

TEOC TeleEngineering Emergency Operations Center

terrain analysis (NATO, joint)—Collection, analysis, evaluation, and interpretation of geographic information on the natural and man-made features of the terrain, combined with other relevant factors, to predict the effect of the terrain on military operations. (FM 5-105/JP 1-02)

***terrain expert** Person who demonstrates skills and knowledge in rendering geospatial engineering to the tactical and operational situation to leverage the battlespace; experts understand the limits/ capabilities of GI&S and can integrate them into MDMP. (FM 3-34)

terrain intelligence (joint)—Intelligence on the military significance of natural and man-made characteristics of an area. (JP 1-02)

***terrain reinforcement** The development of terrain using obstacles to degrade enemy mobility or to enhance friendly survivability through the construction of fighting positions and cover. (FM 3-34)

terrain study (joint)—An analysis and interpretation of natural and man-made features of an area, their effects on military operations, and the effect of weather and climate on these features. (JP 1-02)

TF task force

theater support contractors (joint)—Contract personnel hire in, and operating in, a specific operational are (JP 4-07/JP 1-02)

TIH toxic industrial hazard—(NATO)—The hazards resulting from the release, by any means, of toxic industrial materials, resulting in contamination or irradiation of personnel or the environment area or any particular object. (FM 3-34.500)

TLP troop leading procedure

TO theater of operation

TOA table of allowances

TOC tactical operations center

TOE table of organization and equipment

topo topographic

***topographic engineering** (joint)—Those engineering tasks that provide geospatial information and services to commanders and staffs across the range of military operations; these tasks include terrain analyses, terrain visualization, digitized terrain products, nonstandard map products, and baseline survey data. See also GI&S. (JP 3-34/JP 1-02) (Replaced in Army doctrine by geospatial engineering [FM 3-34])

topographic map (joint)—A map that presents the vertical position of features in measurable form, as well as their horizontal position. (FM 3-25.26/ FM 101-5-1/JP 1-02)

topography (joint)—The configuration of the earth's physical surface, to include its relief and natural and man-made features; topography addresses both dry land and the sea floor (underwater topography). It also refers to mapping or describing these features. (JP 4-01/JP 1-02)

TP	transportation
TPFDD	time-phased force and deployment data
tph	ton(s) per hour
TRADOC	United States Army Training and Doctrine Command
trl	trailer
TRANSCOM	Transportation Command
trk	truck
TSC	theater support command
TTP	tactics, techniques and procedures
TUAV	tactical unmanned aerial vehicle
UAV	unmanned aerial vehicle
UCT	underwater-construction team
UIC	unit identification code
UJTL	universal joint task list—(joint)—A menu of capabilities (mission-derived tasks with associated conditions and standards, such as the tools) that may be selected by a joint force commander to accomplish the assigned mission; once identified as essential to mission accomplishment, the tasks are reflected with the command joint METL. (JP 1-02)
UMT	unit maintenance team
UN	United Nations
UO	urban operations—Offense, defense, stability, and support operations conducted in a topographical complex and adjacent natural terrain where man-made construction and high population density are the dominant features. (FM 3-0)
US	United States
USACE	United States Army Corps of Engineers
USACOM	United States Atlantic Command
USAEHSC	United States Army Engineering and Housing Support Center
USAES	United States Army Engineer School
USAFEG	United States Army Facility Engineer Group
USAID	United States Agency for International Development
USAMC	United States Army Materiel Command
USAMEDCOM	United States Army Medical Command
USAR	United States Army Reserve
USARC	United States Army Reserve Center or Command
USAREUR	United States Army, Europe
USARPAC	United States Army, Pacific

USARSO United States Army Forces Southern Command—The MACOM with Central and South America and the Caribbean as its AOR; it is the Army component of USSOUTHCOM and is one of five components of the command, which when combined, work together as a team to accomplish USSOUTHCOM's theater engagement missions.

USC United States Code

USDA United States Department of Agriculture

USFK United States Forces Korea

USG United States Government

USGS United States Geological Survey

USJFCOM United States Joint Forces Command

USPACOM United States Pacific Command

USSOUTHCOM United States Southern Command

USSS United States Secret Service

USTRANSCOM United States Transportation Command

UTC unit type code

util utility

UXO unexploded ordnance—(NATO, joint)—Explosive ordnance which has been primed, fused, armed, or otherwise prepared for action, and which has been fired, dropped, launched, projected, or placed in such a manner as to constitute a hazard to operations, installations, personnel, or materiel, and remains unexploded either by malfunction or design or for any other cause. (FM 4-30.11/FM 101-5-1/JP 3-15/JP 1-02)

VA Virginia

VDD visualize, describe, direct

WAN wide-area network

WO warning order

whl wheel

WIN Warfighter Information Network

WES Waterways Experiment Station

WMD weapons of mass destruction

WO warning order

wtr water

XO executive officer

ZOC zone of control

ZOP zone of penetration

ZOS zone of seperation

Bibliography

AAP-4. *NATO Standardization Agreements and Allied Publications.* December 1991.

AAP-6 (2002). *NATO Glossary of Terms and Definitions (English and French).* December 2002.

AAP-15 (F). *NATO Glossary of Abbreviations Used in NATO Documents and Publications.* December 2002.

AFDD 2-4. *Combat Support.* 22 November 1999.

AFDD 2-4.4. *Bases, Infrastructure, and Facilities.* 13 November 1999.

AFDD42. *Air Force Doctrine Document 42 Civil Engineering.* 1 December 1994.

AJP 3.12. *Joint Engineering.* This is a new manual that is in draft and fills a doctrinal gap.

American-British-Canadian-Australian Handbook. Coalition Operations Handbook. 01 November 2001.

AR 71-9. *Materiel Requirements.* 30 April 1997.AR 200-1. *Environmental Protection and Enhancement.* 21 February 1997.

AR 310-50. *Authorized Abbreviations, Brevity Codes, and Acronyms.* 15 November 1985.

AR 500-60. *Disaster Relief.* 1 August 1981.

AR 600-8-105. *Military Orders.* 28 October 1994.

AR 700-4. *Logistic Assistance.* 20 September 2002.

AR 715-9. *Contractors Accompanying the Force.* 29 October 1999.

Army Homeland Security (HLS) Strategic Planning Guidance. This soon to be published document is in draft and fills a doctrinal gap.

CJCSI 3010.02A. *Joint Vision Implementation Master Plan (JIMP).* 15 April 2001.

CJCSI 3121.01A. *Standing Rules of Engagement for US Forces.* 1 September 1999.

CJCSI 3122.03A. *Joint Operation Planning and Execution System Volume II; Planning Formats and Guidance.* 27 November 2000.

CJCSI 3207.01. *Military Support to Humanitarian Demining Operations.* 1 March 1999.

CJCSM 3500.04C. *Universal Joint Task List.* 1 July 2002.

DA Memo 10-1. *Executive Agent Responsibilities Assigned to the Secretary of the Army.* 15 January 1997.

DOD Civil Disturbance Plan, Annex C (Concept of Operations), Appendix 8 (Special Instructions), 15 Feb. 1991; modified by Director of Military Support message 161639Z Jul. 96, Subject: Changes to DOD Civil Disturbance Plan--"Garden Plot."

DODD 3025.1. *Military Support to Civil Authorities (MSCA).* 15 January 1993.

DODD 3025.15. *Military Assistance to Civil Authorities.* 18 February 1997.

DODD 4270.34. *Host Nation-Funded Construction Programs.* 19 October 1982.

DODD 4270.36. *DOD Emergency, Contingency, and other Unprogrammed Construction Projects.* 17 May 1997.

DODD 4270.5. *Military Construction Responsibilities.* 2 March 1982.

DODD 5100.1. *Functions of the Department of Defense and Its Major Components.* 1 August 2002.

DODD 5100.46. *Foreign Disaster Relief.* 4 September 1975.

DODD 6050.7. *Environmental Effects Abroad of Major DOD Actions.* 31 March 1979.

DODI 4715.5. *Management of Environmental Compliance at Overseas Installations.* 22 April 1996.

EO 12114. *Environmental Effects Abroad of Major Federal Actions.* 4 January 1979.

EO 13148. *Greening the Government through Leadership in Environmental Management.* 21 April 2000.

FM 1 (100-1). *The Army.* 14 June 2001.

FM 3-0 (100-5). *Operations.* 14 June 2001.

FM 3-06 (90-10). *Urban Operations.* 1 June 2003.

FM 3-06.11 (90-10-1). *Combined Arms Operations in Urban Terrain.* 28 February 2002.

FM 3-07 (100-20). *Stability Operations and Support Operations.* 20 February 2003.

FM 3-07.2. *Force Protection.* This is a new manual that is in draft and fills a doctrinal gap.

FM 3-21.31 (7-32). *The Stryker Brigade Combat Team.* 13 March 2003.

FM 3-34.2 (90-13-1). *Combined-Arms Breaching Operations.* 31 August 2000. With change 3 dated 11 October 2002.

FM 3-34.221. *Engineer Operations: Stryker Brigade Combat Team.* This is a new manual that is in draft and fills a doctrinal gap.

FM 3-34.230 (5-105). *Topographic Operations.* 3 August 2000.

FM 3-34.331 (5-232). *Topographic Surveying.* 16 January 2001.

FM 3-34.343 (5-446). *Military Nonstandard Fixed Bridging.* 12 February 2002.

FM 3-34.471 (5-420). *Plumbing, Pipefitting, and Sewerage.* 31 August 2001.

FM 3-50. *Smoke Operations.* 04 December 1990. With change 1 dated 11 September 1996. Will be revised as FM 3-11.50.

FM 3-55. *Intelligence, Surveillance, and Reconnaissance Operations.* This is a new manual that is in draft and fills a doctrinal gap.

FM 3-90. *Tactics.* 4 July 2001.

FM 3-100.4. *Environmental Considerations in Military Operations.* 15 June 2000. With change 1 dated 11 May 2001. This is in the process of revision to FM 3-34.500.

FM 3-100.21 (100-21). *Contractors on the Battlefield.* 3 January 2003.

FM 3-100.38 (100-38). *UXO Multiservice Procedures for Unexploded Ordnance Operations.* 23 August 2001.

FM 3-100.81. *Multinational Operations Handbook.* This is a new manual that is in draft and fills a doctrinal gap.

FM 4-0 (100-10). *Combat Service Support.* 29 August 2003

FM 4-01.30 (55-10). *Movement Control.* 1 September 2003.

FM 4-93.4 (63-4). *Theater Support Command.* 15 April 2003.

FM 5-7-30. *Brigade Engineer Combat Operations (Airborne, Air Assault, Light).* 28 December 1994. Will be revised as FM 3-34.223.

FM 5-10. *The Engineer Platoon.* 3 October 1995. Will be revised as FM 3-34.220.

FM 5-33. *Terrain Analysis.* 11 July 1990. With change 1 dated 8 September 1992. Will be revised as FM 3-34.330.

FM 5-71-2. *Armored Task-Force Engineer Combat Operations.* 28 June 1996. With change 2 dated 4 September 1997. Will be revised as FM 3-34.222.

FM 5-71-3. *Brigade Engineer Combat Operations (Armored).* 03 October 1995. With change 1 dated 13 November 1997. Will be revised as FM 3-34.224.

FM 5-71-100. *Division Engineer Combat Operations.* 22 April 1993. Will be revised as FM 3-34.225.

FM 5-100-15. *Corps Engineer Operations.* 6 June 1995. Will be revised as FM 3-34.226.

FM 5-102. *Countermobility.* 15 March 1985. Will be integrated into FM 3-34.2.

FM 5-103. *Survivability.* 10 June 1985. Will be revised as FM 3-34.112.

FM 5-104. *General Engineering.* 12 November 1986. Will be revised as FM 3-34.250.

FM 5-116. *Engineer Operations: Echelons Above Corps.* 09 February 1999. Will be revised as FM 3-34.211.

FM 5-125. *Rigging Techniques, Procedures and Applications.* 3 October 1995. With change 1 dated 23 February 2001. Will be revised as FM 3-34.311.

FM 5-134. *Pile Construction.* 18 April 1985. Will be revised as FM 3-34.334.

FM 5-170. *Engineer Reconnaissance.* 5 May 1998. With change 1 dated 13 July 1998. Will be revised as FM 3-34.212.

FM 5-212. *Medium Girder Bridge.* 07 February 1989. Will be revised as FM 3-34.342.

FM 5-233. *Construction Surveying.* 4 January 1985. Will be revised as FM 3-34.352.

FM 5-410. *Military Soils Engineering.* 23 December 1992. With change 1 dated 4 June 1997. Will be revised as FM 3-34.353.

FM 5-412. *Project Management.* 13 June 1994. Will be revised as FM 3-34.350.

FM 5-415. *Firefighting Operations.* 09 February 1999. Will be revised as FM 3-34.480.

FM 5-422. *Engineer Prime Power Operations.* 7 May 1993. Will be revised as FM 3-34.483.

FM 5-424. *Theater of Operations Electrical Systems.* 25 June 1997. Will be revised as FM 3-34.472.

FM 5-426. *Carpentry.* 3 October 1995. Will be revised as FM 3-34.473.

FM 5-428. *Concrete and Masonry.* 18 June 1998. Will be revised as FM 3-34.474.

FM 5-430-00-1. *Planning and Design of Roads, Airfields, and Heliports in the Theater of Operations – Road Design. Volume I.* 26 August 1994. Will be revised as FM 3-34.461.

FM 5-430-00-2. *Planning and Design of Roads, Airfields, and Heliports in the Theater of Operations – Airfield and Heliport Design. Volume II.* 29 September 1994. Will be revised as FM 3-34.462.

FM 5-434. *Earthmoving Operations.* 15 June 2000. Will be revised as FM 3-34.464.

FM 5-436. *Paving and Surfacing Operations.* 28 April 2000. Will be revised as FM 3-34.466.

FM 5-472. *Materials Testing.* 27 October 1999. With change 2 dated 1 July 2001. Will be revised as FM 3-34.354.

FM 5-480. *Port Construction and Repair.* 12 December 1990. Will be revised as FM 3-34.484.

FM 5-482. *Military Petroleum Pipeline Systems.* 26 August 1994. Will be revised as FM 3-34.485.

FM 5-484. *Multiservice Procedures for Well-Drilling Operations.* 8 March 1994. Will be revised as FM 3-34.481.

FM 5-490. *Engineer Diving Operations.* 31 March 1992. Will be revised as FM 3-34.216.

FM 5-499. *Hydraulics.* 1 August 1997. With change 1 dated 12 December 2001. Will be revised as FM 3-34.467.

FM 5-553. *General Drafting.* 6 January 1984. Will be revised as FM 3-34.351.

FM 6-0 (101-5). *Mission Command: Command and Control of Army Forces.* 11 August 2003.

FM 7-0 (25-100). *Training the Force*. 22 October 2002.

FM 7-1 (25-101). *Battle Focused Training*. 15 September 2003.

FM 7-15. *The Army Universal Task List*. 17 December 2002.

FM 20-3. *Camouflage, Concealment and Decoys*. 30 August 1999. Will be revised as FM 3-58.1.

FM 20-11. *Military Diving*. 20 January 1999. With change 1 dated 14 April 2000. Will be revised as FM 3-34.12.

FM 20-32. *Mine/Countermine Operations*. 29 May 1998. With change 3 dated 01 October 2003. Will be revised as FM 3-34.32.

FM 21-10. *Field Hygiene and Sanitation*. 21 June 2000. Will be revised as FM 4-25.10.

FM 21-31. *Topographic Symbols*. 19 June 1961. With change 1 dated 31 December 1968. Will be revised as FM 3-34.31.

FM 34-1. *Intelligence and Electronic Warfare Operations*. 27 September 1994. Will be revised as FM 2-0.

FM 34-130. *Intelligence Preparation of the Battlefield*. 8 July 1994. This is in the process of revision to FM 2-01.3.

FM 90-7. *Combined-Arms Obstacle Integration*. 29 September 1994. With change 1 dated 10 April 2003. Will be revised as FM 3-34.1, incorporating FM 5-102.

FM 90-13. *River-Crossing Operations*. 26 January 1998. Will be revised as FM 3-34.13.

FM 100-6. *Information Operations*. 27 August1966. This is in the process of revision to FM 3-13.

FM 100-7. *Decisive Force: The Army in Theater Operations*. 31 May 1991. This is in the process of revision to FM 3-93, Strategic Army.

FM 100-8. *The Army in Multinational Operations*. 24 November 1994. This is in the process of revision to FM 3-16.

FM 100-10-2. *Contracting Support on the Battlefield*. 4 August 1999. Will be revised as FM 4-100.2.

FM 100-11. *Force Integration*. 15 January 1998. Will be revised as FM 3-100.11.

FM 100-14. *Risk Management*. 23 April 1998. Will be revised as FM 5-19.

FM 100-17. *Mobilization, Deployment, Redeployment, Demobilization*. 28 October 1992. This is in the process of revision to FM 3-35.

FM 100-17-3. Rece*ption, Staging, Onward Movement, and Integration*. 17 March 1999. Will be revised as FM 4-01.8.

FM 100-19. *Domestic Support Operations*. 1 July 1993. Will be revised as FM 3-07.7.

FM 100-23-1. *HA—Multiservice Procedures for Humanitarian Assistance Operations*. 31 October 1994. Will be revised as FM 3-07.6.

FM 101-5. *Staff Organization and Operations*. 31 May 1997. Will be revised as

FM 5-0, Army Planning and Orders Production.

FM 101-5-1. *Operational Terms and Graphics.* 30 September 1977. This is in the process of revision to FM 1-02.

FM 101-5-2. *US Army Report and Message Formats.* 29 June 1999. Will be revised as FM 6-99.2.

FMFM 13-7. *MAGTF Breaching Operations.* 13 June 1995. Currently under revision and will be renumbered as MCWP 3-17.3.

FMFM 13-8. *MAGTF Explosive Ordnance Disposal.* 20 December 1993. Currently under revision and will be renumbered as MCWP 3-17.2.

JP 0-2. *Unified Action Armed Forces (UNAAF).* 10 July 2001.

JP 1. *Joint Warfare of the Armed Forces of the United States.* 14 November 2000.

JP 1-02. *Department of Defense Dictionary of Military and Associated Terms.* 12 April 2001.

JP 2-03. *Joint Tactics, Techniques, and Procedures for Geospatial Information and Services Support to Joint Operations.* 31 March 1999.

JP 3-0. *Doctrine for Joint Operations.* 10 September 2001.

JP 3-07. *Joint Doctrine for Military Operations Other Than War.* 16 June 1995.

JP 3-07.1. *Joint Tactics, Techniques, and Procedures for Foreign Internal Defense (FID).* 26 June 1996. Currently under revision.

JP 3-07.3. *Joint Tactics, Techniques, and Procedures for Peace Operations.* 12 February 1999.

JP 3-07.5. *Joint Tactics, Techniques, and Procedures for Noncombatant Evacuation Operations.* 30 September 1997.

JP 3-07.7. *Doctrine for Civil Support.* This is a new manual that is in draft and fills a doctrinal gap.

JP 3-08. *Interagency Coordination During Joint Operations*, 2 volumes. 9 October 2002. Currently under revision.

JP 3-10.1. *Joint Tactics, Techniques, and Procedures for Base Defense.* 23 July 1996.

JP 3-15. *Joint Doctrine for Barriers, Obstacles, and Mine Warfare.* 24 February 1999.

JP 3-16. *Joint Doctrine for Multinational Operations.* 5 April 2000.

JP 3-26. *Joint Doctrine for Homeland Defense.* This is a new manual that is in draft and fills a doctrinal gap.

JP 3-33. *Joint Force Capabilities.* 13 October 1999.

JP 3-34. *Engineer Doctrine for Joint Operations.* 5 July 2000.

JP 3-35. *Joint Doctrine for Deployment and Redeployment Operations.* 7 September 1999.

JP 3-55. *Joint Doctrine for Intelligence, Surveillance, and Reconnaissance (ISR) Operations.* This is a new manual that is in draft and fills a doctrinal gap.

JP 4-0. *Doctrine for Logistic Support of Joint Operations.* 6 April 2000.

JP 4-01.3. *Joint Tactics, Techniques, and Procedures for Movement Control.* 9 April 2002.

JP 4-01.8. *Joint Tactics, Techniques, and Procedures for Joint Reception, Staging, Onward Movement, and Integration.* 13 June 2000.

JP 4-04. *Joint Doctrine for Civil Engineering Support.* 27 September 2001.

JP 4-05. *Joint Doctrine for Mobilization Planning.* 22 June 1995.

JP 4-05.1. *Joint Tactics, Techniques, and Procedures for Manpower Mobilization and Demobilizations Operations: Reserve Component (RC) Callup.* 11 November 1998.

JP 5-0. *Doctrine for Planning Joint Operations.* 13 April 1995. Currently under revision.

JP 5-00.1. *Joint Doctrine for Campaign Planning.* 25 January 2002.

JP 5-00.2. *Joint Task Force Planning Guidance and Procedures.* 13 January 1999.

MCWP 3-17. *Engineering Operations.* 14 February 2000.

MCWP 3-17B. *Engineer Forms and Reports.* 3 October 1997.

MCWP 3-35.3. *Military Operations on Urbanized Terrain (MOUT).* 26 April 1998.

NWP 4-11. *Environmental Protection.* 1 March 1999.

OPNAVINST 3501.93D, *Projected Operational Environmental (POE) and Required Operational Capabilities (ROC) for Naval Beach Groups and Their Elements.* 19 June 2001.

OPNAVINST 5450.46K, *Naval Construction Force Policy.* 25 May 1999.

STANAG 2014/QSTAG 506. *Operations Orders, Warning Orders, and Administrative/Logistics Orders.* 27 April 1994.

STANAG 2020/QSTAG 510. *Operational Situation Reports.* 14 July 1995.

STANAG 2437. *Allied Joint Operations Doctrine* (AJP-1). 15 December 1994.

STANAG 3680. *NATO Glossary of Terms and Definitions* (AAP-6). 3 October 1986.

TC 3-34.489. *The Soldier and the Environment.* 8 May 2001. With change 1 dated 26 October 2001.

TC 5-150. *Engineer Qualification Tables.* 16 June 1998.

TC 5-210. *Military Float Bridging Equipment.* 27 December 1988. Will be revised as FM 3-34.340.

TC 5-230. *Army Geospatial Guide for Commanders and Planners.* This is a new manual that is in draft and fills a doctrinal gap.

TC 5-340. *Air Base Damage Repair (Pavement Repair).* 27 December 1988. Will be revised as FM 3-34.463.

Title 10. *Armed Forces.*

Title 10. *Burden sharing.*

Title 32. *National Guard.*

TM 5-235. *Special Surveys.* 18 September 1964. Will be revised as FM 3-34.335.

TM 5-240. *Compilation and Color Separation of Topographic Maps.* 15 June 1971. Will be revised and combined with FM 5-243 as FM 3-34.332, Cartographic Photography.

TM 5-243. *Cartographic Aerial Photography.* 2 January 1970. Will be revised and combined with TM 5-240 as FM 3-34.332, Cartographic Photography.

TM 5-245. *Offset Photolithography and Map Reproduction.* 21 July 1970. Will be revised as FM 3-34.333.

TM 5-323. *Military Standardization Handbook: Generator Sets, Electrical Measurement.* 30 August 1962. Will be revised as FM 3-34.476.

TM 5-349. *Arctic Construction.* 19 February 1962. Will be revised as FM 3-34.486.

TM 5-432. *Pits and Quarries.* 15 December 1967. Will be revised as FM 3-34.465.

TM 5-545. *Geology.* 8 July 1971. Will be revised as FM 3-34.360.

TRADOC PAM 525-3-25. *The United States Army Objective Force Maneuver Support Operational Concept.* This yet to be published document is critical to understanding the future Army and the impact on doctrine.

Quadrennial Defense Review. 30 September 2001.

White Paper. *Offshore Force Projection Operations and Enabling Systems: Deployment of the IBCT.* 1 June 2001.

Index

N

O

Made in the USA
Monee, IL
05 November 2026

67810725R00361